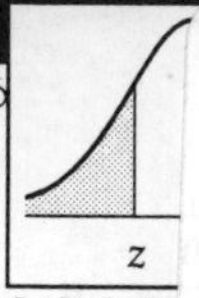

STATISTICS
MAKING SENSE OF DATA

William F. Stout
Kenneth J. Travers
John Marden
University of Illinois at Urbana-Champaign

With contributions from

Amy Goodwin
Chad Schafer
Richard Stalmack
David Travers
Joe Vittitow

Möbius Communications, Ltd.
Rantoul, Illinois

William Stout dedicates this book to his wife, Barbara, for her unflagging emotional and intellectual support.

Kenneth Travers dedicates this book to his wife, Janny, in loving appreciation for her commitment to the importance of life-long learning.

John Marden dedicates this book to his mother, to Ann, and to Phil.

MÖBIUS

COMPLIMENTARY

Statistics: Making Sense of Data

This book is published by Möbius Communications, Ltd.,
221 South Maplewood, Rantoul, Ill., 61866,
telephone 1-800-MOBIUS5.

Design and composition: Publication Services, Inc.

Cover design: Donald Waller
Illustrations: Duane Gillogly

Printed in the United States of America.

Library of Congress Cataloging-in-Publication Data

Stout, William F., 1940-
Statistics : making sense of data / William F. Stout, Kenneth J. Travers, John Marden ; with contributions from Amy Goodwin ... [et al.].
p. cm.
Includes index.
ISBN 1-891304-00-3
1. Statistics. I. Travers, Kenneth J. II. Marden, John I. III. Goodwin, Amy. IV. Title .
QA276.12.S76 1997
519.5--dc21 97-38000
CIP

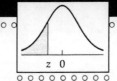

CONTENTS

About the Authors ix

Preface xi

PART I Describing Data 1

1 Exploring Data 2

1.1 What Is the Science of Statistics? 5

1.2 Stems and Leaves 13

1.3 Circle Graphs and Line Plots 19

1.4 Frequency Histograms 24

1.5 Relative Frequency Histograms and Polygons 31

1.6 Misusing Statistics 37

Chapter Review Exercises 41

Computer Exercises 45

2 Centers and Spreads 56

2.1 Centers of Data 58

2.2 Some Properties of Centers 69

2.3 Spread (Variation) in Data 71

2.4 Deviation Values 79

2.5 Mean Absolute Deviation 83

2.6 Variance 86

2.7 Standard Deviation 88

Chapter Review Exercises 92

Computer Exercises 95

3 **Relationships** 102

3.1 Scatter Plots 105

3.2 Graphs of Linear Relationships 111

3.3 Linear Regression: An Informal Visual Approach 116

3.4 Least Mean Error Linear Regression 122

3.5 Least Squares Linear Regression 130

3.6 Covariance, Correlation, and the Least Squares Regression Line 138

3.7 Relation and Causation 153

Chapter Review Exercises 156

Computer Exercises 162

PART II **Statistical Modeling** 173

4 **Modeling Expected Values** 174

4.1 The Expected Value of a Statistic in a Random Experiment 176

4.2 More Five-Step Method Examples 192

4.3 Independence 199

4.4 Random Walks (Optional) 202

Chapter Review Exercises 207

Computer Exercises 212

5 **Probability** 222

5.1 Experimental Probability 224

5.2 Estimating Probabilities 229

5.3 Box Models 235

5.4 Independent Events 241

5.5 Probability as a Theoretical Concept 246

5.6 Complementary Events 254

5.7 Sampling with or without Replacement 258

5.8 The Birthday Problem (Optional) 264

5.9 Conditional Probability 268

Chapter Review Exercises 271

Computer Exercises 277

6	**Making Decisions**	284
6.1	Samples and Populations	286
6.2	Using Samples to Decide on Hypotheses about Populations	291
6.3	The Median Test	304
	Chapter Review Exercises	309
	Computer Exercises	313
7	**Chi-Square Testing**	316
7.1	Is the Die Fair?	318
7.2	How Big a Difference Makes a Difference?	321
7.3	The Chi-Square Statistic	325
7.4	Six Steps to Chi-Square	329
7.5	Smooth Chi-Square Curves	335
7.6	Chi-Square Tables	345
7.7	Unequal Expected Frequencies	351
	Chapter Review Exercises	357
	Computer Exercises	365
8	**Probability Distributions**	376
8.1	The Binomial Distribution: Success Counts	379
8.2	The Poisson Distribution: Events at Random (Optional)	393
8.3	Rolling a Fair Die: The Discrete Uniform Distribution	401
8.4	Continuous Distributions	403
8.5	The Continuous Uniform Distribution	408
8.6	The Normal Distribution	413
8.7	Using the Normal-Curve Table	421
8.8	Applications of the Normal Curve	427
8.9	Other Continuous Distributions (Optional)	433
	Chapter Review Exercises	437
	Computer Exercises	441
9	**Measurement**	446
9.1	Measurements Vary	448
9.2	Measuring Carefully	452

9.3 Standard Error 456

9.4 Systematic Error (Bias) 461

9.5 Accuracy 463

9.6 An Application of Measurement: The Acceleration Due to Gravity 467

Chapter Review Exercises 470

Computer Exercises 473

PART III Estimation and Hypothesis Testing 475

10 Estimation 476

10.1 Parameters and Statistics 478

10.2 Point Estimates and Confidence Intervals 481

10.3 Standard Errors of Estimates 488

10.4 Central Limit Theorem 491

10.5 Confidence Intervals for the Population Mean 496

10.6 Confidence Intervals for the Population Proportion 501

10.7 Confidence Intervals for the Difference between Two Population Means 503

10.8 Point Estimates for the Population Variance 505

10.9 Point Estimates for the Population Median 509

Chapter Review Exercises 511

Computer Exercises 513

11 Hypothesis Testing 522

11.1 A Bootstrap Hypothesis Test of the Population Mean 524

11.2 The *z* Test of the Population Mean 535

11.3 Making a Wrong Decision 540

11.4 The Sign Test 542

11.5 The *t* Test 547

11.6 Comparing Two Population Means 550

11.7 The *z* Test for a Hypothesis about a Population Proportion 556

Chapter Review Exercises 559
Computer Exercises 561

12 **Correlation and Regression** 566
12.1 Is the Regression Line Worthwhile? 568
12.2 Does a Straight Line Work? 574
12.3 Nonlinear Relationships 581
12.4 Using the Correlation Coefficient as a Test Statistic 588
12.5 Two Explanatory Variables 591
12.6 The Multiple Correlation Coefficient 595
12.7 Another Test for Trend: Kendall's Tau 597
Chapter Review Exercises 605
Computer Exercises 610

13 **More Probability with Applications** 616
13.1 Some Basic Rules of Probability 618
13.2 Three Important Discrete Probability Models 624
13.3 Conditional Probabilities 632
13.4 Expected Value as a Theoretical Construct 635
13.5 Hypothesis Testing of the Equality of Two Population Proportions: An Application of the Binomial Distribution 643
13.6 Chi-Square Test for Probabilistic Independence 646
Chapter Review Exercises 652
Computer Exercises 657

Appendices 660

A **Computationally Generated Random Digits** 660

B **Random Number Tables** 662

C **Chi-Square Probabilities** 667

D **Linear Interpolation** 669

E **Normal Probabilities** 671

F Student's *t* Probabilities 673

G Binomial Probabilities 674

Glossary 675

Answers to Selected Problems 683

Index 719

The authors wish to thank the publishers of the following materials for granting permission for their use:

Table 1.1 and the tables in Examples 2.3 and 2.4: *USA Today*, http:www.usatoday.com/olympics.olyfront.htm.

Tables in Exercises 4 and 5 of Section 2.3: *Champaign-Urbana (Illinois) News-Gazette*, Oct. 27, 1996.

Extract in Exercise 3 of Section 3.1: *Weekend Magazine*, Toronto.

Extract in Exercise 4 of Section 3.1: *Champaign-Urbana (Illinois) News-Gazette*, Oct. 21, 1975.

Extract in Chapter Review Exercise 10 of Chapter 3: Healer Inc., http://www.healer-inc.com/a039705a.htm.

Data in Computer Exercise 6 of Chapter 3: David S. Moore and George P. McCabe, *Introduction to the Practice of Statistics* (New York: Freeman, 1989).

Extract in Exercise 4 of Section 6.1: *Education Week*, Jan. 2, 1982.

Table and extract in Exercise 5 of Section 6.1: *Weekend Magazine*, Toronto.

Extract in Exercise 6 of Section 6.1: Associated Press, July 23, 1982.

Extract in Exercise 7 of Section 6.1: 'The Changing Mood of the Nation's Voters, *Family Weekly*, 1980.

Extract in Exercise 8 of Section 6.1: S. Chatterjees, Estimating Wildlife Populations by the Capture-Recapture Method, in *Statistics by Example: Finding Models* (Menlo Park, Calif.: Addison-Wesley, 1973).

Computer Exercise 6 of Chapter 7: David Phillips, Deathday and Birthday: An Unexpected Connection, in Judith M. Tanur et al., eds., *Statistics: A Guide to the Unknown* (San Francisco: Holden-Day, 1972).

Figure E1 in Chapter 9: W. J. Youdin, *Experimentation and Measurement* (Washington, D.C.: National Science Teachers Association, 1972).

Quote appearing in the Key Problem of Chapter 10: Majority Opposes Stadium, Poll Shows, *St. Louis Post Dispatch*, Oct. 12, 1986.

Extract in Exercise 4 of Section 10.1: Walter Sullivan, Ancient Trees Tell a Tale of Past and Future Droughts, *The New York Times*, Sept. 2, 1980.

Computer Exercise 2 of Chapter 12: P. F. Velleman and D. C. Hoaglin, *Applications, Basics, and Computing of Exploratory Data Analysis* (Belmont, Calif.: Wadsworth, 1981).

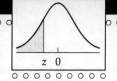

ABOUT THE AUTHORS

William Stout is a professor of Statistics at the University of Illinois at Urbana-Champaign, where he has been on the faculty since 1967. He is an internationally acclaimed researcher in the application of statistics to the fields of educational and psychological measurement. As the founder of the University of Illinois Statistical Laboratory for Educational and Psychological Measurement, he has led the development of a new and widely applicable theory and methodology for improving standardized testing. In particular, his path-breaking research on test bias has had a major impact. Stout is an associate editor of two leading measurement journals: *Psychometrika* and *Journal of Educational and Behavioral Statistics.* He has received extensive grant support from the National Science Foundation, Office of Naval Research, Educational Testing Service, and Law School Admission Council. He was recently awarded the National Council of Measurement in Education prize for the best three-year scientific contribution to educational measurement. Stout has lectured worldwide and was recently the week-long keynote speaker at the Ninth International Netherlands Measurement Workshop. Several times Stout has been on the list of professors voted as outstanding instructors by their students. In the last two years, eight students have received their doctorates in statistics under his guidance. Stout was the codirector of the NSF-funded Illinois Institute for Statistics Education, whose mission was to facilitate quality precollege statistics education nationwide. Stout is the author of over 50 scholarly books and papers. His doctoral degree is in statistics from Purdue University.

Kenneth J. Travers is a professor of Mathematics Education at the University of Illinois at Urbana-Champaign, where he has been on the faculty since 1965. He has been a national postdoctoral fellow at Stanford University and an American Education Research Association Senior Research Fellow at the National Center for Education Statistics in Washington, D.C. He is the author of more than 20 high school and college textbooks on mathematics and mathematics education. For 10 years (1978–87) he served as chairperson of the International Mathematics Committee, a project of the International Association for the Evaluation of Educational Achievement, which was in charge of a widely cited survey of mathematics teaching and learning in twenty countries around the world. He was codirector of the Illinois Institute for Statistics Education. In 1990 he was invited to become a senior-level administrator at the National Science Foundation, where he served for three years, during which time he helped to establish a new unit in the Foundation, the Division for Research, Evaluation, and Dissemination. On his return to the University of Illinois in 1993, he established the University of Illinois Office for Mathematics, Science, and Technology Education, of which he is the director. He holds a doctoral degree in mathematics education from the University of Illinois and bachelor`s and master`s degrees in mathematics and education from the University of British Columbia.

John Marden is a leading internationally known researcher in statistical decision theory, multivariate statistical analysis, and ranking data. He has written numerous papers for prestigious journals, as well as the monograph *Analyzing and Modeling Rank Data.* He is a Fellow of the Institute for Mathematical Statistics and an associate editor for *The Annals of Statistics.* As a faculty member in the Departments of Mathematics (1977–1985) and Statistics (since 1985) at the University of Illinois at Urbana-Champaign, Professor Marden has taught the entire gamut of statistics courses offered, beginning and advanced, applied and theoretical, and has often been the director of the Illinois Statistics Office, the consulting arm of the Department of Statistics. He has been particularly interested in the teaching of statistics to a general audience using the latest computer technology, including the World Wide Web and asynchronous learning networks. He has received a number of grants for the development of innovative technology-assisted methods of teaching statistics: a Summer Teaching Award in 1992 from the University of Illinois College of Liberal Arts and Sciences for developing statistical software; two grants from the Sloan Center for Asynchronous Learning Environments for using conferencing and developing educational statistical computer routines to be used through the World Wide Web; and a grant from the National Science Foundation for establishing a state-of-the-art computer laboratory for students at the University of Illinois. On numerous occasions he has been on the list of professors voted as outstanding instructors by their students. He received his doctoral degree in statistics from the University of Chicago.

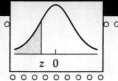

PREFACE

It is the authors' strongly held belief that the most effective way for beginning statistics students to learn statistics well is for them to directly experience its central concepts and standard procedures through closely observing and working with simulated and real data sets. The particular advantage of the use of simulated data is that the student knows precisely the underlying mechanism—that is, the probability model—that is producing the data. Thus the student can see how close the data come, when studied and analyzed with statistical expertise, to revealing their underlying truths. The learning of statistics experientially by direct contact with data is accomplished in this textbook by the presentation of statistical concepts using the five- and six-step simulation methods throughout. The same experiential pedagogical approach was previously developed into a precollege textbook by two of the authors and used in a five-year National Science Foundation Educational Directorate national teacher training grant to William Stout and Kenneth Travers.

A vital complement to the experiential learning of statistical concepts is a clear and understandable presentation of the formal, logical, and deductive aspects of the subject, shorn of all extraneous mathematical baggage. This is unfailingly carried out throughout this text. In most cases this theoretical component is presented after a thorough experiential immersion in the topic of interest. For example, the basic aspects of probability—often a difficult set of principles for beginning students to master—are first presented as empirical phenomena in Chapters 4 and 5 and only then are abstracted and expanded into a deductive body of mathematical propositions and statistical models in Chapters 8 and 13.

To be most effective, an introductory statistics textbook must strike a balance between helping the student learn to reason statistically and exposing the student to the standard body of statistical procedures that are widely used in practice and are judged to constitute statistical literacy. This balance, a central goal of this book, equips a student with both real statistical discernment and the needed familiarity with those elementary statistical procedures so widely used in science, business, and government and so often reported in the media.

We believe that an introductory statistics textbook should both be able to stand alone in educational settings where calculator and computer technology is not being used and also be equipped to make profound and meaningful use of available technology when the instructor wants. This book is designed with these principles in mind. The book is totally self-contained for classrooms in which technology is not available or not desired. However, for those wishing to use it, such technology—in this case Texas Instruments graphing calculators and Windows-based computers—can greatly enhance students' understanding of statistics. A major effort has therefore gone into developing instructional programs for use with TI calculators and Windows-based computers. In particular, the computer software is an enhancement and a customization of software developed over several years by John Marden. A precursor of the textbook software is currently being used in the multiple-section Statistics 100 course at the University of Illinois, which is taken by six hundred to eight hundred students each year. The role of the software provided with the book is to add to a lecture-based statistics course a very efficient laboratory component in which the central concepts are rapidly and clearly experienced first-hand as the student carries out the computer exercises and just plays with the software. The computer exercises constitute a sort of laboratory with manual that guides the student through various data-driven experiences of important statistical concepts. The TI calculator exercises play a similar but more modest role. Throughout these technology exercises, the central role of the technology component is instructional rather than computational.

The textbook breaks new ground in content through its inclusion of bootstrap-based inferential procedures as an important component. The bootstrap approach is one of the most important statistical advances of the past 25 years and is rapidly becoming an essential component of the practicing statistician's toolbox. When presented appropriately, the bootstrap approach is intuitive and hence easily understandable by the beginning statistics student. In keeping with the instructional approach of the book, the bootstrap method is simulation-based, and it seamlessly meshes with the five- and six-step methods used throughout. Most important, it provides the student with a powerful cutting-edge method of statistical inference that, together with large-sample inferential approaches based on the central limit theorem and the analogous result for chi-square statistics—another emphasis of the book—provides the student with enormous statistical inferential power.

The book is written to be accessible to all students having at least a modest exposure to algebra: "intermediate algebra" suffices. Although formulas and graphs appear frequently and students are encouraged to think deeply about what they are learning, the amount of formal mathematical background needed is minimal. The book should work well in any non-calculus-based introductory statistics course at any two-year or four-year college. It can be taught to science and math majors and to humanities and social science majors with equal ease. It presumes no prior exposure to statistics.

Just as our experiential approach to the teaching of statistics makes learning statistics easier and more rewarding for the student, this approach plays the same role for an instructor who is teaching statistics while having a limited background in the subject. In fact, it is a very easy book for an instructor of mathematics or science with limited *statistical* training to learn statistics from while teaching the course.

Several influential nationally circulated reports have stressed the need for new emphases in the teaching of statistics. For example, the widely influential American Statistical Association/Mathematical Association of America Cobb report recommends the teaching of statistics through the heavy use of data and proposes that more emphasis be placed on statistical concepts than on abstract theory. Further, it stresses "active learning." This textbook is tailor-made to address these valid recommendations because of the reasons stated above, and moreover because of the numerous exercises provided, both after each section and as review exercises at the end of each chapter. There are both conventional pencil-and-paper exercises and optional TI and computer exercises. Thus the student has ample opportunity to practice using the concepts learned in the text.

The book begins with three chapters describing how one summarizes data—data being the focus of statistics—graphically and by means of statistical indices. Chapters 4 and 5 then provide a heavily empirical introduction to probability through emphasis on expected value and on probability itself. This is appropriate because probability is the logical underpinning of inferential statistics. Chapters 6 and 7 introduce the process of statistical inference, with the widely used and fascinating topic of chi-square testing appearing unusually early, as contrasted with the typical statistics textbook, in Chapter 7. Then Chapter 8 deepens the student's understanding of probability. The topic of how to achieve accurate measurement in science and industry, and in particular the role of statistics in measurement, is covered in Chapter 9. This important topic of measurement is too often ignored in introductory statistics textbooks. Then Chapters 10, 11, and 12 cover statistical inference in depth, stressing confidence-interval estimation and hypothesis testing. Finally, Chapter 13 provides a formal and rigorous introduction to probability and its most useful modeling applications, while providing some more inferential techniques.

The entire book can easily be covered in a two-semester (or three-quarter) course. If desired, the instructor can omit Chapter 9 or Chapter 13 without loss of continuity. In a one-semester course, Chapters 1–8 and 10–11 can be covered, in which case the instructor is encouraged to monitor the ongoing cognitive progress of the class and set the pace accordingly. In the one-semester course, to avoid undesirable time pressure, the optional sections should likely be omitted, as well as all or selected sections from Sections 1.3, 1.6, 2.5, 6.3, 10.7–10.9, and 11.4–11.6. Moreover, Chapter 7 can be omitted if the instructor wishes greater class time for the important Chapters 10 and 11.

We are confident that you will find the teaching of statistics from this preliminary first edition a rewarding experience. Moreover, the upcoming hardcover edition will be—in part because of instructor input, which will be acknowledged—further improved (it will be available for use in 1998–99). Enabled by the Möbius Web site for this textbook to interact with instructors, we the authors welcome queries and are eager to assist in teaching statistics in this very exciting and effective and experientially based way.

ACKNOWLEDGMENTS

The authors wish to acknowledge their appreciation for the superb and in-depth intellectual and technical support provided by Möbius Communications in the development and preparation of this book, and that provided by Publications Services, Inc., which produced this book for Möbius Communications, in particular the effort of the book's production coordinator, Matt Harris, for his expert management of this complex and challenging endeavor.

Part 1

DESCRIBING DATA

1 *E*XPLORING **D**ATA 2

2 **C**ENTERS AND **S**PREADS 56

3 *R*ELATIONSHIPS 102

1

EXPLORING DATA

Statistical thinking will one day be as necessary for efficient citizenship as the ability to read and write.

H. G. WELLS

OBJECTIVES

After studying this chapter, you will understand the following:

- *The role of statistics*
- *Graphical methods of summarizing data, including stem-and-leaf plots, circle graphs, and line plots*
- *How to use a histogram to summarize the shape of a data set*
- *Graphical descriptions of data that are misleading*

1.1 WHAT IS THE SCIENCE OF STATISTICS? 5

1.2 STEMS AND LEAVES 13

1.3 CIRCLE GRAPHS AND LINE PLOTS 19

1.4 FREQUENCY HISTOGRAMS 24

1.5 RELATIVE FREQUENCY HISTOGRAMS AND POLYGONS 31

1.6 MISUSING STATISTICS 37

CHAPTER REVIEW EXERCISES 41

COMPUTER EXERCISES 45

Will the Real Author Please Stand Up?

A famous TV show in the 1960s had sets of three people each pretending to have the same occupation in an effort to fool a panel of experts while only one person really did have the occupation. After the panel's questioning was over, the experts would predict who the person with the occupation really was, and then the person with the occupation would stand up. The Key Problem is like this.

Finding out who actually wrote a play, a novel, or a will can be a very important problem to solve. For example, when the billionaire Howard Hughes died, several different wills came to light. It was left to the courts to decide which was actually written by Hughes. Interestingly, the court could likely have benefited by consulting a statistician!

Some scholars doubt that William Shakespeare really did write all those plays and poems. Persons named as possibly having done some of the writing include Sir Walter Raleigh and Sir Francis Bacon.

In the history of the United States, there is a famous problem about authorship. It has to do with the *Federalist Papers,* a set of papers written anonymously between 1787 and 1788 by Alexander Hamilton, John Jay, and James Madison to persuade people in the state of New York to ratify the Constitution. The papers contain many details about the writing of the U.S. Constitution. The authorship of 70 of the papers is agreed on, but there are 12 about which there is disagreement as to whether they were written by Hamilton or Madison.

By using words known as *markers,* researchers use statistical methods to help decide the authorship of those papers in dispute. A marker is a set of words that can be interchangeably used with no change in meaning. The words *while* and *whilst* have been used as marker words in the case of the *Federalist Papers.* In the 14 papers that are known to have been written by Madison, *whilst* is used in 8 and *while* is never used (see Figure 1.1*a*). On the other hand, in the 48 papers written by Hamilton, *while* is used in 15 and *whilst* is never used (see Figure 1.1*b*).

In the 12 papers of disputed authorship, *whilst* was frequently used in 5 and *while* was not used at all. Therefore, the researchers concluded, those 5 papers were written by Madison. This reasoning is deductive, based on the presumption that Hamilton never used *whilst,* preferring *while,* and Madison often used *whilst,* but never *while.* Almost always, the author identification process is not this easy. More typically, both authors would be shown in their writings to use both marker words, with Madison, say, using *whilst* more frequently and

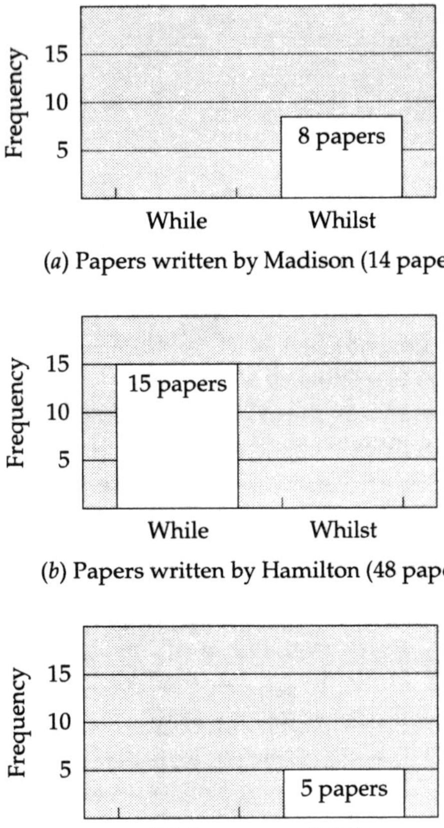

(*a*) Papers written by Madison (14 papers)

(*b*) Papers written by Hamilton (48 papers)

(*c*) Unknown (12 papers)

Figure 1.1 **Word frequencies used to determine authorship of selected *Federalist Papers*.**

Hamilton using *while* more frequently. Then the reasoning would have to be statistical in nature. For example, it would be based on whether the observed usage rate of *while* (versus *whilst*) in a disputed work is sufficiently more like the rate at which Madison used *while*. In that case it would be asserted that the statistical evidence is strong that Madison was the author. The ideas of such statistical inference are the central focus of this book. The science of statistics is essential. Statistical reasoning enables us to make valid inferences that are totally hidden from the statistically uninformed observer of the data. Author identification problems are of this character.

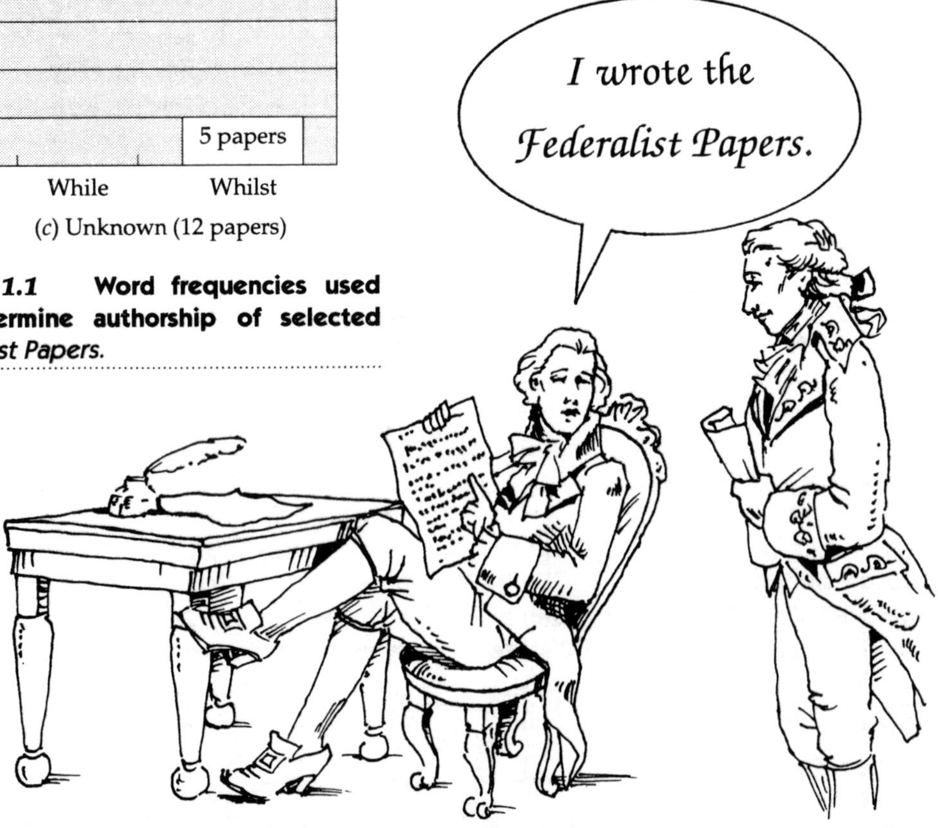

1.1 WHAT IS THE SCIENCE OF STATISTICS?

Considered as a scholarly or scientific discipline, statistics is the science of gathering and describing data and the subsequent drawing of conclusions (inferences) from data. In this book we will study the basic ideas in statistics and learn how to use these ideas to help interpret numerical data in the world around us.

However, to the average person, statistics are merely numerical information (often called *data*) about aspects of the world around us. Indeed, the word comes from *state:* originally, *statistics* meant such numerical descriptions of the affairs of state, that is, of government. **Statistics**—that is, numerical information—can help us learn about world hunger, the likely consequences of a weather system that is headed our way, and so on. From this viewpoint, one piece of data or a single number computed from data (like the average temperature on July 23, 1998, in 20 cities) that somehow informs us about the unknown world around us is called by statisticians and lay people alike a **statistic.** Surely, you've heard someone say, "Isn't that an interesting statistic!"

The daily newspaper is filled with statistics, often presented in clever ways. This numerical information can be about sports, health, or science, among other things. Here are some examples, either of statistics as numerical information or of statistics as inferential reasoning, from data about the unknown character of things.

Winter Olympic Games: CBS, which broadcast the 1994 Winter Olympic Games, held in Lillehammer, Norway, posted an average A.C. Nielsen rating of 27.8% of all U.S. television households for its prime-time telecasts. That was up 49% from CBS's ratings for the 1992 winter games in Albertville, France. Here the 27.8% and 49% are statistics of interest.

More Math Means More Money: One of the most extensive studies ever made of the careers of high school graduates confirms the conventional wisdom that those who take more mathematics courses earn considerably more in the marketplace (*Education Week,* Jan. 11, 1989, p. 9). Superficially it would thus appear from the study that taking more math courses should increase one's earning potential. But it may be that other characteristics of those taking lots of math, such as the motivation to take on a difficult challenge, are the major causes of their higher incomes. For this reason, taking more math courses might not increase one's earning potential. Indeed, it is dangerous to assume that when paired quantities are statistically associated—in that they are large or small together—an increase in one quantity *causes* an increase in the other. Such reasoning is an example of statistics as inference, although unfortunately in this case one of invalid statistical inference. One intent of this book is to help you learn how to engage in valid statistical inference.

Avoiding a Cold? Don't Get Stressed Out! Survey research carried out in England indicates that persons under stress have a greater tendency to catch a cold than those who are not stressed ("Psychological Stress and Susceptibility to the Common Cold," *New England Journal of Medicine,* Aug. 29, 1991, pp. 606–612). Provided the experimental work that led to this conclusion was done properly, concluding that stress increases susceptibility to colds may be an example of valid statistical inference.

Headaches Run in the Family: If you suffer from migraine headaches, don't blame your stars—but perhaps blame your genes. Migraine headaches may be a hereditary condition. Data show that if both father and mother have had them, there is a 75% chance that the son or daughter does, too. If only one parent has had migraine headaches, data show that the chance of the condition occurring is 50%. Even having distant relatives with migraine headaches increases the risk of having them by about 20% over the general population rate. Of course, concluding from the data that heredity causes migraine headaches is risky. For example, stress in the family environment could really be causing migraine headaches to occur more frequently in all family members. (From *Headway,* vol. 3, Dec. 93.)

The Dangers of Opening One's Own Drink of Milk or Juice: Research supported by a British juice company found that 50,000 people in Great Britain seek hospital treatment every year from injuries incurred while struggling to open milk and juice cartons. (From *The Medical Post,* Sept. 13, 1994.)

New Statistic Forecasts Danger from Ultraviolet Rays: Based on a careful study of available data, in the summer of 1994 the National Weather Service introduced an index to forecast the amount of ultraviolet light that will hit the Earth's surface at noon the next day. The scale goes from minimal (0 to 2) to very high (10 to 15). The ratings are based on upper atmosphere ozone levels. Ozone blocks ultraviolet radiation, the part of solar energy that causes sunburn. Thus this index, or statistic, which is computed from upper atmospheric ozone data readings, helps us assess the danger or sunburn or skin cancer.

All these news items deal with topics of general interest. Each contains numerical information (that is, *statistics*) or presents statistical inferences drawn (rightly or wrongly!) from numerical information. Consider some questions about which these statistically based discussions may help inform us:

- *Entertainment:* The popularity of TV coverage (Nielsen ratings) of the 1994 Winter Olympics contrasted with the 1992 Winter Olympics
- *Health:* How dangerous it is to expose your skin to the sun on a hot summer's day, and the relationship between stress and susceptibility to colds
- *Economics:* Whether the study of mathematics may increase one's earning power

Let's return to considering the science of statistics. We said that statistics is the science of gathering, describing, and drawing conclusions from numerical data. It is important to study the field of statistics so that we can draw conclusions about or make better decisions about numerical information that is reported to us, especially in the media. For example, we can sometimes draw our own statistically informed conclusions based on information (data) provided and then compare our conclusions with those of media writers. When we read the newspaper we often need to make decisions about the data-based information that is there. Some illustrative questions are

- Will being in the sun for two hours produce a severe sunburn on a particular day?
- If I take a minor in math, am I likely to obtain a better-paying job?
- Were the Winter Olympics of 1994 more interesting to Americans than the Winter Olympics of 1992?
- Am I more likely than the average person to have migraine headaches if my stepmother has them? If my mother has them?

In this book *statistics* as numerical information to be described and as the science of drawing conclusions from data are both very important. This book will therefore present you with many actual data sets to be described as well as present the basic ideas and techniques of gathering, describing, and analyzing data in order that you will be able to draw sound conclusions that are not obvious to the casual observer of the data.

Next we consider two interesting data sets. For now, we will simply treat them as opportunities to extract information from tabulated data, without applying any sophisticated statistical inferential reasoning to help in this process. Indeed, one important statistical skill is the ability to look at a table of numbers and extract useful information from it by thoughtfully considering the data. The two examples that follow illustrate this.

Table 1.1 **Medals Won in 1996 Summer Olympics**

		Total	Gold	Silver	Bronze
1.	United States	101	44	32	25
2.	Germany	65	20	18	27
3.	Russia	63	26	21	16
4.	China	50	16	22	12
5.	Australia	41	9	9	23
6.	France	37	15	7	15
7.	Italy	35	13	10	12
8.	South Korea	27	7	15	5
9.	Cuba	25	9	8	8
10.	Ukraine	23	9	2	12
11.	Canada	22	3	11	8
12.	Hungary	21	7	4	10
13.	Rumania	20	4	7	9
14.	Netherlands	19	4	5	10
15.	Poland	17	7	5	5
16.	Spain	17	5	6	6
17.	Bulgaria	15	3	7	5
18.	Brazil	15	3	3	9
19.	Britain	15	1	8	6
20.	Belarus	15	1	6	8
21.	Japan	14	3	6	5
22.	Czech Republic	11	4	3	4

Example 1.1

1996 Summer Olympics Champs

The numbers of medals (gold, silver, and bronze) won by the top 22 countries in the 1996 Summer Olympics (held in Atlanta) are given in Table 1.1. We see that the United States, Germany, and Russia won the most medals. What interesting questions does this table answer or suggest? For example, is it usually true that the gold medal ranking of countries is the same as the overall ranking? Is there a strong association between population size and medals ranking?

Example 1.2

World's 10 Largest Cities

The populations of the 10 most populated cities in the world are given in Table 1.2. We see that Tokyo/Yokohama is the most populated city in the world, with 27.2 million people, while Rio de Janeiro and Buenos Aires tie for the least populated (in this list) with 11.7 million people each.

Table 1.2 **Populations of World's 10 Largest Cites**

City	1991 population (millions)
Tokyo/Yokohama	27.2
Mexico City	20.9
São Paulo	18.7
Seoul	16.8
New York	14.6
Osaka/Kobe/Kyoto	13.7
Bombay	12.1
Calcutta	11.9
Rio de Janeiro	11.7
Buenos Aires	11.7

There is another way of looking at how many people live in a city: by how crowded it is. We can consider the city's density—for example, the number of persons per square mile of the city. To find the density, we will need to know the city's area. Table 1.3 gives the area (in square miles) of the world's 10 largest cities.

We now use the information in Table 1.3 to find the population density of each city.

$$\text{Density} = \text{number of persons per square mile} = \frac{\text{population}}{\text{area of city in square miles}}$$

For example,

$$\text{Population of Tokyo/Yokohama} = 27{,}200{,}000$$

$$\text{Area of Tokyo/Yokohama} = 1089 \text{ square miles}$$

Therefore,

$$\text{Density of Tokyo/Yokohama} = \frac{27{,}200{,}000}{1089} = 25{,}018 \text{ persons per square mile}$$

Table 1.3 **Populations and Areas of World's 10 Largest Cities**

City	1991 population (millions)	Area (square miles)
Tokyo/Yokohama	27.2	1089
Mexico City	20.9	522
São Paulo	18.7	451
Seoul	16.8	342
New York	14.6	1274
Osaka/Kobe/Kyoto	13.7	495
Bombay	12.1	95
Calcutta	11.9	209
Rio de Janeiro	11.7	260
Buenos Aires	11.7	535

Table 1.4 Population Densities of World's 10 Largest Cities

City	1991 population (millions)	Area (square miles)	Density (population per square mile)
Tokyo/Yokohama	27.2	1089	25,018
Mexico City	20.9	522	40,036
Sao Paulo	18.7	451	41,465
Seoul	16.8	342	49,099
New York	14.6	1274	11,479
Osaka/Kobe/Kyoto	13.7	495	28,024
Bombay	12.1	95	127,378
Calcutta	11.9	209	56,928
Rio de Janeiro	11.7	260	44,953
Buenos Aires	11.7	535	21,788

Table 1.5 Population Densities, Ordered

City	1991 population (millions)	Area (square miles)	Density (population per square mile)
Bombay	12.1	95	127,378
Calcutta	11.9	209	56,928
Seoul	16.8	342	49,099
Rio de Janeiro	11.7	260	44,953
Sao Paulo	18.7	451	41,465
Mexico City	20.9	522	40,036
Osaka/Kobe/Kyoto	13.7	495	28,024
Tokyo/Yokohama	27.2	1089	25,018
Buenos Aires	11.7	535	21,788
New York	14.6	1274	11,479

Table 1.4 gives the population density of each of the 10 cities. Table 1.5 orders the 10 cities according to their population density, from high to low. Now the picture for the cities has changed. Of the 10, Bombay is by far the densest city, with over 127,000 people per square mile—more than five times as crowded as Tokyo/Yokohama. New York, on the other hand, is the least dense of the 10, with about 11,000 persons per square mile. Thus we have brought out an interesting fact: it may be informative to view city population from the viewpoint of population density rather than total population. For example, as a quality-of-life index to be used in a public health study, density may be the more informative statistic to use. This example illustrates an important point: we often rescale or develop new indices (statistics) from the original data because those new indices are more useful or informative.

In closing this section, we note that the techniques of describing data in ways easy to capture the essence of the information in the data are often called **descriptive statistics.** By contrast, the techniques used to draw conclusions from data are often called **inferential statistics.** Good inferential statistics often occurs because it was preceded by a good descriptive statistical analysis of a data set. This textbook presents both aspects of statistics. The first systematic technique of descriptive statistics will be presented in the next section, namely, a graphical approach called **stems and leaves.**

SECTION 1.1 EXERCISES

1. Listed below are the medal standings of countries that participated in the 1996 Summer Olympics. The countries are ranked in order of total number of medals won.

	Total	Gold	Silver	Bronze
1. United States	101	44	32	25
2. Germany	65	20	18	27
3. Russia	63	26	21	16
4. China	50	16	22	12
5. Australia	41	9	9	23
6. France	37	15	7	15
7. Italy	35	13	10	12
8. South Korea	27	7	15	5
9. Cuba	25	9	8	8
10. Ukraine	23	9	2	12
11. Canada	22	3	11	8
12. Hungary	21	7	4	10
13. Rumania	20	4	7	9
14. Netherlands	19	4	5	10
15. Poland	17	7	5	5
Spain	17	5	6	6
16. Bulgaria	15	3	7	5
Brazil	15	3	3	9
Britain	15	1	8	6
Belarus	15	1	6	8
17. Japan	14	3	6	5
18. Czech Republic	11	4	3	4

Reference:
http://www.usatoday.com/olympics/olyfront.htm

a. Rank the countries by number of gold medals.
b. Find three countries whose position in the rankings goes up.
c. Find three countries whose position in the rankings goes down.

2. In Example 1.2 we should not conclude that Bombay is the most densely populated city in the world, because we only considered the most highly populated cities. Here are data on other large cities:

City	1991 population (millions)	Area (square miles)
Manila	10.2	188
Jakarta	9.9	76
Lagos	8.0	56
Hong Kong	5.7	23
Ho Chi Minh City	3.7	31

a. Calculate the population density for the cities above. (Round your answers to the nearest person.)
b. Construct a table combining these cities with the cities in Table 1.5.
c. What are the five most densely populated cities in your new table?

3. The density of New York State in 1990 was 361 persons per square mile. But the density of New York City (according to Table 1.5) is over

11,000 persons per square mile. Explain why the density of New York State is dramatically lower than the density of New York City.

4. *Infant mortality* refers to the number of infants that die for every 1000 births. *Life expectancy* is the number of years, on average, that a person can expect to live. Here are the infant mortality and the life expectancy in the United States for people born in the years 1920–1990:

	Infant mortality	Life expectancy
1920	86	54
1930	65	60
1940	47	63
1950	29	68
1960	26	70
1970	—	71
1980	13	74
1990	9	76

a. The infant mortality rate has decreased over this time period, and the life expectancy has increased. Give three reasons for these trends.
b. Predict the life expectancy for a person born in the year 2000.
c. The data for infant mortality rates for the decade 1970 are missing. Predict the value.

5. In a survey, teenagers were asked to respond to this statement: "My parents would be happy if I got a job at a fast-food restaurant."

Response	Percentage
Very true	21
Somewhat true	37
Somewhat untrue	20
Very untrue	19
Don't know	3

a. Overall, would you say that parents tend to be happy about their teenagers working in fast-food restaurants?
b. Do you think that the results would depend on the region of the country? Give at least two reasons to support your answer.

6. A national survey of Canadians asked about their level of education and their level of participation in physical activity. "Active" persons were defined as those who take part in three or more hours of physical activity per week for nine months or more during the year.

Amount of education	Percentage who are active
Elementary school only	41
Some secondary school	53
Secondary school or more	56
Certificate or diploma	58
University degree	63

According to this survey, fewer than one-half of the persons with only elementary school education were physically active, while nearly two-thirds of those with university degrees were physically active. Give at least two reasons why people with more education tend to be more physically active.

7. The states having the highest density of millionaires were reported as follows:

State	Number of millionaires per 1000 persons
Idaho	27
Maine	8
North Dakota	7
Nebraska	7
Minnesota	6
Indiana	5
Wisconsin	4
Iowa	4
New Jersey	4
Connecticut	3

a. The population of Idaho is approximately 1 million. Approximately how many millionaires live in Idaho?

b. From this table, can you conclude which state has the largest number of millionaires? Why or why not?

8. Here are 1986 infant mortality rates for five regions of the world. What factors could account for the differences in infant mortality rates?

Sub-Saharan Africa	132
South Asia	178
Near East and North Africa	152
Latin America and the Caribbean	95
East and Southeast Asia	104

1.2 STEMS AND LEAVES

Think back on the preceding example. We dealt with a list of cities giving their populations and areas. Usually the science of statistics does not concern itself with identifying the individuals that were measured—in our example, we might have a list of city areas or populations, but the names of the cities might not even appear. In statistics we are not interested in such questions as *which* city is the largest, the most populous, or the most densely populated. Such questions may be important or interesting, but they are dealt with outside of statistics. In statistics we are interested in the general aspects of the measured data. Are the population densities of the world's largest cities all more or less the same, or do they vary widely? What is the general shape, or distribution, of the city population densities? Is there some value of population density that we can consider typical or representative for the most populated cities?

In this book we will learn a variety of descriptive and inferential statistical techniques, some based on computing interesting statistics, such as the average population among the most populated cities, and others based on graphical techniques. A thorough statistical analysis usually includes both the computation of interesting statistics and the production of informative graphs. Such analysis is usually partly descriptive of the data and partly inferential, in that conclusions concerning the data are usually drawn.

The stem-and-leaf approach is a graphical method used to describe the general shape, or distribution, of the data. It is distinguished by the fact that the value of every piece of data used to construct the graph can in fact be read off the graph.

When we have unorganized pieces of numerical information, it is very helpful (and sometimes necessary) to organize them so that the reader can quickly see certain aspects of the data. Table 1.6 provides the number of stories in 14 of the tallest buildings in the United States and Canada. These data are ordered alphabetically by city. Thus, the table is not designed to enable us to see important patterns in the data. For example, if we want to

Table 1.6 **Notable Buildings of the United States and Canada**

City	Building	Number of stories
Atlanta	Peachtree Center	63
Austin	One American Center	32
Boston	John Hancock Tower	60
Calgary	Petro Canada Center	52
Chicago	Sears Tower	110
Cleveland	Society Center	57
Detroit	Westin Hotel	71
Edmonton	Manulife Place	36
Honolulu	Imperial Plaza	40
Los Angeles	First Interstate World Center	73
New York City	World Trade Center I	110
Seattle	Columbia Seafirst Center	76
Toronto	First Canada Place	72
Vancouver	Royal Center Tower	36

see at a glance the largest number and the smallest number of stories, we would have to reorder the data, as we did with the city population data in Section 1.1. Nor can we easily see from such a table how the building heights are grouped. Are they tightly clustered around one value, or are they spread out over a wide range of values? Is one height typical? What is the general shape of the data? We cannot easily tell from the table.

There is another way of organizing data that is easy to do, and it makes the data easier to understand and discuss. It is the **stem-and-leaf plot,** one of the standard graphical techniques used in statistics. The stem-and-leaf plot will be seen below to be a special kind of *histogram,* which is perhaps the most widely used graphical display of tables of numbers, such as city populations or building heights. So it is important to realize that introducing the stem-and-leaf plot is this textbook's way of introducing histograms. The stem-and-leaf plot's main virtue is that no information about the data is lost while it carries out its desired role of showing the statistician the shape of the data, that is, how frequently the data occur in various ranges of values. This easy way of organizing data was developed by the well-known American statistician John Tukey of Princeton University.

Here is how the stem-and-leaf plot works for the above data set. The first step is to group the data by 10s, 100s, or 1000s, and so on, so that we obtain a reasonable number of "stems," say, 5 to 15 of them. Thus in this case we see from the data in Table 1.6 that we can group them by 10s. That is, we can group the 30s together, the 40s together, and so on, using a list such as the following. We start with a 1, which stands for the group of buildings having from 10 to 19 stories. The 2 stands for the group from 20 to 29 stories high. These numbers are used to form the first column in the stem-and-leaf plot below. These numbers that stand for 10, 20, 30, and so on, are called *stems.* Now we start writing down the data on the number of stories in the buildings. Let's start with these two buildings:

Peachtree Center, Atlanta	63 stories
One American Center, Austin	32 stories

We list those two data points as follows:

Stem	Leaf
1	
2	
3	2
4	
5	
6	3
7	
8	
9	
10	
11	
12	

For 63, we have separated the 6, a stem (which in this table stands for 60), from 3, which is called a *leaf.* In order to read this table, we need a key. Here is a key to this table:

Key: "6 3" in the above stem-and-leaf plot stands for 63.

The key tells us how to read the table. Here the key reminds us that in this table the stems are 10s and the leaves are units. When you make a stem-and-leaf table, you should always provide a key.

We now add the rest of the data. The completed stem-and-leaf plot is presented in Table 1.7.

A stem-and-leaf plot has two very attractive properties. First, it contains *all* the numerical data that were in the original table. That is, we can write down exactly each data point from the stem-and-leaf diagram. In some other kinds of plots the numbers that were used to make the plot are not recoverable from the plot—if you want to read the actual numbers, you must go back to the original list of data (which may not be possible if, for example, you have come across the graph in a publication). Second, and

Table 1.7 Stem-and-Leaf Plot of Number of Stories of Notable Buildings

Stem	Leaf
1	
2	
3	2,6,6
4	0
5	2,7
6	3,0
7	1,3,2,6
8	
9	
10	
11	0,0
12	

Key: "6 3" stands for 63.

powerfully, we now see the general centering, spread, and shape of the data *at a glance.* For example, they seem to center around 50 stories, and they are mostly spread between 30 and 70 stories but have two extreme values (later to be called *outliers*) at 110 stories.

We sometimes need to use certain tricks to construct stem-and-leaf plots for special sets of data. Consider the following data.

Example 1.3 Newly Minted Coins

A sample of 25 pennies from the U.S. mint were weighed on a balance. Here are the results, in centigrams ($\frac{1}{100}$ of a gram):

301	304	305	302	312
314	311	316	320	314
308	309	310	310	313
316	317	312	313	311
310	307	309	312	317

A stem-and-leaf can be used to display these weights in a useful manner. As in the example above, we can group the data by 10s, using 300–309, 310–319, and 320–329. Thus the stems become 30, 31, 32. See Table 1.8.

We see that the range of measurements is rather small—from a low weight of 301 centigrams to a high weight of 320 centigrams. Thus, there is a problem with this table. The data are scrunched together with only three stems, so much so that

Table 1.8 **Stem-and-Leaf Plot of Weights of Pennies**

Stem	Leaf
30	1,8,4,9,7,5,9,2
31	4,6,0,1,7,6,0,2,0,3,2,2,4,3,1,7
32	0

Key: "30 1" stands for 301.

we find it hard to get a good idea about the general shape of the data, for example. We can overcome this difficulty by stretching out the table. But first, let's put the leaves of the table in order. If time permits, that is usually a desirable step, even if we don't plan to split the stem intervals.

Stem	Leaf
30	1,2,4,5,7,8,9,9
31	0,0,0,1,1,2,2,2,3,3,4,4,6,6,7,7
32	0

We can divide each interval into two parts. (Clearly, the only other choice is to split the interval 300–309 into five parts: 300–301, 302–303, and so on. This seems to create too many stems.) For example, for the first row we can write the following:

Stem	Leaf
30	1,2,4
30*	5,7,8,9,9

The asterisk (*) means that the interval is continued from the previous row; that is, we now have two rows with the same stem. So the top row includes values from 300 to 304, and the second row includes the remaining values, in the interval from 305 to 309. Table 1.9 is the expanded table. We now get a much better look at the spread of the data, and in particular we find that the quite narrow interval from 310 to 314 contains about one-half of all the weights (12 out of 25). We also see a pattern, very typical in data sets, in which many of the data pile up in the middle, providing a somewhat bell-shaped stem-and-leaf plot. Note that this bell-shaped nature of the data was not evident from Table 1.8.

Some standard statistical computer programming packages for practitioners have a stem-and-leaf program. Clearly, the user has to decide on the stem intervals. We will not get into a detailed discussion of such judgments,

Table 1.9 **Expanded Stem-and-Leaf Plot of Data of Table 1.8**

Stem	Leaf
30	1,2,4
30*	5,7,8,9,9
31	0,0,0,1,1,2,2,2,3,3,4,4
31*	6,6,7,7
32	0
32*	

Key: "32 0" stands for 320.

but some of the exercises that follow help with this. The important thing is to be familiar with this powerful and often used tool.

SECTION 1.2 EXERCISES

1. The following scores were obtained by 30 students on a final exam in statistics. Construct a stem-and-leaf plot of these data.

51	46	31	35	37	51	56	51	43	48
33	42	37	27	57	65	36	37	55	42
33	49	31	46	50	57	52	35	38	47

2. The following weights of a group of rats used in an agricultural research project are reported to the nearest gram:

206	202	200	215	191	193	196	202
204	191	190	215	188	196	190	182
205	192	194	201	207	205	203	211
198	206	203	192	215	195	210	216
206	208	190	197	210	220	220	211

   a. What stems would you use for recording these data?
   b. Construct a stem-and-leaf plot of these weights using only five stems.
   c. Now construct a stem-and-leaf plot of the weights using 10 stems. (*Hint:* Look at Example 1.3.) Which plot gives you a better picture of the data?

3. The stem-and-leaf plot that follows gives stopping distances (in meters) of 10 test cars going 50 kilometers per hour when the brakes were applied. What are the original data from which the table was made?

Stem	Leaf
6	0,1,4,4,8
7	1,3,5
8	0,2

4. The following stem-and-leaf plot gives final scores on a statistics test for 20 students. From the table, write down each student's score.

Stem	Leaf
7	2,3,4,5,9
8	0,2,3,4,4,5,7,8
9	0,1,2,3,5,6,6

5. Using the stem-and-leaf plot of Exercise 4, answer the following questions:
   a. What was the lowest score obtained?
   b. What was the highest score obtained?
   c. How far apart were the lowest and highest scores? (This statistic is called the *range* of the data.)

6. The following table gives the high temperatures (°F) for 20 cities on April 3, 1997.

Asheville, North Carolina	72
Champaign, Illinois	68
Indianapolis, Indiana	69
Abilene, Texas	67
Los Cruces, New Mexico	72
Los Angeles, California	71
Billings, Montana	56
Chicago, Illinois	65
Albany, New York	53
Charleston, South Carolina	77
Miami, Florida	77
Birmingham, Alabama	74
Iowa City, Iowa	63
Detroit, Michigan	58
Molokai, Hawaii	77
St. Louis, Missouri	76
Lincoln, Nebraska	63
Boston, Massachusetts	53
Baltimore, Maryland	66
Boise, Idaho	60

a. Construct a stem-and-leaf plot of the high temperatures.
b. How far apart were the lowest and highest temperatures for these cities?

7. Make a stem-and-leaf plot for the following data:

152	387	482	627	109	182
523	610	240	267	395	

(*Hint:* Make the stems to represent the 100s place.)

8. Refer to Table 1.5 on population densities.
a. What stems would you use to make a stem-and-leaf plot?
b. Draw a stem-and-leaf plot for the data.
c. What does this plot tell you about what the data looks like?

1.3 CIRCLE GRAPHS AND LINE PLOTS

When we describe data, we can take advantage of the adage that a picture is worth a thousand words. We consider three kinds of graphs used to visually communicate ideas from data: circle graphs, line plots, and bar graphs (histograms). In this section we discuss circle graphs and line plots. In the next section we introduce histograms.

Usually data consist of measurements of something, such as people's heights, corn crop yields (per acre), annual exports (in dollars) of countries, and populations of cities. Another important type of data besides measurement data is categorical data. Here each individual (person, farm, country, and so on) is classified into one of a small number of categories instead of being measured in some manner. For example, populations of people can be classified by gender (female, male), religious preference (Christian, Jewish, Muslim, or other, in the United States), type of living area (urban, suburban, rural), and so on. Sometimes the categories are created from measurements, as when sociologists classify us into socioeconomic classes (upper, middle, and lower class) on the basis of our (measured) incomes.

The proportion of individuals in each category can be very important. Although one could graphically display this information using bar graphs, a very natural way is to use circle graphs.

Circle Graphs

A **circle graph,** sometimes called a *pie chart,* shows the relative proportions for categorical data. Such a graphical representation is useful when it is important for the reader to compare the proportions of the various categories.

Example 1.4

Math Help

In a national survey in the United States, eighth-grade students were asked if their parents were interested in helping them do mathematics. Here are the results:

"My parents are interested in helping me do mathematics."	
Agree	63%
Undecided	21%
Disagree	16%

Source: F. J. Crosswhite, et al., *Second International Study of Mathematics* (Champaign, Ill: Stipes, 1986), p. 390.

These results are shown in the circle graph in Figure 1.2.

It is easy to construct a circle graph from a set of percentages that add to 100%. Each percentage is converted to an angle by multiplying by

$$\frac{360 \text{ degrees}}{100\%} = 3.6$$

For example,

$$\begin{aligned} 63 \times 3.6 &= 227 \text{ degrees} \\ 21 \times 3.6 &= 76 \text{ degrees} \\ 16 \times 3.6 &= 57 \text{ degrees} \end{aligned}$$

We have rounded down once to make the angles sum to exactly 360 degrees. Then manually, with a protractor and compass, or with appropriate software it is easy to draw the circle graph.

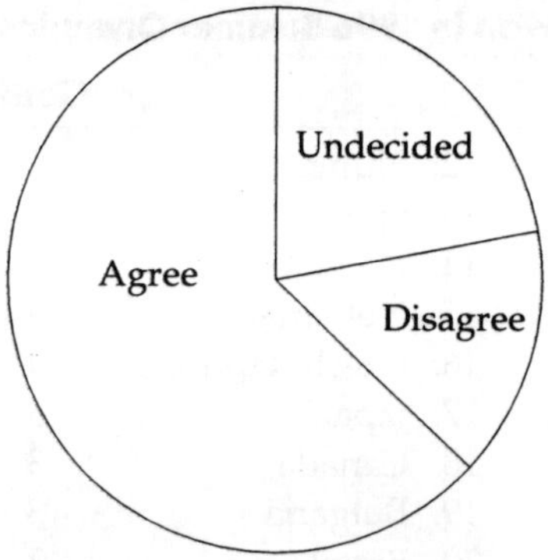

"My parents are interested in helping me do mathematics."

Figure 1.2 **Circle graph of data on perceived parent interest in helping with math homework.**

Line Plots

The ordered nature of a set of numbers sometimes needs to be emphasized and visually displayed. For example, questions about the largest, smallest, and middle (median) data points are important in statistics. A very natural way to display data to emphasize their order is in a **line plot.** As you will see, a line plot also informs us about the general shape of the data, although often not as effectively as a well-designed histogram, as will be discussed in Section 1.4

Consider the data set in the following example.

Example 1.5 **The 1996 Summer Olympics**

The number of gold medals won in the 1996 Summer Olympics are given in Table 1.10. We can plot these data on a *number line* as follows. The mark × is placed on the number line to correspond to each data point in the table. For example, the 3 gold medals for Canada and 13 for Italy are plotted below.

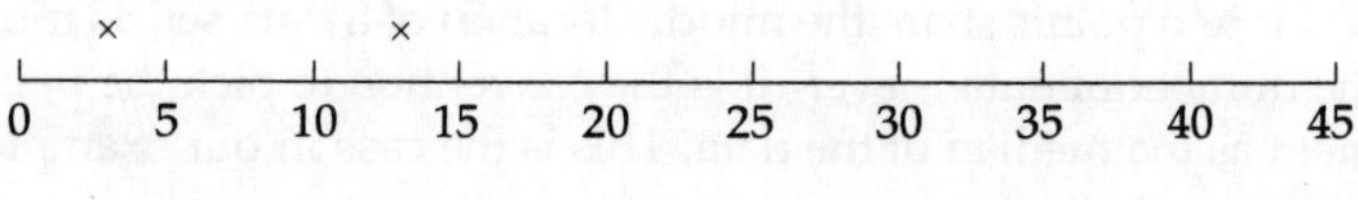

Table 1.10 **Gold Medals Won in 1996 Summer Olympics**

	Gold		Gold
1. United States	44	12. Hungary	7
2. Russia	26	13. Spain	5
3. Germany	20	14. Rumania	4
4. China	16	15. Netherlands	4
5. France	15	16. Czech Republic	4
6. Italy	13	17. Japan	3
7. Ukraine	9	18. Canada	3
8. Australia	9	19. Bulgaria	3
9. Cuba	9	20. Brazil	3
10. South Korea	7	21. Britain	1
11. Poland	7	22. Belarus	1

All the data are now shown in the complete line plot below.

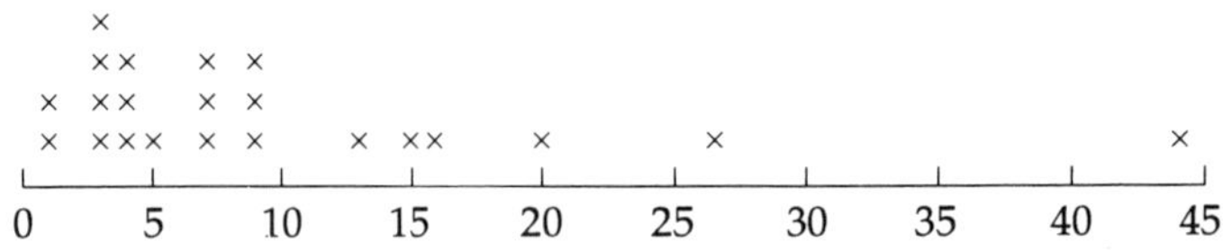

Here are some statements we can quickly make about the order of the Olympics data, based on the line plot above. The smallest and the largest number of gold medals are easily obtained:

Smallest number: Two countries won 1 gold medal each.

Largest number: One country won 44 gold medals.

The number of medals *ranged* from the smallest number (1 medal) to a largest number (44 medals). Thus, the **range** is the *difference* between the largest and smallest numbers in a table (or a line graph) of numbers. That is,

$$\text{Largest number} - \text{smallest number} = 44 - 1 = 43 \text{ medals}$$

The middle point of a set of data is called the **median.** That is, the median is the number on the line plot such that the same number of data points are above it as below it. Two countries are in the middle of the set of data (that is, at the midpoint of the data): countries 11 and 12 (Poland and Hungary), each with seven gold medals. The first 11 countries in the line plot won seven or fewer gold medals, and the next 11 won seven or more gold medals. Thus 7 is the median.

If two points share the middle location of a data set, as must be the case when the number of data is even, it is the convention to pick the point halfway between them as the median of the data. This is the case in our example, since there are 22 countries. The middle two countries both had 7, and the midpoint between 7 and 7 is of course 7. If the middle two countries had 7 and 9 gold medals, then we would have selected 8 as the median, according to the convention.

SECTION 1.3 EXERCISES

1. In a national survey, men were asked why they exercise. Here are the reasons they gave:

Reason	Percentage responding
Health	51
Stress relief	25
Weight loss	20
Other	4

Draw a circle graph of these data.

2. Toss a coin 30 times and record the numbers of heads and tails. Display your result as a circle graph. (*Hint:* To draw a circle graph, you will need to change your results into percentages.)

3. Draw a line graph for the high temperatures in Exercise 6 of Section 1.2. Use the graph to answer the following questions:
   a. What are the highest and lowest temperature?
   b. What is the range of the data?
   c. What do the center and spread of the data tell you about the weather on this day?

4. Draw a line graph for the weights of rats used in agricultural research in Exercise 2 of Section 1.2. Use the graph to answer the following questions:
   a. What are the highest and lowest weights?
   b. What is the range of the data?
   c. Find the median, or middle number, of the data.

Graphing Calculator Exercises

5. Table 1.6 shows notable buildings of the United States and Canada. Enter the city names in alphabetical order into L1 and the number of stories into L2. Plot a line graph in statplot 1 (see Section 1.3 in the TI Graphing Calculator Supplement). Set up the statplot as shown. Use TRACE to read from the calculator which notable building has the most stories and which has the fewest.

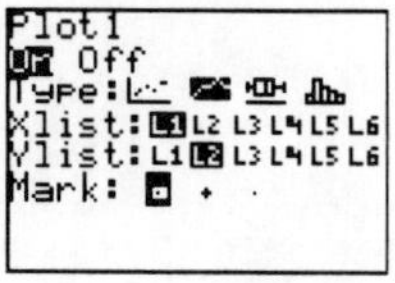

6. Draw a line plot of the data given in Table 1.10. Instead of the country name, enter into L1 its number in the table (for the United States, enter 1; for Russia, enter 2; and so on), and enter the number of gold metals won into L2. Use TRACE to read the number of gold metals won by each country.

1.4 FREQUENCY HISTOGRAMS

In statistics, a bar graph is called a **frequency histogram,** or just **histogram.** One axis of the graph tells the scale of the measurements involved, and the other axis shows how many times each of the measurements occurred. That is, it gives the number of data points, or **frequency,** of the observations in a certain range. Thus, when statisticians say "frequency," they simply mean the number of data points. Since it is the frequency of the data occurring in various locations over the range of the data that determines the overall shape of a data set, the concept of frequency is very useful and important. The histogram is the most common way to graphically display data. It loses some information (the exact value of each data point, that is, the leaf), but it is extremely informative about centering, spread, and shape. These are the three aspects of data we usually care most about.

Our first example of a histogram is the one presented in Figure 1.3. The amount of tar in a cigarette (measured in milligrams) was measured for 24 brands of cigarettes. The graph shows the number (frequency) of those 24 cigarette brands that contain given amounts of tar. For example, the leftmost bar represents the number of cigarette brands that contain

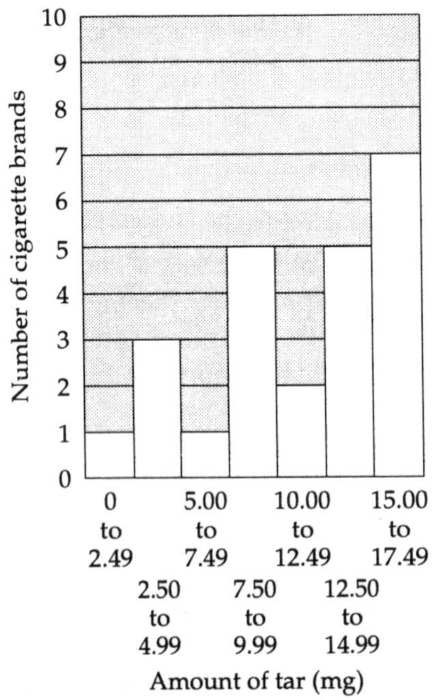

Figure 1.3 **Histogram of tar content of selected brands of cigarettes.**

Table 1.11 **Frequency Table of Tar (mg) per Cigarette**

Interval	Frequency (f)
0–2.49	1
2.50–4.99	3
5.00–7.49	1
7.50–9.99	5
10.00–12.49	2
12.50–14.99	5
15.00–17.49	7
Total	24

from 0 to 2.49—or, roughly 0 to $2\frac{1}{2}$—milligrams of tar per cigarette. That bar reaches to a height of 1, meaning that one brand has a tar content in that range. We read the other bars, corresponding to tar ranges $2\frac{1}{2}$ to 5, 5 to $7\frac{1}{2}$, and so on, in a similar manner.

How did we construct this histogram? The first step was to take the "raw" data, or original 24 pieces of data, and construct a frequency table. Here are the raw data:

1.1	4.0	4.5	4.9	7.2	7.5	7.9	8.3
8.9	9.0	11.6	12.1	12.5	12.9	13.3	14.1
14.7	15.0	15.0	15.3	15.8	16.2	16.8	17.3

From these data and using intervals of width 2.50, we obtain the needed frequency table, shown in Table 1.11. Then we simply convert this frequency table into Figure 1.3 in the obvious way.

What does the histogram of Figure 1.3 tell about the data—that is, tell us about tar content? We see that although some of the 24 brands have relatively low amounts of tar, more of the brands are near the high end of the range of tar contents measured. Further, we see that none of the brands has a tar content greater than 17.49 mg. Thus the graph shows the frequencies of the data fully in various intervals, and this shows us the overall shape of the data.

We might ask why there is no value exceeding 17.49 mg. Does that value represent the natural tar content of tobacco, and does the shape of the graph (high to the right, low to the left) arise from the fact that some manufacturers remove tar from their tobacco? To answer such questions, we must leave statistics; we might ask a chemist, for example.

What if, instead of using the ranges—or *intervals*—0 to $2\frac{1}{2}$, $2\frac{1}{2}$ to 5, and so on, we had used the broader ranges 0 to 5, 5 to 10, and so on? Of course we could construct a new frequency table, starting from the original data,

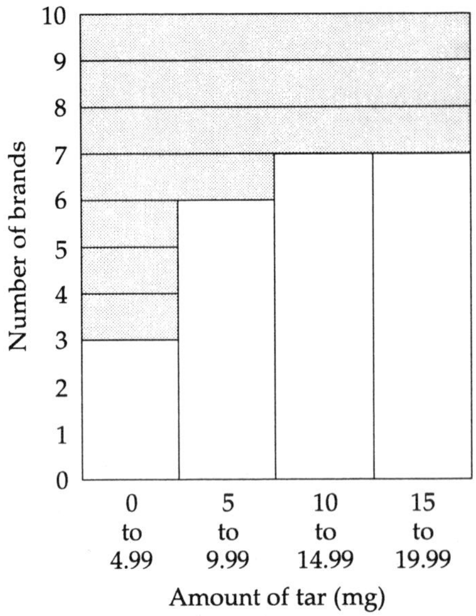

Figure 1.4 **Histogram derived from that of Figure 1.3 by combining bars.**

as we did above. But we can now more easily work directly from Figure 1.3. Looking at our first group, it is clear that the number of cigarette brands whose cigarettes contain between 0 and 4.99 mg of tar is 3. Similarly, the number of brands having a tar content from 5 to 9.99 mg is 6. We continue in this manner and create the histogram in Figure 1.4.

Have we lost or gained anything by graphing the data this way? We see that the rectangles still increase in height as we move from low tar content to high tar content, so our new graph still communicates the fact that more of the 24 brands have higher tar content than lower. But the heights increase more gradually than in Figure 1.3. And the two rightmost rectangles in Figure 1.4 are of the same height, meaning that just as many of the brands have a tar content between 10 and 15 mg as between 15 mg and 20 mg. Judging from Figure 1.4 alone, we would not be able to say whether, within the range from 10 mg to 20 mg, more of the cigarette brands had higher tar content. All this suggests that by consolidating the bars of Figure 1.3—that is, by combining the intervals—we have lost information that we would prefer to keep. For these data we would do better to leave the graph as it is in Figure 1.3.

It is also possible to graph a greater number of bars than is useful. For example, is the histogram of Figure 1.5 very useful in displaying the general shape of the data set it was constructed from? If we choose wider bars, we

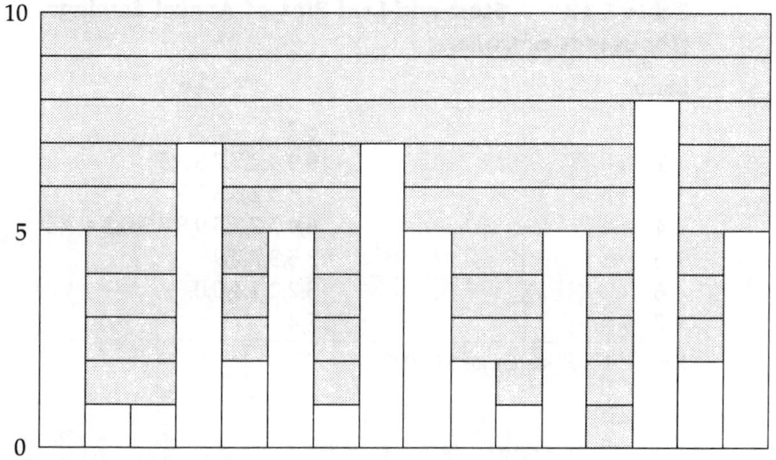

Figure 1.5 **A histogram with too many bars.**

obtain the histogram of Figure 1.6. Thus we discover that the shape of the histogram is rather flat over the range of the data—a very useful piece of information. Clearly, Figure 1.5 hid from us this important fact about the approximate flatness in the shape of the data.

As a general rule of thumb, 5 to 15 bars are usually appropriate. Too few fail to display the shape of the data, as we saw above. Too many lead to a wide variation of bar heights that hides the general shape of the data,

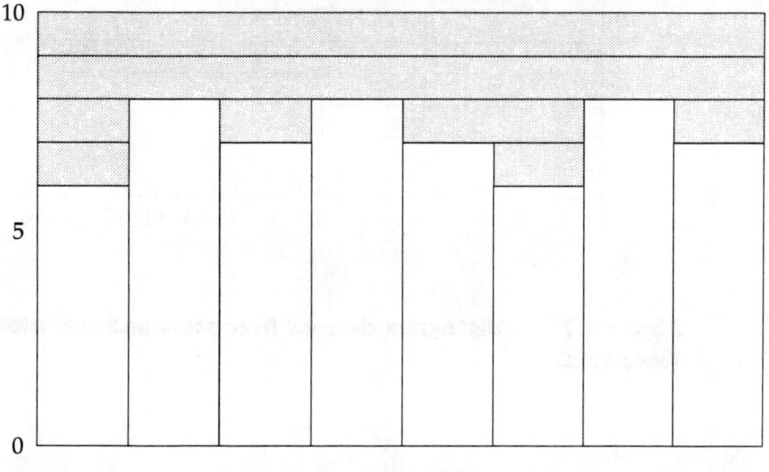

Figure 1.6 **Histogram derived from that of Figure 1.5 by combining bars.**

Table 1.12 **Stem-and-Leaf Plot of Annual Earnings (Thousands of Dollars)**

Stem	Leaf
1	6,7
2	9,9,9,2,7,7,8,3,5
3	4,5,1,2,6,6,3,9
4	6,9,7,7,7,5,9,8,6,3,0,3,4,8,3
5	3,5,3,7,7,9
6	3,2,2,4,4,0,0
7	7,4

Key: "1 6" stands for $16,000.

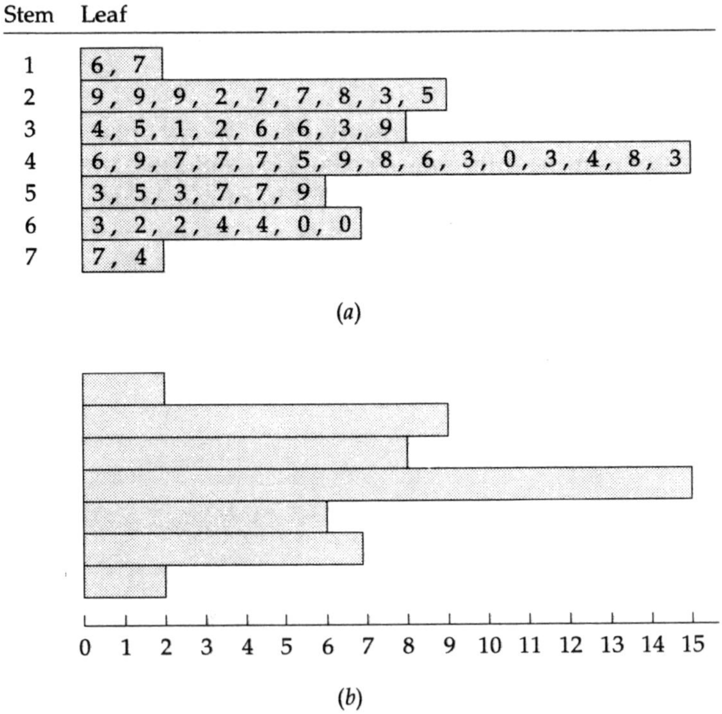

Figure 1.7 **Histogram derived from stem-and-leaf plot of Table 1.12.**

also as we saw above. Clearly, the more data points there are, the more bars are appropriate. For example, 200 data points could allow 15 bars, whereas 30 data points might be best displayed with 5 to 6 bars. This decision is a judgment call on the part of the user; rigid rules about this are not appropriate.

Histograms can be readily made from stem-and-leaf tables, which are basically just a special kind of histogram. Let's look at the stem-and-leaf table of annual earnings to the nearest $1000 in Table 1.12. By simply enclosing each row (set of leaves) in a bar, we have a respectable histogram. See Figure 1.7*a*. We can also count how many leaves there are for each stem and redraw our histogram showing the frequencies (how many leaves there are) along one axis of the graph. See Figure 1.7*b*. It is more common to use vertical bars instead of horizontal bars in bar graphs. In that case, our histogram of the annual earnings looks like the one in Figure 1.8. Note that the horizontal axis scale (salaries in $1000s) has been included, too; we put

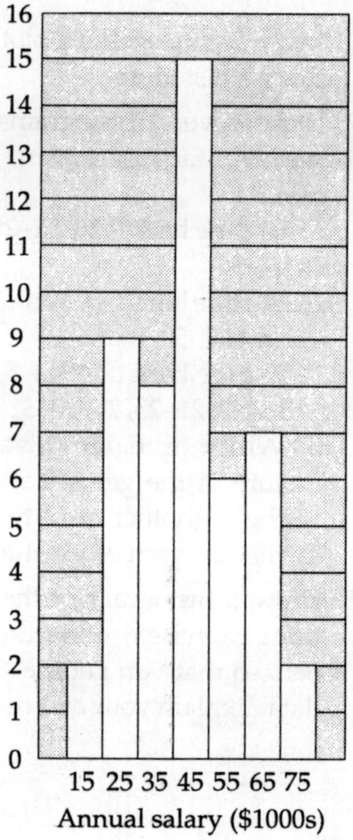

Figure 1.8 **Histogram of Figure 1.7 put in usual position.**

in the scale of the horizontal axis so that it is clear which income each bar represents. Thus, as stated above, a stem-and-leaf plot is really a special case of a histogram.

SECTION 1.4 EXERCISES

1. What is one main disadvantage in using frequency histograms, or, for short, histograms, to represent data? What are two advantages in using histograms to represent data?
2. Draw a histogram of the data given in the stem-and-leaf plot below. The data (from Exercise 4 of Section 1.2) are the scores of 20 students on a statistics test.

Stem	Leaf
7	2,3,4,5,9
8	0,2,3,4,4,5,7,8
9	0,1,2,3,5,6,6

3. Draw the stem-and-leaf plot in Exercise 2 with six stems instead of three. Draw a histogram of the data. Does the addition of more stems give a clearer picture of the data?
4. Draw a histogram of the data in Exercise 2 of Section 1.2 using five bars: 180–189, 190–199, and so on. Now draw a separate histogram with each bar split in half: 180–185, 186–190, and so on. Which histogram gives you a clearer picture of the data?
5. Toss a coin 30 times and keep track of the numbers of heads and tails. Construct a histogram to show your results. (*Hint:* Your histogram will have only two bars.) If you perform the experiment again, would you expect to get the same results? Why or why not?
6. Fingerprints can be classified according to the number of ridges between "loops" in the patterns. The number of ridges is called the *ridge count* for a particular person. Suppose we have the following ridge counts for the fingerprints of 20 people:

189	181	205	210	198
207	201	185	188	192
186	189	192	194	215
205	213	207	213	220

Draw a stem-and-leaf plot of these data. Then draw a histogram.

7. Make several histograms of the total gold medals standings from Exercise 1 of Section 1.1.
   a. Use the bars 1–10, 11–20, 21–30, 31–40, and 41–50.
   b. Use the bars 1–5, 6–10, 11–20, 21–30, 31–40, and 41–50.
   c. Use the bars 1, 2, 3, 4, 5, 6, 7, 8, 9, 10, 11–15, 16–20, 21–25, 25–30, 31–40, and 41–50.
   d. Which histogram gives you the best picture of the general shape of the data? Is there another way to assign the bars so that the picture is better?
8. Draw a histogram of the high temperatures from Exercise 6 of Section 1.2, choosing the bars so that you get the "best" picture of the data. Explain your choice of bars.

Graphing Calculator Exercises

9. Draw a histogram of the data given in the stem-and-leaf plot below. Enter values into data table L1 (see Section 1.1 in the TI Graphing Calculator Supplement). Plot the graph in statplot 1 (see Section 1.4 in the TI Graphing

Calculator Supplement). Set up the statplot and the window as shown:

Stem	Leaf
7	3,5,2,4,9
8	7,0,4,4,5,3,2,8
9	6,3,6,5,2,0,1

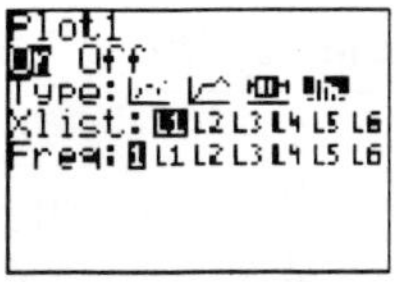

10. Table 1.6 shows notable buildings of the United States and Canada. Enter the city names in alphabetical order into L1. L2 should contain the corresponding number of stories. Plot the graph using statplot 1. Set up the statplot as shown. Use TRACE and read from the calculator which city's notable building has the greatest number of stories. What has the smallest number?

11. Draw a histogram of the data given in Table 1.10. Instead of the country name, enter into L1 its number in the table (for the United States, enter 1; for Russia enter 2; and so on), and enter the corresponding number of gold metals won into L2. Use TRACE to read the number of gold metals won by each country.

1.5 RELATIVE FREQUENCY HISTOGRAMS AND POLYGONS

As we have seen, the histogram is a powerful graphical tool for showing the general shape, or distribution, of a data set. Consider the frequency table of 30 exam scores given by Table 1.13. We first note that we can rescale these frequencies relative to the total number of data points, 30, by dividing each frequency by 30. Doing so converts them to proportions. In this textbook such proportions are thought of as probabilities, and in fact they are called **experimental probabilities** and are denoted by P. We will study them in detail in Chapter 5. Table 1.14 augments Table 1.13 by displaying the experimental probabilities in the column headed by P. Note that these experimental probabilities add to 1.

It is very useful to use these experimental probabilities to construct a histogram in which the total area of all the rectangles is 1. Such a probability-oriented histogram is called a **relative frequency histogram.** The relative

Table 1.13 **Frequency Table of Exam Scores**

Score	Frequency (f)
75	1
76	0
77	2
78	3
79	4
80	11
81	5
82	3
83	1
Total	30

Table 1.14 **Proportions of Persons Obtaining Various Exam Scores, Used for Figure 1.9**

Score	f	P
75	1	0.03
76	0	0
77	2	0.07
78	3	0.10
79	4	0.13
80	11	0.37
81	5	0.17
82	3	0.10
83	1	0.03
Total	30	1.00

frequency histogram for Table 1.14 is shown in Figure 1.9. The sum of the areas of all the rectangles is 1, as planned, because the P's sum to 1 and each rectangle is of width 1 and height P.

A slight modification of this relative frequency histogram, called a **relative frequency polygon,** is often used in this textbook. We obtain the relative frequency polygon from the relative frequency histogram simply by drawing line segments joining the middles of the tops of the rectangles of the relative frequency histogram. To see this, first place a dot in the middle of the top of each rectangle in Figure 1.9, thus obtaining Figure 1.10. Now join these dots, forming the relative frequency polygon given in Figure 1.11.

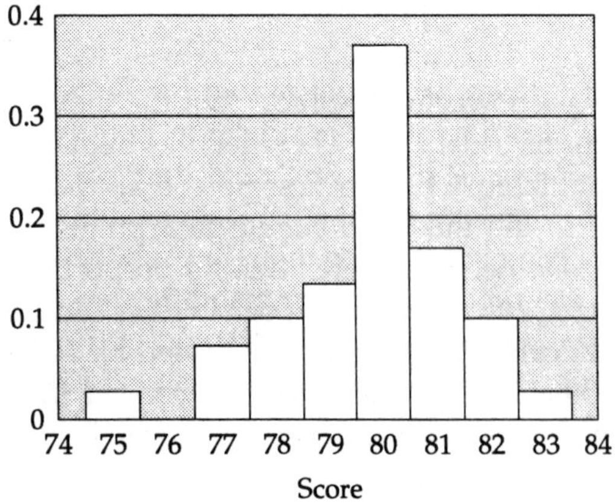

Figure 1.9 **Relative frequency histogram derived from data in Table 1.14.**

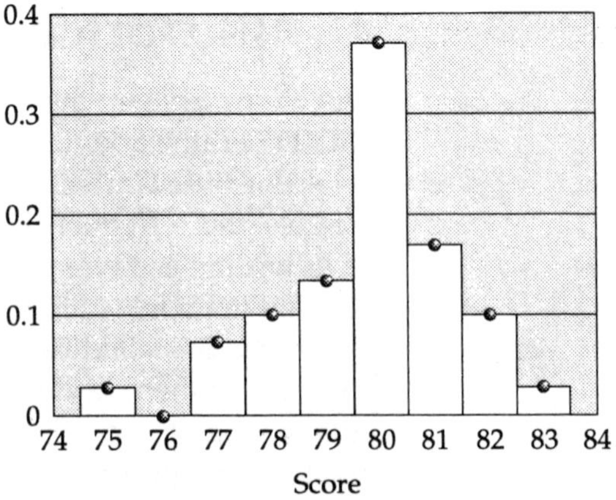

Figure 1.10 **Relative frequency histogram showing midpoints of bars.**

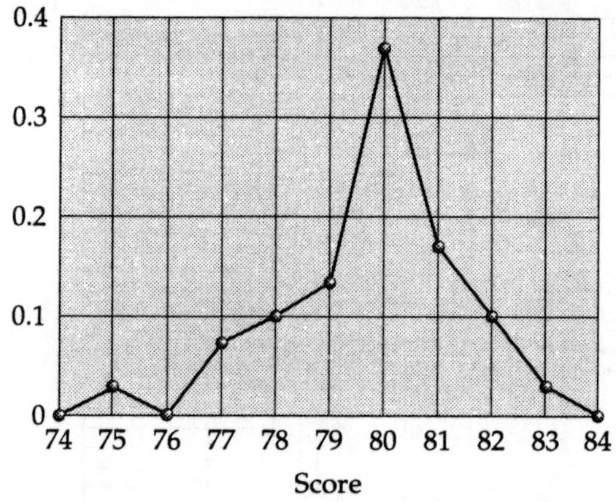

Figure 1.11 **Relative frequency polygon constructed by joining midpoints of histogram bars of Figure 1.10.**

The area under a relative frequency polygon will be close to 1, because the area of the relative frequency histogram it was constructed from is exactly 1.

As another example, consider the stem-and-leaf plot of the weights of 50 men in Table 1.15. First we form the frequency table, including the experimental probabilities, as shown in Table 1.16. Now let's construct the relative frequency histogram and associated relative frequency polygon. We are tempted to construct rectangles of heights given by the P's. But the bases of these rectangles are each going to be of width 10. For example, the first rectangle will start at 130 and end at 140. Thus in order for the areas of the rectangles to add to 1, we need to make the height of each rectangle $P/10$,

Table 1.15 **Stem-and-Leaf Plot of Weights of 50 Men**

Stem	Leaf
13	8,9,9
14	2,9,8,2
15	5,3,6,0,8,3,0,7,1,0
16	7,4,5,4,8,4,9,0,2
17	2,0,9,7,5,8,1,5
18	2,8,7,5,9,2,4,2,1,7,5,2
19	8,6
20	2,3

Key: "13 8" stands for 138 pounds.

Table 1.16 **Experimental Probabilities of Weight Data from Table 1.15, Used for Figures 1.12 and 1.13**

Stem	Leaf	Frequency	P
13	8,9,9	3	0.06
14	2,9,8,2	4	0.08
15	5,3,6,0,8,3,0,7,1,0	10	0.20
16	7,4,5,4,8,4,9,0,2	9	0.18
17	2,0,9,7,5,8,1,5	8	0.16
18	2,8,7,5,9,2,4,2,1,7,5,2	12	0.24
19	8,6	2	0.04
20	2,3	2	0.04
Total		50	1.00

Key: "13 8" stands for 138 pounds.

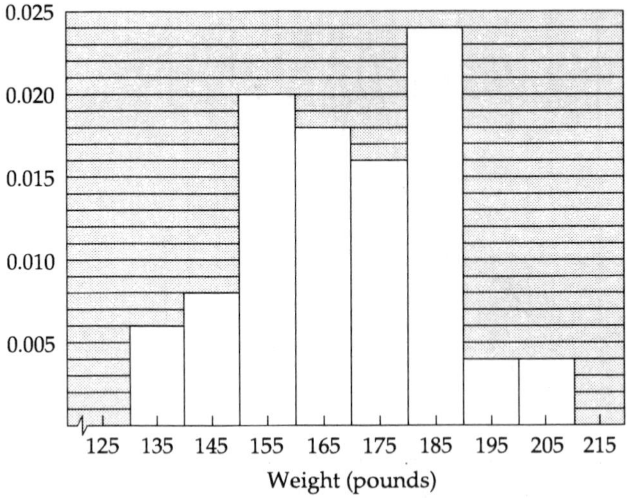

Figure 1.12 **Relative frequency histogram of 50 men's weights derived from Table 1.16.**

because each rectangle has

$$\text{Area} = \text{base} \times \text{height} = 10 \times \frac{P}{10} = P$$

Hence the sum of the areas of the rectangles is the sum of the P's, which is 1. Our relative frequency histogram is given in Figure 1.12.

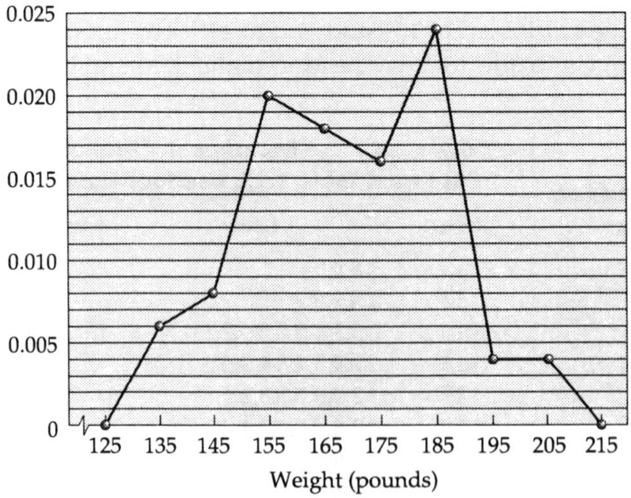

Figure 1.13 **Relative frequency polygon constructed from relative frequency histogram of Figure 1.12.**

Now it is easy to construct the relative frequency polygon, given in Figure 1.13.

SECTION 1.5 EXERCISES

1. This table shows the number of children per household for 50 households interviewed in a survey.

Number of children	f	Proportion
0	3	
1	8	
2	26	
3	10	
4	2	
5	0	
6	1	
Total	50	

   a. Complete the third column to show the proportion of families with each number of children.
   b. Draw a relative frequency histogram of the data.
   c. From your graph, find the proportion of families having two or fewer children.

2. The following gives the number of telephone calls of various durations (to the nearest minute), received by a doctor's office.

Length of call (minutes)	Number of calls
1	18
2	29
3	23
4	15
5	8
6	2
7	1
8	3
9	1

   a. Construct a relative frequency histogram of the data.
   b. From your graph, find (i) the proportion of telephone calls lasting 3 minutes or less, (ii) the most frequent, or common, length of telephone call, and (iii) the proportion of telephone calls lasting more than 6 minutes.

3. A six-sided die is rolled 120 times, and the following outcomes are obtained:

Outcome	f	Proportion
1	15	
2	21	
3	23	
4	19	
5	17	
6	25	
Total	120	

   a. Draw a relative frequency histogram of the data.
   b. From the graph, find (i) the proportion of rolls that give 3 or less, (ii) the proportion of rolls that give 5 or more, and (iii) the proportion of rolls that give an even number.

4. Consider the data in Exercise 2 of Section 1.2.
   a. Draw a relative frequency histogram of the data using the five bars 180–189, 190–199, and so on. (*Hint:* Do not forget to divide each bar's proportion by 10, the number of units in each bar.)
   b. Draw a relative frequency polygon for the data in part (a) in order to obtain a graph of the general slope of the data.

5. Draw a relative frequency histogram of the numbers of gold medals given in Exercise 1 of Section 1.1. Use the bars 1–5, 5–10, 11–20, 21–30, 31–40, and 41–50. (*Hint:* Do not forget to divide each bar's proportion by the number of units in the bar.)

6. Consider the data in Exercise 4 of Section 1.2.
   a. Draw a relative frequency histogram of the data using the bars 70–79, 80–89, and 90–99.
   b. Draw a relative frequency polygon of the data in part (a).

Graphing Calculator Exercises

7. The table below shows the number of children in each of 50 households. Use the program CUMFREQ to find proportion (experimental probability of a particular outcome), cumulative frequency (number less than or equal to a particular outcome), and cumulative proportion (experimental probability of an outcome being less than or equal to a particular outcome). To start the program, press [PRGM] and scroll down to find "CUMFREQ." Then press [ENTER]. The program prompts you for the total number of outcomes or scores (seven in this case: from 0 to 6). It then prompts you to enter the score for each outcome and its corresponding frequency. When the data for seven outcomes are entered, the program will return "Done." The proportion, cumulative frequency, and cumulative proportion will be listed in L3, L4, and L5, respectively. View by pressing [STAT] [edit]. Note: You will discover later in the book that cumulative proportions are very useful.

Number of children	Frequency
0	3
1	8
2	26
3	10
4	2
5	0
6	1

a. Plot the cumulative frequency (L4) versus the number of children per household (L1). Set up the statplot as shown below. Use [TRACE] to find the number of households with two or fewer children. How many have three or fewer?

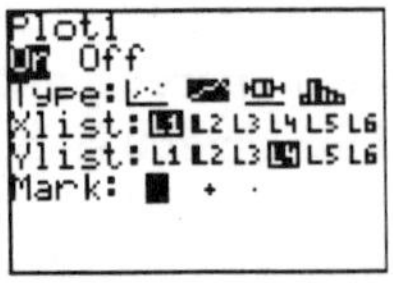

b. Plot the proportion (L3) versus the number of children per household (L1). Use TRACE to find the experimental probability (proportion) of families with two children. What is the experimental probability of there being three children?

c. Plot the number of children per household (L1) versus the experimental probability (cumulative proportion) (L5). Use TRACE to find the cumulative proportion of families with two or fewer children. What is the experimental probability (cumulative proportion) of there being four or fewer children?

8. The following data are the result of rolling a six-sided die 120 times. Use the program CUMFREQ to calculate the proportion, cumulative frequency, and cumulative proportion.

Outcome	Frequency
1	15
2	21
3	23
4	19
5	17
6	25

a. Draw a cumulative proportion (L5) versus outcome (L1) line plot for these data. Find the following:
   (i) The experimental probability of rolls resulting in a 2 or less
   (ii) The experimental probability of rolls resulting in a 4 or less
   (iii) The experimental probability of rolls resulting in a number greater than 4

9. Use the proportions (experimental probabilities) for the outcomes of rolling a six-sided die 120 times in Exercise 8 to answer the questions below. To help answer these questions, run CUMFREQ and plot the graph of cumulative frequency (L5) versus outcome (L1).
a. *P*(3 or less) (that is, the experimental probability of 3 or less)
b. *P*(5 or less)
c. *P*(more than 3)

1.6 MISUSING STATISTICS

Many years ago there was a popular book titled *How to Lie with Statistics.** People enjoyed the book and, what is more important, learned to be wary

*By Darrell Huff (New York: Norton, 1954).

of the uncritical acceptance of statistics as they are often presented to us in the media and by government and industry. To be sure, it is possible to use statistics to give misleading or wrong impressions. Indeed, either intentionally or unintentionally, statistics in books, reports, and the media are often quite misleading. We will sometimes give examples of how statistics might be misused, so that you can be misled less often when you encounter statistics in newspapers or magazines or on radio or television and be better able to interpret them correctly.

Many of the misuses of statistics are achieved by graphical distortions, but there are other ways as well. For now, we focus mainly on graphical misuses.

Example 1.6

The Top 10 Millionaire States

The New York Times of June 28, 1979, reported on the number of millionaires in the United States. The report first gave the numbers of millionaires per 1000 inhabitants for each state. When viewed this way, the top 10 states are as shown in Table 1.17. After looking at this report, one natural reaction is to remark that Idaho is a wealthy state—look at all its millionaires! However, it must be noted that many of these states also have a relatively small population.

Therefore, data about the state-by-state distribution of millionaires can be obtained in a different way and can present a rather different picture. The number of millionaires, reported state by state, can be given as a frequency table in Table 1.18, for the new top 10 list. Notice that the number of states appearing in both lists is rather small: only New Jersey, Indiana, Idaho, and Minnesota appear twice, and

Table 1.17 **Frequencies of Millionaires in Selected States**

State	Frequency per 1000
Idaho	26.7
Maine	7.7
North Dakota	6.9
Nebraska	6.6
Minnesota	5.7
Indiana	4.6
Wisconsin	4.1
Iowa	4.0
New Jersey	3.6
Connecticut	3.4

Table 1.18 **Numbers of Millionaires in Selected States**

State	Number of millionaires
New York	51,031
California	33,509
Illinois	31,131
Ohio	27,607
Florida	26,647
New Jersey	26,565
Indiana	24,345
Idaho	23,797
Minnesota	22,873
Texas	21,051

they appear toward the end of the second list. New York, the state with the most millionaires, is in fact not even among the top 10 states of Table 1.17.

When data are presented, it is important to see how they have been constructed. As the millionaire example shows, data about a topic can be constructed very differently to tell different stories. If not careful, one can be misled in this manner.

Example 1.7 The Closing Gas Stations

We have seen in this chapter that a picture, or graph, can be very helpful. But a graph also can be used to convey different messages with the same data. An August 22, 1977, magazine story on closing gas stations stated that the number of stations in the United States had dropped from 226,000 in 1973 to 180,000 in 1977. Depending on the impressions we want to give, we could use different versions of a graph. Figure 1.14 presents the frequency of gas stations in operation between 1973 and 1977. Figure 1.14*a* presents a picture of relative stability by starting the vertical scale at 0—the usual method of graphing. After all, only a moderate decline is seen in a period of four years. Figure 1.14*b*, on the other hand, obtained by breaking the graph between 180,000 and 0, suggests a dizzying decline in the number of service stations in the United States. The data are the same, but the change of scale in the vertical axis of (*b*) presents a rather different picture, unless one is very careful in interpreting this graph.

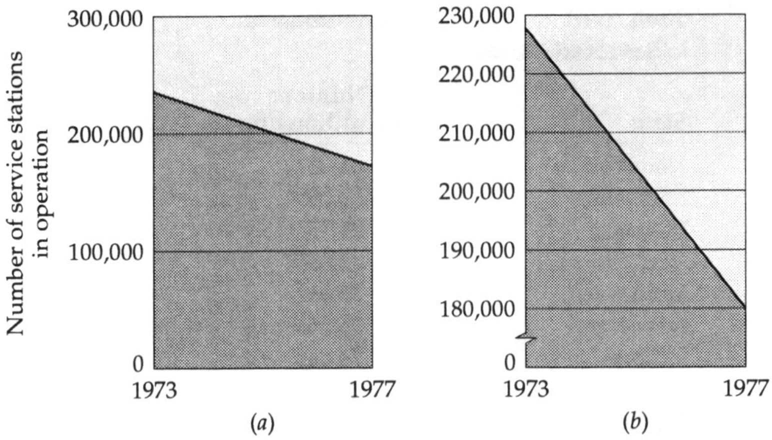

Figure 1.14 **Graphs of data on closing gas stations.**

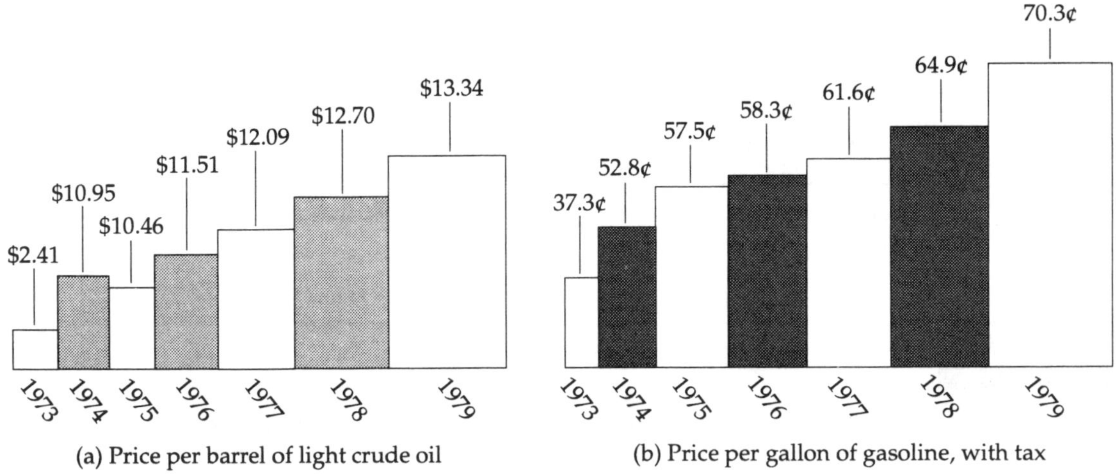

Figure 1.15 **Graphs of data on rising prices of crude oil and gasoline.**

Example 1.8 **Seeing Is Believing**

Graphs can be used (or misused) in another way to create the impression intended by the author or artist. In 1979 the rising prices of crude oil from OPEC countries and the corresponding price increases of gasoline in the United States were much in the news. One artist illustrated the increases by means of two graphs, represented in Figure 1.15.

Figure 1.15*a* shows a series of successively larger columns representing the price of light crude oil leaving Saudi Arabia. The smallest, of dimensions approximately 0.5 cm by 0.5 cm in the original illustration, is labeled with the price $2.41 per barrel, which was the price of light crude on January 1, 1973. The largest, of dimensions approximately 2.0 cm by 3.5 cm in the original illustration, is labeled with the price $13.34 per barrel, the price on January 1, 1979.

Figure 1.15*b* shows a series of columns representing the price of gasoline at the pump. The smallest, of dimensions approximately 0.5 cm by 1.5 cm, is labeled with the price 37.3¢ per gallon, which was the average price of gasoline in the United States in 1973. The largest, of dimensions approximately 2.0 cm by 5.0 cm, is labeled with the monthly averages of price per gallon for three months in 1979.

Although when the illustrations appeared, there was great concern about the price inflation of gasoline, there remains the question of how accurately those graphs represent the price inflation. The column for the cost of oil in 1979, for example, is—in terms of area—almost 14 times as large as the one for 1973, while the actual price increase was by a factor of about $5\frac{1}{2}$. An increase of 550% is indeed whopping, but it is still much less than the increase suggested by the relative areas of the graphs.

What evidently happened is that the concern of the artist was focused upon the heights of the figures representing the price increase. The ratio of 0.6 cm (representing 1973) to 3.2 cm (representing 1979) is 5.3, a sufficiently accurate depiction of the price increase. Unfortunately, when we observe such a figure, we tend to compare areas and not heights, as the artist intended.

SECTION 1.6 EXERCISES

1. Conduct an analysis of the graph representing the price increase of gasoline at the pump similar to the one carried out in this section for the price of oil.
2. Find examples from newspapers, magazines, and similar materials of presentations of statistics that are subject to the kinds of misinterpretation presented in this section.
3. A very famous example of the misuse of statistics took place in the 1932 presidential election. Statisticians predicted that the Republican candidate, Alfred Landon, would win the election by a wide margin. However, the Democratic candidate, Franklin D. Roosevelt, won by a large margin. Find why their prediction was so far off.

CHAPTER REVIEW EXERCISES

1. The distance from the earth to the sun is called the *astronomical unit* (A.U.). Here are the values for the astronomical unit as obtained by astronomers in the period from 1895 to 1961:

Source and date	A.U. (millions of miles)
Newcomb, 1895	93.28
Hinks, 1901	92.83
Noteboom, 1921	92.91
Spencer Jones, 1928	92.87
Spencer Jones, 1931	93.00
Witt, 1933	92.91
Adams, 1941	92.84
Brower, 1950	93.00 (rounded from 92.977)
Rabe, 1950	92.91 (rounded from 92.9148)
Millstone Hill, 1958	92.87 (rounded from 92.874)
Jodrell Bank, 1959	92.88 (rounded from 92.876)
S.T.L., 1960	92.93 (rounded from 92.9251)
Jodrell Bank, 1961	92.96 (rounded from 92.960)
Cal. Tech., 1961	92.96 (rounded from 92.956)
Soviets, 1961	92.81 (rounded from 92.813)

What would you consider to be a good estimate of how far it is from the earth to the sun, based on this information? Why do you think more decimal places are reported in the more recent measurements?

2. The ages of the first 36 U.S. presidents at inauguration and death are given in the table below. Make a stem-and-leaf plot of the ages at inauguration and the ages

President	Age at inauguration	Age at death
1. Washington	57	67
2. Adams	61	90
3. Jefferson	57	83
4. Madison	57	85
5. Monroe	58	73
6. Adams	57	80
7. Jackson	57	78
8. Van Buren	54	79
9. Harrison	68	68
10. Tyler	51	71
11. Polk	49	53
12. Taylor	64	65
13. Fillmore	50	74
14. Pierce	48	64
15. Buchanan	65	77
16. Lincoln	52	56
17. Johnson	56	66
18. Grant	46	63

President	Age at inauguration	Age at death
19. Hayes	54	70
20. Garfield	49	49
21. Arthur	50	57
22. Cleveland	47	71
23. Harrison	55	67
24. Cleveland	55	71
25. McKinley	54	58
26. Roosevelt	42	60
27. Taft	51	72
28. Wilson	56	67
29. Harding	55	57
30. Coolidge	51	60
31. Hoover	54	90
32. Roosevelt	51	63
33. Truman	60	88
34. Eisenhower	62	78
35. Kennedy	43	46
36. Johnson	55	64

at death. For which data is there more variation—age at inauguration or age at death?

3. Using the information in Exercise 2, construct a new set of data: the difference in age between a president's age at death and his age at inauguration. For example, for George Washington, the number is 67 − 57 = 10 years.

   Construct a stem-and-leaf plot of the new data. What is the most frequent number of years a president lived after inauguration? What is the longest amount of time? What is the shortest amount of time?

4. Construct a stem-and-leaf plot from the following data, which are the times in hours that 60 batteries lasted before needing to be recharged. (*Hint:* For 63.2, use a stem of 63 and a leaf of 2. Don't forget to give a key at the end of the table to explain how to read the table.)

63.2	71.0	81.6	68.5	76.2	82.3
70.3	84.6	65.9	74.7	74.2	79.2
63.2	89.7	78.4	86.8	59.0	77.9
50.3	55.5	74.3	60.6	62.1	74.9
63.6	63.2	79.4	57.6	61.4	81.8
71.3	64.1	64.8	64.9	71.8	64.0
74.5	51.8	64.6	75.2	64.4	70.9
74.2	83.3	64.1	77.5	54.1	71.6
63.4	55.2	64.2	66.2	73.4	61.1
60.1	65.3	85.8	68.1	74.5	66.3

5. Draw a histogram of the two data sets in Exercise 2.
6. Draw a histogram of the data set in Exercise 3.
7. Draw a histogram of the data set in Exercise 4.
8. Draw a line graph for each of the following two sets of data:
   a. 25, 30, 43, 51, 65, 70, 87, 91, 96, 100, 115
   b. 50, 53, 58, 63, 67, 70, 72, 78, 80, 83, 90

   For each set of data, find the median and the range. What is the difference between the two sets of data?
9. Toss a die 100 times and record the number of 1s, 2s, and so on. Draw a circle graph of your results.
10. One hundred rolls of a die were made, and the results were as recorded below.

Number thrown:	1	2	3	4	5	6
Frequency:	13	22	10	16	19	20

    a. Draw a relative frequency histogram of the data.
    b. Use the histogram from part (a) to answer the following questions.
       i. P(3 or higher)
       ii. P(odd number)
       iii. P(2 or lower)

11. Draw a relative frequency histogram of the high-temperature data in Exercise 6 of Section 1.2. Choose the number and width of the bars to give the best picture of the data.
12. Draw a relative frequency histogram of the silver and bronze medal standings in Exercise 1 of Section 1.1. Use the bars 1–5, 6–10, 11–15, 16–20, 21–25, 26–30, and 31–35. Explain any differences between the histograms of the two sets of data.

1

COMPUTER EXERCISES

1. THE DATA COMPUTER PROGRAM

The computer problems for the first two chapters use the Data program. We first show how to begin.

a. The Data program is started by double-clicking on the Data icon. Where you find this icon on your particular computer depends on the local experts. Ask them where to find it. It should look like this:

After you double-click with the mouse on the icon, a screen should appear with the words "No Data Loaded Yet." That is fine.

b. Notice the *menu* choices at the top, as in the picture:

The main menu of interest is the Data menu. That one contains all the relevant options, such as loading the data set and analyzing the data.

c. Click on the word "Data" to see the options:

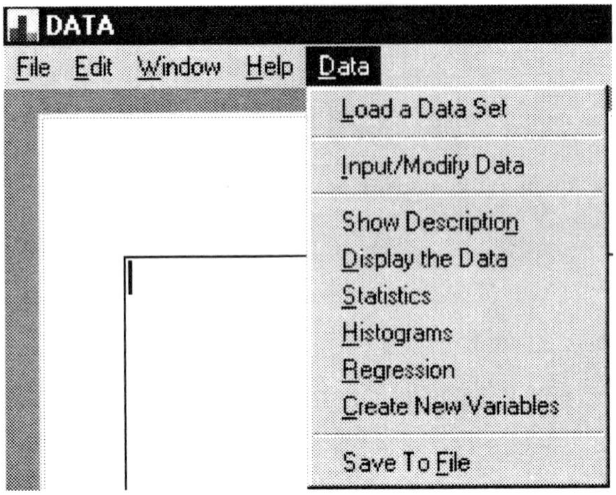

d. The first action to take is to "Load a Data Set." Click on that choice. A *dialog box* will appear with a list of the available data sets:

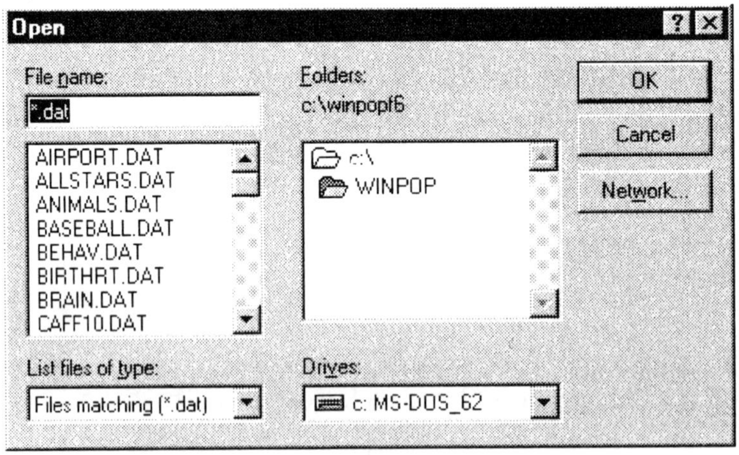

Your computer might have a slightly different list.

e. Choose the data set you wish to use. For example, click on "BIRTHRT.DAT," and then press OK. A screen like the one below will appear that briefly describes the data.

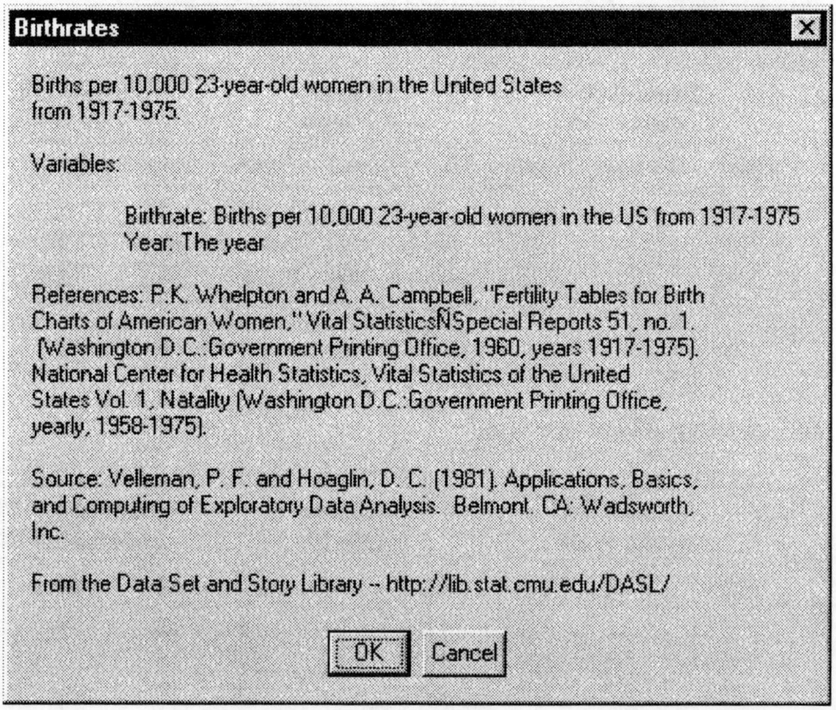

f. If you wish to use this data set, press OK. Otherwise, you can press Cancel to get back to the list of data sets. When you press OK, the program will load that data set. It may take a little time. Once the data set is loaded, the screen will have the title of the data set where "No Data Loaded Yet" used to be, and a description will appear in the main part of the screen. Now you can try other options in the Data menu to analyze the data, or load a different data set. The problems in this and later chapters will guide you through these options.

g. When you have completed your work in the program, press the "exit fish":

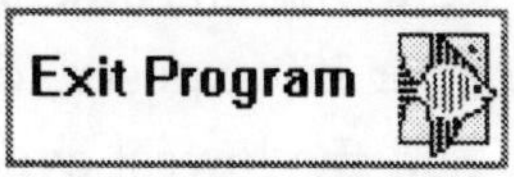

2. MYSTERY

The tables below display the data from a certain disaster. People are classified by sex, economic status (I = high, II = middle, III = low, other = ?), and age (adult or child). Look over the tables, and then answer the questions below.

By economic status and sex:

Economic status	Population exposed to risk			Number of deaths			Deaths per 100 exposed to risk		
	Male	Female	Both	Male	Female	Both	Male	Female	Both
I (high)	180	145	325	118	4	122	65	3	37
II	179	106	285	154	13	167	87	12	59
III	510	196	706	422	106	528	83	54	73
Other	862	23	885	670	3	673	78	13	76
Total	1731	470	2201	1364	126	1490	80	27	67

By economic status and age:

Economic status	Population exposed to risk			Number of deaths			Deaths per 100 exposed to risk		
	Adult	Child	Both	Adult	Child	Both	Adult	Child	Both
I (high)	319	6	325	122	0	122	38	0	37
II	261	24	285	167	0	167	64	0	59
III	627	79	706	476	52	528	76	66	73
Other	885	0	885	673	0	673	76	–	76
Total	2092	109	2201	1438	52	1490	69	48	67

Source: Data in both tables are from Robert J. MacG. Dawson, *Journal of Statistics Education* 3, no. 3 (1995).

a. Look at the "Population exposed to risk" parts of the tables.
   i. What percentage of the total was from the "other" economic status?
   ii. What is the percentage of females among those in economic status group I? Group II? Group III? The others? What trend do you see?
   iii. What is the percentage of children among those in economic status group I? Group II? Group III? The others? What trend do you see?

b. Now look at the "Deaths per 100 exposed to risk" parts of the tables.
   i. Which status group had the lowest overall death rate? Which had the highest?
   ii. Is the overall death rate higher among males or among females?
   iii. Compare the death rates of males and females for economic group I; for economic group II; for economic group III. What difference do you see among the groups when comparing these male/female rates?

iv. Look at the death rates for children in economic groups I, II, and III. (There are no children among the others.) What do you notice?

c. Guess what caused all these deaths. Explain what evidence you used to make your guess.

d. Who do you think the "Others" were? Why?

3. AIRLINE ARRIVALS

The table below gives the number of arrivals and number of arrivals on time, as well as the totals for each airline, in 1987.

	Alaska Airlines			*America West Airlines*		
Destination	**Number of arrivals**	**Number on time**	**Percentage on time**	**Number of arrivals**	**Number on time**	**Percentage on time**
Seattle	2146	1841		262	201	
Phoenix	233	221		5255	4840	
Total	2379	2062	94.8%	5517	5041	91.4%

Data taken from Arnold Barnett, "How Numbers Are Tricking You," *MIT's Technology Review*, October 1994.

a. Find the percentage of on-time arrivals to Phoenix for Alaska Airlines and for America West Airlines. Which airline had a better on-time percentage?

b. Find the percentage of on-time arrivals to Seattle for Alaska Airlines and for America West Airlines. Which airline had a better on-time percentage?

c. Which airline had a better on-time percentage, combining all flights to Seattle and Phoenix?

d. What is the overall percentage of on-time arrivals to Seattle, combining the two airlines?

e. What is the overall percentage of on-time arrivals to Phoenix, combining the two airlines?

f. How do you explain the answer to (c), given the answers to (a) and (b)? Isn't there a contradiction?

4. CALIFORNIA PEOPLE I

Go through the steps discussed in Exercise 1, but load the data set CAPEOPLE.DAT. It contains the gender (1 = male, 2 = female), years of education, and annual income for a sample of 258 people from California. This problem uses the "Histograms" option in the Data menu, which enables us to construct relative frequency histograms. Choosing Histograms yields the following screen:

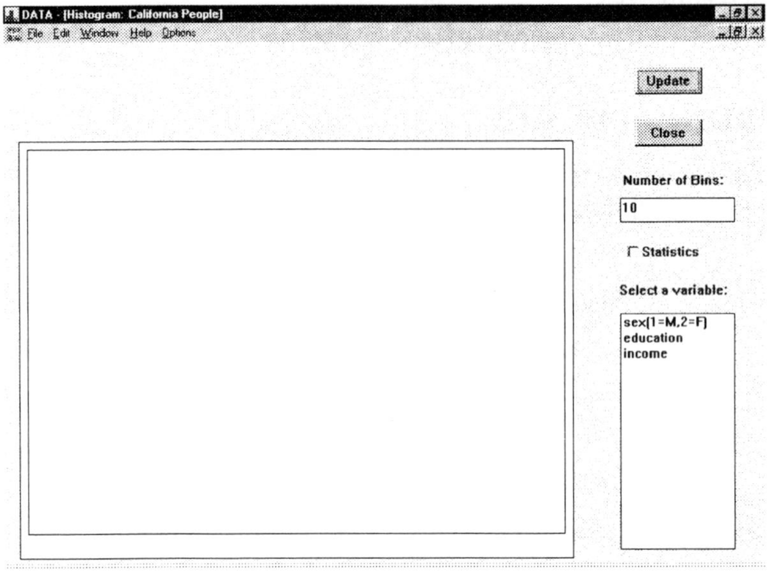

a. Create the histogram for years of education using 19 bins by doing the following:

   i. Choose the "education" item from the list headed "Select a variable."

   ii. Type the number 19 in place of the 10 that is in the box below "Number of Bins."

   iii. Press the Update button.

   You should see a histogram with 19 bars, one bar for each number of years of education from 0 to 18 ("18" represents people with 18 or more years).

   Explain what has caused the three tallest bars. (For which years are the bars highest? What do those years correspond to?) What are the fourth and fifth tallest bars?

b. Create the histogram for income. Try different values for the number of bins (clicking on Update after each change). What are the main features of the data? (Are there a few low incomes and many around $100,000? Or are most bunched within some range of values, with a few very large incomes? If so, between what values do most of them lie?)

c. Create the histogram for sex. How many bars are there? Are there more men or more women?

d. Next is to compare the educational levels of men with those of women. To do this, we need to create a "Split Histogram":

i. While still in the Histogram screen, go to the Options menu at the top and click, so that you see the following choices:

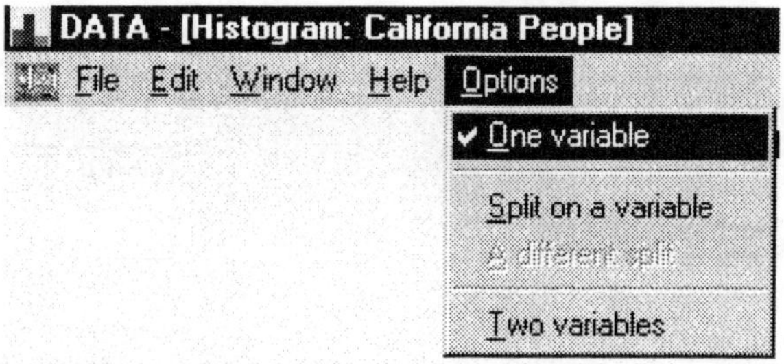

Click on the "Split on a variable" choice. A new screen will pop up that looks like the figure below:

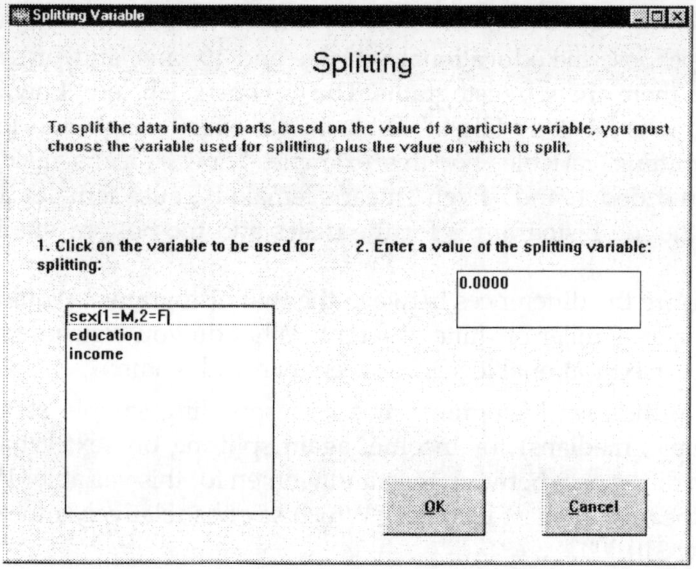

ii. This screen allows you to split the data into two parts according to whether a variable is above or below a certain value. You have to choose the variable on which to split (in box 1 on the screen) and the value on which to split it (in box 2). In this example, we wish to split on sex. The two values are 1 for men and 2 for women, so the splitting value is 1.5. (It could be anything between 1 and 2.) You can either double-click on "sex," in which case 1.5 will appear as the splitting value, or you can single-click on sex and type in 1.5 yourself. Either way, the relevant part of the screen should look like the next picture:

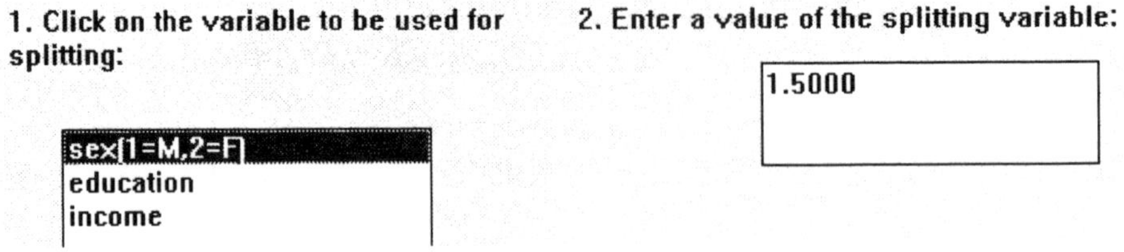

iii. Click on the OK button. (If the splitting value you choose is outside of the range of values of the variable you choose, the program may fail. If that happens, just go to the File menu and choose "Exit." That will close everything. Then you can start from the beginning.) Clicking on the OK button should bring you back to the histogram screen.

Now choose the educational variable and 19 bins, and press Update. This time there are two histograms. The top one is labeled "Low," and the bottom one is labeled "High." These labels refer to high and low levels of the splitting variable. For this example, "Low" means male because males are coded 1, and "High" means female because females are coded 2. Thus the top histogram is for the males and the bottom one is for the females.

What are the differences between the two histograms? Would you say they are quite similar or quite different? What do you conclude about the difference in educational levels between men and women?

e. Create the split histograms and corresponding sample statistics (for example, medians) for Income, again splitting on sex. What are the main differences between men and women for this variable? Which sex has the very highest incomes? Which sex has relatively more incomes below $10,000?

f. Do you think the difference in incomes can be "explained" by the differences in education? (There is no right answer for this question, but do consider whether the differences in education and the differences in income are about the same, or, rather, some differences are more pronounced.)

g. Press the Close button to close the Histogram screen and return to the opening screen. Now you are ready for the next exercise.

5. STATES' ELEVATIONS

The data set STATES.DAT contains the highest elevation and lowest elevation, in feet above (or below) sea level, for the 50 states. Load this data set into the Data program.

a. Go to the Histogram screen and create the histogram for the highest elevations. (Use 25 or 30 bins.) Describe the main features of the histogram (shape, centering, spread, and so on).

b. Create the histogram for the lowest elevations. Describe its main features. What appears to be the most common low elevation? Why?

c. Now close the Histogram screen. The "flatness" of a state is the difference between the highest and lowest points. Use the Display option in the Data menu to find the flatness of Illinois. (A spreadsheet should appear containing the data. You can find the highest and lowest elevations for the states, and then the flatness by subtraction.) What is Illinois' flatness?

d. Which is the flattest state?

e. Try to guess the five flattest states. Then use the spreadsheet to find which five states are really the flattest. What do they have in common? Do they include Kansas? Nebraska? Iowa? Illinois?

6. CLASS DATA I

This problem uses data collected from 180 students enrolled in STAT 100, a beginning statistics course at the University of Illinois that uses a version of this software in its lab sections. The data are in CLASS96.DAT. Load that data set into the Data program.

a. Go to the histogram screen, and choose "Split on a variable" from the Options menu. Select Gender as the variable, and make the splitting value 0.5. (This time, men are coded as 0 and women as 1.) Create the histograms for height. (Use 19 bins.)

   i. In general, are men taller than women, or women taller than men?

   ii. Is there some overlap, or are all men taller than all women?

   iii. Are the shapes of the two histograms similar or quite different? If different, in what ways?

b. Create the histograms for weight.
   i. In general, are men heavier than women, or women heavier than men?
   ii. Is there some overlap, or are all men heavier than all women?
   iii. Are the shapes of the two histograms similar or quite different? If different, in what ways?

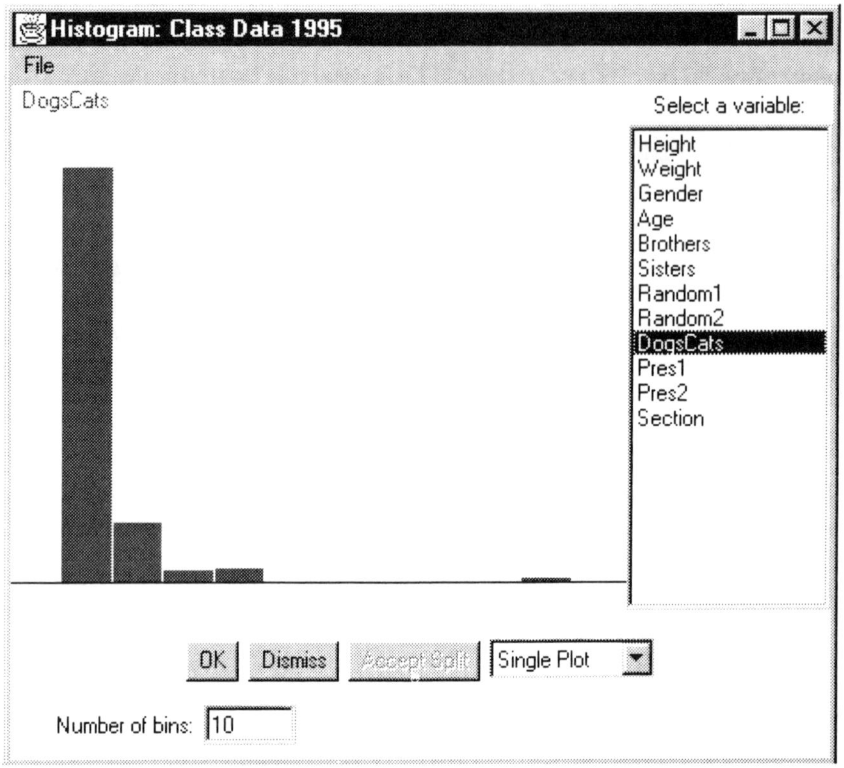

c. Create the histograms for the "DogsCats" variable. Use 41 bins. This variable gives the students' responses to the question of how many dogs and cats they have owned.
   i. In general, have men or women owned more dogs and cats?
   ii. Of the people who have owned 20 or more, do they tend to be men or women, or is it about equal?
   iii. What are the three highest numbers of dogs and cats owned? Do you think these numbers are exact? Why or why not?
   iv. Choose a lower number of bars to get a better representation of the overall shape of the sample.

7. NBA ALLSTARS I

The ALLSTARS.DAT data set contains the heights in inches and in centimeters of 10 famous basketball players. Load that data set into the Data program.

a. Choose "Display the data" from the Data menu. Find the ratio "# centimeters"/"# inches" for a few of the players. What values do you get? Do all players have essentially the same ratio? Why?

b. Create the histogram for the players' heights in inches and the histogram for their heights in centimeters. (Do these separately.) Do they look similar?

c. What do you notice? Are you surprised?

2

CENTERS AND SPREADS

2.1 CENTERS OF DATA 58

2.2 SOME PROPERTIES OF CENTERS 69

2.3 SPREAD (VARIATION) IN DATA 71

2.4 DEVIATION VALUES 79

2.5 MEAN ABSOLUTE DEVIATION 83

2.6 VARIANCE 86

2.7 STANDARD DEVIATION 88

CHAPTER REVIEW EXERCISES 92

COMPUTER EXERCISES 95

Society suffers from the curse of the arithmetic mean. Such compression necessarily ignores the fact that most measurements involve a spread of values around the mean—arising from real differences and random fluctuations.

R. E. Beard

OBJECTIVES

After studying this chapter, you will understand the following:

- *The importance of assessing the center and the spread of a data set*
- *The two most important measures of center: the mean and the median*
- *Two important measures of spread: the interquartile range and the mean absolute deviation*
- *The standard deviation, the most widely used measure of spread*
- *How to read and construct box plots*

Key Problem

Box plots

Consider the "average" monthly high temperatures (in degrees Fahrenheit) for Seattle, Washington, and Salt Lake City, Utah, for January through December. Here each monthly *average* high is defined in a complex but appropriate way as follows (think it through to be sure you understand it). First, consider a given month in a given year for one of the cities (July 1989 in Salt Lake City, say). The daily highs for the month (there are 31 such values in July) are averaged. This monthly average high is computed for each of 30 years. Then these 30 monthly average highs are averaged, producing 92.2 for July in Salt Lake City. Note that we have averaged twice—once over the days of the month and then over 30 years.

Salt Lake City: 36.4, 37.8, 43.6, 61.3, 71.9, 82.8, 92.2, 89.4, 79.2, 66.1, 52.2, 50.8

Seattle: 45.0, 45.1, 52.7, 57.2, 63.9, 69.9, 75.2, 75.2, 69.3, 59.7, 50.5, 49.5

The problem is to find a way to represent the data that is graphical and yet presents statistics showing where the data center and how widely they spread out. A histogram would partially do this, but it would not provide such statistics.

The data are presented in Figure 2.1 in graphs called *box plots.* In the box plot, the center line locates the median (the second quartile, actually), the ends of the box locate the first and third quartiles, and the extended lines, or *whiskers,* locate the minimum and maximum data points.

Such box plots, along with stem-and-leaf plots, are widely used in descriptive statistics. In this chapter we will use these graphs to graphically represent medians, quartiles, and the largest and smallest data values in order to better understand data sets.

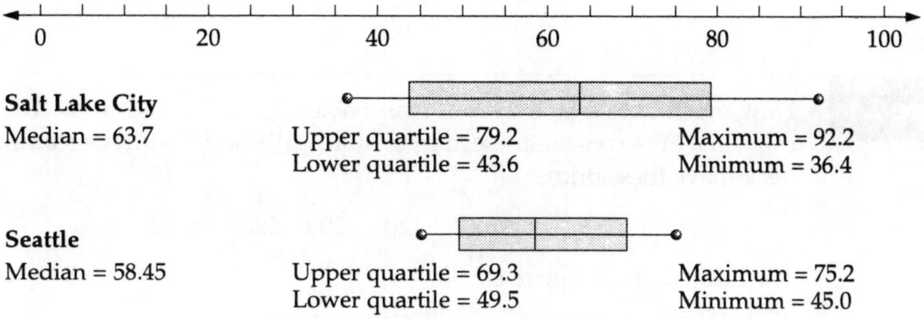

Figure 2.1 **Box-plot comparison of two temperature data sets.**

2.1 CENTERS OF DATA

One of the first things we usually want to know about a set of data, such as salaries among a group of people, prices of laptop computers among the many brands and models available, or the number of home runs in a season by all the major league ball players, is, About how much? About how many? That is, we want to know what is a typical, or central, value for a data set, which may in fact display a lot of variation among its data values. So, for example, we could ask

About how much do students earn at their part-time jobs?

About how much does a decent, but reasonably priced, laptop computer cost?

Let's look at one situation a little more closely. Suppose we have measured the heights of a group of school children. We might have done this, for example, as part of a study of how nutrition affects growth, or to learn whether girls and boys grow at the same rates. We see that the heights we have measured vary greatly; yet we believe it makes sense to speak of a typical or representative height of the children we have measured. In other words, we want a number that we can consider to be the *center* of the data.

So what number do we consider to be the center? We have a choice. The center of such a set of data can be defined in different ways. In this chapter we consider three kinds of center values. Each measure of the center of a data set is useful in somewhat different ways.

Mean

The most common definition of the center value is the **mean**. This is the familiar arithmetic average, so *mean* is just another term used in statistics to refer to the average. The mean of a set of data is easy to find. Just add all the numbers and divide by how many of these numbers there are.

Example 2.1

How much does a tape deck of reasonable quality typically cost? To find out, we call some area consumer electronics stores. We find that five recommended tape decks have these prices:

$100 $120 $200 $200 $450

The mean (average) price is

$$\frac{\$100 + \$120 + \$200 + \$200 + \$450}{5} = \frac{\$1070}{5} = \$214$$

The procedure for finding the mean of a set of data: As we stated above, to find the average, or mean, of a set of data, add up the data and divide by the number of data there are in the set. We can be more concise in our definition by giving a rule, or formula. (The sole purpose of mathematical formulas is to give mathematical rules concisely and without ambiguity. If we relied only on words, both conciseness and precision would often suffer.)

If a set consists of data values X and there are N such numbers, then the mean is $\overline{X}$ (read "X-bar"), and we have

$$\overline{X} = \frac{\text{sum of X}}{N}$$

The mean has many useful mathematical properties that appeal to statisticians. Statisticians have thus depended heavily on the mean as a measure of the center of a data set. However, as we shall see, it can be misleading in some applications, so statisticians have been increasingly using a second index of center, the *median.*

Median

Example 2.2 reintroduces the **median,** or **median value,** which was briefly introduced in Chapter 1 as a natural outgrowth of the use of line plots, which order data from smallest to largest value. The median is often used by statisticians. The median, or median value, is a value such that just as many data values are greater than that value as are less than it.

Example 2.2

Paul wants to know how well he did on his statistics quiz relative to the other 24 students. The instructor showed the students a table giving all the scores (see Table 2.1). Paul got a 62.

Table 2.1 **Student Scores on a Quiz**

Stem	Leaf
3	8
4	0, 1, 1, 2, 3, 5
5	0, 0, 1, 2, 5
6	2, 4, 5, 6, 6, 7
7	1, 2, 8
8	2, 7, 7
9	0

Key: "3 8" stands for a test score of 38 points.

What does Paul know about how he is doing in the course, relative to the rest of the class? Observe that 12 students got a score lower than 62 and 12 got a score higher than 62. Thus Paul's score is exactly the middle—that is, the median—score. Thus Paul knows his rank is exactly in the middle.

Let's consider another example.

Example 2.3

Olympic Medals

The total numbers of medals won by the top 21 countries in the 1996 Summer Olympics were as follows:

		Total			Total
1.	United States	101	12.	Hungary	21
2.	Germany	65	13.	Rumania	20
3.	Russia	63	14.	Netherlands	19
4.	China	50	15.	Poland	17
5.	Australia	41	16.	Spain	17
6.	France	37	17.	Bulgaria	15
7.	Italy	35	18.	Brazil	15
8.	South Korea	27	19.	Britain	15
9.	Cuba	25	20.	Belarus	15
10.	Ukraine	23	21.	Japan	14
11.	Canada	22			

To find the median when the number of data points is at all large, it is easiest to work from a ranked list of data, as in the above table or as is provided by the line plots of Chapter 1. Here the countries are ranked, from highest to lowest, by the total number of medals received (lowest to highest would work just as well).

There are 21 countries in the list. So by counting 10 countries down from the top (or up from the bottom), we find that Canada is in the middle of the list, with 22 medals. Looking at it another way, we see from the list that the same number of countries are above Canada (won more than 22 medals) as below Canada (won fewer than 22 medals). The median is therefore 22 medals.

Note that if the United States had won only 66 medals, the median would still be 22 medals. However, by contrast, the mean would be quite different. Thus the median is **robust** in the sense that it is not influenced much (in fact it is not influenced at all) by a large change in a single data point in the data set. Such robustness for a statistic computed from a data set is quite desirable, especially if we do not trust the accuracy of each data point. In addition, it seems appropriate that the definition of the middle of a data set should not be greatly dependent on a single data value.

Example 2.4

A Slight Complication

This table shows the number of gold medals won by every country that won a gold medal in the 1996 Summer Olympics.

		Gold			Gold
1.	United States	44	12.	Hungary	7
2.	Russia	26	13.	Spain	5
3.	Germany	20	14.	Rumania	4
4.	China	16	15.	Netherlands	4
5.	France	15	16.	Japan	3
6.	Italy	13	17.	Canada	3
7.	Ukraine	9	18.	Bulgaria	3
8.	Cuba	9	19.	Brazil	3
9.	Australia	9	20.	Britain	1
10.	South Korea	7	21.	Belarus	1
11.	Poland	7			

South Korea, Poland, and Hungary share the middle score of 7 gold medals. Thus, more than one data point can be a median point, as we already saw in Chapter 1. This causes no problems as long as one is aware that ties for the median can happen. In summary, the median number of gold medals (for the top countries) is 7, shared by three countries.

We now consider another example, one that introduces another slight complication, namely, that there is no single middle number when there are an even number of data in the set.

According to the definition of the median, just as many of the values in the data set lie above the median value as lie below it. If the number of data in the set is even, such a value would have to fall somewhere between the two middle data values in the set (write out lists of numbers of different lengths, both even and odd, to convince yourself that this is so). In such a case the convention is that we take as the median the value midway between the two middle values—or, which works out to be the same, the average of the two middle values. The next example illustrates this.

Example 2.5

A Further Complication

Sue is looking for a used car and finds six cars within her budget that interest her. The prices are as follows:

$1500 $1650 $1700 $1900 $1925 $1950

($1700 $1900: Middle values)

The middle two numbers are $1700 and $1900. So in order to find the median price, we find the average of the middle two numbers $1700 and $1900, which is $1800. Note that the mean of the six numbers can be found to be $1771 (we have rounded to the nearest dollar). This illustrates what you may have already conjectured, namely, that the mean and median are usually not the same.

Example 2.6

More Tall Buildings

Let's take another look at data on some of the tallest buildings in the United States and Canada given in Chapter 1. Here we have taken only the 10 tallest from the list in Table 1.6. What is the median number of stories of those 10 buildings?

City	Building	Number of stories
Chicago	Sears Tower	110
New York City	World Trade Center I	110
Seattle	Columbia Seafirst Center	76
Los Angeles	First Interstate World Center	73
Toronto	First Canada Place	72
Detroit	Westin Hotel	71
Atlanta	Peachtree Center	63
Boston	John Hancock Tower	60
Cleveland	Society Center	57
Calgary	Petro Canada Center	52

There are 10 data in the table. Counting five down from the top of the list gives 72 stories (First Canada Place in Toronto). Counting five up from the bottom gives 71 stories (the Westin Hotel in Detroit). To obtain the median, we split the difference between 72 and 71 by averaging. So we have

$$\text{Median number of stories} = \frac{71 + 72}{2} = 71.5$$

The procedure for finding the median of a set of data: To find the median of a set of values, first arrange them from high to low (or low to high, as is done in a line plot). Then:

If there are an odd number of data, find the value such that just as many data are above it as below it.

If there are an even number of data, find the average of the middle two values.

Mode

The **mode** (or **modal value**) of a set of data is another kind of center. Suppose we are looking for a suit of clothes. Glancing around the shop, we notice a lot of price tags with $200 on them. We might be tempted to view $200 as the center of the data, in some sense. We are giving our impression of the typical price of a suit based on seeing lots of price tags for $200. This brings us to the idea of the mode, or the modal value. The mode is the most frequently occurring value in a set of data.

Even though the mode is often used by nonstatisticians, statisticians in fact make little use of it in describing data or in making statistical inferences. Unfortunately, it is a textbook tradition, which we are reluctantly following here, to include the mode as a measure of the center of a data set. Usually data occur most frequently near their center as measured by the mean or the median, and in these cases the mode is close to those other measures of the center. However, there sometimes occur oddly distributed data sets that have areas of high occurrence away from the center. In such cases the mode is not representative of the center of the data. The next example illustrates this.

Example 2.7

The table below shows the prices of 10 used bicycles. What is the modal, or most common, price?

Stem	Leaf
9	0
10	0, 5, 0
11	0, 5
12	
13	0, 0, 0, 5

Key: "9 0" stands for $90.

Solution

We see that more bicycles are $130 than any other price (four bicycles cost $130). So the modal price is $130. Yet this price is certainly not in any sense at the center of the distribution.

As remarked above, it sometimes happens that a data set has concentrations of values away from its center. Let's consider such a data set. Table 2.2 provides the lowest temperatures (in degrees Fahrenheit) recorded for February 1–5, 1996, at locations in California and Colorado.

First we construct a stem-and-leaf plot (a line plot would be equally useful).

Stem	Leaf
−3	6
−2	0, 0, 0, 1, 6
−1	0, 6, 6, 8, 8
−0	4, 8
0	
1	2, 9
2	
3	
4	0, 1, 2, 6, 6, 6, 7, 7, 7, 9, 9
5	0, 1

Key: "−3 6" stands for −36 °F.

Table 2.2 **Temperatures (°F) in California and Colorado**

California		Colorado	
San Luis Obispo	41	Akron	−16
Santa Ana	49	Alamosa	−10
Santa Barbara	46	Boulder	−8
Santa Cruz	47	Brighton	−20
Santa Maria	47	Colorado Springs	−18
Santa Monica	50	Denver	−16
Shelter Cove	46	Eagle	−21
Stockton	47	Fort Collins	−20
Tahoe Valley	19	Fort Morgan	−26
Thermal	42	Fraser	−36
Torrance	51	Glenwood Springs	−4
Ukiah	40	Grand Junction	12
Vallejo	46	Grand Lake	−18
Ventura	49	Greeley	−20

There are two clumps in these data. One concentration of temperatures is roughly in the range from 10 to 20 degrees below zero. The other group is above 40 degrees. The median here is +15.5, showing us that very few data appear near the center of this data set. If a set of data has two areas of high concentration, it is called **bimodal.** Bimodal data are uncommon and are usually caused by the blending of two different populations, such as California temperatures with Colorado temperatures, as in the example.

SECTION 2.1 EXERCISES

1. A stem-and-leaf plot of the annual earnings of 50 people appears below. From this plot, find the following:
   a. The mean and median earnings. Compare the two values.
   b. Find the mode of the annual earnings.
   c. How does the mode compare with the other two measures of central tendency?

Stem	Leaf
1	6, 7, 8
2	2, 3, 5, 7, 7, 7, 7
3	1, 2, 3, 4, 5, 6, 6, 9
4	0, 3, 3, 3, 4, 5, 6, 6, 7, 7, 7, 8, 9, 9
5	3, 3, 5, 7, 7, 9, 9
6	0, 0, 2, 2, 3, 4, 4, 6, 8
7	4, 7

Key: "1 6" stands for $16,000.

2. The numbers of medals won by the top 21 countries during the 1996 Summer Olympics are shown in the following table:

	Total	Gold	Silver	Bronze
United States	101	44	32	25
Germany	65	20	18	27
Russia	63	26	21	16
China	50	16	22	12
Australia	41	9	9	23
France	37	15	7	15
Italy	35	13	10	12
South Korea	27	7	15	5
Cuba	25	9	8	8
Ukraine	23	9	2	12
Canada	22	3	11	8
Hungary	21	7	4	10
Rumania	20	4	7	9
Netherlands	19	4	5	10
Poland	17	7	5	5
Spain	17	5	6	6
Bulgaria	15	3	7	5
Brazil	15	3	3	9
Britain	15	1	8	6
Belarus	15	1	6	8
Japan	14	3	6	5

a. Find the median number of silver medals won and of bronze medals won, and name the countries winning those numbers of medals.
b. Find the mean number of silver and bronze medals won.
c. Compare the mean and the median of the number of silver and bronze medals won. Which measure of central tendency—the mean or the median—do you prefer for these data? Explain your answer.

3. Here are the salaries of the Chicago Cubs for the 1996 season. The figures are in millions of dollars.

Player	Salary ($ millions)
Ryne Sandberg	5.975
Mark Grace	4.400
Randy Myers	3.583
Jose Guzman	3.500
Mike Morgan	3.375
Sammy Sosa	2.950
Steve Buechele	2.550
Shawon Dunston	2.375
Dan Plesac	1.700
Glenallen Hill	1.000
Willie Wilson	0.700
Jose Bautista	0.695
Rick Wilkins	0.350
Shawn Boskie	0.300
Derrick May	0.300
Mark Parent	0.250
Anthony Young	0.230
Rey Sanchez	0.225
Frank Castillo	0.225
Willie Banks	0.190
Karl Rhodes	0.145
Jim Bullinger	0.145
Steve Trachsel	0.112
Eddie Zambrano	0.112
Jose Hernandez	0.112
Jessie Hollins	0.109
Blaise Ilsley	0.109

a. Find the mean and the median salaries.
b. Explain the difference between the two measures of central tendency.
c. Which one is the "better" statistic to use to report the "average" salary for the Cubs?

4. The number of campsites in each of five campgrounds in Yoho National Park is as follows:

Campground	Number of campsites
Kicking Horse	92
Hoodoo Creek	106
Chancellor Peak	64
Takakkaw Falls	35
Lake O'Hara	30

What is the mean number of campsites per campground?

5. The number of calories in a serving of various kinds of cheese is given by the following table:

Cheese type	Number of calories
American	106
Cream	99
Feta	75
Monterey	106
Ricotta (whole milk)	216
Swiss	107

What is the mean number of calories per serving?

6. Here are NCAA final men's basketball scores for the indicated years:

1939	Oregon	46	Ohio State	33
1940	Indiana	60	Kansas	42
1941	Wisconsin	39	Washington State	34
1942	Stanford	53	Dartmouth	38
1943	Wyoming	46	Georgetown	34
1944	Utah	42	Dartmouth	40
1945	Oklahoma State	49	NYU	45
1946	Oklahoma State	43	North Carolina	40
1947	Holy Cross	58	Oklahoma	47
1948	Kentucky	58	Baylor	42
1949	Kentucky	46	Oklahoma State	36
1950	CCNY	71	Bradley	68
1959	California	71	West Virginia	70
1960	Ohio State	75	California	55
1961	Cincinnati	70	Ohio State	65
1962	Cincinnati	71	Ohio State	59
1963	Loyola-Chicago	60	Cincinnati	58
1964	UCLA	98	Duke	83
1965	UCLA	91	Michigan	80
1966	Texas-El Paso	72	Kentucky	65
1967	UCLA	79	Dayton	64
1968	UCLA	78	North Carolina	55
1969	UCLA	92	Purdue	72
1970	UCLA	80	Jacksonville	69
1984	Georgetown	84	Houston	75
1985	Villanova	66	Georgetown	64
1986	Louisville	72	Duke	69
1987	Indiana	74	Syracuse	73
1988	Kansas	83	Oklahoma	79
1989	Michigan	80	Seton Hall	79
1990	UNLV	103	Duke	73
1991	Duke	72	Kansas	65
1992	Duke	71	Michigan	51
1993	North Carolina	77	Michigan	71

a. Make a stem-and-leaf plot of the winning scores for each of these three groups of years: 1939–1950, 1959–1970, 1984–1993.
b. Comment on the patterns you see. Explain using means or medians.

7. The numbers of medals won by the top countries in the 1994 Winter Olympics are as follows:

Country	Gold	Silver	Bronze	Total
Norway	10	11	5	26
Germany	9	7	8	24
Russia	11	8	4	23
Italy	7	5	8	20
United States	6	5	2	13
Canada	3	6	4	13
Switzerland	3	4	2	9
Austria	2	3	4	9
South Korea	4	1	1	6
Finland	0	1	5	6
Japan	1	2	2	5
France	0	1	4	5
Netherlands	0	1	3	4
Sweden	2	1	0	3
Kazakhstan	1	2	0	3
China	0	1	2	3
Slovenia	0	0	3	3
Ukraine	1	0	1	2
Belarus	0	2	0	2
Britain	0	0	2	2
Uzbekistan	1	0	0	1
Australia	0	0	1	1

a. What is the modal number of total medals won? Which countries won this number of medals? Interpret this statistic.
b. Find the mean, median, and mode of the number of gold medals won. Which of these is a better measure of the number of gold medals won?

8. Here are the numbers of earthquakes that were recorded as 7 or greater on the Richter scale during the years 1900–1909 and 1980–1989:

Year	Number of earthquakes	Year	Number of earthquakes
1900	13	1980	18
1901	14	1981	14
1902	8	1982	10
1903	10	1983	15
1904	16	1984	8
1905	26	1985	15
1906	32	1986	6
1907	27	1987	11
1908	18	1988	8
1909	32	1989	7

a. Make a stem-and-leaf plot of these two sets of data.
b. Compare them using means or medians.

9. The word *mode* is related to a French word meaning fashion. In what way can the mode of a set of numbers be thought of as "fashionable"?

Graphing Calculator Exercises

10. The city council of McBride is investigating the efficiency of its fire department. The times (in minutes) taken by the fire department to respond to 15 alarms are given below. Enter the data into data list L1 (see Section 1.2 of the TI Graphing Calculator Supplement). Find the mean and the median response times (see Section 1.5 of the TI Graphing Calculator Supplement).

1 3 2 2 1 9 4 6 1 10 1 4 5 10 1

11. Toss three coins 10 times. Write down the number of heads obtained on each toss. Then enter these data into data list L1 and find the mean and the median number of heads obtained in 10 tosses of three coins.

12. The following data are the midrange prices (1993), in thousands of dollars, of 10 different cars. Enter the following price data into L1 using a data table.

Model	Midrange price ($ thousands)	Model	Midrange price ($ thousands)
Acura Legend	33.9	Infinity Q45	47.9
BMW 535i	30.0	Lexus SC300	35.2
Chevy Lumina	15.9	Mercedes-Benz 300E	61.9
Ford Taurus	20.2	Mitsubishi Diamante	26.1
Hyundai Sonata	13.9	Nissan Maxima	21.5

a. What is the lowest price for a car?
b. What is the range of the prices?

c. Calculate the mean and the median prices of the cars. By comparing the mean and the median, comment on the selection of cars in the table.

2.2 SOME PROPERTIES OF CENTERS

Here are the hourly wages of five workers:

$4.50 $5.00 $5.50 $6.00 $7.00

This set of wages has a median of $5.50 and a mean of $5.60.

Now let's see what happens if *one of the workers* gets a large pay increase. We change $7.00 to $15.00, getting

$4.50 $5.00 $5.50 $6.00 $15.00

This set of wages has a median of $5.50 and a mean of $7.00.

Notice first that the median remains at $5.50. One-half of the remaining four workers still earn more and one-half still earn less than $5.50. But the mean wage in the second example has increased so much that it is larger than the wages of four of the five workers. In the second data set, which measure of center—the median or the mean—seems more representative of the typical worker's salary? Clearly one person's salary (the $15 salary) should not play an overly influential role. Hence the median, which is robust against such extreme values, seems more meaningful than the mean. We see here that the mean is strongly influenced by a single unusually large value in the set. It also can be influenced by an unusually small value in the set. The median is not influenced by very large or small values.

We thus say that the median is a more robust measure than the mean. It resists undue influences of individual scores in the set. Since statisticians and the general populace both want the measure of center that we use not to be overly influenced by a single extreme value, we sometimes choose the median because it is so robust against this unwanted influence. Again, if we want the measure of center to be typical of the data, then a statistic like the mean, which can be controlled by a single data point, can be undesirable. Thus the median is a *robust measure of the center* of a data set and as such is often used.

Though not robust, the mean has many properties that make it a very important measure of central tendency in statistics. We will consider only one of these properties here, a very simple one.

Recovery of the Sum of the Data

It is sometimes important to regain information that may not have been given in a statistical summary of the data. The next example presents a particularly simple case of this.

Example 2.8

Suppose we are told that the mean value of scholarships given to 10 people was \$450. How much money was awarded for scholarships?

Solution

We remember the rule for finding the mean of a set of numbers. Let $\overline{S}$ stand for the mean value of a scholarship:

$$\overline{S} = \frac{\text{sum of all scholarships}}{\text{total number of scholarships}}$$

So we can say that

$$\begin{aligned}\text{Sum of all scholarships} &= \overline{S} \times \text{total number}\\ &= \$450 \times 10 = \$4500\end{aligned}$$

A total of \$4500 was awarded in scholarships.

The median does not have this desirable property. Knowing the median does not enable us to compute the sum of the data.

SECTION 2.2 EXERCISES

1. Find the mean and median of the following sets of data:
   a. 10, 15, 20, 25
   b. 10, 15, 20, 2500

   How have the mean and median been affected by replacing 25 in set (a) with 2500 in set (b)?
2. Find the mean and median of the following sets of data:
   a. 100, 110, 120, 130, 140
   b. 0, 110, 120, 130, 140

   How have the mean and median been affected by reducing 100 in set (a) to 0 in set (b)?
3. Ron mowed lawns last summer. His weekly earnings for seven weeks are as follows:

   \$22.00 \$30.00 \$5.50 \$17.50
   \$32.50 \$28.00 \$30.00

   a. Find the mean and median values of his earnings.
   b. Suppose the \$5.50 amount should be \$15.50. Without doing any calculations, what happens to his mean and median earnings?
4. In Exercise 3, find the mode of Ron's weekly earnings. Which of the three measures of central tendency presents a better picture of Ron's earning power last summer?
5. A dispute over salaries has arisen in a print shop between the workers and the owner. The owner believes she should include her \$50,000 salary when reporting the mean salary at the shop. The workers do not agree. Here are the workers' salaries:

   \$15,000 \$16,500 \$17,000 \$17,000 \$20,000

   How might the owner solve the dispute and still include her salary in the computations?
6. An outpatient clinic reported that the mean charges per patient were \$60.50 last month, and that a total of 110 patients were served by doctors during that month. What was the total of the doctors' charges for that month?
7. Mary got a 75 and an 80 on two tests this semester. If a grade of B requires a mean of at least 80, what must she get on her next test in order to get a B average?

8. If Mary (in the preceding exercise) were able to persuade her teacher to average grades using medians instead of means, what minimum grade would she need on her next test to have a B average?
9. Here are the heights (in inches) of 25 twenty-year-old men.

70	71	70	68	70	69	72	72	73	72
74	72	71	70	76	73	74	71	70	73
69	66	75	72	71					

a. Find the mean, median, and mode of these data.
b. Which measure of central tendency is more appropriate for reporting the "average" height of the 25 men? Explain your answer.

2.3 SPREAD (VARIATION) IN DATA

So far in this chapter we have discussed three measures of the center of a group of data. Of these three, the mean and the median are more useful than the mode. Both the mean and the median give a good idea of the center of a group of data—they enable us to quantify and communicate where the center is. Further, these two measures of center have different properties that make one or the other more useful for a particular purpose or a particular set of data.

Now we need a way to measure the amount of variation in a group of data. After all, if someone states that the center of a group of data lies at such and such a value, we would naturally ask whether the data are for the most part close to that value, or whether they instead vary quite widely.

Measures of variation are of great practical importance. Suppose two groups of students take a standardized test scaled from 0 to 100. Everybody in the range from 85 to 89 receives a scholarship of $1000, and everybody scoring 90 or above receives a full scholarship. Here are the data from the two groups (A and B):

Group A

81 61 85 84 73 68 75 77 68 81 83 68 68 74 75 68 82 69 74 60 67 76 91 85 69
73 56 92 70 81 75 84 90 84 88 85 84 59 64 65 63 61 80 78 83 72 77 89 81 60

Group B

81 77 75 76 77 82 78 79 73 82 83 79 80 77 76 84 81 81 82 81 84 79 76 79 82
81 82 80 78 72 78 80 80 80 77 82 83 78 78 76 84 82 80 78 80 81 80 79 78 81

Consider the histograms of the two groups, shown in Figure 2.2. According to this test, the first group (A) is less talented on average, having a mean score of 75.1 while the second group (B) averages 79.4. However, the less talented (on average) group A is in fact much more variable. Hence

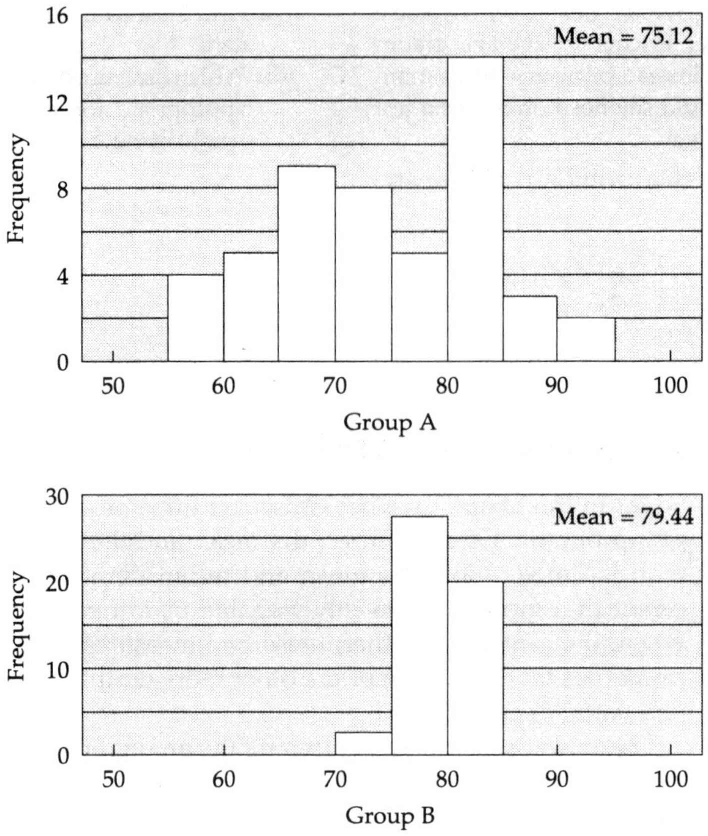

Figure 2.2 **Histogram comparison of two test score data sets.**

it is this group A that gets all the scholarships: three partial and two full. Thus variability matters here. Indeed, variability often plays a role in which group gets a higher proportion of such scholarships as the National Merit Scholarships.

Table 2.3 shows the number of medals (gold, silver, and bronze) won by the 22 countries in the 1994 Winter Olympics (held in Lillehammer, Norway) that won medals. We see that the countries differ a great deal in how many medals they won—from a high of 26 for Norway to a low of 1 for Australia (other countries, of course, won no medals at all). Let's see how we can show these differences in the number of medals won—that is, let's see how we can describe the *variation* in these data.

A box plot of data provides good information about data spread. First, we draw a line plot of the data, as shown below. From the line plot it is easy to see the total variation in the data—from a high of 26 medals to a low of 1

Table 2.3 **1994 Winter Olympics Medal Totals**

	Total	
Norway	26	
Germany	24	
Russia	23	
Italy	20	
United States	13	
Canada	13	
Switzerland	9	
Austria	9	
South Korea	6	
Finland	6	
Japan	5	← Median
France	5	
Netherlands	4	
Sweden	3	
Kazakhstan	3	
China	3	
Slovenia	3	
Ukraine	2	
Belarus	2	
Britain	2	
Uzbekistan	1	
Australia	1	

medal. As was mentioned in Chapter 1, the difference between the high and low values in a set of data is called the *range*. The range here is 26 − 1 = 25.

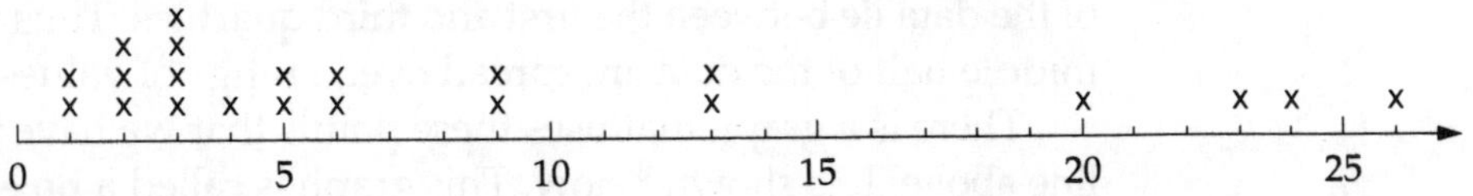

Now, we show the median, or halfway point of the data. Since there are 22 countries, we know the median is between the 11th and the 12th, which are Japan and France, each with 5 medals. So the median number of medals won was 5. The median is shown on the line below.

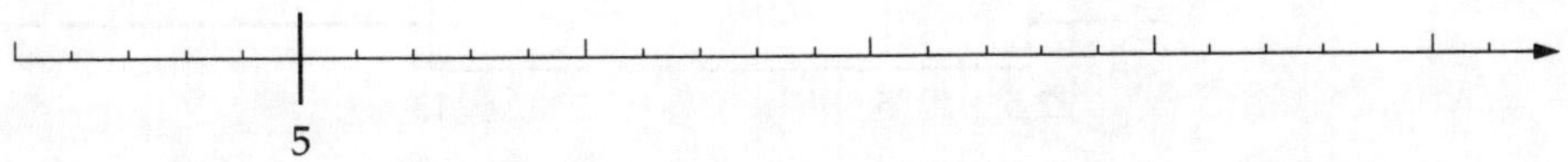

Two more points are also of interest in helping to summarize how much variation there is in the data: the **first quartile** and the **third quartile.** *Quartile* means quarter. The first quartile tells the number below which one-quarter of the data are found. In our example, 5 countries are fewer than $\frac{1}{4}$, and 6 countries are more than $\frac{1}{4}$. The dividing point between the bottom five countries and the sixth is between 2 and 3, and for simplicity we say that the first quartile is 2.5 (halfway between 2 and 3). The third quartile divides the top one-quarter of the countries from the rest of the group. In this case the third quartile is 13. The first and third quartiles are marked on the line below. Note that by the definition of the quartile, the median of 5 is the second quartile. A *box-and-whisker plot,* defined below, actually plots all three quartiles.

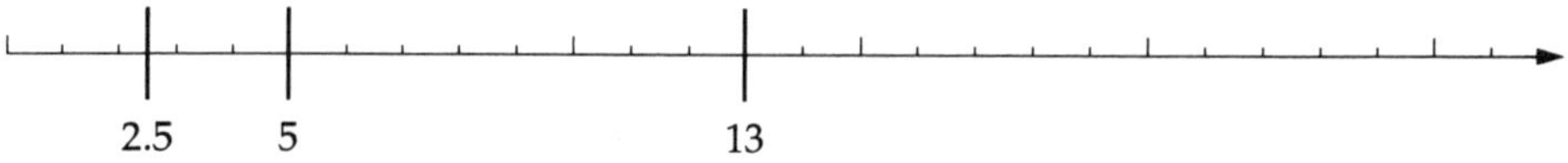

The difference between the third quartile and the first quartile, called the **interquartile range,** is a useful measure of the amount of variation in a set of data. The justification of its use is that besides being a good measure of the amount of variation in a data set, it is not strongly influenced by the location of a small number of extreme points. That is, like the median it is robust against the influence of a small proportion of the data. For these data we have

$$\begin{aligned}\text{Interquartile range} &= \text{third quartile} - \text{first quartile}\\ &= 13 - 2.5 = 10.5\end{aligned}$$

How do we interpret the interquartile range? We know that the middle half of the data lie between the first and third quartiles. Thus we can say that the middle half of the data are spread over a range of values 10.5 medals wide.

There is a graph that uses these points that we have just marked on the line above. It is shown below. This graph is called a **box-and-whisker plot,** or sometimes just a **box plot.** Here we have included the original body of data to provide an idea of how well the box plot represents it:

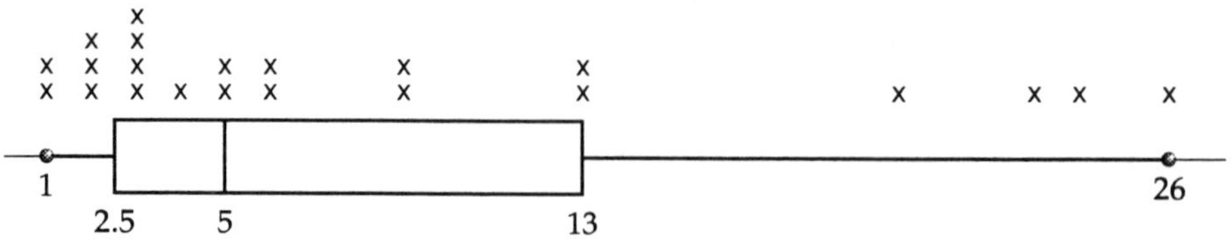

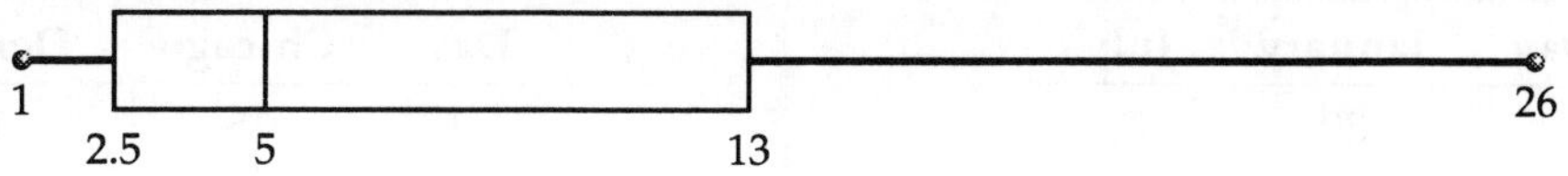

Figure 2.3 **Box plot of 1994 Winter Olympics medals data.**

The box joins the first and third quartiles and shows the median of the data. The whiskers extend from each end of the box to the highest and lowest data values, thereby marking the range of the data. The box includes the middle one-half of the data. In our example, the whiskers extend from 1 to 26 (the range of the data). The box encloses the middle one-half of the countries. That is, one-half of the countries won between 2.5 and 13 medals. The final form of the box plot, as you will see it from now on, is presented in Figure 2.3.

Sometimes such plots will also display the mean (by an asterisk, *), to be contrasted with the median. Many statistical computer program packages include both stem-and-leaf plotting and box-and-whisker plotting.

Now return to the Key Problem and consider its box-and-whisker plots as given in Figure 2.1. (Could you construct these box-and-whisker plots?) Note how informative it is to contrast these two very different populations, both in their centering and in their spread. The general shapes and the smallest, largest, median, first quartile, and third quartile values are an intrinsic part of this graphical representation.

SECTION 2.3 EXERCISES

1. Refer to the city mileages (miles per gallon) for selected 1993 automobiles given in Table 2.4.
   a. Find the mean and median mileages for these automobiles.
   b. Find the first and third quartiles of the mileage data.
   c. Find the interquartile range of these data.
   d. Draw a box plot of these data.
2. The following table gives the daily maximum high temperatures for Flagstaff, Arizona, in January and July 1995.

Day	January	July
1	32	74
2	42	77
3	37	77
4	33	80
5	33	84
6	32	87
7	34	86
8	40	88
9	45	87
10	50	85
11	39	84
12	36	80
13	46	75
14	51	77
15	39	82
16	30	72
17	29	78
18	36	69
19	35	80
20	40	81
21	29	82
22	33	80
23	41	84
24	45	86
25	38	88
26	32	90
27	36	91
28	39	94
29	35	91
30	41	89
31	58	86

a. Find the mean and median temperatures for January and July.
b. Find the interquartile range for January and for July.
c. Draw a box plot for the January data and for the July data.

3. The peak wind gusts (in miles per hour) at Chicago's O'Hare International Airport and Denver's Stapleton Airport for February 1995 are given below.

Day	Chicago	Denver
1	15	32
2	15	39
3	30	15
4	30	27
5	26	17
6	18	16
7	23	14
8	21	14
9	27	32
10	34	26
11	32	20
12	21	12
13	26	11
14	24	38
15	25	14
16	12	17
17	17	22
18	26	36
19	22	17
20	33	20
21	33	13
22	18	18
23	34	12
24	24	14
25	24	17
26	25	25
27	15	26
28	17	13

a. Find the mean and median of the peak wind gusts for Chicago and Denver.
b. Find the interquartile range of the peak wind gusts for Chicago and Denver.
c. Draw a box plot of the data for each city.
d. Comment on Chicago's reputation as "the windy city."

4. The following table of U.S. weather information appeared in the *Champaign-Urbana News-Gazette* on Sunday, October 27, 1996.

National Temperatures

Temperatures on left indicate the previous day's high and overnight low to 7:31 p.m. CDT. Today's forecast temperatures and tomorrow's temperatures and outlook are in the right columns.

	Yesterday			Today		Tomorrow		
	Hi	Lo	Prc	Hi	Lo	Hi	Lo	Otlk
Albany, N.Y.	65	32		64	37	57	46	cdy
Albequerque	40	29	.01	49	33	50	36	cdy
Amarillo	69	47		45	35	55	34	cdy
Anchorage	20	11		20	8	23	7	clr
Asheville	63	53		74	54	74	56	cdy
Atlanta	75	60	.01	75	64	78	64	cdy
Atlantic City	65	46		66	55	67	54	cdy
Austin	86	69	.01	83	69	80	69	cdy
Baltimore	71	50		70	52	70	52	cdy
Billings	34	31	.17	48	18	60	34	clr
Birmingham	71	61	1.02	78	66	76	60	cdy
Bismarck	38	37	.92	41	30	55	30	cdy
Boise	54	35		52	27	55	32	clr
Boston	67	47		59	49	58	53	cdy
Brownsville	87	76	.01	87	75	86	75	cdy
Buffalo	66	45		65	53	50	45	clr
Burlington, Vt.	59	35		62	38	51	45	cdy
Casper	m	30	.94	39	15	51	26	cdy
Chrlstn, SC	78	52		82	62	82	64	cdy
Chrlstn, WV	62	50	.15	76	57	64	49	cdy
Chrltte, NC	73	59		78	59	80	59	cdy
Cheyenne	33	33	.18	41	20	53	25	cdy
Chicago	68	55		68	58	57	38	cdy
Cincinnati	61	51	.32	74	58	64	46	clr
Cleveland	68	46		72	58	53	45	cdy
Colmbia, SC	76	54		79	58	82	60	cdy
Columbus, Ohio	62	52	.12	74	57	58	43	clr
Concord, N.H.	65	32		61	31	56	42	rain
Dallas–Ft W.	80	67		77	70	71	59	rain
Dayton	61	53	.14	72	56	61	43	clr
Denver	42	33	.32	45	24	54	28	cdy
Des Moines	75	57		63	44	62	33	clr
Detroit	60	48		69	55	54	40	cdy
Duluth	53	45	.02	48	41	45	25	cdy
Evansville	63	55	.65	75	58	67	47	cdy
Fairbanks	−5	−24		10	−15	17	0	sn
Fargo	54	39	.10	43	35	53	23	cdy
Flagstaff	30	18	.13	34	23	34	18	sn
Gnd Rapids	63	43		64	54	53	37	cdy
Great Falls	41	23		64	54	53	37	cdy
Grnsboro, NC	67	56	.01	76	53	78	55	cdy
HartfrdSpgfld	66	m		65	36	58	48	cdy
Helena	41	33		47	14	57	25	cdy
Honolulu	88	71		88	72	88	73	clr
Houston	87	66	.06	81	76	81	74	cdy
Indianapolis	62	54	.08	72	57	60	43	cdy
Jacksn, Miss.	82	66	.74	82	68	80	67	cdy
Jacksonville	85	62		85	66	83	64	cdy
Juneau	m	38		44	39	33	28	sn
Kansas City	76	63		61	51	63	39	cdy
Las Vegas	59	37		62	47	63	46	cdy
Little Rock	79	61	.07	78	65	70	58	cdy
Los Angeles	72	55		74	58	73	59	clr
Louisville	65	52	.28	76	56	64	48	clr
Lubbock	71	40		45	42	50	35	cdy
Memphis	77	62	1.00	80	63	74	55	cdy
Miami Beach	86	76		86	76	86	75	cdy
Midland, Tx.	79	44		57	49	57	41	cdy
Milwaukee	62	53		65	53	54	38	cdy
Mpls–St Paul	66	54		54	45	51	29	cdy
Nashville	71	57	.15	77	62	70	52	cdy
New Orleans	84	71	.04	85	74	85	72	cdy
New York C.	65	52		68	56	63	55	cdy
Norfolk, Va.	70	57		72	55	74	60	cdy
North Platte	61	47	.03	49	26	60	23	cdy
Okla City	79	64	.08	67	60	63	50	rain
Omaha	77	57		55	39	61	30	cdy
Orlando	88	66		88	67	86	68	cdy
Philadelphia	70	m		66	50	66	53	clr
Phoenix	63	51		64	50	70	51	rain
Pittsburgh	68	47		71	55	59	48	cdy
Portland, Me.	62	36		57	35	54	44	rain
Portland, Ore	53	38		59	36	62	41	cdy
Providence	66	39		63	40	59	50	cdy
Ralgh-Durham	75	55		77	55	79	58	cdy
Rapid City	41	31	1.49	43	23	49	25	cdy
Reno	40	32	.01	50	24	57	21	cdy
Richmond	68	52	.02	72	52	70	53	cdy
Sacramento	70	48		71	46	74	45	clr
St Louis	67	58		71	58	63	46	cdy
Salt Lake C.	42	33	.19	52	29	55	34	cdy
San Antonio	88	64		84	70	81	70	cdy
San Diego	67	m		69	55	68	56	clr
S Francisco	65	49		75	52	73	52	clr
S Juan, PR	m	77		85	75	86	75	rain
Santa Fe	m	m		45	28	46	31	sn
St Ste Marie	61	42		55	50	45	35	cdy
Seattle	51	42		51	39	54	43	rain
Shreveport	75	65	.04	80	70	80	68	cdy
Sioux Falls	73	51	.01	51	42	59	29	clr
Spokane	47	28		46	29	50	35	rain
Syracuse	67	34		69	44	52	49	cdy
Tampa–St Ptr	88	71		89	71	86	68	cdy
Topeka	78	61		59	50	61	38	cdy
Tucson	48	47	1.57	56	40	66	47	rain
Tulsa	78	65		65	60	57	50	rain
Washing., DC	72	56		73	55	71	55	cdy
Wichita	76	61		65	52	62	45	cdy
Wilkes-Barre	69	38		63	46	57	47	clr
Wilmngtn, De	68	46		66	49	65	53	clr

National temperature extremes for Saturday: high 95 at Alice, Texas; low 10 at Monticello, Utah

a. Find the high and low temperatures for these U.S. cities as reported for October 26, 1996:

Atlantic City, N.J.

Bismarck, N.D.

Dayton, Ohio

Des Moines, Iowa

Flagstaff, Ariz.

Honolulu, Hawaii

Orlando, Fla.

Reno, Nev.

Sacramento, Calif.

Seattle, Wash.

b. Draw a box plot for the high temperatures in these cities.

c. Draw a box plot for the low temperatures in these cities.

5. The following table of international weather information appeared in the *Champaign-Urbana News-Gazette* on Sunday, October 27, 1996.

a. Find the high and low temperatures for these international cities as reported for October 26, 1996:

Athens

Barbados

Berlin

Cairo

Dublin

Paris

Rio de Janeiro

Rome

Seoul

Tokyo

b. Draw a box plot for the high temperatures in these cities.

c. Draw a box plot for the low temperatures in these cities.

World Temperatures

Temperatures, conditions from midnight to midnight previous day.

	Hi	Lo	Wthr
Athens	60	48	cdy
Barbados	86	75	cdy
Berlin	51	42	cdy
Bermuda	77	69	rain
B' Aires	80	62	clr
Cairo	77	60	clr
Dublin	55	41	clr
Havana	86	69	clr
Hong Kng	78	73	cdy
Jerusalem	62	50	rain
London	60	41	cdy
Madrid	75	50	clr
Manila	87	71	cdy
Mexico C.	77	53	clr
Montreal	51	33	clr
Moscow	33	29	cdy
Nassau	86	71	clr
Paris	62	48	cdy
Rio	89	64	rain
Rome	71	42	cdy
Seoul	62	50	clr
Tokyo	77	62	cdy

6. *Money* magazine investigated the work done by income tax preparation specialists. They sent the same 1040 income tax report to 50 tax preparers. Below are values of the calculated income tax due, in thousands of dollars, for the same tax form.

12.5	14.7	16.0	16.6	16.7
16.8	16.9	17.3	17.3	17.5
18.4	18.9	19.1	19.1	19.2
19.7	21.1	21.8	21.9	22.6
22.7	22.7	22.7	22.7	22.8
22.9	22.9	22.9	22.9	23.1
23.3	23.4	23.4	23.5	23.5
23.5	23.5	23.6	23.7	24.0
24.0	24.0	23.2	24.4	24.5
25.0	25.3	25.7	25.9	35.8

a. Find the mean and median amount of tax due.
b. Find the interquartile range of the amount of tax due.
c. Draw a box plot of the amount of tax due.
d. Suppose a person with this income tax report was audited by the Internal Revenue Service. How could he or she use this study to his or her advantage?

7. Here are the densities (in grams per milliliter) of U.S. pennies produced before and after 1983.
a. Draw a box plot of the pre-1983 and the post-1983 pennies.
b. What can you say about the two groups of pennies? What could explain the changes you see?

Pre-1983 density (g/ml)	Post-1983 density (g/ml)	Pre-1983 density (g/ml)	Post-1983 density (g/ml)
8.95	8.04	8.61	8.09
8.82	7.84	8.97	8.17
9.32	7.95	8.93	8.07
8.74	7.67	8.91	8.05
9.19	8.10	9.02	7.99
8.59	8.08	9.10	8.18
8.90	8.06	9.56	8.16
8.96	8.03	9.08	7.89
8.85	7.83	8.95	7.85
9.26	7.90	8.81	8.00
8.73	7.66	9.02	7.96
9.01	8.19	8.77	7.75

Graphing Calculator Exercises

8. Here are the average prices, per night, of first-class hotel rooms in several cities around the world as reported in 1980:

Zurich	$120	Tokyo	$144
Chicago	98	Paris	113
New York	106	London	139
Stockholm	109		

Enter the data into data list L1. Find the mean, median, and range (see Section 1.5 in the TI Graphing Calculator Supplement) of these prices.

9. Here are 1997 prices of a first-class hotel room in the same cities as in Exercise 8.

Zurich	$263	Tokyo	$287
Chicago	290	Paris	479
New York	275	London	393
Stockholm	266		

Do side-by-side box plots of the 1980 and 1997 data.

2.4 DEVIATION VALUES

In Chapter 1 we introduced our first measure of spread, the range. Few use this measure seriously, for it is very nonrobust, being entirely controlled

Table 2.4 **City Mileages of Selected 1993 Cars**

Make and model	City MPG	Difference from average city MPG
Geo Metro	46	46 − 24 = +22
Subaru Justy	33	33 − 24 = +9
Ford Festiva	31	31 − 24 = +7
Dodge Colt	29	29 − 24 = +5
Mitsubishi Mirage	29	29 − 24 = +5
Hyundai Scoupe	26	26 − 24 = +2
Mazda 626	26	26 − 24 = +2
Acura Integra	25	25 − 24 = +1
Chevrolet Lumina	21	21 − 24 = −3
Volvo 240	21	21 − 24 = −3
Buick Riviera	19	19 − 24 = −5
Pontiac Firebird	19	19 − 24 = −5
Lincoln Continental	17	17 − 24 = −7
Cadillac Seville	16	16 − 24 = −8
Chevrolet Astro	15	15 − 24 = −9
Ford Aerostar	15	15 − 24 = −9
Average MPG	24.25	

by the two extreme values of the data set. In the section preceding this one we showed how to present a clearer idea of the spread of a set of data by means of a graphical box plot. We also discussed another measure of the spread: the statistic called the *interquartile range.* The middle half of the data are spread over a range of values whose width is given by the interquartile range.

In Sections 2.5 through 2.7 we will discuss three more measures of spread: the mean absolute deviation, the variance, and the standard deviation. These measures are of great importance in statistics, and we will utilize them throughout this book. Before discussing them, however, we must introduce the notion of the *deviation* of a given data point.

Table 2.4 shows the gasoline mileage for city driving of selected 1993-model cars. The average (mean) mileage for these cars is 24 miles per gallon (after rounding 24.25 to 24 to simplify computations). Here is how to read the last column of this table. A Geo Metro gets 46 miles per gallon (MPG) in city driving. This is 22 MPG *above* the average of 24 MPG. So we can say that the mileage of a Geo Metro *deviates* from the average car mileage by $46 - 24 = +22$ MPG. Similarly, we note from the table that a Lincoln Continental gets 17 MPG in city driving. This is 7 MPG *below* the average of 24 MPG. So we can say that the mileage of a Lincoln Continental deviates from the average car mileage by $17 - 24 = -7$ MPG.

The number that tells us a value's distance away from the average, or mean, is called the **deviation value.** The deviation value of the Geo is 22 MPG. The deviation value of the Continental is −7 MPG.

Notice that since we also know that the average mileage for these cars is 24 MPG, we can recalculate the original value (in this case, the city MPG) from the deviation value.

Make and model	Deviation value	City MPG
Geo Metro	+22 MPG	24 + 22 = 46 MPG
Lincoln Continental	−7 MPG	24 − 7 = 17 MPG

How to compute the deviation value: Suppose we have a set of data and their mean. The deviation value of a given data value is the difference between that data value and the mean.

A deviation value is also sometimes called a *residual value,* or just a *residual,* by statisticians because it is what is left over after the influence of the average is subtracted out. We now consider an example.

Example 2.9

Let's consider the quiz scores of five students given in Table 2.5.

Deviation scores make it easy for the students to compare their performances on the quiz with those of the rest of the class.

Bill: Deviation score = 2.
"I was 2 points above the average (mean)."

Mary: Deviation score = −1.
"I was 1 point below the average."

Fred: Deviation score = 0.

Table 2.5 **Quiz Scores for Five Students**

Student	Score	Deviation score
Bill	9	9 − 7 = 2
Mary	6	6 − 7 = −1
Fred	7	7 − 7 = 0
Tom	4	4 − 7 = −3
Bea	9	9 − 7 = 2
Total	35	
Average (mean) score = 35/5 = 7.0		

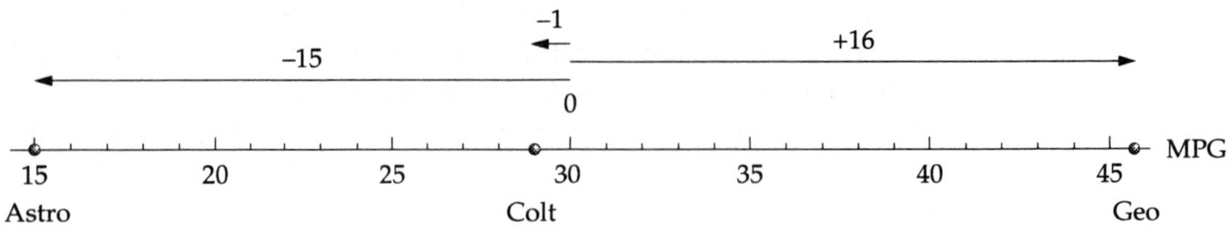

Figure 2.4 **Deviation values for MPG of three cars that average 30 MPG.**

Deviation values can be shown in a graph to help describe the amount of variation in data. If the data display only random fluctuations about the average rather than showing some systematic influence, then the residuals will be scattered around 0 in a way that shows no obvious pattern, such as tending to be positive for the women and negative for the men in the case of the above quiz scores.

Here are the data for three of the cars listed in Table 2.4.

Make and model	City MPG	Deviation value
Geo Metro	46	16
Dodge Colt	29	−1
Chevrolet Astro	15	−15
Mean	30	0

Figure 2.4 shows these deviation values on a number line.

SECTION 2.4 EXERCISES

1. The following were the high temperatures on June 20, 1996, in five cities:

City	°F
Atlanta	81
Bismarck	60
Great Falls	66
Miami	85
El Paso	96

   a. Find the mean high temperature for these cities.
   b. Find the deviation temperature for each city.
   Which city was closest to the mean temperature?

2. Refer to the data in Exercise 1, Section 2.1.
   a. Find the deviation values of the annual earnings of the 50 people.
   b. Which earnings value is closest to the mean? Which is farthest from the mean?

3. Refer to the data in Exercise 3, Section 2.1.
   a. Find the deviation values for the 10 most highly paid Chicago Cubs.
   b. Which Cubs player has the largest deviation income?

4. Refer to the data in Exercise 5, Section 2.1.
   a. Find the deviation values for the number of calories in the various kinds of cheeses listed.
   b. Which cheese has the largest deviation value? Which cheese has the smallest deviation value?

Graphing Calculator Exercises

5. Table 2.4 lists the city MPG values for 16 different cars. Find the deviation value for each vehicle by entering the data (MPG) into L1 in the data table. Find the mean of the MPG data. Then, for each vehicle, subtract the mean from its MPG (see Section 1.5 in the TI Graphing Calculator Supplement). Which vehicle has the smallest deviation value, and which has the largest? Interpret these deviation data.

6. The table below shows the outcomes of rolling a fair six-sided die 120 times. Use program DEV to find and store the deviations in L2. First enter the frequencies into L1. The program prompts for the total number of outcomes (six in this example) and returns with the mean and tells you that the deviation values are in L2. View the table STAT [edit]. Which outcome had the largest deviation value, and which had the smallest deviation value? Interpret these results.

Outcome	Frequency
1	15
2	21
3	23
4	19
5	17
6	25

2.5 MEAN ABSOLUTE DEVIATION

The mean, or average, of the deviation values can be useful in describing how much a set of data varies. Let's consider an example. Jayne's record in shooting baskets for early in the season and late in the season is shown in Table 2.6. During six games early in the season, she was getting the number of baskets out of 10 throws shown on the left. During six games late in the season, her record was as shown on the right.

We change each of these scores to deviation values (Table 2.7). In the early season, we see that Jayne's shooting record was somewhat irregular. Sometimes she was as low as four points below her average of five baskets. Other times she was above average by the same amount. We can find the

Table 2.6 **Number of Baskets out of 10 Throws**

Game	Early in season	Late in season
1	3	5
2	7	7
3	5	6
4	9	5
5	1	7
6	5	6
Totals	30	36
Mean	30/6 = 5.0	36/6 = 6.0

average distance she was from her average value by adding all the deviation values (the vertical bars, which stand for *absolute value,* say to ignore the minus signs) and finding their mean. Early in the season:

$$\text{Mean (absolute) deviation} = \frac{|-2| + |2| + |0| + |4| + |-4| + |0|}{6} = \frac{12}{6} = 2.0$$

Late in the season:

$$\text{Mean deviation} = \frac{|-1| + |1| + |0| + |1| + |1| + |0|}{6} = \frac{4}{6} \approx 0.67$$

We have described our observations statistically. Early in the season the scores varied considerably. On the average, they were two points from the mean. Late in the season Jayne's shooting improved slightly on average and also became less variable. She consistently shot around her late-season mean of six baskets in 10 throws, averaging a mean elevation of only two-thirds of a basket (per 10 throws) away from that mean.

Table 2.7 **Deviation Values for the Number of Baskets in Table 2.6**

Game	Early in season	Late in season
1	3 − 5 = −2	5 − 6 = −1
2	7 − 5 = 2	7 − 6 = 1
3	5 − 5 = 0	6 − 6 = 0
4	9 − 5 = 4	5 − 6 = −1
5	1 − 5 = −4	7 − 6 = 1
6	5 − 5 = 0	6 − 6 = 0

The procedure for finding the mean absolute deviation:

1. Find the mean of the scores.
2. Change all the scores to deviation scores.
3. Find the mean (average) of all the deviation scores, ignoring the minus signs.

SECTION 2.5 EXERCISES

1. Find the mean absolute deviation of this set of quiz scores:

Bill	3
Jane	7
Mary	6
Pat	5
Phil	9

2. Here are wind speeds observed in five cities, in miles per hour:

Juneau	8.4
Chicago	10.3
Boston	12.5
Nashville	8.0
Miami	9.2

Find the mean wind speed, median wind speed, range of wind speeds, and mean absolute deviation of wind speeds.

3. Find the mean absolute deviation of the car mileage data given in the table in Table 2.4. Refer to Exercise 1 in Section 2.3.
4. Find the mean absolute deviation of the number of gold medals won in the 1994 Winter Olympics given in Exercise 7 in Section 2.1.

Graphing Calculator Exercises

5. A scientist recorded the seismic vibrations produced by waterfalls. The measurements are given below. Enter the vibrations into L1 in the data table.

Lower Yellowstone	5
Yosemite	3
Canadian Niagara	6
American Niagara	8
Upper Yellowstone	9
Gullfoss (lower)	6
Firehole	19
Godafoss	21
Gullfoss (upper)	40
Fort Greely	40

a. Use the program DEV to find the deviation scores.
b. Find the mean absolute deviation of these scores. Sum the absolute value of L1. Use

2ND [list] [math] [sum] ENTER 2ND [abs] 2ND [L1] ENTER

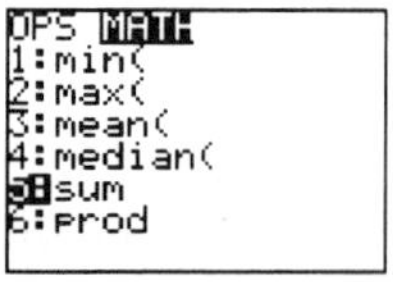

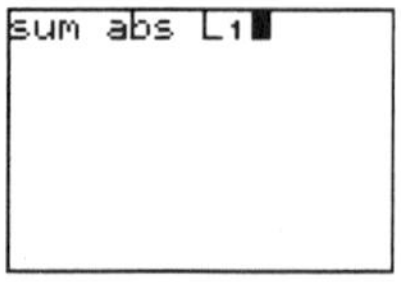

Then divide the result by the number of data points.

6. Enter the data from Table 2.4 (city mileage of selected 1993 cars) into L1. Use the program DEV and find the mean absolute deviation.

7. Use data from Table 2.2 and the program DEV to answer the following:
a. Find the mean absolute deviation of the California temperature data.
b. Find the mean absolute deviation of the Colorado temperature data.
c. Which state has locations reported here that have the least extreme temperatures? Explain your answer.

2.6 VARIANCE

So far we have seen several ways to tell how much a set of data varies: range, interquartile range, and mean absolute deviation. Another important measure of the variation in a set of data is **variance.** In practice, the variance is used as a measure of the spread of data much more often than the others. The reason is that the variance has many very useful mathematical properties not shared by other equally or even more plausible measures of data spread.

We will show how the variance is found by using the data on Jayne's shooting given in Table 2.8. We find the variance by squaring each deviation value and finding the mean (average) of the squares. Early in the season:

$$\text{Variance} = \frac{(-2)^2 + 2^2 + 0^2 + 4^2 + (-4)^2 + 0^2}{6} = \frac{40}{6} \approx 6.67$$

The variance of the early-season values is 6.67. That is, the average (mean) square distance of the values from their mean was 6.67.

Late in the season:

$$\text{Variance} = \frac{(-1)^2 + 1^2 + 0^2 + (-1)^2 + 1^2 + 0^2}{6} = \frac{4}{6} \approx 0.67$$

Table 2.8 **Number of Baskets out of 10 Throws**

	Early in season	Deviation value	Square of deviation value	Late in season	Deviation value	Square of deviation value
	3	−2	4	5	−1	1
	7	2	4	7	1	1
	5	0	0	6	0	0
	9	4	16	5	−1	1
	1	−4	16	7	1	1
	5	0	0	6	0	0
Total	30	0	40	36	0	4

The variance among the late-season values is much smaller than that of the early-season values. This shows, as does the mean deviation, that Jayne's shooting was much more consistent late in the season.

SECTION 2.6 EXERCISES

1. Find the variance of the quiz scores of Exercise 1 in Section 2.5.
2. Find the variance of the wind speeds of Exercise 2 in Section 2.5.
3. What is the variance of the following set of numbers? Explain.

   5 5 5 5

4. Four students made the following numbers of spelling errors in an essay.

   12 7 5 4

   What is the variance of the number of errors?
5. The four students in Exercise 4 were asked to rewrite their essays. The numbers of spelling errors made in the rewritten essays were

   8 3 1 0

   Compare the mean and variance of the number of errors made in the original essays with the number of errors made in the rewritten essays.
6. Here are the normal temperatures at Camelot and Eldorado during the year. Find the mean and variance of the temperatures for each city and compare the values.

	Camelot	Eldorado
January	29	15
February	30	20
March	35	22
April	40	30
May	42	45
June	58	58
July	60	78
August	59	77
September	50	60
October	42	58
November	38	32
December	30	20

7. The manufacturers of resistors for electric circuits have put in bids to a television company. Their prices are comparable, so the television company purchases 10 resistors of each brand and tests them. The resistors are marked 100 ohms. Each is measured for resistance, and the following values are found. Which company, Circuits R Us or Electronics Superstore, would you recommend as the supplier for the television company?

Circuits R Us	Electronics Superstore
95	99
91	103
105	101
103	98
95	99
87	102
106	103
105	97
79	101
103	98

Graphing Calculator Exercises

8. Enter the 1994 Winter Olympics medal totals from Table 2.3 into L1. Run the program DEV. Sum the square of each deviation:

2ND [list] [math] [sum] ENTER 2ND [L1]° x^2 ENTER

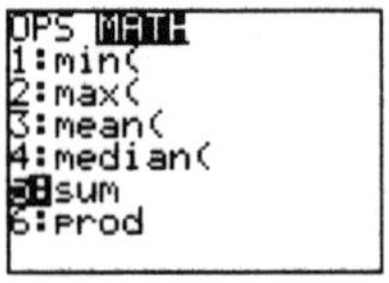

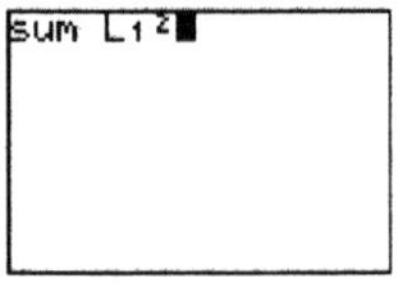

Then divide the sum by the number of data points. The resulting average (mean) is the variance of the number of medals won.

9. Find the variance of the car mileages in Table 2.4.

10. Refer to the temperature data in Table 2.2.
a. Find the variance of the California temperatures.
b. Find the variance of the Colorado temperatures.
c. Which state has locations reported here that have the least extreme temperatures? Explain your answer. Would you have expected an answer different from that of Exercise 7 in Section 2.5?

2.7 STANDARD DEVIATION

Let's first review the ways we can describe the spread, or variation, in a set of data.

1. **Range**: the distance between the smallest and the largest value. (This measure is not often used.)
2. **Interquartile range:** the difference between the first quartile and the third quartile. This measure tells the difference between the smallest and the largest value in the middle one-half of the data. (Recall that this measure is quite robust.)
3. **Mean absolute deviation:** the average (mean) difference between the data values (in absolute terms, that is, ignoring the plus and minus signs) and the mean of the set of data. (This measure is quite intuitive.)
4. **Variance:** the average (mean) squared difference between the data values and the mean score of the set of data. (This measure is less intuitive than the mean absolute deviation, but it possesses nice mathematical properties.)

Recall that the mean (the average) has the same units as the data it is computed from. If the data are in inches, their mean is in inches; if the data are city MPG values, their mean is a city MPG value; if the data are in baskets per 10 throws, their mean is in baskets per 10 throws. Similarly, the deviation value, which is the difference between a data point and the mean and expresses how far off that data value is from the mean, is in the same units as the data. Finally, the mean absolute deviation, which is the average of the absolute values of the deviation values, is also in the same units as the original data.

The variance, in contrast, is not in the same units as the data. Because it is the average of the squares of the deviation values, its units are squares of the units of the data. If the original data are in inches, the variance is in inches squared, or square inches; if the data are city MPG values, the variance is a city MPG value squared.

Recall that the mean absolute deviation expresses how far off the data are, on average, from the mean of the data. The interpretation of the variance is less clear, because its units are different from those of the original measurements, and we cannot compare numbers having different units. For this reason we often use another measure of the spread of data: the *standard deviation*.

The standard deviation, sometimes abbreviated SD, is the square root of the variance. This change in the variance rescales the variance to put it on the *same* scale as the data and the other measures of spread. So finding the standard deviation undoes the effect of distorting the scale of the data by squaring the deviation values. That is, the standard deviation is a new measure of the average difference of values from their mean. It differs from the average distance produced by the mean absolute deviation.

The standard deviation has many important statistical properties, some of which we will study later in this book.

Example 2.10

Consider Jayne's basketball scoring record again.

Early in the season:

$$\text{Variance} = 6.7$$
$$\text{Standard deviation} = \sqrt{6.7} \approx 2.6$$

Late in the season:

$$\text{Variance} = 0.67$$
$$\text{Standard deviation} = \sqrt{0.67} \approx 0.81$$

Just as the variance of Jayne's late-season values is less than the variance of her early-season values, so is the standard deviation of her late-season values less than the standard deviation of her early-season values. The standard deviation of Jayne's early-season values, 2.6, is again a kind of average difference between Jayne's baskets-per-10-throws values and their average. Taking the square root undoes the effect of squaring the deviation scores, which was done to find the variance. In this new way of thinking about the average, we can say that, on the average, Jayne's shooting record was about 2.6 from her mean of 5 baskets in every 10 throws. Late in the season, she was shooting much closer to her average value, or mean (which was now 6), since the standard deviation went down to 0.81.

Example 2.11

Using Table 2.9, find the standard deviation of city mileage for the three cars.

Solution

We know that the standard deviation is given by the square root of the variance. From Table 2.9,

$$\text{Square root of variance} = \sqrt{160.7} = 12.7$$

Table 2.9 **Deviation Values for MPG**

Make and model	City MPG	Deviation MPG value	Deviation MPG value squared
Geo Metro	46	16	256
Dodge Colt	29	−1	1
Chevrolet Astro	15	−15	225
Mean	30	0	482/3 = 160.7

It is instructive to compare the standard deviation of these mileages with their mean deviation. The mean absolute deviation is 10.7 miles per gallon. This is seen by averaging the magnitudes of the deviations from Table 2.9: $(16 + 1 + 15)/3 = 10.7$. We interpret this by saying that on average, the cars differed from the mean mileage by 10.7 miles per gallon. (Two got less than the average of 30 MPG, and one got more than 30 MPG.) Similarly, we interpret the standard deviation of 12.7 miles per gallon by saying that "on average," the cars differed from the mean mileage by 12.7 miles per gallon. But this time the average is determined by squaring how much each car deviated (differed) from the average MPG, averaging these squared deviation scores, and then taking the square root. Although these two statistics are on the same scale, they will almost always differ in value.

As this suggests, these two measures of variation will in fact often be rather close. The mean absolute deviation is more robust against a small proportion of unusually large or small numbers. The square of a large deviation is huge, thus tending to inflate the variance and its directly derived standard deviation over the mean absolute deviation.

In summary, if you wish to use statistics with more mathematical properties, you are likely to prefer the mean and the standard deviation. However, if robustness is important (and many statisticians now insist that it is essential), then you would choose the median and either the interquartile range (of the box plot) or the mean absolute deviation.

SECTION 2.7 EXERCISES

1. The number of accidents occurring in each of five weeks on a busy freeway are given below. Find the variance and standard deviation of these data.

   4 0 6 10 5

2. The number of defective valves found in each of four batches of 1000 in a machine shop are given below. Find the range, mean deviation, variance, and standard deviation of these data.

   2 4 0 10

3. Refer to the quiz scores of Exercise 1 in Section 2.5. Find the standard deviation of the quiz scores.

4. Find the variance and standard deviation for each of the following data sets:

   10 15 20 25
   10 15 20 2500

   How are the variance and standard deviation affected by changing the 25 in the first data set to a 2500 in the second data set?

5. Find the mean absolute deviation of the two data sets in Exercise 4. Which do you think is a more robust measure of variation—the mean deviation or the standard deviation? Explain your answer. (*Hint:* Reread the discussion of the robustness of statistics in Section 2.2.)

6. The following data are the mean ages (in months) and the standard deviations of students who were tested in an international study of mathematics.
   a. In which country were the students the closest in age?
   b. In which country were the students the farthest apart in age?
   c. Explain your answers to parts (a) and (b).

Country	Mean	Standard deviation
Belgium (Flemish)	171	8.0
Belgium (French)	174	11.3
Canada (B.C.)	168	6.0
France	170	8.3
Hungary	171	13.4
Japan	162	3.5
New Zealand	168	5.4
Nigeria	200	37.7
Scotland	168	4.3
Swaziland	188	22.5
United States	170	6.0

Source: *Second International Mathematics Study*, 1987.

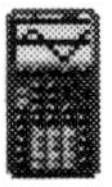

Graphing Calculator Exercises

7. Refer to Exercise 8 in Section 2.6. Find the standard deviation of the numbers of medals won. Recall that the standard deviation is the square root of the variance.

8. Refer to Exercise 9 in Section 2.6. Find the standard deviation of the car mileages. Keeping in mind that this number (the standard deviation) is in miles per gallon, give a meaning to the value of the standard deviation that you have found.

9. Use the California and Colorado temperatures provided in Table 2.2.
   a. Find the standard deviation of the California temperatures.
   b. Find the standard deviation of the Colorado temperatures.
   c. Which state has locations reported here that have the least extreme temperatures? Compare your answer with the one you found in Exercise 10 in Section 2.6.

CHAPTER REVIEW EXERCISES

1. The heights (in inches) of the five starting members of a basketball team are listed below. Find the mean and median height.

   71 75 78 81 84

2. Suppose the shortest player is replaced by a 73-inch-tall player. What is the new mean and median height? Why does the mean change?

3. A sample of 25 light bulbs was taken from an assembly line during each of five weeks and tested. The number of defectives per week were 1, 3, 0, 0, and 7. Which measure of central tendency would make the number of defective light bulbs appear low? Which measure of central tendency would make the number of defective light bulbs appear high?
4. In the 1997 NBA playoffs, Michael Jordan of the Chicago Bulls scored 29, 55, and 28 points in three games for an average of 37.33 points per game against the Washington Bullets. In five games against the Atlanta Hawks, he scored 34, 27, 21, 27, and 24 points for an average of 26.6 points per game. Can you conclude that Michael Jordan's scoring pace was lower in the games against the Hawks? Explain your answer.
5. Draw a box plot for each of the following two sets of data.
   a. 25, 50, 55, 60, 65, 70, 75, 80, 85, 90, 115
   b. 45, 50, 55, 60, 65, 70, 75, 80, 85, 90, 95

   What is the main difference between the two sets of data?
6. Calculate the interquartile range and the range for both sets of data in Exercise 5. Which measure of spread—the interquartile range or the range—is more robust?
7. Find the median, range, and interquartile range of the high-temperature data from Exercise 6 of Section 1.2. Draw a box plot of the temperatures.
8. In Exercise 8 of Section 2.1, the number of earthquakes 7.0 or higher on the Richter scale are recorded for the years 1900–1909 and 1980–1989. Draw a box plot for each data set. Explain the differences in the data using the box plots.
9. Calculate the mean absolute deviation for the following data:

   50 49 54 60 79 65 45 62

10. For the earthquakes data in Exercise 8, calculate the mean absolute deviation for both sets of data. Does it appear that the two data sets have the same spread? Explain your answer.
11. Below is a list of the numbers of home runs hit by the home run leaders of the American League during the years 1951–1965.

    33 52 43 32 37 52 42 42 42 40 61 48 45 49 32

    Calculate the mean, variance, and standard deviation of the data.
12. See Exercise 11. Roger Maris of the New York Yankees hit 61 home runs in 1961 to set a baseball record that still stands (as of the 1997 baseball season). Given this information, which measures of central tendency and spread would you report for these data? Explain your answer.
13. The data set below is the yearly corn yield in bushels per acre for six Midwestern states for the years 1890–1927. Use at least two of the following statistics to

describe this data set: mean, median, mode, range, box plot, variance, standard deviation, mean absolute deviation.

24.5	33.7	27.9	27.5	21.7	31.9	36.8	29.9	30.2	32.0
34.0	19.4	36.0	30.2	32.4	36.4	36.9	31.5	30.5	32.3
34.9	30.1	36.9	26.8	30.5	33.3	29.7	35.0	29.9	
35.2	38.3	35.2	35.5	36.7	26.8	38.0	31.7	32.6	

14. The data set below is the "PICK 3" numbers for the State of New Jersey for 50 days in 1976. Use at least two of the following statistics to describe this data set: mean, median, mode, range, box plot, variance, standard deviation, mean absolute deviation.

810	156	140	542	507	972	431	981	865	499
020	123	356	015	011	160	507	779	286	268
698	640	136	854	069	199	413	192	602	987
112	245	174	913	828	539	434	357	178	198
406	079	034	089	257	662	524	809	527	257

2

COMPUTER EXERCISES

1. CALIFORNIA PEOPLE II

Recall the CAPEOPLE.DAT data from Computer Exercise 1 in Chapter 1. Load it into the Data program again. This exercise will look at the averages, medians, and standard deviations. Choose Histograms from the Data menu. In the Histogram screen, check the Statistics box by clicking in the box to the left of "Statistics" so that it looks like this:

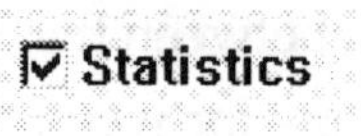

Now when you create a histogram, the statistics for the variable(s) will appear in a table at the top.

a. Select the income variable, select about 25 bins, and press Update. What are the average and the median? Which is larger? By how much? Why are these measures so different if they are both measuring "typical" incomes?

b. Go to the Options menu, select "Split on a variable," and split on sex. (Be sure that 1.5 is in the splitting value box before pressing OK.) Select the education variable with 19 bins, and press Update. What are the average educational levels for men and women? What is the difference between the averages? What are the median levels for men and women? What is the difference between the medians?

c. Now select the income variable, and press Update. What are the average incomes for men and women? What is the difference? What are the median incomes for men and women? What is the difference?

d. Are the differences between men and women greater for educational level or for income? Now do you think the difference in incomes can be "explained" by the differences in education?

2. CLASS DATA II

Recall the STAT 100 class data from Computer Exercise 6 in Chapter 1. (Close the histogram screen if you are still there.) Load the CLASS96.DATA data set, and go to relative frequency histograms. Make sure that the Statistics box is checked. Choose "Split on a variable" from the Options menu, and split on gender. (The splitting value should be 0.5.)

a. Use 19 bins, select the height variable, and click on Update. Find the difference in average height and the difference in median height between men and women. Are these differences about the same?

b. Look at the two histograms for heights. Do they seem to have similar spreads, or is one much more spread out than the other? What are the standard deviations of the men's and women's heights? Are they close?

c. Now choose the weight variable, and then click on Update. Find the difference in average weight and the difference in median weight between men and women. Why is the difference between the averages larger than the difference between the medians?

d. Look at the two histograms for weights. Do they seem to have similar spreads, or is one much more spread out than the other? What are the standard deviations of the men's and women's weights? Are they close? Which is more spread out, the men's weights or the women's weights? Why?

e. Now go to the Options menu and choose "One variable." Choose 41 bins, the DogsCats variable, and Update. Which is larger, the average or the median? Why?

3. MORE CLASS DATA: GRADES

The GRADES96.DAT data set has the grades from another STAT100 class at the University of Illinois. Load this data set, and then go to the histograms.

a. Choose "Split on a variable" from the Options menu, and split on gender again. Be sure the "Statistics" box is checked. Find the differences in averages between males and females on all the variables (except gender).

b. For the class, do women perform better at some tasks (among labs, homework, quizzes, tests, and finals) than men, and worse at others? Is there a pattern? Or do men and women perform fairly equally?

c. What percentage of the class is female? (You can find the numbers of men and women in the class by looking at the "n = ??" part in the little table of statistics.) Is this percentage more than or less than the overall percentage of females at the University of Illinois? (Overall, about 54% of the students are male and 46% are female.)

d. Go to the Options menu, and choose "A Different Split." Double-click on "Labs." The number in the box on the right should be 94.4450. Click on OK. Now look at the histograms for HW. The top histogram is for the people who did not do well on the labs; the bottom one is for the people who did very well. What is the difference between the averages? How do the histograms differ? Find the difference between averages, keeping this same split but looking at the other variables, Quiz, Tests, Final, and Total.

e. Go to the Options menu, and choose "A Different Split," but this time double-click on "HW." Click on OK, and find the differences in averages for Tests, Final, and Total.

f. Do the same again, but split on "Quiz."

g. Roughly, how many points better do people who do well on the labs, homework, and quizzes do on the tests, final, and total than people who do not do well on the labs, homework, and quizzes?

h. What lessons can be learned from these data? Does performing well on the homework and labs cause good performance on the exams, or might there be other factors? What other factors?

4. ANIMALS I

Load the data set ANIMALS.DAT. It contains the typical body weights in kilograms and brain weights in grams of 27 animal species.

a. Go to the histogram screen, select the Statistics box, set the number of bins to 100, choose the body weight variable, and click on Update. Write down the mean and median body weights. Are they about the same? If not, why not?

b. Select brain weights, and then Update. Write down the mean and median brain weights.

c. One animal had an especially large body weight. Which animal was it? To find out, close the histogram screen, and then select "Input/Modify Data" from the Data menu. A new screen will appear. Click on the little circle next to "Use current data set: Body and Brain Weights," so that it looks like this:

Now click on OK. A spreadsheet with the data should appear. Look down the body weight column to find the largest value. (It may be that some values are "#######" or something similar. That occurs when the number is too large to fit in the space. You can expand the column. First, place the cursor between the two columns at the top as in the indicated place in the next figure:

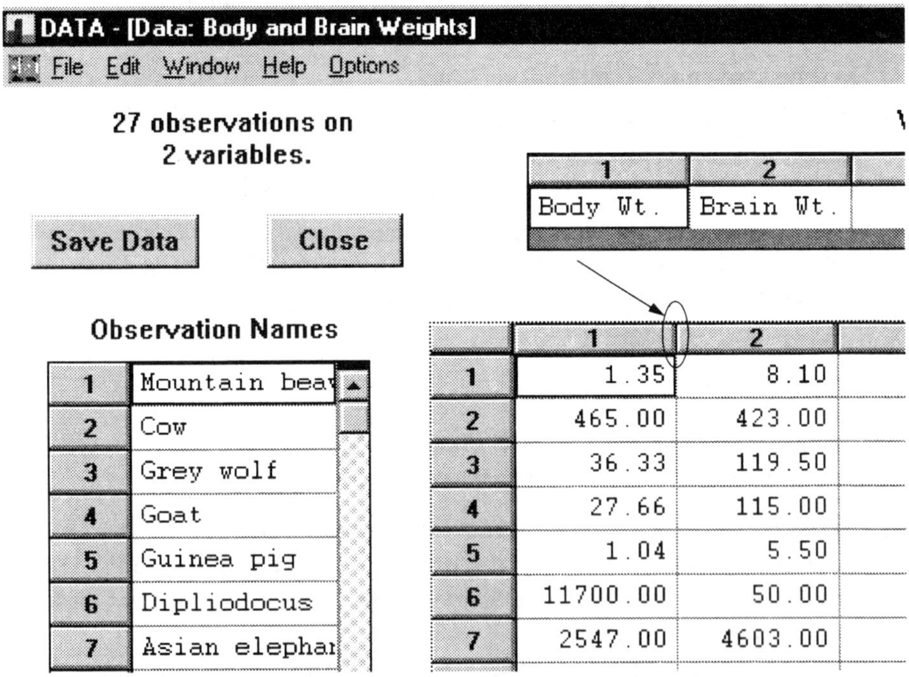

Then move the cursor slightly to the right while holding down the left mouse button.) What is the largest body weight? Which row is it in? (For example, row 4 has the goat, which weighs 27.66 kilograms.)

What is the animal? (Look for the same row in the Observation Names column.)

d. While still in this same screen, select "Delete Observation" from the Options menu. A new box will pop up, asking for the number of the observation to delete:

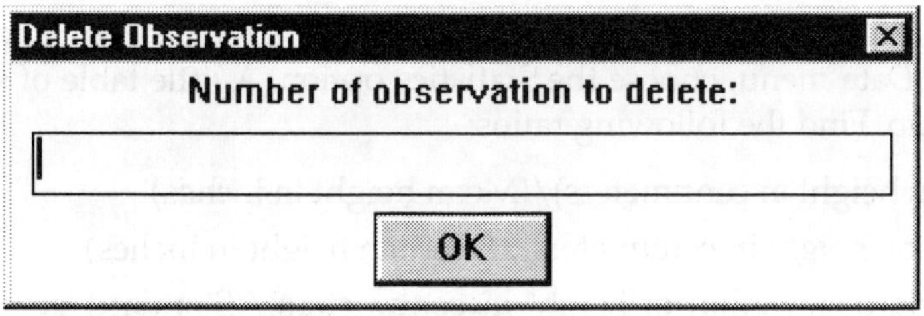

Type in the row number of the largest animal, and then click on OK. That animal will be deleted. Now click on the Close button. Another box will appear asking whether to save the data. Click on "No."

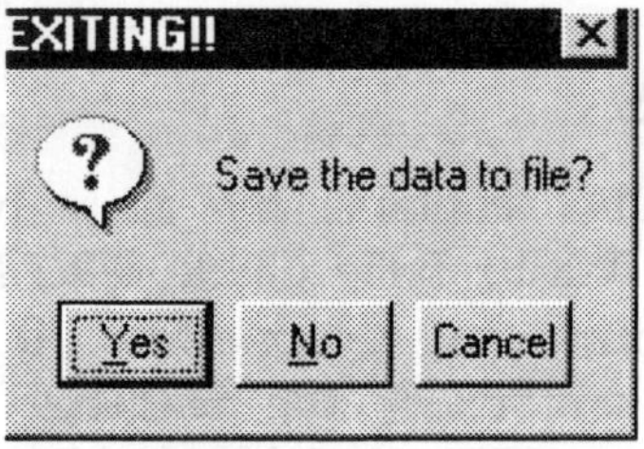

Now you should be back at the beginning screen. As before, go to the histogram screen, select the Statistics box, set the number of bins to 100, choose the body weight variable, and click on Update. Write down the new mean and median body weights (without that one big animal).

e. Select brain weights, and then Update. Write down the new average and median brain weights.

f. Find the difference in the mean body weight with and without that biggest animal. Did the average go up or down? Find the difference in the median body weight with and without that animal. Did the median go up or down? Which was most affected by deleting the largest animal—the mean or the median? Or were they affected to about the same degree? Why?

g. Find the difference in the mean brain weight with and without that biggest animal. Did the average go up or down? Find the difference in the median brain weight with and without that animal. Did the median go up or down? Which was most affected by deleting the largest animal—the mean or the median? Or were they affected to about the same degree? Why?

5. NBA ALLSTARS II

Load the ALLSTARS.DAT data (see Computer Exercise 7 of Chapter 1). From the Data menu, choose the Statistics option. A little table of statistics will pop up. Find the following ratios:

a. (Mean height in centimeters)/(Mean height in inches)
b. (Median height in centimeters)/(Median height in inches)
c. (Standard deviation of height in centimeters)/(Standard deviation of height in inches)
d. (Minimum height in centimeters)/(Minimum height in inches)
e. (Maximum height in centimeters)/(Maximum height in inches)

Are these ratios all about the same? Why?

6. TEMPERATURES

The data set TEMPS.DAT contains the daily high temperatures for 1987 at six sites: Olga, Wash.; Belle Glade, Fla.; Mexia, Tex.; Mayville, N.D.; Minneapolis, Minn.; and Hoopeston, Ill. Choose Histograms from the data menu, and check the Statistics box. Go to the Options menu and choose "Two variables." Now there are two lists of variables. You choose one variable from each list and select Update to see the two corresponding histograms.

a. Select Mayville and Minneapolis in the variable lists, type in 25 bins, and then click on Update. Are their histograms similar or quite different? Are the averages similar? Are the spreads (standard deviations) similar? Are the extremes similar? Are the histograms bell-shaped, or do there seem to be two clusters of temperatures? If so, what do those clusters represent?
b. Now select Olga and Minneapolis in the variable lists, and select Update. Looking at the histograms, do both cities have approximately the same mean temperature? Does the mean tell the whole story? Which city seems to have more extreme temperatures? What are the standard deviations of the temperatures? Are the shapes of the histograms different? In what way?

c. Now select Olga and Belle Glade in the variable lists, and select Update. Compare these two cities on their average temperatures, their standard deviations, and the shapes of the histograms.

d. Do the same as in part (c), but compare Mexia and Belle Glade in the variable lists and select Update. Compare these two cities on their average temperatures, their standard deviations, and the shapes of the histograms.

e. Histograms that have just one peak are called *unimodal* (because they have just one mode), and those with two peaks (or clusters) are called *bimodal.* Bimodal histograms are often mixtures of two different populations, such as men's and women's heights. For each of the six cities, decide whether it has a high or moderate average temperature, a large or small spread, and a unimodal or bimodal shape:

	Average	**Spread**	**Shape**
Olga, Wash.			
Belle Glade, Fla.	High	Small	Unimodal
Mexia, Tex.	High	Large	Bimodal
Mayville, N.D.			
Minneapolis, Minn.			
Hoopston, Ill.			

Can you group some of the cities by the characteristics given in the table? What do these cities have in common (other than similar temperature distributions)?

3

RELATIONSHIPS

One has to draw the line somewhere.

OBJECTIVES

After studying this chapter, you will understand the following:

- *The scatter plot as a graph of two-variable data*
- *How to find the line that best fits a set of two-variable data by means of the least mean error and the least squares method*
- *Covariance as a measure of the association of two variables*
- *Correlation as a measure of how well a straight line fits two-variable data*

3.1 SCATTER PLOTS   105
3.2 GRAPHS OF LINEAR RELATIONSHIPS   111
3.3 LINEAR REGRESSION: AN INFORMAL VISUAL APPROACH   116
3.4 LEAST MEAN ERROR LINEAR REGRESSION   122
3.5 LEAST SQUARES LINEAR REGRESSION   130
3.6 COVARIANCE, CORRELATION, AND THE LEAST SQUARES REGRESSION LINE   138
3.7 RELATION AND CAUSATION   153

CHAPTER REVIEW EXERCISES   156

COMPUTER EXERCISES   162

Key Problem

Building a Linear Model of Data

Do heavier cars get poorer mileage? We might think so, but how can we be more certain? Table 3.1 gives the weights and mileages of 16 car models. Going through this table of data as it stands, it's hard to tell how the two variables—weight and mileage—are related. So we draw a graph to see if there is an underlying pattern in the data.

Figure 3.1 is a plot of the data. Each point in the plot represents one of the cars from Table 3.1. For example, the point that is on the far left of the graph (and highest on the plot) is the Geo Metro. From the table we see that the Geo Metro weighs about 1700 pounds and gets 46 miles per gallon in city driving. Make sure that you see that the point (1700, 46), which represents the Geo Metro, is in about the right place in the plot.

Table 3.1 **Auto Weights and Mileages**

Car	Weight (pounds)	City MPG
Acura Integra	2705	25
Buick LeSabre	3470	19
Cadillac Seville	3935	16
Chevrolet Lumina	3195	21
Chevrolet Astro	4025	15
Dodge Colt	2270	29
Ford Festiva	1845	31
Ford Aerostar	3735	15
Geo Metro	1695	46
Hyundai Scoupe	2285	26
Lincoln Continental	3695	17
Mazda 626	2970	26
Mitsubishi Mirage	2295	29
Pontiac Firebird	3240	19
Subaru Justy	2045	33
Volvo 240	2985	21
Mean	2994	**24.2**

Source: *Journal of Statistical Education* (1993) online data sets.

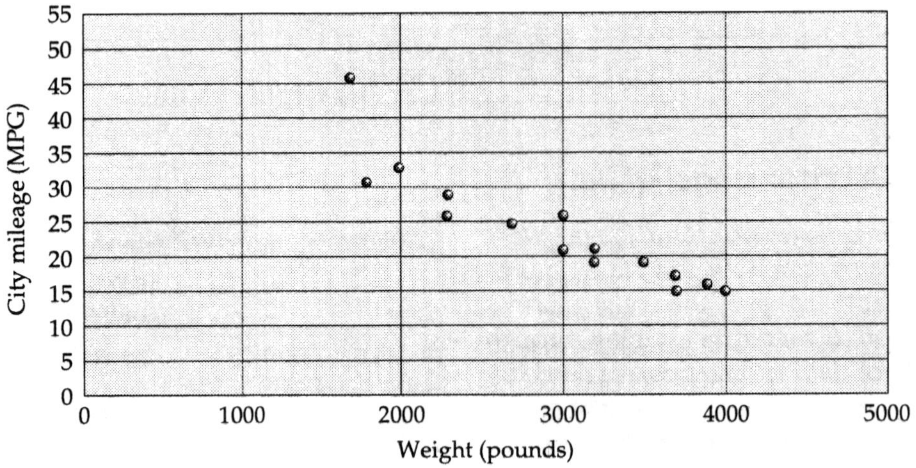

Figure 3.1 **Weight versus miles per gallon for city driving for selected cars.**

We can now look at the plot of the data and ask whether there appears to be a relationship between the weight of a car and its mileage. Indeed, we see in the plot that *generally,* the heavier a car is, the lower its mileage is. Our Key Problem is to find out how to express such relationships between two variables and to measure the strength of the linear relationship.

3.1 SCATTER PLOTS

So far in this book, we have been dealing with descriptive statistics used to summarize certain characteristics of the data, such as the mean and the standard deviation, for only one variable at a time. Such data could be the daily rainfall amounts in Seattle, the daily high temperatures in Phoenix, or the weights of members of the college football team, or any data for which each unit (day, person) yields one data value. That is, only one measurement is taken of each member chosen to form the data set. But often we want to know how two measures are related for the chosen members (farms, people, cities, and so on).

For example, we might want to know the relationship between the number of hours per day spent on homework and performance on final examinations: if I do more homework, can I expect to do better on the final exam in this course? We might want to know how the number of cigarettes smoked per week and illness are related: if I decrease the number of cigarettes per week that I smoke, does this also decrease the chance that I will get lung cancer? Fertilizer and crop yield provide another example: if I increase the amount of fertilizer that I put on this year's corn crop, can I expect an increase in the number of bushels per acre?

Plots of data like Figure 3.1 can help us see if there is a relationship between two variables. (Later in the chapter we will discuss the very complex and important issue of when the existence of a relationship between two variables can be interpreted as one variable's *causing* the other variable's behavior.) Such graphs are called **scatter plots.** Each point in the plot represents values for two variables, one for the X axis and one for the Y axis.

If the number of points in the data set is large and if the underlying relationship is linear, except for "error" or "noise," then the shape of the cloud of points in the scatter plot is usually roughly like a tipped ellipse. See Figure 3.2, where city gas mileage as related to weight is given for 92 cars.

Here, especially if one ignores the three very heavy cars, the cloud of points looks like a long and fairly narrow ellipse. If, as a class project, you collected the weights and heights of 100 adult men, the resulting scatter plot should be quite elliptical in shape. Such a plot of 100 (height, weight) pairs is shown in Figure 3.3. We will discover that obtaining an elliptically shaped scatter plot is a a good thing in that it legitimizes use of the standard technique for relating the two variables, called *least squares linear regression.*

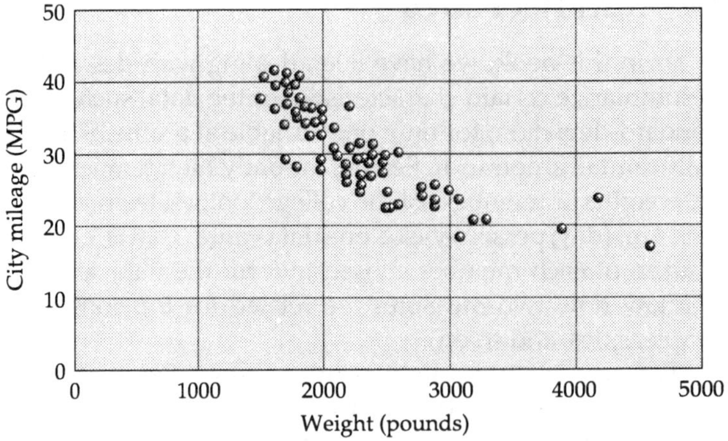

Figure 3.2 **City mileages of 92 selected cars versus their weights.**

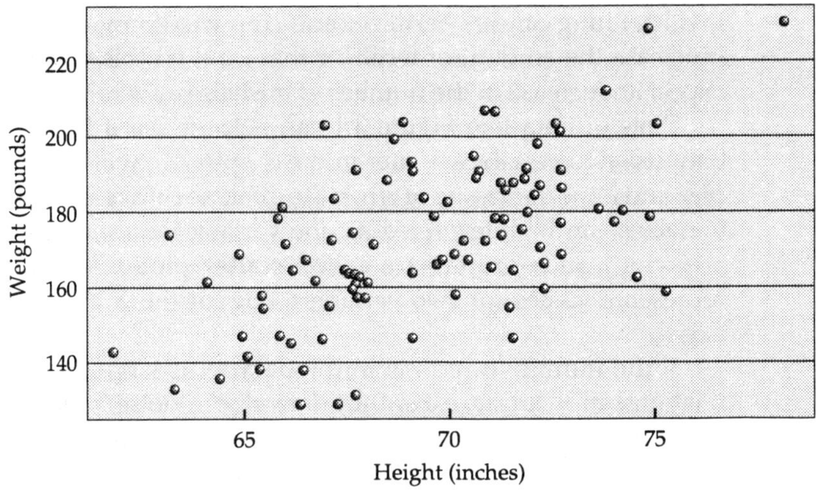

Figure 3.3 **Weights of 100 men versus their heights.**

Statistics and Human Growth

The relationship between a person's height and weight depends on many factors, including race, nationality, sex, nutrition, and age. However, an equation has been found that relates height and weight for children and youth in several countries. The equation is

$$\log w = 0.02h + 0.76$$

where w is weight in pounds, h is height in inches, and $\log w$ is the logarithm of w. (Logarithms are expressed in terms of powers of a base. For example, $\log 100 = 2$, since $10^2 = 100$. We are using base 10.)

Here is a summary of conditions under which the height/weight relationship ($\log w = 0.02h + 0.76$) has so far been found to hold reasonably well:

Race: White, Black, Chinese (in the West Indies)

Country: United Kingdom, Ghana, France, Katanga, West Indies, Canada

Time: 1880–1970, approximately

Age: 2–18 years

Sex: Male (2–18 years) and female (2–13 years)

Socioeconomic class: Various classes in United Kingdom, France, and Canada

The investigation of conditions is still continuing. However, the conditions are so varied that some have already called the equation $\log w = 0.02h + 0.76$ a "law-like relationship" for human growth patterns. Notice that the United States is not referred to in the list. Would you expect the relationship to hold for the United States as well?

SECTION 3.1 EXERCISES

1. What pair of variables are of interest when each of these questions is asked?
   a. Do children who view a lot of television do poorly in school work?
   b. Does a copper bar expand when heated?
   c. Do people who receive more vaccinations become sick less often?
   d. Do tires with higher inflation (more air in them) give better mileage than those with lower inflation?
2. Professor Eron of the University of Illinois did a study of the television viewing of 875 third-grade children. The following excerpt reports his conclusions.

   > There is a strong positive relationship between the violence rating of favorite programs, whether reported by mothers or fathers, and aggression of boys as rated by their peers in the classroom.... There were no significant relationships when TV habits of girls were reported either by mothers or fathers. (*Journal of Abnormal and Social Psychology*, vol. 67, 1963, p. 195)

   What pairs of variables are of interest in this study?
3. The following extract reports on a relationship between variables. What are the variables, and what is the relationship? (*Hint:* Pick what pair of variables you like.)

   > As people get older there is a marked decline in sports involvement—except for viewing sports on television. The average time devoted to active sports per week is 5.1 hours for those aged 18 to 24 compared with 1.7 hours for those 25 and over. But the time spent watching TV sports increases somewhat for all those over 35.

4. Identify three sets of variables whose relationships are discussed in the following extract.

    During the months immediately following the 1974 gasoline shortage it was noted that deaths from vehicular accidents were on a significant decline for the first time in decades. The phenomenon was hailed at the time as one of the few benefits from the fuel crunch.

    More recently notice has been paid to other statistics from that period which indicate a drop in deaths from causes other than highway accidents, but which some researchers believe could have a tie-in to reduced driving habits. San Francisco County, for one, noted a 13 percent decline in deaths from all causes during a three-month period at the height of the gasoline shortage.

    A decrease of almost 33 percent in deaths from chronic lung disease was noted, along with a 16 percent decline in deaths from cardiovascular diseases. It is possible, of course, that a combination of factors brought about these unexpected results.

    But a number of driving-related causes also have been suggested. Less stress from driving less, fewer pollutants in the air, more walking and simply more opportunity to relax at home are among suggested possibilities. No definite conclusions have been reached, but the correlations are intriguing. (*Champaign-Urbana News-Gazette*, October 21, 1975)

5. Figure 3.2 shows the city miles per gallon for a large number of 1993-model cars of various weights.
    a. According to the graph, what is the best city mileage that a car gets?
    b. What is the worst (lowest) city mileage?
    c. What is the approximate average city mileage among all of these cars?
    d. According to the graph, as the weight of a car increases, what tends to happen to its mileage?

6. Figure E1 reports the percentage of people favoring cuts in federal aid for college students versus those people's incomes. For example, of those whose income is less than $15,000 a year, 31% favored cuts in federal aid.

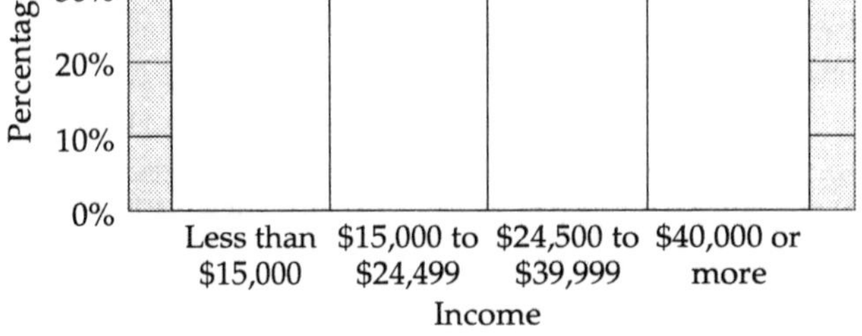

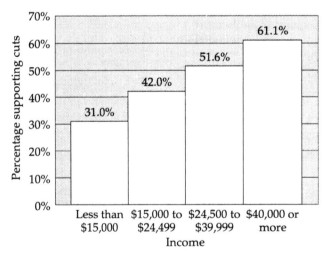

Figure E1 **Percentage of people supporting cuts in federal aid for college students, by income.**

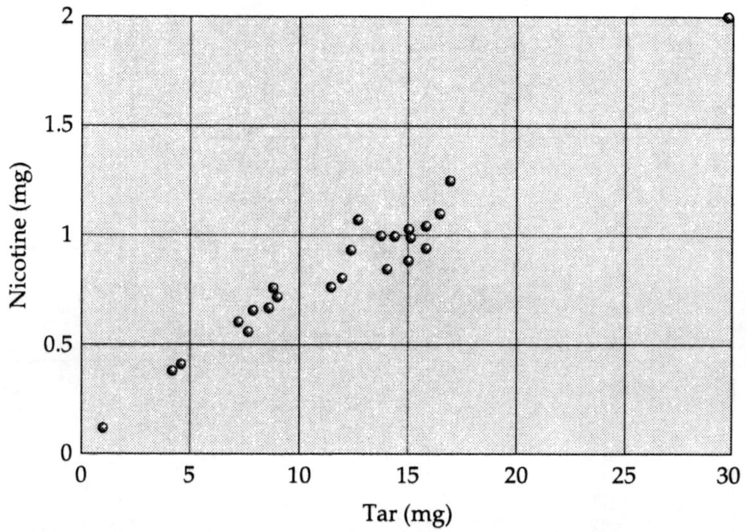

Figure E2 **Nicotine content versus tar content for 25 cigarette brands.**

a. What two variables are shown in this graph?
b. What relationship between these variables is suggested?

7. Find an article in a newspaper or a magazine that deals with a relationship between two variables. What are those variables? What relationship between those two variables is presented or discussed?

8. Figure E2 shows the amount of nicotine (in milligrams) and tar (in milligrams) for 25 brands of cigarettes.
   a. As the amount of tar in a cigarette increases, what tends to happen to its nicotine?
   b. What is the range in nicotine content for the 25 brands of cigarettes?

9. Figure E3 shows the city and highway mileage (miles per gallon) for 92 cars (they are 1993 models).
   a. According to the graph, what is the general relationship between the city mileage and highway mileage for the cars?
   b. Compare the range in city mileage with the range in highway mileage for these cars.

10. Figure E4 shows the infant mortality rate (deaths per 1000 of population for babies less than one year old) and the female life expectancy in years for 97 countries.
    a. What is the range in infant mortality for these countries?
    b. Estimate the median infant mortality rate for these countries.
    c. What is the range in female life expectancy for these countries?
    d. Estimate the median female life expectancy.
    e. What is the general relationship between infant mortality and female life expectancy?
    f. Does this relationship make sense? Explain.

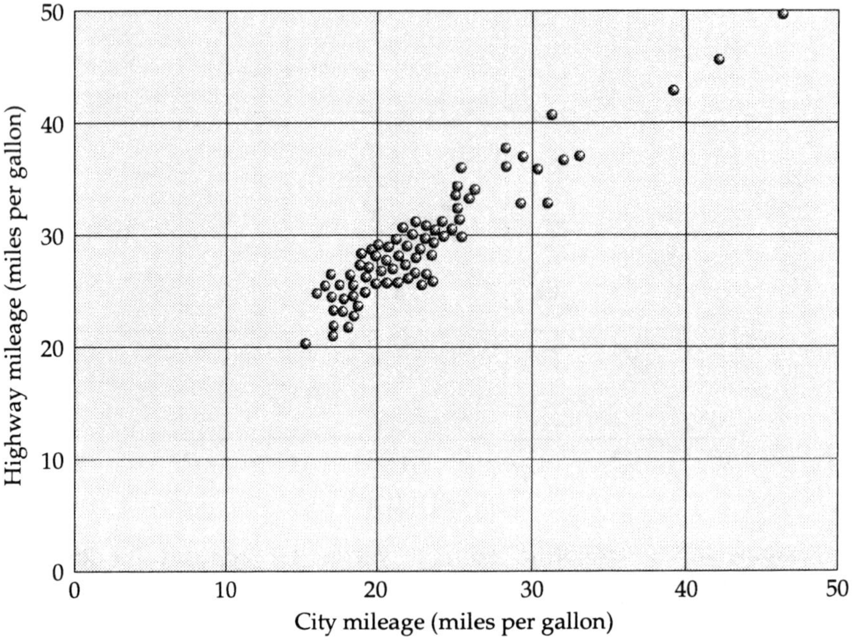

Figure E3 **Highway mileage versus city mileage for 92 cars.**

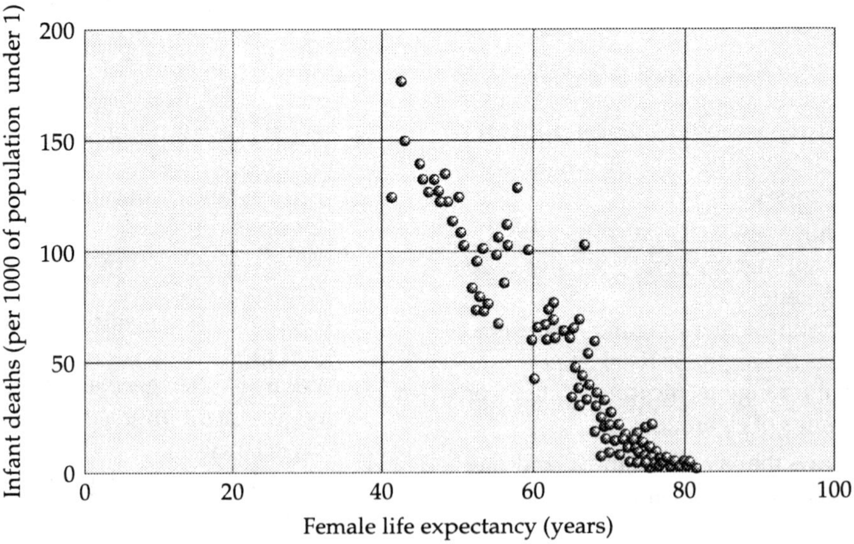

Figure E4 **Infant death rate versus female life expectancy.**

Graphing Calculator Exercise

11. The following data show a relationship between the number of people per television set and the life expectancy for various countries.

Country	Life expectancy	Number of people per television set
Argentina	70.5	4.0
Canada	76.5	1.7
Egypt	60.5	15.0
India	57.5	44.0
Iran	64.5	23.0
Japan	79.0	1.8
Morocco	64.5	21.0
South Africa	64.0	11.0
Sudan	53.0	23.0
United States	75.5	1.3

Make a scatter plot of these data (see Section 1.3 of the TI Graphing Calculator Supplement). Use [TRACE] to help answer these questions:

a. Which country has the lowest life expectancy?
b. Which country has the largest number of persons per television set?
c. Generally, can you see a relationship between life expectancy and number of persons per television set? Would you claim that the number of people per television set has a causal influence on life expectancy?

3.2 GRAPHS OF LINEAR RELATIONSHIPS

Many, but not all, relationships between two variables are approximately linear (straight lines). We consider one such case. The following data were obtained from an experiment that you could easily do yourself. A statistics class collected cylindrical objects, such as coffee cans, juice cans, and a rolled oats container. They then measured the circumference (distance around) and the diameter (distance across) of each object and prepared a table like Table 3.2. They used string to measure the diameters. The purpose of this experiment was to approximately discover the basic relationship between diameter and circumference, namely that circumference $= \pi \times$ diameter.

The class prepared a graph to show how the data are related, as shown in Figure 3.4. As you can see in the graph, the data roughly lie along a straight line. Next they looked for a rule that would fit the data they found. They knew that a straight line has an equation that looks like

$$Y = mX + C$$

Table 3.2 **Diameters and Circumferences of Selected Cylindrical Objects**

Object	Diameter (centimeters)	Circumference (centimeters)
Orange juice can	3.0	10.0
Coffee can (small)	5.0	16.0
Tomato juice can	10.8	32.5
Coffee can (large)	13.0	40.0
Rolled oats container	10.0	32.3
Soup can	6.8	21.0
Candle	4.5	18.0

where m is the slope and C is the Y intercept. So they tried drawing a line through the data in such a way that it would pass as close to the points as possible. Once they found such a line, they could write its equation, since they could find its slope and intercept (approximately) from the graph.

Many of the students had ideas about how to do this. We will discuss a few different approaches in various parts of this book. We report here only one of the suggestions. One student took a piece of thread and held it taut over the data points in such a way that it was possible to see where the "best" line should be drawn. When the thread appeared to pass closest to the points, she held the thread as a guide to draw in that best straight line. Note

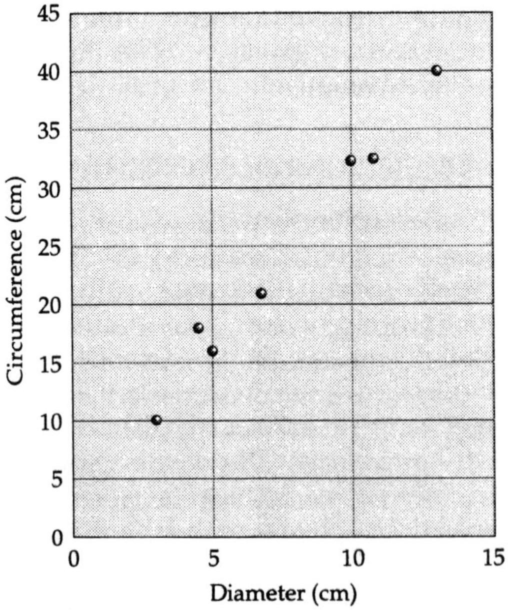

***Figure* 3.4 Diameters and circumferences of selected cylindrical objects.**

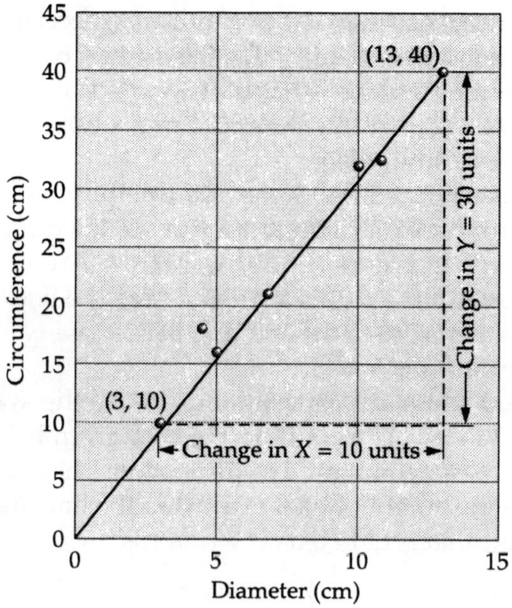

Figure 3.5 **An ad hoc approach to fitting a line to data.**

that she made this judgment totally informally and subjectively, without any rule or criterion to guide her. She made the line go through the origin because it is clear that a diameter close to 0 produces a circumference close to 0. Figure 3.5 shows the resulting line. This approach is *not* acceptable statistically, but we will consider it in order to explain some important concepts about fitting a straight line to data.

Once the line was drawn in, she estimated the slope of the line and the intercept. She noted the following from the graph. In a change in X of 10 units, the corresponding change in Y was 30 units. Hence she estimated that the slope of the "best" line was $30/10 = 3.0$. Recall that the intercept is 0 because the line goes through the origin. Thus her proposed line is

$$Y = 3.0X + 0$$

or

$$\text{Circumference} = 3.0 \times \text{diameter}$$

You may know from other mathematics courses that the circumference of a circle is π times its diameter, and the value of π is approximately 3.1416. Therefore, this experiment has provided a way of estimating the value of π. The class was concerned, however, that this method was quite crude and the slope 3 was likely not very accurate. Indeed, there are two sources of serious error. The measuring of the cylindrical objects may have been quite inaccurate. Even more fundamentally, the method of choosing the line was totally drawn out of thin air—that is, it was "ad hoc." The line produced by it might have a slope quite different from a careful choice of the "best fitting" line derived from theory.

You may have noticed that some of the points were rather far from the line that was found. A major reason for this is the difficulty of measuring accurately around and across cylindrical objects. You may think of other reasons why some of the points are away from the line. However, most of the points are rather close to the line, and in this case we were able to find a line that fits the data fairly well.

The standard statistical convention is to use the symbol Y to refer to actual data, and to use Y' for a fit to the data given by a proposed line or other predictive *model* equation. The distinction is helpful because some (or many, or all) points might not fall exactly on the line drawn. Therefore the equation of a line that is fit to data is written as

$$Y' = mX + C$$

Again, the value of Y' may not be the same as the actual observed value of Y. For example, when $X = 5$ centimeters we have

$$Y' = 3.0 \times 5 = 15.0 \text{ centimeters}$$

In the class experiment, however, a can with a diameter of 5 centimeters was found to have a circumference of 16.0 centimeters. The *error of estimation* between the actual measured value and the value estimated by the line in this case is

$$Y - Y' = 16.0 - 15.0 = 1.0 \text{ centimeter}$$

This deviation from the fitted line is often called a **residual.** It is the amount that remains unexplained by the estimated rule used to "fit" the data. Note that if we use the correct equation, $y' = \pi \times 5$, then $Y' \approx 17$ results and there is still estimater error. This demonstrates that the measuring process is quite inaccurate.

SECTION 3.2 EXERCISES

1. The following data were collected at a slot-racing track on a slot car going full speed.
   a. Draw a graph to show the relationship between time and distance. Then use the informal thread method to draw a line that best fits the data.

Racing time (minutes)	Distance traveled (meters)
2	520
3	770
4	1050
6	1560

b. Find a rule to relate racing time and distance.
c. What is the error (or residual) when you use your rule to estimate the distance traveled for each of the four racing times?

2. A psychologist is interested in the relationship between athletic ability and popularity. The following table gives physical education scores (*X*) and popularity ratings (*Y*) of 10 high school boys. A rating of 10 means "very popular."
   a. Draw a graph of these data and then informally find a rule that estimates popularity (Y') when athletic ability (*X*) is given.
   b. What kind of relationship does this rule imply between the two variables?

Athletic ability (X)	Popularity (Y)
8	7
4	5
7	8
7	6
3	4
5	8
9	8
7	10
6	6
8	7

3. This table records the time required to cook a turkey. Plot the data and, using your informal visual judgment, find a rule that gives the cooking time for a turkey for a given weight.

Weight (pounds)	Cooking time (hours)
5	3.5
7	4.0
10	4.5
14	5.5
18	6.75
22	8.5

4. Taxi fares recorded for five trips are given below.
   a. Plot the data and find a rule that gives the fare in terms of the length of the trip in miles.
   b. According to the rule you found in (a), what is the estimated fare for each of the five trip lengths?
   c. Using part (b), find the residual for each of the five trip lengths.

Length of trip (miles)	Total fare
2	$1.20
3	1.30
4.5	1.45
7	1.80
10	2.00

5. The table below gives the length (from nose to tail) and weight of eight laboratory mice. Plot the data and find a rule to estimate weight knowing the length of a mouse.

Length (centimeters) X	Weight (grams) Y
16	32
15	26
20	40
13	27
15	30
17	38
16	34
21	43

6. If a rule fit to a set of data turns out to produce estimates that are all exactly the same as the original observations,
   a. What would the residuals necessarily all be?
   b. What would the scatter plot necessarily look like?

3.3 LINEAR REGRESSION: AN INFORMAL VISUAL APPROACH

We have already used the *idea* of linear regression. Now we will use the name **linear regression,** which was given to this idea by statisticians almost one hundred years ago. *Linear* means that the relationship is in a straight line. *Regression* means going back to a lower level. Linear regression is the statistical method of fitting a straight line to data. In Chapter 12 we will learn more about why this word *regression* is appropriate.

Table 3.3 provides the number of weeks nine people spent in a weight-loss program and the amount of weight, in pounds, that they lost. The data are plotted in Figure 3.6. To find a relationship between time spent in the weight-loss program and amount of weight loss achieved, we try to fit a line to these data by following the four steps listed below. We will call finding a straight line by this method the *visual method of linear regression.* It is really an augmentation of the student's approach in Section 3.2. This informal approach is not used in real statistical practice, but it is in fact very useful in helping us become familiar with the problem of finding a "best fitting" line to a scatter plot of data such as Figure 3.6. The equation of the line we are looking for is called the *regression equation.*

The heart of the method is to choose the line through $(\overline{X}, \overline{Y})$ that seems by visual inspection to best fit the data. Augmenting the student's approach in the previous section by forcing the line through $(\overline{X}, \overline{Y})$ seems very reasonable because when the X value is $\overline{X}$, then it is very plausible to predict the Y value

***Table* 3.3 Weight Loss versus Number of Weeks in Program**

	Number of weeks in program (X)	Total weight loss (Y)
	12	19
	4	18
	8	8
	4	16
	12	26
	7	12
	5	12
	9	29
	14	31
Total	75	171
Mean	$\overline{X} = 8.33$	$\overline{Y} = 19.0$

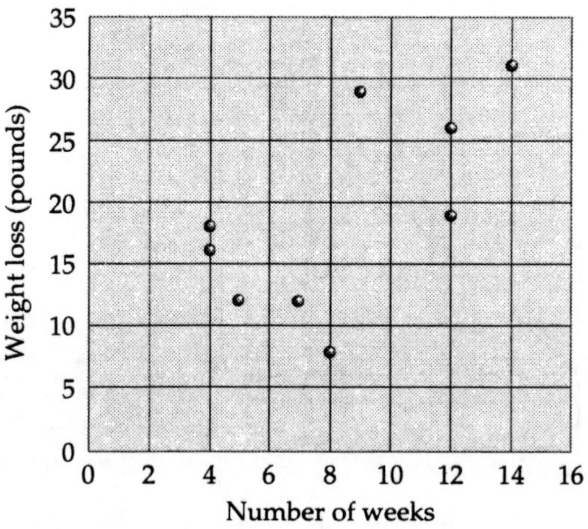

Figure 3.6 **Weight loss versus number of weeks in program for nine people in weight-loss program.**

to be $\overline{Y}$. Indeed, the most commonly used statistical method, the method of least squares, which is discussed later in this chapter and in Chapter 12, *forces* the line to pass through $(\overline{X}, \overline{Y})$.

Step 1: Since the regression line will pass through point $(\overline{X}, \overline{Y})$, find $\overline{X}$ and $\overline{Y}$ and draw in $(\overline{X}, \overline{Y})$, as indicated in Figure 3.7 by the point (8.33, 19.0).

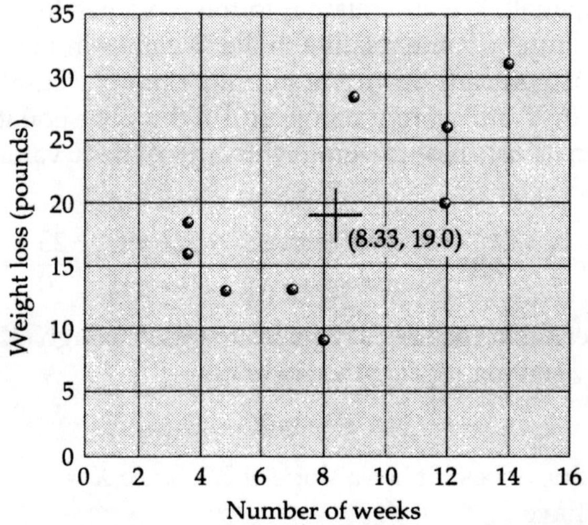

Figure 3.7 **Plot of $(\overline{X}, \overline{Y})$ for weight-loss data.**

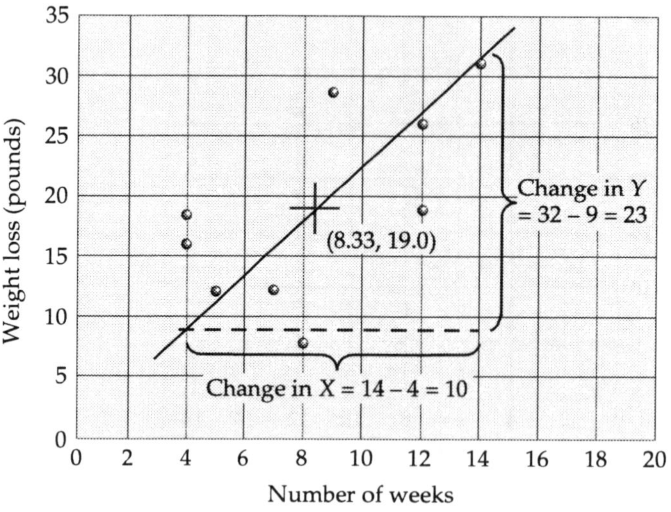

Figure 3.8 **A first attempt to fit a line through ($\overline{X}, \overline{Y}$) and "closest" to the data.**

Step 2: Using a thread, string, or transparent rule as a guide, draw in the line that appears to best fit the data (Figure 3.8). Of course this judgment is subjective: your classmates would all draw slightly different lines.

Step 3: Measure the slope of the line from the graph (graph paper helps in this). See Figure 3.8. Two points (14, 32) and (4, 9) from the line are chosen. It is up to you which two points you choose, but you ought to choose one point with a small X value relative to the X values in the data, and one point with a large X value relative to the X values in the data. The reason for this is that you can determine the slope more accurately if you use a large change in X and a large change in Y in the slope equation. Your two X values chosen *do not* have to agree with any of the X values of actual data points.

$$m = \text{slope} = \frac{\text{change in } Y}{\text{change in } X} = \frac{32 - 9}{14 - 4} = \frac{23}{10} = 2.3$$

Step 4: Recall that the line is required to pass through ($\overline{X}, \overline{Y}$). We find the Y intercept by solving for C in the equation

$$\overline{Y} = m\overline{X} + C$$

Suppose the line of best fit has a slope of 2.3. Since $\overline{X} = 8.33$, $\overline{Y} = 19.0$, and $m = 2.3$, we have

$$19 = 2.3(8.33) + C$$

So

$$C = 19 - 19.17 = -0.17$$

Therefore the estimated regression equation is

$$Y' = 2.3X - 0.17$$

Now we can measure how closely the line fits the points by computing the **mean absolute error,** often called the **mean error.** To do this, use the equation

$$Y' = 2.3X - 0.17$$

obtained from the four-step visual method to estimate the weight loss for a person based on how long he or she spends in the program. See Figure 3.9. For example, consider the person who spent 4 weeks in the program and lost 18 pounds. We predict his weight loss to be

$$\begin{aligned} Y' &= 2.3(4) - 0.17 \\ &= 9.2 - 0.17 = 9.03 \text{ pounds} \end{aligned}$$

So the prediction error for his weight loss is

$$Y - Y' = 18 - 9.03 = 8.97 \text{ pounds}$$

This prediction error is a positive number. In Figure 3.9 you can see that the point representing this person is about 9 pounds *above* the regression line.

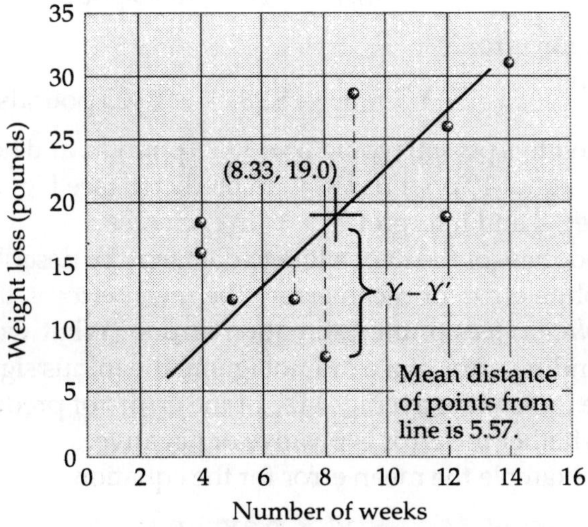

Figure 3.9 **Prediction error for $Y' = 2.3X - 0.17$.**

Table 3.4 Mean Error for Regression Equation $Y' = 2.3X - 0.17$

X	Y	Y'	$Y - Y'$	$\|Y - Y'\|$
12	19	27.43	−8.43	8.43
4	18	9.03	8.97	8.97
8	8	18.23	−10.23	10.23
4	16	9.03	6.97	6.97
12	26	27.43	−1.43	1.43
7	12	15.93	−3.93	3.93
5	12	11.33	.67	.67
9	29	20.53	8.47	8.47
14	31	32.03	−1.03	1.03
				50.13

Mean error = 50.13/9 = 5.57

This person lost about 9 pounds *more* than we predicted by the regression line. Note that there was another person in the course for 4 weeks who in fact lost 16 pounds. This is an illustration of the obvious fact that length of time in the program does not exactly control the amount of weight lost.

Now we find the prediction error for the person who was in the program for 8 weeks. His predicted weight loss is

$$
\begin{aligned}
Y' &= 2.3(8) - 0.17 \\
&= 18.4 - 0.17 = 18.23 \text{ pounds}
\end{aligned}
$$

The prediction error is

$$Y - Y' = 8 - 18.23 = -10.23 \text{ pounds}$$

We predicted this person would lose 18.23 pounds, and he actually lost only 8 pounds—about 10 pounds *less* than predicted (see Figure 3.9). You should check Table 3.4 and interpret $Y - Y'$ in each case.

The goodness of the fit of a line to data may be described in terms of the mean absolute error, or mean error. The mean error is the mean (average) of the *absolute values* of the estimation errors. (What would the mean be, within rounding error, if we did not ignore the minus signs? It would be 0.) That is, we look at the average size of the errors of prediction, deliberately ignoring whether the error is positive or negative.

In this example the mean error for the equation

$$Y' = 2.3X - 0.17$$

is found for the data of Table 3.4 to be 5.57.

SECTION 3.3 EXERCISES

1. Find the equation of the line that passes through the following pairs of points.
   a. (0, 0), (4, 6)
   b. (1, 3), (10, 15)
   c. (5, 3), (1, 2)
   d. (10, 10), (6, 8)
2. The following table is used by paperhangers to estimate the number of rolls, R, of wallpaper required to do a room, given the distance, P (in feet), around the room. (This table is for a 9-foot ceiling height.)
   a. Use the informal visual approach to find a rule that fits these data.
   b. Find the mean error of the resulting line.

Distance around room P	Number of rolls of wallpaper R
30	9
40	12
50	15
60	18
70	21
80	24
90	27

3. Refer to the cylinder circumferences and diameters in Table 3.2.
   a. Use the informal visual approach to find a rule to fit the data in Table 3.2.
   b. Find the mean error of the line found in (a).
4. The following table shows the average weight of men aged 20–24, for different heights.
   a. Fit the data using the informal visual approach, with height as the X variable and weight as the Y variable.

Height (inches)	Weight (pounds)
62	133
64	142
66	151
68	160
70	170
72	179
74	189

   b. Find the mean error of the line that you obtained in (a).
5. These data are final exam scores obtained by 12 students in their senior year of high school mathematics and their first year of college mathematics.
   a. Use the informal visual approach to estimate college mathematics scores given high school mathematics scores.
   b. Find the mean error of your chosen line.

High school final exam score	Final exam score in first year of college
13	23
27	28
18	29
17	27
21	29
26	26
28	31
19	20
23	19
7	18
21	26
19	30

Graphing Calculator Exercises

6. The following table, from Exercise 3 of Section 3.2, gives the time required to cook a turkey as a function of its weight. Make a scatter plot

Weight (pounds)	Cooking time (hours)
5	3.5
7	4.0
10	4.5
14	5.5
18	6.75
22	8.5

of these data. Find a possible linear equation to fit these data (see Section 1.6 in the TI Graphing Calculator Supplement.)

7. Enter the data of Table 3.1, from the Key Problem. Find the equation of a line that could be used to predict the gas mileage of a car on the basis of its weight.

3.4 LEAST MEAN ERROR LINEAR REGRESSION

The mean error for a regression line tells us how far, on the average, the data points are from that line. For the line of Section 3.3, the data points giving the actual weight losses are, on average, 5.57 vertical units from this line of predicted weight losses.

We will now see how well two other regression lines fit these same data, one with a slope of 1.5 and the other with a slope of 2.5 (see Figure 3.10). Tables 3.5 and 3.6 show how we calculate the mean error for each of the two lines. (Check the first-row computations.) If we draw a graph that shows the mean error produced by each of the three lines (Table 3.7), it looks like Figure 3.11.

With the help of a computer or calculator we can calculate the mean error for several lines with different slopes (Table 3.8) and fill in the graph some more (Figure 3.12). From the graph and the table it appears as if the mean error reaches its smallest value for some line with a slope near 1.9. With the help of our computer, we can calculate the mean error for lines with slopes close to 1.9 (Table 3.9) if more accuracy is desired. With these data, the graph can be filled in even more (Figure 3.13). From Table 3.9 we can state that the slope of the regression line that gives the least mean error is, to the nearest hundredth, 1.91. Note that this line was required to pass through $(\overline{X}, \overline{Y})$.

Without the help of a computer (or at least a calculator), this trial-and-error method of finding the slope of the best-fitting line can be time-consuming. If you do not use a calculator or computer, you should concentrate on the meaning of the best-fitting line without doing a lot of

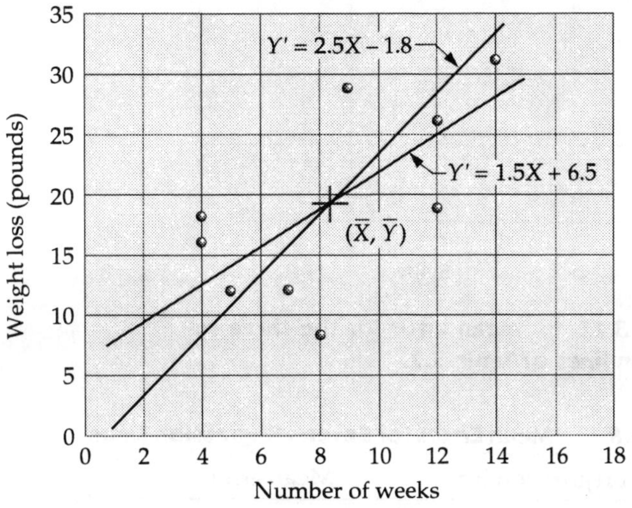

Figure 3.10 **Two other regression lines fit to the weight-loss data.**

Table 3.5 **Mean Error for Regression Equation $Y' = 1.5X + 6.5$**

X	Y	Y'	$Y - Y'$	$\lvert Y - Y' \rvert$
12	19	24.5	− 5.5	5.5
4	18	12.5	5.5	5.5
8	8	18.5	−10.5	10.5
4	16	12.5	3.5	3.5
12	26	24.5	1.5	1.5
7	12	17.0	− 5.0	5.0
5	12	14.0	− 2.0	2.0
9	29	20.0	9.0	9.0
14	31	27.5	3.5	3.5
				46.0

Mean error = 46.0/9 = 5.11

Table 3.6 **Mean Error for Regression Equation $Y' = 2.5X - 1.8$**

X	Y	Y'	$Y - Y'$	$\lvert Y - Y' \rvert$
12	19	28.17	− 9.17	9.17
4	18	8.17	9.83	9.83
8	8	18.17	−10.17	10.17
4	16	8.17	7.83	7.83
12	26	28.17	− 2.17	2.17
7	12	15.67	− 3.67	3.67
5	12	10.67	1.33	1.33
9	29	20.67	8.33	8.33
14	31	33.17	− 2.17	2.17
				54.67

Mean error = 54.67/9 = 6.07

Table 3.7 **Mean Errors for Three Regression Lines**

Slope of regression line	Mean error
1.5	5.11
2.3	5.57
2.5	6.07

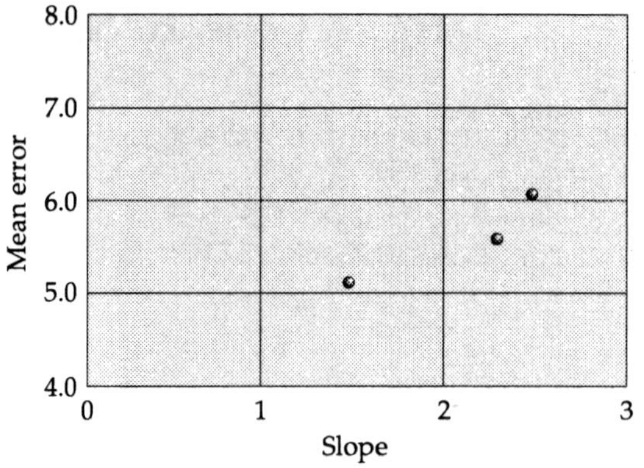

Figure 3.11 **Mean error for the three regression lines of Table 3.7.**

Table 3.8 **Mean Errors for Several Regression Lines**

Slope of regression line	Mean error	
1.0	5.26	
1.1	5.23	
1.2	5.20	
1.3	5.17	
1.4	5.14	
1.5	5.11	
1.6	5.08	
1.7	5.05	
1.8	5.02	
1.9	4.99	← Least error
2.0	5.04	
2.3	5.57	
2.5	6.07	
3.0	7.33	

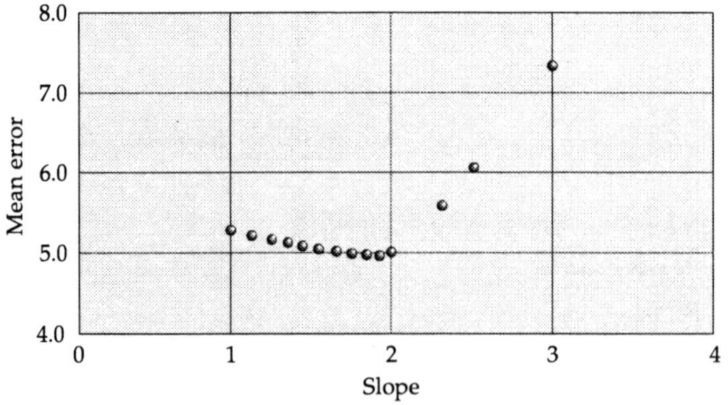

Figure 3.12 **Mean error for several regression lines.**

Table 3.9 **Magnified Mean Errors for Regression Lines with Slopes near 1.9**

Slope of regression line	Mean error	
1.80	5.022	
1.81	5.019	
1.82	5.016	
1.83	5.013	
1.84	5.010	
1.85	5.007	
1.86	5.004	
1.87	5.001	
1.88	4.998	
1.89	4.995	
1.90	4.992	
1.91	4.990	← Least error
1.92	4.995	
1.93	5.000	
1.94	5.005	
1.95	5.011	
1.96	5.016	
1.97	5.021	
1.98	5.026	
1.99	5.031	
2.00	5.037	

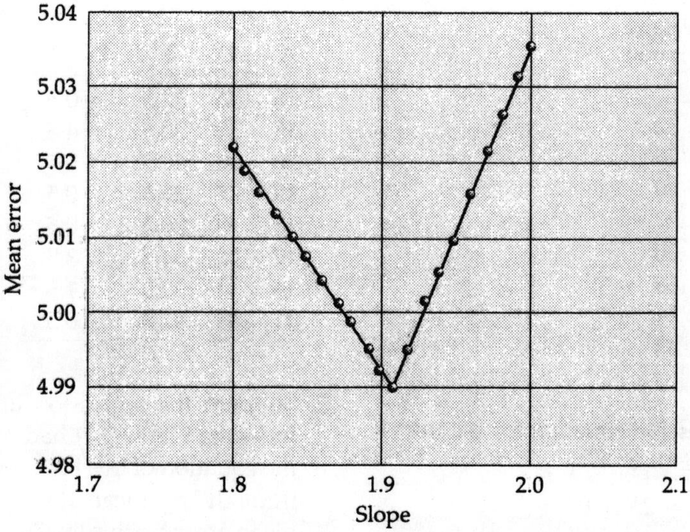

Figure 3.13 **Mean error for many regression lines with slopes near 1.9.**

calculations. Note that if desired we could find the slope to the nearest thousandth or even to greater accuracy. That is, for all practical purposes we can find by our trial-and-error method the line that is best-fitting in the sense that it passes through $(\overline{X}, \overline{Y})$ and minimizes the mean error.

Practicing statisticians apply a very similar method that differs from our method only by dropping the requirement that the line pass through $(\overline{X}, \overline{Y})$. The method simply finds the line that minimizes the mean error. This best-fitting line will usually come very close to $(\overline{X}, \overline{Y})$ but will usually not pass through it. The line found by this new method is usually close to the line found by the method used here, that of finding the line passing through $(\overline{X}, \overline{Y})$ that minimizes the mean error. When the criterion used to select a best line is to minimize the mean error, the method is robust. In fact it is more robust than the widely used method of least squares, discussed below. The method used here and the more advanced method not requiring the line to pass through $(\overline{X}, \overline{Y})$ both use the mean error, and hence both are robust. When you wish to use a robust method, you should use our method of seeking the line that minimizes the mean error and passes through (X, Y).

SECTION 3.4 EXERCISES

1. The table below gives the length (from nose to tail) and weights of eight laboratory mice. Plot the data and the mean data point $(\overline{X}, \overline{Y})$.

Length (centimeters) X	Weight (grams) Y
16	32
15	26
20	40
13	27
15	30
17	38
16	34
21	43

2. Suppose the regression equation for the mice data in Exercise 1 is

$$Y' = 2X + 0.5$$

The table below shows the estimated weight in grams of each mouse, given its length in centimeters, using the regression equation

$$Y' = 2X + 0.5$$

Find the mean error for this equation.

X	Y	Y'	$Y - Y'$
16	32	32.5	−0.5
15	26	30.5	−4.5
20	40	40.5	−0.5
13	27	26.5	0.5
15	30	30.5	−0.5
17	38	34.5	3.5
16	34	32.5	1.5
21	43	42.5	0.5

3. Suppose the regression line for the weight-loss data (Table 3.3) had a slope of 1.5. Find the equation of this line, assuming that it goes through the mean data point. What is the mean error for this line?

4. The data in the table below are the final examination scores obtained by 12 students in their senior year of high school mathematics and their first year of college mathematics. Plot the data and the mean data point. Draw in your estimated regression line using a trial-and-error process. What is the equation of the line you found?

High school final exam score X	Final exam score in first year of college Y
13	23
27	28
18	29
17	27
21	29
16	26
28	31
19	20
23	19
7	18
21	26
19	30

5. Suppose the regression equation for the final exam data (Exercise 4) is

$$Y' = 0.4X + 17.7$$

Use this equation to estimate a student's score on the final examination in first-year college mathematics, given that student's score on the final exam in the senior year of high school mathematics. Prepare a mean error table like Tables 3.4–3.6. What is the mean error for this equation?

6. Suppose the regression equation for the weight-loss data of Table 3.3 is

$$Y' = 2.5X - 0.2$$

Find the mean error for this equation.

7. The following table gives the prediction errors for the mice data of Exercises 1 and 2, but this time for the equation

$$Y' = X + 17.13$$

Regression equation: $Y' = X + 17.13$

X	Y	Y'	$Y - Y'$	$\lvert Y - Y'\rvert$
16	32	33.13	−1.13	1.13
15	26	32.13	−6.13	6.13
20	40	37.13	2.87	2.87
13	27	30.13	−3.13	3.13
15	30	32.13	−2.13	2.13
17	38	34.13	3.87	3.87
16	34	33.13	0.87	0.87
21	43	38.13	4.87	4.87
			Total	25.00

a. What is the mean error for this equation?
b. Which of the two regression equations for the mice data—that of Exercise 2 or the one above—is a better fit of the data, in terms of mean error?

$$Y' = 2X + 0.5 \quad \text{or} \quad Y' = X + 17.13$$

Why?

8. The table below gives the prediction errors for a line having a slope of 1.9. What is the mean error for this line? Compare this result with the value you get by using Figure 3.13.

Regression equation: $Y' = 1.9X + 3.16$

X	Y	Y'	$Y - Y'$	$\lvert Y - Y'\rvert$
12	19	25.97	− 6.97	6.97
4	18	10.77	7.23	7.23
8	8	18.37	−10.37	10.37
4	16	10.77	5.23	5.23
12	26	25.97	0.03	0.03
7	12	16.47	− 4.47	4.47
5	12	12.67	− 0.67	0.67
9	29	20.27	8.73	8.73
14	31	29.77	1.23	1.23

9. Ten students took a spelling test (X) and a reading test (Y) and got the scores in the table below.
   a. Use the regression equation

   $$Y' = 2X - 3.8$$

   to estimate each student's reading score knowing his or her spelling score.
   b. What is the mean error of this equation?

Spelling X	Reading Y	Spelling X	Reading Y
12	20	6	14
10	12	6	6
10	18	5	7
8	10	4	3
7	12	2	1

10. The table below estimates reading scores (Y'), given spelling scores (X), for the 10 students in Exercise 9 using the equation

    $$Y' = 2.3X - 5.8$$

    a. What is the mean error for this equation?
    b. Which line is better, according to the least mean error criterion? This one, or the one in Exercise 9?

X	Y	Y'	$Y - Y'$	$\lvert Y - Y' \rvert$
12	20	21.8	−1.8	1.8
10	12	17.2	−5.2	5.2
10	18	17.2	0.8	0.8
8	10	12.6	−2.6	2.6
7	12	10.3	1.7	1.7
6	14	8.0	6.0	6.0
6	6	8.0	−2.0	2.0
5	7	5.7	1.3	1.3
4	3	3.4	−0.4	0.4
2	1	−1.2	2.2	2.2

11. The table below gives the mean error for the mice data of Exercise 1 when regression lines having slopes from 1.0 to 3.0 are used. What is the slope of the regression line that best fits the data, according to this table? What is the equation of the best-fitting line?

Slope	Mean error
1.0	3.12
1.2	2.75
1.4	2.37
1.6	2.00
1.8	1.62
2.0	1.50
2.2	1.56
2.4	1.80
2.6	2.07
2.8	2.35
3.0	2.65

12. Using the best-fitting line that you found in Exercise 11, estimate the weight of mice having these lengths:
    a. 15 cm
    b. 18 cm
    c. 20 cm

Graphing Calculator Exercises

13. Refer to the lengths and weights of eight mice given in Exercise 1. Enter the data into a data table using L1 for length (cm) and L2 for weight (gm). Find Y' using the equation $Y' = 2X + 0.5$ (see Section 1.5 of the TI Graphing Calculator Supplement). Subtract L3 from L2 and store the difference in L4 ($Y - Y'$). Sum the absolute values of L4 and divide by the number of values to find the mean error.

14. Exercise 4 gives the examination scores obtained by 12 students in their senior year of high school mathematics and their first year of college mathematics. Enter those data into a data table. Plot the data as a scatter plot. Also plot the mean data point ($\overline{X}, \overline{Y}$). Draw in your estimation for the line of best fit. Use this equation to find the mean error.

15. Find a line that fits the data from the Key Problem by trying various lines through ($\overline{X}, \overline{Y}$) and comparing their mean errors. Set up the data table with L1 containing the weight in 100s of pounds and L2 containing the city MPG).

16. Which of the regression equations below for the mice data (see Exercise 13) is a better fit of the data? Explain why.

$$Y_1' = 2X + 0.5 \quad \text{or} \quad Y_2' = X + 17.13$$

Do each equation separately. Find Y' (See Section 1.5 of the TI Graphing Calculator Supplement). Subtract L3 from L2. Find the absolute value of L3 and store in L4. Sum L4 and divide by the number of data points. Repeat for Y_2'. Explain why one equation is a better estimate than the other.

17. Refer to the following data regarding crickets. X is related to how fast crickets move their wings. Y is the ambient temperature. Enter the number of pulses per second (L1) and the temperature (L2). Using the program LEASTERR, find the least mean error for these data. Use starting slope = 2.5, step = 0.1; calculations = 10, and sample size = 15. View the slopes and the corresponding mean error by viewing L4 and L5 in table mode ([STAT] [edit]). Plot a line graph and use the [TRACE] function. See the statplot setup below.

X Number of pulses per second	Y Temperature (°F)
20	89
16	72
20	93
18	84
17	81
16	75
15	70
17	82
15	69
16	83
15	80
17	83
16	81
17	84
14	76

18. Ten students took a spelling test (X) and a reading test (Y) and got the scores presented in Table 3.10 (see below). Use these data, and use starting slope $= 1.9$, step $= 0.1$, calculations $= 10$, and number of data points $= 10$. Run the program LEASTERR, and find the least mean error to two decimal places. *Hint:* Use the slope that produced the least mean error as the starting slope and use a step of 0.01

3.5 LEAST SQUARES LINEAR REGRESSION

Another way of describing how well a line fits a set of data is to *square* the prediction errors $Y - Y'$ and find the mean (average) of all the squared errors. This should remind you of how you found the variance of a set of data in Chapter 2. You found the deviation values (difference of each score from the mean), squared them, and averaged the squares. In the case of the least squares method we find the mean of the squared distance of the Y data values from a proposed regression line. That is, the difference between finding the mean of the squared errors for a proposed regression line and finding the variance of the Y's is that in the case of the squared errors we take the deviation between Y and the line instead of the deviation between Y and $\bar{Y}$. Thus if we instead found the mean of the squared distances of the Y's from $\bar{Y}$, this larger quantity would be the variance of the Y's.

Consider the scores of 10 students in a spelling test (X) and in a reading test (Y) given in Table 3.10. The reading test scores are plotted against the

Table 3.10 **Spelling and Reading Scores of 10 Students**

Spelling score X:	12	10	10	8	7	6	6	5	4	2
Reading score Y:	20	12	18	10	12	14	6	7	3	1

spelling test scores in Figure 3.14. We want to find the equation of a line that best fits these data when we consider the average of the squared vertical distances of the data points from this line.

To begin our search, we try a line with a slope of 1.5. This is a rather arbitrary slope to start with. (You may wish to start with some other slope.) Since this line must go through $(\overline{X}, \overline{Y})$, we can find its equation. We have $\overline{Y} = 1.5\overline{X} + C$. But $\overline{X} = 7.0$ and $\overline{Y} = 10.3$. So we have $C = 10.3 - (1.9)(7.0) = -0.2$, and the equation of this line is

$$Y' = 1.5X - 0.2$$

Table 3.11 gives the squares of the deviation scores, and the mean (average) squared deviation, for the line whose equation is $Y' = 1.5X - 0.20$.

We will now do some trial-and-error exploration to see if we can improve on the fit of a line to these data. Let's try a line with a slope of 2.0. Table 3.12 gives the results for the equation $Y' = 2.0X - 3.7$.

We find the mean square error for the regression line $Y' = 2.0X - 3.7$ of Table 3.12 to be 7.01. From the viewpoint of choosing a line with as small a mean square error as possible, this regression line fits the data a little better than the previous one, since the mean (average) square error has been reduced from 7.51 to 7.01. We are looking for the line that best fits the data when we are judging fit in terms of the mean (average) square error. So far

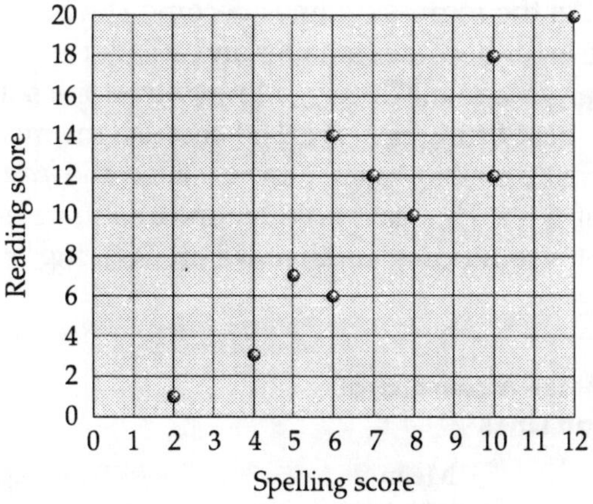

Figure 3.14 **Reading versus spelling scores for 10 students.**

Table 3.11 **Mean Square Error Regression Equation $Y' = 1.5X - 0.2$**

X	Y	Y'	$\|Y - Y'\|$	$(Y - Y')^2$
12	20	17.8	2.2	4.84
10	12	14.8	2.8	7.84
10	18	14.8	3.2	10.24
8	10	11.8	1.8	3.24
7	12	10.3	1.7	2.89
6	14	8.8	5.2	27.04
6	6	8.8	2.8	7.84
5	7	7.3	0.3	0.09
4	3	5.8	2.8	7.84
2	1	2.8	1.8	3.24
Total				75.10

Mean square error = 7.510/10 = 7.51

Table 3.12 **Mean Square Error for Regression Equation $Y' = 2.0X - 3.7$**

X	Y	Y'	$\|Y - Y'\|$	$(Y - Y')^2$
12	20	20.3	0.3	0.09
10	12	16.3	4.3	18.49
10	18	16.3	1.7	2.89
8	10	12.3	2.3	5.29
7	12	10.3	1.7	2.89
6	14	8.3	5.7	32.49
6	6	8.3	2.3	5.29
5	7	6.3	0.7	0.49
4	3	4.3	1.3	1.69
2	1	0.3	0.7	0.49
Total				70.10

Mean square error = 7.01

we have found the mean square errors for two regression lines, one with a slope of 1.5 and an intercept of −0.2, and one with a slope of 2.0 and an intercept of −3.7. They are shown in Table 3.13.

With the help of a computer or a calculator, we can find the mean square errors for many more regression lines with slopes ranging from 1.5 to 2.5. In each case the line passes through $(\overline{X}, \overline{Y})$ (we can show theoretically that this must be the case for the best-fitting least squares line, that is, that the best-fitting least squares line in fact always passes through $(\overline{X}, \overline{Y})$). Table 3.14 and the graph of the values in the table (Figure 3.15) help give us an idea of what is taking place. As the regression lines become steeper than a slope of 1.5, the mean square error becomes smaller and smaller—for a while. Then this error becomes larger again. Clearly, a line with slope somewhere between 1.7 and 1.9 is the best-fitting line. It is *best* because the mean square error for this line is the least possible. Such a line is called the *least squares best-fitting line* for the set of data. The least possible mean square error is denoted by S_e^2, the subscript e reminding us that we are looking at the mean square *error*.

Table 3.13 **Mean Square Error for Two Regression Lines**

Slope of regression line	Mean square error
1.5	7.510
2.0	7.010

Table 3.14 **Mean Square Error for Several Regression Lines**

Slope	Mean square error	
1.5	7.51	
1.6	7.07	
1.7	6.80	
1.8	6.70	← Least mean square error
1.9	6.77	
2.0	7.01	
2.1	7.41	
2.2	7.98	
2.3	8.72	
2.4	9.63	
2.5	10.71	

The **least squares best-fitting line** for a set of data in two variables is the regression line that has the smallest possible mean square error for estimating one variable from the other. Often textbooks will use minimization of the root mean square error (RMSE) as the criterion for deciding which line fits the data best. This is just the square root of the mean square error, so both

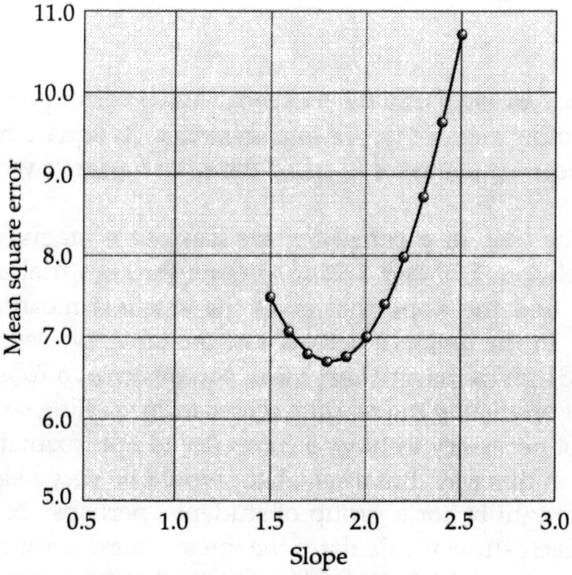

Figure 3.15 **Mean square error for several regression lines for reading and spelling score data.**

Table 3.15 **Mean Square Errors for Lines with Slopes near 1.80**

Slope	Mean square error	
1.70	6.806	
1.71	6.788	
1.72	6.772	
1.73	6.758	
1.74	6.745	
1.75	6.735	
1.76	6.725	
1.77	6.718	
1.78	6.712	
1.79	6.708	
1.80	6.706	Least mean
1.81	6.705	← square error
1.82	6.706	(error variance)
1.83	6.708	
1.84	6.713	
1.85	6.719	
1.86	6.726	
1.87	6.735	
1.88	6.746	
1.89	6.759	
1.90	6.774	

the mean square error and the root mean square error produce the *same* line: if we minimize a quantity, we also minimize its square root! This smallest possible mean square error is called the **error variance** when Y is estimated from X.

With the help of a computer, we can put a magnifying glass on the interval between 1.70 and 1.90 to attempt through trial and error to more accurately find the slope that gives the smallest mean square error. See Table 3.15. To the limits of accuracy of the table, we see that a line with a slope of 1.81 gives the smallest mean square error, 6.705. That is, the error variance of predicting the reading scores from spelling scores is 6.705.

It is not necessary to have a computer to approximately find the best-fitting line in this way. But a calculator would be very helpful! A good way to do this might be for a group of students, perhaps six or eight, to work together. Each student calculates the mean square error for one regression line passing through $(\bar{X}, \bar{Y})$ with a specified slope (see also the calculator exercises for this section). However, with a large number of data, this is a big job, too. In the next section we will give a formula that computes the slope of the least squares best-fitting regression line directly from the data and thus allows us to bypass our instructive but time-consuming trial-and-error method of finding the least squares best-fitting regression line.

Which Method for Finding the Best Regression Line?

We have now three methods for finding the regression line. In general, each approach will produce a different line, so you need to decide which method you prefer. If you need just a quick and general idea of what the best regression line is, then the visual method of Section 3.3 is adequate. The method used most often in standard statistical practice is to find the line that makes the mean square error as small as possible (the method of least squares). The statistical properties of this approach, which are presented in advanced statistic courses, make this method attractive. But in this approach a table of slopes and errors like Table 3.14 or 3.15 is not typically used. Instead, the formulas given in Section 3.6 below (or very similar formulas) are employed. We favor your using formulas, but we recommend that you use the slopes approach until you are confident you understand the way the least squares method works. The formulas are then just a shortcut for finding the exact slope of the line you are seeking.

In certain cases you should really find the smallest mean error regression line instead of the smallest mean square error regression line because of its greater robustness against undue influence from one or two data points. (Imagine that a measurement error has caused one data point to be way off from where it should be. Certainly, if you suspect that some of the data points are likely measured badly, then having a robust method would be necessary.) If you are not told which method to use, use your own judgment or preference. However, if accuracy is at all an issue, *do not* use the visual method.

SECTION 3.5 EXERCISES

1. The following table again provides the lengths (from nose to tail) and weights of eight laboratory mice. The regression equation

$$Y' = 1.5X + 8.81$$

was used to estimate the weight of a mouse given its length.

a. What is the mean square error for estimating weight using this equation? (X = length in centimeters, Y = weight in grams.)

X	Y	Y'	$\lvert Y - Y'\rvert$	$(Y - Y')^2$
16	32	32.81	0.81	0.6561
15	26	31.31	5.31	28.1961
10	40	38.81	1.19	1.4161
13	27	28.31	1.31	1.7161
15	30	31.31	1.31	1.7161
17	38	34.31	3.69	13.6161
16	34	32.81	1.19	1.4161
21	43	40.31	2.69	7.2361

b. The regression equation

$$Y' = 1.9X + 2.16$$

is used in the table below to estimate the weight of a mouse given its length. What is the mean square error for this equation?

X	Y	Y'	$\lvert Y - Y'\rvert$	$(Y - Y')^2$
16	32	32.56	0.56	0.3136
15	26	30.66	4.66	21.7156
20	40	40.16	0.16	0.0256
13	27	26.86	0.14	0.0196
15	30	30.66	0.66	0.4356
17	38	34.46	3.54	12.5316
16	34	32.56	1.44	2.0736
21	43	42.06	0.94	0.8836

c. Which of the two equations

$$Y' = 1.5X + 8.81$$

$$Y' = 1.9X + 2.16$$

fits the mice data better, in terms of mean square error?

2. Ten students took a spelling test (X) and a reading test (Y) and got the scores in the table below.

a. The regression equation

$$Y' = X + 3.3$$

was used to estimate a student's reading score given his or her spelling score. What is the mean square error for this equation?

X	Y	Y'	$\lvert Y - Y'\rvert$	$(Y - Y')^2$
12	20	15.3	4.7	22.09
10	12	13.3	1.3	1.69
10	18	13.3	4.7	22.09
8	10	11.3	1.3	1.69
7	12	10.3	1.7	2.89
6	14	9.3	4.7	22.09
6	6	9.3	3.3	10.89
5	7	8.3	1.3	1.69
4	3	7.3	4.3	18.49
2	1	5.3	4.3	18.49

b. Suppose the regression equation

$$Y' = 1.5X - 0.2$$

was used in part (a) to estimate reading scores from spelling scores. Complete the table below, and find the mean square error for this equation.

X	Y	Y'	$\lvert Y - Y'\rvert$	$(Y - Y')^2$
12	20	17.8	2.2	
10	12	14.8	2.8	
10	18	14.8	3.2	
8	10	11.8	1.8	
7	12	10.3	1.7	
6	14	8.8	5.2	
6	6	8.8	2.8	
5	7	7.3	0.3	
4	3	5.8	2.8	
2	1	2.8	1.8	

c. Of the two regression equations

$$Y' = 1.5X - 0.2$$

$$Y' = X + 3.3$$

which fits the spelling test and reading data best? Why?

3. The table below gives the mean square errors for the spelling and reading data for regression equations having different slopes from 1.0 to 2.3. According to this table, what is the slope of the line that best fits these data? What is the equation of this line?

Slope	Mean square error	Slope	Mean square error
1.0	12.21	1.7	6.80
1.1	10.93	1.8	6.70
1.2	9.82	1.9	6.77
1.3	8.88	2.0	7.01
1.4	8.11	2.1	7.41
1.5	7.51	2.2	7.98
1.6	7.07	2.3	8.72

4. Refer back to the weight-loss data of Table 3.3. The following table gives the mean square errors for the weight-loss data for regression equations having different slopes. According to this table, what is the slope of the best-fitting line for these data? Find the equation of this line.

Slope	Mean square error
1.0	36.00
1.1	35.14
1.2	34.53
1.3	34.16
1.4	34.04
1.5	34.16
1.6	34.53
1.7	35.14
1.8	36.00
1.9	37.10
2.0	38.44

5. The following table gives the mean square error (for the mice data; see Exercise 1) for estimation equations having several different slopes from 1.5 to 2.5. What slope gives the least mean square error? What is the equation of the regression line for the mice data? What is the error variance for the weights of the mice?

Slope	Mean square error
1.5	6.99
1.6	6.24
1.7	5.62
1.8	5.10
1.9	4.71
2.0	4.49
2.1	4.37
2.2	4.38
2.3	4.50
2.4	4.74
2.5	5.11

6. What is the estimated weight, using the equation found in Exercise 5, for mice having these lengths?
   a. 18 centimeters
   b. 22 centimeters
   c. 26 centimeters

Graphing Calculator Exercises

7. Refer to the mouse length (X) and weight (Y) data of Exercise 1. Which of the two equations fits the mouse data the best according to the least squares criterion of Section 3.5? (Enter X values in L1 and Y values in L2.)

$$Y_1' = 1.5X + 8.81 \quad \text{or} \quad Y_2' = 1.9X + 2.16$$

Do each equation separately. Find Y'. Subtract L3 from L2. Square L3 and store in L4. Sum L4 and divide by the number of data points. Repeat for Y_2'. Explain why one equation is better than the other.

8. Ten students took a spelling test and a reading test and got the scores of Table 3.10. Enter the data into a data table: spelling into L1 and reading into L2. Run the program ERRVAR (see Section 1.8 in the TI Graphing Calculator Supplement). Use starting slope = 1.0, step = 0.1, calculations = 10, and sample size = 10. View the slopes and the corresponding mean absolute error by viewing L4 and L5 in table mode (STAT [edit]). Plot a line graph; see the statistical plot setup on the following page.

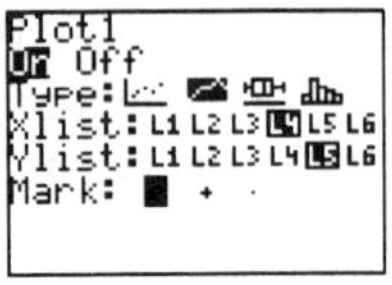

Find the least square error for these data by using [TRACE].

9. Refer to Table 3.3, which shows the number of weeks nine people attended a weight-loss program and the corresponding total weight they lost. Enter the weeks they attended into L1 and the weight lost into L2. Use the program ERRVAR to find the slope, to two decimal places, of the line that has the least square error.

3.6 COVARIANCE, CORRELATION, AND THE LEAST SQUARES REGRESSION LINE

You no doubt have concluded by now that it can be a long and complicated business to find the least squares best-fitting line for a set of data by our trial-and-error approach. There is a shortcut you can use, now that you understand what the method is about. This shortcut is helped along by a new measure—**covariance**—that is important to study for its own sake because it is a basic statistical concept. We learned about variance in Chapter 2. Variance is a measure of the amount that a set of data varies about its mean, or average. Covariance, another measure of variation, tells how closely two variables change, or vary, in a linear (straight line) manner with respect to one another. (Think of the variables *co*operating.)

In this chapter we have seen examples of how two-variable data vary. If the slope of the regression line is positive, then as one variable increases, so does the other. Such data have a *positive covariance.*

Figure 3.16 is a graph of the data given in Table 3.16 on the possible harmful effects on our health from pollution due to nuclear reactors. It shows the amount of exposure of eight Oregon counties to radioactive waste from the Atomic Energy Commission's plant in Hanford, Washington, and the rates of death due to cancer in those counties. It appears from the graph that as the exposure to radiation increases, so does the cancer death rate.

If the slope of the regression line is negative, then as one variable increases, the other decreases. Such data have a *negative covariance.* Consider Figure 3.1, in the Key Problem for this chapter. As the weight of the cars increases, the mileage decreases.

Now let's see how to calculate covariance. We use the Key Problem as an example.

Step 1: Find the means of the X values (car weight) and Y values (car MPG). To avoid having to work with large numbers, we will express the car

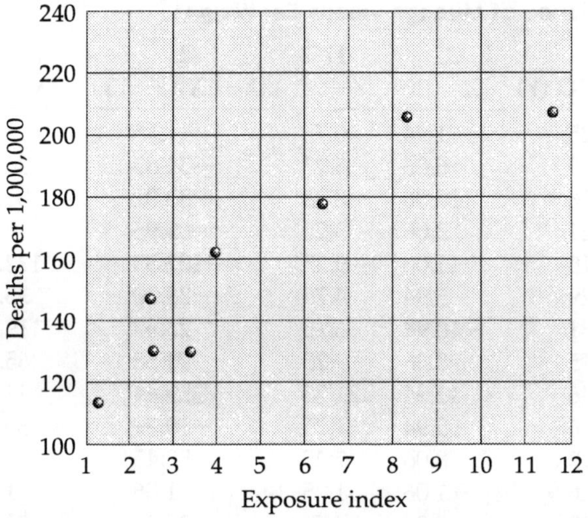

Figure 3.16 **Cancer death rate versus exposure to radiation.**

Table 3.16 **Cancer Mortality in Eight Oregon Counties versus Exposure to Radioactive Waste**

County	Index of exposure	Cancer mortality (per 1,000,000 persons)	County	Index of exposure	Cancer mortality (per 1,000,000 persons)
Umatilla	2.49	147.1	Hood River	3.83	162.3
Morrow	2.57	130.1	Multnomah	11.64	207.5
Gilliam	3.41	129.9	Columbia	6.41	177.9
Wasco	1.25	113.5	Clatsop	8.34	210.3

weights in hundreds of pounds, not pounds. See column A in Table 3.17. As is seen in Table 3.17,

$$\overline{X} = \text{mean of } X = \frac{463}{16} = 28.90$$

$$\overline{Y} = \text{mean of } Y = \frac{388}{16} = 24.3$$

Step 2: Find the deviation for each X value and each Y value, that is, $X - \overline{X}$ and $Y - \overline{Y}$. (See Chapter 2.)

Step 3: For each data pair X and Y, find the product of the X deviation and Y deviation. That is, find $(X - \overline{X})(Y - \overline{Y})$.

Step 4: Find the sum of the products from step 3. From Table 3.17 (column E), this sum is -857.75.

Table 3.17 **Finding Covariance of Mileage versus Car Weight**

	A Weight (X)	B City MPG (Y)	C $X - \bar{X}$	D $Y - \bar{Y}$	E $(X - \bar{X})(Y - \bar{Y})$	F $(X - \bar{X})^2$
	27	25	−1.94	0.75	−1.45	3.75
	35	19	6.06	−5.25	−31.83	36.75
	39	16	10.06	−8.25	−83.02	101.25
	32	21	3.06	−3.25	−9.95	9.38
	40	15	11.06	−9.25	−102.33	122.38
	23	29	−5.94	4.75	−28.20	35.25
	18	31	−10.94	6.75	73.83	119.63
	37	15	8.06	−9.25	−74.58	65.00
	17	46	−11.94	21.75	−259.64	142.50
	23	26	−5.94	1.75	−10.39	35.25
	37	17	8.06	−7.25	−58.45	65
	30	26	1.06	1.75	1.86	1.13
	23	29	−5.94	4.75	−28.20	35.25
	32	19	3.06	−5.25	−16.08	9.38
	20	33	−8.94	8.75	−78.20	79.88
	30	21	1.06	−3.25	−3.45	1.13
Sum	463	388	0.00	0.00	−857.75	862.94
Mean	28.94	24.25			−53.61	53.93

Step 5: Find the average of the products from step 3. This average is

$$\frac{-857.75}{16} = -53.61$$

The covariance of X and Y is this average of the products of the deviations. In this case the covariance is −53.61. Many statisticians denote the covariance by S_{XY}. We can say that since the covariance of the data is negative, the slope of the least squares best-fitting regression line for these data is negative. That is, as the car weight (X) increases, the mileage (Y) tends to decrease. Thus the sign of the covariance tells us the sign of the slope of the least squares regression line.

The Slope of the Least Squares Regression Line

In addition to the covariance, we also need to calculate the variance of X, since we will need that in our formula to find the slope of the least squares best-fitting line for the data. Refer to Chapter 2 for a reminder of how to calculate the variance of a set of data. To find the variance of the X values, we add a column to our covariance table (see Table 3.17, column F). This gives us the squares of the X deviation values. As we recall from Chapter 2,

$$\text{Variance of car weight } X = \frac{\text{sum of squares of deviations of } X}{\text{number of scores}}$$

$$= \frac{862.9}{16}$$

$$= 53.9$$

It is shown in more advanced statistics courses that the slope of the least squares best-fitting line is found as follows:

$$\text{Slope} = \frac{\text{covariance of } X \text{ and } Y}{\text{variance of } X}$$

For our car data this formula gives

$$\text{Slope} = \frac{-53.61}{53.93} = -0.99$$

Recall that the least squares line passes through $(\overline{X}, \overline{Y})$. Hence, we now have an exact formula for the regression line:

$$Y' = mX + C$$

with the slope

$$m = \frac{\text{Covariance of } X \text{ and } Y}{\text{Variance of } X}$$

So we have

$$Y' = -0.99X + C$$

But since the least squares line goes through $(\overline{X}, \overline{Y})$, we have

$$\overline{Y} = -0.99\overline{X} + C$$

And we know that

$$\overline{X} = 28.9 \quad \text{and} \quad \overline{Y} = 24.3$$

So

$$C = 24.3 - (-0.99)(28.9)$$

$$= 52.9$$

Therefore the equation of the least squares best-fitting line for the car data is

$$Y' = -0.99X + 52.9$$

Correlation

We have three ways of finding a line that fits well or even best fits a set of data. But how good is this fit, for a given set of data? The goodness of fit is seen by examining the estimation errors for Y values, given X values. Another way of describing the goodness of fit of data to a line is to ask how strongly the X and Y values are related. If they are strongly related, Y values can be predicted from X values very accurately (with little estimation error). If the X and Y scores are only weakly related, Y is predicted from X with considerable error. If the scatter plot is the typically occurring elliptical cloud as in Figure 3.3, then if the cloud is long and narrow the fit will be good, whereas the more close to circular the cloud is, the weaker the fit will be.

The Pearson Correlation Coefficient

Around 1900, the statistician Karl Pearson invented a statistic to describe the strength of the linear relationship between two variables. It is called the **Pearson correlation coefficient** and is given the label r. This correlation coefficient is defined so that it has values between -1 and 1, inclusive, and so that the sign of r (positive or negative) is the same as the sign of the slope of the least squares best-fitting line for the data.

Some examples will illustrate.

Example 3.1 **Strong Positive Linear Relationship**

Figure 3.17 shows scores on midterm (X) and final (Y) exams for a class of students. These scores are strongly related, and the slope of the least squares best-fitting line for the data is positive. The correlation between X and Y when calculated is close to 1.0.

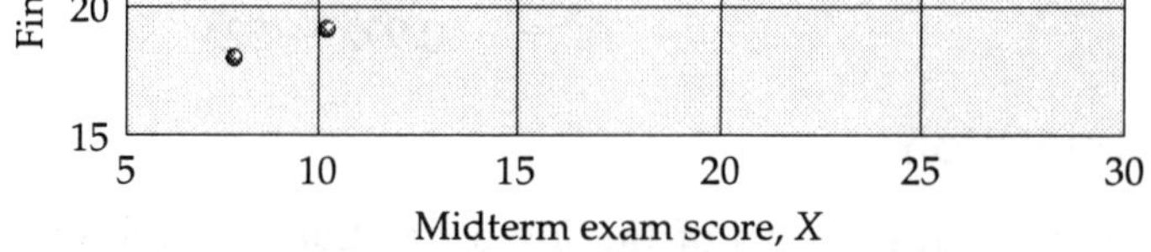

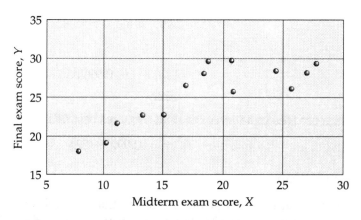

Figure 3.17 **Final exam versus midterm scores.**

Example 3.2 Weak Positive Linear Relationship

Figure 3.18 shows an index of interest in mathematics (X) and final mathematics test score (Y) for a class of students. These scores are weakly related, and the slope of the regression line is positive. This is called a weak, positive relationship between X and Y. The correlation between X and Y is positive, but fairly close to zero.

Example 3.3 No Linear Relationship

Figure 3.19 shows an index of student popularity (X) and mathematics test score (Y) for a class of students. There appears to be no relationship between these two variables. The slope of the best-fitting line is close to zero.

Example 3.4 Weak Negative Linear Relationship

Figure 3.20 shows an index of anxiety about mathematics (X) and final mathematics score (Y). These scores are weakly related, and the slope of the regression line is negative. This is called a weak, negative relationship between X and Y. The correlation between X and Y is negative and fairly close to zero.

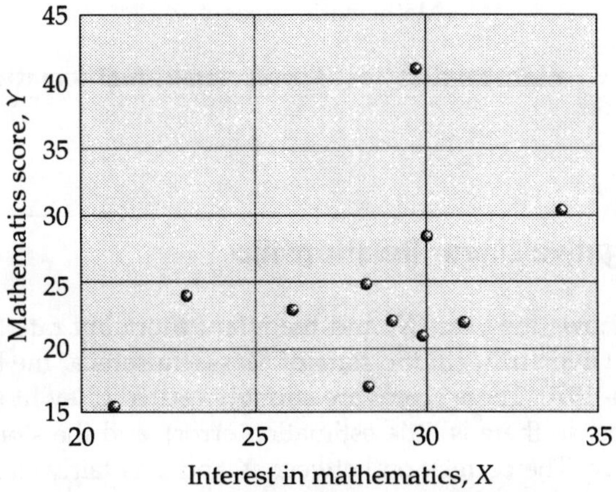

Figure 3.18 **Mathematics achievement versus interest in mathematics.**

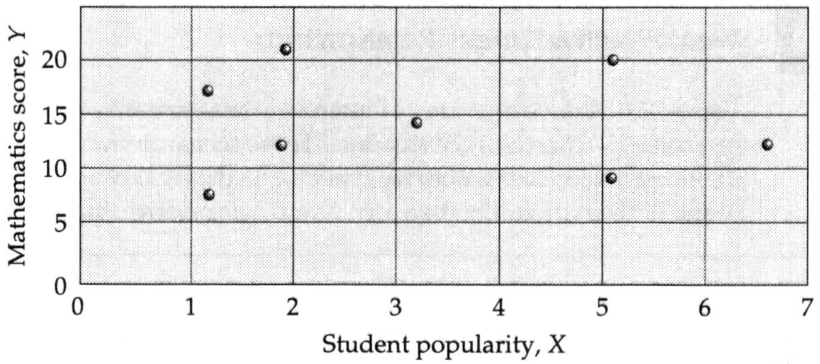

Figure 3.19 **Mathematics test scores versus student popularity.**

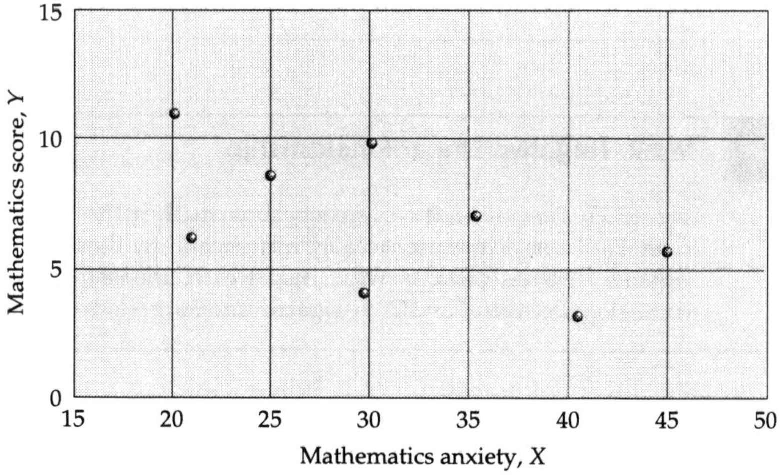

Figure 3.20 **Mathematics test scores versus mathematics anxiety.**

Example 3.5

Strong Negative Linear Relationship

Figure 3.21 shows the year (X) and the infant mortality rate (Y = deaths in one year per 1000 live births) in the state of Massachusetts at the beginning of the 13 decades 1850–1970. These scores are strongly related (they fit rather closely to the regression line, so there is little estimation error), and the slope of the regression line is negative. The correlation between X and Y is fairly close to -1.0. But it is interesting to note that using a nonlinear curve might be a better choice. This issue is taken up in Chapter 12.

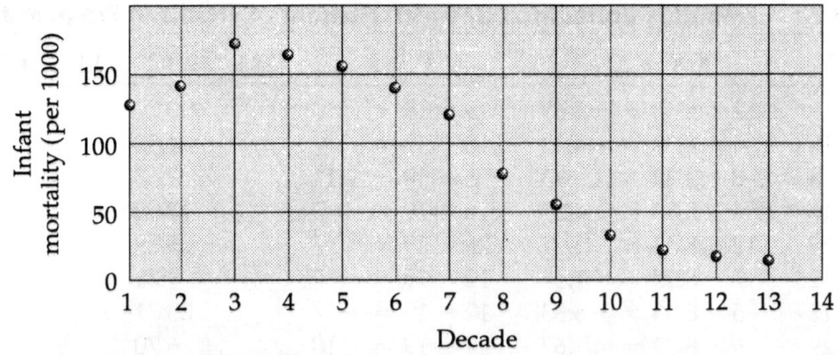

Figure 3.21 **Infant mortality rates for 13 decades.**

How to Calculate the Correlation Coefficient *r*

There are several ways to calculate r. Here is one:

$$r = \frac{\text{covariance of X and } Y}{S_X \cdot S_Y}$$

That is, the correlation between X and Y is the covariance of X and Y divided by the product of the standard deviation of X, denoted S_X, and the standard deviation of Y, denoted S_Y.

A table is helpful in our calculations. This table is not much different from the one used to find covariance.

Example 3.6

Find the correlation between the number of weeks spent in a weight-loss program (X) and the weight lost by participants (Y) as given in Table 3.3.

Solution

We follow six steps, using data from Table 3.18. Compare this table with Table 3.17.

Step 1: Find the means of X and Y. We see from Table 3.18 that

$$\text{Mean of } X = \overline{X} = \frac{75}{9} = 8.33$$

$$\text{Mean of } Y = \overline{Y} = \frac{171}{9} = 19.0$$

Step 2: Find the deviation value for each X and each Y (that is, $X - \overline{X}$ and $Y - \overline{Y}$).

Table 3.18 **Finding Correlation between Number of Weeks in Program and Weight Loss**

X	Y	$X - \overline{X}$	$Y - \overline{Y}$	$(X - \overline{X})(Y - \overline{Y})$	$(X - \overline{X})^2$	$(Y - \overline{Y})^2$
12	19	12 − 8.33 = 3.67	19 − 19 = 0	0	13.47	0
4	18	4 − 8.33 = −4.33	18 − 19 = − 1	4.33	18.75	1
8	8	8 − 8.33 = − .33	8 − 19 = −11	3.63	.11	121
4	16	4 − 8.33 = −4.33	16 − 19 = − 3	12.99	18.75	9
12	26	12 − 8.33 = 3.67	26 − 19 = 7	25.69	13.47	49
7	12	7 − 8.33 = −1.33	12 − 19 = − 7	9.31	1.77	49
5	12	5 − 8.33 = −3.33	12 − 19 = − 7	23.31	11.09	49
9	29	9 − 8.33 = .67	29 − 19 = 10	6.70	.45	100
14	31	14 − 8.33 = 5.67	31 − 19 = 12	68.04	32.15	144
75	171			154.00	110.01	522

Step 3: Find the squares of the deviation values for X and Y. Then find the sum of the squares for each and divide by the number of values, N. This gives the variance of X and the variance of Y.

$$\text{Variance of } X = \frac{\text{sum of } (X - \overline{X})^2}{N}$$

$$= \frac{110.01}{9} = 12.22$$

$$\text{Variance of } Y = \frac{\text{sum of } (Y - \overline{Y})^2}{N}$$

$$= \frac{522}{9} = 58.0$$

Step 4: Find the square root of the variance of X and the variance of Y. This gives the standard deviation of X and the standard deviation of Y.

$$\text{Standard deviation of } X = S_X = \sqrt{\text{variance of } X}$$

$$= \sqrt{12.22} = 3.50$$

$$\text{Standard deviation of } Y = S_Y = \sqrt{\text{variance of } Y}$$

$$= \sqrt{58.0} = 7.62$$

Step 5: Find the product of the deviation values, add them up, and divide by the number of values N.

$$\text{Covariance of } X \text{ and } Y = \frac{\text{sum of } (X - \overline{X})(Y - \overline{Y})}{N}$$

$$= \frac{154.0}{9} = 17.11$$

Step 6: Find the correlation between X and Y.

$$r = \frac{\text{covariance } X \text{ and } Y}{S_X \cdot S_Y} = \frac{17.11}{(3.50)(7.62)} = 0.64$$

Finding Correlation from the Slope

We have another way to calculate the correlation r between two variables if we happen to know the slope of the least squares best-fitting line for the data. Recall that we can find this slope—call it m—from a table of slopes and errors (if we have a computer to help us) or we can calculate it using a formula already given above, that is,

$$m = \frac{\text{covariance of } X \text{ and } Y}{\text{variance of } X}$$

Once we know the slope, m, of the line that predicts Y scores from X scores, we find r by using the equation in step 6 as follows:

$$r = \frac{m \cdot S_X}{S_Y}$$

It is interesting to note that this equation can be written

$$m = r\frac{S_Y}{S_X}$$

Thus, if r is small in magnitude (that is, the fit is bad), then we choose a small slope m. The result is that we do not let the value of X have much influence on the prediction Y'. In fact, if $r = 0$, we let X have no influence. Then the least squares regression line becomes

$$Y = \overline{Y}$$

That is, the prediction is $\overline{Y}$ regardless of the X value, as is appropriate. For small r note that the slope of the best-fitting least squares regression line "regresses" back to 0.

Example 3.7

Weight-Loss Data

We need to find the slope of the least squares best-fitting line for the weight-loss data of Table 3.3. See Table 3.18. We calculate the variance of X as $110.0/9 = 12.22$ and the covariance of X and Y as $154.0/9 = 17.11$. If we use the above formula, we get a slope of

$$m = \frac{\text{covariance of X and Y}}{\text{variance of X}} = \frac{17.11}{12.22} = 1.40$$

Since we know that

$$S_X = 3.50$$

and

$$S_Y = 7.62$$

we have

$$r = \frac{(1.40)(3.50)}{7.62} = 0.64$$

Correlation: How Good Is the Fit?

Remember that the least squares best-fitting line gives the smallest mean square estimation error of all possible lines fitting the data. This smallest error is usually called the *error,* or *residual, variance* of Y and is sometimes denoted S_e^2. This error variance—the error after the line is fit to the data—is less than the variation of Y in the data about $\overline{Y}$. That is, it can be shown that

$$S_e^2 \leq S_Y^2$$

We can use the error variance S_e^2 and the ordinary Y variance S_Y^2 (the variance we studied in Chapter 2) as another way to calculate r, the correlation between X and Y. Here is the reason for this. If the error variance is small compared with the variance of Y, there is little estimation error. In other words, the regression line fits the data well. If the error variance is almost as large as the variance of Y, there are large estimation errors. The regression line is not very good at estimating Y values from X values. So a small error variance relative to the Y variance means a large correlation (close to $+1$ or -1). A large error variance relative to the Y variance means a small correlation (close to 0).

We can therefore look at the ratio of the error variance of Y to the regular variance of Y, or

$$\frac{S_e^2}{S_Y^2}$$

as a measure of goodness of fit of a regression line. In fact, when this ratio is subtracted from 1, it can be shown that the result is exactly equal to the square of the correlation between X and Y; that is,

$$r^2 = 1 - \frac{S_e^2}{S_Y^2}$$

Thus the closer r^2 is to 1, the better the fit of the line. If r^2 is close to 0, the estimate error is large. We can then find the magnitude of r by

$$|r| = \sqrt{1 - \frac{S_e^2}{S_Y^2}}$$

Example 3.8 **Spelling and Reading Data**

Recall the spelling and reading scores of 10 students given in Table 3.10. We found in Table 3.15 that the least mean square error was 6.705. This is the error variance for the reading scores. The variance of the reading scores is

$$S_Y^2 = 34.21$$

So

$$r^2 = 1 - \frac{S_e^2}{S_Y^2} = 1 - \frac{6.71}{34.21}$$
$$= 1 - 0.196 = 0.804$$

We now take the square root of 0.804:

$$r = \pm 0.90$$

How can we decide whether the correlation is positive or negative? The correlation has the same sign as the covariance and the slope. So we can look at the slope of the regression line, which in most cases can be easily seen by looking at the scatter plot, or we can find the covariance of X and Y. In this example we can also look at Table 3.15, which tells us that the slope of the regression line is 1.81. We therefore have

$$r = +0.90$$

SECTION 3.6 EXERCISES

1. a. Find the covariance of the mice data of Exercise 1 of Section 3.1. Also find the variance of X, the lengths in centimeters of the mice.
   b. Use the slope formula of Section 3.6 to find the slope of the best-fitting line for the mice data. Then compare this value with that given in the table of slopes and errors for the mice data of Exercise 5 of Section 3.5.

2. a. Find the covariance of the mathematics exam and anxiety data in the following table. This covariance should be negative. Explain why. Also find the variance of the X scores.

	Anxiety score X	Exam score Y
Robert	15	37
Juan	20	33
Margo	23	5
Debi	30	31
Phil	34	40
Jo	34	35
Harry	35	32
Mark	36	30
Polly	38	14
Jim	40	9
Pete	42	20
Mary	44	24
Sue	46	18
Roger	49	21
Tim	50	13
Jose	53	14
Sue	59	7
John	60	9
Jill	62	7
Jon	63	12

b. Find the slope of the least squares best-fitting line for the anxiety data using the formula. Compare with the slope of the line from the following table. They should be very close.

Slope	Mean absolute error	Mean square error	
−1.00	7.84	112.66	
−0.95	7.42	104.47	
−0.90	7.00	97.20	
−0.85	6.68	90.84	
−0.80	6.36	85.40	
−0.75	6.10	80.87	
−0.70	5.87	77.26	
−0.65	5.67	74.56	
−0.60	5.64	72.77	
−0.55	5.84	71.90	← Least mean
−0.50	6.12	71.94	← square error
−0.45	6.40	72.90	
−0.40	6.69	74.77	
−0.35	7.08	77.55	
−0.30	7.46	81.25	
−0.25	7.85	85.87	
−0.20	8.24	91.39	

3. a. The data in the table below are the number of push-ups that could be done by a sample of 12 male instructors at Howard Community College (the 38-year-old teacher was the track coach). Find the covariance of these data. Also find the variance of X, the age of the teachers doing the push-ups.

Age	Number of push-ups	Age	Number of push-ups
21	10	22	9
25	8	27	6
22	11	44	4
28	6	48	3
30	7	35	8
38	15	48	5

b. Find the slope of the least squares best-fitting line for the push-ups data using the formula of Section 3.6 with the information found in part (a).

4. Use the slope of the least squares best-fitting line for the push-up data that you found in Exercise 3. Find the equation of the regression line with this slope. Then use the regression line to estimate the number of push-ups that can be done by teachers of each of the following ages:
   a. 26 years
   b. 39 years
   c. 50 years

   What assumptions are you making when you make these estimates?

5. Figure E5 shows graphs of different collections of data in two variables. Tell, by inspection, whether the correlation between the two variables is

   Strong positive
   Weak positive
   Weak negative
   Strong negative

6. Tell whether you would expect the correlation between the two variables in each of these sets of data to be strong positive, weak positive, weak negative, or strong negative.
   a. Height and weight of children from ages 4 to 12 years
   b. Time spent studying for an exam and score on that exam

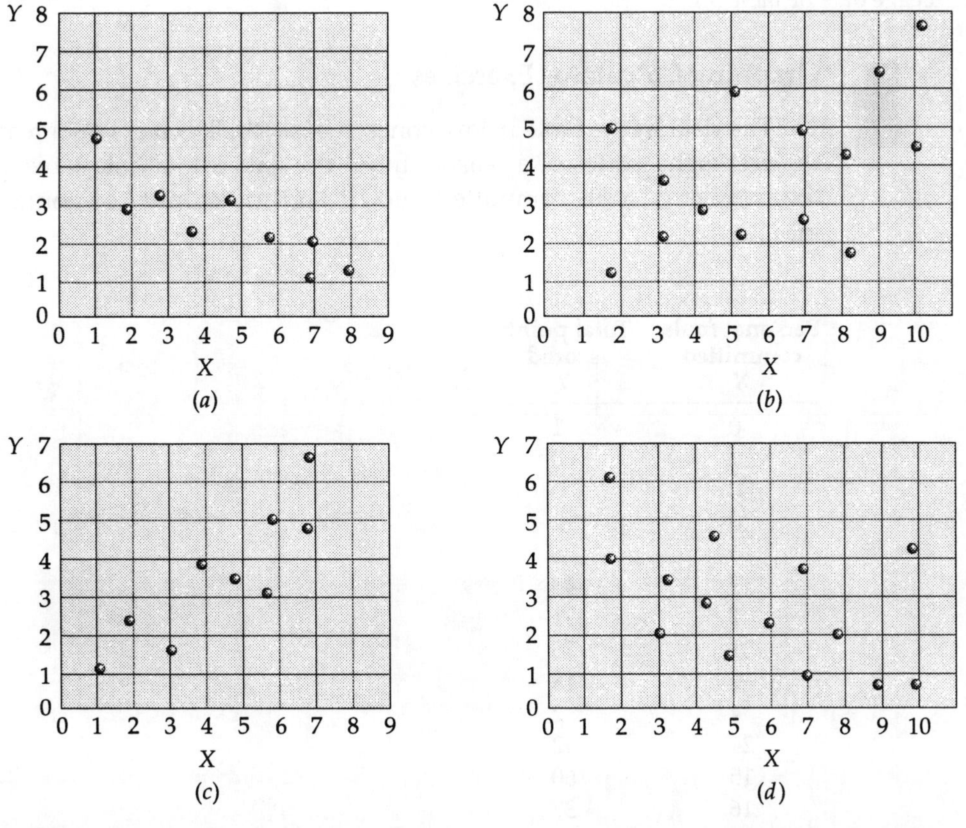

Figure E5

c. Attitude toward mathematics (whether you like it or not) and final grade in mathematics
d. Popularity in school or college and academic achievement
e. Price of a watch and how many minutes it gains or loses per week
f. Age of a car and number of trouble-free miles per month
g. Pressure applied to a gas (such as hydrogen) and volume that the gas occupies
h. Air pressure kept in automobile tires and mileage obtained from the tires

Use the covariance and standard deviation method of this section to find the correlation in Exercises 7 through 12.

7. Find the correlation between the car weight and mileage data of Table 3.1.
8. Find the correlation between age and number of push-ups as given in Exercise 3.
9. Find the correlation between the length and body weight of mice as given in Exercise 1 of Section 3.4.
10. Find the correlation between radiation in the environment and cancer rate, as given in Figure 3.16.
11. The error variance for the problem introduced in Exercise 4 is 8.48. Use the error variance to find the correlation between age and number of push-ups. Is this correlation positive or negative? Why?
12. Find the correlation between length and body weight of mice using the slope method. Compare with your answer obtained in Exercise 9 of this section.

Graphing Calculator Exercises

13. The data in the table below concern basketball players on a team over the first eight games of a season. Input the data into a data table, placing the number of fouls committed into L1 and the number of points scored into L2.

Personal fouls committed X	Total points scored Y
0	1
0	0
0	0
0	2
6	6
17	48
7	21
31	75
9	18
4	2
24	42
15	60
16	37
0	3

a. Make a scatter plot.
b. Estimate the line of best fit (see Section 1.6 of the TI Graphing Calculator Supplement).
c. Find the covariance between X and Y (see Section 1.9 of the TI Graphing Calculator Supplement) and the standard deviation of X and Y. What is the correlation between X and Y? (See Section 1.12 of the TI Graphing Calculator Supplement.) Refer to Sections 2.6 and 2.7 for how to calculate variance and standard deviation.

14. In the exercises of Sections 3.4 and 3.5 you calculated the slope with the least mean absolute error and least mean square error for the spelling and reading data of Table 3.10. Now calculate the slope of the least squares best-fitting line by using the covariance of X and Y and the variance of X, as explained in Section 3.6. Compare the three values for slope that you obtained. (Note that the trial-and-error least squares line of Section 3.5 should have a slope quite close to the best-fitting least squares line found here.)

15. Enter the data from Exercise 2 on mathematics anxiety versus exam scores. Enter anxiety score (X) into L1 and exam score (Y) into L2. Find the covariance, and explain your result. Also find the variance of the X scores. Now calculate the slope of the least squares best-fitting line. What is the relationship between the sign of the covariance and the slope of this line?

3.7 RELATION AND CAUSATION

In this chapter we have seen many examples of data in which there was found to be a strong relationship between the two variables. It seems reasonable in this case to use the data to conclude that the aging process *causes* a person to possess less ability at physical tasks such as running, jogging, or playing football.

However, the existence of a relationship between two variables does not prove that one causes the other. Suppose, for example, we were to collect data on shoe size and score on a 40-point, 12th-grade mathematics test administered to male students from grades 4 through 12. We would expect the data to look something like Figure 3.22. That is, we would expect that people who get better scores on the test will wear bigger shoes. But this is far from saying that a knowledge of mathematics *causes* big feet (or vice versa)! In this, there is another factor to take into account—the aging of the student, which causes both mathematics scores to improve and feet to grow larger.

There are many examples of the perils of interpreting correlations between variables as indicating causality. The radiation and cancer data in Figure 3.16 illustrate a relationship between the amount of radiation in the

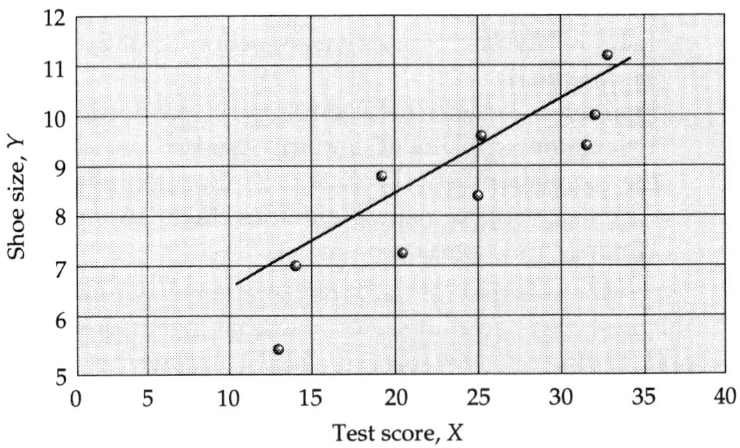

Figure 3.22 **Shoe sizes versus test scores of nine students.**

environment and the rate of cancer. But the data do not *prove* (though they strongly suggest) that the radiation is the cause of the cancer. It could be that there are other, as yet unknown, factors that enter into the picture, such as nonradioactive pollutants.

One of the best-known examples of the confusion about correlation and cause regards smoking and cancer. For the past 50 years or more, research projects have produced data showing a strong positive correlation between the amount a person smokes and the person's chance of getting cancer. On one side of the debate, medical associations and consumer groups have claimed that smoking causes cancer. On the other side—notably the tobacco industry—it has been argued that although smoking and cancer are statistically related, smoking has not been established as the cause of cancer. Recently, however, a study reported in *Science* has isolated a medical causal link in humans. Further, carefully designed studies on other mammals have shown strong evidence of causality in these species, which likely applies to humans too.

SECTION 3.7 EXERCISES

1. The data in the following table were reported by Dr. Al Shulte of the Pontiac, Michigan, school system. The data are the number of personal fouls committed (X) and the total number of points scored (Y) by the junior varsity basketball team at Waterford-Kettering High School in Michigan.
   a. Draw a graph of these data.
   b. Suppose the slope of the best-fitting line for the points scored and fouls committed is 2.5. Assuming the regression line goes through the mean data point, what is the equation of this line?

Player	Personal fouls committed X	Total points scored Y
Bogert	0	1
Bone	0	0
Campbell	0	0
Forbes	0	2
Godoshian	6	6
Graham	17	48
Madill	7	21
Manning	31	75
McGrath	9	18
Nutter	4	2
Nyberg	24	42
Shipman	15	60
Spencer	16	37
Watson	0	3

c. Find the covariance between *X* and *Y* and the standard deviation of *X* and *Y*. What is the correlation between *X* and *Y*?

d. If these data are interpreted causally, what are the implications for coaching?

2. It is often the case that two variables are correlated, and the simplest explanation—that a change in one variable causes the other to change—is often the correct one. But we must *not* leap to this conclusion on the basis of the statistical correlation alone. In many cases causality, at least in its simplest form, is not the explanation. It may be that the two variables are unrelated but are both being strongly influenced by a third one. Or the correlation may be sheer coincidence—nothing more than a statistical accident.

Often, the person on the street assigns causality to a variable when skepticism would be wise. For example, the fact that the number 9 was winner three times at a roulette wheel does not make it any more likely that a 9 will come up on the next spin of the wheel.

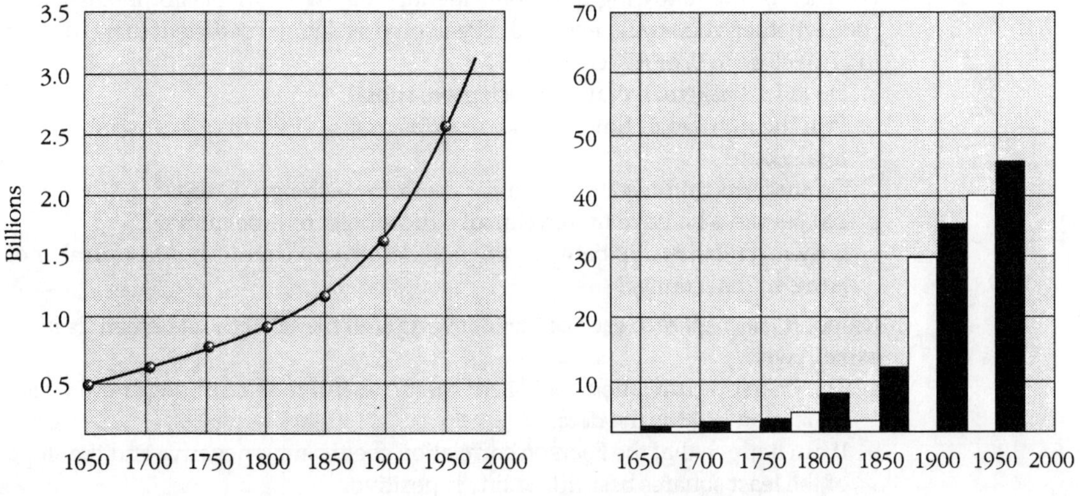

(*a*) The increase in human population over the past 300 years.

(*b*) The number of exterminated mammal forms (white bars) and bird forms (black bars) eliminated over the past 300 years. Each bar represents a 50-year period.

Figure E6

Consider the following examples of correlated variables and judge whether causality is likely. If you think it is, note carefully your reasons, and make sure that these reasons are based on your assumptions and knowledge about the situation and not on the statistical correlation.

a. Over the years there has been an increase in the suicide rate and in the availability of handguns.
b. In a certain country there has been an increase in poor physical fitness and a decrease in the life expectancy.
c. There has been an increase in alcohol consumption and an increase in traffic fatalities in City X.
d. The rate of cigar smoking and the incidence of cancer in the mouth has gone up.
e. The more college courses one takes, the higher ones starting salary is upon graduation, on average.
f. The more cars a household owns, the better is the health of its members.

3. Find a newspaper clipping or quote an item from radio or television news that uses data in two variables in such a way as to imply a causal relationship between them.
4. What correlations are evident from the graphs shown in Figure E6? Is there a causal relationship, do you think, between one variable and another? Explain. (Note: Relationships between variables need not be linear, that is, a straight line.)

CHAPTER REVIEW EXERCISES

1. For each of the following questions, identify the two variables of interest, and state whether you would expect the Pearson correlation coefficient to be (i) close to 1, (ii) close to 0, or (iii) close to -1.
   a. Do taller basketball players block more shots?
   b. Will the number of times a student skips a class effect his or her score on the final exam?
   c. Do students from wealthier families perform better on IQ tests?
   d. Do people who have more years of education earn more money?
   e. Is there a relationship between the population of a county and the number of farms in that county?
2. Respond *true* or *false* to each of the following statements. If the statement is false, explain why.
   a. The Pearson correlation coefficient can be useful in determining how well a regression line fits the data.
   b. It can happen that the Pearson correlation coefficient is negative and the slope of the least squares best-fitting line is positive.
   c. The mean error for a regression line must be positive.
   d. The method of least squares is the most widely used and most accurate method for finding regression lines, and it should always be used if a computer or calculator is available.
3. Consider the following verbal (X) and quantitative (Y) scores on the SAT from five students:

Student	X	Y	$X - \overline{X}$	$Y - \overline{Y}$	$(X - \overline{X})(Y - \overline{Y})$	$(X - \overline{X})^2$	$(Y - \overline{Y})^2$
1	670	710					
2	550	500					
3	720	620					
4	410	490					
5	520	560					
Sum	2870	2880					

A researcher wants to answer this question: "Is there a relationship between verbal and quantitative scores on the SAT?" For statistical evidence, the researcher wants to calculate the Pearson correlation coefficient (r) for the above data. The following exercise will take you through the six steps of calculating r.

a. Complete the table above.
b. Using the information from the table you just formed, find the variance of Y, the variance of X, the standard deviation of X, and the standard deviation of Y.
c. What is the covariance between X and Y? Again, all the calculations you need should have been performed in filling in the table.
d. Using the results from parts (b) and (c), find the correlation between X and Y.
e. What kind of relationship between mathematical and verbal scores on the SAT does this value of r imply?

4. What is the relationship between the length and width of covers of books? Find nine books, and measure the length and width of the cover for each.
   a. Do you think there will be a strong relationship between the length and width of covers of books? Why or why not?
   b. Construct a scatter plot of your nine observations. On this plot, draw the regression line using the informal visual method.
   c. What is the slope of your regression line? What is the equation of your regression line?
   d. Using the regression equation, calculate Y' for each of the points. What is the mean error?
   e. How accurate was your prediction from part (a)?
5. The following table is a list of 10 teams and their total number of runs scored and total number of home runs hit from the 1996 season.
   a. The regression line obtained by the method of least squares is $Y' = 2X + 455$. Which of the 10 observations has the largest squared error? Which has the smallest?
   b. True or false: If instead the method of least mean error were used to find the regression line, the regression equation would come out exactly the same. Justify your answer.
   c. A team not on the list, the Atlanta Braves, hit 197 home runs. Predict how many runs they scored.

Team	Number of home runs (X)	Total number of runs (Y)	Y′	(Y − Y′)²
Colorado Rockies	221	961	897	4096
Pittsburgh Pirates	138	776	731	2025
Los Angeles Dodgers	150	703	755	2704
St. Louis Cardinals	142	759	739	400
Cincinnati Reds	191	778	837	3481
Chicago Cubs	175	772	805	1089
Florida Marlins	150	688	755	4489
San Diego Padres	147	771	749	484
New York Mets	147	746	749	9
Houston Astros	129	753	713	1600

6. In Exercise 5 we found the least squares best-fitting line for the baseball team data.
   a. What is the mean squared error for the regression equation given in part (a) of Exercise 5?
   b. Recall that the mean squared error of the best-fitting least squares regression line is sometimes denoted as S_e^2. The variance of the number of runs scored, or S_Y^2, is 4878.01. What is the value of r^2?
   c. The square root of r^2 will of course give you r, except that it will not give you the sign. What information can be used to find the sign of r?
7. Using the data from Exercise 5, draw a scatter plot and fit a regression line using the method of median fit.
   a. What is the equation of this regression line? Does it seem to be somewhat similar or dissimilar to the equation found in part (a) of Exercise 5 using the method of least squares?
   b. Comparing these two regression equations (method of least squares and method of median fit), which would give you a larger mean squared error? How do you know this? (*Hint:* This should require no computation.)
8. Recall this equation:

$$\text{Slope from method of least squares} = \frac{\text{covariance between } X \text{ and } Y}{\text{variance of } X}$$

   a. Using this equation, find the covariance between the number of runs scored and the number of home runs hit from Exercise 5. Recall that the number of home runs is the X variable in this case.
   b. Again using the above equation, go back to Exercise 3 and find the slope of the regression line from the method of least squares for the SAT data.
9. Find a newspaper article that cites a study in which two variables are related to each other.
   a. What are the two variables that are being related?

b. Just because there is a relationship between these two variables, does that mean that one is causing the other? Do you feel that the authors of this article took that fact into consideration?

10. Consider the following section from a published health study:

    Among the current smokers who had high fish intake, the number of cigarettes smoked per day had no effect on the incidence of heart disease and the death rates from heart problems. As seen with lung diseases, the consumption of fish seems to protect smokers from the harmful effects of smoking on the heart and blood vessels. It has been suggested that fish oils alter some metabolic processes that result in beneficial effects including more dilatation (or relaxation) of blood vessels, less platelet adhesiveness, reduced inflammatory response to the injury caused by smoking, lower triglyceride and fibrinogen levels, and lower blood pressure (especially in patients who have mildly elevated blood pressure). (Source: http://www.healer-inc.com/a039705a.htm)

    a. What are the two variables being compared in this study?
    b. Could you conclude from this study that if you were a smoker and you ate a lot of fish, you are definitely at a reduced chance of suffering from heart disease? Why or why not?

11. In this chapter, three methods of finding a regression line were discussed: the informal visual method, the method of least squares, and the method of least mean error. For each of the following statements, tell which of the methods it applies to.
    a. The slope of the line found using this method is related the Pearson correlation coefficient.
    b. With this method, the equation for the regression line can be found easily without the use of a calculator or computer.
    c. This method is not considered robust—that is, it can be unduly influenced by one or two of the points.
    d. Using this method, the mean error of the regression equation is less than the mean error of any other regression line that passes through $(\overline{X}, \overline{Y})$.

12. The following is a list of the sizes of nine diamonds (in carats) and their prices:

Size (X)	Price (Y)	Size (X)	Price (Y)
0.17	$353	0.16	$342
0.16	$328	0.15	$322
0.17	$350	0.19	$485
0.18	$325	0.21	$483
0.25	$642		

Source: http://www2.ncsu.edu/ncsu/pams/stat/info/jse/datasets.index.html

a. Draw a scatter plot showing the relationship between these two variables.
b. Find a regression line using the informal visual method. What is the equation for the line?
c. You find a diamond in a store of size 0.23 carat, and the salesperson is going to sell it to you for \$500. Is it a good deal, according to the regression line found above?
d. Which of the nine diamonds listed above was the worst deal for the person who bought it, according to your regression line?

13. What is the covariance between the size of a diamond and its price, using the data from Exercise 12?

14. The chapter discusses two methods for finding the slope of the least squares best-fitting line. Explain both of the methods briefly.

15. The table below gives the weight (X) of eight adult men and the amount of weight they are able to lift (Y):

Man	X	Y	$X - \overline{X}$	$Y - \overline{Y}$	$(X - \overline{X})(Y - \overline{Y})$	$(Y - \overline{Y})^2$
1	150	172	−21	−12	252	144
2	174	210	3	26	78	676
3	163	159	−8	−25	200	625
4	167	175	−4	−9	36	81
5	210	200	39	16	624	256
6	189	220	18	36	648	1296
7	140	150	−31	−34	1054	1156
8	175	185	4	1	4	1
Sum	1368	1472	0	0	2896	4235

a. What is the covariance between X and Y?
b. What is the slope of the least squares best-fitting regression line?
c. Using the fact that the line must pass through (171, 184), determine the equation for the least squares best-fitting regression line.
d. If another male 10th-grade student weighs 156 pounds, what would be the best prediction as to how much he will be able to lift?

16. Figure E7 is a scatter plot of four data points. Copy the plot onto graph paper and try to fit a regression line to the data minimizing the mean error. The line must pass through the point $(\overline{X}, \overline{Y}) = (2.5, 8)$, which is marked on the graph.
a. Obtain an estimate of the mean error for your graph.
b. Determine the equation of the line that you drew on the scatter plot. (*Hint:* Estimate the slope of the line you drew and use the fact that the line must pass through the point (2.5, 8).)

17. Repeat Exercise 16, but this time try to fit a line to the data points that will minimize the mean square error instead of the mean error. Recall that single points will have more influence in least squares regression.

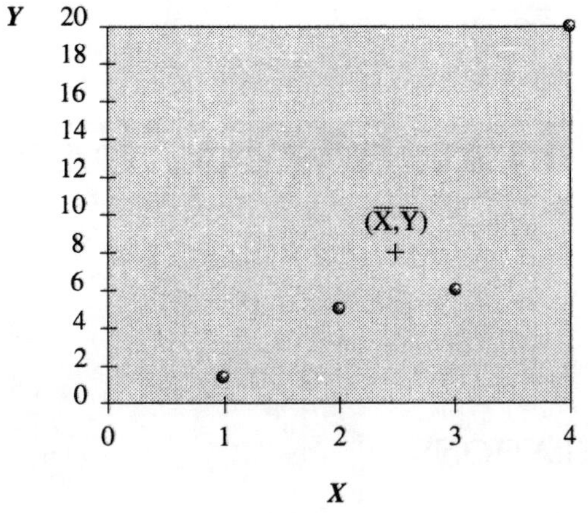

Figure E7

18. The following table shows the return and risk for selected investments during the period 1960–1990. (The risk variable is a quantitative measurement of the amount of risk for that type of investment.)

	Return (*X*)	Risk (*Y*)
Common stock	9.8	15.8
Long-term corporate bonds	6.9	11.1
Long-term treasury bonds	6.4	10.9
Short-term treasury bonds	6.4	2.9
Residential real estate	10.6	10.7
Farm real estate	9.9	8.5
Business real estate	8.7	4.9

a. What sort of relationship would you expect between return and risk for investments? Why?
b. Find the covariance between return and risk.
c. What is the correlation between return and risk?
d. Does your answer agree with what you guessed in part (a)?

3

COMPUTER EXERCISES

1. GUESSING CORRELATIONS

This problem uses the Correlations program. You start it by clicking on the following icon:

Correlations

The program will present you with a screen. To get started, click on the "New Plots" button. Four scatter plots will appear, as will a list of correlation coefficients along the table of circles at the right:

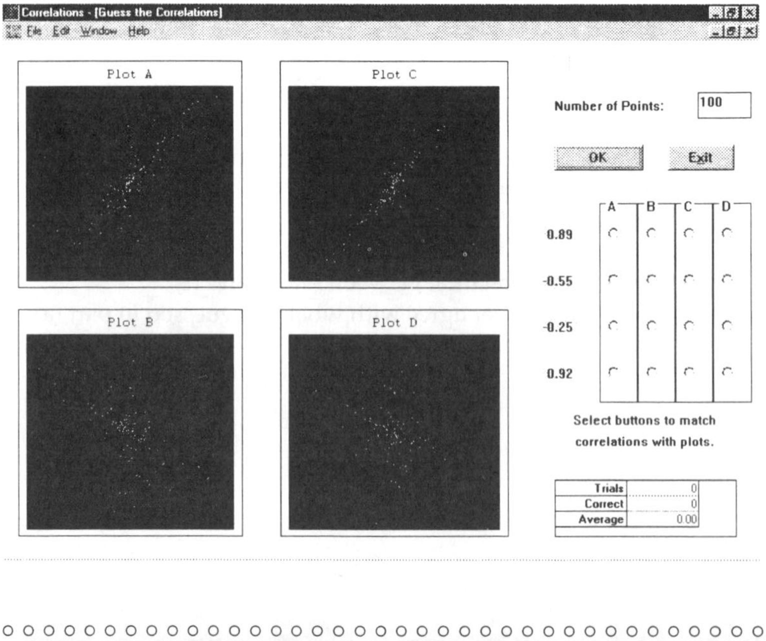

Each time you do it, the plots will be different, so yours will be different from this one. The plots are labeled A, B, C, D, as are the columns of the table. Your job is to match the plots with their correlation coefficients by clicking in the appropriate circles. In this example, it looks as if plots A and C have the strongest positive correlations, with C a little stronger. So we guess that plot C has correlation 0.92 and plot A has correlation 0.89. The other two plots have negative correlations. Plot B looks a little tighter than plot D, so we guess that plot B has correlation −0.55 and plot D has −0.25. After we click in the appropriate boxes, the table will look like this:

	A	B	C	D
0.89	●	○	○	○
-0.55	○	●	○	○
-0.25	○	○	○	●
0.92	○	○	●	○

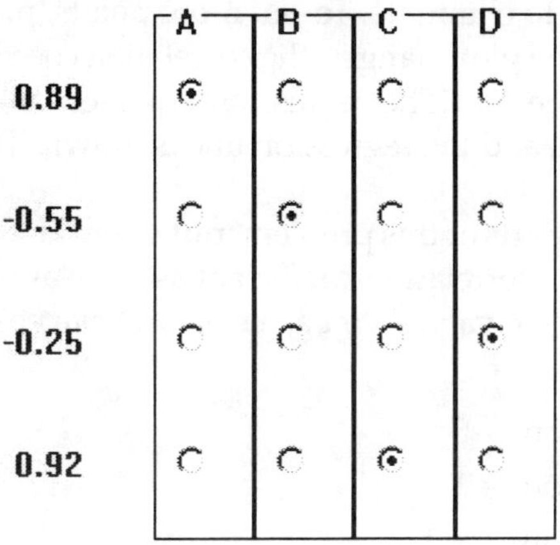

Now click on the OK button. (It used to be the "New Plots" button.) It will check the answers. In this case, we guessed all of them correctly! It is not always easy. Sometimes it is almost impossible. That is the way it is supposed to be.

To get another set of plots, press "New Plots" again. Notice that in the lower right there is a table that keeps track of how many times you have played and the average number you have guessed correctly.

a. Practice the guessing many times. When you feel confident, do the guessing five times and report your average number correct. Do you think you were unlucky in the plots you got? Or did you guess everything correctly?

b. Briefly describe the strategy you used in formulating your guesses.

c. In general, is it easier to distinguish a difference of 0.05 when the correlations are over 0.9 (say, 0.92 and 0.97), or when the correlations are small (say, 0.06 and 0.11)?

d. Try guessing again, but first set the number of points to 2. Is it easier or harder with 2 points than with 100 points? Why?

2. PUTTING POINTS I

Start the Put Points program by clicking on its icon:

Put Points

By clicking the mouse on the chart area, you will create data points. If you want to erase a point, first select the "Erase Points" button, and then click on the point you wish to eliminate. To put more points, press the "Put Points" button. Each time the plot changes, the correlation coefficient is recalculated and presented at the top. The regression equation and root mean square error are also given, and the regression line is drawn. This problem focuses on the correlation.

For each of the parts of this problem, put down at least 10 points in such a way as to obtain a correlation coefficient as close as you can to the target value. Write down the values you achieve, and sketch the scatter plot you create.

a. Target correlation $= 0$
b. Target correlation $= 1$
c. Target correlation $= -1$
d. Target correlation $= 0.5$
e. Target correlation $= -0.5$
f. Target correlation $= 0.3$
g. Target correlation $= -0.8$

Which targets are easiest to achieve? What strategy did you use in putting down the points?

3. DECATHLON

Start the Data program. (Go back to Chapter 1 to recall how to use the program.) Load DECATH.DAT. This file has data on the performances in the decathlon of the best decathlon athletes in the world. The decathlon is a 10-event athletic contest consisting of the 100-meter, 400-meter, and 1500-meter runs, the 110-meter high hurdles, the javelin and discus throws, the shot put, the pole vault, the high jump, and the broad jump.

For the three runs (100 meter, 400 meter, and 1500 meter) and the hurdles, the scores are times, so it is better to have a small number (a faster time). In the other events, it is better to have a large number.

a. For each of the following pairs of variables, guess whether their correlation is positive or negative.
   i. 100-meter run and 400-meter run
   ii. 100-meter run and 1500-meter run
   iii. Discus throw and javelin throw
   iv. Long (or broad) jump and high jump
   v. High jump and pole vault
   vi. 100-meter run and high jump
   vii. Discus throw and 100-meter run

b. After guessing, make the "Regression" choice in the Data menu. A new screen will appear that allows you to create scatter plots. Click on the box next to "Regression Line" in the upper right. That will make sure that the correlation coefficient is displayed. For each of the above pairs, choose the first variable as the "X Variable" and the second as the "Y Variable," and then click on Update. The next picture shows the result of picking "100M" as *X* and "400M" as *Y*:

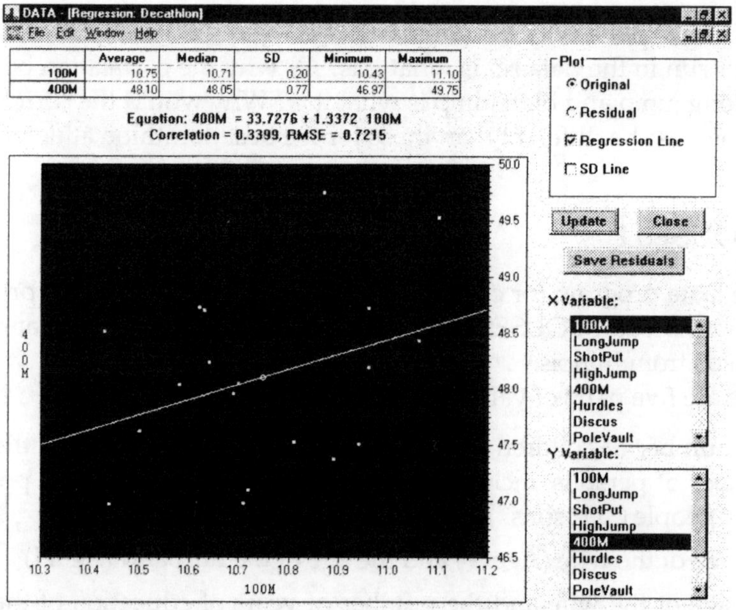

Over the scatter plot, you will see the correlation coefficient. In this case it is 0.3399. For the same pairs of variables, find the correlations:
   i. 100-meter run and 400-meter run (0.3399)
   ii. 100-meter run and 1500-meter run

iii. Discus throw and javelin throw
iv. Long (or broad) jump and high jump
v. High jump and pole vault
vi. 100-meter run and high jump
vii. Discus throw and 100-meter run

How good were your guesses in part (a)?

c. Suppose everyone in a typical statistics class participated in a decathlon. Imagine looking at the data. Do you think the correlations in the following pairs would be positive or negative?
   i. 100-meter run and 400-meter run
   ii. 100-meter run and 1500-meter run
   iii. Discus throw and javelin throw
   iv. Long (or broad) jump and high jump
   v. High jump and pole vault

   Give reasons for your answers. (We would guess that they would all be fairly positive.)

d. Can you explain why the correlation between the 100-meter and 1500-meter run in the data set is what it is? Or why the correlation between the long jump and high jump is what it is? Why would the correlations for the class be different than those for the best decathlon athletes?

4. SPOUSES I

Start the Data program (or close the regression program from the previous exercise) and load SPOUSES.DAT. The data pertain to 177 husband-wife households from Illinois.

Here are five pairs of variables:

- The number of children living in the household ("Kids") and the total number of persons, including children, the two parents, and possibly other people ("Persons")
- The age of the wife (AgeW) and the age of the husband (AgeH)
- The age of the wife and the number of years of education of the wife (EducationW)
- The number of years of education of the wife and the number of years of education of the husband (EducationH)
- The income of the wife (IncomeW) and the income of the husband (IncomeH)

a. Put the five pairs in order from what you would guess is the most weakly correlated pair to the most strongly correlated pair. Do this before creating any scatter plots. (A strong negative correlation is stronger than a weak positive correlation; for example, -0.75 is stronger than $+0.35$.)

b. Now choose the Regression option from the Data menu. Create the scatter plots of the pairs of variables above. Be sure the Regression Line box is checked, so that the correlation coefficient will appear above the plot. What are the correlation coefficients for the five pairs?

c. How good were your guesses in part (a)?

d. Find the correlation coefficient between the age of the wife and the number of children. Does this correlation suggest that if a woman adopts a child, she will become younger? Why or why not?

5. DRAFT LOTTERY

In 1969 the U.S. government instituted a lottery to assign numbers to young men of draft age. They were to be drafted into the military in the order of their lottery numbers. People with high numbers were less likely to be drafted. The lottery numbers were assigned according to birthdays. There was a drawing in which each day of the year received a different number from 1 to 366. (The drawing included leap year birthdays, since it applied to everyone born in 1950 or earlier who was not older than 26.) Whichever number was assigned to one's birthday was one's number. The data set DRAFT69.DAT contains the data, with the variables being the day of the year and the lottery number. Load the data, and choose "Regression" from the Data menu.

a. Create the plot using Day as the "X Variable" and Lottery Number as the "Y Variable." (Don't check the Regression Line box yet.) Do you see any pattern? What would you guess is the correlation?

b. Now check the Regression Line box, and click on Update. What is the correlation coefficient?

c. Some people criticized the lottery because it seemed that people born later in the year tended to receive lower numbers. Do you think the lottery was fair? (This question is purposely ambiguous. Actually, there are several questions: (i) Was the assignment of lottery numbers to birthdays totally random? (ii) Does it matter? (iii) Is it politically good to draft people randomly?)

d. President Bill Clinton was born on August 19. What was his lottery number? Use the Display Data choice in the Data menu to find it. Don't forget to close the Regression screen first.)

e. The data set DRAFT69M.DAT has the same data, except that the lottery numbers were averaged over each month. Load that data set, and find the correlation coefficient. Is there a clear trend? What is the correlation?

f. Is the correlation coefficient obtained in part (e) higher or lower than that obtained in part (b)? Can you explain why?

g. The data sets DRAFT70.DAT and DRAFT70M.DAT contain similar data, but for the draft lottery in 1970 conducted for people born in 1951. The procedure for assigning numbers was supposed to be better this time. Find the correlations for these data sets. Are they closer to 0 than before? Which lottery was more random, the 1969 one or the 1970 one?

6. SMOKING AND CANCER

Load the SMOKE.DAT data set. It compares rates of smoking and death from lung cancer for men in 25 occupational groups in England. The "Smoking" variable is 100 times the average number of cigarettes smoked per day by men in each occupational group divided by the overall average, so that numbers over 100 indicate more smoking than average for that group. The "Mortality" variable similarly measures the death rate of men in each group, normalized so that numbers over 100 indicate more deaths due to lung cancer than average for that group. (Data are from David S. Moore and George P. McCabe, *Introduction to the Practice of Statistics* [New York: Freeman, 1989].)

Choose Regression from the Data menu.

a. Find the correlation coefficient between Smoking and Mortality.

b. Clicking on the actual data points will cause the name of the occupational group to appear. Click on the points. What do the groups with low smoking and lung cancer rates have in common? Which group has the highest cancer rate?

c. Does the answer in part (a) prove that smoking causes lung cancer? Or could there be other factors?

d. Do you think smoking causes lung cancer?

7. PUTTING POINTS II

As in Exercise 2, this problem uses the Put Points program. Click on its icon. This time, the goal is to put down at least 10 points in such a way as to obtain the slope of the regression line as close as you can to the target values listed below. Write down the slopes you achieve, and sketch the scatter plots you create.

a. Target slope $= 0$
b. Target slope $= 1$
c. Target slope $= -1$
d. Target slope $= 2$
e. Target slope $= -2$
f. Target slope $= 0.25$
g. Target slope $= 4$

8. CLASS DATA III

This exercise uses the data from a STAT 100 class described in Computer Exercise 6 of Chapter 1 and Computer Exercise 2 of Chapter 2.

a. Imagine the scatter plot with X = height and Y = weight. What do you guess is the slope of the regression line, approximately?
   i. Less than 0
   ii. Less than 1, but more than 0
   iii. Around 2
   iv. Around 5
   v. Around 30

   Hint: To come up with a guess, think of two heights in inches, H_1 and H_2. How much heavier, on average, is someone who is H_2 inches tall than someone who is H_1 inches tall? Divide that difference in weight in pounds by the difference in inches, $H_2 - H_1$. That gives a guess of the slope. Try a few pairs H_1, H_2.
b. Start the Data program, load CLASS96.DAT, choose Regression from the Data menu, and click on the Regression Line box. Choose height as the X variable and weight as the Y variable, and then click on Update. What is the correlation coefficient?
c. What is the slope? How good was your guess in part (a)?
d. One person in the class was on the University of Illinois football team. Click on the point you think represents him. What is the point's name?
e. The "RMSE" value given beside the correlation above the scatter plot is the square root of the error variance—that is, RMSE $= S_e$. What is the RMSE for this regression line? What is S_e^2?
f. Check the following formulas:
   i. Look at the formula that relates the slope m and the correlation coefficient r:

$$r = \frac{mS_X}{S_Y}$$

The table at the top of the regression screen contains S_X and S_Y, the standard deviations of X and Y, respectively. Calculate mS_X/S_Y and compare it with r. Are they essentially the same?

ii. Look at the formula that relates the correlation coefficient to the variances of Y and of the error:

$$|r| = \sqrt{1 - \frac{S_e^2}{S_Y^2}}$$

Calculate the square root in the equation. Does it equal $|r|$, essentially?

g. Click on the Save Residuals button. The residuals are the errors, $Y - Y'$. A new variable called "R(3.2)" will be created. (The "3" refers to the weight, which is the third variable in the list, and the "2" refers to the height, which is the second variable in the list.) Select the variable of errors, R(3.2), as the Y variable. (You will have to scroll down the list to find it.)

i. What is the correlation coefficient?

ii. What is the slope?

iii. What is the average of the errors?

iv. What is the standard deviation of the errors?

v. What is RMSE $= S_e$? How does it compare with the standard deviation of the errors? How does it compare with the RMSE for the regression with $Y =$ weights found in part (e)?

vi. Who has the largest error?

9. NBA ALLSTARS III

Recall the ALLSTARS.DAT data set from Computer Exercise 7 of Chapter 1 and Computer Exercise 5 of Chapter 2.

a. Imagine that X is the height in inches and Y is the height in centimeters. What do you think the slope of the regression line is?

b. What do you think the value of the correlation coefficient is?

c. What do you think is the value of the error variance, S_e^2? (Recall that S_e^2 is the square of the RMSE.)

d. Now load the data set into Data, choose Regression from the Data menu, check the Regression Line box, and create the regression with $X =$ "height #in" and $Y =$ "height #cm." What do you notice about the points?

e. What is the slope? Were you correct in part (a)?
f. What is the correlation coefficient? Were you correct in part (b)?
g. What is the error variance? Were you correct in part (c)?

10. SPOUSES II

Recall the SPOUSES.DAT data set used in Exercise 4. Load this data set into the Data program. Choose Regression from the Data menu. Check the Regression Line box.

a. Find the correlation coefficient and slope for X = education of wife and Y = income of wife.
b. Find the correlation coefficient and slope for X = education of husband and Y = income of husband.
c. Of the two correlation coefficients in parts (a) and (b), which is higher? What does this mean?
d. Of the two slopes in parts (a) and (b), which is higher? What does this mean? Press "Save Residuals." A variable containing the errors will be created, called "R(4.3)." Close the regression screen, and select Histograms from the Data menu. Create the histogram for R(4.3) using 20 bins. Are there more large positive errors or large negative errors? What does this mean?
e. Close the histogram screen, and choose "Display the Data" from the Data menu. Look for the observations corresponding to the two or three largest positive errors R(4.3). Find the wife's and husband's ages for each observation. Do the same for the observations with the largest negative residuals. What do you notice? What does this say about our culture?

11. ANIMALS II

Go back to the ANIMALS.DAT data considered in Computer Exercise 4 of Chapter 2. Load the data into the Data program. Let X = body weight and Y = brain weight.

a. Guess whether the correlation between body weight and brain weight will be positive or negative.
b. Choose the Regression option from the Data menu. Create the regression with X = body weight and Y = brain weight, making sure the Regression Line box is checked. What is the correlation coefficient? What is the slope? Were you correct in part (a)?

c. Click on the three points farthest to the right. Which animals are they? What do they have in common?

d. Click on the three points closest to the top. Which animals are they?

e. Close the regression screen, and go to Input/Modify Data from the Data menu. Choose the "Use current data set: Body and Brain Weights" circle, and click on OK. Now from Options, choose "Delete Observation" to get rid of one of the three animals from part (c). (Look in the Observation Names column to see the numbers for the animals you wish to delete.) Do it twice more, each time getting rid of another one of those three animals. When they are all gone, select "Close," and click on "No" when the program asks if you want to save the data. Go back to the regression screen, and create the regression with X = body weight and Y = brain weight, making sure the Regression Line box is checked, as before. Does the scatter plot look different than before you deleted the three animals? In what way?

f. What is the correlation coefficient now? Is it positive?

g. Click on "Save Residuals," and then make the plot X = body weight and Y = R(2.1) (= errors). Click on the five points having positive errors. Which animals are they? Do you see several points representing a single type of animal?

Part 2

STATISTICAL MODELING

4 *M*ODELING EXPECTED VALUES 174

5 *P*ROBABILITY 222

6 *M*AKING DECISIONS 284

7 *C*HI-SQUARE TESTING 316

8 *P*ROBABILITY DISTRIBUTIONS 376

9 *M*EASUREMENT 446

4

MODELING EXPECTED VALUES

Measure, measure, measure. Measure again and again.

Galileo

OBJECTIVES

After studying this chapter, you will understand the following:

- *The expected value as a measure of the typical size of a statistic*
- *The five-step simulation method for estimating an expected value*
- *How to use a random number table*
- *Independence of outcomes in a probability experiment*
- *The "random walk" of the physical sciences*

4.1 THE EXPECTED VALUE OF A STATISTIC IN A RANDOM EXPERIMENT 176

4.2 MORE FIVE-STEP METHOD EXAMPLES 192

4.3 INDEPENDENCE 199

4.4 RANDOM WALKS 202

CHAPTER REVIEW EXERCISES 207

COMPUTER EXERCISES 212

The Hermits' Epidemic

Six hermits live on an otherwise deserted island. An infectious disease strikes the island. None of the hermits has ever contracted the disease before. Once a person contracts the disease, he becomes infectious and also becomes immune to the disease in the future. Suppose one of the hermits gets the disease. While infectious, he randomly visits one of the other hermits. The visited hermit, while infectious, then visits another hermit at random. If the visited hermit has not had the disease, he gets it and is infectious. The disease is transmitted in this manner until an infectious hermit visits an immune hermit (who has already had the disease), at which time the disease dies out. Assuming this pattern of visits, how many hermits, on average, can be expected to get the disease?

4.1 THE EXPECTED VALUE OF A STATISTIC IN A RANDOM EXPERIMENT

Many occurrences in the world around us involve random or chance outcomes. Often what we observe around us daily is best viewed as being the result, at least partially, of chance or randomness—that is, the result of a random "experiment," as statisticians say. We think of the weather, the stock market, the number of three-point shots Michael Jordan makes in a game, the number of traffic lights that are red as we drive to campus, the time when the instructor arrives in class, and so on, in this way. Moreover, we are often interested in the number or amount of a chance quantity (that is, a statistic) being observed. We have the intuitive notion that there is a typical, or average, value, called the *expected value* for this chance quantity. For example, if a coin is tossed 10 times we "expect" five heads. The expected value is a value seen as more typical or central in the sense that for any one occurrence, such as the number of red traffic lights experienced tomorrow, the observed value will fall reasonably close to the expected value. Here are some examples of questions whose answers are expected values:

About how much precipitation is expected next year in the Kansas wheat belt?

About how many people in your town are expected to get the flu this winter?

About how many three-point shots is Michael Jordan expected to make in the next Chicago Bulls game?

About how much of an increase is expected in the value of a particular stock in the next two weeks?

The numerical answers to these questions are surely useful in future decision making: Will irrigation be required in Kansas during the next wheat-growing season? Should you plan to take a flu shot next winter? What defense strategy should be used against Michael Jordan in the next game?

Even if we cannot predict the actual numerical value of a future trial, knowing a sort of average or typical value to be expected can be very useful. Let's consider the case of rainfall in the Kansas wheat belt. Of course, after the year is over, if we wish to know the actual precipitation for the year, we can check with the Kansas State Agricultural Department. However, suppose we wish to *predict* the precipitation for next year. If the average annual precipitation over the past 10 years was 22.3 inches, we could predict that the typical or expected value will be 22.3 inches. Our assumption here is that the future, except for random noise, should be like the past.

In fact, one of the most important ideas in science is that we can predict future or unseen occurrences from past ones, making combined use of data and a scientific model. For example, predicting the Standard & Poor's stock index one year from now requires economic modeling and appropriate economic data. In this chapter we will see how random models and data can help us estimate the expected value of chance quantities that we observe. In particular we will see how such estimates can be accurately obtained by designing and carrying out experiments.

The Cereal Box Problem

Suppose "Tripl Crisp Cereal" is running a promotion on its cereal by including one of six different colored pens in each box.

Fire Orange	Brilliant Yellow	Baby Blue
Shell Pink	Rosie Red	Grassy Green

Assume that when you buy a box of cereal your chances of getting any of the six colored pens are equal. About how many boxes of cereal would you expect to have to buy to get a complete set of all six colored pens?

The first thing to notice is that this problem involves chance. What is the *smallest* number of boxes of cereal that you would have to buy? Clearly, six. It is possible (but not very likely!) that each box purchased would have a different colored pen. In that case six boxes would do it. On the other hand, what can we say about the largest number of boxes that you might have to buy? It is possible (but again, not too likely) that you could buy 1000 or even 10,000 boxes and not get that last color of pen! So for any one person the answer is at least six boxes, but it could be an arbitrarily large number of boxes.

How can we find the expected number of boxes needed in our cereal box problem? There is a theoretical answer, but it requires sophisticated mathematical reasoning. An informal approach would be to actually go on a shopping trip and start repeatedly buying cereal boxes until we get a pen of each color. We could record our purchases as in Table 4.1. Here is how Table 4.1 works. For each cereal box, a tally mark (/) shows what color of pen was obtained. Table 4.1 shows that so far, in shopping trip 1, three boxes have been purchased, with one red and two green pens.

Table 4.1 **Partial Results of One Cereal-Buying Shopping Trip**

Shopping trip	*Outcomes (pens obtained)*						Number of boxes of cereal
	Orange	Yellow	Blue	Pink	Red	Green	
1					/	//	3

Table 4.2 **Results of One Cereal-Buying Shopping Trip**

Shopping trip	*Outcomes (pens obtained)*						Number of boxes of cereal
	Orange	Yellow	Blue	Pink	Red	Green	
1	///	/	//	////	~~////~~ //	////	21

A shopping trip ends when we have a complete set of all six colors. Table 4.2 shows the results of shopping trip 1 completed. We see that we bought 21 boxes of cereal before we had a complete set of six colors. We also see, for example, that we have three orange pens and two blue pens.

We now have one estimate of how many boxes of cereal are typically needed to get a complete set of six pens: 21 boxes. But we could have been lucky, or unlucky. With this single shopping trip we have no way of knowing whether 21 is unusually large, or unusually small—or in fact close to the typical or expected number that we seek. In order to get a better overall estimate of the typical number of boxes needed, we can go on several more shopping trips and then find the average number of boxes of cereal that were purchased. Indeed, if we carry out enough trips, the average will get as close to the theoretical expected value as we wish. Collecting lots of data can therefore substitute for a mathematical derivation of the unknown expected value!

It is not practical to solve this problem in this way (actually going out and buying cereal). It is too time-consuming, too expensive, and not sensible—unless, of course, we really like Tripl Crisp Cereal! Another approach to solving this problem is to do a simulation experiment as a substitute for the real experiment of actually going shopping for cereal. In particular, we can use a physical simulation model—a fair, six-sided die—to substitute for the process of purchasing boxes of cereal.

Doing a Simulation Experiment to Solve the Cereal Box Problem

There are six different pens to collect. Assume that when we buy a box of Tripl Crisp, our chances of getting any of the six colors of pen are equal. So we can use a familiar six-sided die as a physical model for buying pens. Let each side of the die correspond to one of the six pen colors.

1 = orange	2 = yellow	3 = blue
4 = pink	5 = red	6 = green

One toss of the die will correspond to the purchase of one box of cereal. Thus the repeated tossing of the six-sided die until all six pen colors appear becomes our substitute physical simulation model for an actual shopping trip. The number of tosses of the die that are made until all six digits appear

Table 4.3 **Outcomes of Five Rolls of a Die**

Shopping trip	Outcomes of rolls of a die						Number of boxes of cereal
	Orange	Yellow	Blue	Pink	Red	Green	
2	/	//	/	/			5

becomes one estimate of the typical number of cereal boxes required to get all colors of pen.

To begin our simulation, we roll a die five times and get these results: 3 2 2 4 1. We record these results in Table 4.3. We continue to roll the die until our shopping trip is completed (that is, until we get all six colors). It takes 16 rolls. We let this count as shopping trip 2, shown in Table 4.4. Think for a moment about shopping trip 1, an actual trip, and shopping trip 2, a pretend (or simulated) trip. Either trip is equally useful or informative. We lose *nothing* by using a die to simulate taking shopping trips. In science one often uses simulation models that capture the essence of a problem, as the die throwing clearly does here.

The Random Number Table

By now you may have concluded that the above procedure, although an enormous improvement over taking actual shopping trips, is still somewhat tedious if we need to simulate many shopping trips (as we will). Fortunately, there is an easier way to simulate random (chance) outcomes than carrying out a physical simulation experiment such as tossing a die. You can use a *table of random digits*. To understand this claim, we first note that the results of rolling a die are really *random numbers* (or digits). Roll a die, say, four times, and write down the result of each roll. Let us say you get 2, 3, 2, 5. This is a set of four random numbers. In this case the numbers have been produced by actually rolling a die. But computers and some calculators can also produce such random numbers and can do so much more rapidly. Such computer-generated random numbers are often presented in a table, and such a table is referred to as a *table of random digits* or *random number table*. Thus we can now replace the physical experiment of rolling a die repeatedly to simulate cereal box shopping trips by a computer-generated random number table. The random number table simulates the die tossing that in turn simulates the cereal box shopping trips.

Table 4.4 **Outcomes of Rolls of a Die to Simulate a Complete Shopping Trip**

Shopping trip	Outcomes of rolls of a die						Number of boxes of cereal
	Orange	Yellow	Blue	Pink	Red	Green	
2	//	///	///	/	~~////~~	//	16

Before we see how to use such a random number table, let's consider one physical way such a random number table can be obtained. Seeing how a random number table can be constructed from a physical model helps us understand its basic characteristics. Suppose we have a box containing 10 tickets that are identical except for being numbered 0, 1, 2, 3, 4, 5, 6, 7, 8, 9. After mixing the tickets thoroughly, we draw one of the tickets, note the number written on it, and return it to the box. We repeat this process many times, being sure to mix the tickets thoroughly between drawings. Note that each digit of this *box model* is just as likely to be drawn as any other digit. Thus we say the digits are *uniformly distributed.* Note also that the digit obtained on one draw is in no way influenced by the digit obtained on the previous (or any other) draw, because the previously drawn digit was returned before the current draw. That is, the digits on different draws are *independent* of each other. We will later see that this independence of draws is central to statistical analyses that are based on many random draws from a population, and we will study independence more carefully. This equal likelihood of the digits and their independence are the two basic characteristics of a table of random numbers, whether generated by a physical box model or by a computer (which is certainly a more mysterious way to obtain random numbers unless one is told what algorithm—set of computing rules—the computer uses).

Suppose as a result of 30 such drawings from the box model the following random digits were obtained:

74102 76512 99800 27493 80217 65241

This could become one row of a random number table. For visual convenience only, they are grouped in sets of five digits each.

A table of random numbers, then, is any table produced by a model that produces digits, such as the digits 0, 1,..., 9, with equal likelihood and with the digits occurring independently of each other. Thus our box model of digits 0, 1,..., 9 will work fine to generate a random number table, provided we draw *with replacement* between draws. That model also makes it easy to understand and experience the true nature of such random numbers. But such tables are typically produced by a computer that follows a particular numerical procedure, or algorithm. (For more discussion about random number generators and one such algorithm—the von Neumann method—see Appendix A.) The advantage of using a computer algorithm is simply the enormous speed at which such random digits can be produced.

Table B.2 in Appendix B is a computer-generated set of random numbers of the digits 1, 2,..., 6 designed to simulate the rolling of a six-sided die. That is, instead of the digits 0, 1,..., 9 being uniformly likely, the digits 1,

2, . . . , 6 are uniformly likely; the digits 7, 8, 9, and 0 do not occur. Here are the first two rows from such a table.

Row									
1	66533	45332	24614	22231	26431	35541	12165	62116	16111
2	61261	22613	26252	14622	32262	33244	34614	13316	41136

To use a random number table, you may start with any row and at any place in that row. But once you pick a place, you should keep moving in a fixed way from that place in the table, from left to right, say, through the rest of the table. In using a random number table, we can move right or left or diagonally, or in whatever direction we wish. The important thing is that we move systematically so that we do not let any human selective (or discretionary) behavior slip in. If that happens, it is unlikely that the numbers selected will be independent, as required.

We can use this table to simulate another shopping trip. That is, we can regard the first five numbers in the first row, 6 6 5 3 3, as the results of rolling a die five times and as indicating the first five pen colors obtained.

Since we are using the rules 1 = orange, 2 = yellow, 3 = blue, 4 = pink, 5 = red, and 6 = green, the pens corresponding to (6, 6, 5, 3, 3) are (green, green, red, blue, blue). This can be the beginning of shopping trip 3, which is shown completed in Table 4.5.

We now go to the next group of five numbers, 4 5 3 3 2, and record the outcomes in Table 4.5. Then we go to the next group, 2 4 6 1 4. In this group we need only the first four numbers, since the 1 gives us the color needed to complete the set. We can draw a loop around these 14 numbers in the table to show that this is a completed shopping trip.

Row									
1	(66533	45332	2461)4	22231	26431	35541	12165	62116	16111
2	61261	22613	26252	14622	32262	33244	34614	13316	41136

We now begin shopping trip 4 with the number 4 and continue in the same way to complete trips 4 and 5. The results are recorded in Table 4.5. The first set of all six colors was obtained in 14 rolls of the die, the second set was obtained in 13 rolls, and the third in 28 rolls. In Table 4.5 three trials were carried out: 14, 13, and 28 boxes were purchased. This gives an average of $(14 + 13 + 28)/3 = 18.3$ boxes. So, based on these three trials, we estimate that 18.3 boxes of cereal will have to be purchased for us to get

Table 4.5 **Outcomes from Using a Random Number Table to Simulate a Complete Shopping Trip**

Shopping trip	*Outcomes of rolls of a die*						Number of boxes of cereal
	Orange	Yellow	Blue	Pink	Red	Green	
3	/	//	////	//	//	///	14
4	//	////	///	//	/	/	13
5	~~////~~ ~~////~~ //	~~////~~	/	/	//	~~////~~ //	28

a complete set of six colored pens. If we add in the two trials we got earlier (21 boxes in Table 4.2 and 16 boxes in Table 4.4), we now have an average of $(21+16+14+13+28)/5 = 18.4$ boxes of cereal. It is interesting to note that by the mathematical argument referred to above, one can derive the actual expected value to be 14.7 ($= 1 + 6/5 + 6/4 + 6/3 + 6/2 + 6$). Our value of 18.4, based on only five trials, is not all that close. Again, if we average over a large number of trials, 10,000 say, we would obtain a number very close to the correct 14.7.

Steps to Solving the Cereal Box Problem

Let's now outline the steps that we are following to obtain a good estimate for the cereal box problem. In each application of the following five-step method, the end result is to obtain, through simulation trials, a good estimate of the true theoretical answer. For practical purposes, having a good estimate is as useful as having the exact theoretical answer. (Of course, when possible it is interesting and informative to understand how the theoretical answer is obtained.) The more simulation trials conducted, the more accuracy we can expect for our estimated answer. These five steps, with only slight modifications, can be used to solve a wide variety of problems involving probabilistic, or chance, situations in which we seek an unknown expected value.

Step 1. Choice of a Model: First we must decide on a model to represent the chance setting of the problem we are trying to solve. For the cereal box problem, we use the familiar (fair) six-sided die to represent pen colors obtained, where

1 = orange	2 = yellow	3 = blue
4 = pink	5 = red	6 = green

As the above discussions make clear, the model could be other than a six-sided die. For example, it could be a box model with tickets 1, 2, . . . ,

6 to be drawn from the box with replacement between draws. Moreover, the actual use of the physical model (die, box, and so on) is unnecessary and can, for example, be replaced by the random number table, Table B.2 of Appendix B, which simulates die tossing by presenting uniformly and independently produced digits from the set $1, 2, \ldots, 6$. Indeed, even this table can be bypassed by a computer program that very rapidly generates the random digits $1, 2, \ldots, 6$. Thus the actual experiment of the shopping trip has been replaced by a simulated die-throwing or box-sampling model whose behavior may be rapidly simulated using random number generation from a table or even more rapidly using a computer. It is always useful to think of a concrete physical simulation model (dice, coins, a box), even if that model is in turn simulated by random numbers. Box models are especially important to us because of their great versatility. In a box model we can have as many tickets with as many numbers (some repeated) as we wish. Hence it is good practice to construct a box model (step 1) for every example involving the five-step method.

Step 2. Definition of a Trial: A trial consists of rolling the die until all six sides are obtained. Each trial represents a completed shopping trip.

Step 3. Numerical Outcome of a Trial: Record the number of rolls of the die (number of cereal boxes purchased) in step 2. The outcome of a trial is often called the *statistic of interest*.

Step 4. Repetition of Trials: Steps 2 and 3 should be repeated many times. As stated above, the larger the number of trials, the better (more reliable) is your estimate of the desired expected value. For many problems, 100 trials provide a reasonably reliable estimate. A simulation of 1000 or 10,000 trials guarantees great accuracy and is extremely easy to carry out on a computer or powerful calculator.

Step 5. Finding the Mean (Average) of the Statistic of Interest: In the cereal box problem the statistic of interest is the number of boxes (as modeled by random number selection, say) that had to be obtained to get a pen of every color. Divide the total number of boxes obtained in all the trials by the total number of trials, thereby obtaining the average number of cereal box purchases per trial.

If enough trials are conducted, this five-step method provides a way of getting an accurate estimate of a desired expected value. The model of step 1 is usually easy to construct (no complicated formulas!) and straightforward to use via steps 2 through 5 to get the desired estimated expected value. This way of solving a problem by setting up a model and then doing a sufficient number of trials to estimate the desired result accurately is used by statisticians and scientists to solve a great variety of difficult problems in nuclear physics, industrial operations research, economics, and many other fields. This approach is sometimes called the *Monte Carlo method,* named after the famous gambling site on the Mediterranean. Many of the problems in this book will be solved by using the Monte Carlo method. The Monte Carlo method is a *simulation method* because it uses dice or some other model (such as a box model) to represent, or simulate, a problem situation. It should be stressed that the physical model (dice, tickets in a box, and so on) can always be replaced by the use of random numbers to rapidly produce a large number of trials, as large as needed for the required accuracy of step 5.

As a statistics student you can carry out the five-step method fairly rapidly by using a 10-digit box model to generate random numbers. You can do so more rapidly by using an already generated random number table. Or by using a computer program that generates random numbers and carries out the five steps itself, you can quickly obtain an expected value based on a very large number of trials. The instructional software included with this book will enable you to do just that.

In the example below we introduce the probability 0.6 of winning a baseball game. What does this mean? The concept of the probability of an event, like winning a baseball game, will be carefully discussed in the next chapter. Here we provide a brief informal discussion of probability. We are already, in fact, familiar with the idea of probability. Every day the weather forecast states the chance of precipitation, usually given as a percentage. One's understanding of probability should begin with an understanding of what a probability such as 0.6 means. (In everyday life we usually express probabilities as percentages; for example, we would say that the baseball team has about a 60% chance of winning each World Series game in the example below. But as you will see, we actually use proportions instead of percentages. Hence, if presented with a probability as a percentage, you should divide it by 100 to convert it to a proportion. Almost universally, practicing statisticians and scientists use proportions to represent probabilities. Hence proportions will be used throughout in this book for probabilities.) Here a probability of 0.6 means that if the team were to play many games—thousands, say—against the other team, its proportion of wins would be very close to 0.6. And the more games played, the more likely it would be that the team's winning proportion is very close to 0.6. Thus 0.6 is a theoretical probability that one gets close to empirically by actually doing a large of number of trials. This empirical reasoning is often called the **frequency interpretation** of probability.

Example 4.1

How Many Games to Win the World Series?

Suppose the National League team is judged to be slightly better than the American League team in a certain baseball season. Let's say that the National League team has a probability of 0.6 of winning each game of the World Series. How can we estimate the expected number of games needed to complete the Series?

In order to win the Series, a team must win four games, so the Series can go on for a maximum of seven games. Assume that the probability of the National League team winning a game is independent of the win/loss record for preceding games. This assumption may be somewhat weak, since psychological factors, team coaching, and so on may play an important role as the Series progresses.

Solution

We will use the five-step method to estimate the expected number of games needed to win the Series as follows:

1. Choice of a Model: Use a six-sided die. To achieve a model having the desired probabilities, we interpret the faces of the die as follows:

$$\begin{aligned} 1,2,3 &: \text{National League team wins game.} \\ 4,5 &: \text{American League team wins game.} \\ 6 &: \text{Ignore the result, and roll die again.} \end{aligned}$$

This represents a probability of 0.6 for a National League win and 0.4 for an American League win. Can you explain why? What other models (such as a box model) could be used to obtain the desired probability?

Right now, it probably seems that we have thrown a trick at you that is rather hard to understand and to apply: we have a probability of 0.6 to simulate, which is not achievable by the toss of a die or coin or some simple physical model. But actually this probability is also easy to simulate by an ordinary 10-digit random number table (or the more concrete box model with 10 tickets having digits 0, 1, ..., 9). For example, clearly we can use the assignment

$$\begin{aligned} 0,1,2,3,4,5 &: \text{win} \\ 6,7,8,9 &: \text{lose} \end{aligned}$$

How would you handle the probabilities $0.3 = \text{win}, 0.1 = \text{tie}, 0.6 = \text{lose}$, as might be the case in a sport in which ties are possible, such as soccer?

What about a probability requiring two significant digits, such as 0.43? The answer in such cases is to look at pairs of random numbers. Thus we have 100 pairs of equally likely random numbers: $00, 01, \ldots, 99$. Clearly we can model a probability of winning of 0.43 by making the assignments

$$\begin{aligned} 00, 01, \ldots, 42 &= \text{win} \\ 43, 44, \ldots, 99 &= \text{lose} \end{aligned}$$

Table 4.6 **Simulations to Determine Number of Games to Win the World Series**

World Series number (trial)	Game number 1	2	3	4	5	6	7	Number of games until win	Winner of Series
1	1 ↔ N	2 ↔ N	1 ↔ N	2 ↔ N				4	N
2	2 ↔ N	1 ↔ N	3 ↔ N	2 ↔ N				4	N
3	2 ↔ N	5 ↔ A	2 ↔ N	1 ↔ N	4 ↔ A	2 ↔ N		6	N
4	2 ↔ N	3 ↔ N	2 ↔ N	2 ↔ N				4	N

A ↔ American League; N ↔ National League; winner of game is as follows: 1, 2, 3 ↔ N; 4, 5 ↔ A; 6 ↔ ignore.

Now a little thought can convince us we can use a random number table to simulate any set of probabilities we wish. (Try 0.15, 0.5, and 0.35 as an exercise.)

2. Definition of a Trial: A trial consists of rolling the die, up to seven times (why seven?), until a series winner is determined. Write down the number obtained on each roll. Beside each number record the winner of the game. The first team to win four games is the winner of the Series.

3. Definition of the Statistic of Interest: Record for each trial the number of "games" needed for the Series to end (that is, the number of games played until one team has won four games).

4. Repetition of Trials: Repeat steps 2 and 3 many times. Here we do four trials to illustrate the process. Instead of actually tossing a die, we used random numbers from Table B.2, row 2, as follows:

Row									
2	61261	22613	26252	14622	32262	33244	34614	13316	41136

The trials are recorded in Table 4.6. Note that the first 10 random numbers not ignored are 12122 13225.

5. Finding the Mean of the Statistic of Interest: Calculate the average number of games required to win the Series. Based on the four trials of step 4, the estimated number of games to win the Series is $(4 + 4 + 6 + 4)/4 = 18/4 = 4.5$ games. Of course, to get an accurate answer for the expected number of games, we would need to do many more trials.

We should keep in mind that even though the National League team is a heavy favorite, it is possible for the American League team to win the Series, even in four games. Indeed, we could use the five-step method to estimate the probability of an American League victory. We will learn how to do that in Chapter 5.

Now consider this slightly different version of the World Series problem. Here the chances of either team winning are balanced.

Example 4.2

Suppose now that the National League and American League teams are evenly matched going into the World Series. That is, each team has a probability of 0.5 of winning a game. Now how many games is the Series expected to last?

Solution

For this problem we can use a particularly simple physical model, namely a fair coin, as follows.

1. Choice of a Model: We use a coin, where

H : National League team wins game.
T : American League team wins game.

(H is for heads, and T is for tails.) Would a box model containing two tickets also work?

2. Definition of a Trial: A trial consists of tossing the coin up to seven times. Write down the outcome (heads or tails) obtained on each toss.

3. Definition of the Statistic of Interest: Record for each trial the number of "games" needed for the Series to end (that is, the number of games played until one team has won four games).

4. Repetition of Trials: Repeat steps 2 and 3 many times. We will use random data from Table B.1, Appendix B, row 1, as follows:

Row									
1	01110	10000	00010	10111	00010	11001	10011	10001	10110

Here 1 denotes a head and 0 denotes a tail. We record the trials in Table 4.7.

5. Finding the Mean of the Statistic of Interest: Calculate the average number of games required to win the Series. From Table 4.7, the estimated number of games to win the Series is $(6 + 4 + 5 + 5)/4 = 20/4 = 5.0$ games.

Table 4.7 **Simulation of Four World Series: Even Contest**

World Series number (trial)	*Game number* 1	2	3	4	5	6	7	Number of games until win	Winner of Series
1	0 ↔ A	1 ↔ N	1 ↔ N	1 ↔ N	0 ↔ A	1 ↔ N		6	A
2	0 ↔ A	0 ↔ A	0 ↔ A	0 ↔ A				4	A
3	0 ↔ A	0 ↔ A	0 ↔ A	1 ↔ N	0 ↔ N			5	A
4	1 ↔ N	0 ↔ A	1 ↔ N	1 ↔ N	1 ↔ N			5	N

A ↔ American League; N ↔ National League; winner of game is as follows: 1 ↔ N; 0 ↔ A.

To clarify these ideas, we close Section 4.1 with more discussion about models of chance and with one more application of the five-step method that uses a large number of trials, as is desirable for accuracy.

The idea of a model is a useful one in mathematics and science. A scientific model is simply any entity (a set of mathematical equations, a replica of a new airplane wing design in a wind tunnel, a computer animation of turbulent ocean currents, and so on) that substitutes for the real object of study. As such, it mimics certain behaviors of the real phenomena (such as the aerodynamic performance of an actual aircraft wing) we wish to study. The vital characteristic of a scientific model is that it must be able to reveal interesting information about the real thing. For example, an engineering firm that is designing a new type of aircraft might first build a model aircraft that incorporates the new features and fly it in a wind tunnel. Similarly, the chemists who discovered the structure of the DNA molecule built many three-dimensional models to help them in their investigation.

Certainly the most common and traditional kind of scientific model is a set of mathematical formulas, like the well-known physics formula $F = ma$ relating the force applied to an object to its acceleration. As already stated, in this book we also use models to help us solve problems involving chance. Today computers are often used to produce highly informative visual models that evolve over time (like a movie) to help explain airplanes undergoing turbulence, black holes in space, major water currents in the Atlantic Ocean, and many other phenomena. In this book our models will be probabilistic models, usually based on coins, dice, and boxes containing numbered tickets, and will often use computer-generated random numbers. For example, a toss of a coin could be used to represent whether a child is a boy or a girl, as in the following problem.

Example 4.3

The Expected Number of Girls in a Three-Child Family

What is the expected number of girls in a three-child family? The answer, 1.5, is probably obvious to you, but the point is to verify this intuition empirically by using the five-step method. One way to solve this problem would be simply to go out and conduct interviews of many three-child families, keeping records of how many girls there are in each family. The results of such a survey might be recorded in a table like Table 4.8. In our survey the average (mean) number of girls is

$$\frac{0 \times 13 + 1 \times 32 + 2 \times 38 + 3 \times 17}{100} = \frac{0 + 32 + 76 + 51}{100} = \frac{159}{100} = 1.59$$

However, instead of actually interviewing families, we can use the approach suggested in this section. We can think of a model of a three-child family. Instead

Table 4.8 **Number of Girls in a Three-Child Family**

Number of girls	Number of families
0	13
1	32
2	38
3	17
Total	100

of actually interviewing families, then, we can toss one coin three times, once for each child, letting heads stand for a girl and tails stand for a boy. Therefore, a coin is a model for a child, and the outcome of tossing a coin represents the child's being a boy or girl. We then conduct a simulation experiment in which we wish to find the average number of heads (or tails) obtained.

We summarize again the five steps to follow in using a Monte Carlo solution to the problem, relating them to our example of the three-child family:

1. Choice of a Model: In our example of the three-child family, an appropriate model is a coin. (You can probably think of other, equally appropriate models. For example, you might roll a die and use this rule: even number = boy, odd number = girl.)

2. Definition of a Trial: A trial consists of tossing the coin three times, once for each child in the family, and counting the number of heads obtained.

3. Definition of the Statistic of Interest: In this case we are interested in the number of heads (number of girls) obtained in each trial.

4. Repetition of Trials: When you do a probabilistic experiment such as tossing coins, your answers will tend to be different from one experiment to another, as we have already seen. Each trial will probably be different from the theoretical value and from other trials, but the average (mean) value will approach closer and closer to the theoretical value as the number of trials is made larger and larger.

The more trials you do, the closer the averaged results of probability experiments will be to the theoretical value being estimated. This fact is exactly why the expected value is of practical importance, namely, that it *is* the number that the average of the statistic of interest over a large number of trials is drawn to. This fact is well known as the *law of averages.* If we carry out 10,000 trials using a computer, then probability theory can be called on to show that our answer will usually be within 0.0075 of the theoretical expected value of 1.5—high accuracy indeed. For now, we assume that 100 trials are sufficient to give us reasonable accuracy in our estimates whenever we use the five-step method.

Returning to our example, we repeat step 2 for a total of 100 trials. That is, the experiment of tossing a coin three times is repeated 100 times. The results of the tosses are in Table 4.9. By *frequency* is meant the number of times the event occurs; here it means the number of times the event occurs in the 100 trials.

Table 4.9 **Three Tosses of a Coin to Simulate the Number of Girls in a Three-Child Family**

Number of heads (girls)	Frequency
0	14
1	34
2	36
3	16
Total	100

5. Finding the Mean of the Statistic of Interest: Add the values of the statistic of interest defined in step 3 and divide by the total number of trials decided on in step 4. For our example the mean number of heads obtained was

$$\frac{0 \times 14 + 1 \times 34 + 2 \times 36 + 3 \times 16}{100} = \frac{154}{100} = 1.54$$

In this experiment the mean number of girls expected in a three-child family is estimated to be 1.54.

This is close to the theoretical expected value of $0.5 \times 3 = 1.5$. In other problems the theoretical value may not be so easy to find, and we will often estimate it using our five-step method.

SECTION 4.1 EXERCISES

For all exercises, unless otherwise noted, use the five-step method and do 25 trials. (More trials are feasible if you are using a calculator or the textbook's optional software.)

1. a. From the table below find the estimated expected number of boxes of cereal needed to get the complete set of six pens.
   b. Using a six-sided die, do another 20 trials. Extend the table to provide the information for the additional trials.
   c. Find the expected number of boxes of cereal needed to be purchased, based on 25 trials.
   d. It can be shown by advanced mathematics (probability theory) that the expected number of boxes required to obtain all six colors, as more and more trials (shopping trips) are carried out, becomes closer and closer to 14.7. Compare your answer in (c) to the theoretical result of 14.7.

	1. Orange	2. Yellow	3. Blue	4. Pink	5. Red	6. Green	Number of boxes of cereal
1	///	/	//	////	~~////~~ //	////	21
2	//	///	///	/	~~////~~	//	16
3	/	//	////	//	//	///	14
4	//	////	///	//	/	/	13
5	~~////~~ //	~~////~~	/	/	//	~~////~~ //	28

2. Refer to Example 4.2. Using the five-step method, estimate the average number of games that will be played before one of the teams (National or American League) wins the World Series. Assume that each team has a 50% chance of winning each game.

3. Suppose that instead of six different colors of pens, Tripl Crisp Cereal decides to use four different colors of pens in its boxes. What is the average number of boxes of cereal one would expect to have to buy in order to get the complete set of four pens? *Hint:* Use a four-sided die or Table B.2 (in Table B.2 you could use digits 1–4 and ignore 5 and 6).

4. A student takes a true-false test by guessing. Assume that his chance of getting a question correct is 50%.
   a. How many answers can he expect to get correct, by guessing, on a 10-question test?
   b. How did you model the probability (step 1 of the five-step method) in this case?
   c. Does your answer agree with what you would expect?

5. Another student takes a multiple-choice test by guessing. Each question has three possible answers.
   a. Predict, using the five-step method, how many questions the student can expect to answer correctly on a 10-question exam.
   b. How did you model the probability (step 1 of the five-step method) in this case?
   c. Does your answer agree with what you would expect?

6. A newly married couple agree that they want to have children. The husband wants to have at least one son, and the wife wants to have at least one daughter. They are interested in finding how many children they can expect to have before they will have one boy and one girl. Assume that the probability of having a boy or a girl is 50%.
   a. How would you model the probability (step 1 of the five-step method) in this case?
   b. Using the model from part (a), describe one complete trial. How would you know when the trial has ended?
   c. Compute, using the five-step method, the number of children this couple can expect to have.

7. In certain regions, the climate is very constant. Assume there is an area where the probability of any rainfall during the day is 0.25 (25%), every day of the year. A newcomer to the area fails to bring an umbrella. If the newcomer buys an umbrella the first day that it rains, how many days can he expect to wait before purchasing an umbrella? (*Hint:* A probability of 0.25 can be modeled by using four sides of a six-sided die: 1 = rain, 2 = no rain, 3 = no rain, 4 = no rain, 5 = do over, 6 = do over.)

8. Suppose 1 out of every 13 cars of a certain make and model will develop engine problems before the warranty expires. How many of these cars can the dealer expect to sell before selling two that will have this problem?
   (*Hint:* When you use the five-step method, note that 1 out of every 13 cards in a deck of cards is an ace. When using a deck of cards, be careful to reshuffle the deck completely after each trial.)

9. **Extra credit:** It may not be reasonable to assume that the colors are uniformly distributed among the boxes of cereal. It could be, for example, that each color is weighted in inverse proportion to its popularity. Suppose the pen colors are distributed among the cereal boxes as shown below. Now what is the average number of boxes of cereal that must be purchased to acquire all the colors?

Color	Proportion with color
Orange	2/6
Yellow	1/6
Blue	1/6
Pink	1/6
Red	1/12
Green	1/12

Graphing Calculator Exercises

The exercises call for 25 trials. Do more if you wish to increase accuracy.

10. Each pack of a brand of bubblegum contains a photo of one of six pop bands. Make a frequency table that records the number of packs of bubblegum you need to buy if you want to collect a complete set of six photos. When you buy a pack of gum, your chances of getting any of the six photos are equal. Use the program GUM to simulate the buying of packs of bubblegum. Do 25 "shopping trips." GUM prompts you for the number of possible outcomes (six in this case, one for each photo) and how many shopping trips you want to carry out (25 in this case). Press ENTER after each response. The number of the trip and the number of packs of gum bought are displayed. Record the result in your table, and press ENTER to obtain the next trial's results. Calculate the expected number of packs of bubblegum needed to obtain a complete set of photos.

11. Refer to Exercise 10, but change the number of photos to eight. Use the program GUM, and make a frequency table for 25 shopping trials. What is the expected number of packs of gum that one would have to buy in order to get a complete set?

12. Examples 4.1 and 4.2 describe the Monte Carlo approach to finding the number of games needed to determine a World Series champion. Use the program WSERIES to simulate playing 25 World Series. WSERIES asks for the probability of Team A winning, and the number of World Series to simulate (25 in this case). Press ENTER after each entry. Do a plot of Team A probabilities versus the expected number of games to win, for each of the following Team A probabilities:

a. 0.6 c. 0.5
b. 0.25 d. 0.8

4.2 MORE FIVE-STEP METHOD EXAMPLES

We have already seen how we can use a six-sided die, a fair coin, or a box of tickets as physical realizations of probability models to solve the problem of estimating the expected value. In this section we will see how still other models can be used with the five-step method to solve expected value problems.

Example 4.4

A certain type of plastic bag that is being tested has been found in the past to burst under a given pressure 25% of the time. About how many bags can be expected to be tested under that pressure until one is found that bursts?

Solution

We need a model for a probability of 0.25 (the probability that a bag bursts). One model would be a four-sided die. Such dice actually exist, having four triangular faces and being pyramidal in shape. But instead of actually tossing such a die, one can obtain data by using random numbers. We can also consider using two fair coins. When we toss a (fair) coin, we are equally likely to get heads (H) or tails (T), so we can use one coin as a model for getting a probability of 0.50. We know that the probability of heads is 0.50 and the probability of tails is 0.50. But how do we get a probability of 0.25? We will use two coins.

When two coins are tossed, there are four possible outcomes: HH, HT, TH, TT. It seems evident that each outcome is equally likely. So, for example, the probability of getting two heads, HH, when tossing two coins is $\frac{1}{4}$, or 0.25. Now, to set up our model, we make the following assignments:

$$\text{HH} = \text{Bag bursts.}$$

$$\text{HT, TH, TT} = \text{Bag does not burst.}$$

Thus we have a physical model that assigns 0.25 to the bag bursting.

Instead of actually tossing coins, we can use random digits from 0 to 9 to simulate the tossing of pairs of coins. How can we do this? Here is row 1 from Table B.3 of random numbers:

Row									
1	32236	12683	41949	91807	57883	65394	35595	39198	75268

Since we are interested in a model that uses two coins, we read pairs of digits from the table. Starting at the left, we have these pairs: (3,2), (2,3), (6,1), and so on. These pairs of data can correspond to tosses of two coins, as specified in step 1 of the five-step method below. These are only suggestions. Can you think of others?

1. Choice of a Model: For each "coin," 0, 2, 4, 6, and 8 (that is, the even numbers) will be considered "heads," and 1, 3, 5, 7, and 9 (that is, the odd numbers) will be considered tails. We therefore have $(3, 2) \leftrightarrow$ TH, $(2, 3) \leftrightarrow$ HT, $(6, 1) \leftrightarrow$ HT, and so on. Thus the model is specified, and moreover our simulation will proceed using random numbers.

2. Definition of a Trial: A trial consists of repeatedly tossing a pair of coins until we get HH, which means the bag bursts for the first time.

3. Defining the Statistic of Interest: The statistic of interest is number of tosses of two coins we make until HH is obtained.

4. Repetition of Trials: Three sample trials are provided in Table 4.10. Only the first trial is given in detail. We used row 1 of Table B.3, given above.

5. Finding the Mean of the Statistic of Interest: The average number of bags tested until one bursts is $(4 + 8 + 10)/3 = 22/3 \approx 7.3$. Of course, to get a reasonably

Table 4.10 **Number of Bags Tested Until One Bursts**

Trial	Outcomes*	Number of bags until one bursts
1	32 ↔ TH 23 ↔ HT 61 ↔ HT 26 ↔ HH	4
2	HT HT TH TT TH HT TT HH	8
3	TH TT TH TT TT TT TT TH TH HH	10

*Trials 2–3 not given in detail.

accurate estimate we need many trials—100, say. The theoretical expected value happens to be 4. Thus 7.3 is rather far off from the actual expected value.

Example 4.5

According to one psychological theory of personality, people can be divided into three personality types: introvert (shy), extrovert (outgoing), or ambivert (balanced between shy and outgoing). These categories are defined in such a way that the general population is approximately equally divided between the three personality types.

Suppose eight pairs of persons are needed for a psychological experiment. If the pairs are formed by chance, about how many pairs will have at least one extrovert?

Solution

We use the five-step method.

1. Choice of a Model: Use random digits from 0 to 9. This is a total of 10 digits, so we will divide them into groups of 3 digits each, as follows:

$$1,2,3 \leftrightarrow \text{introvert ("In")}$$

$$4,5,6 \leftrightarrow \text{extrovert ("Ex")}$$

$$7,8,9 \leftrightarrow \text{ambivert ("Am")}$$

We will ignore 0:

$$0 \leftrightarrow \text{Ignore; go to next digit.}$$

(Can you come up with another scheme that assigns equal probability to each personality trait?)

2. Definition of a Trial: A trial consists of selecting digits in eight groups of two digits. (Satisfy yourself as to why the choice of eight groups of two digits is appropriate.) Use the rule in step 1 to find out if the "person" chosen is introvert, extrovert, or ambivert. To illustrate, we again go to Table B.3, row 1, which has these digits: 32236 12683. The first three pairs of digits give $(3,2) \leftrightarrow$ (In,In), $(2,3) \leftrightarrow$ (In,In), $(6,1) \leftrightarrow$ (Ex,In).

Table 4.11 **Number of Pairs with At Least One Extrovert**

Trial	Outcomes	Number of pairs with at least one extrovert
1	(3, 2) = (In, In) (2, 3) = (In, In) (6, 1) = (Ex, In) (2, 6) = (In, Ex) (8, 3) = (Am, In) (4, 1) = (Ex, In) (9, 4) = (Am, Ex) (9, 9) = (Am, Am)	4
2	(In, Am) (Am, Ex) (Am, Am) (Am, In) (Ex, Ex) (In, Am) (Ex, In) (Ex, Ex)	4
3	(Am, Ex) (In, Am) (In, Am) (Am, Am) (Ex, In) (Ex, Am) (Ex, In) (In, Am)	4
4	(Ex, Ex) (Ex, Am) (In, In) (Am, Am) (Am, Am) (Am, Ex) (Am, Ex) (Am, Ex)	5

3. Definition of the Statistic of Interest: Count the number of pairs with *at least one extrovert.*

4. Repetition of Trials: Repeat steps 2 and 3 many times.

We continue with Table B.3, rows 1 and 2. We record the results of four trials in Table 4.11.

Row									
1	32236	12683	41949	91807	57883	65394	35595	39198	75268
2	40336	50658	32089	78007	58644	73823	62854	31151	64726

5. Finding the Mean of the Statistic of Interest: Calculate the average number of pairs with at least one Ex. From Table 4.11, the average number of pairs with at least one extrovert is $(4 + 4 + 4 + 5)/4 = 17/4 = 4.25$. The mathematics of probability can be used to show that the true expected value is $\frac{40}{9}$, which is rather close to our estimate of 4.25.

Example 4.6 Reservations, Please!

One might find the following notice at an airline counter.

Disclosure Notice—Deliberate Overbooking

Airline flights may be overbooked, and there is a slight chance that a seat will not be available on a flight for which a person has a confirmed reservation. A person denied boarding on a flight is entitled to a compensatory payment.

Consider the following problem. An airplane has 40 seats, and the airline accepts 43 reservations for a certain flight. If the airline's records indicate a no-show rate for this flight of 10 percent, find for this flight (a) the expected number of empty seats and (b) the expected number of ticketed passengers who do not get a seat. (Both (a) and (b) are of interest because often fewer than 40 ticketed passengers will show up for the flight and occasionally over 40 passengers will show up.)

Table 4.12 **Airline Seating with Oversell Strategy**

Trial	Number of zeros in string of 43 digits (no-shows)	Number of persons seated	Number of empty seats	Number of persons not seated
1	5	38	2	0
2	4	39	1	0
3	8	35	5	0

Solution

The procedure we use for solving this problem is, as usual, the five-step method:

1. Choice of a Model: Use a table of random digits.

$$0 = \text{no-show}$$
$$1\text{–}9 = \text{person appears for reservation}$$

2. Definition of a Trial: A trial consists of reading 43 digits, one for each reservation made.

3. Definition of the Statistic of Interest: In the terms of our experiment, we wish to find the number of nonzero digits exceeding 40 (corresponding to those persons not seated on a flight).

4. Repetition of Trials: As an example, we will use rows 2–4 of Table B.3.

Row									
2	40336	50658	32089	78007	58644	73823	62854	31151	64726
3	88795	93736	22189	47004	48304	77410	78871	98387	44647
4	12807	65194	58586	78232	57097	01430	00304	32036	23671

In row 2, read off the first 43 digits (one for each reservation) and count the number of nonzero digits (the number of no-shows). We see that 38 people will be seated for the flight, and there are $40 - 38 = 2$ empty seats. Begin trial 2 with row 3 of Table B.3. Table 4.12 shows the results of three trials.

5. Finding the Mean of the Statistic of Interest: The average number of empty seats in our three trials is $(2+1+5)/3 = 8/3 = 2.6$. The average number of people refused seating is $(0+0+0)/3 = 0$. (If we carried out a large number of trials, this latter average would be greater than 0.)

We now return to the hermit's epidemic as our final example. First reread the Key Problem. Number the hermits $1, 2, \ldots, 6$. The solution to the hermit's epidemic can be obtained by any model in which an infected hermit is equally likely to visit any one of the five other hermits. For example, one could use an ordinary die and ignore the face corresponding to the hermit who is visiting. On each visit, the currently infected hermit visits one of the five other hermits. The trial ends when a visit is made to a previously infected hermit. The statistic of interest is the number of infected hermits (at least two, at most six), which is exactly equal to the total number of visits.

The Key Problem could be changed in a variety of ways to make it more realistic as a model for the spread of disease. For example, the number of hermits on the island could be increased. Suppose there were 100 hermits. Then instead of using five sides of a six-sided die we could use the digits of a random number table in pairs—00, 01, 02, . . . , 99—to designate which hermit is visited each time. We ignore the number of the visiting hermit. For example, suppose hermit 44 is visiting and we select 44. Then since the hermit cannot visit himself, we simply ignore the 44 and select again. We would make a list of which hermits have received a visit and hence are infected. Then we would stop our trial when an already infected hermit is visited a second time.

As stated, the problem assumes that if an uninfected hermit comes into contact with an infected hermit, the disease is transmitted with certainty. But suppose instead that the chance of contracting the disease when in contact with an infected person is 75% instead of 100%. How could the problem be solved now?

Other possible modifications include making the number of days a hermit is infectious to be random, assuming that every member of the hermit society makes a random number of visits each day (that is, comes into contact with a random number of other hermits), and assuming that the disease is initiated with several infectious hermits. Now the model begins to seem fairly realistic in its attempt to simulate the way a disease actually spreads.

Many of the problems of Chapter 4 have theoretical solutions for their expected values, but the hermit problem with the kinds of modifications suggested here quickly becomes a problem for which Monte Carlo simulation is the *only* way to find a solution. Many questions of great practical importance can only be answered using the five-step approach. For example, many problems in industrial operations research, such as complex inventory management, can be solved only through such simulation.

You are encouraged to think up your own modifications to the hermit problem to make it a more realistic model for the spread of a disease in a population. Be creative. But for each complication that you pose, you also need to provide a method for modeling the modified problem!

SECTION 4.2 EXERCISES

1. Change Example 4.4 as follows:
   a. Suppose the plastic bags burst 20 percent of the time (instead of 25 percent). About how many bags can be expected to be tested before one bursts?
   b. Suppose the bags burst only 10 percent of the time. Then how many can be expected to be tested before one bursts?
2. Refer back to Table 4.11.
   a. Use the results from the table to estimate the average number of pairs in which both persons are extroverts.
   b. Do another 21 trials to reach a better estimate of the number of pairs in which both persons are extroverts.
3. In Example 4.6, make the following changes, but still assume that the aircraft has 40 seats. For each part below, find the expected number of empty seats and the expected number of persons *not* getting a seat.
   a. Suppose the airline accepts 45 (instead of 43) reservations for a flight. Assume a no-show rate of 10%.
   b. Suppose the airline accepts 43 reservations but the no-show rate is only 5% (instead of 10%).
4. Suppose, on the average, one out of every eight houses will experience some sort of crime in a certain neighborhood. What would be a way to model the $\frac{1}{8}$ chance using only a coin? (Refer back to how you used a coin in the original plastic bag example. Try to extend that idea.)
5. A certain professor has eight keys, but he never recalls which one fits his office door lock. He tries one key at a time, each time choosing one of the keys at random from his pocket. (That is, he does not remember which keys he has already tried, and they all look the same.) What is the expected number of tries needed for him to find the correct key? Refer to Exercise 4 for a way to model this probability using only a coin.
6. The probability that a baby will be a boy is very close to 0.50 (50%). In 10 three-child families, about how many will consist of at least two boys? About how many will consist of exactly two boys?

Graphing Calculator Exercises

7. The program BURST can be used to provide data for Example 4.4. BURST asks for the burst percentage and the number of trials (25 for this question). Press ENTER after each response. The program returns the number of bags tested until a bag bursts. Record the data in a frequency table, and for the following burst percentages find the average number of bags tested until a bag bursts:
   a. 25%
   b. 50%
   c. 10%

8. Reservations, please! Example 4.6 can be solved using the program OVERBOOK. Enter the number of seats, the number of seats sold, and the

no-show percentage. The program displays the number of seats occupied per flight and then finishes by telling the percentage of flights overbooked. Simulate 25 flights to find the number of seats occupied per flight and the percentage of flights overbooked expected for the following numbers of seats sold and no-show rates:

a. 45 seats sold; 10% no-show rate
b. 43 seats sold; 5% no-show rate
c. 45 seats sold; 20% no-show rate

4.3 INDEPENDENCE

The idea of independence is very important in statistical thinking. We have already used it informally in the discussion of random numbers. We will introduce it by means of examples now and will make our understanding of it more exact in the next chapter.

We say that two outcomes are **independent** if the occurrence of one has no influence on the chance of the occurrence of the other. For example, if you toss a coin and roll a die, the outcomes of the two activities are clearly independent in this sense. (You could, of course, make them dependent by actually fastening the die to the coin. But we assume that you have given the coin a good, honest toss and the die is a fair roll.)

Clearly, tosses of distinct coins are independent. It may not be quite so obvious that individual tosses of the same coin are likewise independent. But when you toss a coin, the outcome of the next toss is not at all influenced by the outcome of the preceding toss. Hence, repeated tosses of the same coin are independent. Because of independence, the model of one coin tossed three times is equivalent to the model of three coins tossed once. The one you would prefer in defining a trial for an application of the five-step method is just a matter of which one you find to be easier or more sensible to use in solving a particular problem.

Consider the independence of outcomes having to do with the weather. We can think of examples in which the occurrence of rain in one city is closely linked to the likelihood of rain in another, and therefore the outcomes are dependent. Consider, for example, the cities Chicago and Milwaukee, which are less than 100 miles apart. Clearly, the occurrences of rainfall in each city will not be independent! Let's instead think of two cities in which the occurrences of rain are independent. For example, Chicago and San Juan, Puerto Rico, may both have a forecast of rain of 50%. But whether it rains in Chicago would seem to be independent of rain occurring in San Juan. Hence the chance of rain in Chicago given that we know it is raining in San Juan is still 50%. We will make the assumption of independence when we carry out the next experiment.

Example 4.7

On 12 days the chance of rain in both Chicago and San Juan has been 50%. On the average, on how many of these days would we expect it to rain in both cities?

Solution

We follow our five steps to solve this problem.

1. Choice of a Model: As a model for this problem, we use two coins, one for each city. The outcomes of the two coin tosses are independent of each other, as are the outcomes of rain (we assume) in Chicago and San Juan. For example,

Penny: Chicago
Nickel: San Juan

In each case, heads means rain and tails means no rain. (Would a box model with tickets 00, 01, 10, 11 work? How?)

2. Definition of a Trial: A trial consists of tossing the penny (for Chicago) and the nickel (for San Juan) a total of 12 times each (once for each of the 12 days mentioned in the problem statement).

3. Definition of the Statistic of Interest: We are asked about the number of days when it rains in both cities. So we want to know how many times we obtain two heads when the two coins are each tossed 12 times. For example, here are the results we obtained by actually tossing a penny and a nickel 12 times each (the outcome of the penny is shown first).

TH TT (HH) (HH) TH TT TH (HH) TH (HH) HT (HH)

Two heads were obtained five times, so for this trial the statistic of interest is 5.

The data may be recorded in a frequency table such as this:

Number of days with rain in both cities	Tally	Frequency
0		
1		
2		
3		
4		
5	/	1
6		
7		
8		
9		
10		
11		
12		

Table 4.13 **Simulated Number of Days with Rain in Both Chicago and San Juan**

Number of days with rain in both cities	Tally	Frequency
0		0
1	~~||||~~ /	6
2	~~||||~~ ////	9
3	~~||||~~ ~~||||~~ ~~||||~~ ///	18
4	~~||||~~ ~~||||~~ /	11
5	////	4
6	/	1
7	/	1
8		0
9		0
10		0
11		0
12		0
		50

Here *frequency* simply means the number of times the event of interest has occurred.

4. Repetition of Trials: We repeat the trial of tossing the penny and nickel 12 times for a total of 50 trials. The results are presented in Table 4.13.

5. Finding the Mean of the Statistic of Interest: The mean number of days (out of 12) when it rained in both cities is the mean number of trials resulting in two heads.

$$\text{Mean} = \frac{6 \times 1 + 9 \times 2 + 18 \times 3 + 11 \times 4 + 4 \times 5 + 1 \times 6 + 1 \times 7}{50}$$

$$= \frac{6 + 18 + 54 + 44 + 20 + 6 + 7}{50} = \frac{155}{50} = 3.1$$

So according to our five-step simulation method we expect rain in both cities an average of 3.1 out of 12 days. This is quite close to the correct expected value of $\frac{1}{4} \times 12 = 3$.

SECTION 4.3 EXERCISES

1. In Example 4.7 we estimated the probability of it raining on the same day in both Chicago, Illinois, and San Juan, Puerto Rico. Because of the great distance between the two cities, we assumed that the chance of it raining in Chicago on a certain day is independent of whether it rains in San Juan.

Consider now two cities that adjoin each other: Champaign and Urbana, Illinois. For practical purposes we can assume that if it rains in one on a certain day, it also rains in the other that day. During a certain week, the chance of rain is predicted to be 50% in each city. Describe how you would estimate the probability of it raining in both cities on a given day. Would this require two coins?

2. Recall Example 4.5 in Section 4.2, where researchers were choosing people for a psychological experiment. There we assumed that $\frac{1}{3}$ of all people are extroverts. Once again, four pairs are drawn for a total of eight people, but this time researchers draw each pair from within a family (that is, two people from each of four families).

   In Exercise 2 in Section 4.2 you estimated the number of pairs out of the eight in which both are extroverts, and your answer should have been around 0.89. In this experiment, the researchers found that the average number of pairs in which both are extroverts is 2.2. How can this difference be explained? What assumptions were we making when working with that example previously?

3. Obviously, the chance of seeing five heads in a row on a fair coin is not very good. A friend of yours is flipping a coin (which you know to have a 50% chance of producing a head) and he promptly gets four heads in a row. Knowing the fact that seeing a head in five straight flips is rare, he argues that the chance that the next one will be a tail is very good, saying that the chance of a tail is much higher than the chance of a head. Does his argument make sense? Why or why not? What do you think is the probability that he will see a head on the next flip?

4. Think of two events that are independent of each other, and two events that are not.

4.4 RANDOM WALKS (OPTIONAL)

Although it is not used in the study of statistics in the rest of this book, a fun probability topic is that of *random walks.*

Imagine yourself in a city in which the streets are laid out in a rectangular grid. You wander about aimlessly from block to block, and when you reach an intersection you choose your direction—north, south, east, or west, all equally likely—at random. You always follow the rectangular grid of streets. Probability theory predicts that if you keep walking in this manner, you will keep returning to the place you started, instead of drifting farther and farther away. (In a *three*-dimensional random walk, in a grid in which one goes north-south, east-west, or up-down, one will eventually drift far away from the origin, never to return again!) This example illustrates the *random walk model* of modern physics, which is the path of a particle moving at random along a rectangular lattice. The random walk model is extensively employed in many areas of physics.

Example 4.8 **One-Dimensional Random Walk**

A man takes a walk, always starting from 0 on the infinite line below. In order to decide whether to go to the left or to the right, he tosses a coin. If the coin falls heads, he walks one block to the right. If the coin falls tails, he walks one block to the left. For example, if he obtained H H T in three tosses, he would go from 0 to 1 to 2 and then back to 1. On average, how far from zero would he expect to be after 10 tosses of a coin?

$$\ldots \quad -7 \quad -6 \quad -5 \quad -4 \quad -3 \quad -2 \quad -1 \quad 0 \quad 1 \quad 2 \quad 3 \quad 4 \quad 5 \quad 6 \quad 7 \quad \ldots$$

Solution

The table below gives the results of 12 trials (random walks) consisting of 10 tosses each.

Random walk number:	1	2	3	4	5	6	7	8	9	10	11	12
Location at end of walk:	2	0	0	−2	2	−4	−4	−2	−4	2	6	−2

The average distance of the stopping point from the starting point is

$$\frac{2+0+0+2+2+4+4+2+4+2+6+2}{12} = \frac{30}{12} = 2.5 \text{ units}$$

We could use a table of random digits to solve this problem, instead of tossing coins. Our correspondence rule could be

Odd digit = Walk one block to right.
Even digit = Walk one block to left.

Example 4.9

Two-Dimensional Random Walk

Think of a rectangular grid of streets on a city map. Let this grid extend infinitely far in all directions. A woman takes a walk, starting at (0, 0). She can walk either north, south, east, or west. In order to decide, she tosses two coins and uses this rule:

HH = one block north
HT = one block south
TH = one block west
TT = one block east

Suppose the two coins are tossed four times, with these results: HT, TT, HH, HT. Satisfy yourself that after these four tosses the woman will be at the location (1, −1).

After four tosses of the two coins, how far can she expect to be from her starting point? We will not solve this problem here, but we will point you in the direction of solving it. The distance from the starting point of the person who has taken the random walk can be calculated by using the Pythagorean theorem, which says that the length d in Figure 4.1 is given by the equation $d^2 = x^2 + y^2$. Here x is the

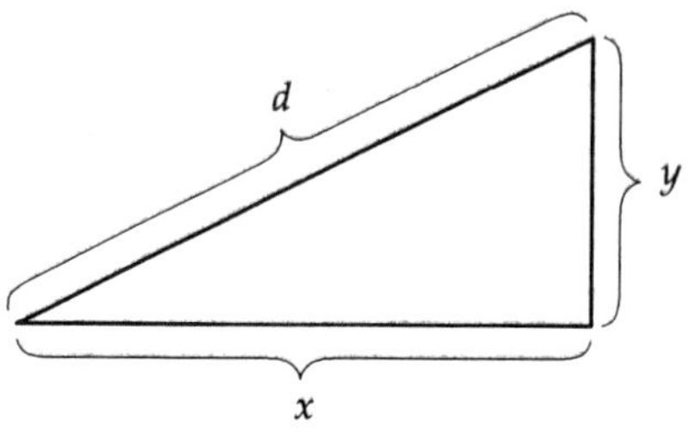

Figure 4.1 **The hypotenuse, *d*, in the right triangle is related to *x* and *y* by $d^2 = x^2 + y^2$.**

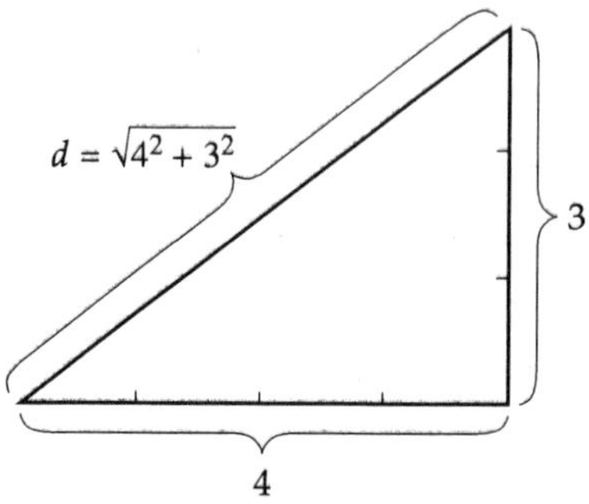

Figure 4.2 **An application of the Pythagorean theorem.**

distance from (0, 0) in the east-west direction, and y is the distance from (0, 0) in the north-south direction. For example, in Figure 4.2, the length of d is 5 units.

How would you use the five-step method to solve this problem?

Interestingly, this idea of a two-dimensional random walk can be used to study heat distribution problems in thermodynamics that were once thought solvable only by differential equations. We consider one example.

Example 4.10 The Temperature of a Point on a Steel Plate

We heat a steel plate by maintaining each edge at the temperature indicated in Figure 4.3. What is the temperature of an arbitrary point P inside the boundaries of the plate after temperatures have settled down to their equilibrium values (that

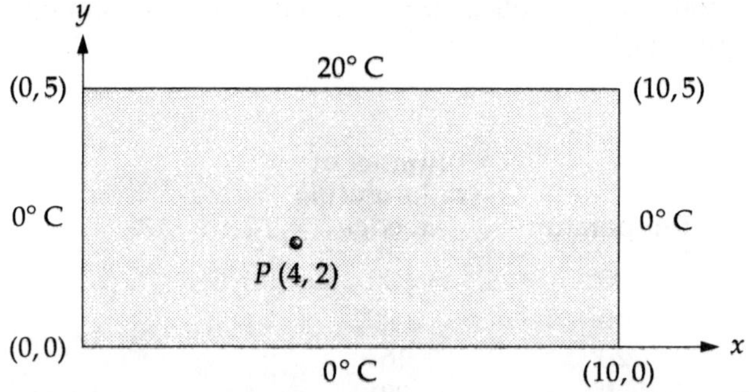

***Figure* 4.3** **Estimating the temperature at the point (4, 2).**

is, no further change of temperature occurs at any location on the plate as time passes)? For example, the equilibrium temperature at (5, 0.25) is likely just slightly higher than 0° Celsius.

Surprisingly (perhaps), advanced probability theory allows us to solve this problem by averaging over many two-dimensional random walks. Starting at P, a walk continues until one of the edges is reached. When an edge is reached, the temperature of that edge is recorded. The walk then resumes at P and is repeated. After a sufficient number of trials (random walks), the average of all the temperatures obtained is taken to be an estimate of the temperature at P.

The table below gives the result of 100 random walks, or trials, estimating the temperature at the point (4, 2). Here the two-dimensional random walk takes a unit step in each direction. For example, six steps to the right from point P would reach the 30° edge.

Edge	Temperature	Number of walks ending at edge
Right	0°	2
Left	0°	7
Top	20°	34
Bottom	0°	57
		100

Then we compute

$$\text{Temperature} = \frac{2 \times 0^\circ + 7 \times 0^\circ + 34 \times 20^\circ + 57 \times 0^\circ}{100} = 6.8^\circ$$

to obtain an estimate of the true temperature at the point (4, 2).

For 500 random walks, done by a computer, the following result was obtained for the point (4, 2).

Edge	Temperature	Number of walks ending at edge
Right	0°	13
Left	0°	42
Top	20°	164
Bottom	0°	281
		500

Then we compute

$$\text{Temperature} = \frac{13 \times 0^\circ + 42 \times 0^\circ + 165 \times 20^\circ + 281 \times 0^\circ}{500} = 6.56^\circ$$

to obtain an estimate of the true temperature at the point (4, 2).

This procedure is an experimental solution to the analytic approach used to solve this problem in advanced "potential theory." The theoretical solution is 6.808 degrees.* Our solution, obtained by averaging numerous random walks, can be shown through advanced mathematical methods to come closer and closer to this theoretical solution as the step size shrinks and it takes more steps to reach an edge (we have a relatively large step size here, and that is likely the reason that the experimental random walk solution of 6.56° is a bit low, even though we simulated 500 random walks). Hence a close approximation to the theoretical answer is obtained when numerous random walks (500, say) of very small step size (0.1, say) are averaged.

SECTION 4.4 EXERCISES

1. Continue the one-dimensional random walk of Example 4.8 for another 18 trials. What is the average stopping position of the man after 10 moves, combining the 30 trials together?
2. Explain how a six-sided die could be used to do the random walk of Exercise 1.
3. Suppose Example 4.8 is changed as follows. Walls are put up at the points −5 and 5. When the person reaches either wall, the walk ends. What is the average number of blocks the person could expect to walk before reaching one of the walls? Base your estimate on 20 trials.
4. As in Example 4.9, take a walk from (0, 0) based on 20 tosses of two coins. What is the distance of the point where you end up

*Obtained by University of Illinois graduate student Ed Malczewski. The analytic solution involves solving Laplace's equation.

from (0, 0)? (*Hint:* Recall that the distance from (0, 0) to the point (x, y) is $d = \sqrt{x^2 + y^2}$.)

5. Look back at Example 4.10, and estimate the temperature at each of the following points, based on 30 random walks for each point.
   a. (1, 3)
   b. (2, 4)
   c. (3, 1)
   d. (8, 1)

6. A gambler begins betting with $20. He has decided beforehand that he will stop gambling if he reaches $50 in winnings, or, obviously, if he runs out of money. The game he is playing is such that on each play he has a 25% chance of gaining $10, a 25% chance of breaking even, and a 50% chance of losing $10.
   a. Explain how this situation can be modeled by a one-dimensional random walk starting at point 20, with walls at 0 and 50. What will the step size be?
   b. On average, how many games will the gambler play before either running out of money or winning $200? Use the five-step method, and do at least 20 trials.
   c. Out of the 20 or more trials you performed in part (b), how many times did the gambler end up losing all his money? Divide this by the number of trials to find the proportion of time the gambler lost all his money.

Graphing Calculator Exercises

7. The program WALK1 simulates a one-dimensional walk. WALK1 prompts for the number of trials and the number of fair coins tossed in a trial (each coin determining a step to the left or right). The program displays the distances from zero for the number of trials and then gives the overall average distance. Do 30 trials for each of the following numbers of coins tossed, and record the average distance obtained.
   a. Toss 10 coins (10-step walk).
   b. Toss 15 coins.

8. Program WALK2 simulates a two-dimensional walk. WALK2 prompts you for the number of trials and the number of tosses of two fair coins in each trial. The program displays the overall average distance from the starting point. Do 30 trials for each of the following:
   a. Toss two coins 10 times.
   b. Toss two coins 15 times.

CHAPTER REVIEW EXERCISES

1. State how you would model a probability of 0.25 (25%) with the following:
   a. A coin
   b. A standard deck of playing cards
   c. A six-sided die
   d. A random number table

2. What are the five steps of the five-step method? State at least one advantage and one disadvantage of the five-step method.
3. A particle begins a one-dimensional random walk at the point 0. The probability that it moves to the right one step (adds one) is 0.25 (25%) at each step. The probability that it moves to the left (subtracts one) is 0.75 (75%). At what point can you expect the particle to be after 10 steps? Perform at least 20 trials to find your estimate.
4. Every day after school, a boy has to walk home, but he never can recall the way to take. Assume that his house is located at the point $(1, 1)$ and his school is at the point $(0, 0)$. The city streets form a perfect grid, with each city block making a square on the grid. If the boy randomly chooses the next direction to walk, with each direction equally likely (a 0.25 chance of his going north, south, east, or west), out of five tries in one week how many times will he make it home after walking 10 blocks or less? Perform at least 20 trials.
5. Repeat Exercise 4, but this time assume that the boy will never choose to go back the direction that he just came. So, now, after his first move, he has three choices at each intersection, each with probability $\frac{1}{3}$. How many times in five tries will he arrive home walking less than 10 blocks? Do at least 20 trials.
6. Your aunt, who is not very willing to take risks, decides to invest in the stock market. She buys some stock, and says that as soon as the stock declines in value for two straight days, she is going to sell it. Assume that the stock has a 40% chance of increasing in value on any day, a 30% chance of keeping the same value, and a 30% chance of losing value. How many days do you expect she will hold the stock before selling? Use the five-step method with at least 20 trials.
7. It is assumed that if a couple has a child, there is a 50-50 chance that the child will be a girl.
   a. Using the five-step method and 25 trials, find the average number of girls in a family with eight children.
   b. How many times out of the 25 trials did you have a family with seven girls and one boy? If an actual family has several girls, what does it tell you about the assumption that each birth is independent from the other?
8. A waiter receives a tip according to the following guidelines:

Good service	\$2.00
Satisfactory service	\$1.00
Poor service	\$0

   The waiter is currently giving good service 30% of the time, satisfactory service 60% of the time, and poor service 10% of the time. If he serves five customers an hour, how much money in tips can he expect to make per hour? Use the five-step method with at least 20 trials.
9. If you were playing cards, how many five-card hands would you expect to have to be dealt before receiving a hand in which at least three cards are of the same suit (the suits are spades, clubs, diamonds, and hearts)?

10. A simple computer weather model makes the assumption that each day is independent of the next. The model is run many times to simulate the month of January in a certain city. The results, in terms of the amount of snowfall per day, are summarized as follows:

Snowfall (inches)	Percentage of days
0	60%
0.5	18%
1	12%
2	7%
3	3%

What is the estimated total snowfall for the entire month of January (31 days)? (*Hint:* This should not require the use of the five-step method.)

11. The owner of a restaurant buys a new set of glasses for his restaurant. She knows that the probability that any day will go by without anyone breaking a glass is 90%. How many days can the owner expect to go without needing to replace a broken glass? Use the five-step method with 25 trials. Use a random number table to model the probability.

12. Refer to Exercise 11. Assume that there is never a day when more than one glass is broken.
    a. What would you guess to be the expected number of days passed before two glasses are broken? What do you base your guess on?
    b. Perform another 25 trials, this time ending each trial when two glasses are broken. What is the average number of days until two glasses are broken? Is it close to what you guessed?

13. A certain basketball player is able to make 25% of his three-point shots and 50% of his two-point shots. Assume that the player has five opportunities to take a shot during a game.
    a. Using the five-step method, estimate how many points the player can expect to score if he only tries three-point shots. Use at least 20 trials.
    b. Again using the five-step method, estimate the number of points the player can expect to score if he only attempts two-point shots.
    c. On the basis of your results, should this player concentrate on two-point or three-point shots?

14. A company believes that a certain percentage of its products will turn out to be defective. For each defective product the company will lose $15, and for each nondefective product the company will earn $10. The management used the five-step method to investigate the amount of money they could expect to earn off of the product, assuming that 100,000 units are sold.

    To model the probability, they used a random number table with numbers 0 through 9. Two numbers next to each other were put together to form a random number between 0 and 99, where 00 was taken to be 0, 01 is 1, and so on. A complete trial consisted of looking at 100 random numbers in this way, and if the number was 00, 01, 02, 03, or 04, the product was assumed to be defective. Fifty trials were performed, with the results as follows:

Number of defectives	Number of trials
0	5
1	14
2	20
3	8
4	3

a. What was the average number of defective units in the 50 trials? Divide this by 100 to get the percentage of units expected to be defective. How many of the total 100,000 units to be sold can be expected to be defective?
b. How much money can the company expect to earn from this product?

15. A child is collecting trading cards. The complete set consists of 10 cards, and there are three cards in each pack of cards sold. Assuming that the 10 cards are randomly distributed among the packs, how many packs should the child expect to have to buy before having the complete set? Use the five-step method to reach an estimate, and use at least 25 trials. (*Hint:* A deck of cards would be a convenient way to model this situation.)

16. Refer to Exercise 15. Instead of wanting the complete set, the child is interested in only one certain card. How many packs should the child expect to buy before getting that card? Use the five-step method again with 25 trials.

17. John is taking a test consisting of 10 multiple-choice questions with four choices for each question. For each question he gets right, he receives four points, but for every one he answers incorrectly, he loses one point.
    a. If John knows no answers to any of the questions and he just guesses, what is his expected score? Again, use the five-step method with 25 trials.
    b. Would John be better off not answering any of the questions—in other words, just leaving the test blank?

18. Refer to Exercise 17. Assume this time that John is able to eliminate one choice out of the four on each question, but he has to guess among the other three choices. Assuming that the scoring structure is the same, what is his expected score in this case?

19. Estimate the number of times you expect you would have to roll a die before one side appears twice. (That is, how many times do you have to roll before something comes up that you have already seen before?) Perform at least 25 trials to estimate this.

20. Every day Sue wakes up to her alarm. Sometimes she hits the snooze bar, and other times she does not. If she hits the snooze bar more than once in a

morning, she will end up being late for class. The following table shows how often Sue hits the snooze bar:

Does not hit it	50% of mornings
Hits it once	30% of mornings
Hits it twice	15% of mornings
Hits it three times	5% of mornings

In a five-day school week, how many times can Sue expect to be late for class in the morning? Perform at least 25 trials.

4

COMPUTER EXERCISES

1. CEREAL BOXES I

The Five Step program is designed to carry out the five-step method, which is used in Chapters 4 and 5 to experimentally estimate theoretical expected values and theoretical probabilities. We will introduce the program through the cereal box problem described in Section 4.1. The picture shows what the opening screen looks like:

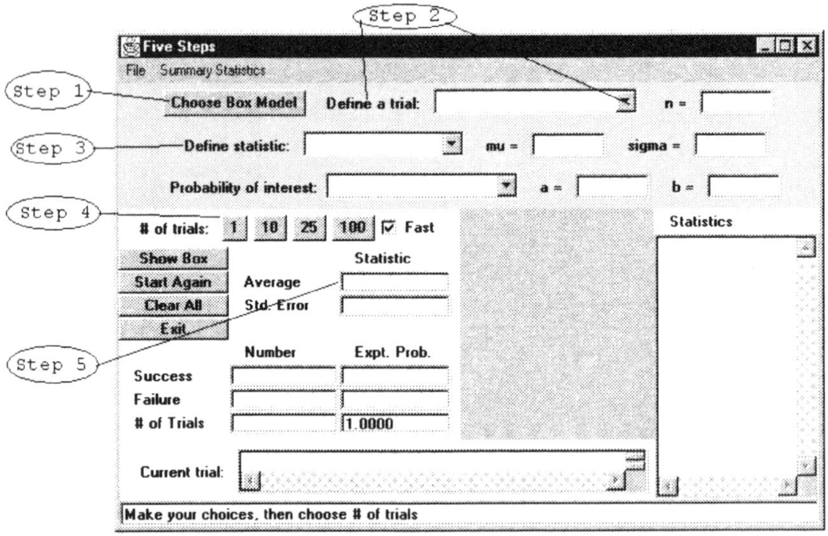

The bubbles labeling the steps are not part of the program, but relate the various parts of the screen to the steps of the five-step method. As in the text, these computer probability problems are approached by carrying out the five steps. The five steps for the cereal box problem are next.

Step 1. Model: The models in this program are given as box models. A box model is simply a box containing tickets having numbers written on them. Every ticket is equally likely to be chosen. In some cases more than one ticket will have the same number (Section 5.3 will discuss box models). One then draws tickets from the box. This step is very important because it specifies the probability model, or *population,* for whatever statistical problem is being studied. The most important step is to decide what tickets should go in the box: how many tickets there are, what value is written on each, and how many of each kind there are. For the cereal box problem, we need six tickets: one with a 1, one with a 2, . . . , and one with a 6. Press the Choose Box Model button to obtain the following screen:

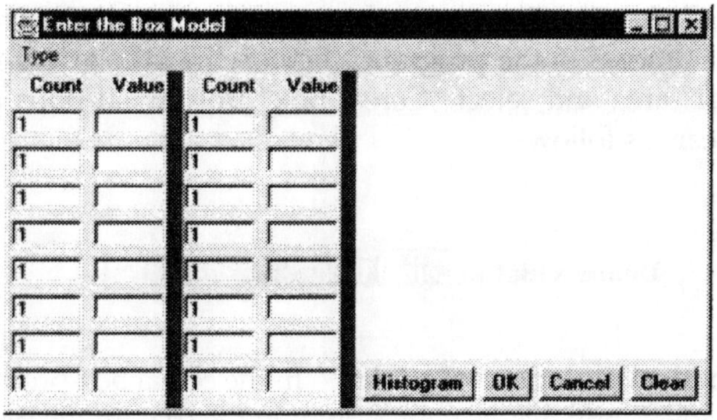

The Value columns should contain the values that are to be written on the tickets, and the Count columns reveal how many tickets with each value there are. For the current problem, we need six values, 1, 2, 3, 4, 5, and 6, each with a count of 1. Counts of 1 are already there, so just fill in each of the values, so that the screen looks like the following:

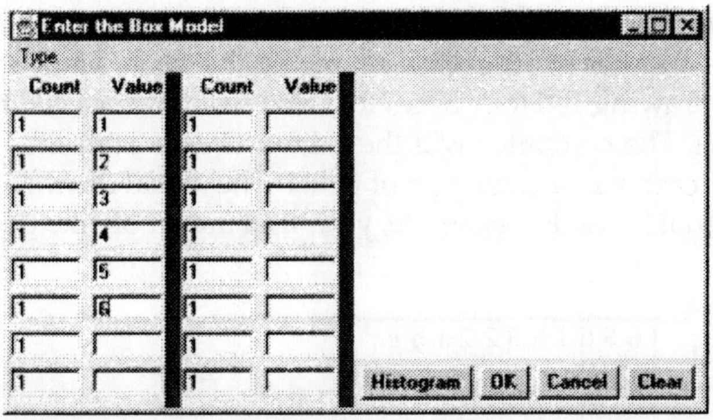

Click on OK when finished. The screen will disappear. Some of the elements of the original screen will disappear, too. That is fine. They will reappear if needed. To see the box, click on the "Show Box" button:

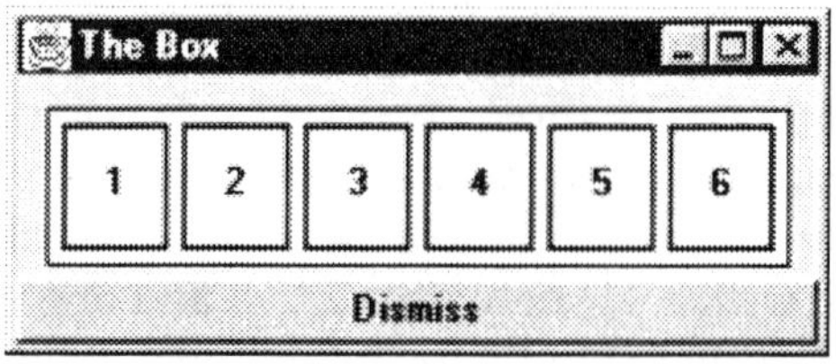

Step 2. Definition of a Trial: A trial consists of drawing tickets from the box *with replacement* until we have drawn at least one of each value. To represent that process in the program, click on the little arrow next to the "Define a trial" area and select "Cereal Box" from what pops up, so that that area appears as follows:

Step 3. Numerical Outcome of a Trial: In the cereal box problem we are interested in how many tickets were drawn in the trial. In the program, the number of draws is called the "sample size." Click on the arrow next to the "Define statistic" area, and select "Sample size," so that it looks like the following:

Step 4. Repetition of Trials: The four buttons labeled 1, 10, 25, and 100 determine the number of times one can repeat the trials. More are possible. For example, drawing 100 four times in a row will produce 400 trials. Press the "1" button. The computer will then draw tickets from the box until it has drawn at least one of each type of ticket. The actual draws are given in the "Current trial" area. For example, your trial may look like this:

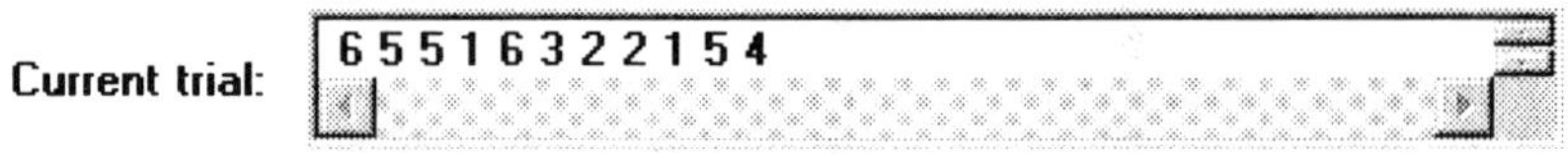

Thus on this trial, the first draw was a 6, the next two were 5, then there was a 1, then another 6, and so on, until on the 11th draw there was the first 4. At that point at least one of each had been drawn, so the trial ended. The numerical outcome (statistic of interest) is the sample size, 11, which appears in the area labeled "Sample sizes" on the right. Your trial's outcome may be 11, but it will probably be different.

One trial is not enough. To perform more trials, just click on one of the other numbered buttons. Suppose you click on the "100" button. The program will run more trials, each time showing the trials (quickly) at the "Current trial" area and writing the actual numerical outcome at the right.

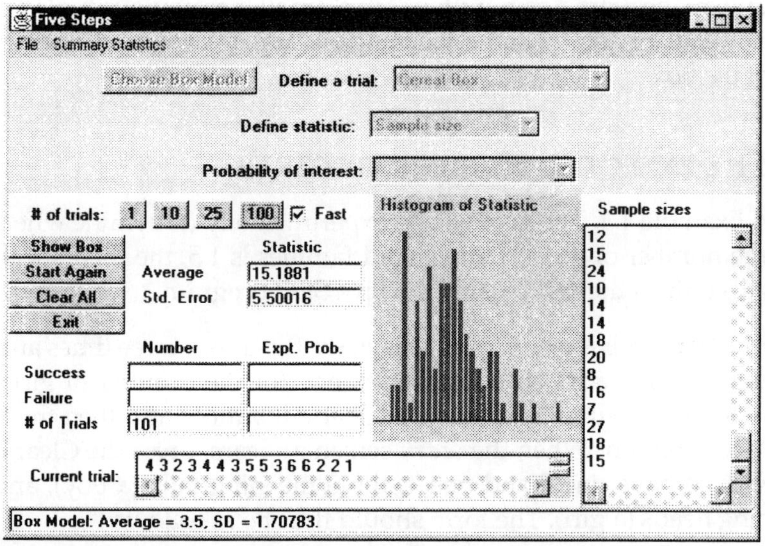

The total number of trials so far has been 101, which is given in the "# of Trials" area. Notice also the histogram. It is constructed of the 101 numerical outcomes—that is, the 101 sample sizes.

Step 5. Finding the Average of the Statistic of Interest: You can take all the 101 sample sizes collected at the right and find the average of them by hand. Fortunately, the computer already found the average, which is in the area labeled "Step 5" in the first figure. From the figure in step 4, we see the average is 15.1881, which means it will take on average about 15 boxes in order to obtain all six colors of pens. Your average will be different, but it should be similar. (Recall that 14.7 is the true theoretical value.)

Besides the average, one can find other summary statistics by going up to the Summary Statistics menu:

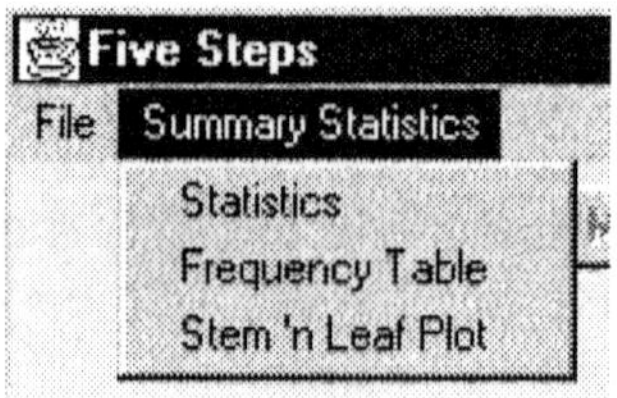

Choosing "Statistics" will cause a screen to pop up having the average, standard deviation, median, interquartile range, and other common statistics for the trials. The other two choices will cause a frequency table and a stem-and-leaf plot to pop up.

More trials can be obtained by clicking on the numbered buttons. Clicking on the Start Again button will clear all the previous trials but leave the choices at the top the same. Clicking on "Clear All" will clear everything except the box.

2. THE EXPECTED NUMBER OF GIRLS

Recall Example 4.3. The goal was to experimentally determine whether the average number of girls in three-child families is 1.5, the theoretical value. We will confirm the result using the Five Step program.

Step 1. Choose a Model: In the text, we flip a coin three times and count the number of heads. That number stands for the number of girls in the family. In the program, click on the Clear All button, and then the Choose Box Model button. When the entry screen appears, press the Clear button. We wish to have one 0 and one 1, the 0 meaning tails (a boy), and the 1 meaning heads (a girl). The form should thus be as follows:

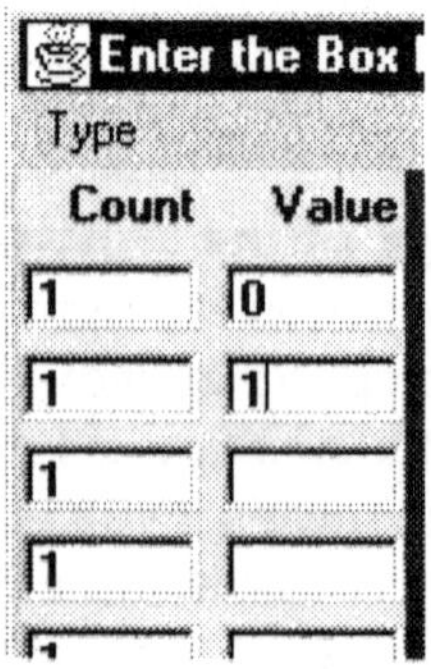

Press OK.

Step 2. Definition of a Trial: A trial is three flips. From the "Define a trial" area, choose "Draw n with replacement." A new little area with "n =" will appear. Type in "3" because that is how many we wish to draw for each trial.

Step 3. Numerical Outcome of a Trial (Statistic of Interest): This time we just want to count how many of the children are girls—that is, how many of the flips are heads, or how many of the draws are 1s. Thus we just want to take the sum of the draws. From the "Define statistic" area, choose "Sum":

Step 4. Repetition of Trials: Press the "1" button. You will see three numbers in the "Current trial" area, each one either a 0 or a 1. The number of 1s, or the sum of the draws, will appear in the list of sums at the right. Press "100" to have the program run through 100 more trials. Press "100" some more times, if you wish.

Step 5. Finding the Average of the Statistic of Interest: In the "Average" area you can see the average number of 1s per trial, which represents the average number of girls per three-child family. Is it close to 1.5?

3. CEREAL BOXES II

It seems intuitive that the more different types of pens there are, the more cereal boxes you need to open before having one of each type. How many more? Does it take on average one more box if there are six pens than if there are five pens? Or is it proportional—that is, would it take twice as many boxes to get all of six pens than to get all of three pens? This problem asks you to use the five-step method to try to discover the general character of the relationship between the number of different pens and the expected number of boxes needed to obtain one of each pen.

a. If there is only one type of pen, then you only have to buy one box of cereal to get it. (Why?) For other numbers, you need to use the five-step method outlined in Section 4.1. Carry out the steps for the box with one

1 and one 2; then for the box with one 1, one 2, and one 3; then for the box with one each of 1, 2, 3, 4; then for the box with one each of 1, 2, 3, 4, 5; and finally for the box with one each of 1, 2, 3, 4, 5, 6. Perform 200 trials for each. (Be sure to press the Start Again button after every 200 trials.) Fill in the following table with the averages of the numbers of cereal boxes needed (that is, the averages of the sample sizes):

Number of pens	Average number of boxes	(Average number of boxes)/(number of pens)
1		
2		
3		
4		
5		
6		

b. Does the average number of boxes needed increase by 1 when the number of pens increases by 1?

c. Fill in the third column in the table by dividing the second column by the first column. This column has the number of boxes needed per pen. Are the numbers in this column approximately the same? Are they increasing? Are they decreasing?

d. Make a scatter plot with X being the number of pens and Y being the average number of boxes needed. Are the points approximately on a straight line, or do you notice some curvature in their pattern?

e. Guess what the average number of boxes needed would be if there were seven different pens. Use the five-step method to estimate this average. How good was your guess?

4. CEREAL BOXES III

Until now, it was assumed that the chance of any color of pen was the same. Often, the chances will be different. Especially if there is a prize attached to collecting all types of pen, a cereal manufacturer might have one color that is much rarer than the other colors. How does this rareness affect the number of boxes needed? This problem will look at the case in which there are four different colors of pen and one pen is rarer then the others. Set up the box in the five-step method so that there are five tickets each of 1, 2, 3, and 4. The picture below shows a part of the screen for entering the box, and the resulting box:

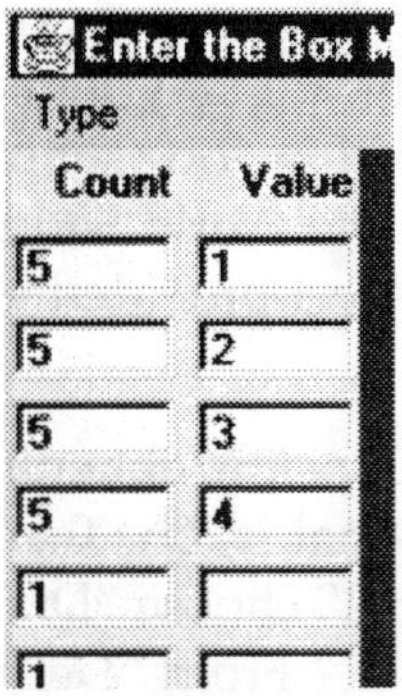

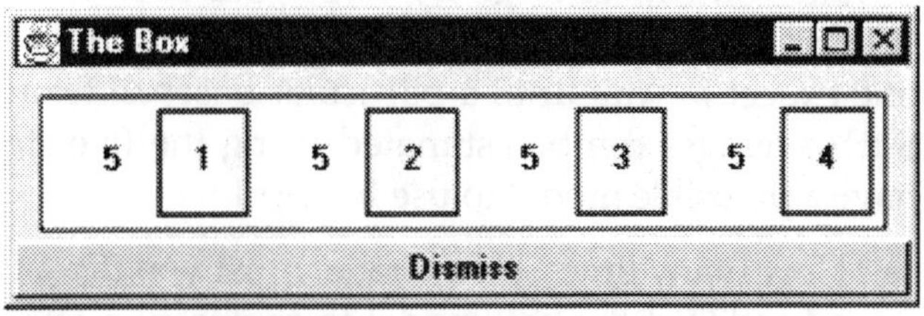

Drawing with replacement from this box is the same as drawing with replacement from the box with one each of 1, 2, 3, 4, because there is a $\frac{1}{4}$ chance of getting any one of the numbers each time.

Set up the cereal box problem as before, with five each of 1, 2, 3, and 4. Run 200 trials. Write down the average number of cereal boxes needed. In addition, go to the Summary Statistics menu, and select "Statistics." Find the maximum sample size. This is the largest number of cereal boxes needed for the 200 trials. Write down the average and maximum in the first row of the table, next to the "5" in the column containing the number of 4s.

Number of 4s	Average number of boxes	Maximum number of boxes
5		
4		
3		
2		
1		

Next, press "Start Again" and change the box model so that there are four 4s instead of five. (Change the "Count" column for the value 4 from 5 to 4.) Run 200 trials of the cereal box problem, and enter the average and maximum in the table in the second row. Do the same three more times: once with three 4s, once with two 4s, and once with one 4.

Do you see a pattern? When there are fewer 4s than the other numbers, does it tend to take more cereal boxes to get one of each pen, or fewer, or about the same number of cereal boxes?

5. WAITING TIMES

We often want to know how long one should expect to wait before something happens. For example, if the Internal Revenue Service randomly chooses a certain number of tax forms to audit each year, how many years should you expect to go by until your first audit? Or how many lottery tickets should you expect to buy before getting a winner? How long should one expect to wait until a particular baseball team wins the World Series? Such averages can be estimated using the five-step method provided we have a plausible model to use in step 1.

a. How many times, on average, must you flip a fair coin before obtaining a head? Set the box model to represent a fair coin—that is, one 0 and one 1, where the 1 indicates heads. From "Define a trial" choose "Draw until first X." Then type "1" in the box labeled "X = ." From "Define statistic" choose "Sample size." The top part of the screen should look like the following:

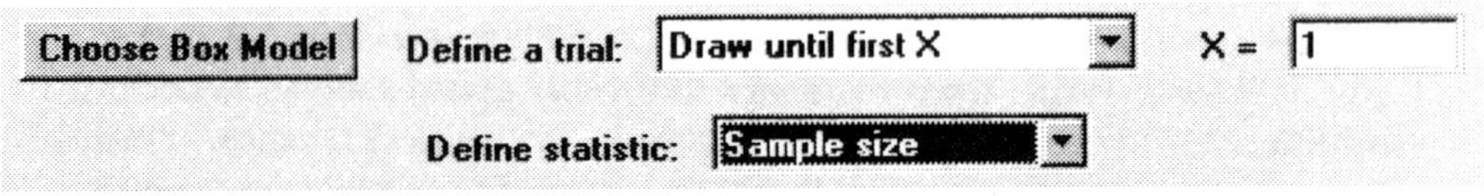

A trial then consists of drawing tickets until the first 1 is drawn. The number of draws needed is the statistic. In the "# of trials" buttons, press "1" a few times, and notice the current draws each time. You will see some 0s and a 1. Often there will be no 0s, just a 1. Press the "100" button a few times. What is the average sample size?

b. How many times, on average, must you roll a fair six-sided die until you see your first 6? (Press "Start Again" if you have not done so already.) Set up the box model so that there is one each of 1, 2, 3, 4, 5, and 6 in the box. Define a trial to be drawing until first X, where X = 6, and define the statistic to be the sample size again. Press the "100" button a few times. What is the average now?

c. You would expect that the rarer an event is, the longer you have to wait until it occurs. How is the rareness related to the average waiting time? Use the five-step method to estimate the average number of draws it takes until the first 1 for boxes with different numbers of 0s but one 1. For each box, run 200 trials, and record the averages in the table at the top of the following page.

d. Make a scatter plot with X equal to the proportion of 1s in the box and Y equal to the average number of draws until the first 1. Are the points roughly on a straight line, or is there some curvature?

e. Make a guess about the average number of draws until the first 1 when there are ten 0s and one 1. Test your guess using 400 trials. Does your guess seem reasonable?

Box	Average number of draws	(Average number of draws)(proportion of 1s)
One 0, one 1		
Two 0s, one 1		
Three 0s, one 1		
Four 0s, one 1		
Five 0s, one 1		
Six 0s, one 1		
Seven 0s, one 1		
Eight 0s, one 1		
Nine 0s, one 1		

f. Now fill in the third column in the table. Are the numbers about the same? If so, what are they approximately equal to? If not, is there a pattern to them?

g. Now create the box with three 0s and two 1s. Make a guess about the average number of draws until the first 1. Test your guess using 400 trials. Does your guess seem reasonable?

h. Can you make a general hypothesis about the relationship between the proportion of 1s in the box and the average number of draws until the first 1?

5

PROBABILITY

It is remarkable that a science which began with the consideration of games of chance should have become the most important object of human knowledge.

Pierre-Simon Laplace (1749–1827)

OBJECTIVES

After studying this chapter, you will understand the following:

- *The five-step method for computing experimental probabilities*
- *The experimental probability $P(E)$ as estimator of the theoretical probability $p(E)$*
- *How to model probability problems using coins, dice, random digits, and box models*
- *The independence of events as an experimental and theoretical concept*
- *The theoretical probability model of equally likely outcomes*
- *The law of complementary events*
- *Sampling from a population with and without replacement*
- *The birthday problem as an illustration of the five-step method*
- *Introductory concepts of conditional probability*

5.1 EXPERIMENTAL PROBABILITY 224

5.2 ESTIMATING PROBABILITIES 229

5.3 BOX MODELS 235

5.4 INDEPENDENT EVENTS 241

5.5 PROBABILITY AS A THEORETICAL CONCEPT 246

5.6 COMPLEMENTARY EVENTS 254

5.7 SAMPLING WITH OR WITHOUT REPLACEMENT 258

5.8 THE BIRTHDAY PROBLEM 264

5.9 CONDITIONAL PROBABILITY 268

CHAPTER REVIEW EXERCISES 271

COMPUTER EXERCISES 277

The Birthday Problem

Consider a roomful of people who can be viewed, for the purposes of this problem, as having been randomly selected from a large population (say, the residents of San Francisco). What is the probability that at least two of them share the same birthday (for example, September 10)? For simplicity we presume that nobody has a February 29 leap-year birthday.

Suppose there are 25 people in the room. Most people's intuition is that the chance must be quite small—say 8%, 10%, or 20%. But this is an example of a common happening in probability problems: often our guess about how probable an event is can be way off. The answer to this problem can be computed using the rules of theoretical probability, but it is a challenging computation. Our Key Problem is whether we can find the answer (approximately) experimentally as we did with expected values in Chapter 4.

Doing statistical inference on a data set begins with building a probability model for how the data were generated. A vital feature of the five-step method introduced in Chapter 4 is that it *requires* us to think carefully and explicitly about our choice of this probability model. Choosing this probability model is always a necessary first step in a statistical analysis. Hence this chapter is an important prelude to doing effective model-based statistical inference. In this chapter we adapt the five-step method of Chapter 4 to provide us with a method of finding approximate values of important theoretical probabilities of probability models experimentally.

5.1 EXPERIMENTAL PROBABILITY

Consider questions such as the following:

If a basketball player shoots 60%, what is the chance that she will get four baskets in her next six tries?

What is the probability that a family of four children will have children of both sexes?

If light bulbs last on average six months, what is the chance a particular light bulb will last nine months?

If I make an airline reservation, what is the chance that, because of overbooking, I will not be able to get a seat on the airplane?

A lot of 50 VCRs contains 10 defectives. If one randomly chooses 10 of the 50 VCRs, what is the probability that none are defective?

We live in a probabilistic world. Weather reports tell of the chances of a hurricane hitting the Florida coast. Public health officials recommend vaccinations to reduce our chances of getting the flu. We send out numerous resumes with the hope of increasing our chances of getting a job. As we grow older, the cost of life insurance increases because the insurance company computes our chances of living for a certain period of time. As the above questions demonstrate, probability is important in a wide variety of fields, including sports, genetics, and manufacturing.

Probability arises when there is uncertainty about the occurrence of events. Games of chance by their very nature involve interesting probability problems, and they were well known in Greece at least 2300 years ago. It was in the Middle Ages when gamblers (and mathematicians and philosophers) began to think about probability in a more systematic way—with the obvious hope of improving their winnings! The serious study of probability apparently began in the seventeenth century when the French nobleman Chevalier de Méré approached the famous mathematician Blaise Pascal with questions about betting strategies in games of chance. Pascal took up the challenge and, together with another famous mathematician, Pierre de Fermat, did the groundwork for the formal rules of probability theory.

There are various interpretations of probability. In this book we use an *experimental*, or *relative frequency*, interpretation. This interpretation simply says that an observed "experimental probability" resulting from many trials (for example, many tosses of a coin) provides a good estimate of the unknown theoretical probability. Indeed this theoretical probability is *understood* to be the resulting experimental probability that would result "in the limit" if we kept doing such trials forever. We illustrate with some examples.

If an automobile insurance company finds that 9 cars out of 1000 randomly chosen insured cars are stolen, then the experimental probability of an insured car being stolen is

$$\frac{9}{1000} = 0.009$$

This seems a good estimate of the true theoretical probability of a car being stolen for the 200,000 policies the company holds.

If a possibly loaded coin is tossed 10 times and heads occur 7 times, then the experimental probability is

$$P(\text{heads}) = \frac{7}{10} = 0.7$$

The experimental probability value of 0.7 is important as an estimate of the theoretical probability of the coin coming up heads. Of course, our intuition correctly tells us we will need many more trials before we can have much confidence in the accuracy of this estimate.

If a die is tossed 500 times and the single dot appears 80 times, then empirically,

$$P(\text{one dot appearing}) = \frac{80}{500} = 0.16$$

This is quite close to the true probability of $\frac{1}{6} \approx 0.1666$.

In each of these examples,

$$P(\text{desired event occurring}) = \frac{\text{number of occurrences of desired event}}{\text{number of repetitions, or trials}}$$

The ratio obtained in this manner is an **experimental probability,** based on the relative frequency (that is, the frequency or count relative to the

Table 5.1 **10,000 Tosses of a Coin**

Number of tosses	Number of heads	Relative frequency (experimental probability)
10	6	0.600
20	6	0.300
30	10	0.333
40	14	0.350
50	22	0.440
100	48	0.480
150	66	0.449
200	88	0.440
300	136	0.453
400	182	0.455
500	235	0.470
600	286	0.477
700	336	0.480
800	384	0.480
900	431	0.479
1000	486	0.486
2000	995	0.498
3000	1498	0.501
4000	2001	0.500
5000	2502	0.500
6000	3009	0.502
7000	3547	0.507
8000	4059	0.507
9000	4541	0.504
10,000	5056	0.506

total number of all trials) of the desired event occurring. Note that by its definition it is a number between 0 and 1. It gives us an *estimate* of the true theoretical probability of the desired event, and one we can trust if the number of trials is large. The concept of the theoretical probability will be discussed later in the chapter.

To see what can happen as the number of repetitions increases, refer to Table 5.1. This table gives the simulated results of tossing a coin 10,000 times. The first row gives the number of heads that appeared in the first 10 tosses, the second row gives the number appearing in the first 20 tosses, and so on. The table is based on the experience of J. E. Kerrich, who, while a prisoner of war in Denmark during World War II, actually did toss a coin 10,000 times and record the results.* Notice the sequence of relative

*J. E. Kerrich, *An Experimental Introduction to the Theory of Probability* (Copenhagen: J. Jorgenson and Co., 1946), p. 14.

frequencies (each an experimental probability of heads). At first they show considerable variability. It is somewhat remarkable that tosses 11–20 were all tails! However, after hundreds of trials, the experimental probabilities, although continuing to vary, become more nearly constant. If the coin were tossed even more times, it seems reasonable to expect that there would be still less fluctuation in the relative frequency of heads. But some variation would still always occur. It also appears that the relative frequency is settling on a number near 0.500, possibly $\frac{1}{2}$ itself. The fact that experimental probabilities stabilize as the number of trials increases is one of the fundamental laws of science. It is sometimes called the *law of statistical regularity*. It is the cornerstone of statistical reasoning, as we will learn. We cannot predict one coin toss, but we *can* predict quite accurately the proportion of heads in 10,000 tosses.

SECTION 5.1 EXERCISES

1. What estimated probability of getting heads based on each of the following numbers of tosses is obtained from the data in Table 5.1?
   a. 100
   b. 500
   c. 1000

   Do you think the coin we tossed to produce Table 5.1 was well balanced? Why or why not?
2. Roll a die 60 times and observe how often you obtain each of the six numbers 1, 2, 3, 4, 5, and 6. Find the experimental probability for each of the following based on the outcomes of your experiment.
   a. *P*(rolling a 1)
   b. *P*(rolling a 2)
   c. *P*(rolling a 3)
   d. *P*(rolling a 4)
   e. *P*(rolling a 5)
   f. *P*(rolling a 6)
3. Shuffle a standard deck of 52 cards and draw one card. Repeat at least 50 times. Estimate the probability of drawing an ace. Is your answer consistent with what you would think is the theoretical probability of drawing an ace?
4. Toss three coins 50 times and keep a tally of the number of heads on each toss of the three. Then use your data to approximate the following probabilities:
   a. *P*(All three are heads)
   b. *P*(Exactly two are heads)
   c. *P*(Exactly one is a head)
   d. *P*(None of the three is a head)
   e. The probabilities from parts (a) and (d) should be fairly equal, and the probabilities from (b) and (c) should be close to equal. Does this make sense? Why is this so? (*Hint:* Getting exactly one head means you got exactly two tails.)
5. A building supplies manager wants to estimate the probability that a certain type of carpet tack will fall point-up when dropped on a smooth wood floor. A worker drops one of the tacks 500 times and finds that it falls point-up 35 times. What is the estimated probability that the tack will fall point-up?
6. Over his career, a basketball player has made 1210 free throws and missed 214. What is his estimated probability of making a free throw?
7. Toss two dice 50 times and record the sum of the dots appearing each time. Then compute the following probabilities:
   a. *P*(The sum of the two dice equals 2)
   b. *P*(The sum of the two dice equals 7)
   c. *P*(The sum of the two dice equals 10)
   d. *P*(The sum of the two dice equals 6)

Graphing Calculator Exercises

8. Simulate rolling a six-sided die 120 times by using the program DICE. Select your choice at the menu; enter the number of rolls (120). The possible outcomes are in the first column, labeled L1, of the displayed table, and the results are in the second column, L2. Make a histogram of the data (see "Making a Histogram" in Section 3 of Appendix G). Set up the statplot and the window as shown and obtain the following probabilities:

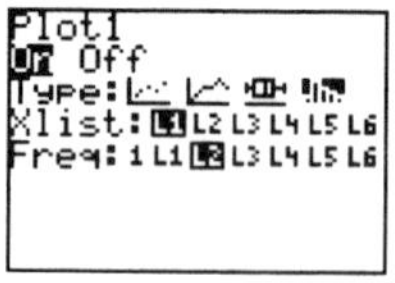

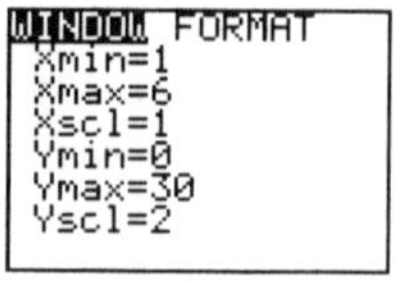

a. *P*(rolling a 1)
b. *P*(rolling a 2)
c. *P*(rolling a 3)
d. *P*(rolling a 4)
e. *P*(rolling a 5)
f. *P*(rolling a 6)

Note: It may be useful to first check the data table (STAT [edit] ENTER) to be sure that the scale in the *Y* axis is large enough.

9. Repeat Exercise 8, but this time run DICE using 180 trials. Compare the experimental probabilities for the six outcomes that you get now with what you got in Exercise 8. What number does each estimated probability seem to approximate?

10. Simulate tossing two coins 50 times by using the program FLIPS. At the prompt enter the number of coins to be tossed (2) and the number of trials (50). In the data table, L2 gives the frequency of heads obtained for each possible outcome listed in L1. View the results in the data table by pressing STAT [edit] ENTER. Obtain the following experimental probabilities:
a. *P*(2 heads)
b. *P*(exactly 1 head)
c. *P*(1 or more heads)
d. *P*(0 heads)

11. Repeat Exercise 10, but this time do 100 trials. Compare the estimated probabilities you get now with those obtained in Exercise 10.

12. Toss five coins 50 times by using the program FLIPS. At the prompt enter the number of coins to be tossed (5) and the number of trials (50). The possible number of heads is in column L1 of the data table, and the

frequency is in column L2. View the results in the data table by pressing [STAT] [edit] [ENTER]. Obtain the following:

a. *P*(All five are heads)
b. *P*(Exactly three are heads)
c. *P*(Three or more are heads)
d. *P*(None of the five is a head)

13. Use the program DICE to simulate rolling two six-sided dice 50 times. The possible sums are in column L1 of the data table, and the resultant frequencies are in column L2. View the data table and obtain the following.

a. *P*(Sum equals 2)
b. *P*(Sum equals 7)
c. *P*(Sum equals 10)
d. *P*(Sum is greater than 6)

5.2 ESTIMATING PROBABILITIES

It seems clear that the approach of estimating expected values in Chapter 4 should also work here in Chapter 5 for estimating probabilities. Consider the following example.

Example 5.1

In a family of three children, what is the probability that *all three* are boys? (Assume that boys and girls have an equal chance of being born.)

Solution

One approach to solving this problem would be to take a survey of a large number of families with three children and find out in what proportion of them all three children are boys. However, it would take a lot of time to collect enough data to provide a good estimate of the probability. Another approach would be to simulate the survey by tossing three coins in succession, using the following representation:

Tails (T) represent "boy is born."

Heads (H) represent "girl is born."

Note that this simulation model assumes an equal chance of a boy or a girl being born. This is approximately true in reality: about 51% of newborns are male.

Suppose now that we toss three coins and get H T H. This is interpreted as follows: the first child is a girl, the second child is a boy, and the third child is a girl. We are assuming not only that boys and girls each have the same chance of being born but also that each succeeding child has an equal chance of being a boy or a girl, regardless of the sex of the previous child.

We can use the five-step procedure for estimating the theoretical probability that in a family of three children, all are boys:

1. Choice of a Model: Use a coin to simulate the sex of a child, where

Heads:	girl
Tails:	boy

2. Definition of a Trial: A trial consists of tossing the coin three times.

3. Definition of a Successful Trial: A successful trial occurs when all three coin tosses fall tails (that is, all three children are boys). Note that from the viewpoint of the five-step method, when we are estimating the theoretical probability of an event, observing whether or not that event has occurred in the trial corresponds to observing the statistic of interest in step 3 of Chapter 4.

4. Repetition of Trials: Do a sufficiently large number of trials (at least 100 is recommended for accuracy). Consider a sample of three trials. Instead of actually tossing coins, we will use Table B.1, the coin-toss random number table, which uses the digits 0 and 1 instead of the symbols T and H. The simulated coin tosses are from row 1 of Table B.1, reproduced here:

01110 10000 00010 10111 00010 11001 10011 10001 10110

The three sample trials are recorded in Table 5.2.

5. Probability Estimate: The estimated (experimental) probability that all three children are boys is

$$P(\text{three boys in a family of three}) = \frac{\text{number of successful trials}}{\text{total number of trials}}$$

Based on only three trials, our estimate for the probability of three boys in a family of three children is

$$P(\text{three boys in a family of three}) = \frac{1}{3} = 0.33$$

To obtain an accurate estimate for this probability, more trials are needed. As an exercise, do 100 trials using three coins, Table B.1, or a computer coin-toss simulation program if you have one. If you use a computer, you can do 1000 or even 10,000 trials in order to achieve extreme accuracy. The true probability is $\frac{1}{8}$ here.

Table 5.2 **Sample Trials to Estimate Probability of Three Boys in a Three-Child Family**

Trial			Success?
1	THH	1 boy	No
2	HTH	1 boy	No
3	TTT	3 boys	Yes

How many trials should one carry out? Above we recommended that 100 trials be performed if accuracy is desired. One hundred trials means that typically—indeed, $\frac{2}{3}$ of the time—the estimated probability will be within 0.05 of the true theoretical probability. If one uses a computer to do the trials, as is possible with the instructional software provided with this textbook, then one can do a large number of trials. For example, doing 400 trials means that one will typically be within 0.025 of the true theoretical probability. Why this is so will become clear in Chapter 10. By doing only 25 trials, one will typically be within 0.1 of the true probability—a level of accuracy useful for many purposes. Sometimes in our examples we will demonstrate the five-step method using many trials to show the high accuracy that is possible; at other times we will use 100, 25, or fewer trials so that the approach is made clear. One always has the option to do more, even without access to computer simulation.

If in step 3 we were to assign a 1 to a successful trial and a 0 to an unsuccessful trial (this in effect being our definition of the statistic of interest for the five-step method, making the third step above the same as that in Chapter 4), then step 5 merely becomes, as it was in Chapter 4, the computation of the average of the statistic of interest. This average is simply the proportion of successes. This remark makes clear that the five-step method for finding a probability is the same as the five-step method of Chapter 4. Indeed our instructional software uses the same approach for estimating a probability as for estimating an expected value.

We now consider another example involving three-child families. You might have guessed (not a good method in general, because our guesses about probabilities are wrong surprisingly often) or even supplied some reasoning to convince yourself that the true probability above was $\frac{1}{8}$, but the answer in the following example is less clear.

Example 5.2

In a family of three children, what is the probability that *at least two* are boys? (Assume that boys and girls have an equal chance of being born.)

Solution

In order to solve this problem, we make only a slight modification of our five-step procedure of Example 5.1.

1. Choice of a Model: Again the coin models the sex of a child:

Heads:	girl (1)
Tails:	boy (0)

2. Definition of a Trial: Again, a trial consists of tossing the coin three times.

***Table 5.3* Sample Trials to Estimate Probability of At Least Two Boys in a Three-Child Family**

Trial			Success?
1	TTT	3 boys	Yes
2	THT	2 boys	Yes
3	HTH	1 boy	No
4	HHT	1 boy	No

3. Definition of a Successful Trial: A successful trial occurs when at least two of the three coin tosses fall tails (that is, at least two children are boys). This is the only change from Example 5.1.

4. Repetition of Trials: Do a sufficiently large number of trials, say, 100.

Here we consider a sample of four trials. The coin tosses are from Table B.1, row 1. *Start where we left off in Example 5.1 (start with the 10th character).*

01110 10000 00010 10111 00010 11001 10011 10001 10110

These four trials result in Table 5.3.

5. Probability estimate: The estimated probability that at least two children are boys is

$$\frac{\text{Number of successful trials}}{\text{Total number of trials}}$$

Based on our four trials, we estimate the probability of there being at least two boys in a family of three children as

$$P(\text{at least two boys in a family of three}) = \frac{2}{4} = 0.5$$

The true theoretical probability is 0.5; we were rather lucky that the experimental probability was exactly correct!

We now consider another example from everyday life, one that does not use a fair coin model for step 1.

Example 5.3

Suppose the first traffic light on your route to class is green for 20 seconds and red for 40 seconds. Thus, the light has a 60-second cycle. What is the probability that you will get *exactly three* green lights on the next four mornings?

Solution

We need to find a model to represent the probability of $\frac{1}{3}$ of the lights being green (that is, 20/60 is the theoretical probability of the lights being green). A coin will not do the job. We could, for example, use a die having six faces. Our five-step procedure for estimating this probability is then as follows.

1. Choice of a Model: Let one or two dots appearing on the die represent a green light (G). That is,

1 or 2 dots: G

Let three to six dots represent no green light (N). That is,

3, 4, 5, or 6 dots: N

2. Definition of a Trial: A trial consists of throwing four dice, one for each morning. (We could also use only one die and throw it four times, once for each morning.)

3. Definition of a Successful Trial: A successful trial occurs when either one or two dots appear on exactly three of the four dice.

4. Repetition of Trials: Do at least 100 trials.

For an example, we consider four trials. The dice rolls are from Table B.2, row 1. It is not good practice to always start with the first random number in row 1. Hence, say, start with the 11th digit.

66533 45332 24614 22231 26431 35541 12165 62116 16111

Our four sample trials yield the results presented in Table 5.4.

5. Probability Estimate: Based on only four trials, our estimate for the probability of getting three green lights on four mornings is

$$P(\text{getting three green lights on four mornings}) = \frac{\text{number of successful trials}}{\text{total number of trials}} = \frac{1}{4} = 0.25$$

Table 5.4 **Estimating the Probability of Three Green Lights on Four Mornings**

Trial			Success?
1	2461	GNNG	No
2	4222	NGGG	Yes
3	3126	NGGN	No
4	4313	NNGN	No

In Chapter 8 and Chapter 13 we will learn how to solve for theoretical probabilities involving the random number of successes for a fixed number of subtrials, as in Example 5.3. Such probabilities are called *binomial probabilities.* Examples 5.1 and 5.2 are also binomial probability problems. In Example 5.3 the theoretical probability is

$$p(\text{three green lights in four trials}) = \frac{8}{81} \approx 0.1$$

Our experimental value of 0.25 is rather far from this theoretical value. The discrepancy confirms our suspicion that four trials are likely to produce a poor estimate of the true theoretical probability.

SECTION 5.2 EXERCISES

Note: Most of the exercises suggest that you use 25 trials in the five-step method. Clearly, using more trials can improve the accuracy, as discussed below.

1. For Example 5.1, perform a total of 25 trials to provide a better estimate of the desired probability that all three children in the family are boys.
2. For Example 5.2, perform a total of 25 trials to provide a better estimate of the desired probability that at least two of the children are boys.
3. Consider this revised version of Example 5.3: The first traffic light on your route to class is green for 20 seconds. It has a 60-second cycle. What is the probability that you will get the green light on at least two of five mornings? Do a total of 25 trials to solve this problem.
4. Suppose you are to take a multiple-choice test made up of five questions. Each question has three alternatives to choose from. Since you do not know any of the answers, you decide to throw a die to get the answer. Perform a total of 25 trials, each trial being one try at the test.
   a. What is the probability that you will answer at least two questions correctly?
   b. If passing requires that you answer at least three correctly, what is the probability that you pass by throwing a die?
5. Estimate the probability of getting at least one ace in a five-card hand from a standard 52-card deck. Perform at least 25 trials.
6. A tire company sends a shipment of 26 tires to a dealer. Because of a problem in the manufacturing process, four of those tires have a serious flaw.
   a. Estimate the probability that the first customer to buy a set of four of these tires receives one of the flawed tires. Assume that the bad tires are randomly scattered throughout the 26. (*Hint:* Think about how a deck of cards could be used.) Perform 25 trials.
   b. Assuming that the first customer did not receive one of the flawed tires, what is the probability that the second customer, who also buys a full set of four, will not receive any of the bad tires? Perform another 25 trials.

Graphing Calculator Exercises

7. For Example 5.1, use the program FLIPS to do 50 trials to provide a better estimate of the probability that in a family of three children all will be boys.

The results may be viewed in the data table provided by FLIPS. Column L1 gives the possible number of boys in a three-child family, and column L2 shows the number of occurrences (that is, the frequency) for each possible outcome.

8. For Example 5.2, use FLIPS to do 50 trials and provide a better estimate of the desired probability that at least two of the three children are boys.

9. Use FLIPS to estimate the probability that in a family of four children,

a. At least two are girls

b. There are exactly three girls

Do 50 trials.

10. Repeat Calculator Exercise 9, but improve on your estimates by using FLIPS to do 100 trials.

5.3 BOX MODELS

So far, we have been able to solve a variety of probability problems by using a model based on a coin or a die. Often, though, it is useful and convenient to have other models to help solve problems. Just as in Chapter 4, where our focus was on the expected value, in problems in which our focus is on probability (the probability of some event), box models of tickets are also very helpful. Recall that a box model is simply a (possibly large) collection of numbered tickets that are each equally likely to be chosen. Often several tickets will have the same number. For example, a box containing tickets numbered $1, \ldots, 6$ is clearly equivalent to a fair die in its functioning as a step 1 model. We will continue to substitute random digits for the actual physical model of ticket sampling from a box; we will do just that in the next example. Being able to use random numbers to obtain trials from a step 1 box model is very useful in obtaining a large number of trials quickly.

Example 5.4

A camera manufacturer finds that 10% (0.10) of the springs used in the shutters it manufactures are defective. Of the next five springs tested by the manufacturer, what is the probability of finding *two or more* defective?

Solution

We need some way of representing a probability of 0.1 for a defective spring. A box with 10 tickets with digits $0, 1, \ldots, 9$ would do. Clearly, choosing one of the digits—0, say—to denote a defective spring will produce the desired physical model of an event with probability $\frac{1}{10}$.

Instead of sampling tickets from an actual box, with replacement between draws, we can use an ordinary 10-digit random number table (which can be thought of as simulating a 10-sided die). The probability that any individual spring is defective, we are told, is 10%, or 0.10, or $\frac{1}{10}$. To represent this probability, we can use one-digit random numbers.

To estimate the probability of finding two or more defective springs in five tested, we use the following five-step procedure.

1. Choice of a Model: Use one-digit random numbers, where

0:	defective spring
1–9:	nondefective spring

2. Definition of a Trial: A trial consists of reading off five consecutive digits from the table.

3. Definition of a Successful Trial: A success occurs whenever two or more of the five digits are 0 (representing defective springs).

4. Repetition of Trials: Do at least 100 trials.

Here we will do five sample trials. We will use Table B.3, rows 1 and 2.

1	32236	12683	41949	91807	57883	65394	35595	39198	75268
2	40336	50658	32089	78007	58644	73823	62854	31151	64726

On the first trial, the following digits are read from the table:

3 2 2 3 6

These correspond to the following: nondefective, nondefective, nondefective, nondefective, nondefective. This trial is not successful, because no defective spring was found. We now carry out the remaining four trials, with the results presented in Table 5.5.

5. Estimated Probability: Based on five trials, an estimate of P(at least two defective springs) is found to be

$$P(\text{at least two defective springs}) = \frac{\text{number of successful trials}}{\text{total number of trials}} = \frac{0}{5} = 0$$

Table 5.5 **Estimating the Probability of Two or More Defectives in Five Springs**

Trial	Digits	Success?
1	32236	No
2	12683	No
3	41949	No
4	91807	No
5	57883	No

Although we suspect (correctly) that the corresponding theoretical probability is small, it is of course not 0. Indeed, the true probability to three digits is 0.081.

Now we consider an example of disease transmission. Although it is a simplistic example, it will serve to introduce the use of statistics in the field of epidemiology. Probability models and statistical inference are vital to epidemiology.

Example 5.5

The chance of contracting strep throat when coming into contact with an infected person is estimated as 0.15. Suppose the four children of a family come into contact with an infected person. What is the chance of *at least one* of the four children getting the disease?

Solution

1. Choice of a Model: Use pairs of random digits, with the following rule:

01 through 15:	Child gets strep throat.
16 through 99, plus 00:	Child does not get strep throat.

(Think of these random digits as substituting for a box having 15 tickets with "strep" written on them and 85 tickets with "no strep" written on them.)

2. Definition of a Trial: Read four pairs of random digits, one for each child in the family.

3. Definition of a Successful Trial: A trial is a success if at least one of the four pairs of digits is in the range 01–15.

4. Repetition of Trials: Repeat for a total of 100 trials.

Six trials are shown in Table 5.6. We used the random digits below (from a table like Table B.3) and read across the table row by row.

Row									
1	69531	54637	06640	35956	26693	27891	06397	70132	29186
2	56905	10986	53970	31729	18700	91782	65398	63865	81835

5. Probability Estimate: From Table 5.6 we get

$$P(\text{At least one child gets strep}) = \frac{4}{6} = 0.67$$

The correct answer is approximately 0.48, so our estimate is rather high.

Table 5.6 **Estimating the Probability of At Least One in Four Children Contracting Strep Throat**

Trial	Digits	Success?
1	69, 53, (15), 46	Yes
2	37, (06), 64, (03)	Yes
3	59, 56, 26, 69	No
4	32, 78, 91, (06)	Yes
5	39, 77, (01), 32	Yes
6	29, 18, 65, 69	No

All of the above examples have concerned trials in which each subtrial had only two outcomes. This is often not the case, as in the next example, in which each subtrial has five possible outcomes.

Example 5.6

A secretary receives telephone calls at a rate described by the following table:

Calls in an hour	Proportion of the time
0	15%
1	35%
2	30%
3	15%
4	5%

Estimate the probability that he will receive *four or more* calls during a two-hour period. That is, consider two subtrials.

Solution

1. Choice of a Model: Using a random number table, assign the digits in the following way:

Calls in an hour	Proportion of the time	Random digits
0	15%	00–14
1	35%	15–49
2	30%	50–79
3	15%	80–94
4	5%	95–99

2. Definition of a Trial: Read two pairs of random digits, one for each of the two hours.

3. Definition of a Successful Trial: Using the table in step 1, convert the random two-digit numbers into a number of calls in each of the two hours. Add these together, and if it is at least 4, then the trial is a success. Otherwise, it is a failure.

4. Repetition of Trials: Repeat for a total of 100 trials. Six are shown below, generated by the random digits from Example 5.5.

Sample trial	Random digits	Calls in hour 1	Calls in hour 2	Total number of calls	Success?
1	69 53	2	2	4	Yes
2	15 46	1	1	2	No
3	37 06	1	0	1	No
4	64 03	2	0	2	No
5	59 56	2	2	4	Yes
6	26 69	1	2	3	No

5. Probability Estimate: There were 28 successes (not shown) in the 100 trials. Thus

$$P(\text{at least four calls in two hours}) = \frac{28}{100} = 0.28$$

Because we did 100 trials, we expect our experimental probability to be relatively close to the theoretical probability. The correct theoretical probability is $p(\text{at least four calls}) = 0.2468$. Thus we *are* fairly close to the true value!

SECTION 5.3 EXERCISES

1. In Example 5.4, estimate the probability that one or more springs in a series of five will be defective.
2. In Example 5.5, suppose strep throat is contracted with a probability of 0.20 (instead of 0.15). What is the probability that at least one of four children in a family becomes infected, assuming that each has come into contact with an infected person?
3. A manufacturer of 60-second film advertises that 95 out of 100 prints will develop. Suppose you buy a pack of 12 and find that two prints do not develop. If the manufacturer's claim is true, what is the probability that two or more prints do not develop in a pack of 12? Based on the company's advertising, do you have reason to complain?
4. A pharmaceutical company knows from previous testing that a certain antibiotic capsule falls below prescribed strength 6% of the time, making the dosage ineffective. What is the probability that a prescription of 20 such capsules will contain two or more with ineffective dosages?
5. The following table gives the proportion of each type of hit Babe Ruth achieved during

the 1928 season:

Type of hit	Percentage of total hits
Single	50
Double	15
Triple	4
Home run	31

If in a given game Ruth had three hits, estimate the probability that he had at least one home run.

6. Using the data from Exercise 5, estimate the probability that Ruth had no singles in a game in which he had exactly two hits.
7. Use the data of Exercise 5. Suppose Ruth had three hits in a particular game. Estimate the probability that he hit exactly one home run and one single.
8. Describe how you could use a random number table to model an event that has a probability of 1 in 1000 of occurring.

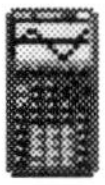

Graphing Calculator Exercises

For this section, use the program LOAD. This program is similar to FLIPS, with the important difference that it simulates the tossing of a coin that is loaded instead of one that is fair. The probability of getting heads with the loaded coin may be specified in 10ths, from 0.1 to 0.9 (so when the probability of getting heads is specified as 0.5, LOAD works the same as FLIPS).

9. A certain type of plastic bag has burst under 15 pounds per square inch of pressure 10% of the time. If a prospective buyer tests seven bags chosen at random, what is the probability that at least one will burst? In this problem we will use as a model a loaded coin with $p(\text{Heads}) = 0.1$ (Why?) Or we could use a box model, with nine 0s and a 1 (to give the stated probability of 0.10), which is sampled from with replacement seven times (once for each bag). Do 50 trials.

10. Suppose an infectious disease is contracted with a probability of 0.20. What is the probability of *at least one* of four children in a family getting the disease, assuming that each comes into contact with an infected person? Do 50 trials.

11. In Example 5.4 we estimated the probability of *two or more* defective springs appearing in five that are tested (10% of the springs are defective). Based on only five trials, the experimental probability was found to be zero. However, we stated that the theoretical probability is 0.081. Do 100 trials and compare your answer with the theoretical one.

12. Paul is in a German-speaking country and has decided to apply for a driver's license. Since he cannot read or write German, he is concerned about the written examination, which consists of 20 true/false questions. He has heard that he needs a score of at least 13 to pass the examination. He has decided to answer the questions by flipping a coin. What are his chances of passing the test? Do 100 trials.

5.4 INDEPENDENT EVENTS

You may wish to review Section 4.3, where independence was introduced. Let's return to our example of rain in San Juan and Chicago from Section 4.3. We assume that whether it rains in one city is unrelated to whether it rains in the other.

Example 5.7

The weather forecast for a certain day is a 60% chance of rain (0.60) for San Juan and a 50% chance of rain (0.50) for Chicago. What is the estimated probability of rain in *both* cities?

Solution

We'll use a table of random numbers to solve this problem.

1. Choice of a Model: A model of probability of 0.60 for San Juan:

1, 2, 3, 4, 5, 6: Rain in San Juan
7, 8, 9, 0: No rain in San Juan

A model of probability of 0.50 for Chicago:

1, 2, 3, 4, 5: Rain in Chicago
6, 7, 8, 9, 0: No rain in Chicago

2. Definition of a Trial: A trial consists of reading two digits (one for each city) from the table and recording them. Note that the first digit in a pair of digits has *no influence* on which digit appears second. In this manner our model captures the independence of rain in San Juan from rain in Chicago. That is, we *are* modeling independence.

3. Definition of a Successful Trial: In the terms of our problem, a successful trial occurs when it rains in both cities—that is, when we get one of the digits 1–6 for San Juan and one of the digits 1–5 for Chicago.

4. Repetition of Trials: Conduct at least 100 trials.

We will conduct five sample trials. Several rows from a random number table are presented in Table 5.7 (a table just like Table B.3) so that you can conduct the 100 trials if you wish. In order to conduct five trials, we will use row 1 as our source

Table 5.7 **Random Digits from 0 to 9**

Row										
1	47914	16071	36031	59991	06587	20041	72258	43406	89002	57786
2	87803	01790	94569	84915	35267	19243	18401	49493	51687	55104
3	73195	84474	62039	01874	30812	08586	81047	11382	49653	72071
4	70782	99421	57972	49804	45883	27459	51242	48024	01517	43432
5	95902	41240	99773	47194	23315	49300	12459	03162	06584	68944

Table 5.8 **Estimating the Probability of Rain in San Juan and Chicago**

Trial	Rain in San Juan (1–6)	Rain in Chicago (1–5)	Success?
1	4 = rain	7 = no rain	No
2	9 = no rain	1 = rain	No
3	4 = rain	1 = rain	Yes
4	6 = rain	0 = no rain	No
5	7 = no rain	1 = rain	No

of random digits. The first 10 digits are 47914 16071 and are paired as follows:

$$4,7 \quad 9,1 \quad 4,1 \quad 6,0 \quad 7,1$$

The results are presented in Table 5.8.

5. Probability Estimate

$$P(\text{rain in both San Juan and Chicago}) = \frac{\text{number of successful trials}}{\text{number of trials}}$$

Therefore, based on our very few trials—only five—we have estimated the probability as

$$P(\text{rain in both San Juan and Chicago}) = \frac{1}{5} = 0.20$$

We are now about to demonstrate an important property, or law, regarding independent events. First, in order to get a more reliable estimate of the probability of rain in both cities in our example, we conducted a total of 100 trials, with the results in Table 5.9. From the table, we find

$$P(\text{rain in both San Juan and Chicago}) = \frac{28}{100} = 0.28$$

Also note from Table 5.9 that

$$P(\text{rain in San Juan}) = \frac{58}{100} = 0.58$$

$$P(\text{rain in Chicago}) = \frac{47}{100} = 0.47$$

Observe the following:

$$P(\text{rain in San Juan}) \times P(\text{rain in Chicago}) = 0.58 \times 0.47 = 0.2726$$

Table 5.9 **Estimating the Probability of Rain in San Juan and Chicago**

Number of trials	*Number of times it rained* San Juan	Chicago	Number of successes*
100	58	47	26

*Rain in both cities occurring simultaneously

which is close to 0.26, the experimental probability of rain occurring in both cities on the same day.

These estimates are consistent with the following important law about experimental probabilities of independent events.

The law of experimental independence

If two events are independent, then the estimated (experimental) probability that they both occur will be approximately equal to the product of the two individual estimated probabilities (assuming a reasonably large number of trials has been conducted).

Suppose the two events are denoted A and B. Then this law states that the probability of both A and B occurring together satisfies

$$P(A \text{ and } B) \approx P(A) \times P(B)$$

For Example 5.7, let

$$A = \text{rain in San Juan}$$
$$B = \text{rain in Chicago}$$

Again, "A and B" simply means that the result of the trial is that both A and B occur. Then according to the law of experimental independence,

$$\begin{aligned} P(A \text{ and } B) &= P(\text{rain in both San Juan and Chicago}) \\ &\approx P(A) \times P(B) \\ &\approx P(\text{rain in San Juan}) \times P(\text{rain in Chicago}) \end{aligned}$$

Recall that from our experiment we found

$$P(\text{rain in both San Juan and Chicago}) = \frac{26}{100} = 0.26$$
$$P(\text{rain in San Juan}) = 0.58$$
$$P(\text{rain in Chicago}) = 0.47$$

So we have confirmed the law:

$$0.58 \times 0.47 = 0.2726 \approx 0.26$$

We can use $P(A) \times P(B)$ as an experimental estimate of the theoretical probability $p(A \text{ and } B)$—a useful thing to do when we do not have data on how often A and B have occurred simultaneously. You will be asked in the exercises to conduct your own experiment to obtain more trials and show that this law provides a good estimate for the probability of two independent events occurring simultaneously.

Later in this chapter you will learn how the theoretical probability of two independent events taking place can be calculated exactly, rather than just estimated, if the theoretical probability of each event is known. (You can probably guess the rule.)

SECTION 5.4 EXERCISES

Use the random digits provided in the tables in Appendix B. Do 50 trials unless told otherwise.

1. For the rain experiment of Example 5.7, conduct another 10 trials beyond the five trials of Table 5.7. Give all the details required. To illustrate, we have done one trial, trial 6, using the next two digits, 3 and 6, in Table 5.7.

Trial	San Juan	Chicago	Success?
6	3—rain	6—no rain	No

You should do trials 7 through 15.

2. Suppose 80 trials of the rain experiment are conducted and there are 26 successes. Further, out of the 80 trials, you get rain in Chicago on 42 trials and rain in San Juan on 50 trials.
   a. What are P(rain in both cities), P(rain in Chicago), and P(rain in San Juan), based on these 80 trials?
   b. What is P(rain in Chicago) $\times$ P(rain in San Juan)? Compare this with P(rain in both cities). Does the law of experimental independence seem to hold?
3. There are two traffic lights on your route to class. Each has a 60-second cycle. Out of those 60 seconds, each light is green for 20 seconds.
   a. Assuming that the lights work independently of each other, conduct an experiment to estimate the probability of getting a green on both lights by finding P(Both lights green).
   b. Look back at your results from part (a), and use the available information about P(first light green) and P(second light green). How well does the law of experimental independence hold?
4. Both engines of a two-engine rocket must fire before it will lift off. The theoretical probability that each engine will fire is 0.75. The engines fire independently.
   a. Conduct an experiment to estimate the probability of a liftoff.
   b. Verify the law of experimental independence.
5. Assume that a two-engine airplane can make a safe flight if one of its engines stops in operation. We want to determine the probability of not having a safe flight (that is, of both engines failing) assuming that the engines operate independently of each other. Each engine has a 0.10 chance of failing in flight.
   a. Since the engines are assumed to be independent of each other, which formula tells us what P(Both engines fail) approximately equals?
   b. Using the formula for P(Both engines fail) you found in part (a), estimate the chance that a flight will not be safe.

6. Suppose automobile accidents are equally likely to occur on each day of the week. What is the probability that if two accidents occur, they will both be on the weekend (that is, on Saturday or Sunday)? Assume that the two accidents are independent of each other.

Graphing Calculator Exercises

For this section, use the program RANDLIST. This program generates lists of random numbers (like those in Table B.3). You specify the number of lists (between two and four) and the length of the lists, up to 99 (usually corresponding to the number of trials desired).

The lists are then used as the basis for defining the model to solve the problem, as well as providing the random data for the associated experiment.

7. See Example 5.7. The weather forecast for a certain day is a 60% (0.60) chance of rain in San Juan and a 50% (0.50) chance of rain in Chicago. What is the estimated probability of rain in both cities? Do 50 trials.

The box model consists of 100 tickets marked as follows:

For San Juan:

Tickets corresponding to random digits 1 to 60 are marked 1 (for rain).

Tickets corresponding to random digits 61 to 100 are marked 0 (for no rain).

For Chicago:

Tickets corresponding to random digits 1 to 50 are marked 1 (for rain).

Tickets corresponding to random digits 51 to 100 are marked 0 (for no rain).

Set the number of lists equal to 2 (one for San Juan and one for Chicago). Set the number of trials equal to 50. We show only 7 trials below. Here L_1 is the simulation of rain in San Juan (1–60) on successive trials, while L_2 is the simulation of rain in Chicago (1–50) on successive trials. L_3 denotes the event of rain in both cities, and is left for you to determine ("yes" or "no").

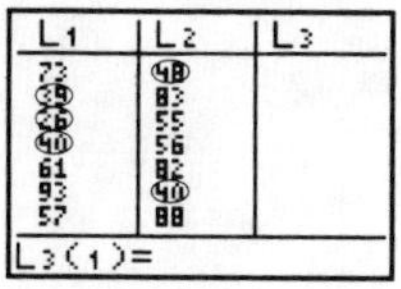

Rain in each of San Juan and Chicago is indicated by a circle. In none of these seven trials did it rain in both cities. So based on these seven trials, $P(\text{Rain in San Juan and Chicago}) = 0$.

Do an additional 43 trials for a better estimate.

8. Both engines of a two-engine rocket must fire before it will lift off. The probability that each engine will fire is 0.85. The engines fire independently. Find the experimental probability of the rocket lifting off. Do 50 trials.

9. The law of experimental independence is stated on page 243. Use RANDLIST to test this law for two events, A and B, whose theoretical probabilities you can choose. Then extend the law to three events, A, B, and C. Do at least 100 trials for each test (you will need to generate new lists, since the TI-82 allows only 99 rows (trials) in each list).

5.5 PROBABILITY AS A THEORETICAL CONCEPT

So far, we have solved probability problems by estimating the required probability after conducting some sort of experiment and collecting data. But probability may be approached from another point of view: in many cases we can calculate the theoretical probability in advance by determining the number of ways a successful outcome can occur and dividing by the total number of possible outcomes in the trial. We have already used this viewpoint in simple cases—for example, when we decided which model to use in a given experiment.

We first need to discuss the idea of *equally likely outcomes*. When the outcomes are equally likely, the method for computing the theoretical probability just proposed in the preceding paragraph is justified.

Equally Likely Outcomes

Many real-world trials or experiments have a finite set of outcomes possible on each trial. In many situations involving probability, all the outcomes are equally likely. That is, if the situation were to be repeated a large number of times, each of the outcomes would occur approximately the same proportion of times. A coin toss, for example, has two equally likely outcomes: heads and tails (see Figure 5.1*a*). A roll of an ordinary die has six equally likely outcomes: 1, 2, 3, 4, 5, and 6 (see Figure 5.1*b*).

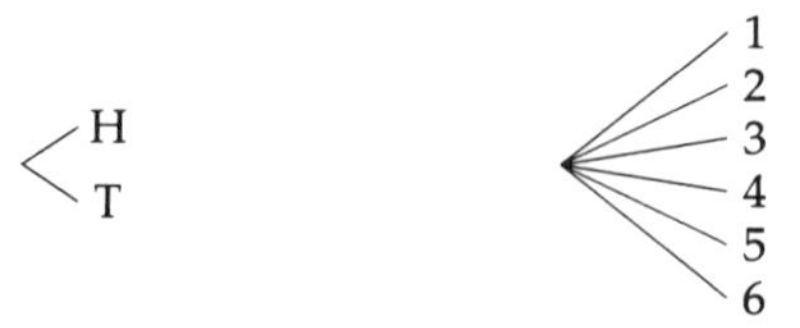

(*a*) Two outcomes of tossing a coin

(*b*) Six outcomes of rolling a die

Figure 5.1 **Possible outcomes of tossing a coin and rolling a die.**

Suppose we seek the theoretical probability of an event, such as the event that a rolled die produces an even number of spots. Then of the possible outcomes (1, 2, 3, 4, 5, 6) we see that there are three ways (2, 4, 6) of obtaining a "successful" outcome. By *successful* we simply mean an outcome that causes the event of interest—namely, the die being even—to occur. For such probability problems in which we are willing to assume equally likely outcomes, the following rule is extremely useful.

Theoretical probability of an event when outcomes are equally likely

Suppose we have a set of N equally likely outcomes, of which S of them are successes in the sense that each "successful" outcome causes the event to occur. Then the theoretical probability of the event is

$$p(\text{event}) = \frac{\text{number of ways of obtaining a successful outcome}}{\text{total number of outcomes}} = \frac{S}{N}$$

For the toss of a coin, we can say

$$p(\text{heads}) = \frac{1}{2} = 0.5$$

For the probability of getting a 1 or a 2 on the roll of a die, we can say

$$p(1 \text{ or } 2) = \frac{2}{6} = \frac{1}{3}$$

because 2 of the possible outcomes (1, 2) cause the event to happen and there are 6 possible outcomes. We did not do an experiment to find this probability. Instead, we used some basic ideas about equally likely outcomes and successful outcomes.

Observe that we use a lowercase p, as in

$$p(\text{heads}) = \frac{1}{2} = 0.5$$

to indicate a theoretical probability. Earlier in this chapter we referred to the coin-tossing experiment of Kerrich and found, based on 10,000 simulated tosses of the coin, that

$$P(\text{heads}) = 0.507$$

The capital P shows that we are talking about an *experimental* probability based on data obtained from a certain number of trials. So P(heads) is an estimate of p(heads). That is, the experimental probability of getting heads is an estimate of the theoretical probability of getting heads on the toss of a fair coin.

For a roll of a die, the probability of getting a 3 is

$$p(3) = \frac{1}{6}$$

As a slightly more complicated example, suppose we want to find the probability of getting an even number when rolling a die, as was briefly mentioned above. According to Figure 5.1*b*, three of the outcomes give even numbers. So

$$p(\text{even number when rolling a die}) = \frac{3}{6} = \frac{1}{2} = 0.5$$

Now let's see how to find the theoretical probability of getting two heads in the toss of two coins. Figure 5.2, called a *tree diagram*, shows the possible outcomes. The tree diagram helps us keep track of the possible outcomes when there are multiple subtrials, as in the tossing of two coins. Along the top branch of the tree, we find the successful outcome HH. Along the bottom branch we have one of the other outcomes, TT, and so on.

From this diagram we count a total of four equally likely outcomes; only one of these is two heads. So the theoretical probability is

$$p(\text{two heads}) = \frac{1}{4} = 0.25$$

If we want the theoretical probability of getting one or more heads when tossing two coins, we can count three outcomes from the tree diagram (Figure 5.3). We find that three of the outcomes give one or more heads. So

$$p(\text{one or more heads}) = \frac{3}{4} = 0.75$$

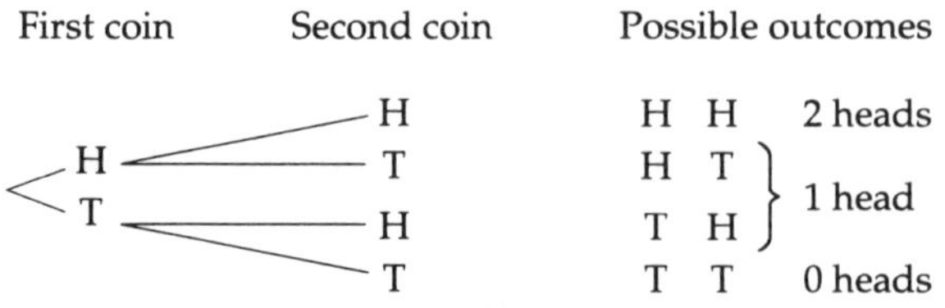

Figure 5.2 **Possible outcomes of tossing two coins, showing outcome of two heads.**

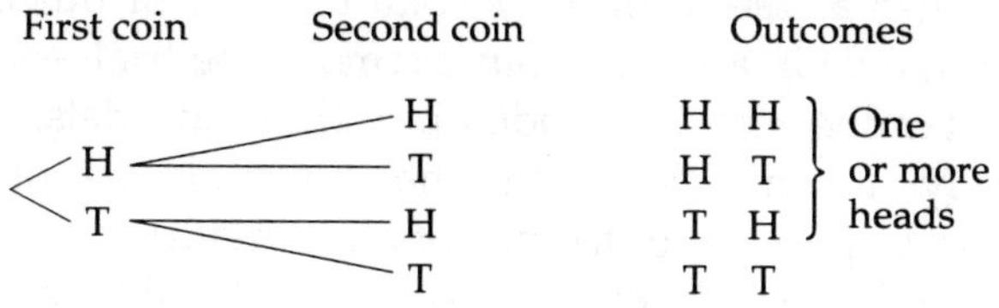

Figure 5.3 **Possible outcomes yielding one or more heads.**

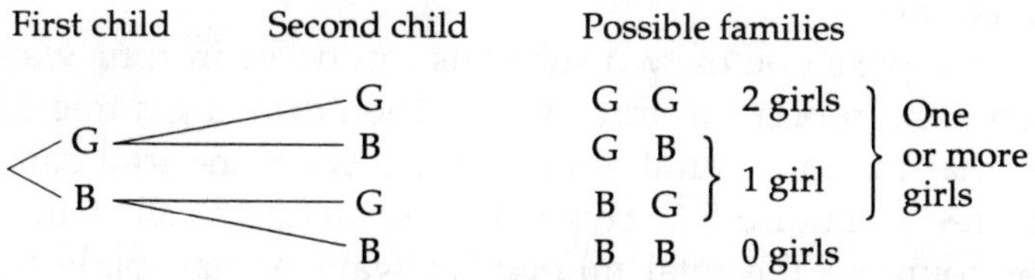

Figure 5.4 **Possible outcomes yielding one or more girls.**

If we want to find the theoretical probability of one or more girls in a two-child family, for example, we can draw a family tree that is just like the diagram for the two coins (see Figure 5.4). Again, counting the number of families with one or more girls, we find three. So

$$p(\text{one or more girls}) = \frac{3}{4} = 0.75$$

This process can be extended. For example, for a three-child family, as we considered in Section 5.2, the number of possible family combinations is found as shown in Figure 5.5. You can see that a tree diagram is not required to carry out the counting process, but it is a very helpful tool.

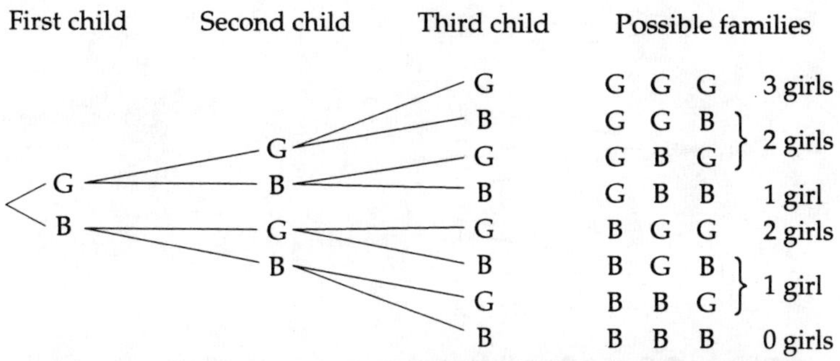

Figure 5.5 **Possible family combinations for three children.**

We can use a rule to find the total number of outcomes (families, in this case) possible in a trial. In our example, the trial—an observation of a three-child family—can be divided into three subtrials, one for each child in the family. To find the total number of trials, we find the product of the numbers of possible outcomes of all individual subtrials. Since in our example each child can be either a boy or a girl and there are three children, we have

$$2 \times 2 \times 2 = 8$$

possible family combinations. Drawing the tree diagram also makes this fact clear.

Suppose one of two subtrials can occur in four ways and the second subtrial can occur in three ways. Then drawing a tree diagram would tell us that the total number of possible ways the trial can occur is 4×3, as we see in Figure 5.6. This rule is sometimes called the *multiplication rule* for counting the total number of ways a trial made up of subtrials can occur.

Returning to the example of the three-child family, we find that the case of three girls happens in one of the eight possible family combinations. So

$$p(\text{three girls}) = \frac{1}{8} = 0.125$$

Some problems involving equally likely outcomes require complex reasoning involving permutations, combinations, and other counting formulas. For example, what is the theoretical probability of being dealt five hearts in a

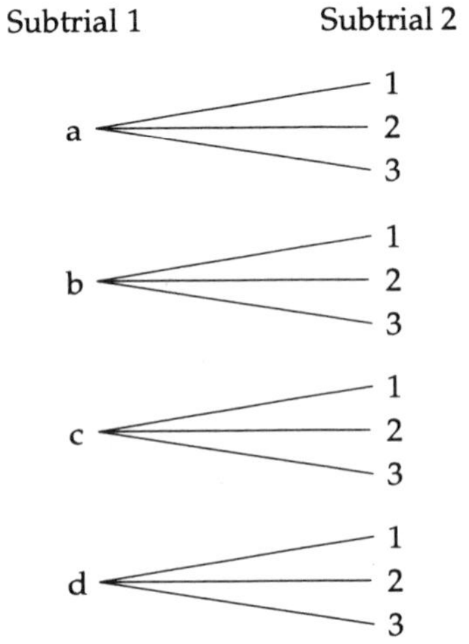

Figure 5.6 **Possible outcomes for subtrials 1 and 2.**

five-card hand from an ordinary deck of cards (a nice hand if one is playing poker)? We would need to study permutations and combinations before being able to answer a problem like this. One important theoretical probability model involving equally likely outcomes—the binomial probability model—will be discussed in Chapter 8.

Theoretical Independence

We have already learned the law of experimental independence, namely that

$$P(A \text{ and } B) \approx P(A) \times P(B)$$

It should not surprise us that this is an approximate empirical stand-in for an analogous law for theoretical probabilities for which exact equality holds.

The law of theoretical independence

If two events are independent, then the theoretical probability that they both occur is exactly equal to the product of the two individual probabilities.

If we denote the two independent events as A and B, then according to this law,

$$p(A \text{ and } B) = p(A) \times p(B)$$

Suppose, for example, we toss a coin and roll a six-sided die. What is the theoretical probability of heads on the coin and a 6 on the die?

Since

$$p(\text{heads}) = \frac{1}{2}$$

and

$$p(6 \text{ on die}) = \frac{1}{6}$$

and since the tossing of a coin and the rolling of a die are independent, we have

$$p(\text{heads and } 6) = \frac{1}{2} \times \frac{1}{6} = \frac{1}{12} \approx 0.08$$

Example 5.8

Suppose we choose a pair of random digits independently, each taken from the digits 0 to 9. What is the theoretical probability of getting the pair 2, 3 (that is, first 2 and then 3)? (Note that this is what sampling two digits from Table B.3 amounts to.)

Solution

We have

$$p(2) = \frac{1}{10}$$

and

$$p(3) = \frac{1}{10}$$

So

$$\begin{aligned} p(2 \text{ and then } 3) &= p(2) \times p(3) \\ &= \frac{1}{10} \times \frac{1}{10} = \frac{1}{100} = 0.01 \end{aligned}$$

This property of independence also applies to more than two events for both estimated and theoretical probabilities.

Example 5.9

Three Independent Events Involving Estimated Probabilities

Suppose we collect weather data on three cities—San Juan, Chicago, and Calcutta—and find these estimated probabilities of rain on a certain day:

$$\begin{aligned} P(\text{rain in San Juan}) &= 0.58 \\ P(\text{rain in Chicago}) &= 0.47 \\ P(\text{rain in Calcutta}) &= 0.34 \end{aligned}$$

We assume that rainfall in each city is independent of rainfall in the other cities. So we estimate the probability of rain simultaneously in all three cities to be

$$\begin{aligned} P(\text{rain in all three cities}) &= P(\text{rain in San Juan}) \\ &\quad \times P(\text{rain in Chicago}) \\ &\quad \times P(\text{rain in Calcutta}) \\ &= 0.58 \times 0.47 \times 0.34 \approx 0.093 \end{aligned}$$

We know that $p(\text{rain in all three cities}) \approx P(\text{rain in all three cities}) \approx 0.093$. Thus we have an experimental estimate of the desired probability.

Let's now consider in a direct manner the theoretical probability of three independent events all occurring together.

Example 5.10 **Three Independent Events Involving Theoretical Probabilities**

What is the theoretical probability of three heads occuring in three tosses of a fair coin?

Solution

For each toss,

$$p(\text{heads}) = \frac{1}{2}$$

Therefore,

$$\begin{aligned} p(\text{three heads}) &= p(\text{heads on first toss}) \\ &\quad \times p(\text{heads on second toss}) \\ &\quad \times p(\text{heads on third toss}) \\ &= \frac{1}{2} \times \frac{1}{2} \times \frac{1}{2} = \frac{1}{8} = 0.125 \end{aligned}$$

Compare this answer with the one obtained earlier in this section for p(three girls) using a tree diagram (see Figure 5.5) and treating the eight outcomes possible in the trial as equally likely. We can thus reach the same correct answer of 0.125 in two different ways, one using the idea of eight equally likely outcomes and the other using the idea of three independent subtrials.

What is the probability of three heads occuring if the coin is loaded with $p(\text{heads}) = 0.6$ on each trial? Here the tree diagram approach fails, and you *must* use the multiplication rule of theoretical independence.

SECTION 5.5 EXERCISES

Unless otherwise noted, the probabilities required in these exercises are theoretical, not experimental.

1. Identify three probability experiments having multiple stages (such as tossing two dice) and equally likely outcomes. Draw a tree diagram for each to justify your choices.
2. A four-sided die is called a *tetrahedral* die. Assume that the possible outcomes of tossing a four-sided die are 1, 2, 3, 4. Under the assumption of equally likely outcomes, find the following:
   a. $p(1)$
   b. $p(2)$
   c. $p(3)$
   d. $p(4)$
   e. p(even number)
   f. p(number less than 4)
3. A ten-spinner is shown below. The arrow is free to spin until it stops in one of the 10 sectors.

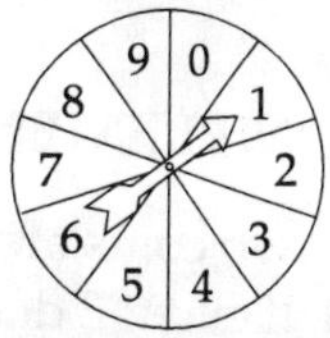

What is the theoretical probability of obtaining each of the digits 0–9? What is the theoretical probability of obtaining an odd digit?

4. Suppose you have 19 classmates and the instructor is about to choose one member of your class to present a problem on the board. What is the theoretical probability that you will be chosen, assuming that each class member has an equal chance of being selected?
5. A regular deck of playing cards has 13 hearts (red), 13 diamonds (red), 13 clubs (black), and 13 spades (black). Suppose you deal the top card from a well-shuffled deck, which is face down. Find the theoretical probabilities for the following:
   a. p(black card)
   b. p(heart)
   c. p(ace)
   d. p(king of diamonds)
6. Suppose a pair of dice are rolled. Find
   a. p(Sum of dots is less than 5)
   b. p(Sum of dots is 7)
   c. p(Sum of dots is greater than 11)

   Hint: Use the following table to see all the possible outcomes of throwing a pair of dice, and the corresponding sum in each case. This works better than the tree diagram in this case.

	Die 2					
Die 1	1	2	3	4	5	6
1	2	3	4	5	6	7
2	3	4	5	6	7	8
3	4	5	6	7	8	9
4	5	6	7	8	9	10
5	6	7	8	9	10	11
6	7	8	9	10	11	12

7. Repeat Exercise 6, but this time find the *experimental* probability of each of the following events, based on the throwing of a pair of dice 50 times. That is, find the following:
   a. P(Sum of dots is less than 5)
   b. P(Sum of dots is 7)
   c. P(Sum of dots is greater than 11)
8. Once again, throw a pair of dice. Using the law of theoretical independence, find the following probabilities. In each case, show what two probabilities you are breaking the probability of the occurrence of both events into.
   a. p(Both dice come up sixes)
   b. p(First die comes up a one, and the second one comes up a two)
   c. p(First die comes up an odd number, and the second one comes up an even number)
   d. p(Each die comes up a value greater than or equal to 5)
9. Assume that when a certain basketball player takes a shot, she is equally likely to make or miss the shot. Also assume each shot she takes is independent of any other shot.
   a. Draw the tree diagram for all the possible outcomes of the first three shots.
   b. What is the probability that she makes the first three shots she takes?
   c. Show how the law of theoretical independence can be used to find the probability in part (b).
   d. Using the tree diagram from part (a), find the probability that she makes at least one of her first three shots.

5.6 COMPLEMENTARY EVENTS

Suppose two coins are tossed, with E denoting the event that two heads are obtained. If event E does *not* take place, then two heads are *not* obtained. (For example, we may get one head and one tail.) These two events are called *complements* of each other. (Remember that when something complements something else, it makes it complete. An event and its complement form the complete set of all possible outcomes of the trial.)

Here is another example. Suppose F is the event that a person makes a reservation for dinner at a restaurant. The complement of F is the event that a person does not make a reservation for dinner.

We can refer to complementary events in this way: if E is some event, then *not* E is the complementary event.

Let E be the event that at least one child is a girl in the three-child family. Then not E, the complement, is the event that no children are girls. Note carefully that the complement is not the event that at least one child is a boy.

Example 5.11

Three coins are tossed 50 times, and the number of times each number of heads is obtained is recorded in Table 5.10. From this table we see that three heads were obtained 9 times out of 50. If E is the event that three heads are obtained, then not E happened

$$50 - 9 = 41 \text{ times}$$

In 41 of the 50 tosses of the three coins, three heads were not obtained 41 times.

In terms of experimental probability, we can say that in the preceding example,

$$P(3 \text{ heads}) = \frac{9}{50} = 0.18$$

and

$$P(\text{not } 3 \text{ heads}) = \frac{41}{50} = 0.82$$

Table 5.10 **Number of Heads Obtained in 50 Tosses of Three Coins**

Number of heads	f
0	7
1	16
2	18
3	9
Total	50

Notice, though, that we can also say that

$$P(\text{not 3 heads}) = 1 - P(\text{3 heads}) = 1 - 0.18 = 0.82$$

This property of complementary events always holds.

Property of complementary events for experimental probability

For any event E, the probability that E will not happen is 1 minus the probability that E will happen. That is,

$$P(\text{not } E) = 1 - P(E)$$

(The claim is that exact equality holds, in contrast with the law of experimental independence.)

In terms of theoretical probability, we have the same rule. Look at the tree diagram for three coins in Figure 5.7. The event E, three heads, occurs in one out of the eight possible outcomes. Therefore not E occurs in seven out of the eight outcomes. So we have

$$p(\text{3 heads}) = \frac{1}{8}$$

and

$$p(\text{not 3 heads}) = 1 - \frac{1}{8} = \frac{7}{8}$$

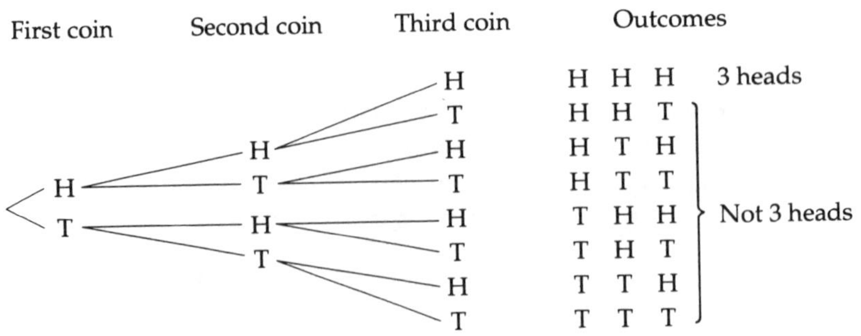

Figure 5.7 **Possible outcomes for "three heads" and "not three heads."**

Property of complementary events for theoretical probability

For any event E, the theoretical probability that E will not happen is

$$p(\text{not } E) = 1 - p(E)$$

This property of complementary events is useful when it is easier to figure out the value of $p(\text{not } E)$ than $p(E)$. For then $p(E) = 1 - p(\text{not } E)$ becomes an easy calculation.

SECTION 5.6 EXERCISES

1. The following table gives the results of tossing two coins 100 times. Let E be the event of no heads occurring. What, in words, is *not* E? From this table, find $P(E)$ and $P(\text{not } E)$.

Number of heads	f
0	20
1	52
2	28

2. Let E be the event of no heads occurring when two fair coins are tossed. What are the theoretical probabilities $p(E)$ and $p(\text{not } E)$?
3. A deck of playing cards is shuffled. Let event A be getting an ace.
   a. Write in words the event not A.
   b. Describe an experiment to find $P(\text{not } A)$.
4. A survey of 60 emergency telephones shows that six do not work. Let W be the event that the phone you come to does not work. What are $p(W)$ and $p(\text{not } W)$? Write out not W in words.
5. Let G be the event that there is at least one girl in a family of four children.
   a. Write out in words the event not G.
   b. Estimate the value of $P(\text{not } G)$ using at least 50 trials.
   c. What is the theoretical probability of $p(\text{not } G)$?
   d. Use your answer from part (c) to find $p(G)$.
6. A student says that her probability of passing Math 122 is 0.8 and that of passing English 100 is 0.6. Assume that these two events are independent. Let A be the event that she passes Math 122, and B be the event that she passes English 100. (Is this likely a good assumption in the sense of being realistic?)
   a. Write out in words the event that both not A and not B occur.
   b. What are $p(\text{not } A)$ and $p(\text{not } B)$?
   c. What is the theoretical probability that she fails both courses?
   d. What is the theoretical probability that she passes both courses?
   e. Why do the probabilities in (c) and (d) not add to 1?
7. A manufacturer of transistors submits a bid on each of four government contracts. A firm will receive the contract if it submits the lowest bid. In the past, this manufacturer's bid has been the low bid 15% of the time. Assume that this continues to be the case, and assume independence from one bid to another.
   a. Conduct an experiment of 50 trials to find the experimental probability that the manufacturer will not receive any of the four contracts.

b. What is the theoretical probability that the manufacturer does not receive any of the contracts?
c. What is the theoretical probability that the manufacturer will receive at least one of the contracts?

8. Each person in a group of three randomly chooses one other in the group as a friend. A person who ends up with no friends by this random choice is called an *isolate*. Use random digits to estimate the following:
   a. The mean number of isolates produced by this process of choosing friends
   b. The probability that no one out of the three is an isolate

   Do 50 trials. A table such as the one that follows, which shows one trial, will help you keep track of the friend-choosing process. An X marks the person's choice.

Person choosing	*Person chosen as friend* 1	2	3
1		X	
2	X		
3	X		

9. During a certain week, there is a 25% chance of rain each day. Assume that whether it rains on one day is independent of whether it rains on any other day.
   a. What is the theoretical probability that it does not rain on any day during the week?
   b. Write out the complement to the event, "It does not rain on any day during the week."
   c. Using parts (a) and (b), find the theoretical probability that it rains at least one day during the week.

5.7 SAMPLING WITH OR WITHOUT REPLACEMENT

We will introduce the issue with an example. Suppose we have four colors of flags. We decide to wave three flags in sequence to signal from a ship to the shore (not a likely approach in this electronic age). Suppose a young girl finds the flags and sends a message at random. She begins by sampling one flag—that is, she chooses the first color she will transmit. Now she has a choice. She can sample from the three remaining flags, or she can replace the first flag and transmit the second color by randomly choosing from the same four flags as the first signal. This choice is the distinction between sampling randomly *without* or *with replacement*. It is a very important distinction in statistics, where we very often sample randomly from a population in order to use properties of the sample to estimate properties of the entire population.

Let's count the number of signals possible if the child is sampling without replacement. Denote the four colors as R (red), O (orange), Y (yellow), and B (blue). We'll use a tree diagram to count all possible outcomes. See Figure 5.8. Notice that 24 signals are possible ($4 \times 3 \times 2 = 24$).

What is the probability of the signal ORY (O followed by R followed by Y)? Since this signal can occur only in one way, we have

$$p(\text{ORY}) = \frac{1}{24}$$

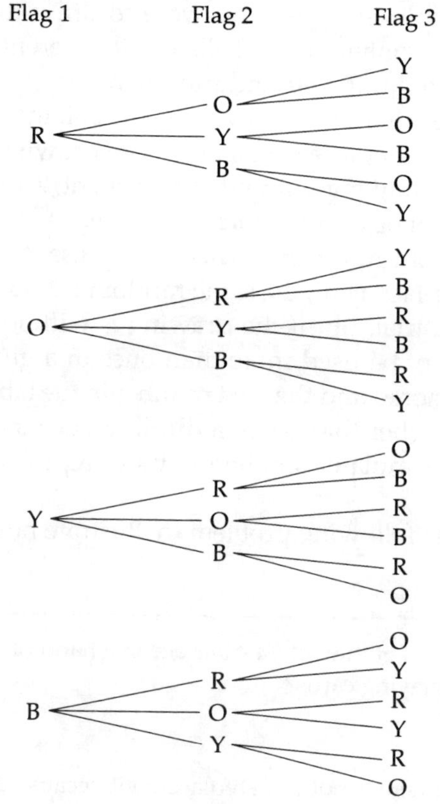

Figure 5.8 **Possible three-flag signals without replacement.**

because all possible sequences (keeping in mind that the sampling is without replacement) are equally likely.

What is the probability of a signal containing no yellow flag? Applying the principle of equally likely outcomes, we see by counting the outcomes in Figure 5.8 having no Y (such as OBR) that

$$p(\text{no Y}) = \frac{6}{24}$$

If instead the child creates signals randomly *with replacement*, the above probabilities will change. Even the total number of possible outcomes will be different: there will be $4 \times 4 \times 4 = 64$ possible signals rather than $4 \times 3 \times 2 = 24$. You can solve the problem of $p(\text{no Y})$ by drawing the tree diagram, analogous to that of Figure 5.8, and counting outcomes having no Y.

As another example, suppose you were to draw three balls with replacement from a jar containing 10 balls of 10 different colors. You must return each drawn ball to the jar and mix thoroughly before drawing the next one. It would be possible to draw the same ball more than once in one trial. If you were *not* to replace each ball before drawing the next one, the drawing would be a drawing without replacement, and in any given trial the same ball could not be drawn more than once.

If a table of random numbers is used to represent the above drawing of three balls with replacement, the same random number may occur more than once in a given trial. But if the drawing is without replacement, the same number may not be used more than once in a given trial. That is, a duplicate must be ignored and the next number in the table must be chosen, and so on until a number that is not a duplicate occurs. Thus we *can* use random numbers to simulate sampling without replacement provided we ignore duplicates.

Now consider the following problem to illustrate how this works.

Example 5.12 What is the probability of getting *at least one* ace in a hand of five cards dealt from a deck of 52 ordinary playing cards?

Solution

The drawing involved here is not with replacement because the same card cannot occur twice in a hand of five cards—it is a drawing without replacement. Let's find an experimental probability as an estimation of the theoretical probability. One way of estimating the probability of getting at least one ace in a hand of five cards would be to actually deal out hands of five cards to find out the number of times one or more aces occur. Another approach would be to use a table of random numbers and follow the five-step procedure.

1. Choice of a Model: Use two-digit random numbers from 01 to 52, inclusive. Ignore all others.

01–04:	ace
05–52:	remaining cards in deck

If the first six digits are 09 75 48, we treat them as 09 48 because 75 is greater than 52 and hence is ignored. Thus we now have 52 equally likely outcomes by this trick of ignoring 53–99. This is a powerful tool for obtaining equally likely probabilities when the number of outcomes is not 10, 100, or 1000, say.

2. Definition of a Trial: A trial consists of reading off five random numbers between 01 and 52, *ignoring duplicates.* That is, if the first six digits in the table are 03 03 27, then we treat this as 03 27 because the second 03 is a duplicate.

Table 5.11 Estimating the Probability of At Least One Ace in Five Cards

Trial	Random numbers	Success?
1	49 29 25 (02) 52	Yes
2	42 45 40 49 07	No
3	37 20 30 38 21	No
4	48 32 07 30 22	No
5	43 49 (04) 26 09	Yes
6	09 38 44 22 36	No
7	39 16 51 19 06	No
8	10 09 49 50 24	No
9	(01) 21 (03) 26 (02)	Yes
10	10 (02) 16 47 13	Yes

3. Definition of a Successful Trial: A successful trial occurs when at least one of the five two-digit random numbers is between 01 and 04, inclusive (that is, when at least one of the numbers obtained represents an ace).

4. Repetition of Trials: Do at least 100 trials.

Suppose 10 trials produced the results listed in Table 5.11. Here we have removed all pairs larger than 52. In 4 of the 10 trials, at least one of the random numbers is less than or equal to 4 (trials 1, 5, 9, and 10). Therefore, in four of the trials we have drawn at least one ace.

5. Probability Estimate:

$$P(\text{at least one ace in hand of five cards}) = \frac{\text{number of successful trials}}{\text{total number of trials}}$$

$$= \frac{4}{10} = 0.4$$

In Chapter 13 we will learn how to solve problems like this theoretically using the *hypergeometric distribution,* which deals with probabilities involving two types of outcomes (like ace or not ace) when the sampling is without replacement.

Step 2 of the above procedure is the one that involved drawing samples without replacement. We had to sometimes search through more than five two-digit random numbers less than or equal to 52 before we got five that were different. When drawing *with replacement,* however, we always take the two-digit random numbers less than or equal to 52 as they come from the table (that is, we do not ignore duplicate numbers).

Why Consider Sampling without Replacement?

One of the most important applications of statistics is the random sampling of real populations, often people, in order to decide, based on the sample, what the population is like. Since it is easier to find theoretical probabilities for the independent subtrials that result from doing such sampling with replacement, one prefers the theory that results when the sampling is with replacement. In real sampling situations, however, it would be silly to allow an individual to be sampled twice. Therefore, as a practical matter, all sampling of populations in real statistical problems is done without replacement.

It can be shown that the two theoretical probabilities of an event computed for sampling with replacement and for sampling without replacement, such as the two differing values for p(At least 60 of 100 sampled people favor legalizing abortion), are in fact almost equal to each other if the size of the population is large relative to the sample size. This is true because if the population is large, the chance of choosing an individual who has already been sampled is very small even if the sampling is with replacement. Thus, excluding this possibility of resampling a person, which is exactly what sampling without replacement does, makes almost no change in a computed sampling probability of interest that is more easily computed assuming sampling with replacement.

Suppose we decide to address a small sample, large population sampling situation with the five-step method because we need to estimate a probability (or an expected value of interest to us). Then the two methods of defining a trial (sampling with replacement and sampling without replacement between subtrials) are, roughly, equally easy to carry out. Therefore we can decide to sample with or without replacement, with the knowledge that either approach is acceptable because the associated theoretical probabilities of a successful trial are so close to each other. By contrast, if we were instead doing a theoretical analysis of the sampling situation to solve for a probability (or an expected value), we would likely assume sampling with replacement because the mathematics needed to do our probability computations is so much easier and the resulting answer is so close to the (often difficult to compute) theoretical answer assuming sampling without replacement.

SECTION 5.7 EXERCISES

Many of these probability problems are difficult to solve theoretically, especially when the sampling is without replacement. You will likely often need to resort to the five-step method.

1. A bag contains five black marbles, four red marbles, and three white marbles. Three marbles are drawn in succession.

a. If each marble is replaced before the next one is drawn, what is the probability that at least one of the three is white? *Hint:* What is the complement of at least one being white?
b. If each marble is not replaced before the next one is drawn, what is the probability that at least one of the three is white?
c. If each marble is not replaced before the next one is drawn, what is the probability that all three are of the same color?

2. A student has to match three terms that she has never seen before with their definitions. If she guesses, what is the probability of her
   a. Getting all three correct?
   b. Getting none correct?
   c. Getting exactly one correct?
   d. Getting exactly two correct?

   Hint: Label the terms 1, 2, 3 for convenience and look at all the different possible orders of 1, 2, 3 that the student can respond with. For example, the response 3, 2, 1 produces one match.

3. A man has 10 keys and exactly one of them fits the lock on his office door. He tries the keys one at a time, never trying any key more than once.
   a. Is this sampling with or without replacement? Why?
   b. What is the probability that he will obtain the correct key within his first three choices?

4. In a small town with only 100 registered voters, there are two candidates running for mayor, Mr. Jones and Mrs. Smith. You know that out of the 100 registered voters, 40 would vote for Mr. Jones and 60 for Mrs. Smith, but not everyone will actually vote. In fact, because of a terrible snowstorm, only 10 of the voters are able to vote. For each of the following questions, assume that the people who show up to vote are a random sample from the population of 100.

   Simulate the election using at least 50 trials, recalling that this is sampling without replacement (since each person can vote only once). Use a random number table, and assign numbers 00 through 39 to Mr. Jones and the rest to Mrs. Smith. Sampling without replacement means that if you see a number you have already seen (within the same trial), you have to throw out that number and go to the next.
   a. What is the probability that Mr. Jones will win the election (that is, receive six or more votes)?
   b. What is the probability that Mrs. Smith will win?
   c. What is the probability that they each receive five votes?
   d. Do you think that the low voter turnout helped Mr. Jones's chances or hurt them? (*Hint:* What would his chances have been if all 100 had voted?)

5. A group of 10 men are choosing teams to play a game of basketball, and they decide to randomly assign people to either team, with five players on each team. Two of the players are taller than the rest, both of them being 6 feet, 5 inches tall. What is the probability that they will end up on the same team?

6. Return to Exercise 5. Two additional people show up wanting to play. The men decide to play two games of three men against three men instead of one game of five men versus five men. What is the probability in this case that the two tall men end up playing on the same team?

7. Recall the example given at the beginning of the section concerning the possible signals made from flags of four colors. Using Figure 5.8, calculate the following probabilities. Recall that this is sampling without replacement.
   a. p(YOR)
   b. p(First flag is red)
   c. p(Last flag is blue)
   d. p(At least one of the signal flags is yellow)

8. Consider now that in the example with the four colored flags the flags are chosen at random *with* replacement. Draw the corresponding tree diagram for this situation and calculate the following probabilities.
   a. p(YOR)
   b. p(Last flag is yellow)
   c. p(Middle flag is blue)
   d. p(None of the three flags is red)

5.8 THE BIRTHDAY PROBLEM (OPTIONAL)

We now take up the Key Problem. It is an enjoyable problem to play around with and provides a nice example in which a theoretical probability analysis is somewhat difficult and the five-step method is quite easy to carry out.

The birthday problem is engaging partly because its answer is very surprising. Informally, it is simply the question whether at least two people in a group (such as students at a lecture) share the same birthday. More formally, for a group of 25 people, say, the birthday problem is the following:

> Among a group of 25 people chosen at random, what is the probability that at least two of them have the same birthday? Disregard the year of birth, and count only 365 possible birthdays (that is, ignore leap-year birthdays, occurring on February 29).

Perhaps you'd like to make your own guess about the correct answer. To understand how to solve this problem experimentally, we will first scale it down to a simpler one.

Example 5.13 The Birth Month Problem

What is the probability that among a group of four people chosen at random, at least two were born in the same month (not necessarily the same year)?

Solution

As we have suggested so often in this book, we could get a start on this problem by gathering data from around us. Let's say we go to groups of four of our classmates and ask them the month in which they were born. We then keep a record of how many of the groups of four have at least two persons born in the same month.

Alternatively, we could use the five-step approach:

1. Choice of a Model: We could use a 12-sided die (dodecahedron). A 12-sided die would give us one side for each month of the year. Alternatively, we could use the random data generated from 12 digits provided in Table B.4. We make the assignments

1: January
2: February

and so on.

2. Definition of a Trial: A trial consists of rolling the die four times, once for each person in the group.

3. Definition of a Successful Trial: A trial is a success if one of the digits is repeated (that is, the same month occurs at least twice).

4. Repetition of Trials: Do at least 100 trials.

As an example, we will do 10 trials. In random digit Table B.4, we have the nine digits 1–9 plus

$$0 \leftrightarrow 10 \qquad E \leftrightarrow 11 \qquad T \leftrightarrow 12$$

Here is the first row:

Row									
1	7840T	9647E	42368	8309T	85516	24907	42T0E	T5120	31E34

The results of the 10 trials are given in Table 5.12.

5. Probability Estimate: The estimated probability of two or more people in a group of four sharing a birth month, based on 10 trials, is

$$P(\text{At least two people share a birth month}) = \frac{\text{number of successful trials}}{\text{total number of trials}}$$

$$= \frac{3}{10} = 0.30$$

The theoretical probability happens to be $990/1728 \approx 0.57$.

Table 5.12 **Estimating the Probability That Two or More People in a Group of Four Share a Birth Month**

Trial	Random numbers	Success?
1	7840	No
2	T964	No
3	7E42	No
4	36(88)	Yes
5	309T	No
6	8(55)1	Yes
7	6249	No
8	0742	No
9	(T)0E(T)	Yes
10	5120	No

The Birthday Problem

We want to find the probability that two or more people in a randomly chosen group of 25 people share a birthday. We use the same approach for solving the birth month problem to solve the birthday problem.

1. Choice of a Model: Assume that each birthday is as likely as any other. (Is this a reasonable assumption?) We will use Table B.3 and read off three-digit random numbers from 001 to 365 and ignore all others. That is, each time three consecutive digits form a number from 001 to 365, we select that number. For example, 72234 40065 yields 223 and then 006 because 7, 44, and 5 are discarded (they would have formed numbers larger than 365). Alternatively, you could look at digits in groups of three and choose only those groups from 001 to 365. Either variation makes each three-digit number from 001 to 365 *equally likely*. The numbers 001–365 correspond to the 365 days of the year (deliberately ignoring leap years, as stated in the Key Problem).

2. Definition of a Trial: A trial consists of reading off a series of 25 three-digit random numbers from 001 to 365, inclusive. Clearly, 001 to 365 are each equally likely and independently chosen for the 25 different subtrials.

3. Definition of a Successful Trial: A trial is a success if at least one duplicate random number is obtained (corresponding to at least two people having the same birthday).

4. Repetition of Trials: Do many trials (at least 100).

Here we will carry out two trials as an example. We will use groups of three digits, giving numbers from 1 to 365, from rows 11–15 of Table B.3.

Row									
11	72234	40065	24052	95658	98335	21125	45364	67989	32451
12	02833	78254	05112	95160	62546	85982	85567	27427	21436
13	20565	20846	63664	72162	75338	04022	77166	83339	99021
14	18090	91089	09799	75883	36480	37067	40933	65634	79883
15	11519	97203	70899	00697	84864	24470	07933	48202	15392

The two trials are recorded in Table 5.13. Here the ignored digits have been removed.

5. Probability Estimate: The estimated probability of sharing a birthday, based on two trials, is

P(At least two people in a group of 25 share a birthday)

$$= \frac{\text{number of successful trials}}{\text{total number of trials}} = \frac{1}{2} = 0.50$$

Table 5.13 **Estimating the Probability That Two or More People in a Group of 25 Share a Birthday**

Trial	Random numbers	Success
1	223 006 240 295 335 211 (254) 364 324 102 337 (254)*	Yes
2	051 129 160 254 285 274 272 143 205 208 216 275 338 040 227 166 333 021 180 091 089 097 336 037 067	No

* To save time, a trial may be ended when a duplicate number (here, 254) is found, because in that case the trial is a success.

This answer is totally unreliable, of course. From 400 trials (using a computer) it was found that P(At least two people share a birthday) $=$ $229/400 = 0.5725$. This is to be compared with the correct theoretical answer p(At least two share a birthday) $= 0.5687$. Thus the error is $0.5725 - 0.5687 =$ 0.0038. Hence we have gotten very close to the true probability, which is startlingly large!

The really surprising thing about the birthday problem is that only 23 people are needed to make the theoretical probability of at least two people sharing the same birthday slightly larger than $\frac{1}{2}$. Most people expect the number of people required to be much larger. In fact, for larger numbers of people than 23, the theoretical probability gets close to 1 quite rapidly as the number grows.

Just to show the power of the method, we did a huge number of trials, namely, 8000. Then we obtained

$$P(\text{At least two people share a birthday}) = \frac{4489}{8000} = 0.5691$$

Thus the error now is $0.5691 - 0.5687 = 0.0004$, close enough for just about any purpose we might have. Thus we have found an answer extremely close to the theoretical value *without* knowing how to solve the problem theoretically. It took the computer a mere couple seconds of computing time to find this answer, too! Thus, if we are willing to do enough trials, we can come as close to the true answer as we wish.

SECTION 5.8 EXERCISES

1. For the birth month problem of Example 5.13, do a total of 25 trials to get a better estimate of the probability that in a group of four people, at least two were born in the same month.
2. Solve the birth month problem for a group of five people instead of four. That is, what is the probability that among a group of five people chosen at random, at least two were born in the same month?

3. What is the largest group of people needed before it is certain that at least two were born in the same month? Why?
4. Our goal is to obtain a better estimate of the probability that in a group of 25 people, at least two were born on the same day of the year.
   a. What would you guess this probability to be?
   b. Perform a total of at least 25 trials to add to the results shown in Table 5.13. What is your estimate of the probability?
   c. Was your guess in part (a) close to your estimate in part (b)?
5. Determine the experimental probability that at least two people in random groups of the following sizes have the same birthday:
   a. 5
   b. 10
   c. 15
   d. 20

   (Your accuracy depends on your stamina, on obtaining the assistance of other class members, or on the use of a computer simulation program.)
6. Draw a graph of the data of Exercise 5. Let the x value be size of the group, and let the y value be the estimated probability of a shared birthday.
7. Suppose each of three people all select a number between 1 and 10. What is the probability that at least two of them choose the same number?
8. Suppose each of six students selects at random an integer between 1 and 52, inclusive. What is the probability that at least two of the students will select the same integer? (You may wish to use a deck of cards to provide the data for this problem, instead of a table of random digits.)

5.9 CONDITIONAL PROBABILITY

Suppose we wish to find a probability when certain information is given or some event is known to have happened. We denote the probability that an event A occurs given that B is true or is known to have occurred by $P(A \mid B)$ or $p(A \mid B)$, and we read this as "the probability of A given B." It is also often referred to as the **conditional probability** of A given B. As usual P denotes an experimental probability and p denotes a theoretical probability. We will discover in Chapter 13 that the concept of conditional probability is an important one and has a close relationship to independence. But for now we merely introduce the idea.

Example 5.14 A carnival spieler shows an audience three cards. One card is red on both sides, one blue on both sides, and one red on one side and blue on the other. Someone in the audience shuffles the cards and places one on the table in such a way that no one knows what color is on the bottom side. The top side is red. The spieler says, "Obviously this is not the blue-blue card. Then it is either the red-red card or the red-blue card." Suppose the spieler will bet even money that it is the red-red card (he will pay you $1 if he is wrong and you will pay him $1 if he is right). Is this a fair bet? (From W. L. Ayres, C. G. Fry, and H. F. S. Jonah, *General College Mathematics*, 3rd ed., McGraw-Hill, 1970, p. 234.)

Solution

Since any of the three cards would be put on the table with equal probability, we will use a die to simulate this problem. We set up the following step-by-step procedure.

1. Choice of a Model: There are six possible ways of choosing and laying a card on the table, because each of the three possibly chosen cards has two faces to choose as the upward face. Let a toss of 1 or 2 represent the red-red card being put on the table, 3 or 4 represent the blue-blue card, and 5 or 6 the red-blue card. Further, if the toss is 5, let us say that the red side is up, and if the toss is a 6 we will say that the blue side is up.

2. Definition of a Trial: Toss the die. Since we are given that the top side is red, we will ignore a toss of 3, 4, or 6, which correspond to a blue side being up. Toss until a 1, 2, or 5 appears. This is one trial of the experiment. It amounts to conditioning on the red side being up.

3. Definition of a Successful Trial: Call the event that the toss is a 1 or 2 a success, because this would correspond to the red-red card being placed on the table.

4. Repetition of Trials: Do at least 100 trials.
The results listed in Table 5.14 were obtained for 10 trials.

5. Probability Estimate: Estimate $P(\text{red-red card} \mid \text{red face up})$ as the number of successes divided by the number of trials. Thus we estimate that

$$P(\text{red-red card} \mid \text{red face up}) = \frac{6}{10} = 0.6$$

Of course, 10 trials is not enough to produce much accuracy. Interestingly, $p(\text{red-red card} \mid \text{red face up}) = 2/3$, in contrast with the intuitive (and incorrect) notion that $p(\text{red-red card} \mid \text{red face up}) = 1/2$. If we ran 400 trials (easy to do with

Table 5.14 **Estimating $p(\text{red-red card} \mid \text{red face up})$**

Trial	Die tosses	Success?
1	2	Yes
2	4, 2	Yes
3	3, 4, 1	Yes
4	5	No
5	6, 2	Yes
6	1	Yes
7	5	No
8	3, 5	No
9	6, 5	No
10	2	Yes

a computer), we would expect to be quite close to the true probability, and close enough to easily decide that 1/2 is not correct, as the spieler in effect claimed with his even-money bet.

Another such puzzling example is the *Let's Make a Deal* television show problem. Behind each of two curtains is a pig, and behind the third is a BMW car. You choose a curtain at random (clearly, $p(\text{win car}) = \frac{1}{3}$). But the announcer reveals a pig behind one of the other curtains before you get to look behind your curtain. He then says that if you wish, you can change the curtain you have chosen. Should you?

Most people would think that changing curtains does not alter your chances of winning. But what is $p(\text{win} \mid \text{change curtain})$? This problem yields the startling result that

$$p(\text{win} \mid \text{change curtain}) = \frac{2}{3}$$

while

$$p(\text{win} \mid \text{do not change curtain}) = \frac{1}{3}$$

Thus you *should* change curtains!

Do you believe this? If not, you could use the five-step method to test it.

SECTION 5.9 EXERCISES

1. Continue the problem of the carnival spieler by doing at least another 30 trials and obtaining a new estimate of the probability that the spieler wins the bet.

2. What is the probability that a family with three children has all boys if you know that at least one of the children is a boy?

3. The student body of a certain college is composed of $\frac{1}{3}$ freshmen, $\frac{1}{4}$ sophomores, $\frac{1}{4}$ juniors, and $\frac{1}{6}$ seniors. Suppose 20%, 50%, 60% and 80% of the freshmen, sophomores, juniors, and seniors, respectively, have their own cars. What is the probability that a student selected at random will have a car?

4. A somewhat absent-minded hiker forgets to bring his insect repellent on 30% of his hikes. The probability of being bitten is 90% if he forgets the repellent, and 20% if he uses the repellent. Perform at least 50 trials simulating whether the hiker is bitten or not.
   a. What is the probability that he will be bitten?
   b. Now look only at the times when the hiker was bitten. What is the total number of times he was bitten? Of these times, how many times did he forget to bring his repellent?

c. Using the information from part (b), calculate the probability that he forgot his repellent given that he was bitten.

5. You throw a pair of dice, one after the other. You are interested in the following events:

A = the sum of the dice is greater than or equal to 6

B = the sum of the dice is greater than or equal to 5

C = the first die comes up 3

D = the second die comes up 1

Perform 50 trials and then use those trials to estimate the following probabilities:

a. $P(A \mid B)$
b. $P(B \mid C)$
c. $P(B \mid A)$
d. $P(B \mid D)$
e. $P(C \mid D)$

6. Simulate the *Let's Make a Deal* example to estimate
a. $P(\text{win car} \mid \text{do not change curtain})$
b. $P(\text{win car} \mid \text{change curtain})$

CHAPTER REVIEW EXERCISES

1. Assume you know that 90% of the entire student population is right-handed and 10% is left-handed. Your class has 20 students, and the room the class is assigned to has 20 desks, all designed for right-handed people.
   a. What is the probability that all the students in your class will be right-handed? Use the law of theoretical independence.
   b. What is the probability that in your class of 20 students there will be at least one left-handed person?
   c. Assume now that there is one left-handed desk in the room. Estimate the probability that there will be a perfect match between the students and their preferred type of desk. Do at least 50 trials using a random number table.
   d. The probabilities in parts (b) and (c) should be different. Why?
2. State each of the following:
   a. The law of theoretical independence
   b. The law of experimental independence
   c. The property of complementary events for theoretical probabilities
3. You are interested in estimating the percentage of married couples in which both husband and wife are Democrats, and the percentage in which both husband and wife are Republicans. You know that in the population in general, 50% of people are Democrats.
   a. You should not assume that the political party of the wife is independent of the party of the husband. Why not?
   b. Since the husband's and wife's political parties are not independent, this situation cannot be simulated by simply flipping two coins, one for the husband and one for the wife. Assume that you know that 75% of the time, the wife will have the same political party as the husband. So to estimate

the desired percentages, begin as follows: flip a coin 25 times, and let heads indicate that the husband is a Democrat, and tails indicate that the husband is a Republican.

c. Now, for each of the husbands in part (b), simulate the political party of the wife. (*Hint:* this is equivalent to performing 25 trials, with each trial having a 75% chance of the outcome "same party" and a 25% chance of "different party.")
d. Using your results, estimate the number of couples in which both husband and wife are Democrats; both Republicans.

4. A friend brings you a coin and claims it is not fair (not equally likely to give heads or tails). He claims that he flipped it five times and saw only one head.
   a. Use the five-step method with 25 trials to estimate the probability of seeing only one head in five flips of a fair coin.
   b. Does your friend have a legitimate claim? Why or why not?
5. Explain the difference between sampling with replacement and sampling without replacement. Give an example of where each method is used.
6. The following table presents the results of a survey asking people whether they have volunteered or not during the past 12 months:

By gender	
Men	45% volunteered
Women	52% volunteered
By employment status	
Full-time	50% volunteered
Part-time	58% volunteered
Not employed	46% volunteered
Retired	40% volunteered

Source: *Champaign-Urbana News-Gazette,* March 6, 1997, B1.

   a. If you ask a woman at random, what is the probability that she has not volunteered during the past 12 months? (Do not use the five-step method to find this answer.)
   b. Estimate the probability that if you ask five retirees (men and women) if they have volunteered in the past year, at least one of them will tell you that they have. Perform at least 50 trials using a random number table.
7. Now we want to calculate the probability in part (b) of Exercise 6 theoretically.
   a. Write out the complement of the event "at least one of the five retirees has volunteered during the past 12 months."
   b. Assume that the retirees are independent of each other with respect to whether or not they have volunteered. Using the law of theoretical independence, write the probability of the event you found in part (a) as the sum of five separate probabilities.

c. According to the formula you found in part (b), what is the theoretical probability of the complement to the event "at least one of the five retirees has volunteered during the past 12 months."
d. Finally, use the property of complementary events for theoretical probability to calculate p(At least one of the five retirees has volunteered).

8. Once again, look at the data in Exercise 6. In a certain neighborhood, the breakdown of employment status (for both men and women) is as follows:

Full-time	65%
Part-time	10%
Not employed	5%
Retired	20%

There are 100 adults living in this area.
a. Using a random number table, estimate the probability that if you draw an individual at random from the neighborhood, that person has volunteered during the past 12 months. (*Hint:* Use four consecutive digits from the random number table. The first two tell you what employment status the person has, and the second two tell you whether or not the person has volunteered.)
b. Using the results from part (a), estimate the probability that given a person volunteers, he or she has a full-time job.
c. Suppose you walk around the neighborhood asking people whether they have volunteered or not. Are you sampling with or without replacement?

9. A group of five friends go to a restaurant where there are 10 choices on the menu. Estimate the probability that none of the five chooses the same item, assuming each chooses one thing.

10. You and some friends begin to draw straws to see who has to go pick up the pizza you ordered. There are a total of 10 people in the group, and there are 10 straws. The loser is whoever draws the one long straw.
a. Before any straws are drawn, what is the theoretical probability that you will be the one who has to go pick up the pizza? Assume that each straw is equally likely to be drawn?
b. Three of your friends draw first, and none of them draws the long straw. What is your probability of being the loser now? In other words, what is p(drawing long straw | first three drew short straws)?

11. The bus is scheduled to arrive at a bus stop at exactly 8:24 A.M. But the bus actually arrives according to the following table:

8:21	5%
8:22	10%
8:23	15%
8:24	35%
8:25	20%
8:26	10%
8:27	5%

You arrive at the stop to catch the bus according to this table:

8:21	5%
8:22	15%
8:23	30%
8:24	20%
8:25	20%
8:26	5%
8:27	5%

a. Do at least 50 trials to estimate your probability of catching the bus. (If you arrive at the same time the bus does, assume that you catch the bus.)

b. Estimate the probability that you will have to wait 3 minutes or more for the bus. (If you arrive at 8:21 and the bus comes at 8:24, you have waited exactly 3 minutes.)

12. Return to the bus data in Exercise 11. Estimate the following conditional probabilities:

a. P(You catch the bus | The bus arrives at 8:24)

b. P(You miss the bus | You arrive at the bus stop at 8:25)

c. P(You are forced to wait three or more minutes | The bus arrives at 8:26)

13. Four football teams are in a league together. Based on the strengths and weaknesses of each team, you deduce the following probabilities:

p(Team A beats Team B) = 0.75
p(Team A beats Team C) = 0.40
p(Team A beats Team D) = 0.50

p(Team B beats Team C) = 0.60
p(Team B beats Team D) = 0.30

p(Team C beats Team D) = 0.40

a. What is p(Team C beats Team A)?

b. Assume that each game is independent of the others and that each team plays each of the other teams twice during the season. Estimate the probability that Team A will finish the season with two or fewer losses (that is, a 4-2 record or better). Do at least 50 trials, with each trial being the six-game season that Team A has to play.

14. Return to the data in Exercise 13. The four teams are to play a single-elimination tournament with the following form:

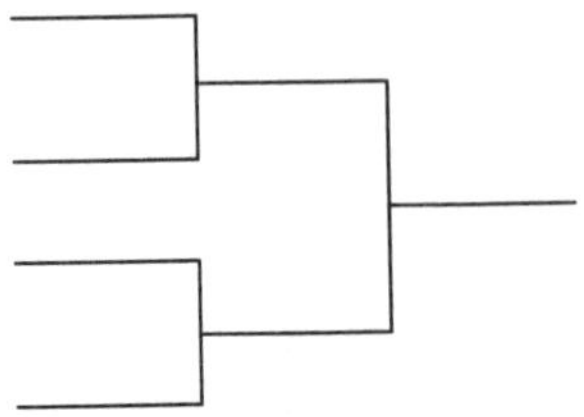

The teams are to be randomly assigned to the brackets.

a. Draw out 25 bracket diagrams such as the one shown above. In each case randomly assign the teams to the four starting positions. Is this sampling with replacement or without replacement?

b. Simulate each of the tournaments using the probabilities given in Exercise 13. What is your estimate for the probability that Team C wins the tournament?

c. Using the 25 simulations you just performed, what is your estimate for the conditional probability *P*(Team C wins the tournament | Team C plays Team A in the first round)?

15. The goal of this exercise is to calculate the theoretical probability in an example similar to the birthday problem. Assume that each of three individuals chooses a number from 1 to 3. We are interested in estimating the probability that at least two of them choose the same number.

a. Draw the tree diagram in this case. There should be 27 equally likely outcomes.

b. Using the tree diagram you found in part (a), what is *p*(Two of them choose the same number)?

c. What is the complement of the event "two of them choose the same number"? State the probability of that complementary event.

16. Jan plays a game in which she spins a wheel with 10 equally likely outcomes, numbered 0 through 9. Whatever number she spins is the amount of money she wins. After her first spin, she gets to decide whether she wants to spin again and try to improve on the amount she has won. If she spins a second time, she wins only the amount she receives on the second spin, and not what she would have received from the first spin.

a. Assume that Jan always decides to spin the second time. Estimate the probability that she will reduce the amount of money she wins on her second spin. Do at least 50 trials. This can be done very easily by means of a a random number table.

b. Estimate the following experimental conditional probabilities:

P(On the second spin she increases her winnings | On first spin she won $3)

P(On the second spin she increases her winnings | On first spin she won $4)

P(On the second spin she increases her winnings | On first spin she won $5)

c. On the basis of the results of part (b), what should Jan's rule be for deciding whether or not to spin the wheel a second time?

17. Look back at Exercise 16. This time we are going to calculate theoretical probabilities based on this situation.

a. Each outcome is equally likely when Jan spins the wheel once. What is the probability that a given spot on the wheel comes up?

b. Each spin is independent of the other, so what is the theoretical probability that a given combination of two numbers comes up? Use the law of theoretical probabilities.

c. Reconsider the probability: p(On the second spin Jan increases the amount she wins | On the first spin she won \$4). Explain why this probability is equal to p(On the second spin she spins a 5 or higher). What is this probability?
d. Calculate the following theoretical probabilities:

p(On the second spin Jan increases the amount she wins | On first spin she won \$5)

p(On the second spin Jan increases the amount she wins | On first spin she won \$6)

18. A box contains six tickets, labeled 0 through 9. What is the probability that if you draw out three of the tickets, the sum of the three is equal to or greater than 20 if you
    a. Draw with replacement?
    b. Draw without replacement?

    Perform at least 50 trials for each part to estimate the probabilities.
19. You survey 100 married couples as to whether they have exercised in the past week. The results are as follows:

    Out of 100 wives, 56 have exercised.

    Out of 100 husbands, 49 have exercised.

    Out of 100 couples, there are 40 in which both the husband and wife have exercised.

    a. Using these data, estimate P(Wife has exercised), P(Husband has exercised), and P(Both husband and wife have exercised).
    b. Does the law of experimental independence seem to hold in this case? Do a calculation to prove your point. Does this seem logical?
20. Write out the complement of each of the following events, but do not use the word *not*:
    a. "There was at least one game where he scored a point."
    b. "At least five of the houses on the block are painted white."
    c. "The first card dealt from the deck of cards is a diamond."
    d. "Exactly 7 of the 10 members of the class passed the final exam."

5

COMPUTER EXERCISES

Before attempting these problems, you should do the computer exercises from Chapter 4 so that the five-step program is familiar to you.

1. EXAMPLE 5.2 REVISITED

In Example 5.2 we are interested in the chance that there are at least two boys in a three-child family. The assumptions are that the chance of a boy is 50% and that the sexes of the children are independent. Start the five-step program. Press the "Choose Box Model" button, and create the box with one 0 (girl) and one 1 (boy). Define a trial to be "Drawing n with replacement," where $n = 3$. Define the statistic to be "Sum" because we want the number of boys, which is the sum of the draws. The objective is to find the proportion of three-child families that have at least two boys, so the probability we are interested in is "Sum ≥ 2." Choose "Stat >= a" from the "Probability of interest" area, and type "2" in the "a = " area. The top part of the screen should be as follows:

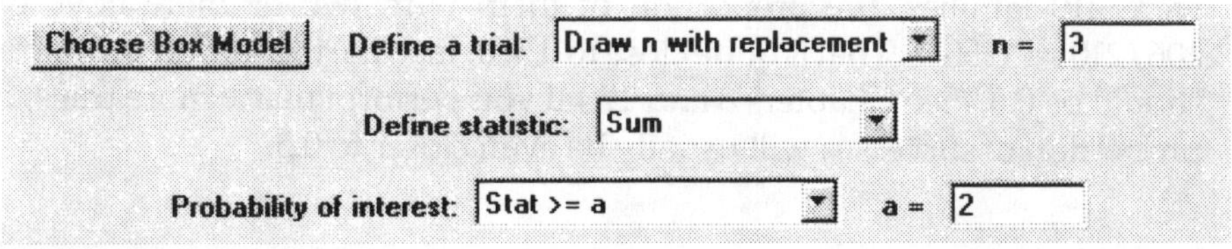

Now press the "1" button in the "# of trials" area. You will see the first family in the "Current Draws" area. The sum of the numbers is the number of boys. If that sum is 2 or 3, then we have a "Success" as defined by the "Probability of interest." If the sum is less than 2 (that is, 0 or 1), then we have a "Failure." In the "Sums" area at the right, the sum will appear, as well as whether it was a success or failure. Press "1" nine more times, each time noticing what happens. Then press "100" a couple of times. You should see a screen somewhat like this:

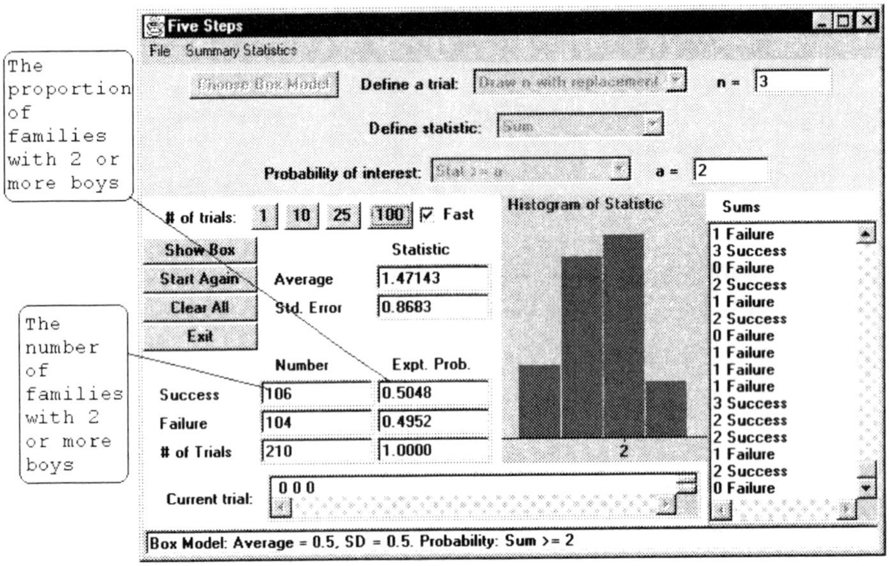

Your numbers will be somewhat different, but the basic idea should be the same. Notice that this time, as was not the case in Chapter 4, there are also numbers in the table just above the "Current trial" area. Here the numbers of successes and failures are counted, as well as the experimental probabilities, labeled "Expt. Prob.," which are the observed proportions of successes and failures. In this particular screen, it turned out that there were 106 successes: 106 families had two or three boys. Because there were 210 families, the proportion of them with two or more boys was $106/210 = 0.5048$. That is quite close to 0.5. In fact, the theoretical proportion should be 0.5. (See Chapters 8 and 12.) If you perform many more trials, the proportion of successes will usually be even closer to 0.5.

2. THE LAW OF AVERAGES

Imagine flipping a fair coin a number of times. The more flips you do, the closer you will come to half heads and half tails, right? It depends on what is meant by "closer." In absolute numbers of heads, or in percentages? Start the five-step program, and set the box to have one 0 and one 1. We will look at three numbers of flips: 10, 40, and 100. We will also look at three probabilities: (a) exactly half the flips yield heads, (b) the number of heads is within five of half the number of flips, and (c) the number of heads is within 10% of half the number of flips. First, make guesses for the following questions, circling either "increases" or "decreases":

- The chance of getting exactly half heads increases/decreases as the number of flips goes up.
- The chance of coming within five heads of exactly half heads increases/decreases as the number of flips goes up.
- The chance of coming within 10% of exactly half heads increases/decreases as the number of flips goes up.

The rest of this exercise will explore these questions.

a. Obtaining Exactly Half Heads: Define the trial to be "Draw n with replacement" with "n = 10." Define the statistic to be "Sum." Set the "Probability of interest" to be "Stat = a" with "a = 5" because 5 is half of 10. Perform 200 trials by pressing "100" twice. Watch the "Sums" area to see which trials give exactly 5 heads and which do not. When the trials are finished, fill in the observed proportion of successes in the "n = 10" row of the following table:

Number of flips	Proportion of trials having exactly half heads
$n = 10$	
$n = 40$	
$n = 100$	

Repeat the process (press "Start Again") but with "n = 40" and "a = 20." Fill in the observed proportion of trials yielding 20 heads in the "n = 40" row. Finally, repeat the process again but with "n = 100" and "a = 50," and fill in the last row.

Does the chance of getting exactly half heads seem to increase or decrease as the number of flips increases?

b. Coming within Five Heads of Exactly One-half Heads: Define the trial to be "Draw n with replacement" with "n = 10," and keep the statistic as "Sum." This time, set the "Probability of interest" to be "a <= Stat <= b" with "a = 0" and "b = 10" because the numbers from 0 to 10 are those within 5 heads of 5. Perform 200 trials by pressing "100" twice. Watch the "Sums" area to see which trials give between 0 and 10 heads. (They all do, of course.) When the trials are finished, fill in the observed proportion of successes (which should be 1.00) in the "n = 10" row of the following table:

Number of flips	Proportion of trials with number of heads within five of half the number of flips
$n = 10$	
$n = 40$	
$n = 100$	

Repeat the process (press "Start Again") but with "n = 40," "a = 15," and "b = 25," because coming within five heads of 20 means getting from 15 to 25 heads. Fill in the observed proportion in the "n = 40" row. Finally, repeat the process again but with "n = 100," "a = 45," and "b = 55," and fill in the last row.

Does the chance of getting within five heads of exactly half heads seem to increase or decrease as the number of flips increases?

c. Coming within 10% of Exactly Half Heads: Define the trial to be "Draw n with replacement" with "n = 10," again keeping the statistic as "Sum." To get within 10% of half heads, we must get within 1 (which is 10% of $n = 10$) of 5. So set the "Probability of interest" to be "a <= Stat <= b" with "a = 4" and "b = 6" because the numbers from 4 to 6 are those within 10% of 5 heads. Perform 200 trials, and fill in the observed proportion of successes in the "$n = 10$" row of the next table:

Number of flips	Proportion of trials yielding within 10% of exactly half heads
$n = 10$	
$n = 40$	
$n = 100$	

Repeat the process but with "n = 40," "a = 16," and "b = 24," because 10% of 40 is 4, so getting within 10% of half heads means getting from 16 to 24 heads. Fill in the observed proportion in the "n = 40" row. Finally, repeat the process again but with "n = 100," "a = 40," and "b = 60," because 10% of 100 is 10. Fill in the last row.

Does the chance of getting within 10% of exactly half heads heads seem to increase or decrease as the number of flips increases?

How well did you guess at the beginning of this exercise? Try to summarize what appears to happen as the number of flips increases.

3. OVERBOOKING

Suppose an airplane is 5% overbooked, and the chance that a person with a reservation shows up is 0.90. What is the probability that at least one person with a reservation will not have a seat? Does it depend on the capacity of the airplane, or is the chance about the same no matter how many seats the plane has?

For this exercise, set the box to have one 0 and nine 1s, so that drawing a "1" from the box indicates the person shows up, and 0 means the person is a no-show. Define the statistic to be "Sum." We will consider three planes, with capacities 20, 100, and 200. Using 200 trials for each plane, estimate the following:

a. For the plane with only 20 seats, the airline sold 5% of 20, or 1, too many seats; that is, it sold 21 tickets. Thus you should define the trial to be drawing 21 with replacement. The probability of interest is the chance more than 20 people show up, so choose "Stat > a" with "a = 20."

b. For the plane with 100 seats, the airline sold 5% of 100, or 5, too many seats; that is, it sold 105 tickets. Define the trial to be drawing 105 with replacement. The probability of interest is the chance more than 100 people show up, so choose "Stat > a" with "a = 100."

c. For the plane with 200 seats, 5% overbooking means at the airline sold 210 tickets. The probability of interest is the chance that "Stat > a" with "a = 200."

In each case, the percentage of overbooking is the same. Which of the following appear to be true?

- The chance of too many people showing up is about the same no matter how big the plane.
- The chance of too many people showing up is larger for a larger plane.
- The chance of too many people showing up is smaller for a larger plane.

4. ROULETTE

Roulette is a gambling game that utilizes a wheel with 38 slots numbered 0, 00, 1, 2, ..., 36. (Notice that "0" and "00" are different.) The wheel is spun, and a small ball careens around the wheel, eventually settling in one of the slots. The player can bet on whether a particular number comes up, or on any of many combinations of numbers. For example, one can bet on the number 12, or bet on all the even numbers (2, 4, 6, ..., 36), or bet on a block of six numbers such as 13,

14, 15, 16, 17, and 18. If your number comes up, or one of the numbers in your combination comes up, you win a certain amount depending on the bet. If none of your numbers comes up, you lose your bet. We will focus on three bets:

- Bet \$1 on the number 12. If 12 comes up, you win \$35; if it does not, you "win" −\$1.
- Bet \$1 on the block of six numbers 13, 14, 15, 16, 17, 18. If one of those comes up, you win \$5. If none does, you win −\$1.
- Bet \$1 on the even numbers. If an even number comes up, you win \$1. If an odd number or 0 or 00 comes up, you win −\$1.

Imagine playing one of these bets 100 times. How much money do you expect to win, or—what is more likely—lose? What is the most you are likely to win or lose? What is the chance that you end up winning money? Do the answers to these questions depend on the type of bet you make? Answering these questions requires a blend of estimating expected values and estimating probabilities.

To address these questions, we will use the five-step program to fill in the following table:

Bet	Average	Minimum	Maximum	Chance > 0
12				
13, 14, 15, 16, 17, 18				
Even numbers				

a. Bet on 12: Set up the box to mimic the amount of money you would get on one play of this bet. That is, there is one chance for a 35, and 37 chances for a −1; hence the box should be as follows:

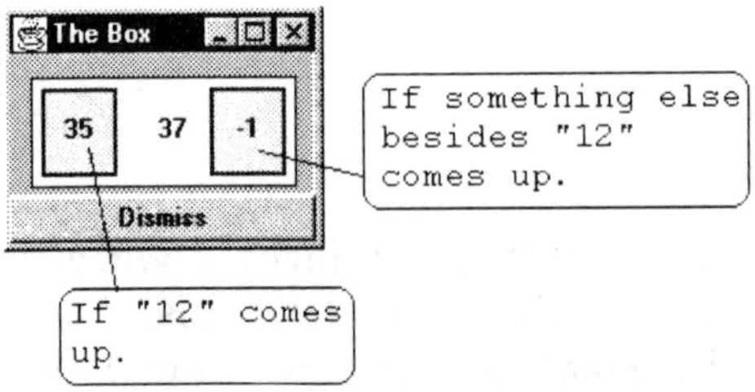

Define the trial to be drawing n with replacement, where $n = 100$. The statistic should be "Sum" (the total amount of money you end up with), and the probability of interest is "Stat > a" with "a = 0."

Now perform 200 of these trials. Fill in the average, minimum, maximum, and the proportion of success ("Chance > 0") in the table for the bet on 12. You can find the minimum and maximum by choosing "Statistics" from the "Summary Statistics" menu at the top. Notice that the average (which is perhaps negative) is our estimated expected value of our winnings.

b. Bet on 13, 14, 15, 16, 17, 18: The only difference between the setup here and that for the previous bet is the box. Choose "Start Again," and then change the box so that there are six 5s (the winners) and 32 −1s (the losers). Do 200 trials, and fill in the table for this bet.

c. Bet on Evens: Now change the box so that there are 18 1s (the winners) and 20 −1s (the losers). Do 200 trials, and fill in the table for this last bet.

The bets range from conservative (betting on evens) to risky (betting on the one number 12). Answer the following questions in light of the results from above:

1. Does the average amount you end up with seem to be positive or negative?
2. Does the average amount you end up with appear to depend on the bet, or are the averages for the three bets about the same?
3. Look at the minima and maxima. Which tend to be larger in absolute value, the maxima or minima? What does this mean?
4. Do the maxima and minima increase in absolute value as the bets become riskier, or are they about the same for the different bets?
5. Does the chance of ending up with more than $0 seem to be larger for riskier bets, smaller for riskier bets, or about the same for all bets?
6. If you had to choose one of these bets to play 100 times, which one would you choose? Why?

6

MAKING DECISIONS

You don't have to eat the whole ox to know the meat is tough.

Samuel Johnson (Boswell)

OBJECTIVES

After studying this chapter, you will understand the following:

- *A random sample as representative of a population*
- *How to use a sample to decide a hypothesis about a population*
- *How to use the six-step method to measure the strength of evidence about a hypothesis*
- *The median test of the hypothesis that two populations have the same center value*

6.1 SAMPLES AND POPULATIONS 286

6.2 USING SAMPLES TO DECIDE ON HYPOTHESES ABOUT POPULATIONS 291

6.3 THE MEDIAN TEST 304

CHAPTER REVIEW EXERCISES 309

COMPUTER EXERCISES 313

Key Problem

The standard drug used in the treatment of a certain disease is known to be effective 60% of the time. An experimental drug is used on 10 patients in a clinical trial and is found to be effective for 8 of them. Is this reliable evidence that the experimental drug has a better than 60% cure rate?

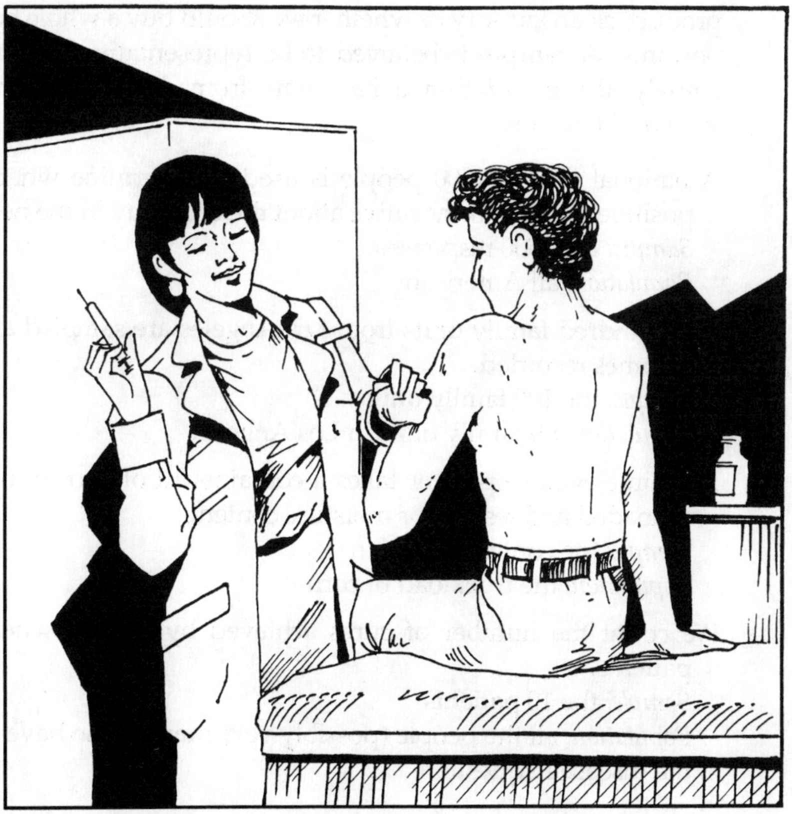

Later in the chapter we will solve the Key Problem. We first need to discuss two major ideas in statistics that are critical to the use of data to make decisions: samples and populations.

6.1 SAMPLES AND POPULATIONS

When we take a sample, such as a sample packet of a new brand of detergent sent in the mail or a whiff of a new brand of perfume in a department store, we are not usually interested in the sample for its own sake. Rather, we are interested whether a full-sized box of the new detergent will repeatedly produce clean laundry or whether we should buy a whole bottle of the new perfume. A **sample** is believed to be representative of something larger, namely, the **population** it has come from. Here are some examples of statistical relevance:

A national poll of 1500 people is used to determine whether people are positive, neutral, or negative about the economy in the next year.
Sample: the 1500 responses
Population: all Americans

One hundred family units from Los Angeles are sampled and their family incomes recorded.
Sample: the 100 family units
Population: all family units in Los Angeles

A grain-elevator operator takes a containerful of corn from a truck being unloaded and tests it for moisture content.
Sample: the container of corn
Population: the truckload of corn

We count the number of cures achieved by a drug when tested on 10 patients.
Sample: the 10 patients
Population: all the people (possibly very many) who have or will contract the disease in question

In Chapter 2 we calculated such values as the mean, median, interquartile range, and standard deviation of sets of data such as car mileages, outdoor temperatures, and number of Olympic medals won by each country. Often the set of data has been a sample taken from a larger population. In such cases we can use our measures of sample data center and spread as estimates of the center and spread of the population. An example will illustrate.

Suppose five people are randomly chosen and are asked their opinion about the taste of a new food product. They rate it from 5 (delicious) to 0 (tasteless)—what psychologists call a Likert scale. Their responses are

$$3 \quad 2 \quad 3 \quad 4 \quad 3$$

The mean response is

$$\frac{3+2+3+4+3}{5} = \frac{15}{5} = 3.0$$

The variance is

$$\frac{(3-3)^2 + (2-3)^2 + (3-3)^2 + (4-3)^2 + (3-3)^2}{5}$$

$$= \frac{0+1+0+1+0}{5} = \frac{2}{5} = 0.40$$

and hence the standard deviation is

$$\sqrt{0.40} \approx 0.63$$

Now suppose we want to use this information to give the food producer some idea *not* about how only five people like the new product, but about how the public, or people in general, will like the product.

We will use these five sample values to *estimate* the opinion of the population that the sample is believed to represent. There is an important idea here, namely, that the sample, if typical of the population, can in fact be very informative about the nature of the population. Let's suppose there are 1,000,000 potential buyers of this new food product. These 1,000,000 people make up the population in which we are interested. We can think of a population mean and variance of the opinions of these 1,000,000 people concerning the new product. To get the population mean, we could in principle add up all 1,000,000 responses (from delicious to tasteless) and divide by 1,000,000. This would be a lot of work, and indeed for many reasons it would be very difficult to do. Similarly, we could in principle find the variance of these 1,000,000 ratings. These numbers, often called *population parameters,* would then tell us exactly the center and the variation about the center of the population.

In statistics we learn how to use samples to estimate properties of populations. We can use the mean of a sample to estimate the mean of the population from which it was taken and the variance of a sample to estimate the variance of the population. Using the sample mean, we thus estimate (with error) where the population's center is. As we will learn later, the variance of the sample will help us in two important ways. First, it estimates how much variation from member to member there is in the population. Second, it helps us assess how accurate the sample mean is in its role of estimating the true population mean.

Table 6.1 **Home-Run Hits by American League Leaders, 1921–1980**

Stem	Leaf
1	
2	2, 4
3	2, 2, 2, 2, 2, 2, 3, 3, 3, 4, 5, 6, 6, 6,
*	7, 7, 7, 7, 9, 9, 9
4	0, 1, 1, 1, 2, 2, 2, 3, 3, 4, 4, 4, 4, 5, 5,
*	6, 6, 6, 6, 6, 7, 8, 8, 9, 9, 9, 9, 9, 9
5	2, 2, 4, 8, 8, 9
6	0, 1
7	

Table 6.1 is a stem-and-leaf plot of the number of home runs scored by each year's home-run leader in the American League for the 60 seasons from 1921 through 1980.

We can find the mean and variance of these 60 numbers in the usual way.

$$\text{Mean number of home runs} = 42.85$$
$$\text{Variance} = 95.26$$

Now let's think of this example from a sampling point of view. The numbers in Table 6.1 make up a population (the population of the numbers of home runs hit by the American League's home-run leaders for every year from 1921 to 1980). Thus the above mean number of home runs and variance become parameters, namely, the population mean and variance. Now suppose instead that we do not have access to all the information in Table 6.1. Suppose we have only a random selection of some of those numbers. For instance, our data might be taken from baseball statistical yearbooks, and we might have only an unsystematic assortment of, say, 10 yearbooks between 1921 and 1980. In that case we would estimate the average number of runs by using the "random" sample of data. This trick of treating a set of data as a random sample when it was actually not randomly sampled is often used in statistical analyses. It is acceptable practice as long as the data sampled are representative of the population. Whether such a data set is representative of the population is very hard to judge. Hence conservative statisticians argue strongly for random sampling and are wary of treating such "samples of convenience" as random samples.

Let's say our sample, of size 10, is as given in Table 6.2. The mean of these 10 numbers is 45.0. This appears to be a reasonably good estimate of the true population mean of the 60 numbers (that is, 42.85).

Table 6.2 **Ten Values from Table 6.1**

Stem	Leaf
2	
3	2, 6, 7
4	1, 3, 6, 7, 9
5	8
6	1

Later in this book (Chapter 12) we will discover mathematical reasons why the mean of a random sample serves very well as an estimator of the mean of the population from which the sample was drawn. By a *random* sample we simply mean that every member of the population has the same chance of being part of the sample. In this sense, the sample should be representative of the population.

Returning to the example, the variance of these 10 numbers is 78.0. This is not a very good estimate of the population variance of 95.26. As we will also see later, this way of finding the variance may not serve very well as an estimator of the population variance when the sample size is small. (We will call it a *biased* estimator of the population variance because, as we will learn, it is consistently too small, on average.) Later we will see how to remove the bias from the sample variance.

As we have said, we often take samples in order to estimate parameters of the population the samples are taken from. Thus we let the mean of a sample, for example, be our estimate of the mean of the population. We do that when the nature of the population (its mean) is unknown. That is, statisticians use numerical indices called statistics to estimate unknown population parameters. Estimation is one major kind of statistical inference.

Here is another way samples can help us answer questions about a population. Sometimes we suppose the mean of a population has a certain value and we then sample the population to confirm or rule out our supposition. We know that every population has some spread and that a given sample may have values somewhat far from the center. But if the sample mean is extremely far—improbably far—from the supposed center, we would conclude that our supposition is incorrect. In that case we have used our sample to rule out our supposition about the population. This kind of statistical inference is called *hypothesis testing* and is introduced in Section 6.2, below.

Estimation and hypothesis testing are the two major kinds of statistical inference. In this chapter we will begin to study hypothesis testing. A wide survey of important topics concerning estimation and hypothesis testing appears in Chapters 7 and 10 through 13.

SECTION 6.1 EXERCISES

1. For each of the following, identify the sample and the population.
   a. A poll of 40 college freshmen is taken to find out the views of the freshman class on lowering the drinking age from 21 to 18.
   b. A few small pieces of rock from a river valley region are analyzed at a laboratory to see how much gold they contain.
   c. Maurice is taking a course in French cooking. He has made a pot of onion soup and tastes a spoonful to see if he has added enough seasoning.
   d. Jerry and Elaine are at the beach. They want to go swimming, so Elaine puts her toe in the water to help decide whether they should go in.
   e. A building inspector wants to see whether the concrete being used is of sufficient quality. She takes a small sample of mix from the truckload every morning for five days and sends it to the laboratory for analysis.
   f. Lisa has symptoms of appendicitis. Her doctor does a blood test to see if her white blood cell count is normal.
2. For the situations in Exercise 1, find the following:
   a. An example of a sample mean (which we label $\overline{X}$). Describe the population mean being estimated.
   b. An example of a sample variance (S^2) that might be obtained. Describe the population variance being estimated.
3. Make up your own example of a population you want information about and the sample you would use to obtain the information.
4. The following paragraph is from *Education Week* (January 12, 1982, p. 8). What is the sample, and what is the population?

   > Most American students today have positive feelings about music, but many teenagers consider art less important than did their counterparts five years ago... according to recently released reports of the National Assessment of Educational Progress.

5. The table and extract that follow are from *Weekend Magazine* ("Armchair Athletes Do More Than Watch"). What is the sample, and what is the population to which the article refers?

Have you actually participated in sports over the last year?

		Those who watch TV sports	Those who don't watch TV sports
Yes	63%	69%	57%
No	35	31	43

> Tomorrow is Grey Cup day and many of us will spend it glued to our television sets. Does the enthusiasm of Canadians for sports extend beyond the armchair? It certainly does. Almost two-thirds of the people surveyed in the *Weekend* Poll say they have participated actively in sports during the last 12 months—and those who watch sports on television are more likely to be active than those who don't. Swimming turns out to be the most popular sport; more than 30 percent of those active in sports have engaged in swimming over the last year. Next in order of preference are cycling (20 percent) and jogging (18 percent).
>
> About 64 percent watch sports on television: hockey is by far the most popular game, followed by football and baseball. Thirty-nine percent have attended a paid-admission sporting event within the last year: again, hockey is the most popular event.
>
> On the average, Canadians spent 6.4 hours per week last year on sports, half of it in active participation. They spend

2.1 hours watching TV sports and 1.1 hours attending sports events. A slight majority of people say the time they devote to active participation has increased more than the time they spend watching sports.

6. Describe the sample and population referred to in the following paragraph.

   College students take more than twice as many courses in the fields of science, mathematics, and engineering as they do in the humanities, a government-sponsored survey indicates. The statistics were based on a survey of course selections at 760 colleges and universities in the United States. (Source: Associated Press, July 23, 1982)

7. Identify the sample and the population in the following extract from *Family Weekly* ("The Changing Mood of the Nation's Voters").

   What America Is Thinking

   More than 130,000 households responded to FAMILY WEEKLY's recent "Timely Issues" survey. Here's a summary of their opinions.

   Should the government provide more health services, to be paid for ultimately by taxes?

   Yes 23.9% No 67.6%
   Not sure 8.1%

   Will marriage fade in importance?

   Yes 27% No 59.7%
   Not sure 12.9%

   Are today's schools adequately preparing children for the future?

   Yes 8.1% No 82.3%
   Not sure 8.5%

8. Identify the sample or samples and population in the following paragraph.

   An environmental group is trying to estimate the number of fish in a lake. A spot in the lake is chosen using a table of random digits. A catch of fish is made there. The fish in the catch are counted, tagged, and then quickly released back into the lake. In a few days, another spot in the lake is chosen (again, using a table of random digits). A second catch of fish is made. The total number of fish caught and the number of tagged fish [are] noted. This information is used to estimate the total number of fish in the lake. (Source: S. Chatterjees, "Estimating Wildlife Populations by the Capture-Recapture Method," in *Statistics by Example: Finding Models*, Addison-Wesley, Menlo Park, California, 1973)

9. Find two examples from newspapers or magazines that involve samples and populations. Identity the samples and populations in each example.

6.2 USING SAMPLES TO DECIDE ON HYPOTHESES ABOUT POPULATIONS

Let's start with an example. A local newspaper editorial for a community of 2500 people suggests, without giving any statistical evidence, that the community favors raising the driving age from 16 to 18. The students of a sociology class decide to conduct a survey to statistically assess this claim. They ask 25 people chosen at random from the local telephone directory whether they favor raising the driving age or not. That is, they have chosen a random sample of size 25. The results are as follows:

Number of persons in favor of raising driving age: 18

Number of persons not in favor (opposed): 7

In such surveys there is usually a "no opinion" category. Here, for simplicity, this category is not included. From these results, should the class conclude that the community is in favor of raising the driving age?

We will solve this problem by taking an approach that may at first seem unusual. Later, as you see how the approach works, you will see how powerful it is.

The class is asking whether the community is in favor of raising the driving age. We start by *supposing* that the community is not in favor of raising the driving age but instead is evenly split on the issue—that one-half of the people in the community are in favor of raising the driving age and one-half are opposed. If the observed data (from the 25-person survey) provide strong evidence against our *null hypothesis* of a 50/50 community split, we can safely reject that hypothesis. In this case, *strong evidence* means an improbably large number of people in the sample favoring raising the driving age. Specifically, we want to know whether the observed 18 is improbably large if in fact the population split is 50/50.

Another way of putting it is as follows. If the community is really split 50/50, what is the probability that in a random sample of 25 people we would find as many as 18 *or even more* who favor raising the driving age? If that probability, which we express as

$$p(\text{18 or more in favor})$$

is quite small, then we find ourselves forced to reject the 50/50 hypothesis. How do we find p(18 or more in favor)? We will use the five-step method to obtain P(18 or more in favor) as an estimate of p(18 or more in favor).

Note that the *smaller* this probability, the *stronger* the evidence to discard the supposition of a 50/50 community split. Be sure not to get confused on this point. It really is logical once you carefully think it through. It is often helpful to check out our intuition here: Doesn't 18 or more out of 25 seem rather unlikely if the community is really split 50/50 on the issue? We will find out shortly.

We need to simulate doing a 25-person survey many times. Thus, as we lay out the five-step method, keep in mind that each trial must be a simulation of 25 responses. We will simulate the situation using a box model designed to model a 50/50 split in the community. We will assess how strongly (if at all) the data of our original 25-person survey argue against this model's being true.

The five-step method requires us in steps 1 and 2 to select the model and to define a trial. An obvious choice of a model is a box of 2500 tickets with 1250 labeled "in favor" and 1250 labeled "opposed." Then a trial would consist of 25 draws without replacement between draws, thus guaranteeing that the same person is never sampled twice. However, recall from Section 5.7 that if the population size is large compared to the sample size, we may instead estimate a probability using sampling with replacement, with the knowledge that our estimate is virtually as good as if we sampled without replacement. Thus it is acceptable to obtain P(18 or more in favor) using sampling with replacement. But then the box model only needs two tickets, one labeled "in favor" and one labeled "opposed." Because we are sampling with replacement, this method creates the 50/50 split on each draw from the box as required. Finally, it is clear that we could dispense with the box altogether and toss a coin instead. We will do that.

We now examine the five steps.

1. Choice of a Model: We use a fair coin to model the 50/50 split.

Heads: In favor
Tails: Opposed

2. Definition of a Trial: A trial consists of tossing the coin 25 times, once for each person in the poll. We record the number of heads—the number of persons in favor of raising the driving age.

3. Definition of a Successful Trial: The trial is a success if 18 *or more* favor raising the driving age.

4. Repetition of Trials: We should do about 100 trials in order to get a good estimate of p(18 or more in favor of raising the driving age). Table 6.3 presents the results of 100 trials.

5. Finding the Probability of Interest: From our table of outcomes (Table 6.3), we see that 18 or more heads (in 25 tosses of a coin) were obtained 3 times in 100 trials. So,

$$P(\text{18 or more in favor}) = \frac{3}{100} = 0.03$$

Let's think about our result. Just by chance, we obtained 18 or more heads in 25 tosses of a coin 3 times out of 100. In the terms of our problem, this is like getting 18 or more people who favor raising the driving age in a sample of 25 people *even though the community is evenly split on the issue.* On the basis of our experiment, we conclude that getting 18 or more people in a sample of 25 happens *very rarely* in a 50/50 population, because in our 100 trials, it happened only 3 times.

Table 6.3 **Number of Heads in 100 Trials of Tossing a Fair Coin 25 Times**

Number of heads obtained in 25 tosses	Frequency	
0	0	
1	0	
2	0	
3	0	
4	0	
5	0	
6	0	
7	1	
8	2	
9	7	
10	9	
11	14	
12	18	
13	15	
14	11	
15	8	
16	7	
17	5	
18	2	18 or more in favor
19	1	
20	0	
21	0	
22	0	
23	0	
24	0	
25	0	

From these results, the class *should* conclude that the community favors raising the driving age.

Another important idea emerges here. Suppose the 50/50 split holds in the community. Then the data *can* deceive us by chance. For example, 19 heads did actually occur once in the simulation of a 50/50 population. That is, we might erroneously conclude, if chance acted against us and produced an 18 or 19, that the community favors raising the driving age when in fact it *is* split 50/50.

In Chapter 13 we will learn how to compute p(at least 18 for a 50/50 split). Then doing a simulation will not be necessary. But the experience of doing one is quite instructive here in helping us understand how hypothesis testing actually works.

Statistics and the Stars

People have studied the stars for untold centuries. But only within the last one hundred years have the relatively modern concepts of statistics and probability been applied to astronomy. As we look at the stars, it may seem that they are scattered at random in the heavens. Closer observation, however, reveals groups of stars—double stars and sometimes chains of four or five that form a glittering necklace. Have such arrangements occurred by chance?

An astronomer in the eighteenth century, John Michell, calculated the probability that the star pairings he found could have occurred by chance under the hypothesis of a random scattering of stars. He obtained such small probabilities that he concluded that some stars must be physically linked together. This conclusion was later verified by the astronomer William Herschel.

Such calculations of unlikelihood, or improbability, have been applied to other configurations of stars and galaxies as part of a continuing quest for more knowledge about the origins of the universe. This is exactly the kind of reasoning we used in the example of the driving-age survey.

Example 6.1

Let's revisit the Key Problem. The standard drug used in the treatment of a certain disease is known to be effective 60% of the time. An experimental drug is used on 10 patients in a clinical trial and is found to be effective for 8 of them. Is this reliable evidence that the experimental drug has a better than 60% cure rate?

Solution

The fact that there were 8 cures out of 10 treatments does *suggest* the possibility that the experimental drug is better. The previous cure rate of 0.6 leads us to expect that on average there should be 6 cures out of 10 treatments. But it also seems quite likely that a drug with a true cure rate of 0.6 might easily produce 8 out of 10 on occasion. So we ask whether achieving 8 cures is reliable evidence of real improvement. We want to know whether there has been improvement in the actual, or real, underlying cure rate—what the experimental drug would achieve if it were used thousands and thousands of times. By *real improvement* we would mean, then, that for the experimental drug the actual, or real, cure rate has increased above 0.6.

Again, it seems likely that a drug with a true probability of 0.6 might easily produce 8 out of 10 cures. If we are not willing, then, to claim that getting 2 more cures than the expected 6 cures is strong evidence of improvement, how many cures *would* we require before stating with confidence that the experimental drug is better? Do we insist that we need to get 9 out of 10, or even a perfect 10 out of 10 before we are convinced? In other words, how big an improvement do we require before we are willing to say, "Yes, that *does* indicate real improvement"?

From a statistical point of view, the logic that we use to answer this question goes along these lines:

a. We make the null hypothesis assumption that the experimental drug has *not* improved from the standard 60% cure rate (just as we assumed earlier that 50% of the community favored raising the driving age).

b. Under this assumption, we ask how likely it is that we would get 8 cures in 10 treatments or do even better. (We admit that it is entirely possible to get 8 cures out of 10 with a cure rate of 0.6. Indeed, it is even possible—but not very likely—to get 10 cures out of 10.)

c. We then do a five-step experiment (because we do not yet know how to directly compute the theoretical probability) to estimate the probability of getting 8 or more cures out of 10, *under the assumption that the experimental drug is not better* (that is, it remains at 0.6). We are interested in 8 or more cures, because getting more than 8 cures would be even stronger evidence of improvement.

d. If this probability is large (what *large* means will be defined later!), we conclude with a statement such as, "It is not unlikely (that is, the estimated probability is large) to get 8 cures out of 10 if a drug's cure rate is 0.6. So we have no reliable evidence that the experimental drug is better."

   We are taking the position that we will accept the claim that the experimental drug is better only if there is strong evidence to support it. Clearly, 8 or more out of 10 provides some evidence. But the key issue is whether this evidence is strong!

e. If this probability is small (a nominal value for *small* will also be defined later), we conclude with a statement such as, "It is unusual (the probability is small) to get cures 8 out of 10 treatments with a cure rate of 0.6. Thus, we are doubtful that the experimental drug has a cure rate of 0.6." Hence we are willing to conclude there has been improvement, knowing, of course, that there is still some chance we have been misled by the data.

We can use our familiar five-step procedure to solve this problem.

1. Choice of a model: We need to estimate the probability of observing 8 or more cures out of 10 treatments, assuming a cure rate of 0.6. We can use a box model of 10 digits (0–9) where

$$1, 2, 3, 4, 5, 6 \leftrightarrow \text{Cure}$$
$$7, 8, 9, 0 \leftrightarrow \text{No cure}$$

For expediency, as usual, we substitute drawing random numbers for our physical model, which in this case is a box model.

2. Definition of a Trial: Read 10 random digits (one for each patient).

3. Definition of a successful trial: Count the number of "cures" (that is, digits 1 through 6, inclusive) achieved in the 10 treatments. Define the trial as successful if 8, 9, or 10 cures occur.

Table 6.4 **Estimating *p*(8 or more cures out of 10 attempts) Using 25 Trials**

Trial	Random digits		Number of cures	Success?
1	32236	12683	9	Yes
2	41949	91807	4	No
3	57883	65394	6	No
4	35595	39198	6	No
5	75268	40336	7	No
6	50658	32089	5	No
7	78007	58644	4	No
8	73823	62854	7	No
9	31151	64726	9	Yes
10	88795	93736	4	No
11	22189	47004	5	No
12	48304	77410	5	No
13	78871	98387	2	No
14	44647	12807	6	No
15	65194	58586	7	No
16	78232	57097	4	No
17	01430	00304	5	No
18	32036	23671	8	Yes
19	62932	99837	5	No
20	20160	27792	5	No
21	37090	62165	6	No
22	11172	66827	7	No
23	39830	04587	4	No
24	64810	25649	7	No
25	56530	94864	6	No

4. Repetition of trials: We do 25 trials (100 would be much better). The results appear in Table 6.4.

5. Finding the probability of interest: We wish to estimate the probability of getting 8 or more cures out of 10 attempts:

$$p(\text{getting 8 or more cures out of 10 attempts})$$

Using the results of step 4, given in Table 6.4, the corresponding experimental probability is

$$P(\text{getting 8 or more cures out of 10 attempts}) = \frac{3}{25} = 0.12$$

based on 25 trials.

Can this be judged a rare event? Statisticians have made their minds up long ago on what criteria to use. Almost universally, 0.05 is used as the value that the probability must be less than before we are willing to declare the observed data rare enough to discard the null hypothesis (here that the new drug has a cure rate of 0.6). Occasionally 0.1 is used, and occasionally 0.01 is used.

We note that the actual theoretical probability can be shown to be about 0.17, even bigger than the estimated 0.12. Even using 0.12, we simply cannot claim to have strong evidence that the new drug is better. (It *may* be, but the data do not make the case strongly enough to claim so.) Suppose the manufacturer of the experimental drug claims we should have allowed 50 treatments to allow for the possibility of stronger evidence. Do you think this is a good idea?

We do not have reliable evidence that the experimental drug is better than the standard drug, so we withhold judgment. That is, we do not conclude that the experimental drug is *not* an improvement. We merely withhold judgment. This illustrates an underlying truth in science. It is easier to disprove hypotheses, theories, and so on, than to confirm them. Thus science progresses very slowly in its acceptance of new theories and the corresponding rejection of old ones.

As we have seen numerous times in this textbook, the first step in doing a probability experiment empirically to solve a problem is to decide which physical model to use to produce the outcomes with the desired probabilities. Examples of such models include a coin, a die, or a box of tickets. Also, we often use random digits that we get from a table, a calculator, or a computer to stand in for the physical model.

In statistical decision making, it is important to choose a model that is likely to describe the mechanism producing the observed data. That is, we seldom interpret data without a viewpoint (scientific model) that makes certain assumptions that we are willing to accept as truth about the nature of the data-producing mechanism. That is, certain assumptions are made about the nature of the data being considered.

Consider our example above. If we are looking at a drug trial, we might assume independence among patients and assign a particular parameter value to the probability of a cure for each randomly chosen patient. This is really our scientific model, which in fact states those things we are willing to assume before we allow ourselves to be influenced by the data. Then, as in the above example, we look at data in an effort to either disprove or confirm the model. Thus statistical decisions result from the combination of two sources: the model used *and* the data collected.

Example 6.2

Rob claims he has the power of mind over matter (psychokinesis). He claims he can make a coin come up heads several times in a row. Mary is skeptical and challenges him to demonstrate his powers. He then tosses a coin repeatedly, getting

H H H H H T. Mary says, "See, you failed on the sixth toss." Rob says, "Well, I don't claim to have perfect control over the coin; I merely produce improbably long strings of heads in a row."

We need to make a decision. Does Rob have the power he claims, or could H H H H H T have occurred just by chance for a perfectly fair coin? How long a string of consecutive heads is needed for us to be persuaded that the fair coin null hypothesis is implausible?

Solution

We make our decision in the same way as in the above examples. We hypothesize a model that denies Rob's claim, and we ask how strong the evidence from the data is that this model is wrong. Our model, then, is that of a fair coin repeatedly tossed. That is, we *assume* that Rob has no power over the coin's outcome. Then we estimate how likely it is that he would get five or more heads in a row before the first tail (at least five heads before the first tail).

If such an event occurs fairly often by chance, we will conclude that he has no influence on the coin. That is, although the data *hint* that Rob may have an influence on the coin tossing, his performance is not good enough to provide conclusive evidence that he has actually increased the chance of a head above 50%. If we instead find that such a result (five or more heads in a row) was unlikely to obtain from a fair coin, we would conclude that he does have an influence on the coin.

As discussed above, this sort of argument is powerful and is used repeatedly by statisticians to answer yes/no questions from data. An interesting analogy is that of a trial by jury. A person charged with a crime is presumed innocent until proven guilty. That is, under the jury system, the model assumed is that of *innocence.* If the members of the jury are not presented with enough evidence to persuade them of the defendant's guilt, they pronounce the person innocent, which corresponds to our failing to reject the null hypothesis. Other individuals looking at the evidence may conclude otherwise (and indeed, the defendant may in fact be guilty). But the jury's decision means that they did not have sufficient evidence to assert that the defendant is guilty. So, to give the accused the benefit of the doubt, innocence is presumed.

Similarly, we are not willing to conclude that Rob influences coin tossing unless the five (or more) heads in a row provides convincing evidence that the assumed model of a fair coin is false. Is five or more heads in a row convincing evidence? We use our familiar five steps to help us make the decision. Now, however, we add a sixth step, that of making our decision. Let's see how we can decide if Rob has ESP.

1. Choice of a model: We assume Rob has no influence, so a good model to use is a fair coin.

2. Definition of a trial: A trial consists of tossing the coin until the first tail occurs. We record the number of heads (possibly 0) in a row before the first tail.

3. Definition of a successful trial: The trial is a success if the number of heads occurring before a tail is five or more.

Table 6.5 P(5 or more heads in a row)

Number of heads in a row	Frequency
0	48
1	27
2	11
3	6
4	4
5	2
6	0
7	1
8	1
	100

4. Repetition of trials: We do 100 trials. The results appear in Table 6.5.

5. Finding the probability of interest: This empirical probability comes from the results of step 4. We are told that Rob got 5 heads in a row. So we want to determine

$$P(\text{five or more heads in a row})$$

We see from Table 6.5 that the probability of getting five or more heads in a row is estimated as

$$\frac{4}{100} = 0.04$$

These 100 trials suggest that obtaining five or more consecutive heads is unusual. Rob's coin tossing is convincing evidence that something other than the chance behavior of a fair coin is going on.

Our decision-making procedure involves a sixth step added to the usual five-step method.

6. Making a decision about the proposed model: If *P*(obtained statistic or greater) is *large,* then we decide that the hypothesized model is acceptable. (That is, the obtained statistic may be regarded as a chance event likely to have been produced by the model.) If *P*(obtained statistic or greater) is as small as 0.05 or smaller, then we conclude that the model is incorrect and we reject it as a source of the data. In our case, the occurrence of five heads in a row thus leads to rejection of fair coin tossing (and *maybe* that Rob has a special influence on the coin).

Suppose, instead, that Rob had gotten three heads in a row. Without knowing anything about statistics, we have the feeling that this is not very unusual for a fair coin. But now we have a way of testing our hunch. We see from our frequency table that getting three (or more) heads happened 14 out of 100 trials. Therefore, we would conclude that the evidence is weak that Rob is influencing the coin and that it is likely just an ordinary fair coin toss.

Large and *small* are defined differently in different circumstances. For now, we simply use the following rule, which is usually preferred in statistics: A *large probability* is any probability greater than 0.05. A *small probability* is any probability less than or equal to 0.05. Such a "small" probability is our standard for rejecting a null hypothesis.

Example 6.3

A certain disease is contagious for 30% (0.30) of the people exposed to the disease. A new vaccine is given to 20 people exposed to the disease, and only 4 of them become infected. Is the new vaccine effective?

Solution

As the null hypothesis that we wish to challenge, we assume that the vaccine is not effective. In that case the probability of infection for a vaccinated person is the same as for everybody else—30%. So we need a model for a probability of infection of 0.3 for each person in a 20-person trial. Then we will count the number of trials in which out of 20 persons, 4 persons *or fewer* were infected. Note that an important change has occurred. Here we will be estimating p(observed statistic's value or less) to make our decision. Note that until now we have always estimated p(observed statistic's value or *more*). You should be alert to this possibility—namely, that some null hypotheses are rejected because the statistic of interest is improbably *small*.

1. Choice of a Model: We will use a 10-ticket box model, where

1, 2, 3: Infected
4, 5, 6, 7, 8, 9, 0: Not infected

2. Definition of a Trial: A trial consists of doing 20 draws with replacement from the box, one for each person given the vaccine.

3. Definition of a Successful Trial: Record the number of persons infected in each trial. Count as successful a trial in which four *or fewer* people become infected.

4. Repetition of Trials: We do 100 trials. (The results are in Table 6.6).

5. Finding the Probability of Interest: We see from Table 6.6 that four or fewer people out of 20 became infected in 21 out of the 100 trials. So we have

$$P(4 \textit{ or fewer} \text{ persons infected}) = 0.21$$

6. Decision: According to the result of step 5 (0.21), it is not very unusual that four or fewer people become infected by this disease out of 20. We have *no* conclusive evidence that the vaccine is effective.

Table 6.6 **P(4 or fewer persons infected)**

Number of persons infected out of 20	Frequency	
0	0	4 or fewer persons
1	1	
2	6	
3	7	
4	7	
5	14	
6	24	
7	15	
8	13	
9	6	
10	5	
11	2	
12	0	
13	0	
14	0	
15	0	
16	0	
17	0	
18	0	
19	0	
20	0	
	100	

SECTION 6.2 EXERCISES

1. Recall the problem of deciding whether the community is equally split or whether it favors raising the driving age. Our solution led to Table 6.3, the table of outcomes of tossing a coin 25 times. Using Table 6.3, find the following experimental probabilities:
   a. *P*(16 or more heads)
   b. *P*(18 or more heads)
   c. *P*(20 or more heads)
   d. *P*(12 or fewer heads)
   e. *P*(9 or fewer heads)
   f. *P*(5 or fewer heads)
2. In a sample of 25 college freshmen, 15 people voted in favor of lowering the drinking age from 21 to 18. Can you conclude that the freshman class would be in favor of lowering the drinking age? Use the five-step method and Table 6.3 to determine your answer.
3. Suppose the cure rate for a certain drug is 60%. Scientists claim they have found a new drug having a higher cure rate than the older drug. To test their claim, you give the new drug to 10 infected people. Nine of them are cured. Use the five-step method and random numbers from Table 6.4—but with a different step 1 model—to decide if the new drug is really more effective than the old drug.
4. In Exercise 3, suppose instead that the cure rate for the old drug is 70%. Use the five-step method and the random numbers from a table like Table 6.4—but with a different step

1 model—to decide if the new drug is really more effective than the old drug.

5. Suppose 5 out of 25 executives in a large company are women. Use the five-step method and Table 6.3 to determine whether this company has a gender bias at the executive level.
6. In Exercise 5, what is the lowest number of women executives the company could have and *not* display statistically strong evidence of a gender bias? Use the five-step method and Table 6.3 to determine your answer.
7. In a large city with a nonwhite population of 60%, there have been numerous complaints about racial bias in the jury system. On a recent jury there were four nonwhites among the 12 jurors. The following table presents the results of 100 simulations made using the five-step method as modeled in Example 6.1. Use the table to determine if the complaints of racial bias are valid.

Number of nonwhite jurors	Frequency
0	0
1	0
2	0
3	0
4	2
5	8
6	17
7	29
8	20
9	13
10	10
11	1
12	0

8. Voters are believed to be evenly split on a constitutional amendment. A telephone poll of 20 voters results in 14 being in favor of the amendment. Should it be concluded, on the basis of this poll, that the community favors the amendment? Use the five-step method and the table below to determine your answer.

Number of heads	Frequency	Number of heads	Frequency
0	0	11	17
1	0	12	12
2	0	13	17
3	1	14	2
4	1	15	1
5	3	16	0
6	2	17	0
7	6	18	1
8	9	19	0
9	12	20	0
10	16		

9. Suppose the poll in Exercise 8 had shown that 7 out of 20 people were in favor of the amendment. Should it be concluded, on the basis of this poll, that the community is opposed to the amendment? Use the five-step method and the table in Exercise 8 to determine your answer.
10. Records from several years ago show that in a certain community, 50% of the people owned their homes. A survey of 20 families, just taken, shows that 15 own their homes. Is this reliable evidence that there has been an increase in home ownership in the community? Use the five-step method and the table in Exercise 8 to determine your answer.

Graphing Calculator Exercises

The exercises in this section make use of the program FLIPS. The program asks for the number of (fair) coins to be flipped and the number of times you want to flip them. For large numbers of coins (20 or more) or flips (50 or more), the program may take 60 seconds or more to execute. View the results in a data table (STAT [edit]). The number of heads (0 through N, where N is the number of coins tossed) is in L1, and the frequency (number of times each outcome was obtained) is in L2. Do 50 trials unless otherwise indicated.

11. Simulate flipping 10 coins 50 times. Use the results to find the following:
a. P(3 or more heads)
b. P(4 or fewer heads)
c. P(6 or more heads)
d. P(8 or fewer heads)
e. P(9 or more heads)

12. Arlene claims that she can tell the difference between margarine and butter. Suppose she cannot tell and only guesses. If she is given 10 pieces of bread with either margarine or butter, what is the likelihood that simply by guessing, Arlene will succeed in telling the difference eight or more times?

13. A jury consisting of three women and nine men is selected. The lawyer for the defense believes there was sex bias (in favor of men) in selecting the jury. What is the probability that nine or more men are chosen if there is no bias (that is, men and women are selected with equal likelihood)?

6.3 THE MEDIAN TEST

The median can be used to help make decisions about the equality of the centers of two populations. The procedure presented below is often used in statistics.

Example 6.4

Suppose there is a new way of teaching spelling in grade schools. A teacher has two classes of students, and the students were assigned randomly to the two classes. She teaches one of the classes using the new method, A. She teaches the other class using her usual method, B. Then she gives both classes the same spelling test. These scores are obtained:

Method A:	10	10	10	12	15	17	17	19	20	22	25	26
Method B:	6	7	8	8	12	16	19	19	22			

We want to decide whether method A is better than method B. That is, did the students get a higher average (median) test score with method A than with method B?

Solution

It is required that the classes have comparable spelling ability before being taught by the two methods. Here, as in actual research, the researcher achieves such comparable groups by assigning people (or objects) to the groups at random, thus "averaging out" differences between the people or objects of the two groups.

At first glance it appears that the method A students are doing better. One way to statistically assess this impression is to compare all 21 students' performances with the median of the 21 obtained scores. If the two methods are equally effective, we would expect *roughly* the same proportion of students in each class to be above the common median. The median of this combined set of scores is 16. Now we write a + for each score that is above the median score of 16. Since the scores were given in order, it is easy to see that we have the following:

Method A: + + + + + + +
Method B: + + +

But from a statistical point of view, we recognize that the two groups of students could get different scores on the same test just by chance. Just as a fair coin can easily produce seven heads, the method B students could have done worse on this test just by chance. Thus we need to do a formal six-step analysis by which we will see how likely it is that the scores differ this much by chance. Let's follow our six steps to make our decision.

1. Choice of a model: We require a model that randomly distributes 12 of the 21 numerical scores given above to the method A classroom with the remaining going to the method B classroom. In this manner we obtain a null hypothesis model that presumes there is no influence of teaching method on spelling test performance. Then we can use this null hypothesis model to assess whether the seven pluses and three pluses actually obtained occur rarely or not under the null hypothesis. More precisely, if our null hypothesis model only rarely produces as big a difference between the two groups as appeared in the original test, we would conclude that the model is wrong and hence that method A is better.

We will redistribute the above test scores at random to the students taught by the two methods. One way to do this is to use a box model with the 21 scores as tickets.

2. Definition of a trial: Draw 12 tickets at random *without replacement* to be the scores for group A. The remaining tickets give the scores for group B. The median of the combined group of 21 scores is still 16, since we still have the same 21 scores.

3. Definition of a Successful Trial: Find the number of scores in group A that are greater than 16, and record this number. Define the trial as successful if there are seven or more such group A scores.

4. Repetition of trials: We do 50 trials (100 would be better). Table 6.7 shows the results of 50 trials.

5. Finding the Probability of Interest: We find in Table 6.7 that seven *or more* scores in group A were greater than the median of the combined group in 15 of the 50 trials. Thus,

$$P(7 \text{ or more}) = \frac{15}{50} = 0.30$$

Table 6.7 **P(7 or more group A scores larger than combined median)**

Number of scores greater than 16	Frequency	
0	0	
1	0	
2	0	
3	1	
4	8	
5	10	
6	16	
7	10	7 or more
8	5	
9	0	
10	0	
11	0	
12	0	
	50	

6. Decision: Getting seven or more scores in group A that are larger than the median for the combined group happens often by chance, namely, 30% of the time. Because getting seven or more scores in group A larger than the median is *not* a rare event, we decide that the model chosen in step 1 could be appropriate. The difference in scores between the two groups observed in the original test could easily have happened just by chance. Therefore we *do not* have strong evidence that method A is better, and our initial impression that method A might be better has not been proven by the data.

The problem of deciding whether two populations being sampled from are the same or different in their centering, spread, or some other characteristic is a very important one in statistics. It is often called the *two-sample problem.*

SECTION 6.3 EXERCISES

1. A new feed for cattle is being developed. It is tried out in the laboratory with a group of cattle called group I, or the experimental group. Group II, or the control group, continues to use the old feed. The following weight gains (in pounds) are recorded:

Group I	Group II	Group I	Group II
104	62	204	109
109	83	209	109
127	90	266	205
143	101	277	
187	106		

Can you conclude that the new feed is effective in producing weight gain? Use the five-step method and the table below to determine your answer.

Number above median	Frequency
0	0
1	0
2	2
3	17
4	39
5	32
6	9
7	1
8	0
9	0

2. Two groups of rats run a maze under the influence of two different drugs. The following error scores are recorded:

Group A:	14	12	8	9	14	17	14
Group B:	12	8	11	7	14	8	8

Do the drugs have different effects on the error scores made by the rats? Use the five-step method and the table below to determine your answer.

Number above median	Frequency
0	0
1	1
2	11
3	42
4	33
5	13
6	0
7	0

3. Two groups, one consisting of farmers and the other of union members, were given an attitude test on the use of food stamps. Below are their scores. The following values of the attitude index were obtained:

Farmers	Union members
12	28
10	26
10	25
9	24
8	24
7	19
7	18
6	10
4	9
2	6

Do the groups differ in their attitudes? Use the five-step method and the table below to determine your answer.

Number above median	Frequency
0	0
1	0
2	7
3	23
4	36
5	23
6	10
7	1
8	0
9	0
10	0

4. To test the effectiveness of the new weight-loss drug, Redux, 40 women were split into two groups: the control group, group A, who took

a placebo drug, and the experimental group, group B, who took the real drug, Redux. The amount of weight lost by each member of the groups over a six-month time period is noted below:

Group A	Group B
3	5
4	5
5	7
6	7
7	10
10	10
11	10
12	10
15	15
18	18
19	20
20	24
23	28
24	29
25	38
30	42
33	44
38	49
40	50
42	55

Do the groups differ in the amount of weight lost? Use the five-step method and the table below to determine your answer.

Number above median	Frequency
0	0
1	0
2	0
3	0
4	0
5	0
6	3
7	4
8	11
9	27
10	15
11	19
12	11
13	5
14	4
15	1
16	0
17	0
18	0
19	0
20	0

Graphing Calculator Exercises

For this section, use the program LOAD. This program is similar to FLIPS, with the important difference that it simulates the tossing of a coin that is loaded instead of one that is fair. The probability of getting heads with the loaded coin may be specified in 10ths, from 0.1 to 0.9 (so when the probability of getting heads is specified as 0.5, LOAD works the same as FLIPS). Do 50 trials unless otherwise indicated.

5. Consider the drug trials of the Key Problem. Suppose there are 10 cures in 12 trials with the new drug. Determine whether the new drug is more effective (that is, has a cure rate better than 0.6).

6. A certain medication cures a skin rash in 8 out of 10 applications, according to medical research. A new medication is tried on 20 patients

and cures 17 of them. Is this evidence that the new medication is more effective than the old?

7. Repeat Exercise 6, but this time suppose the new medication cures 19 out of 20 cases.

8. A certain die is believed to be loaded in favor of even numbers. You roll the die 50 times and observe an even number 35 times and an odd number 15 times. Decide whether the die is loaded.

CHAPTER REVIEW EXERCISES

1. Gallup takes a sample of 2000 people to determine the percentage of people in the nation who would vote for a certain presidential candidate. Identify the population the sample was taken from.
2. A mayoral candidate wants to determine what percentage of people in the city would vote for her in the upcoming election. She assigns you the task of finding a sample of the city's residents. What characteristics should the population for your sample have?
3. A sample of 25 residents of the city was taken for the mayoral candidate in Exercise 2. Eighteen of them would vote for the candidate. Should the candidate feel confident about being elected mayor? Use the five-step method and Table 6.3 to determine your answer.
4. Josh went to a basketball clinic over the summer to improve on his three-point shooting. Before camp, out of 10 three-point shots, Josh would make 1, on average. Now, out of 10 three-point shots, Josh makes 3. Has Josh's three-point shooting improved? Use the five-step method and the random numbers from Table B.3 to determine your answer.
5. Consider Exercise 4. Suppose Josh used to make 1 out of 10 three-point shots on average, as in Exercise 4, but now makes 6 out of 20 after attending the clinic. Has Josh's three-point shooting improved? Use the five-step method and the random numbers from Table B.3 to determine your answer. Is your answer the same as in Exercise 4? (*Hint:* Use Table B.3 and combine the first 20 random numbers, combine the second 20 random numbers, etc.)
6. A biologist believes she has developed a method for irradiating the genes of fruit flies so that the ratio of males to females in the offspring will be smaller (there will be more females than males). In the first 20 of the resulting offspring after the treatment, 13 are females and 7 are males. Does the method the biologist used work? Use the five-step method and the table below to determine your answer.

Number of heads	Frequency	Number of heads	Frequency
0	0	11	17
1	0	12	12
2	0	13	17
3	1	14	2
4	1	15	1
5	3	16	0
6	2	17	0
7	6	18	1
8	9	19	0
9	12	20	0
10	16		

7. Engineers for General Motors want to increase the fuel economy of a new car by changing its type of engine. To test their theory, 16 cars were driven and the fuel economy (in miles per gallon) was calculated for each car. Here are the results:

Old engine:	19.8	18.8	20.9	17.6	19.4	20.8	18.3	20.7
New engine:	21.5	19.9	22.6	18.2	19.7	20.4	18.1	20.8

Do the cars with the new engines have better fuel economy (have higher miles per gallon readings)? Use the five-step method and the table below to determine your answer.

Number above median	Frequency
0	0
1	0
2	6
3	25
4	37
5	22
6	8
7	2
8	0

8. To test a new method of teaching calculus, 40 students were split into two groups. After completing the class, the two groups took the same final exam. Here are the results:

Old method:	40	61	65	66	68	71	74	74	74	78
	81	83	86	86	90	91	95	96	98	98
New method:	35	65	70	71	78	79	79	79	80	80
	83	83	85	87	88	92	96	96	98	100

Does the new method improve student learning? Use the five-step method and the table below to determine your answer.

Number above median	Frequency
0	0
1	0
2	0
3	0
4	0
5	0
6	1
7	3
8	12
9	28
10	20
11	20
12	9
13	5
14	2
15	0
16	0
17	0
18	0
19	0
20	0

6

COMPUTER EXERCISES

1. RAISING THE DRIVING AGE

Recall the study described in Section 6.2 in which 25 people were asked whether they were in favor of raising the driving age. It turned out that 18 out of 25, or 72%, were in favor. The question was whether that is evidence that over half of the entire community was in favor. Start the Five Step program, and perform the first five steps as in the text: (1) The box model has one 0 and one 1. (2) A trial is drawing 25 with replacement. (3) The statistic is "Sum," and the probability of interest is the probability that the statistic is greater than or equal to 18. (4) Perform 400 trials. (5) Find the proportion of successes.

Now step 6 is to make a decision: Based on the result in step 5, do you think the box with one 0 and one 1 is plausible? What do you conclude about the community's attitude toward raising the driving age?

2. TEACHING METHODS

Look again at Example 6.4. The question is whether the scores from using the new teaching method (A) are higher than those for the old method (B) by chance, or whether method A is actually superior. In the example, 7 of the 12 method A scores were above the median of 16. If the students were assigned to the methods randomly, and both methods are equally good, then what is the chance that 7 or more of the students using method A would be above the median?

To address the problem, set up a box model with the 21 scores, but code them as 1 if they are over 16, and 0 if they are not. It happens that 10 are over 16, so the box should have ten 1s and eleven 0s. Now under the null hypothesis model that the methods are equally good, the method A people can be modeled drawing 12 randomly from this box, because all these students have the same chance of exceeding the median, and counting how many scores are above the median (that is, counting the number of 1s). One thing we have to be careful of this time is to note that we should draw *without* replacement. That is, after drawing a ticket, we do not put it back into the box because we do not want to choose the same person more than once. Now define the trial to be "Draw n w/o replacement" with "n = 12." The statistic is "Stat >= a" with "a = 7." Now perform 200 trials. What do you get as the experimental probability of getting 7 or more? Does this value lead you to conclude that method A is superior?

3. MALE/FEMALE RATIO

Recall the class data in Computer Exercise 3 of Chapter 2. In the STAT 100 class, there were 102 women and 63 men, so the percentage of women was about 62%. At the University of Illinois there are about 46% women and 54% men overall. Thus the percentage of women in STAT 100 is quite a bit higher than in the university as a whole. Could this be just due to chance, or is there a real tendency for women to take this course? To test this hypothesis, set up the null hypothesis step 1 model that the 165 people in the class can be modeled by 165 draws with replacement from a box in which there are 46 1s (the women) and 54 0s (the men). What is the chance of choosing 102 or more women this way?

Set up the box with 46 1s and 54 0s. A trial is to draw 165 with replacement. The statistic is "Sum," and the probability of interest is the chance that the statistic is greater than or equal to 102. Perform 100 trials.

a. What is the average number of women for these trials? How close is that to 102?

b. What is the proportion of trials in which the number of women was 102 or more? Do you think this box yields a plausible model for the enrollment in this course?

c. What is the maximum number of women for these trials? (Look in the "Statistics" choice of the Summary Statistics menu.) How close is that number to 102?

d. Does it appear that women are more likely to take this course than the overall percentage at the university would indicate?

4. DRAFT LOTTERY

Reread the draft lottery example in Computer Exercise 5 of Chapter 3. Start the Data program. (See the computer exercises in Chapter 1 to refresh yourself about the Data program, if necessary.) Load the DRAFT69M.DAT data set, which has the average lottery numbers for the twelve months.

Choose "Display the Data" from the Data menu. Look at the average lottery numbers. It looks as if the averages for earlier months are higher than for later months. Is this because of chance, or is there something wrong with the randomization process?

a. Find the median of these twelve averages.

b. Look at the average lottery numbers for the first six months of the year. How many of them are above the median found in part 1? Call this number *k*.

c. Next, we will use the Five Step program to see what the chance is that in the first six months, *k* or more averages are above the median, assuming the null hypothesis. In this case, the null hypothesis is that the averages are equally likely to be allocated to the months in any order. The box thus has six 1s (representing averages above the median) and six 0s (representing averages below the median). A trial is to choose six without replacement. The statistic is still the sum, and the probability of interest is the chance that the statistic is greater than or equal to *k*. You have to supply the value of *k* from part (b).

   Perform 200 trials. What proportion had *k* or more averages above the median?

d. What do you conclude from part (c) about the randomization in the lottery? (i) It appears that there is a definite bias toward higher numbers being assigned to earlier months, or (ii) the observed bias could be just due to chance.

Now load DRAFT70M.DAT, and repeat the above steps.

e. Find the median of the twelve averages.

f. How many of the averages are above the median found in part (e)? Call this number *k*.

g. Perform the same steps as in part (c), but with the new value of *k*. What proportion had *k* or more averages above the median?

h. What do you conclude from part (g) about the randomization in this 1970 lottery? (i) It appears that there is a definite bias toward higher numbers being assigned to earlier months, (ii) it appears that there is a definite bias toward higher numbers being assigned to later months, or (iii) there does not appear to be any bias either way.

7

CHI-SQUARE TESTING

I cannot believe that God plays dice with the world

Albert Einstein

OBJECTIVES

After studying this chapter, you will understand the following:

- *The hypothesis of a fair die*
- *The chi-square statistic for assessing the hypothesis of a fair die*
- *How to use the six-step method to test a hypothesis using a chi-square statistic*
- *How to use a smooth chi-square curve in place of the six-step method*
- *Chi-square testing of unequal hypothetical probabilities*

7.1 IS THE DIE FAIR? 318

7.2 HOW BIG A DIFFERENCE MAKES A DIFFERENCE? 321

7.3 THE CHI-SQUARE STATISTIC 325

7.4 SIX STEPS TO CHI-SQUARE 329

7.5 SMOOTH CHI-SQUARE CURVES 335

7.6 CHI-SQUARE TABLES 345

7.7 UNEQUAL EXPECTED FREQUENCIES 351

CHAPTER REVIEW EXERCISES 357

COMPUTER EXERCISES 365

Key Problem

Is the Die Fair?

For thousands of years people have used dice to help them make decisions. Some mathematics instructors have been known to use a die to decide who are to turn in their homework! So a natural question to ask for a particular die is, Is it fair?

Suppose we roll a six-sided die 60 times and obtain the outcomes in Table 7.1. We see from the table, for example, that 1 was obtained seven times, 2 was obtained nine times, and so on. Do we think this die is fair?

Table 7.1 **Sixty Rolls of a Six-Sided Die**

Outcome	f	P
1	7	$7/60 = 0.12$
2	9	$9/60 = 0.15$
3	14	$14/60 = 0.23$
4	7	$7/60 = 0.12$
5	18	$18/60 = 0.30$
6	5	$5/60 = 0.08$
Total	60	1.00

7.1 IS THE DIE FAIR?

Let's address the Key Problem. It is a particular kind of decision-making problem, the topic of the previous chapter. We want to test the null hypothesis that the data were produced by a fair die. Although some of the focus of this chapter is on variations of the fairness of a die question, we will in fact discover that a very large number of fascinating and important real problems can be solved with the methods of this chapter. Indeed, the methods of this chapter can be used to test whether a given data set is well modeled by a particular many-sided loaded die. Thus, we require neither six categories nor equal probabilities for the categories of the null hypothesis model being tested. Consider two illustrations. Suppose a frequency table provides the number of automobile accidents for each of the seven days of the week for a large American city. The question of whether auto accidents are equally likely on each day of the week is equivalent to the question of whether a seven-sided die is fair. As a second example, Mendel's theory of genetics, which revolutionized the life sciences, predicts for a famous pea-plant breeding experiment that four kinds of peas will occur with probabilities $\frac{1}{16}$, $\frac{3}{16}$, $\frac{3}{16}$, and $\frac{9}{16}$. One could use the methods of this chapter to assess how well this four-sided loaded die model fits a frequency table of the distribution of 250 pea plants over the four categories. Statisticians refer to the set of probabilities associated with the faces of a loaded die as a *multinomial* set of probabilities.

Let's think about what sort of outcome frequencies we would expect from a fair die. A natural first question to ask is, What is meant by "fair"? One definition is the following: a fair die is one for which, when it is rolled, each side has an equal probability, $\frac{1}{6}$, of appearing. This definition is based on theoretical probabilities. From an experimental probabilistic point of view, we could say that a fair die is one for which, when it is rolled many times, the experimental probabilities of the faces appearing become closer and closer to each other. Note that the theoretical probability $\frac{1}{6}$ is an abstraction. It tells us the proportion of times each side occurs if we throw a fair die infinitely often. But we cannot ever toss a die infinitely often.

In terms of Table 7.1, what does the above definition mean? It means that since we have rolled the die 60 times, the proportion of times we obtained each side (that is, the experimental probability) should be close to $\frac{1}{6}$ (or 0.16666..., which rounds to 0.167) if the die is fair.

According to Table 7.1, we got a 1 seven times in 60 rolls of the die. In Table 7.1 and often in this chapter f denotes *frequency*, that is, the number of times the outcome occurs. So

$$P(1) = \frac{7}{60} = 0.12 \text{ (rounded)}$$

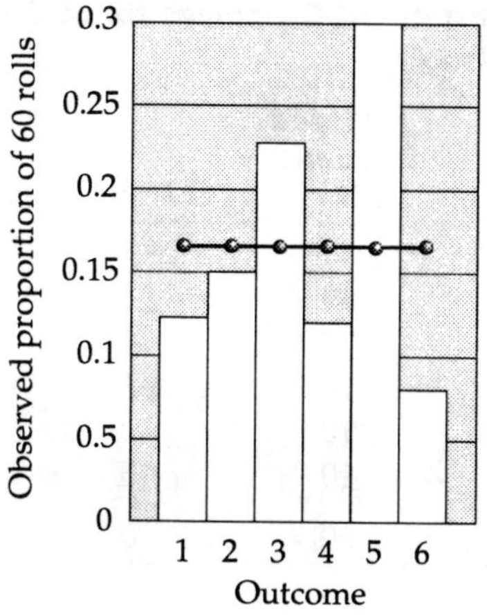

Figure 7.1 **Comparison of experimental and theoretical probabilities for a fair die.**

as compared with $p(1) = \frac{1}{6} = 0.167$ (if it is fair). Similarly $P(2)$ can be compared with $p(2) = \frac{1}{6} = 0.167$, and so on.

The relative frequency histogram (a concept discussed in Section 1.5) of Figure 7.1 presents the observed proportions (experimental probabilities) obtained in the 60 rolls of the die in Table 7.1. Note that the areas of the rectangles do add to 1, as required for a relative frequency histogram. The horizontal line is the expected proportion (theoretical probability) of outcomes for a fair die. An off-the-cuff response to Figure 7.1 might be something like, "We got a one only 7 times, but we got a five 18 times. It seems reasonable to get each outcome about 10 times. Eighteen times seems rather large. So it looks as if the die is *not* fair." But of course, just as when we toss a fair coin 60 times, we do not necessarily get 30 heads and 30 tails, when we roll a six-sided die 60 times we do not necessarily get 10 1s, 10 2s, and so on, *even if the die is fair*. So, as discussed previously, we expect some deviation from the 10 1s, 10 2s, and so on, for a die that is absolutely fair. So the question whether the die of Table 7.1 is fair is still unanswered.

We can start to get a statistical answer to this question by comparing the actual (or obtained) outcomes of the 60 rolls of the die with what we would expect from a fair die. We do this by expanding Table 7.1 as shown in Table 7.2. Let's discuss Table 7.2 column by column.

Outcome of die: Which side fell up?

Obtained frequency (O): This is the obtained frequency of the given outcome of the die in 60 rolls—that is, the number of times that side of the die appeared.

Table 7.2 **Obtained and Expected Outcomes for 60 Rolls of a Six-Sided Die, Assuming It Is Fair**

Outcome of die	Obtained frequency (O)	Expected frequency (E)	$O - E$	$\lvert O - E \rvert$
1	7	10	−3	3
2	9	10	−1	1
3	14	10	4	4
4	7	10	−3	3
5	18	10	8	8
6	5	10	−5	5
Total	60	60	0	24

Expected frequency (E): This is the expected frequency of the given outcome of the die in 60 rolls, *if the die is fair*—that is, the number of times, for a fair die, each side is expected to appear. For this table, $E = 10$ in every case. One convincing argument that $E = 10$ is correct is that if all sides have the same theoretical probability, then the expected number of occurrences of each side among the 60 rolls must be equal. So each side must be expected to occur $60/6 = 10$ times.

$O - E$: This is the difference, or deviation, between the obtained and the expected frequency of the outcome. For example, in row 1 we see that $O - E = 7 - 10 = -3$. This tells us that when the die was tossed 60 times, the 1 appeared three fewer times than expected for a fair die. On the other hand, in row 5 we see that $O - E = 18 - 10 = 8$. That is, 5 was obtained eight more times than expected.

$|O - E|$: This is the absolute value of $O - E$. That is, we ignore whether $O - E$ is positive or negative. For example, for row 6, $O - E = -5$ and $|O - E| = 5$.

Why ignore the signs? One reason can be inferred from Table 7.2. The positive and negative values in column $O - E$ cancel each other out, adding up to zero. This sum is thus not very informative about deviations of the data from the model! However, the sum of column $|O - E|$, which we see from the table is equal to 24, is informative. It tells us that in the 60 rolls of a fair die, the outcomes differed from what was expected by a total difference of 24.

We now need a way to help us decide if the total difference of 24 or more is an unusual one, or a difference that can be obtained rather often, by chance, with a die that is fair. This is clearly a decision-making (hypothesis testing) problem like those of Chapter 6. We create a new statistic, D, that we define as follows:

$$D = \text{sum of } |O - E|$$

D is the total of the absolute deviations of the observed from the estimated frequencies and is the sum of the $|O - E|$ entries in Table 7.2

We now have a way to help us find out whether a die is fair. We repeatedly compute a table like Table 7.2 using a fair die (the null hypothesis model) and thus repeatedly find the value of D. If $D = 24$ is not unusually large compared with the typical values obtained for a fair die, this is an indication that the die is quite possibly fair. But if $D = 24$ is relatively large, this suggests that the die is not fair. We leave it to the next section to help us know what a "small" or a "large" value of D is with respect to providing strong or weak evidence of an unfair die—that is, with respect to providing strong or weak evidence that the null hypothesis should be rejected.

7.2 HOW BIG A DIFFERENCE MAKES A DIFFERENCE?

We were left in Section 7.1 with the question of how to decide if a calculated value of the statistic D is "small" or "large." That is, we need a way of deciding whether an observed value of the D statistic is so large that it would happen only seldom by chance for rolls of a fair die.

Suppose we have a six-sided die that we know is fair (for example, the manufacturer has carefully constructed it to be a fair die). We roll it 60 times and obtain the results shown in Table 7.3. Let's look at the sum of $|O - E|$, which is 20. Of the 60 rolls of this fair die, the number of observed outcomes differed from what was expected by a total of 20.

We will repeat these 60 rolls of the fair die many times, and each time we will calculate a value of D. We will have as a result a frequency table of D values obtainable from 60 rolls of a fair die, and from this frequency table we will be able to determine the experimental probability of observing

Table 7.3 **Results of 60 Rolls of a Six-Sided Die Known to Be Fair**

Outcome	Obtained frequency (O)	Expected frequency (E)	$O - E$	$\|O - E\|$
1	4	10	−6	6
2	6	10	−4	4
3	11	10	1	1
4	10	10	0	0
5	15	10	5	5
6	14	10	4	4
Total	60	60	0	20

a D value of 24 or greater in the case of a fair die. If a D value of 24 is unusually high for a fair die, we would conclude that the die of Table 7.2, which yielded a D value of 24, is not fair. In summary, we would like to find $P(D \geq 24)$ for 60 rolls of a fair six-sided die, thus estimating the theoretical probability $p(D \geq 24)$.

We will take the six-step approach of Chapter 6:

1. Choice of a Model: We take a six-sided die that we know to be fair.

2. Definition of a Trial: We roll the die 60 times, and we record the outcomes of the trial in a table like Table 7.3.

3. Definition of a Successful Trial: We calculate D for each trial. Count as successful a trial in which $D \geq 24$.

4. Repetition of Trials: We do a moderately large number of trials, say, 30 (100 would be better). The results of 30 trials are presented in Table 7.4.

5. Probability Estimate: We estimate the theoretical probability of a successful trial—that is, of getting a D that is 24 or greater—using the results of our experiment. According to Table 7.4, the largest value of D obtained in the 30 trials was 20. In our 30 trials, then, we did not get a value of D that was 24 or larger. Therefore,

$$P(D \text{ is greater than or equal to } 24) = 0$$

6. Decision: We found that, on the basis of 30 trials, the experimental probability of D being greater than or equal to 24 in 60 rolls of a fair die is zero. Recall from Chapter 6 that the convention is to consider as unusual any event whose probability is 0.05 or less. Since a probability of zero is less than our criterion of 0.05 for an unusual event, we conclude that it is unlikely to get a D that is greater than or equal to 24, if the die is fair. So

Table 7.4 **Frequency Table for *D* Statistic**

D	f
4	1
6	2
8	3
10	5
12	5
14	6
16	3
18	3
20	2
Total	30

we conclude that the die of Table 7.2 is *not fair*. That is, we reject the null hypothesis model of a fair die. Of course, if we are to be able to trust our probability estimate and hence our decision, we should have done more trials of the 60-roll die experiment, say, 100 trials.

SECTION 7.2 EXERCISES

Answer these questions using the D statistic and the required table of trials.

1. Here are the results of rolling a six-sided die 60 times. Calculate D and decide whether the die is fair.

Outcome	f
1	4
2	17
3	14
4	6
5	18
6	1
Total	60

   Use Table 7.4 or create a new table by using the six-step hypothesis-testing method and doing many trials.

2. Suppose we roll a six-sided die that we assume is fair. How many times would we expect each side to occur if we roll the die
   a. 150 times?
   b. 300 times?
   c. 600 times?

3. Nancy and Pete go through one page of a telephone book and write down the last digit of 50 telephone numbers. Here are their data:

Digit:	0	1	2	3	4	5	6	7	8	9
f:	1	6	3	2	5	8	2	10	8	5

   Prepare a table of obtained and expected outcomes like Table 7.3, and find the value of D. Do you think the telephone book was a good source of random numbers? Explain. The following table gives the results of 30 simulations of 50 random digits and their associated D's.

D	f
4	1
6	0
8	2
10	1
12	2
14	7
16	7
18	2
20	4
22	3
24	1

4. A breakfast cereal company features a special offer by including one of four differently colored ballpoint pens in a box. In a shopping trip that resulted in 20 boxes of cereal, the following numbers of pens were obtained. Do you think that the company is distributing the pens in equal numbers of colors, or are some colors more likely to be obtained than others? What is the value of D?

Color	f
Blue	8
Yellow	4
Red	3
White	5
Total	20

Graphing Calculator Exercises

5. Use the program RANDOM, which rolls a fair die. The program prompts for the number of sides of the die and the number of times the die is to be rolled. The observed results will be in column L1, and the expected results will be in L2 and will be retrieved by the sequence [STAT] [edit] [ENTER].

Use the program RANDOM to simulate the rolling of a six-sided die

a. 150 times

b. 300 times

c. 600 times

Compare the expected and obtained results for these simulated rolls of a fair die. Then calculate D for each set of data.

6. A six-sided die was rolled 60 times. The observed outcomes are given below. Complete the table to give the outcomes expected if this is a fair die. From this experiment, calculate D (see Section 1.13 of the TI Graphing Calculator Supplement). Do you think that the die is fair ?

Side	Observed	Expected
1	8	
2	10	
3	16	
4	7	
5	8	
6	11	

Now, using RANDOM, generate additional sets of outcomes for 60 rolls of a *fair* six-sided die. For each set of outcomes, calculate D. Prepare a frequency distribution of the values of D, and use it to help you decide whether the six-sided die for which the data are given above (for 60 rolls of the die) is a fair die.

7. James used his phone book to generate the 50 random numbers shown in the table below. Enter the observed frequencies into L1 and the expected frequencies into L2. Calculate D. Do you think that his phone book is a good source of random numbers?

Digit	Observed	Expected	Digit	Observed	Expected
0	1	5	5	8	5
1	6	5	6	2	5
2	3	5	7	10	5
3	2	5	8	8	5
4	5	5	9	5	5

7.3 THE CHI-SQUARE STATISTIC

Calculating D for a set of outcomes of an experiment is a convenient way of telling how far away the results are from what was expected. However, a statistically better (and the usual) way of describing such results is to calculate a value called the **chi-square statistic,** written χ^2. The symbol χ is the Greek letter chi, pronounced to rhyme with *sky*. The chi-square statistic is very commonly used in practice. It is very important in the life sciences, for example.

Let's see how the chi-square statistic is computed. Table 7.5, called a *chi-square frequency table*, shows both the D statistic (the total of the $|O - E|$ column) and the chi-square statistic (the total of the $(O - E)^2/E$ column) for the same data, 60 rolls of a six-sided die. (In practice, a chi-square table does not show the value of the D statistic.) The value of χ^2 is given by

$$\chi^2 = \text{sum of } \frac{(O - E)^2}{E}$$

Therefore,

$$\chi^2 = \frac{3^2}{10} + \frac{1^2}{10} + \frac{4^2}{10} + \frac{3^2}{10} + \frac{8^2}{10} + \frac{5^2}{10}$$

$$= \frac{9}{10} + \frac{1}{10} + \frac{16}{10} + \frac{9}{10} + \frac{64}{10} + \frac{25}{10} = 12.4$$

There are similarities, but also important differences, in how we calculate D and χ^2. First, instead of taking the absolute value $|O - E|$ (that is, the magnitude of $O-E$), we square $O-E$, obtaining the square of the magnitude of $O-E$. Second, we then divide each value of $(O-E)^2$ by its corresponding expected value E.

Why divide by E? This is a natural question to raise. It can be partially justified in this way: Dividing each $(O-E)^2$ by E helps us make comparisons

Table 7.5 **Chi-Square Frequency Table**

Outcome	Obtained frequency (O)	Expected frequency (E)	$O - E$	$\|O - E\|$	$(O - E)^2$	$(O - E)^2/E$
1	7	10	−3	3	9	0.9
2	9	10	−1	1	1	0.1
3	14	10	4	4	16	1.6
4	7	10	−3	3	9	0.9
5	18	10	8	8	64	6.4
6	5	10	−5	5	25	2.5
Total	60	60	0	38		12.4

concerning whether an $(O - E)^2$ value is unusually large when the expected values differ from application to application. That is, dividing by E is a way to appropriately *scale* the $(O - E)^2$ values in the table. For just as we expect a tall person to weigh more than a short person, we expect an experiment involving more rolls of a die to produce larger values of $(O - E)^2$ than an experiment with fewer rolls. Hence, dividing by E (which is proportional to the number of rolls) is a way to appropriately rescale the values of $(O - E)^2$, just as computing the ratio weight/height seems like a good way to adjust for the influence of height on a person's weight.

Now let's ask how often the expected and observed outcomes will differ enough by chance for a fair six-sided die to produce χ^2 values as large as or larger than our obtained 12.4. To answer this question, we repeat many trials (30 here) of the experiment of rolling a die 60 times, computing χ^2 for each set of rolls (that is, for each trial). We then prepare a frequency table of χ^2 values, as shown in Table 7.6, in much the same way as we produced a frequency table of D values when estimating $P(D \geq 24)$ in the previous section. We find a stem-and-leaf plot handy for this task. From Table 7.6 we see that a χ^2 of 12.4 or larger was obtained in 1 of the 30 trials. Therefore, rounded to two significant figures,

$$P(\chi^2 \geq 12.4) = \frac{1}{30} \approx 0.03$$

We estimate that getting a χ^2 of 12.4 or larger, for a fair die, happens very rarely—about 3 times in 100. Again, we use our criterion of 0.05 as the

Table 7.6 **χ^2 Values Obtained from 30 Trials of Rolling a Fair Six-Sided Die 60 Times**

Stem	Leaf	f	
0	6, 8	2	
1	0, 4, 6	3	
2	2, 2	2	
3	6, 8	2	
4	0, 0, 2, 4	4	
5	0, 0, 0, 6, 8, 8	6	
6	0, 4, 4, 6	4	
7	0, 0, 0, 8	4	
8		0	
9	2	1	
10	6	1	
11		0	
12	4	1	$\chi^2 = 12.4$
		30	

Key: "1 0, 4, 6" means 1.0, 1.4, 1.6.

probability of an unusual event. Since 0.03 is less than 0.05, we conclude that the die whose data are in Table 7.5 is *not* fair. Note that we have really just used our six-step decision-making method to evaluate whether $\chi^2 \geq 12.4$ is unusual, as we did earlier for $D \geq 24$.

SECTION 7.3 EXERCISES

1. Consider the experiment of 60 rolls of a fair six-sided die. For each trial the χ^2 statistic is computed. Use Table 7.6 to estimate the indicated probabilities:
   a. $P(\chi^2 \geq 2.2)$
   b. $P(\chi^2 \geq 1.6)$
   c. $P(\chi^2 \geq 5.0)$
   d. $P(\chi^2 \geq 7.8)$
   e. $P(\chi^2 \geq 10.0)$
2. A six-sided die was rolled 60 times with these results:

Outcome	Obtained number	Expected number
1	8	10
2	7	10
3	13	10
4	11	10
5	15	10
6	6	10
	60	60

   a. Calculate chi-square for these data.
   b. Using Table 7.6, estimate the probability of getting a chi-square as large as or larger than the value you obtained in part (a). Do you think the die used in part (a) is fair? Why or why not?
3. In a colored pen offer (Exercise 4 of Section 7.2), this time involving six pen colors, these numbers of pens were obtained from 90 cereal boxes:

Color of pen	Number obtained
Sky blue	17
Passionate pink	31
Deep purple	7
Burnt orange	10
Boring brown	9
Anemic ash	16

Use the chi-square frequency table below to help you decide whether the manufacturer distributed equal numbers of pen colors, or whether it provided more pens of some colors than of others.

Stem-and-Leaf Plot for χ^2

Obtained from 50 trials of rolling a fair six-sided die. Each trial consists of 90 rolls.

Stem	Leaf	*f*
0	5, 7, 8	3
1	1, 3, 5, 6, 7	5
2	3, 5, 7, 7, 8	5
3	1, 1, 1, 2, 2, 3, 5, 6, 6, 6, 7, 7, 9	13
4	1, 3, 4, 4, 5, 5, 6, 9	8
5	1, 2, 3, 6, 6, 6, 7	7
6	0, 3, 3, 9	4
7	2, 3, 5, 9	4
8		0
9		0
10	3	1
		50

Key: "1 1" *stands for 1.1.*

4. During a busy day in a large city, 90 traffic tickets were issued at six locations:

Location	Number of tickets given
A	12
B	7
C	21
D	15
E	11
F	24
Total	90

Use the chi-square frequency table of Exercise 3 to help you decide if the tickets are being given with equal likelihood at each location. (Assume that all locations have the same amount of traffic.)

5. The table below gives the outcomes of rolling a four-sided die 40 times.

Outcome	Expected number	Obtained number
1	10	12
2	10	7
3	10	14
4	10	7
Total	40	40

a. Calculate chi-square for these data.
b. A fair four-sided die was rolled for a total of 50 sets of 40 rolls each. Chi-square was calculated for each set of 40 rolls, and the results are shown in the following table. Use this table to decide if the die rolled in part (a) is fair.

Stem-and-Leaf Plot for χ^2

Obtained from 50 trials of rolling a fair four-sided die. Each trial consists of 40 rolls.

Stem	Leaf	*f*
0	0, 4, 4, 6, 6, 6, 6	7
1	0, 0, 0, 0, 2, 4, 4, 4, 4, 6, 8, 8	12
2	0, 0, 2, 2, 4, 5, 5, 5, 5, 5	10
3	0, 2, 4, 4, 4, 6, 6, 8, 8	9
4	0, 0, 0, 2, 6, 6	6
5	0, 2, 4	3
6	2	1
7	2	1
8		0
9		0
10	0	1
Total		50

Key: "7 2" represents 7.2.

6. The table below tells how many of 40 persons prefer each of four kinds of orange juice. Find *D* and chi-square for these data. Then use the chi-square frequency table of Exercise 5 to help you decide if there is convincing evidence that some kinds of orange juice are preferred over others.

Kind of orange juice	Number of persons preferring
Fresh	13
Freeze-dried	11
Frozen	8
Canned	8
Total	40

Graphing Calculator Exercises

7. Refer to Exercise 5 of Section 7.2. Calculate chi-square for each of those sets of data (rolls of a fair six-sided die). See Section 1.14 of the TI Graphing Calculator Supplement.

8. See Exercise 6 of Section 7.2. The following data are obtained by 60 rolls of a six-sided die:

Side	Observed	Expected	Side	Observed	Expected
1	8	10	4	8	10
2	10	10	5	8	10
3	16	10	6	11	10

Calculate chi-square for these data and decide whether the die is fair. Compare your decision with what you decided when using the D statistic.

9. Refer to James's phone-book data in Exercise 7 of Section 7.2. Calculate chi-square for these data. Now decide whether the phone book is a good source of random numbers.

7.4 SIX STEPS TO CHI-SQUARE

In Chapter 6 and earlier in this chapter (Section 7.2) we learned and used six steps to decision making. Now let's more systematically see how they can be used with chi-square. Also, let's choose a more applications-oriented example than the die problem.

Example 7.1

Accidents at Irongate

The Irongate Foundry has kept records of on-the-job accidents for many years. Accidents are reported according to which hour of the shift they happen. Table 7.7 shows the accident report. Both the union and management at the foundry want

Table 7.7 **Accident Report for Irongate Foundry**

Hour of shift	Number of accidents
1	19
2	17
3	15
4	24
5	20
6	26
7	22
8	25
Total	168

to know whether there is reliable evidence that accidents are more likely to happen during one hour of the shift than another.

Solution

The table below shows a value of chi-square of 5.1 calculated from the data in Table 7.7.

	Number of accidents		
Hour of shift	Expected (E)	Obtained (O)	$(O - E)^2$
1	21	19	$(-2)^2 = 4$
2	21	17	$(-4)^2 = 16$
3	21	15	$(-6)^2 = 36$
4	21	24	$(3)^2 = 9$
5	21	20	$(-1)^2 = 1$
6	21	26	$(5)^2 = 25$
7	21	22	$(1)^2 = 1$
8	21	25	$(4)^2 = 16$
Total	168	168	

$$\chi^2 = \frac{4}{21} + \frac{16}{21} + \frac{36}{21} + \frac{9}{21} + \frac{1}{21} + \frac{25}{21} + \frac{1}{21} + \frac{16}{21} = \frac{108}{21} \approx 5.1$$

In order to determine whether there is statistical evidence that accidents are more likely to happen during some hours of the shift rather than others, we *assume* the null hypothesis model that accidents are equally likely to happen during each hour. So our model is a fair eight-sided, or octahedral, die.

1. Choice of a Model: Use an eight-sided fair die, and let each side correspond to a different hour of the shift. With such a model, we *expect* one-eighth of the accidents to happen during each hour. That is, the expected number of accidents is $\frac{1}{8} \times 168 = 21$.

Now we must determine whether values of chi-square as large as or larger than the one obtained from the above Irongate data ($\chi^2 = 5.1$) occur rarely, or often. To do so, we go to the next steps.

2. Definition of a Trial: A trial consists of rolling a fair eight-sided die 168 times, once for each accident reported in Table 7.7.

3. Definition of a Successful Trial: For each trial we calculate chi-square. Count as successful a trial in which $\chi^2 \geq 5.1$.

Here are the results for the first trial. The table on the following page yields a calculated chi-square of 3.6—not a successful trial.

4. Repetition of Trials: We perform a total of 50 trials of rolling an octahedral die 168 times and record the chi-square statistic values in Table 7.8.

	Number of accidents		
Hour of shift	Expected (E)	Obtained (O)	$(O - E)^2$
1	21	20	1
2	21	24	9
3	21	23	4
4	21	18	9
5	21	15	36
6	21	22	1
7	21	21	0
8	21	25	16
Total	168	168	

$$\chi^2 = \frac{1}{21} + \frac{9}{21} + \frac{4}{21} + \frac{9}{21} + \frac{36}{21} + \frac{1}{21} + \frac{0}{21} + \frac{16}{21} = \frac{76}{21} \approx 3.6$$

Table 7.8 **χ^2 Values Obtained from 50 Trials of Rolling a Fair Eight-Sided Die 168 Times**

Stem	Leaf	f
0		0
1	5, 7, 8	3
2	7	1
3	0, 4, 6, 8	4
4	1, 2, 3, 3, 3, 4, 6, 6, 8	9
5	0, 0, 1, 2, 3, 8	6
6	1, 4, 5, 5, 6, 6, 9	7
7	0, 2, 6, 9	4
8		0
9	0, 0, 4, 6, 9, 9	6
10	0, 9	2
11	2, 6	2
12	7	1
13	5, 5, 7	3
14		0
15		0
16	2	1
17		0
18	0	1
Total		50

Key: "12 7" stands for 12.7.

5. Probability Estimate: From our original data in Table 7.7 we obtained a chi-square of 5.1. From Table 7.8 we see that a chi-square of 5.1 or greater (a successful trial) was obtained 31 times in 50 trials. So

$$P(\chi^2 \geq 5.1) = \frac{31}{50} = 0.62$$

6. Decision: We found in step 5 that a chi-square of 5.1 or more occurred rather often (over one-half of the time) by chance for an absolutely fair die (that is, for a model in which accidents are equally likely each hour). We thus have no evidence to rule out the model of the fair die. So we conclude that there is no statistical evidence of some hours of the shift being more accident-prone than others. The observed differences in occurrence of accidents at the different hours of a shift are typical of what we might expect because of ordinary chance variation.

Note: You should remind yourself of what decision we would have made if the estimated probability we found in step 5 had been small (less than 0.05). In that case we would have concluded that the obtained statistic happens only rarely by chance for a fair die. So the model we chose in step 1 would not be acceptable—the results we obtained from our original data would not likely have been produced by the model. (In the terms of the problem, a small probability for the obtained chi-square—not what we found, to be sure—would have been evidence of some hours of the shift being more likely to have accidents than others.)

Example 7.2 Animal Bites

In 1967 the weekly numbers of animal bites reported to the Chicago Board of Health for part of October and November were as follows:

Week ending	Number of animal bites
October 26	268
November 2	189
November 9	199
Total	656

Do you think that the weekly variation in the number of bites reported is due to chance alone? Use the chi-square table of areas to help you decide.

Solution

We assume a null hypothesis in which each of the three weeks is equally likely to have animal bites reported. With such a model, we would expect one-third of the

Table 7.9 **χ^2 for Data on Animal Bites**

	Number of bites		
Week ending	**Expected***	**Reported**	**(Reported − Expected)^2**
October 26	218.7	268	$(268 - 218.7)^2 = (49.3)^2$
November 2	218.7	189	$(189 - 218.7)^2 = (-29.7)^2$
November 9	218.7	199	$(199 - 218.7)^2 = (-19.7)^2$
	656.1	656	

$$\chi^2 = \frac{(49.3)^2}{218.7} + \frac{(-29.7)^2}{218.7} + \frac{(-19.7)^2}{218.7} = 16.9$$

* Expected values do not exactly add to 656 because of rounding.

bites to be reported each week. So the *expected* number of bites is

$$\frac{1}{3} \times 656 \approx 218.7$$

Now we can compute the chi-square statistic for these data. See Table 7.9. We need to find out if the chance of getting $\chi_2^2 \geq 16.9$ is large or small under the assumption that bites are equally possible in each of the three weeks.

1. Choice of a Model: We can use a box model with tickets 1, 2, 3 in it, corresponding to the three weeks, drawing with replacement.

2. Definition of a Trial: A trial would consist of doing 656 draws with replacement from the box, one for each bite reported. We would record how many times each of the three tickets was drawn. That is, we would find out how many of the 656 bites were reported each week, by chance, in one trial.

3. Definition of a Successful Trial: The statistic of interest is chi-square. We calculate chi-square for each trial. A trial is a success if $\chi^2 \geq 16.9$.

4. Repetition of Trials: We do 50 trials. The results are tabulated in the frequency table of Table 7.10.

5. Probability Estimate: We seek $P(\chi^2 \geq 16.9)$. So from Table 7.10, $P(\chi^2 \geq 16.9) = 0$.

Table 7.10 **Frequency Table for χ^2**

Interval	f
0– 0.99	9
1– 1.99	17
2– 2.99	12
3– 3.99	7
4– 7.99	4
8–16.99	1
≥ 17	0

6. Decision: We found that the experimental probability of getting a chi-square greater than or equal to 16.9 is 0. This is much smaller than 0.05, so we conclude that the differences in number of bites reported to the Chicago Board of Health are *not* due to chance alone. It may be the task of someone to try to find out *why* the numbers reported are different from week to week. Can you think of any reasons?

SECTION 7.4 EXERCISES

1. The table below is a stem-and-leaf table for chi-square based on 50 trials of drawing from a random number table. Use this table to find these probabilities:
   a. $P(\chi^2 \geq 6.0)$
   b. $P(\chi^2 \geq 4.8)$

Stem-and-Leaf Table for χ^2

Obtained by 50 trials of drawing from a random number table. Each trial consists of 100 draws.

Stem	Leaf	f
0		0
1	2	1
2	8	1
3	2, 4, 8	3
4	0, 4, 8	3
5	0, 4, 6, 8	4
6	0, 4, 4	3
7	0, 0, 4, 6, 6	5
8	2, 2, 4, 4	4
9	0, 0, 0, 2, 4, 6	6
10	0, 4, 8, 8, 8	5
11	0, 0, 2, 2, 6	5
12	0, 4, 4, 4, 6, 8, 8	7
13		0
14		0
15	2, 4, 8	3
Total		50

Key: "1 2" represents 1.2.

Use the six-step method to solve Exercises 2 through 4.

2. The statistics class at Johnson Community College rolled an octahedral die 168 times and got the results below. Use Table 7.8 to decide whether their die was loaded.

Outcome	Obtained f
1	29
2	22
3	18
4	19
5	20
6	23
7	12
8	25
Total	168

3. The statistics class at Buffalo Grove College interviewed 100 students at random and asked them to give their favorite number between 0 and 9. Here are the results. Use the table in Exercise 1 to decide whether the students prefer some numbers over others.

Outcome	f
0	5
1	3
2	11
3	10
4	19
5	9
6	11
7	15
8	13
9	4
Total	100

4. A clothing store stocks men's ties that are identical except that they are in four different colors. After 40 sales, inventory shows the following purchases. Use the table of Exercise 5(b) of Section 7.3 to decide whether some colors are preferred over others.

Color of tie	Number sold
Amber	7
Blue	9
Orange	14
Maroon	10
Total	40

5. The last digit of each of 100 telephone numbers was taken (in order) from one page of a telephone book. The frequencies were as follows:

Digit	f
0	3
1	8
2	15
3	14
4	10
5	7
6	8
7	9
8	11
9	15
Total	100

Calculate chi-square for these data. Do you think the telephone book is a good source of random data? Why or why not? *Hint:* Use the table in Exercise 1.

7.5 SMOOTH CHI-SQUARE CURVES

We know from Chapter 5 that the way to obtain good accuracy in our probability estimates is to have a large number of trials in the five- or six-step method. Table 7.11 is a frequency table for the chi-square values obtained in 100 trials, each trial consisting of 90 rolls of a fair six-sided die. It further shows the evaluation of $P(\chi^2 \geq 9.0) = 9/100$.

Any time we have a table of frequencies (number of occurrences) for numerical outcomes, we can graph the relative frequency histogram of the numerical outcomes. If we then join the midpoints of the tops of the rectangles of the relative frequency histogram by straight lines, we get the relative frequency polygon, which gives the general shape of the distribution of these observed proportions. These graphical constructions can be reviewed in Section 1.5.

Consider $P(\chi^2 \geq 9.0)$ as evaluated in Table 7.11. The value of $P(\chi^2 \geq 9.0)$ is precisely the area of the histogram to the right of 9.0 in the relative frequency histogram in Figure 7.2, which was constructed from Table 7.11. Indeed, $P(\chi^2 \geq a)$ for any choice of a can be evaluated as the area to the right of a in the relative frequency histogram. Clearly,

$$P(\chi^2 \geq 9.0) = 0.04 + 0.02 + 0.02 + 0.01 = 0.09$$

which is the area of the shaded region to the right of 9.0 in the histogram.

Figure 7.2 also gives the relative frequency polygon of the chi-square values from the 100 trials given in Table 7.11. In each interval of chi-square values in Table 7.11, we have used the midpoint (for example, 1.5 for

Table 7.11 **Relative Frequency χ^2 Table (100 Trials)**

χ^2	Frequency f	Relative frequency
0.00– 0.99	0	0
1.00– 1.99	9	0.09
2.00– 2.99	19	0.19
3.00– 3.99	14	0.14
4.00– 4.99	13	0.13
5.00– 5.99	11	0.11
6.00– 6.99	7	0.07
7.00– 7.99	10	0.10
8.00– 8.99	8	0.08
9.00– 9.99	4	0.04
10.00–10.99	0	0.00
11.00–11.99	2	0.02
12.00–12.99	0 $\}$ $\chi^2 \geq 9.0$	0.00
13.00–13.99	2	0.02
14.00–14.99	0	0.00
15.00–15.99	0	0.00
16.00–16.99	1	0.01
Mean χ^2	4.79	
$P(\chi^2 \geq 9.0)$	$\frac{9}{100} = 0.09$	

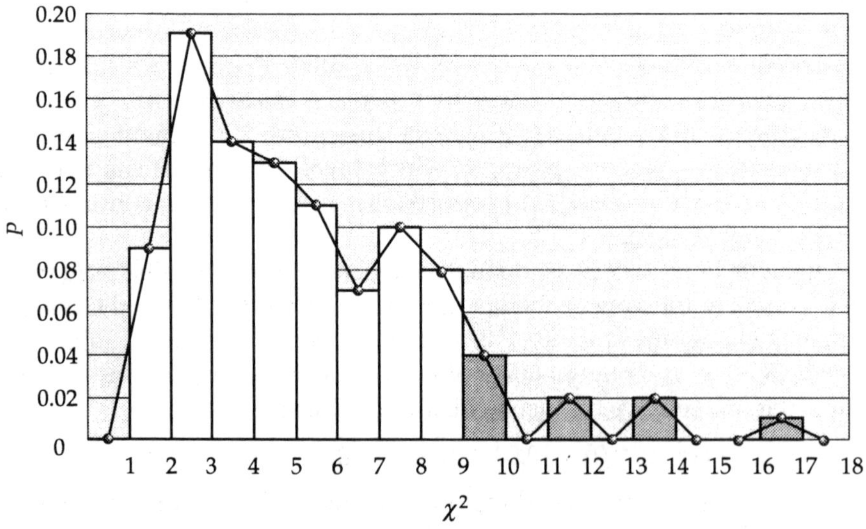

Figure 7.2 **Graph of the chi-square values reported in Table 7.11 (100 trials).**

the 1.00–1.99 interval) as the horizontal coordinate. This relative frequency polygon of chi-square values gives us a visual idea of what are large (unusual) chi-square values for 90 rolls of a fair six-sided die.

As we conduct more and more trials, it is a very important fact that the relative frequency polygon of chi-square values will become closer and closer to a particular smooth curve. To show this, we have added to Table 7.11 a further 1000 trials and a further 10,000 trials to produce Table 7.12. In Figure 7.3 the 100-trial, 1000-trial, and 10,000-trial relative frequency polygons and the smooth curve that results as the trials increase to an arbitrary large number are all displayed. Note in particular how the 10,000-trial relative frequency polygon is much closer to the in-the-limit smooth curve than the 100-trial relative frequency polygon. This smooth curve is called a **chi-square density.**

Table 7.12 **χ^2 Values Obtained from Large Numbers of Trials of 90 Rolls of a Fair Six-Sided Die**

	Relative frequency			
Interval	**100 trials**	**1000 trials**	**10,000 trials**	**Chi-square density**
0– 0.99	0.03	0.033	0.0390	0.0366
1– 1.99	0.12	0.104	0.1013	0.1120
2– 2.99	0.16	0.163	0.1605	0.1476
3– 3.99	0.13	0.152	0.1431	0.1491
4– 4.99	0.10	0.117	0.1417	0.1323
5– 5.99	0.16	0.104	0.1101	0.1087
6– 6.99	0.06	0.102	0.0855	0.0848
7– 7.99	0.08	0.076	0.0641	0.0639
8– 8.99	0.05	0.043	0.0498	0.0468
9– 9.99	0.03	0.038	0.0341	0.0335
10–10.99	0.03	0.026	0.0239	0.0237
11–11.99	0.01	0.014	0.0170	0.0165
12–12.99	0.02	0.010	0.0117	0.0113
13–13.99	0.00	0.004	0.0052	0.0077
14–14.99	0.02	0.003	0.0037	0.0052
15–15.99	0.00	0.004	0.0023	0.0035
16–16.99	0.00	0.001	0.0028	0.0023
17–17.99	0.00	0.005	0.0013	0.0015
18–18.99	0.00	0.001	0.0010	0.0010
19–19.99	0.00	0.000	0.0008	0.0007
20–20.99	0.00	0.000	0.0005	0.0004
21–21.99	0.00	0.000	0.0003	0.0003
22–22.99	0.00	0.000	0.0000	0.0002
23–23.99	0.00	0.000	0.0001	0.0001
24–24.99	0.00	0.000	0.0000	0.0001

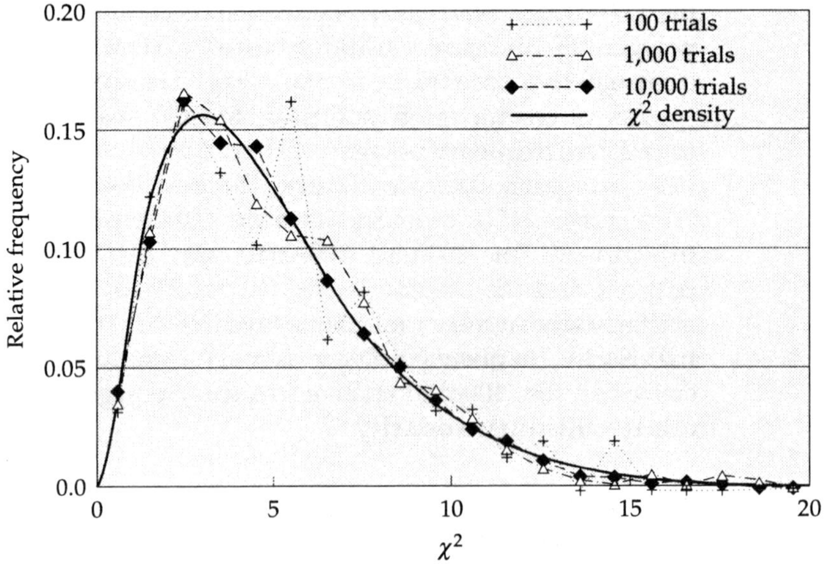

Figure 7.3 **Relative frequency polygons for 100, 1000, and 10,000 trials, and chi-square density.**

Suppose a chi-square hypothesis-testing problem concerning 90 rolls of a possibly loaded six-sided die produces a chi-square value of 9.0. Then as we understand from the previous sections, we would like to evaluate $p(\chi^2 \geq 9.0)$ in order to make a decision. In the previous section we learned how to do the six-step method in order to obtain $P(\chi^2 \geq 9.0)$ as an estimate of $p(\chi^2 \geq 9.0)$. Now we have just learned that $P(\chi^2 \geq 9.0)$ is a certain area of the relative frequency histogram. Moreover, we have also just learned (recall the 10,000-trial case) that this histogram and its associated relative frequency polygon are in fact well approximated by the chi-square density of Figure 7.3. Thus we have another way to approximately evaluate $p(\chi^2 \geq 9.0)$, one that avoids simulation and the six-step model *entirely*. Namely, we simply seek the area to the right of 9.0 under the chi-square density.

Figure 7.4 shows this estimated probability as the area under the chi-square density to the right of 9.0. You might recognize this as an integration problem of calculus. You will see in the next section that there are tables that give the theoretical probability for exceeding a given value of chi-square corresponding to rolls of a die with a given number of sides and, more generally, for other problems requiring chi-square probabilities.

How good is this approximation of $p(\chi^2 \geq a)$ provided by the chi-square density? Table 7.13 shows empirical chi-square frequencies for 100 trials of rolling a six-sided die, for 30, 60, and 90 rolls per trial. Figure 7.5 shows the three estimated smooth curves that result. These curves have been

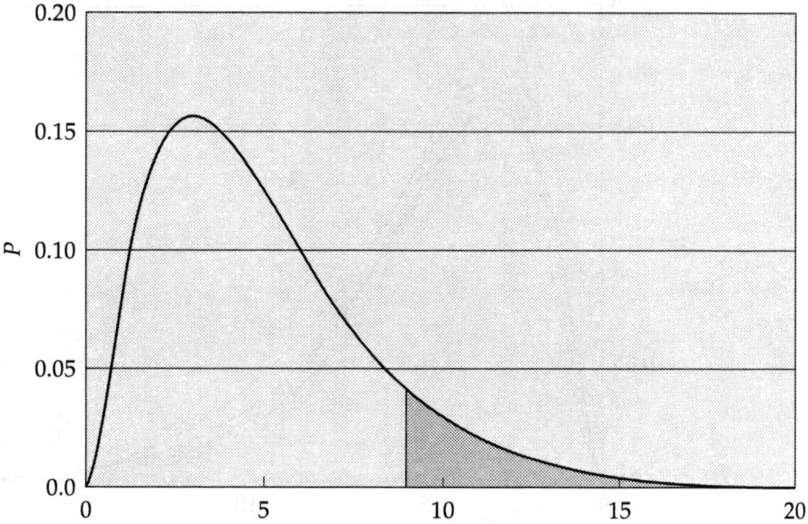

Figure 7.4 **Estimated probability $P(\chi^2 \geq 9.0)$.**

Table 7.13 **Empirical Chi-Square Frequency *f* (100 Trials of a Fair Six-Sided Die)**

χ^2	30 rolls per trial	60 rolls per trial	90 rolls per trial
0.00– 0.99	2	5	3
1.00– 1.99	13	8	16
2.00– 2.99	20	15	13
3.00– 3.99	12	17	18
4.00– 4.99	11	9	9
5.00– 5.99	8	10	11
6.00– 6.99	17	14	6
7.00– 7.99	3	5	6
8.00– 8.99	5	7	7
9.00– 9.99	3	5	2
10.00–10.99	3	2	2
11.00–11.99	2	1	3
12.00–12.99	0	1	1
13.00–13.99	1	0	1
14.00–14.99	0	0	0
15.00–15.99	0	0	1
16.00–16.99	0	1	1
Mean χ^2	4.70	4.94	5.01

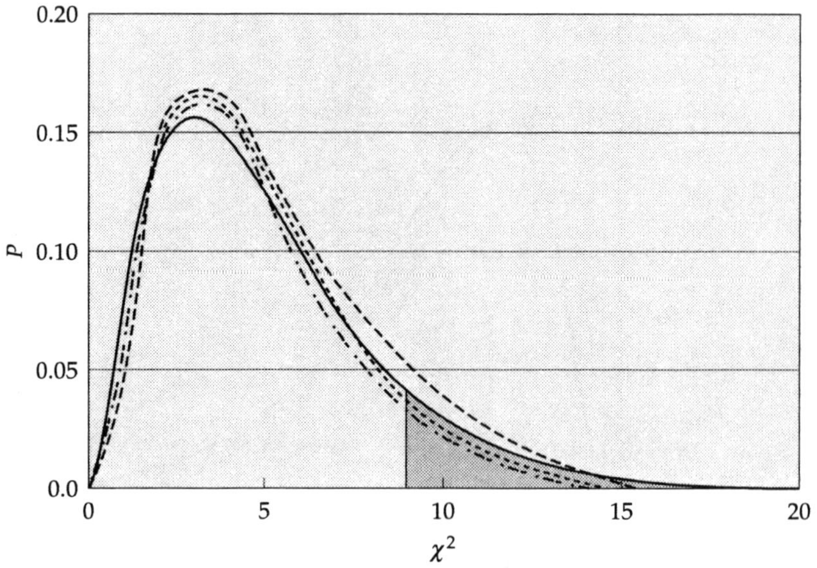

Figure 7.5

informally estimated by a method similar to the visual method of Chapter 3 for doing linear regression. A least squares approach analogous to that of Section 3.5 for linear regression would also be a possible way of obtaining such estimated smooth curves; however, such a calculation would be an unnecessary diversion here. The key point to note is that all three curves are close to each other and close to the theoretical chi-square density, which is also drawn in Figure 7.5. Hence, it is intuitively clear that whether we have 30, 60, or 90 rolls, we can use areas under the chi-square density to solve for probabilities of the form $p(\chi^2 \geq a)$; the value of chi-square is not affected much by the total number of rolls of the die involved in calculating the statistic. Whether we want to compute $p(\chi^2 \geq 9.0; 30 \text{ rolls})$, $p(\chi^2 \geq 9.0; 60 \text{ rolls})$, or $p(\chi^2 \geq 9.0; 90 \text{ rolls})$, the area to the right of 9.0 under the chi-square density of Figure 7.4 provides a good approximation. That is, we simply use the theoretical curve discussed above.

The D statistic does not have this crucial property of not being affected by the number of rolls of the die, nor does it have other important properties of chi-square. This is why statisticians usually prefer to use chi-square rather than D. For now, we assume that areas under a smooth chi-square curve (as tabulated below) can be used to decide whether an observed chi-square value is strong evidence that a given model is unlikely to have produced the data, provided enough data (die throws in this example) are available that every expected frequency is 5 or greater. For example, this criterion of a frequency of 5 or greater holds for Table 7.5. This will be our convention for deciding whether one is allowed to use areas under the chi-square density instead of using the six-step method.

Degrees of Freedom

We have now discovered that we can compute chi-square probabilities such as $p(\chi^2 \geq 9)$, which are useful in hypothesis testing, by using areas under a chi-square density, namely, that of Figure 7.4. Moreover, we have learned that we can use this curve regardless of the number of die tosses per trial provided they are not too few (because of our rule that the expected frequency be at least 5). But this was all for a six-sided die. Is it the *same* theoretical curve regardless of the number of sides on the die of the null hypothesis model? Recall that the model in the animal-bite example was in effect a three-sided fair die. Because the distribution of the actual chi-square statistics in simulations are known to be close to the theoretical curve, we can look at chi-square frequency tables for fair dice of different numbers of sides in an effort to find out whether we get the same chi-square density in the limit as we got in Figure 7.4 or whether we need a whole family of curves (or their tabulated areas).

We therefore now explore the effect on the chi-square values of the number of sides of the fair die being used. We carry out 50 simulated trials of rolling a fair die. Each trial consists of 60 rolls of a die. For the first experiment we use a four-sided die (Table 7.14), for the second experiment we use a six-sided die (Table 7.15), and for the third experiment we use a

Table 7.14 **Table of χ^2 for a Four-Sided Die**

Fifty trials. Each trial consists of rolling the die 60 times.

Stem	Leaf
0	1, 3, 4, 4, 7, 7, 9, 9, 9, 9
1	1, 2, 2, 2, 3, 3, 5, 7, 7, 7, 7
2	0, 0, 0, 3, 3, 3, 4, 5, 8, 8, 8, 9
3	1, 2, 3, 3, 5, 5, 6, 6
4	1, 4, 4, 7, 8
5	7
6	5
7	
8	
9	9
10	3

Mean chi-square = 2.7
Standard deviation = 2.1

Key: "10 3" stands for 10.3.

10-sided die (Table 7.16). Notice first that as the number of sides on the die increases, so does the mean chi-square value. Notice also that the standard deviation of chi-square increases as well. That is, as the number of sides of the die increases, the chi-square values also tend to become more spread out.

Thus Tables 7.14 through 7.16 suggest that the number of sides of the fair die (that is, the number of possible outcomes of the die) has an effect on the size of the chi-square statistics produced. Hence if we are to be able to use areas under a smooth curve analogous to the curve in Figure 7.4, we will need a whole family of curves, one for each number of sides, or outcomes, of the die. This *number of possible outcomes* has to do with the idea of *degrees of freedom* (sometimes abbreviated *df*). We can look at the idea of degrees of freedom in terms of tables of outcomes for rolling a die.

Tables 7.17 through 7.19 each give results for one trial of rolling a fair die. The die in Table 7.17 is four-sided, that in Table 7.18 is six-sided, and that in Table 7.19 is 10-sided. Each trial consists of 60 rolls. Table 7.17 presents the four possible outcomes for that die. But once we know how many times a 1, a 2, and a 3 were obtained ($12 + 17 + 14 = 43$), we know exactly how many 4s were obtained (17), since the total number of rolls was 60. We therefore say that Table 7.17 has three degrees of freedom—one less than the number of sides of the die.

Table 7.15 **Table of χ^2 for a Six-Sided Die**

Fifty trials. Each trial consists of rolling the die 60 times.

Stem	Leaf
0	4, 8
1	0, 4, 6, 8
2	0, 4, 6, 8, 8
3	0, 2, 2, 2, 2, 4, 4, 6, 6, 8
4	0, 0, 2, 2, 4, 4, 6, 6, 6, 6
5	2, 4, 4, 6, 6, 8, 8
6	2, 4, 4
7	0, 4, 8
8	2, 4, 8
9	6
10	
11	2
12	
13	6

Mean chi-square = 4.7
Standard deviation = 2.8

Key: "13 6" stands for 13.6.

Table 7.16 **Table of χ^2 for a 10-Sided Die**

Fifty trials. Each trial consists of rolling the die 60 times.

Stem	Leaf
0	
1	7
2	7, 7
3	3, 3
4	0, 0, 3, 7, 7
5	0, 0, 3, 3
6	3, 3, 7
7	0, 0, 0, 0, 0, 3, 7, 7, 7
8	0, 0, 0, 7, 7
9	0, 7
10	0, 3, 3, 3, 7, 7
11	0
12	0, 3
13	0, 3
14	0, 3
15	0
16	3
17	0
18	0

Mean chi-square = 8.4
Standard deviation = 4.0

Key: "18 0" stands for 18.0.

Table 7.17 **Outcomes of Rolling a Four-Sided Die**

Outcome	Number expected	Number obtained	
1	15	12	3 degrees of freedom (outcomes 1–3)
2	15	17	
3	15	14	
4	15	___	← 17 more needed to make 60 observations
Total	60	60	

Table 7.18 **Outcomes of Rolling a Six-Sided Die**

Outcome	Number expected	Number obtained	
1	10	9	5 degrees of freedom (outcomes 1–5)
2	10	14	
3	10	11	
4	10	7	
5	10	7	
6	10	___	← 12 needed to make 60 observations
Total	60	60	

Table 7.19 **Outcomes of Rolling a 10-Sided Die**

Outcome	Number expected	Number obtained	
1	6	8	9 degrees of freedom
2	6	2	
3	6	10	
4	6	9	
5	6	12	
6	6	5	
7	6	5	
8	6	1	
9	6	4	
0	6		← 4 needed to make 60 observations
Total	60	60	

Similarly, in Table 7.18, once we know how many 1s, 2s, 3s, 4s, and 5s were obtained, we know how many 6s were obtained. We say that Table 7.18 has five degrees of freedom.

Since the number of degrees of freedom (the number of sides on the die minus 1) affects the size of the chi-square statistics produced by tables of die-rolling outcomes, it is important to know how many degrees of freedom are associated with a given chi-square. Therefore, we indicate that a chi-square has, for example, three degrees of freedom by writing

$$\chi_3^2$$

When we wish to find a probability of a certain chi-square, we must first know its number of degrees of freedom. The number of degrees of freedom will tell us which theoretical chi-square density to use to find areas. We study this in the next section.

SECTION 7.5 EXERCISES

1. Find the estimated probabilities below. You must first decide which table of values to use: Table 7.14, Table 7.15, or Table 7.16.
   a. $P(\chi_3^2 \geq 5.7)$
   b. $P(\chi_3^2 \geq 9.9)$
2. Find the estimated probabilities below. You must first decide which table of values to use: Table 7.14, Table 7.15, or Table 7.16.
   a. $P(\chi_5^2 \geq 6.0)$
   b. $P(\chi_5^2 \geq 8.0)$
3. Find the estimated probabilities below. You must first decide which table of values to use: Table 7.14, Table 7.15, or Table 7.16.
   a. $P(\chi_5^2 \geq 11.2)$
   b. $P(\chi_5^2 \geq 9.6)$
4. Find the estimated probabilities below. You must first decide which table of values to use: Table 7.14, Table 7.15, or Table 7.16.
   a. $P(\chi_9^2 \geq 10.0)$
   b. $P(\chi_9^2 \geq 15.0)$

5. A six-sided die was rolled 90 times and the following outcomes were obtained.

Outcome	f
1	17
2	13
3	17
4	10
5	16
6	17

a. Calculate chi-square for these data.
b. How many degrees of freedom are associated with this chi-square? Explain.
c. What is the experimental probability of obtaining a chi-square as large as or larger than the value you obtained in part (a), assuming that a fair die was used? [That is, for whatever the value ν you obtained in part (a), what is $P(\chi^2 \geq \nu)$?]
d. Do you believe the die used in this exercise was a fair die? Why or why not?

6. You can use chi-square to solve the Key Problem for this chapter. How many degrees of freedom does that chi-square have? Explain.
7. Sue uses her telephone book as a source of random digits. She takes the first 300 telephone numbers in the book, in order, and writes down the last digit of each number. If she uses a chi-square to help her decide if the digits are random, how many degrees of freedom will the chi-square have? Explain.
8. A record store stocks albums for the top 20 artists. The manager of the store wants to determine whether the artists are equally popular (in terms of sales of albums) or whether some artists are definitely favored over others. If a chi-square is used, how many degrees of freedom will it have? Explain.

7.6 CHI-SQUARE TABLES

It turns out that the number of degrees of freedom completely determines which chi-square density we use to compute chi-square probabilities with. The chi-square density of Figure 7.4 is a chi-square density having five degrees of freedom, which would be used to compute $p(\chi_5^2 \geq 9.0)$, for example. If a chi-square problem puts data into 10 categories (that is, the null hypothesis is a 10-sided die), we need to use a chi-square density having $10 - 1 = 9$ degrees of freedom. In Figure 7.6 we display chi-square densities having 3, 5, and 9 degrees of freedom, corresponding to the chi-square testing of a 4-sided, a 6-sided, and a 10-sided fair die. The argument above that helped convince us that the chi-square density with five degrees of freedom works well for testing the hypothesis of a six-sided fair die could just as easily be made for a die of any number of sides. Further, the null hypothesis can be one that specifies a loaded die—that is, we can also use chi-square density areas when the die is hypothesized to be loaded in a specified way.

We now have a way to find the probability of a chi-square as large as or larger than a given value that is more accurate than using experimental probabilities obtained from the six-step method: we find the approximate area under a chi-square density. Tables have been produced that provide

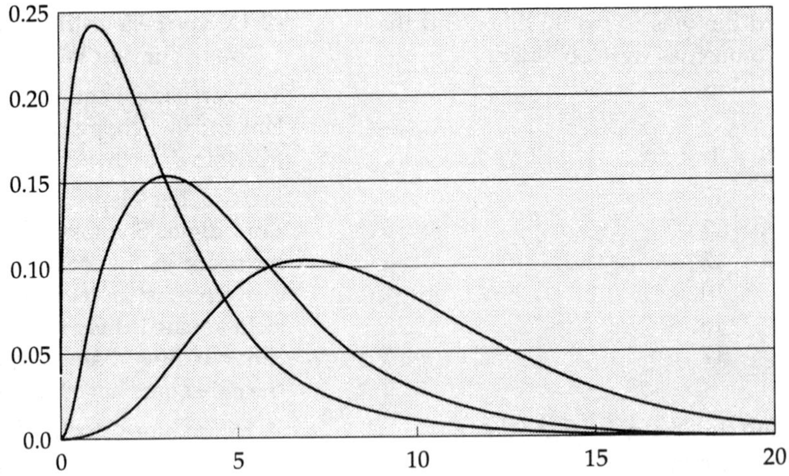

Figure 7.6 **Chi-square densities of three, five, and nine degrees of freedom.**

theoretical probabilities determined from areas under the theoretical chi-square densities.

Now let us consider an example.

Example 7.3

Find $p(\chi_5^2 \geq 1.6)$.

Solution

Note that no mention of a die is made. But we are dealing with the χ^2 curve that would result from a fair six-sided die, which produces $6 - 1 = 5$ degrees of freedom.

We use a table of areas under the chi-square density. The unique theoretical smooth curve for a chi-square with five degrees of freedom is shown in Figure 7.7. The shaded area is the theoretical probability in question.

Table 7.20 gives theoretical areas under the chi-square density. It is the result of integral calculus calculations used to find areas. There are rows for numbers of degrees of freedom from 1 through 5 because, as we have already discovered, different numbers of degrees of freedom produce different distributions of χ^2 values. Each row is associated with a problem for a particular number of degrees of freedom. For example, row 5 would be our choice for a χ^2 problem for a fair six-sided die. An expanded version of this table is in Appendix C.

Look at row 5, which gives areas for various regions under the curve in Figure 7.4. Look along row 5 until you find 1.6. (Actually 1.61 is in the table, but this is 1.6 when rounded to the nearest 10th.) Now look at the top of the column in which 1.6 is located. You find 0.90, which is the area of the shaded part in Figure 7.6. (The

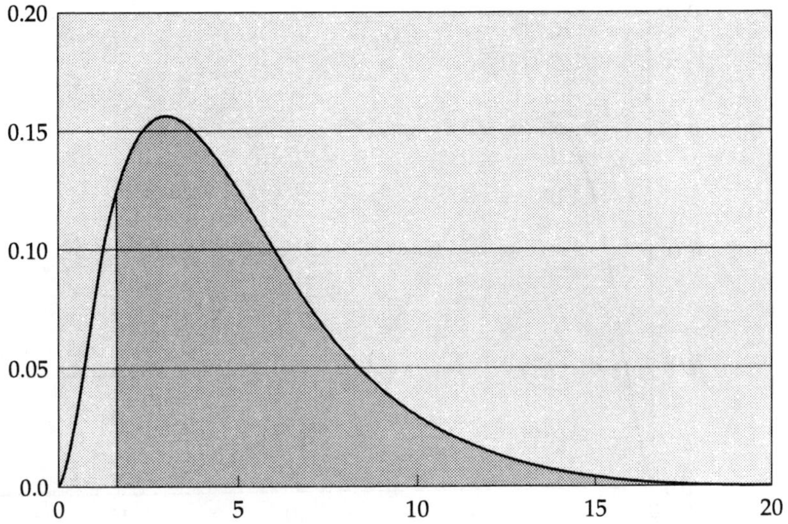

Figure 7.7 **Chi-square density with five degrees of freedom.**

Table 7.20 **Areas under the Chi-Square Density**

Degrees of freedom	*Probability (area) to the right of* 0.99	0.95	0.90	0.50	0.20	0.10	0.05	0.01	0.001
1	0.00	0.00	0.02	0.45	1.64	2.71	3.84	6.63	10.83
2	0.02	0.10	0.21	1.39	3.22	4.61	5.99	9.21	13.82
3	0.11	0.35	0.58	2.37	4.64	6.25	7.81	11.34	16.27
4	0.30	0.71	1.06	3.36	5.99	7.78	9.49	13.28	18.47
5	0.55	1.15	1.61	4.35	7.29	9.24	11.07	15.09	20.52

total area under the curve is 1.0.) So we have

$$p(\chi_5^2 \geq 1.6) \approx 0.90$$

Example 7.4

Find $p(\chi_5^2 \geq 8.0)$.

Solution

Figure 7.8 shows the area we are trying to find. Again we use Table 7.20 and refer to the row for five degrees of freedom. The value of 8.0 does not appear in this row. We have, instead, 7.29 and 9.24. The probability (area) given for 7.29 is 0.20 (see the top of the column). The probability given for 9.24 is 0.10. So the probability for 8.0 is between 0.20 and 0.10.

Here is what we can do to get an estimation of this area. The value 8.0 lies at a certain proportion of the distance between 7.29 to 9.24. The corresponding value

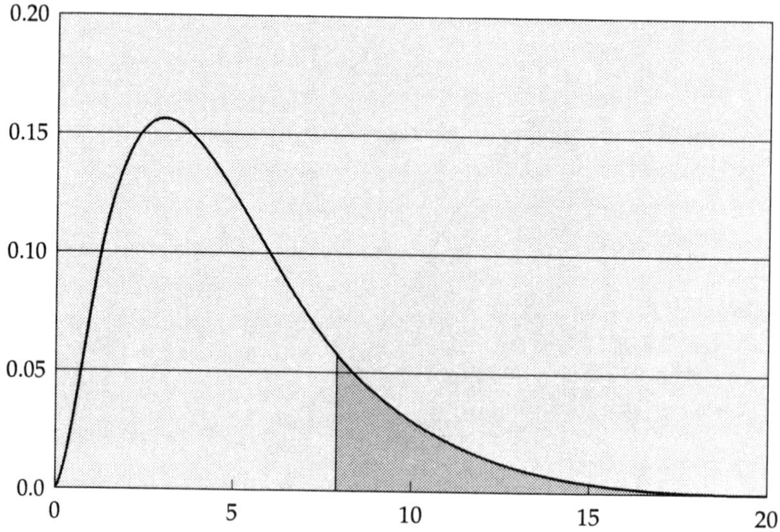

Figure 7.8 $p(\chi^2 \geq 8.0)$.

of area that we are looking for lies at the same proportion of the distance from 0.20 to 0.10. Before we do any calculating, then, we can estimate that 8.0 is roughly half-way between 7.29 and 9.24. So we estimate that the area (probability) we are seeking is roughly half-way between 2.0 and 1.0, or

$$p(\chi_5^2 \geq 8.0) \approx 0.15$$

For many purposes this crudely estimated probability is adequate. However, for a better approximation, use the following rule, called **linear interpolation.** It proceeds as follows.

a. Find at what proportion of the distance from 7.29 to 9.24 the value of 8.0 is located.

$$\text{Distance from 7.29 to 9.24} = 9.24 - 7.29 = 1.95$$

$$\text{Distance from 7.29 to 8.00} = 8.00 - 7.29 = 0.71$$

Calculate the value of the fraction

$$\frac{\text{Distance between 7.29 and 8.00}}{\text{Distance between 7.29 and 9.24}} = \frac{8.00 - 7.29}{9.24 - 7.29} = \frac{0.71}{1.95} = 0.36$$

b. Then find the difference between the probability (area) for 7.29 and that for 9.24:

$$0.20 - 0.10 = 0.10$$

c. From step (a) we know that the desired value lies a fraction of 0.36 of the distance from 0.20 to 0.10. (Because we went from 7.29 to 9.24, we must go from 20, which corresponds to 7.29, to 10, which corresponds to 9.24.) So the area (probability) value is a distance $0.36 \times 0.10 = 0.036$ away from 0.20. Therefore

$$\text{Desired probability} = 0.20 - 0.036 = 0.164 \approx 0.16$$

or

$$p(\chi_5^2 \geq 8.0) \approx 0.16$$

Appendix D has further discussion and formulas regarding linear interpolation.

We now return to Example 7.2 to solve it using the theoretical chi-square area tables instead of the six-step method. It will require use of a chi-square density with a different number of degrees of freedom than 5.

Example 7.5 **Animal Bites Revisited**

Consider Example 7.2. Use the chi-square table of areas to help you decide whether the weekly variation in the number of bites reported is due to chance alone.

Solution

Because it is a three-sided-die null hypothesis model, the number of degrees of freedom is $3 - 1 = 2$. Recall that $\chi^2 = 16.9$ was obtained for the data. Thus we will reject the null hypothesis or not, depending on the value of $p(\chi_2^2 \geq 16.9)$.

We suspect we will reject it because we got $P(\chi^2 \geq 16.9) = 0$ earlier. We use Table 7.20. We go to row 2 since this chi-square has two degrees of freedom (as we noted at the beginning of the solution). A chi-square of 16.9 does not appear in row 2. The largest value in row 2 is 13.82. Because 16.9 is greater than 13.82, we know that $p(\chi_2^2 \geq 16.9)$ is smaller than 0.001. With a better table we could estimate what the probability actually is, but we do not need to know this to reach our decision.

Decision: We found that the probability of getting a chi-square greater than or equal to 16.9 is less than 0.001. That is, the chance is less than 1 out of 1000 (0.001) that a chi-square of 16.9 would be obtained by chance. This is a very small probability (much smaller than 0.05), so we strongly reject the null hypothesis of a three-sided fair die. Thus we conclude that the differences in number of bites reported to the Chicago Board of Health are *not* due to chance alone.

SECTION 7.6 EXERCISES

For each of the following, find the estimated probability (as the capital P indicates) from Table 7.15 and then find the theoretical probability from Table 7.20.

1. $P(\chi_5^2 \geq 4.4)$
2. $P(\chi_5^2 \geq 7.4)$
3. $P(\chi_5^2 \geq 9.2)$
4. $P(\chi_5^2 \geq 11.2)$

For Exercises 5 and 6, find the estimated probability from Table 7.14 and then the corresponding theoretical probability from Table 7.20.

5. $P(\chi_3^2 \geq 4.6)$
6. $P(\chi_3^2 \geq 6.5)$
7. Find these probabilities. Use the expanded table of chi-square areas in Appendix C.
   a. $p(\chi_4^2 \geq 9.5)$
   b. $p(\chi_4^2 \geq 13.3)$
8. Find these probabilities. Use the expanded table of chi-square areas in Appendix C.
   a. $p(\chi_7^2 \geq 12.0)$
   b. $p(\chi_7^2 \geq 14.1)$
9. Find these probabilities. Use the expanded table of chi-square areas in Appendix C.
   a. $p(\chi_{10}^2 \geq 12.0)$
   b. $p(\chi_{10}^2 \geq 20.0)$
10. Find these probabilities. Use the expanded table of chi-square areas in Appendix C.
    a. $p(\chi_{20}^2 \geq 18.0)$
    b. $p(\chi_{20}^2 \geq 40.0)$

For Exercises 11 through 14, use the chi-square table of areas in Appendix C.

11. The statistics class at Forest View College rolled a die and obtained the following results:

Outcome	f
1	26
2	7
3	17
4	19
5	14
6	13
	96

Calculate chi-square for this experiment. How likely are you to get a chi-square value as large as or larger than the one obtained here, by chance, if the die they used was fair? Do you believe the die used in this question was fair? Explain.

12. The statistics class at Central College interviewed 380 students (at random, we assume) and asked them for their favorite number between zero and nine. Here are the results:

Outcome	f
0	0
1	12
2	27
3	56
4	37
5	40
6	58
7	103
8	31
9	16
	380

Calculate χ^2 for these results. Do you think that the Central students prefer some numbers over others? Why?

13. The number of eighths in the fraction on a closing stock price for low-priced stocks from the American Stock Exchange is shown below. Why are some fractions more common than others? Explain.

Fraction	Frequency
0	60
1/8	30
2/8 (1/4)	29
3/8	27
4/8 (1/2)	47
5/8	49
6/8 (3/4)	39
7/8	38
	319

14. In a study of rounding errors made by shoppers when they estimate the cost of articles, 174 cash register slips were examined and the last digit of the price registered on the slip was recorded. Here are the results. Would you conclude that the costs of articles are more likely to have some last digits rather than others?

Last digit	Frequency
0	11
1	10
2	10
3	34
4	7
5	25
6	4
7	18
8	6
9	49
	174

7.7 UNEQUAL EXPECTED FREQUENCIES

The chi-square statistic can be used in a remarkable variety of ways to solve vital problems for science, medicine, and industry. Indeed, it played a key role in confirming Mendel's theory of genetics, a triumph of modern science.

Recall from Section 7.3 that a basic pattern for a chi-square table is the following:

Outcome Expected frequency (E) Obtained frequency (O) $(O - E)^2$

This applies, for example, to the rolling of a die. If the die is fair, all outcomes are expected to occur an equal number of times, so the expected frequencies of the different outcomes are all equal. So far in this chapter, all but one of our chi-square problems have involved equal expected frequencies. Now we will see how chi-square can be used to solve problems in which the expected frequencies are unequal, thus greatly widening the scope of problems we can attack using the chi-square hypothesis-testing approach. Note once again that a chi-square test asks whether the null hypothesis model of a certain loaded die fits the data.

Example 7.6

In a certain city, the racial makeup of the population is as follows:

Whites	40%
Blacks	35%
Others	25%

The racial composition of the city council is as follows:

Whites	26
Blacks	10
Others	9
	45

How proportionally representative of the racial composition of the city is the makeup of the council? Another way of asking this question is this: Does there appear to be *bias* in the makeup of the council that favors, for instance, whites, or is the makeup such that it would occur rather often in a random sample? We will use the chi-square test for this problem.

Solution

What is the null hypothesis under which we find the expected number of outcomes? Under the assumption of proportional representation, the expected numbers of council members of the three racial categories are 18.00, 15.75, and 11.75 (that is, $0.4 \times 45, 0.35 \times 45, 0.25 \times 45$). Thus our model under the null hypothesis is that of a loaded three-sided die with probabilities (0.4, 0.35, 0.25). The chi-square table of outcomes for the city council data under this assumption is as follows:

Outcome	Expected (E)	Obtained (O)	$O - E$	$(O - E)^2$
Whites	18.00	26	8	64.00
Blacks	15.75	10	−5.75	33.06
Others	11.25	9	−2.25	5.06

The chi-square value for the data can now be obtained as follows:

$$\chi^2 = \frac{64}{18} + \frac{33.06}{15.75} + \frac{5.06}{11.25}$$

$$= 3.56 + 210 + 0.45 = 6.1$$

Now we will use the theoretical chi-square density to decide whether this value is improbably high. The chi-square table (Appendix C) will give us the probability of getting a chi-square as large as 6.1 or larger from the loaded three-sided die (0.4, 0.35, 0.25) model defined above. There are three rows in our table of outcomes (corresponding to three possible outcomes of the die), so once any two rows are known, the third row is fixed. The table of outcomes therefore has two degrees of freedom. The chi-square table with two degrees of freedom tells us that the probability of a chi-square equal to 6.1 or more is approximately 0.05, leading us to the conclusion that we should reject the null hypothesis of proportional representation of the three groups. Note that our problem is an important social problem, rather than a mere die-tossing problem.

Example 7.7

A psychology professor teaches several large classes of introductory psychology and assigns grades by aiming at the following proportions: A, 5%; B, 20%; C, 50%; D, 20%; F, 5%; except for some randomness about these percentages. At the end of the term, the professor claimed to have assigned the 200 students letter grades for the course according to Table 7.21.

Does the professor's assignment of grades conform to these percentages? Or do the proportions of students getting the various letter grades differ so much from these proportions that we doubt the professor's model?

Solution

We set up the chi-square frequency table of outcomes in the usual way.

Outcome (grade)	Expected number of grades (E)	Obtained number of grades (O)	$O - E$
A	5% of 200 = 10	25	15
B	20% of 200 = 40	48	8
C	50% of 200 = 100	105	5
D	20% of 200 = 40	18	−22
F	5% of 200 = 10	4	−6
	200	200	

We calculate chi-square with the familiar formula:

$$\begin{aligned}\chi^2 &= \text{sum of } \frac{(O-E)^2}{E} \\ &= \frac{(15)^2}{10} + \frac{(8)^2}{40} + \frac{(5)^2}{100} + \frac{(-22)^2}{40} + \frac{(-6)^2}{10} \\ &= \frac{225}{10} + \frac{64}{40} + \frac{25}{100} + \frac{484}{10} + \frac{36}{10} \\ &= 22.5 + 1.6 + 0.25 + 12.1 + 3.6 \\ &= 40.05\end{aligned}$$

Table 7.21

Grade	Number of students getting grade
A	25
B	48
C	105
D	18
F	4
Total	200

This appears to be a rather large chi-square. But let's check it out in our chi-square table (Appendix C). Since there are five categories in our table (five different letter grades assigned), this chi-square has four degrees of freedom. We go to the table and see that a chi-square as large as 40.05 occurs very seldom by chance (less than 0.001 of the time). So the difference between the number of grades we expect for the 5 : 20 : 50 : 20 : 5 scheme and what was given by the psychology professor is too large to expect by chance very often. We conclude, rather, that the letter grades are *not* proportionally distributed as claimed. (It looks as if she gives more As and Bs and fewer Ds and Fs than would be expected if she were grading the way she claimed.)

We have shown in this section how the chi-square statistic can be used to make conclusions about outcomes even though the expected numbers of observations of the possible outcomes are not equal. One restriction to keep in mind is that the expected number of observations of each outcome should not be smaller than about 5.

Let's note how general and powerful the chi-square approach is. If a problem leads to any hypothesized set of probabilities for a finite number of outcomes (for example, $p(1) = \frac{1}{2}$, $p(2) = \frac{1}{3}$, $p(3) = \frac{1}{6}$) and we can collect frequency data from this experiment by doing enough trials that (number of trials) × (smallest probability) is at least 5, then we can use chi-square to test the null hypothesis "goodness of fit" of the actual data to the hypothesized model. Because so many applications specify such a finite set of probabilities, chi-square goodness of fit testing has an enormously wide scope of application. In Chapter 13 we will take up a totally different type of chi-square testing in which one tests whether two attributes are independent or associated, such as income level and political preference.

SECTION 7.7 EXERCISES

1. Is it true, as is often assumed, that people prefer the flakiest (lightest) pie crust? An experiment was done at Cornell University to test this assumption. Thirty people were involved. Each person was blindfolded and asked to state the order of preference for three pieces of pastry—one light (L), one medium (M), and one heavy (H)—which were presented in random order.

   There are six different ways in which the order of preference could be given, as shown below. Also given is the number of persons choosing each of the six.

Preference ordering	Number of persons choosing (obtained frequencies)
LMH	10
LHM	3
MLH	3
MHL	8
HLM	1
HML	5
	30

   a. Under the assumption that each of the six orderings is preferred equally, what is the expected number of persons choosing each?

b. Test the assumption of equal preference using chi-square.

c. A model was developed under the assumption of preference for lighter pastry. According to this model, the probability of each of the six outcomes being chosen is as follows:

Preference ordering	Probability of outcome under assumption of preference for light pastry
LMH	0.23
LHM	0.18
MLH	0.18
MHL	0.15
HLM	0.15
HML	0.11

Now what is the expected number of outcomes (that is, expected number of persons preferring each ordering of pie crusts)?

d. Test the assumption of preference for light pastry, using the model given in (c).

2. In the study of genetics, scientists are interested in determining how characteristics of living things are inherited from one generation to another. The fundamental principles of heredity were discovered by Gregor Johann Mendel (1822–1884) when he proved by crossbreeding garden peas that there are definite patterns in the way characteristics such as size, shape, and surface texture are passed on from generation to generation.

One of Mendel's classic experiments involved crossbreeding two kinds of pea plants, one with round yellow seeds and the other with wrinkled green seeds. Here are the results that should have been obtained, according to his law of heredity, and the results actually obtained. Did the seed-growing experiment support or cast doubt on his genetic theory?

Type of seed	Expected proportion	Expected number	Obtained number
Round and yellow	$\frac{9}{16}$	312	314
Wrinkled and yellow	$\frac{3}{16}$	104	101
Round and green	$\frac{3}{16}$	104	108
Wrinkled and green	$\frac{1}{16}$	35	32
		555	555

3. A biological experiment with flowers yielded the following results. The expected values were based on a genetic theory of inheritance. Did the results support the theory?

Characteristic of flower	Expected f	Observed f
AB	180	164
Ab	60	78
aB	60	65
ab	20	13

4. A public opinion poll claims that 60% of the voters favor Mr. Alpha, 30% favor Mr. Beta, and 10% favor Ms. Gamma. The actual election produced these votes:

Candidate	Number of votes
Alpha	503
Beta	115
Gamma	35

Does it appear that the poll was accurate?

5. A census in a certain city shows the racial composition to be 30% white, 55% black, and 15% others. The city council has the following

racial composition (number of persons of each race):

White	20
Black	20
Other	10

How representative of the racial composition of the city is the racial composition of the city council?

6. A typist prepared a manuscript of 167 pages, of which 103 contained no errors. Forty-five pages had one typing error each. Sixteen pages had two errors each. Three pages had three errors. None had more than three errors. The table summarizes this information.

Number of typing errors	Number of such pages (observed outcomes)
0	103
1	45
2	16
3 or more	3
Total	167

The probabilistic model called the *Poisson model* has been used to predict the number of typing errors per page. Using this model, the following probabilities are calculated:

Number of typing errors	Probability
0	0.59
1	0.31
2	0.08
3 or more	0.02

a. Calculate the expected number of pages having 0, 1, 2, and so on, errors, according to this model.
b. Use the chi-square to test whether this model is an appropriate one.

7. A probabilistic model has been used to predict occurrences of heavy rainstorms. The following table shows the number of heavy rainstorms reported by 330 weather stations during a one-year period. (For example, 102 stations reported no heavy rainstorms, 114 reported one heavy rainstorm, and so on.) The third column shows the probabilities of the corresponding numbers of heavy rainstorms according to the model being used. Use chi-square to test how well the model predicts the occurrence of heavy rainstorms.

Number of heavy rainstorms	Number of stations reporting	Probability
0	102	0.301
1	114	0.361
2	74	0.216
3	28	0.086
4 or more	12	0.036

8. Geologists sometimes want to know the composition of pebbles in a stream (for example, what fraction of the pebbles are quartzite). R. Flemal, in 1967, gathered 100 samples of 10 pebbles each from the Gros Ventre River in Wyoming. The table below shows the results. The right-hand column gives the expected number of samples according to a binomial probability model. Do the chi-square test to see whether the pebbles can reasonably be thought to be distributed according to the model.

Number of quartzite pebbles	Observed frequency	Expected frequency
10	6	9.1
9	25	24.7
8	31	30.0
7	28	21.7
6	9	10.3
5	0	3.3
4 or less	1	0.9
Total	100	100.0

CHAPTER REVIEW EXERCISES

1. At a restaurant there are six choices on the menu. On a certain day, the items are ordered the following numbers of times:

Grilled Chicken Sandwich	12
Patty Melt	14
Steakburger	6
Garden Salad	5
Fish Sandwich	11
Grilled Steak	12

Were some items preferred over others? Use a chi-square test.

2. Calculate the following probabilities from the table in Appendix C:
   a. $p(\chi_5^2 \geq 14.3)$
   b. $p(\chi_6^2 \geq 20.1)$
   c. $p(\chi_3^2 \geq 4.3)$
   d. $p(\chi_8^2 \geq 22.9)$

3. The paper waste produced in the United States is broken down by type in the following way (percentages are of total weight):

Newspapers	21%
Books/magazines	14%
Office paper	10%
Corrugated cardboard	28%
Mixed paper	27%

Your college conducts an experiment in which it collects all of its paper waste for one day. It measures out to 48.1 tons.
   a. What is the expected amount of each of the five types of paper in that 48.1 tons?
   b. The actual composition of the 48.1 tons is as follows:

Newspapers	11.0 tons
Books/magazines	7.1 tons
Office paper	3.2 tons
Corrugated cardboard	12.9 tons
Mixed paper	13.9 tons

Does the waste collected from your college follow the same distribution as the national figures?

4. Explain intuitively why the number of "degrees of freedom" is one less than the number of possible outcomes.

5. A weather model predicts the following for a certain month in your area:

   Fifteen days will have no rain.

   Ten days will have rain, but less than one inch.

   Six days will have more than an inch of rain.

   The rainfall during the month is looked back on after it is over, and the actual results were as follows:

   Thirteen days had no rain.

   Eleven days had rain, but less than an inch.

   Seven days had more than one inch of rain.

   a. How many degrees of freedom are there in this case?
   b. What is the chi-square statistic?
   c. Did the model predict well what the actual rainfall would be?

6. For each of the following probabilities, give both the estimated value using the appropriate table from the chapter, and the theoretical value from the tables in the back of the book.
   a. $P(\chi_5^2 \geq 5.3)$ (Use Table 7.15 for the estimate.)
   b. $P(\chi_3^2 \geq 4.2)$ (Use Table 7.14 for the estimate.)
   c. $P(\chi_9^2 \geq 15.2)$ (Use the table from Exercise 1 of Section 7.4 for the estimate.)

7. The following data give the average amount of time spent by the typical college freshman on various activities during a weekday.

Activity	Number of hours
Class attendance	3
Studying	3
Job	0.25
Leisure	3
Social activities	2.25
Travel (between classes)	1
Eating	1.5
Grooming	1
Resting and sleeping	6.5
Recreation	1.5
Other	1

Source: David Desmond and David Glenwick, "Time Budgeting Practices of College Students," *Journal of College Student Personnel*, vol. 28, no. 4 (1987), pp. 318–323.

   a. Write out the estimated number of hours you believe you spend on each of these activities on a typical weekday.

b. Do a chi-square test to determine if your schedule matches that of the average college freshman.

8. In 1000 poker hands, you would expect (based on theoretical probabilities) to see the following hands these numbers of times:

Poker hand	Expected number in 1000 hands
Nothing	502
One pair	423
Two pairs	48
Three of a kind	21
Other	7

You and your friends play games of poker over an extended period of time, and in 1000 hands you see the following distribution:

Poker hand	Observed number in 1000 hands
Nothing	488
One pair	438
Two pairs	51
Three of a kind	16
Other	7

a. How many degrees of freedom are there in this case?
b. Did your results follow the expected distribution? Use the chi-square test.

9. Explain why we would be more likely to conclude that a die is loaded the larger the chi-square statistic is.

10. In 1987 the racial makeup of college campuses was as follows:

White	79%
Hispanic	5%
Native American/Native Alaskan	1%
African American	9%
Asian/Pacific Islander	4%
Nonresident	2%

The total number of bachelor degrees awarded, broken down by racial categories, was as follows:

Group	Number of bachelor degrees (thousands)
White	841
Hispanic	27
Native American/Native Alaskan	4
African American	57
Asian/Pacific Islander	33
Nonresident	29

Source: U.S. Department of Education. Cited in QEM, *Education that Works: An Action Plan for the Education of Minorities*, MIT, Cambridge, Mass., 1990.

a. Out of the total of 991,000 degrees awarded, how many could be expected in each of the racial groups, if you assume they follow the same distribution as the student population?
b. Do a chi-square test to determine whether the actual number of bachelor degrees awarded has the same distribution as the student population.

11. Forty times you shuffle a deck of cards and draw out one card.
    a. How many times do you expect each of the four suits (spades, clubs, diamonds, and hearts) to show up?
    b. This test was performed, and the actual results were as follows:

Hearts	12 times
Diamonds	7 times
Clubs	11 times
Spades	10 times

Do a chi-square test to determine if the cards were being drawn randomly from a fair deck.

12. Consider the following two sets of data:

Set A

	Expected	Observed
Red	100	105
Blue	100	97
Green	100	98

Set B

	Expected	Observed
Red	5	10
Blue	5	2
Green	5	3

a. By just looking at the data, do you think the observed values follow the same distribution as the expected values in the case of data set A? What about data set B?
b. Calculate

$$(\text{Observed red} - \text{expected red})^2 + (\text{Observed blue} - \text{expected blue})^2 + (\text{Observed green} - \text{expected green})^2$$

for both data sets A and B.

c. Using your answers from parts (a) and (b), explain why dividing by the expected number is important in the calculation of the chi-square statistic.

13. A clothing store kept track of how often certain sizes of clothing were bought during a day. The results were as follows:

Size	Number of articles sold
Small	32
Medium	40
Large	21
Extra large	15

a. If you believed that each size is equally likely to be sold, how many of each size would you have expected to be sold?

b. Using the chi-square statistic, test the belief that each size is equally likely to be sold.

14. A researcher is interested in determining whether infants are more likely to be born on certain days of the year than on others (ignoring leap years). If the researcher gathered the data for the number of births on every day of the year, what would be the number of degrees of freedom in the chi-square statistic?

15. a. Calculate the following probabilities using the table in Appendix C:

$p(\chi_2^2 \geq 10.2)$
$p(\chi_4^2 \geq 10.2)$
$p(\chi_6^2 \geq 10.2)$

b. What trend do you see in your results to part (a)? That is, as the number of degrees of freedom goes up, what happens to the probability of having a chi-square statistic greater than 10.2?

c. If you had an experiment in which there were 100 degrees of freedom and your chi-square statistic was 10.2, would you conclude that the observed and expected distributions were the same? Explain how you reached your answer. (*Hint:* You should need neither calculations nor tables to solve this.)

16. You make your monthly budget as follows:

Rent	23%
Food	20%
Bills	7%
Taxes	15%
Auto	15%
Savings	10%
Other	10%

You earn \$1500.00 each month.

a. How much do you expect to spend in each of the categories?
b. At the end of a month, you look back on how much you spent on everything, and you see this:

Rent	$345	Auto	$320
Food	220	Savings	150
Bills	110	Other	125
Taxes	230		

Did your spending fit your budget very well? Do a chi-square test to test this.

17. In a class of 197 students, the professor looks at the first number of every student's social security number. The results were as follows:

Digit:	0	1	2	3	4	5	6	7	8	9
Frequency:	0	2	2	179	7	7	0	0	0	0

Was the first digit of the social security number a good source of random numbers? Perform a chi-square test.

18. Recall Exercise 17. This time the professor looks at the last number of every student's social security number. Now the results were as follows:

Digit:	0	1	2	3	4	5	6	7	8	9
Frequency:	25	23	13	16	16	20	26	22	16	20

Was the last digit of the social security number a good source of random numbers? Perform a chi-square test.

19. In the 1968 presidential election the popular vote was divided as follows:

Richard Nixon	31,770,237 votes
Hubert Humphrey	21,270,533 votes
George Wallace	9,906,141 votes

a. What percentage of the vote did each candidate win?
b. The actual winner of the election is determined by the votes of the electoral college. There are 538 members. If the members' votes reflect how the people voted, how many of the electoral votes should each candidate have won?
c. The real results were as follows:

Richard Nixon	301 electoral votes
Hubert Humphrey	191 electoral votes
George Wallace	46 electoral votes

Test the expected versus the actual results using the chi-square test.

20. A car dealer believes that each of the five colors of a certain car is equally likely to be chosen by the customer. At the end of a month, these are the results the dealer sees:

Color	Frequency
Red	21
Blue	10
White	15
Black	17
Light taupe	7

Use a chi-square statistic to test whether the dealer is correct that each color is equally likely to be chosen.

COMPUTER EXERCISES

The program Chi-Square is a separate program that enables you to perform the six steps on chi-square data tables as given in this chapter. We will illustrate its use on two of the examples in the text.

1. EXAMPLE 7.1 REVISITED

Reread Example 7.1, concerning the number of on-the-job accidents by hour of shift.

a. Choose the model, and then decide what the expected number of accidents for each hour should be. In this example, the expected value is 21 for each hour. Type "21" in the first eight spaces in the Expected column. To obtain the chi-square value from the data, we also have to fill in the Obtained column with the actual data. The screen should look as follows:

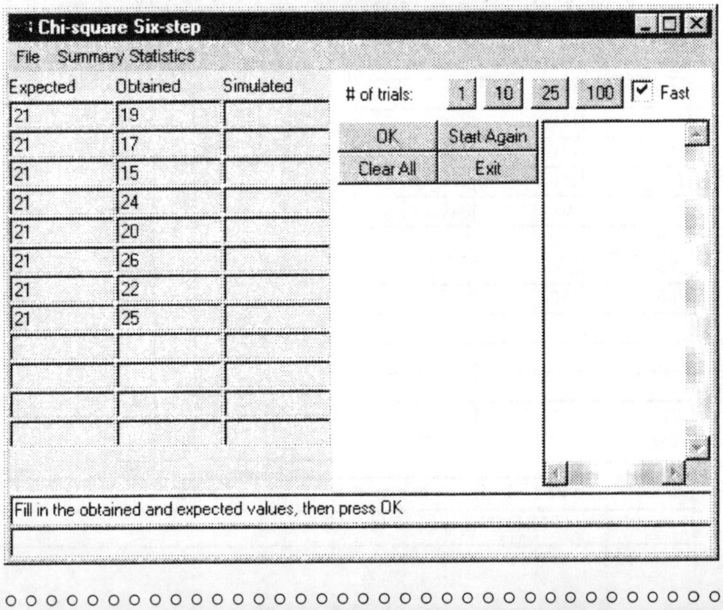

Press the OK button. The program will add up the columns to make sure the obtained and expected counts have the same sum, in this case 168. The second-last line will then present the obtained chi-square (5.14), as well as the number of degrees of freedom:

Obtained Chi-square = 5.14. 8-outcome experiment. Choose the # of trials.

b. An eight-sided fair die will represent the step 1 null hypothesis that accidents are equally likely to happen during each hour. A trial consists of rolling the die 168 times and counting how many of each number appear. The program will simulate this process.

c. Similarly, after each trial, the program will calculate the chi-square statistic from this simulated trial.

Press the "1" button. The program will fill in the third column with a simulated trial and calculate the chi-square. The chi-square value will be printed at the right, with the notation "** OVER" if this value is larger than that found from the data. In the screen below, it happened that the chi-square for this trial was 7.43, which exceeds the chi-square obtained from the data.

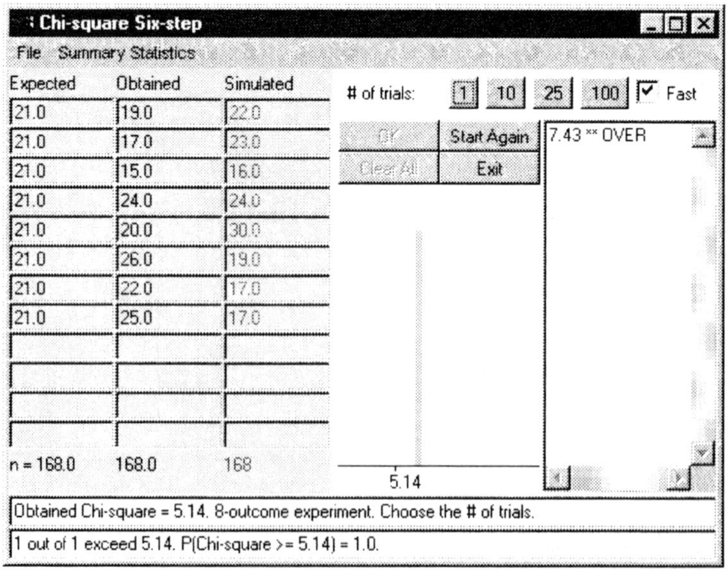

d. Repeat the trials 100 more times by pressing the "100" button. The program will simulate 100 more trials, each time finding a new chi-square:

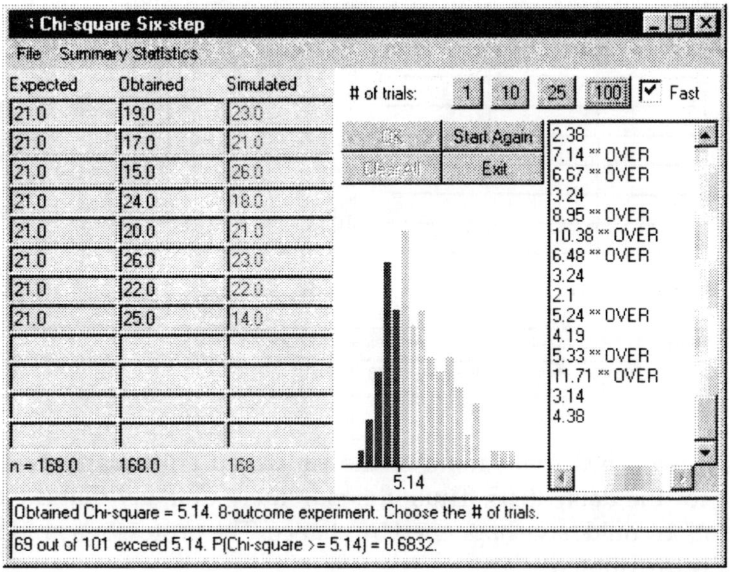

e. The probability of finding a chi-square equal to or greater than that obtained from the data can then be estimated from the trials. The last line of text at the bottom of the screen states the number of trials in which the simulated chi-square exceeded 5.14. For the simulations above, it turned out that 65 out of the 101 trials had a chi-square equal to or larger than 5.14, yielding an experimental probability of $69/101 = 0.6832$. (This number is close to that in the example, which was $31/50 = 0.62$.)

f. Again, the decision is that there is no reason to suspect differences in the shifts. The data look very similar to what you would expect from randomly rolling a die.

2. EXAMPLE 7.7 REVISITED

Recall the null hypothesis model in Example 7.7, which looked at the grades of 200 people in a psychology class. The process follows the same steps as above. Press "Start Again" and then "Clear Values" to erase the previous example from the screen. Fill in the expected and obtained counts, and then press OK. This is what the screen looked like after 500 trials:

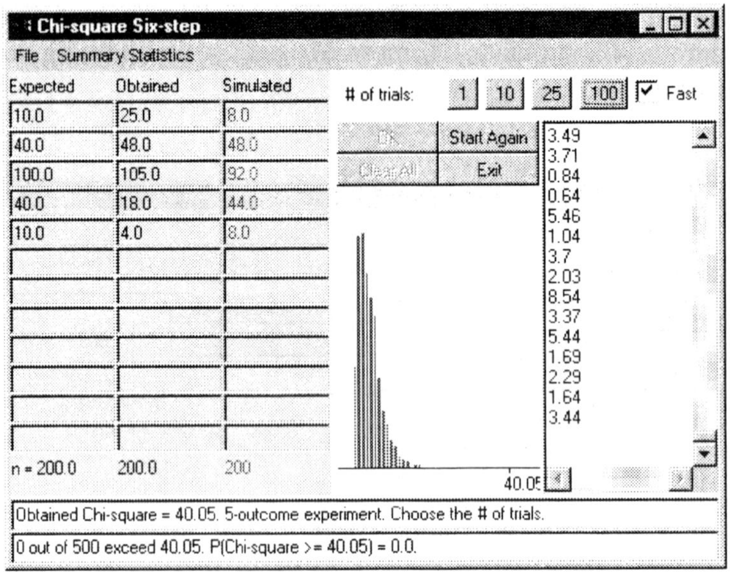

Even with 500 trials, none of the simulated chi-squares was as high as that for the data, 40.05. Thus it is extremely unlikely that the obtained distribution could have happened by chance. The professor's grades are not normally distributed.

3. ILLINOIS STATE LOTTERY

One lottery game the State of Illinois conducts is called "Pick 3." You choose three numbers from the digits 0, 1, 2, . . . , 9. (You can choose a number more than once.) The lottery randomly chooses three table-tennis balls numbered 0, 1, 2, . . . , 9 with replacement. Various types of matches will win you various amounts of money. Pick 3 is run twice daily. We will just look at the results for the evening draws. In 1996, $3 \times 365 = 1095$ numbers were drawn in the evening Pick 3 draws. The table below shows how many of each digit were drawn:

Digit	Number drawn	Digit	Number drawn
0	121	5	107
1	104	6	114
2	107	7	105
3	118	8	107
4	98	9	114

Notice that the numbers are not equal; for example, 121 0s were drawn, but only 98 4s. Is there evidence that some digits are more likely than others, or can the differences be ascribed to chance? To answer that question, set up and run the chi-square program. The expected count for each digit is 109.5. (Why?) A fractional expected value is acceptable. The obtained counts are given in the table. Type the expected and obtained counts in the appropriate areas, and then press OK to make sure everything is OK.

a. What is the obtained chi-square value? What is the number of degrees of freedom?

b. Press the "100" button. (This program may take a while.) What proportion of the simulated chi-squares are equal to or greater than the one obtained from the data?

c. What do you conclude: Are some digits more likely than others, or can the differences be ascribed to chance?

4. MENDEL'S EXPERIMENTS I

Gregor Mendel (1822–1884), an Augustinian monk, conducted a number of experiments on the hybridization of peas that led him to conjecture a number of important genetic truths. Many of the experiments involved two types of plants—for example, one type having round seeds and one having wrinkled seeds. The two types were crossed, yielding the *hybrid generation* of plants, all of whom had round seeds. He termed the round seeds the *dominant* characteristic, and the wrinkled seeds the *recessive* characteristic. Next, members of this hybrid generation were crossed, producing the *first generation* of offspring of the hybrids. Interestingly, among this first generation, some had round seeds and some wrinkled seeds. Thus in the plants of the hybrid generation there was some of the wrinkled seed potential, although none of the seeds were wrinkled. In the first generation from the hybrids there were about three times as many round seeds as wrinkled seeds, so the round seeds still dominated, but not completely as in the hybrid generation. Specifically, the 253 plants in the first generation yielded 7324 seeds, of which 5474 were round and 1850 wrinkled, for a proportion of 0.7474 round.

The table below summarizes seven of Mendel's experiments. These results can be found in his paper "Experiments in Plant Hybridization."* In

*The original paper, in German, was published in 1866 as "Versuche über Pflanzenhybriden" in *Verhandlungen des naturforschenden Vereines in Brünn, Bd. IV für das Jahr 1865,* Abhandlungen, 3-47. This English version can be found at the MendelWeb Internet site (http://www.netspace.org/MendelWeb/). The 1901 translation is by William Bateson, with corrections and changes by Roger Blumberg.

the first two experiments the combined seeds from many plants (253 and 258 plants, respectively) were counted, whereas in the other experiments the numbers of plants were counted.

Dominant character (A)	Recessive character (a)	Total in first generation	Number of dominant in first generation	Number of recessive in first generation
Round seeds	Wrinkled seeds	7324	5474	1850
Yellow albumen	Green albumen	8023	6022	2001
Gray-brown seed-coats	White seed-coats	929	705	224
Inflated pods	Constricted pods	1181	882	299
Green pods	Yellow pods	580	428	152
Flowers along stem	Flowers at end of stem	858	651	207
Long stem	Short stem	1064	787	277

Question a: For each experiment, calculate the proportion of individuals in the first generation from the hybrids that had the dominant character. Are all the proportions close to 0.75?

Genetic research since Mendel's time has explained his data as follows. Each of the characters studied is determined by a *gene*, and each gene has two *alleles*, the dominant one and the recessive one. Thus the first experiment studied the gene that determines the form of the seed, and it has the two alleles "round" and "wrinkled"; in the second experiment the gene was the one that determines the color of the albumen and the alleles were "yellow" and "green"; and so on.

To speak generally, suppose A represents the dominant allele for a particular gene, and a represents the recessive allele. Each plant has two representatives of the allele in its genetic makeup, so it can have two of the dominant alleles, AA; two of the recessive alleles, aa; or one of each, Aa. If the plant's genetic makeup is AA or Aa, then its actual appearance is A, which is what is meant by saying that A is the dominant allele. The only way the plant could have appearance a is if the genetic makeup is aa.

When two plants are crossed, each offspring receives one allele from each parent. If a parent is Aa, then the offspring has a 50% chance of receiving the A and a 50% chance of receiving the a.

Does this theory explain Mendel's observations? Mendel produced the hybrid generation by crossing AA plants with aa plants. All the offspring received an A from one parent and an a from the other and were therefore Aa, which appeared as A:

	Genetic makeup	Appearance
Parent 1	AA	A
Parent 2	aa	a
Offspring (hybrid)	Aa	A

This table explains why the hybrid generation all had the dominant appearance: although half the alleles were recessive, they were always paired with a dominant allele, so that the recessive allele was hidden. No statistical reasoning is involved here.

But the next step was to cross members of the hybrid generation. They all had the Aa genetic makeup. Each offspring randomly received one allele from each parent. The table below shows every possible genetic makeup of the offspring in the first generation from the hybrids and the probability p of the makeup:

	From parent 2	
From parent 1	$A\ (p = \frac{1}{2})$	$a\ (p = \frac{1}{2})$
$A\ (p = \frac{1}{2})$	$AA\ (p = \frac{1}{4})$	$Aa\ (p = \frac{1}{4})$
$a\ (p = \frac{1}{2})$	$Aa\ (p = \frac{1}{4})$	$aa\ (p = \frac{1}{4})$

Thus, using theoretical independence, $p(AA) = p(A)p(A) = \frac{1}{2} \times \frac{1}{2} = \frac{1}{4}$ is the probability that a given offspring will receive an A from both parents. The chance is $(\frac{1}{2} \times \frac{1}{2}) + (\frac{1}{2} \times \frac{1}{2}) = \frac{1}{2}$ that the offspring will receive an A from one parent and an a from the other, because $p(Aa) = p(A)p(a) = \frac{1}{2} \times \frac{1}{2}$, $p(aA) = p(A)p(a) = \frac{1}{2} \times \frac{1}{2}$, and as we will see in Chapter 13, we can add the probabilities of the two disjoint ways of having one a and one A. Finally, the chance is $\frac{1}{2} \times \frac{1}{2} = \frac{1}{4}$ that a given offspring will receive an a from both parents. Because the appearance of the offspring will be A if the genetic makeup is AA or Aa, and the appearance will be a only if the genetic makeup is aa, the chance of the offspring having appearance A is $\frac{3}{4}$ since $p(\text{at least one } A) = 1 - p(aa)$.

Question b: Do Mendel's data fit the theory? That is, is the proportion of individuals in the first generation from the hybrids that have the dominant appearance 0.75? The proportion given by the data is not exactly 0.75, but it is close. Use the chi-square six-step program to see whether the departure from 0.75 can be ascribed to chance. Use 100 trials for each experiment. For each experiment, the two outcomes are "Appearance is A" and

"Appearance is a." The obtained data can be modeled as the number of draws, with replacement, from a box with three A's and one a. Thus the expected counts are $\frac{3}{4}$ × (number of individuals) and $\frac{1}{4}$ × (number of individuals) for the two outcomes. For example, the first experiment has 7324 individuals (seeds), so one expects $\frac{3}{4} \times 7324 = 5493$ round seeds (A) and $\frac{1}{4} \times 7324 = 1831$ wrinkled seeds (a). The observed numbers were 5474 round and 1850 wrinkled. After you enter these numbers and press OK, the screen should look like the following:

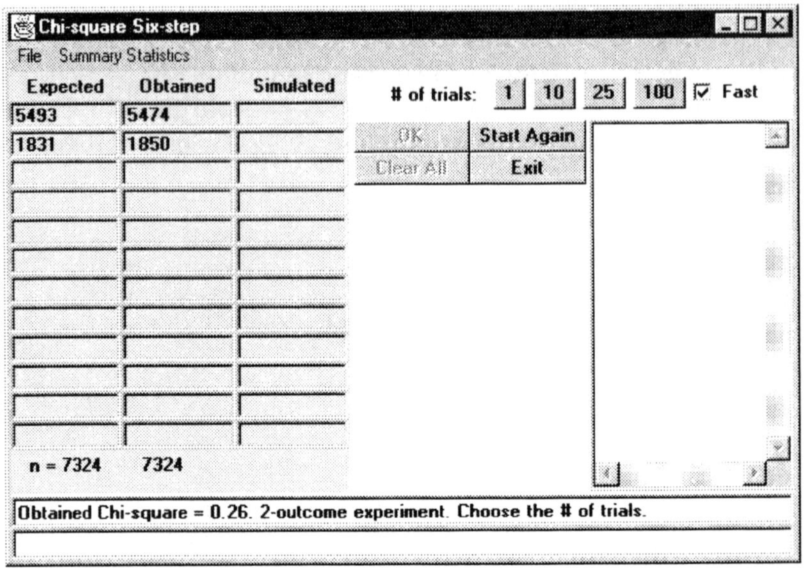

Now press "100" trials. Because there are quite a few individuals in the data set, this program may take a while to run. Note down the value of $P(\chi^2 \geq \text{obtained } \chi^2)$.

Repeat the process for the other six experiments, each time figuring out the correct expected counts. What are the values of $P(\chi^2 \geq \text{obtained } \chi^2)$? Do they suggest that the data fit the model well? Or is there something wrong?

5. MENDEL'S EXPERIMENTS II

Another of Mendel's experiments considered two observable characteristics at once. Hybrids were created by crossing plants that had round yellow seeds with plants that had wrinkled green seeds. The hybrids all had round yellow seeds, both round and yellow being dominant. These hybrids were then crossed, producing 15 plants with 556 seeds of the following kinds:

Appearance	Number
Round yellow seeds	315
Wrinkled yellow seeds	101
Round green seeds	108
Wrinkled green seeds	32

(This is one of the most famous data sets in science.)

The genetic theory to explain these data is a little more complicated than before. We start with two genes, in this case one for the form of the seed and one for the color of the seed. The form has the dominant allele A = round and the recessive allele a = wrinkled. The color has the dominant allele B = yellow and the recessive allele b = green. Each parent has two alleles for each gene. Each offspring receives one form allele and one color allele from each parent. For the original plants, the two types were $AABB$ and $aabb$. Each offspring in the hybrid generation received AB from one parent and ab from the other, so it was $AaBb$, which means it appeared round and green. Now imagine crossing two plants from this hybrid generation. Each offspring has four possibilities for alleles from each parent: AB, Ab, aB, and ab. Combining the possibilities from the two parents, the offspring have the following possibilities for their genetic makeup:

	From parent 2			
From parent 1	***AB***	***Ab***	***aB***	***ab***
AB	*AABB*	*AABb*	*AaBB*	*AaBb*
Ab	*AABb*	*AAbb*	*AaBb*	*Aabb*
aB	*AaBB*	*AaBb*	*aaBB*	*aaBb*
ab	*AaBb*	*Aabb*	*aaBb*	*aabb*

Translating the genetic makeups to the appearance, we find the following:

	From parent 2			
From parent 1	***AB***	***Ab***	***aB***	***ab***
AB	Round yellow	Round yellow	Round yellow	Round yellow
Ab	Round yellow	Round green	Round yellow	Round green
aB	Round yellow	Round yellow	Wrinkled yellow	Wrinkled yellow
ab	Round yellow	Round green	Wrinkled yellow	Wrinkled green

a. In the above table, count how many (out of the 16 cells) of the offspring possibilities are of each type of appearance. Assuming that the possible alleles (outcomes) are equally likely and that the alleles received from one parent are independent of those received from the other, each of the 16 four-allele combinations is equally likely, so the chance of each type of appearance is just the number of combinations yielding that appearance divided by 16. Fill in the following table:

Appearance	Number of cells	Theoretical probability (number of cells/16)
Round yellow seeds	9	0.5625
Wrinkled yellow seeds		
Round green seeds		
Wrinkled green seeds		

b. Using the above table, find the expected number of each type of seed out of 556 seeds. Use the chi-square six-step program to test whether the observed data could come from 556 draws with replacement from a box with proportions of round yellow, round green, wrinkled yellow, and wrinkled green seeds as given in the table. Perform 100 simulations. What is $P(\chi^2 \geq \text{obtained } \chi^2)$? Does the model fit the data adequately?

6. POSTPONING DEATH

In the chapter "Deathday and Birthday: An Unexpected Connection," David Phillips investigates whether people who are coming to the end of their life are likely to try to stay alive until their birthday if it is coming up soon.* He assembled data on months of birth and death for a collection of 348 famous Americans. It appeared that people were less likely to die in the month before their birth month, and more likely to die in the three months after their birthday. The table on the following page summarizes the data.

If people die randomly relative to their birthday, you would expect about 1/12th to die in each of the twelve indicated months. This is our step 1 null hypothesis. (It will not be exactly 1/12th because the months do not have exactly the same number of days, but the approximation is not bad.)

*In J. M. Tanur et al. (eds.), *Statistics: A Guide to the Unknown* (San Francisco: Holden-Day, 1972), pp. 52–65.

Death month	Number of people
6 months before birth month	24
5 months before birth month	31
4 months before birth month	20
3 months before birth month	23
2 months before birth month	34
1 month before birth month	16
Birth month	26
1 month after birth month	36
2 months after birth month	37
3 months after birth month	41
4 months after birth month	26
5 months after birth month	34

a. If people do die randomly, about now many (out of 348) should die in each of the above months?

b. Does it appear that fewer did die than expected during the month before their death month?

c. Is the difference noted in part (b) just due to chance, or is there real evidence that people postpone their death? Perform a chi-square test using the chi-square six-step program. Use 100 trials. We carry out the test because as statisticians we cannot accept our informal judgments that 16 is too small and 36 too large to fit the null hypothesis of uniformly distributed death months.

8

PROBABILITY DISTRIBUTIONS

You have a very serious disease. Of ten persons who get this disease only one survives. But do not worry. It is lucky you came to me for I've recently had nine patients with this disease and they all died of it.

G. Polya

OBJECTIVES

After studying this chapter you will understand the following:

- *Discrete and continuous theoretical probability distributions*
- *The binomial distribution of the number of successes*
- *The Poisson distribution of the number of random occurrences*
- *The continuous uniform distribution of a randomly occuring point*
- *The normal distribution (bell-shaped curve)*
- *How to compute normal probabilities using a standard normal table*

8.1 THE BINOMIAL DISTRIBUTION: SUCCESS COUNTS 379

8.2 THE POISSON DISTRIBUTION: EVENTS AT RANDOM 393

8.3 ROLLING A FAIR DIE: THE DISCRETE UNIFORM DISTRIBUTION 401

8.4 CONTINUOUS DISTRIBUTIONS 403

8.5 THE CONTINUOUS UNIFORM DISTRIBUTION 408

8.6 THE NORMAL DISTRIBUTION 413

8.7 USING THE NORMAL-CURVE TABLE 421

8.8 APPLICATIONS OF THE NORMAL CURVE 427

8.9 OTHER CONTINUOUS DISTRIBUTIONS 433

CHAPTER REVIEW EXERCISES 437

COMPUTER EXERCISES 441

Key Problem

In the 1898 book *Das Gesetz der Kleinen Zahlen (The Law of Small Numbers),* Dr. L. von Bortkewitsch summarized some data on 14 Prussian military units over the 20 years from 1875 to 1894, producing 280 observations. For each unit and year, he presented the number of unfortunate soldiers who died from blows by horses, presumably kicks. In most of the units, no one was kicked to death by horses. The worst were two units that had four such fatalities (unit 11 in 1880, and unit 14 in 1882). The table shows the number of units that had 0, 1, 2, 3, or 4 of these deaths:

Number of soldiers kicked to death	Number of units
0	144
1	91
2	32
3	11
4	2
5 or more	0
	280

Thus, in 144 cases there were no deaths, 91 cases had one, and so on.

Is there a regular pattern to these numbers? How can such data be modeled?

Throughout the book, repeatable random experiments using physical models such as coins, dice, or boxes of numbered tickets (and often using random numbers to rapidly simulate these models) have been used to generate statistics. In the case of a box model, we can think of the tickets (possibly a large number of them) as a representation of the actual population being sampled from. In the case of a die, we can think of the set of all possible throws of the die (infinitely many) as the conceptual population being sampled from. It is conceptual because we can never see all of the population, because the die can continue to be thrown forever. Note that a number is associated with each population member and that therefore randomly sampling once produces a statistic. Because the sampling is random, the statistic is random and has a probability law, called a distribution, to describe its behavior.

In statistics, we are concerned with obtaining information about the probability law that describes sampling from the population associated with random experiments—both simulated and real-world experiments. As suggested above, the population may actually exist or may exist only conceptually. Sampling from either type of population can be physically represented by a box model: sampling without replacement usually models the experiment of randomly sampling from a real population (such as registered voters, each of whom we would want to sample at most once), and sampling with replacement usually models the experiment of sampling from a conceptual population (such as throws of a die).

We refer to the probability law for sampling a statistic from a population as the **distribution** of the statistic. In the specific case of a box model, the distribution is the probability law associated with randomly drawing one numbered ticket from the box. For example, if the experiment is to throw a fair die (equivalently, to draw randomly once from the box model containing the numbered tickets 1, 2, 3, 4, 5, 6), then the distribution is given by

value:	1	2	3	4	5	6
p(value):	1/6	1/6	1/6	1/6	1/6	1/6

We have learned that by using the five-step method we can estimate the distribution of a possibly unfair die (knowing it is not necessarily the one tabulated above) by throwing the die 100 times, say. Then the relative frequency histogram, that is, the observed proportion of 1s, 2s, and so on, can be used to estimate $p(1)$, $p(2)$, and so on.

Statisticians rely on a number of useful distributions. The purpose of this chapter is to present certain basic distributions that statisticians often use to model probability experiments and to do statistical analyses.

Another example of a distribution is that of chi-square, discussed in the previous chapter. One benefit of the chi-square distribution is that the relative frequency histogram (or the associated relative frequency polygon) of chi-square statistics can be approximated by a particular smooth curve, called a chi square *density*. We can therefore use a table of probabilities for this chi-square distribution to bypass the six-step method when doing a chi-square test, obtaining from tables the probability of a chi-square statistic being larger than the observed chi-square value. From this viewpoint, the distribution (the smooth curve) tells how likely it is that one random observation of a statistic exceeds a specified value. Such a probability is determined by computing the area under the density lying to the right of the specified value. This is a *continuous* distribution (recall Figure 7.4), whereas the above fair die distribution is *discrete*. Both types are very important in statistics.

In summary, by the *distribution* of a statistic we mean the set of the theoretical probabilities of its possible values in the discrete case, and a density in the continuous case. In Section 6.1 we discussed the concepts of sample and population. As discussed above, the distribution gives the theoretical probability of each possible value of the statistic resulting from sampling from a population. If the population is a real physical population, such as the heights of adult males, we use a distribution (possibly a density) whose shape approximates that of the relative frequency histogram of the population. For example, the probability law for sampling one adult male's height is given by the famous bell-shaped curve (called a normal density), because the histogram of all the population's heights is bell shaped. Thus the distribution is determined by the histogram of the population from which the statistic (height, say) is randomly sampled.

We will first discuss the discrete distribution associated with yes/no or success/failure situations, such as coin tossing or medical drug testing.

8.1 THE BINOMIAL DISTRIBUTION: SUCCESS COUNTS

The simplest random experiments focus on only two possible results.

Buy a lottery ticket: you either win or lose.

Flip a coin: it comes up heads or tails.

Randomly choose a person from the United States: does the person have confidence in the public schools? Yes or no.

Deal a 5-card hand from a typical 52-card deck: all the cards are from the same suit (a flush), or at least two suits appear.

A person is given a flu shot: the person either does or does not contract the flu.

A couple has a baby: it is either a girl or a boy.

On a given day in a given city, it either snows or does not snow.

The statistic in such an experiment is either 0 or 1. One has to decide which result is the 1 and which is the 0. For example, if your lottery ticket wins, you record 1, but if you lose, you record 0. Often the outcome associated with the 1 is called a "success," especially if the context suggests that one outcome is good and the other is bad.

Suppose you buy one lottery ticket every day for a week. To be specific, each ticket is an Illinois Lottery Instant Game ticket. Each ticket has a 25% chance of winning (the amount won is usually fairly small). What might the data look like? Here is a typical week:

Day	Win or lose	Statistic
Monday	Win	1
Tuesday	Lose	0
Wednesday	Lose	0
Thursday	Lose	0
Friday	Lose	0
Saturday	Win	1
Sunday	Lose	0

Monday and Saturday happened to have winners. The other days lost.

Instead of looking at the above table as representing seven small experiments, consider it to represent one larger experiment that consists of buying seven lottery tickets and counting the number of winners. (How would you build a box model of seven purchases of a lottery ticket with a 25% chance of winning for each purchased ticket? Hint: you can sample seven times with replacement from the box model.) The statistic of interest for the week is the number of winners (you do not care on which days you won, just on how many days you won). In the example there are two winners. Notice that the statistic for the number of winners for the week is just the sum of the statistics for the seven days:

$$1 + 0 + 0 + 0 + 0 + 1 + 0 = 2$$

That is, adding up the 1s counts the number of successes in the larger experiment.

What is the chance of having no winning ticket in a week? Of having three winners? Or (best of all) of having all seven winners? That is, what is the distribution of this sum? Recall that the term *distribution* simply means the specification of the probabilities for the various possible values of the statistic of interest, in this case 0, 1,..., 7. The answer is the **binomial**

distribution. *Bi* means two, *nom* means name. The idea is that each day there are two possible results: win or lose.

For a random experiment in the real world to yield a statistic with a binomial distribution, several conditions have to be satisfied:

1. There is a fixed number of individual trials.
2. Each trial yields a statistic of either 0 or 1.
3. The chance of an individual trial yielding a 1 is the same for each trial.
4. The trials are independent. That is, the 1s and 0s that occurred on prior trials have no influence on whether the current trial produces a 1 or a 0. (It may be helpful to review the discussion on independence in Section 4.3.)

If the four conditions are satisfied, the sum of the individual statistics (the number of successes) has a binomial distribution. Later in the chapter we will learn how to compute binomial probabilities. We refer to an experiment satisfying the four conditions as a *binomial experiment.*

The lottery example appears to satisfy the conditions very well. Condition 1: there are exactly seven trials planned. Condition 2: each trial is either a winner, in which case the statistic is 1, or a loser, in which case the statistic is 0. Condition 3: the chance of winning (getting a 1) on any one day is 25%. Condition 4, independence, supposes that whether you win one day does not have any effect on whether you win on any of the other days. Theoretically that should be true.

Flipping a coin 25 times and counting the number of heads is also well modeled by a binomial distribution. (Ask yourself why the four required conditions hold.) Some other examples follow.

Polling: The Gallup organization conducted a survey during May 28–29, 1996, to assess the public's confidence in various institutions. It interviewed 1019 adults. These people were chosen randomly in such a way that everyone in the population had an equal chance of being chosen. For each person, a 1 was recorded if the person expressed confidence in the public schools, and otherwise a 0 was recorded. (More specifically, a 1 means the person had a "great deal" or "quite a lot" of confidence, while a 0 means the person had only "some" confidence, "very little" confidence, or "no" confidence or did not know.) The number of 1s was 387; that is, 387 of the 1019 people had confidence in the public schools. That is about 38%.

Think about the four conditions for a binomial experiment. First, the number of individual trials is 1019, the number of people interviewed. Did the Gallup Organization fix that number in advance, or actually interview a random number of people? Probably the number was not exactly fixed, but for practical purposes it can probably be considered so. Its polls often have around 1000 respondents, but because of the vagaries of finding people at home, having working telephones, and so on, the actual number randomly fluctuates a bit, depending on how lucky the polling organization is in reaching the people it intends to sample. Second, each trial indeed yields a 1 or a 0.

Third, what is the chance that a given chosen person has confidence in the public schools? Here we have to be careful about where the randomness is. Imagine the entire population of adults in the United States. Some have confidence in the public schools; some do not. For argument, suppose 40% do have confidence. Then the chance that a randomly chosen person is one of those with confidence in the public schools is 40%, or $\frac{2}{5}$. This chance is exactly the same for the first person chosen, the second chosen, and so on, so condition 3 is satisfied. This last statement is not intuitive for many students. It becomes clearer if one realizes that the $\frac{2}{5}$ chance of the second person having confidence in the public schools is figured *without* knowledge of the first person's response. To make this clear, consider a simple example of two people buying milk in the case in which one of the five bottles in the grocery cooler is spoiled. The chance that the first person buys the spoiled milk is of course $\frac{1}{5}$. But the chance for the second person is *also* $\frac{1}{5}$, because the result from the first person is not known. For example, do you care whether you are the first or the second purchaser if as the second purchaser you have no knowledge about the first person's purchase?

Were people chosen independently? In a practical sense, yes. Thus the resulting 1s and 0s of the 1019 trials will be independent as required. There could be a slight problem of dependency between people selected because a person chosen once cannot be chosen again, which is to say that the population being drawn from changes slightly from trial to trial. For example, if a lot of 1s have occurred, then the chance of getting a 1 in the current trial with so many people with 1s already selected may be less. But the population of the United States is so much larger than 1019 that the chance of getting a 1 on the current trial is for all practical purposes uninfluenced by the results of previous trials. For all practical purposes, then, independence holds.

The result of this poll does appear to be reasonably well modeled by a binomial distribution, although there are a few subtleties to suggest caution about our conclusion.

Drawing cards: This example uses just the four aces and four kings from a regular deck. Shuffle those eight cards well, and choose five off the top. Count the number of aces you draw. Is this number binomial? The first condition holds: there are five individual trials. The second holds, too. Each draw is either 1, if it is an ace, or 0, if it is a king. The third condition is true, because the chance of an ace being drawn is $\frac{1}{2}$ for any of the cards. (Remember that you are not considering the result of previous trials when you say the probability of an ace in the third trial, say, is $\frac{1}{2}$.) The fourth condition is false. The chance that the second card is an ace is not independent of the first card. If the first card is an ace, the chance that the second is an ace is $\frac{3}{7}$, because three aces and four kings are left. If the first card is a king, the chance that the second card is an ace is $\frac{4}{7}$. What is the chance that the fifth card is an ace if the first four cards are aces? What is the chance that the fifth card is an ace if the first four cards are kings? Moreover, $\frac{3}{7}$ is a lot less than $\frac{1}{2}$. So, as was not the case in the above polling example, where the dependence was slight and hence ignorable, here it is substantial.

The number of aces should not be modeled as a binomial.

Children: For a randomly selected family, is the number of girls among the children binomial? Certainly, each child is either a girl or boy, and the chance that any given child is a girl is about 49% (generally slightly more boys are born than girls). The genders of the children in a family are probably reasonably independent, especially if we discount the occurrence of twins. But condition 1 may not be valid: in some cultures many parents want to have at least one son, so the number of children is not fixed, because they keep having children until they have at least one son. Therefore the number of girls may not be binomial. Of course, if we only considered three-child families, the binomial distribution would work well.

Snow: Each day of the year, it either does or does not snow in Detroit. Is the number of days it snows during a given year binomial? No. The first two conditions are valid, but the third is not. The chance of snow on winter days is obviously much higher than on summer days. What about condition 4? If it is snowing on one day, then clearly the chance of snow on the next day is higher. Thus condition 4 fails too!

Binomial Probabilities

By knowing the number of individual trials and the chance of a 1 on an individual trial, it is possible to find the chance of obtaining a given number of 1s, or successes, in a binomial experiment. For example, we will simulate the above lottery example. There are seven individual trials, and each has a $\frac{1}{4}$ chance of yielding a 1. We generate seven individual trials 100 times, and each time we find the number of winners. Table 8.1 contains the number of times we obtained each possible sum (for example, 10 simulated weeks had no winner, and 37 had one winner), the proportion (experimental probability) of seven-trial experiments in which we obtained each sum, and the theoretical probabilities.

Table 8.1 **p(number of winners) for Success Probability $\frac{1}{4}$**

Number of winners	Experimental frequency	Experimental probability	Theoretical probability
0	10	0.10	0.1335
1	37	0.37	0.3115
2	32	0.32	0.3115
3	17	0.17	0.1730
4	3	0.03	0.0577
5	1	0.01	0.0115
6	0	0.00	0.0013
7	0	0.00	0.0000

In the next section we will show how to figure out the theoretical probabilities that are given in the last column. There is a good chance that there will be one or two winners over the course of a week, but it is very unlikely that there will be more than four winners. Note that the observed probabilities are reasonably close to the theoretical ones. If instead of simulating 100 weeks we simulated 400 or 1000, the experimental probabilities would be even closer to the theoretical probabilities.

Although the estimate of the chance of getting seven winners rounds to 0, it is possible. The probability of doing so is easy to figure out. Since each individual day has a $\frac{1}{4}$ chance of winning, and the seven days are independent, the chance that all seven are winners is

$$\left(\frac{1}{4}\right)\left(\frac{1}{4}\right)\left(\frac{1}{4}\right)\left(\frac{1}{4}\right)\left(\frac{1}{4}\right)\left(\frac{1}{4}\right)\left(\frac{1}{4}\right) = \frac{1}{16{,}384} = 0.000061$$

(recalling from Section 4.3 the fact that we multiply probabilities to determine the probability of events occurring together when the events are independent). With this very small probability, which is about 1 in 16,380, we would expect to wait on average over 315 years (16,380 weeks) before having an entire week of winners.

Figuring Out Theoretical Binomial Probabilities

The approach to finding the chance of there being seven winners in a trial can be further developed to find the chance of there being any particular number of winners. There are three steps:

1. Write down all the possible sequences of individual trial outcomes that combine to produce the desired number of winners.
2. Find the probability of each of those sequences.
3. Sum the probabilities found in step 2.

Let's start with the simple example of flipping a fair coin twice, which is a simple case of a binomial experiment. The number of heads could be 0, 1, or 2. For each of those numbers, we follow the above steps.

Number of heads = 0: Only one sequence yields 0 heads: both flips are tails. Hence step 1 yields

tails	tails	0 0

The chance that both are tails is $\left(\frac{1}{2}\right)\left(\frac{1}{2}\right)$, because the coin is fair and the flips are independent. So step 2 yields

Probability of (tails,tails)

$$= \text{(probability of tails on first flip)(probability of tails on second flip)}$$

$$= \left(\frac{1}{2}\right)\left(\frac{1}{2}\right) = \frac{1}{4}$$

Step 3 is to sum the probabilities, and since there is only one, the answer is $\frac{1}{4}$.

Number of heads = 1: For step 1 we need the ways to get one heads and one tails. There are two sequences:

heads	tails	1 0
tails	heads	0 1

The chance of each of those two sequences can be figured out as in the previous case, so step 2 gives us

Probability of (heads,tails)

$$= \text{(probability of heads on first flip)(probability of tails on second flip)}$$

$$= \left(\frac{1}{2}\right)\left(\frac{1}{2}\right) = \frac{1}{4}$$

and

Probability of (tails,heads)

$$= \text{(probability of tails on first flip)(probability of heads on second flip)}$$

$$= \left(\frac{1}{2}\right)\left(\frac{1}{2}\right) = \frac{1}{4}$$

Now add the probabilities to complete step 3:

$$\frac{1}{4} + \frac{1}{4} = \frac{1}{2}$$

Number of heads = 2: Now both flips must be heads. Step 1 gives

heads heads 1 1

Step 2 gives

Probability of (heads,heads)

$$= \text{(probability of heads on first flip)(probability of heads on second flip)}$$

$$= \left(\frac{1}{2}\right)\left(\frac{1}{2}\right) = \frac{1}{4}$$

In step 3 we get a sum of $\frac{1}{4}$.

The distribution just derived is binomial with two individual trials in which the chance of a 1 (success) on any individual trial is $\frac{1}{2}$. Here is the summary:

Number of successes	Theoretical probability
0	$\frac{1}{4} = 0.25$
1	$\frac{1}{2} = 0.50$
2	$\frac{1}{4} = 0.25$

Notice that the probabilities add up to 1. They should, because we have accounted for all possible outcomes—you have to get either 0, 1, or 2 heads. This intuitive idea that the sum of the probabilities for all of the distinct possible outcomes adds to one will be made rigorous in Chapter 13. But here we can say that when we list the probabilities of all the distinct possible outcomes of a probability experiment, these probabilities must sum to one. As a simple example, the sum of the six die probabilities of $\frac{1}{6}$ is 1.

Now we tackle a slightly more complicated example. Imagine playing the Illinois Instant Lottery game three times. Recall that the chance of a single ticket being a winner is $\frac{1}{4}$. The possible values of the statistic, the number of wins in three tries, are 0, 1, 2, and 3. We will go through the steps to find the chances.

Number of wins = 0: Step 1: The only sequence is

lose lose lose 0 0 0

The chance of a single ticket winning is $\frac{1}{4}$, so the chance of losing is $1 - \frac{1}{4} = \frac{3}{4}$ (because the complement of winning is losing). Step 2: The chance of three losers in a row is

$$\left(\frac{3}{4}\right)\left(\frac{3}{4}\right)\left(\frac{3}{4}\right) = \frac{27}{64} = 0.421875$$

Step 3: The answer is $\frac{27}{64}$.

Number of wins = 1: Of the three tickets, one is a winner and two are losers. The winning one could either be the first, second, or third ticket. So for step 1 we have the three possible sequences:

Win	Lose	Lose	1 0 0
Lose	Win	Lose	0 1 0
Lose	Lose	Win	0 0 1

Step 2 finds the chance of each sequence by multiplication:

Chance of (win,lose,lose)

$$= (\text{chance of win on first}) \times (\text{chance of loss on second}) \times (\text{chance of loss on third})$$

$$= \left(\frac{1}{4}\right)\left(\frac{3}{4}\right)\left(\frac{3}{4}\right) = \frac{9}{64} = 0.140625$$

Similarly,

$$\text{Chance of (lose,win,lose)} = \left(\frac{3}{4}\right)\left(\frac{1}{4}\right)\left(\frac{3}{4}\right) = \frac{9}{64} = 0.140625$$

$$\text{Chance of (lose,lose,win)} = \left(\frac{3}{4}\right)\left(\frac{3}{4}\right)\left(\frac{1}{4}\right) = \frac{9}{64} = 0.140625$$

Notice that each sequence has the same probability, $\frac{9}{64}$. Step 3 adds the three probabilities:

$$\frac{9}{64} + \frac{9}{64} + \frac{9}{64} = \frac{27}{64} = 0.421875$$

Number of wins = 2: Now there are two wins and one loss. Step 1:

Win	Win	Lose	1 1 0
Win	Lose	Win	1 0 1
Lose	Win	Win	0 1 1

Step 2:

Chance of (win,win,lose)

$$= (\text{chance of win on first}) \times (\text{chance of win on second}) \times (\text{chance of loss on third})$$

$$= \left(\frac{1}{4}\right)\left(\frac{1}{4}\right)\left(\frac{3}{4}\right) = \frac{3}{64} = 0.046875$$

Continuing:

$$\text{Chance of (win,lose,win)} = \left(\frac{1}{4}\right)\left(\frac{3}{4}\right)\left(\frac{1}{4}\right) = \frac{3}{64} = 0.046875$$

$$\text{Chance of (lose,win,win)} = \left(\frac{3}{4}\right)\left(\frac{1}{4}\right)\left(\frac{1}{4}\right) = \frac{3}{64} = 0.046875$$

Again, each sequence has the same probability, in this case $\frac{3}{64}$. Step 3 is to add them up:

$$\frac{3}{64} + \frac{3}{64} + \frac{3}{64} = \frac{9}{64} = 0.140625$$

Number of wins = 3: There is only one sequence with three wins. Step 1:

Win Win Win 1 1 1

Step 2:

$$\text{Chance of (win,win,win)} = \left(\frac{1}{4}\right)\left(\frac{1}{4}\right)\left(\frac{1}{4}\right) = \frac{1}{64} = 0.015625$$

Step 3: The answer is then $\frac{1}{64}$. There is not a very good chance that all three tickets will win.

Now we can summarize the binomial distribution with three trials and with the chance of a 1 on an individual trial being $\frac{1}{4}$:

Number of successes	Theoretical probability
0	27/64 = 0.421875
1	27/64 = 0.421875
2	9/64 = 0.140625
3	1/64 = 0.015625

Add up the probabilities. Do you obtain 1?

When the number of individual trials is large, writing out all the possibilities is tedious, but in principle it can be done. Table 8.1 showed the theoretical probabilities for seven trials for the case in which the probability of a 1 on a single trial is 0.25. Section 13.7 revisits the topic of the binomial distribution and gives a compact formula for theoretical binomial probabilities in general. For your convenience, an incomplete table of binomial probabilities based on this formula appears in Appendix G.

Means and Standard Deviations

Statisticians and probabilists have studied the properties of many specific distributions, including the binomial. An advantage to recognizing that the random experiment (perhaps sampling from a population) yields a statistic with a known probability distribution is that many important characteristics of the experiment are immediately available. For example, theoretical probabilities of interest can often be obtained through tables or formulas (rather than the laborious five-step method). In addition, there is usually a simple expression for the theoretical mean and the theoretical standard deviation, the most widely used measures of the center and spread of a distribution. Concerning the binomial distribution, in Chapter 10 confidence intervals to estimate the probability of a 1 (success) with a built-in measure of estimation accuracy and in Chapter 11 hypothesis tests to carry out decision making concerning the probability of a 1 are developed. Chapters 10 and 11 also deal with confidence intervals and hypothesis testing for the means and standard deviations of other distributions. Below we discuss the theoretical mean and standard deviation of a binomial.

In advanced statistics books that utilize calculus, general formulas for the mean and standard deviation of a distribution are given that involve either integral calculus or complex summation operations. Perhaps more important than having the ability to compute such means and standard deviations, however, is having a clear understanding of their meaning and usefulness. Our five-step method and the notion of sampling from a distribution provide this. First, a distribution always describes the probability law for a statistic that results from one trial of a random experiment. Often the experiment is the random sampling of an object from a population, such as the sampling of one tree from all the trees in a particular forest. Now imagine 100 trials of the experiment, possibly conducted using the five-step method in which the distribution is expressed by a particular physical box model. As we learned in Chapter 4, which discussed expected value, the average of these 100 sampled statistics from the given distribution should be close to a fixed number. Indeed, if more and more trials were carried out, the experimental average, or estimated expected value, would approach closer and closer to a fixed theoretical quantity. This quantity is the theoretical mean (sometimes called the *population mean*). Thus the five-step method can be used to approximately find this theoretical mean of a distribution when we have a physical model or a random number table to simulate sampling from the distribution. Similarly, the experimental standard deviation of a large number of statistics sampled from a distribution provides a good approximation of the distribution's standard deviation (also called the *theoretical standard deviation* or the *population standard deviation*). Thus the real meaning of the theoretical mean and standard deviation is that they indicate where the population (real or conceptual) is centered and how spread out it is. Moreover, these important aspects of the population can be found approximately by sampling from the population.

If you flip a fair coin 10 times, you would expect about five heads. In any binomial experiment, you would expect the proportion of 1s to be equal to the chance any individual trial is a 1 times the number of trials. This conjecture is correct and leads to the following convenient formula: In a binomial experiment, the theoretical mean number of 1s, or successes, is

$$\text{(Number of trials)} \times \text{(probability of a 1 on an individual trial)}$$

In the lottery example above, in which you buy a lottery ticket on each of seven days of the week and the chance of winning on any given day is $\frac{1}{4}$,

$$\begin{aligned}\text{Theoretical mean number of wins} \\ &= \text{(Number of days)} \times \text{(probability of winning on a given day)} \\ &= 7 \times \left(\frac{1}{4}\right) = \frac{7}{4} = 1.75\end{aligned}$$

You expect 1.75 winning tickets in a week. What can this mean? You know you cannot win exactly one and three-quarters times; each day you either win or lose, and thus in one week you win a number of times given by an integer. It is important to realize that *expected* means that after repeating the experiment over and over, the average number of wins per week would be around 1.75. (Recall Chapter 4, where we found such expected values empirically by the five-step method. Here we are simply learning the formula for the expected number of successes in a binomial experiment, which allows us to bypass the five-step method to precisely find the expected number of successes in a binomial experiment, which the five-step method only approximates.)

In the 100 five-step method experiments summarized in Table 8.1, the actual average is 1.69:

$$\frac{(10 \times 0) + (37 \times 1) + (32 \times 2) + (17 \times 3) + (3 \times 4) + (5 \times 1)}{100} = \frac{169}{100} = 1.69$$

This experimental mean, 1.69, is fairly close to the theoretical mean, 1.75, that it estimates.

How far do the actual values from the individual experiments typically differ from 1.75? In most weeks the number of winners is 1, 2, or 3, so you expect to be fairly close to the theoretical mean. How close can be measured by the standard deviation because the standard deviation is a measure of the typical spread one expects of an observed statistic. That is, being off by up to one standard deviation from the theoretical mean is a common occurrence. Indeed, obtaining a sampled statistic that is very close to the theoretical mean is quite unusual: variation is always present in random experiments! Crudely, we expect to be off from the mean by about one standard deviation, on average. For the data in Table 8.1, the sample, or experimental, standard deviation (the square root of the variance of the data) is

$$\text{square root}\left\{\frac{1}{100}[10(0-1.69)^2+37(1-1.69)^2+32(2-1.69)^2+ 17(3-1.69)^2+3(4-1.69)^2+(5-1.69)^2]\right\}$$

$$=\sqrt{\frac{105.39}{100}}=1.0267\approx 1$$

We express this by stating that the numbers of winners per week averaged 1.69, plus or minus about 1. By "plus or minus about 1" we mean that the typical deviation from 1.69 is about 1 *in magnitude* and can be in either direction.

The theoretical standard deviation is the standard deviation you would obtain if you ran the experiment over and over forever. If the experiment is one of sampling from a population, the theoretical standard deviation can be thought of as the population standard deviation. The theoretical standard deviation for a binomial has a convenient, if not particularly intuitive, formula: In a binomial experiment, the theoretical standard deviation of the number of 1s is

square root[(number of trials) × (probability of a 1 on an individual trial) × (probability of a 0 on an individual trial)]

In the lottery example,

Theoretical standard deviation of the number of wins
= square root[(number of days)(probability of winning on a given day) × (probability of losing on a given day)]

$$=\sqrt{7\times\left(\frac{1}{4}\right)\times\left(\frac{3}{4}\right)}=\sqrt{\frac{21}{16}}=\sqrt{1.3125}=1.1456$$

It is close to the experimental value of 1.0267—not surprising, because with 100 trials we anticipate that the experimental or sample value will be close to its theoretical or population value. (This is of course our justification for using the five-step method with a reasonably large number of trials.)

Flip a fair coin 100 times. You expect 50 heads $\left(100\times\frac{1}{2}\right)$, and the theoretical standard deviation is

$$\sqrt{100\times\left(\frac{1}{2}\right)\times\left(\frac{1}{2}\right)}=\sqrt{25}=5$$

Thus it is not unusual to be off by up to 5 heads or so; that is, any number between 45 and 55 is reasonable. In fact, an important rule of thumb is that about $\frac{2}{3}$ of the time we expect to be off from the theoretical mean by less than one standard deviation. In this precise sense, the size of the theoretical standard deviation gives us a quantitative interpretation of the "typical" variability in whatever is being observed—here the number of heads in 100 tosses. You would be surprised to get as few as 20 or 30, or as many as 70 or 80.

If you flip a fair coin 1000 times, you would anticipate that the number of heads would be around 500. How far off is reasonable? More than or less than 5? The answer lies in the standard deviation:

$$\sqrt{1000 \times \left(\frac{1}{2}\right) \times \left(\frac{1}{2}\right)} = \sqrt{250} = 15.81$$

Now from 485 to 515 is a likely number of heads—to be precise, an occurrence that should happen about $\frac{2}{3}$ of the time if we toss a coin 1000 times. So the typical spread with 1000 flips is about three times that with 100 flips. The more flips, the larger the standard deviation. However, an interesting fact is that as a percentage of the number of flips, the standard deviation is actually less with 1000: 15.81 out of 1000 is 1.581%, while 5 out of 100 is 5%. Later this observation will be seen to be involved with the law of averages.

SECTION 8.1 EXERCISES

1. A baseball team has won 60% of its games to this point in the season. Over the weekend it will play three games. Go through the four conditions for a binomial experiment. Why is the binomial distribution not appropriate for the number of games won on the weekend?
2. On a multiple-choice quiz, Joan guesses on each of five questions. There are four possible answers for each question.
   a. What is the probability that Joan guesses the first question correctly?
   b. Go through the four conditions for a binomial experiment. Verify that the number of correct answers out of five follows the binomial distribution.
   c. What is the probability that Joan guesses the correct answer on all five questions?
3. Return to the situation presented in Exercise 2.
   a. Write out the possible quiz outcomes in which Joan correctly guesses exactly two of the five questions. There should be 10 total possible outcomes. To get you started, here is one of the outcomes:

Question number	Guess
1	Right
2	Right
3	Wrong
4	Wrong
5	Wrong

b. Now, next to each of the outcomes, write the probability of seeing that particular outcome.
c. Using the table you have created, find the probability that Joan gets exactly two of the five questions correct on the quiz.

4. Give an example of a situation in which the binomial distribution could be used.

5. On each day of a certain week, there is a 20% chance that the bus will reach the bus stop late. Assume that each day is independent of the next.
   a. What is the probability that the bus reaches the bus stop on time on a certain day?
   b. What is the probability that the bus reaches the bus stop on time every day in a five-day work week?
   c. What is the probability that the bus reaches the bus stop on time exactly four days in the five-day work week?
   d. What is the theoretical mean for the number of times the bus will reach the bus stop on time in a five-day work week? What is the standard deviation?

6. For each of the following binomial distributions, give the theoretical mean and the theoretical standard deviation:
   a. $n = 10$; probability of a 1 is 0.90.
   b. $n = 22$; probability of a 1 is 0.35.
   c. $n = 14$; probability of a 1 is 0.12.
   d. $n = 7$; probability of a 1 is 0.47.

7. There are 100 poker chips in a bowl, 65 of which are blue and 35 red. You are to draw out a chip 50 times. After each draw, you replace the chip into the bowl (that is, there are 100 chips in the bowl prior to each draw). *Hint:* Does replacement make this a binomial probability problem?
   a. What is the theoretical expected value of the number of blue chips you will draw in 50 draws?
   b. What is the theoretical standard deviation of the number of blue chips you will draw in 50 draws?
   c. For each of the following values, give a subjective answer as to whether it is likely or not to draw that many blue chips in 50 draws: 18 or fewer; 29 or fewer. Base your decisions solely on the theoretical expected value and standard deviation you found in parts (a) and (b). Justify your answers.

8. You roll a fair, six-sided die six times. Calculate the following probabilities by writing out all the possible outcomes that result in the stated event and summing those probabilities.
   a. The probability that exactly one of the six tosses will come up a five
   b. The probability that exactly five of the six tosses will come up an even number
   c. The probability that all of the tosses will come up an odd number

9. A shopper goes to the grocery store and purchases a bag of 10 apples. Based on previous experience, he knows that 5% of the apples of this brand will be bruised and inedible. Assume that the apples are randomly assigned to the bag. What is the probability that in his bag he has either no bad apple or only one bad apple? (*Hint:* The probability that there is either one bad apple or none equals the sum of the probability that there is no bad apple and the probability that there is one bad apple.)

8.2 THE POISSON DISTRIBUTION: EVENTS AT RANDOM (OPTIONAL)

Imagine standing on a bridge over a not-too-busy rural interstate and counting the number of cars that pass under it in an hour. Can this number be modeled as a binomial? Each car that passes can be counted as a 1. But how many trials are there? In particular, is there a fixed number of trials, as required for the binomial distribution to be valid? This question requires us to define a population of cars being sampled from and to consider that a 1 (or success) occurs if the sampled car passes under the bridge. Is this population the total number of cars in the world, each car being an individual trial, counting 1 if it passes under the bridge and 0 if it does not? Or should we just consider cars in the United States, or those in the particular state or county the interstate is located in, or just those on the roadway? We seem to have a problem deciding on what the population is, what it means to sample from the population, what the fixed number of trials should be, and what the probability of success on each trial is. The binomial distribution just does not apply!

A similar example is that of a Geiger counter counting the number of radioactive particles being emitted from some substance, such as a mineral, within a minute. Perhaps the total number of trials can be taken as equal to the number of particles in the physical object (the piece of mineral), each giving a 1 if it is emitted and 0 if not?

Trying to consider these experiments as binomial experiments is a bit difficult. If there is a very large but unspecified number of trials and if for each one there is a very small but unspecified chance of a 1, another distribution, the **Poisson distribution,** is likely to model the data quite well and will be easy to work with.

Like the binomial distribution, the Poisson distribution counts the number of occurrences (perhaps "successes") of something, but as is not the case with the binomial, there is no theoretical maximum number of occurrences when the Poisson is used. Instead, one counts the number of occurrences during a preset amount of time. The number of occurrences will have a Poisson distribution if the following conditions are satisfied:

1. The probability of an occurrence within a small time interval of fixed length is the same regardless of when the time interval begins. Thus the rate of occurrence is constant over time.
2. Whether there is an occurrence within a given interval is independent of whether there is an occurrence in another nonoverlapping interval (even if the intervals are close to each other). Thus past occurrences do not influence the likelihood of current occurrences.
3. If we take a very small time interval, it is extremely unlikely that there will be two or more occurrences in the same small time interval.

Sometimes these three assumptions are loosely characterized as implying that the times of the occurrences are totally random.

Consider the car example. We are interested in the number of cars passing in one hour. Compared with a one-hour interval, a one-second interval can be considered small. Condition 1 means that the chance that a car passes under the bridge during any one-second period is the same for every one-second period in the hour. This condition would be violated if, for example, half-way through the hour an athletic event at a nearby high school ended, so that in the second half of the hour there are many more cars than during the first half. Condition 2 means that a car passing during a particular second has no effect on the chance that one passes during the next (or any other) second. (That would not be the case if you are counting regularly scheduled buses stopping at a bus stop. Suppose a bus is scheduled to come once every 20 minutes. If a bus has just passed, then for the next several minutes it would be very unlikely that another one would come.) The third condition is valid for a small time interval because the cars are not attached to each other and hence multiple occurrences are impossible. If instead of counting cars, you counted the number of people going by in cars, condition 3 would not hold, because there could be several people in one car. If we are confident that the three assumptions hold, we can use the Poisson probability law.

Let's return to the Key Problem. For a given military unit, it may be reasonable to consider using the Poisson distribution for the number of deaths in that unit in the year. Since the chance of being kicked to death seems likely to be about the same for each day during the year, condition 1 seems fairly reasonable. It seems that a soldier's being kicked to death is independent of another soldier's being kicked to death, so condition 2 seems reasonable. Condition 3 is also probably satisfied, unless these horses are prone to go on rampages and kick several soldiers to death at once. Thus we suspect the data may be well modeled by the Poisson distribution.

The data in Table 8.2 are based on the 280 military unit and year combinations (14 units in each of the 20 years of the study). For simplicity we will consider them to be 280 different units, each observed for one year. For each unit, the number of soldiers kicked to death within the year

Table 8.2 **Prussian Soldier Data**

Number of soldiers kicked to death	Number of units	Experimental probability
0	144	0.514
1	91	0.325
2	32	0.114
3	11	0.039
4	2	0.007
5 or more	0	0
	280	

was recorded. Since there are data on what happened for 280 different units, we can get a pretty good empirical idea of what the probabilities are for 0 death per unit, 1 death per unit, and so on. On average, only 0.7 soldier per unit died in this manner per year, which one supposes would be somewhat comforting to a Prussian soldier if the units all have very large numbers of soldiers.

The theoretical Poisson probabilities are based on a parameter that is the theoretical mean number of counts in the time interval, here 1 year. That is, the mean number of counts (per unit) is a *parameter* of the distribution, just as for the binomial the number of trials and the probability of a 1 on a trial are both parameters. Thus any formula for a Poisson probability will require the user to have selected a value for the theoretical mean. The actual formula for computing a Poisson probability, such as p(3 deaths), is given in Section 13.2. Table 8.3 gives theoretical probabilities for a few possible values of the mean; each column is the distribution for a particular choice of theoretical mean. The ellipsis dots (...) below the 16th row indicate that the probabilities for 17, 18, and so on are so small that they can be considered equal to zero. Figure 8.1 shows these theoretical probabilities as histograms. Each column of Table 8.3 provides one such histogram.

Table 8.3 **Theoretical Poisson Probabilities**

	Mean			
Number of occurrences	0.1	0.7	1.0	5.0
0	0.9048	0.4966	0.3679	0.0067
1	0.0905	0.3476	0.3679	0.0337
2	0.0045	0.1217	0.1839	0.0842
3	0.0002	0.0284	0.0613	0.1404
4	0.0000	0.0050	0.0153	0.1755
5	0.0000	0.0007	0.0031	0.1755
6	0.0000	0.0001	0.0005	0.1462
7	0.0000	0.0000	0.0001	0.1044
8	0.0000	0.0000	0.0000	0.0653
9	0.0000	0.0000	0.0000	0.0363
10	0.0000	0.0000	0.0000	0.0181
11	0.0000	0.0000	0.0000	0.0082
12	0.0000	0.0000	0.0000	0.0034
13	0.0000	0.0000	0.0000	0.0013
14	0.0000	0.0000	0.0000	0.0005
15	0.0000	0.0000	0.0000	0.0002
16	0.0000	0.0000	0.0000	0.0000
...	...	...	...	...

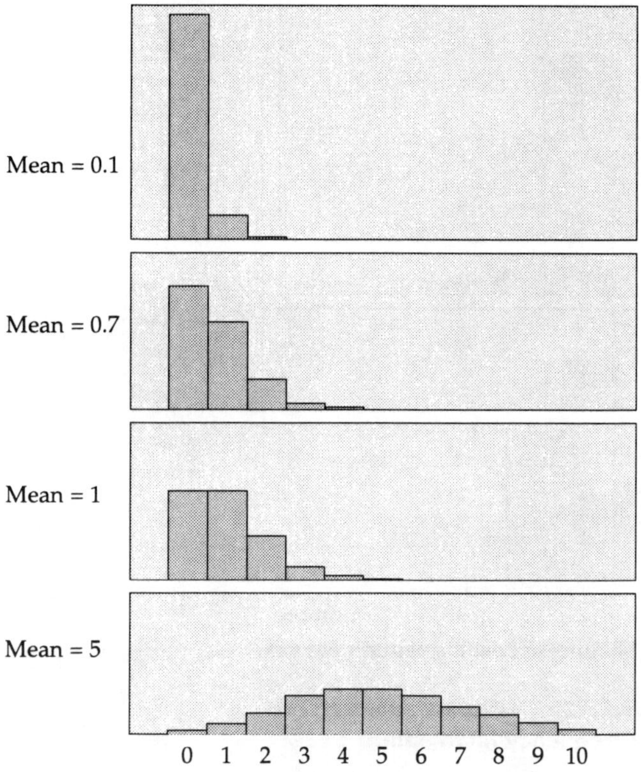

Figure 8.1 **Poisson probability histograms.**

The standard deviation for a Poisson distribution is easy to compute:

$$\text{Theoretical standard deviation} = \sqrt{\text{theoretical mean}}$$

Return to the soldiers' data. Compare the theoretical probabilities for a mean of 0.7 (our estimate of the theoretical mean) in Table 8.3 to the observed proportions in Table 8.2. They are not exactly the same, but they are fairly close. See the upper histogram in Figure 8.2 for the relative frequency histogram of the data. The height of each rectangle is the observed proportion of the data with the specified number of deaths. It appears as if the data are reasonably Poisson-like because the theoretical probability histogram is so closely approximated by the relative frequency histogram. You could check this judgment out with a chi-square test, couldn't you?

For another example, consider the number of runs the Oakland Athletics baseball team scored in each of the 144 games they played in 1995. The question is whether the number of runs per game is reasonably modeled as Poisson. Thus the time interval here is one game. The average number

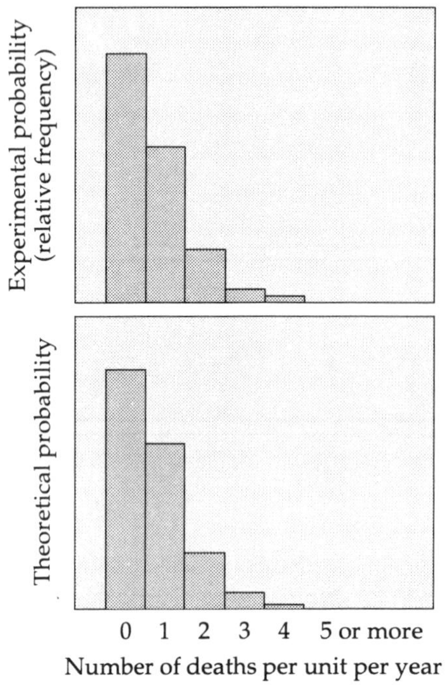

Figure 8.2 **Prussian soldiers' relative frequency histogram.**

of runs per game they scored was 5.07. In Table 8.4 we have, for each possible number of runs, the number of games in which the Athletics scored that number of runs, as well as the observed proportion of games (that is, the experimental probability of that number of runs per game) and the theoretical Poisson probability of that number of runs assuming a theoretical mean of 5.07. Figure 8.3 has the corresponding relative frequency histogram constructed from these data and the theoretical probability histogram.

The last two columns in Table 8.4 show some fairly large discrepancies. For example, the Athletics were shut out (had 0 runs) eight times, or about 5.6% of the time. The Poisson distribution predicts shutouts only 0.6% of the time. They scored four runs 12.5% of the time and five runs 7.6% of the time, while the Poisson distribution predicts over 17% for four runs and also for five runs. What other discrepancies are there? You should try to explain how this situation might violate the three conditions for the Poisson distribution, and explain why the Poisson therefore does not model these data well. Again, one could do a chi-square test if one wished, expecting to reject the hypothesis that a Poisson distribution fits the data well.

Table 8.4 **Number of Runs Scored per Game by Oakland Athletics in 1995**

Number of runs	Number of games	Proportion of games	Theoretical probability
0	8	0.0556	0.0063
1	14	0.0972	0.0319
2	14	0.0972	0.0808
3	14	0.0972	0.1365
4	18	0.1250	0.1730
5	11	0.0764	0.1754
6	20	0.1389	0.1482
7	13	0.0903	0.1073
8	14	0.0972	0.0680
9	5	0.0347	0.0383
10	2	0.0139	0.0194
11	6	0.0417	0.0089
12	1	0.0069	0.0038
13	3	0.0208	0.0015
14	1	0.0069	0.0005
15 and over	0	0.0000	0.0003

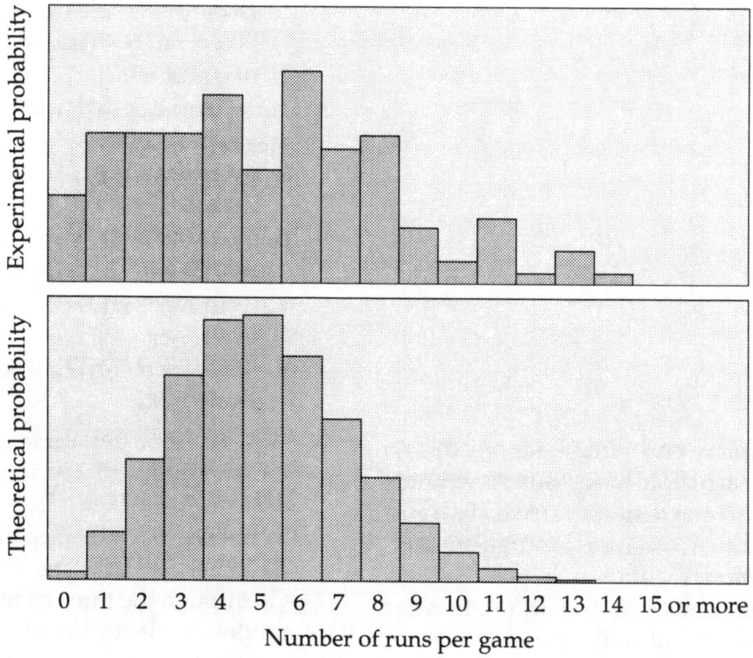

Figure 8.3 **Oakland Athletics' histograms.**

SECTION 8.2 EXERCISES

1. Go through the three conditions for a Poisson random variable, and give possible reasons why in the example of the Oakland Athletics baseball team the data do not seem to follow the Poisson distribution.

2. Rebecca works as a greeter at a grocery store. She theorizes that during the 2:00 to 4:00 time period, customers arrive at the rate of two per minute. She is willing to make the three Poisson assumptions.

   She keeps track of the number of customers who arrive each minute during a 15-minute period. Here are her results:

Minute	Number of customers
1	2
2	0
3	4
4	5
5	1
6	0
7	2
8	3
9	1
10	2
11	3
12	4
13	2
14	1
15	0

   a. In how many one-minute periods did no customer arrive? In how many one-minute periods did one customer arrive? Two customers? Three customers? Four customers? Five customers?
   b. Using the answers from part (a), find the experimental probability that no customer arrives in a one-minute period. Calculate such a probability for every number of customers up to 5.
   c. According to the Poisson distribution with a theoretical mean of 2, the theoretical probabilities of the six outcomes are as follows:

Number of customers in a minute	Theoretical probability
0	0.135
1	0.271
2	0.271
3	0.180
4	0.090
5	0.036

   Create two histograms, one showing the experimental probabilities, and the other showing the theoretical probabilities.
   d. Compare the two histograms you created. Does this evidence seem to support Rebecca's hypothesis concerning the arrival of customers?

3. Using Table 8.3, determine the following theoretical probabilities:
   a. The probability of 0 occurrences when the mean is 0.7
   b. The probability of 3 occurrences when the mean is 5
   c. The probability of 11 occurrences when the mean is 5
   d. The probability of 1 occurrence when the mean is 0.1

4. At a fast-food restaurant, the arrival of customers during the period from 11:00 A.M. to 2:00 P.M. is tracked. The owner is interested in finding out whether or not the arrival of customers follows the Poisson distribution. Go through the three conditions for a Poisson random variable. Which of those three is being violated in this case? Explain.

5. State the relationship between the theoretical standard deviation and the theoretical mean

of a Poisson random variable. What is the theoretical standard deviation in the case of the Prussian army data?

6. The following table shows probabilities obtained in an experiment. The table also has the corresponding probabilities under the Poisson distribution with a mean of 3.4.

Number of occurrences	Experimental probability	Theoretical probability
0	0.053	0.033
1	0.153	0.113
2	0.138	0.193
3	0.103	0.219
4	0.225	0.186
5	0.214	0.126
6	0.068	0.072
7	0.047	0.035

a. Make a histogram of the experimental probabilities, and a histogram of the theoretical probabilities.

b. Compare the two histograms. Do the obtained data seem to follow the Poisson distribution with mean 3.4?

7. Reconsider the data given in Exercise 6. Recall that in Chapter 7 we discussed the chi-square test, a test used to determined whether data follow a certain distribution. Assume that in the experiment described Exercise 6, the obtained probabilities were actually based on 1000 observations.

a. Reconstruct the results of the experiment, giving the number of outcomes for each of the values. For example, there were $1000 \times 0.053 = 53$ outcomes of 0.

b. Using the theoretical probabilities, calculate how many of the 1000 observations are expected for each of the values. For example, we would expect $1000 \times 0.033 = 33$ outcomes of 0.

c. Calculate the value of the chi-square statistic for these data.

d. What are your conclusions based on the chi-square test?

8.3 ROLLING A FAIR DIE: THE DISCRETE UNIFORM DISTRIBUTION

If you roll a fair six-sided die, as has been done often in this book, you are equally likely to get 1, 2, 3, 4, 5, or 6. There is a name for such a distribution: the **discrete uniform distribution** from 1 to 6. *Discrete* means that the possible values are specific values that can be listed. (The opposite of *discrete* is *continuous*, which we discuss in Section 8.5 for a uniform distribution.) *Uniform* means that each of the values is equally likely. Since there are six values in the case of a normal die, the theoretical probability of any one of them appearing is $\frac{1}{6}$. A 32-sided die, if it is fair, will yield a discrete uniform distribution with outcomes ranging from 1 to 32. It is not necessary to start with 1. For example, a five-sided die could have the numbers $-2, -1, 0, 1, 2$. Its distribution will be a discrete uniform distribution ranging from -2 to 2. One needs only to specify the lowest and the highest integer. Also, use of a "die" here is unnecessary. A die is merely a tangible physical object to help us imagine generating the data. The essence of the *discrete uniform distribution* is a sequence of equally spaced integers each having the same theoretical probability.

The theoretical mean is just halfway between the low value and the high value:

$$\text{Theoretical mean} = \frac{\text{low value} + \text{high value}}{2}$$

For example, the theoretical mean for a fair six-sided die is (low value + high value)/2 = (1 + 6)/2 = 3.5.

The formula for the theoretical standard deviation is not very intuitive. For completeness, we give it:

$$\text{Theoretical standard deviation} = \sqrt{\frac{(\text{high value} - \text{low value} + 1)^2 - 1}{12}}$$

For the six-sided die, this standard deviation is

$$\text{Standard deviation} = \sqrt{\frac{(6 - 1 + 1)^2 - 1}{12}} = \sqrt{\frac{6^2 - 1}{12}}$$

$$= \sqrt{\frac{36 - 1}{12}} = 1.7078$$

Thus there will be quite a bit of variability from the average of 3.5 in that the typical variation is roughly from 2 to 5.

SECTION 8.3 EXERCISES

1. For each of the following discrete uniform distributions, give the value of the theoretical mean and the theoretical standard deviation.
   a. High value = 10, low value = 1
   b. High value = 8, low value = 4
   c. High value = 5, low value = −5
   d. High value = 15, low value = 6
2. Repeat Exercise 1 for these values:
   a. High value = −5, low value = −9
   b. High value = 21, low value = 10
   c. High value = 0, low value = −10
   d. High value = 32, low value = 25
3. You and three of your friends agree to meet for lunch every day. You begin to keep track of how many of your friends beat you to the meeting place each day. You suspect that this random variable has a uniform distribution.
   a. What are the possible values of the random variable?
   b. Assuming that this random variable does follow the uniform distribution, what are its theoretical mean and standard deviation?
   c. The actual data are as follows: Out of a total of 50 meetings...

   There were 10 times when none of your friends were there when you arrived.

   There were 13 times when one of your friends was there when you arrived.

   There were 15 times when two of your friends were there when you arrived.

   There were 12 times when all three of your friends were there when you arrived.

   Use the chi-square test as described in Chapter 7 to determine whether these observed data do indeed follow the uniform distribution.

8.4 CONTINUOUS DISTRIBUTIONS

The binomial and Poisson distributions are both based on counting; the discrete uniform distribution deals with a list of consecutive integers. These are all discrete distributions. It is common to find data based on measurements instead of counts: for example, mountain heights, light bulb lifetimes, incomes, agricultural yields, and so on. In many measurements, we may be able to specify the smallest and largest possible values, but any value in between those, not just integers, may be possible. For example, a lightbulb's lifetime (the length of time it is continuously on before breaking) cannot be less than 0 hours, although if it is defective it may last exactly 0 hours. There is some maximum length of time—if not 1000 hours, then maybe 10,000 hours or more. But any value between 0 and some large maximum is possible. It may last exactly one hour, or half an hour, but it is more likely to last a less rounded-off time, such as 134.398459987 hours. Of course, in practice one cannot find the exact lifetime, but instead finds it to the nearest minute, second, or maybe 10th of a second. In any case, the actual number of possible values is immense; theoretically, it is infinite if one ignores the fact that all measurements are rounded off (for example, a length might be rounded to the nearest $\frac{1}{1000}$ of an inch).

Moreover, measurements are often best viewed as random, in which case it is necessary to develop probability distributions to describe them. For example, we might be randomly sampling measurements of a real population, such as heights of adult females, in which case height must be assigned a continuous probability distribution. Or we might be randomly sampling from a potentially infinitely repeatable experimental process, such as assessing the lifetime of lightbulbs as they come off the production line, in which case the lifetime must be described by a continuous probability distribution.

Statisticians have developed continuous distributions that model these measurement-based situations well. We have already seen a case of a continuous distribution in Chapter 7, namely, the smooth chi-square density. Saying that the data are continuous simply means that one cannot list all the possible values because, in principle, there are infinitely many (imagine infinite accuracy in the measuring process). Try listing all the numbers between 0 and 1, for example. You might start with 0, but what is next? 0.1? Why not 0.01? Or 0.0000983762763782? Or 0.000000000000001? There *is* no number that comes next after 0. Mathematicians call such a set of numbers a *continuum*.

Here are some contrasting examples of continuous and discrete situations: A die is discrete; a spinner is continuous. A digital clock is discrete;

a regular analog clock, with hands, is continuous. A digital calculator is discrete; a slide rule is continuous. Is the channel tuner on a car radio discrete or continuous? It depends on whether you use a knob to move a dial or buttons to select specific frequencies.

How does one specify probabilities for a continuous distribution? Theoretically, the probability of any particular value is 0. Imagine shooting at a target. What is the chance you hit exactly the point that is the exact center? Zero! You might get very close—within $\frac{1}{8}$ inch, say, or $\frac{1}{16}$ inch—but it is very unlikely that you would get within one millionth of an inch, or a billionth, and so on. So we have a difficulty: every possible value has a zero probability, but one of the values has to appear. Thus we cannot specify the distribution by specifying the probability of each possible value of the statistic, as in the die example, because each such probability is 0. This type of paradox does not bother mathematicians, but statisticians have to deal with reality and hence need a way around it. Recall the continuous chi-square distribution. We find a probability by computing an approximate area under the chi-square density. Now we consider why this area approach is appropriate.

Even though each number in the range of a statistic having a continuous distribution must have probability 0, nonetheless it is clear that some values are more likely (probable) for some intervals than others. Thus we seek a mechanism for assigning probabilities to *intervals* instead of assigning them to specific values, as in the discrete case.

For example, if all numbers between 0 and 1 are equally likely to be selected, the probability of selecting a number between 0 and $\frac{1}{2}$ is $\frac{1}{2}$, because half of the numbers between 0 and 1 are between 0 and $\frac{1}{2}$. In the more general case of nonuniform continuous distributions, we cannot assign a probability to be proportional to interval length but rather as an area covering the interval under a density curve, as we did with the chi-square density.

To see how this is done, consider a relative frequency histogram for continuous data. For example, in Figure 8.9, which appears later in the chapter, we have the relative frequency histograms for heights of women sampled from a population. Recall that the total area of the relative frequency histogram rectangles is 1. In Figure 8.9 the relative frequency histogram has superimposed over it an approximating bell-shaped density (called a Gaussian, or normal, density; its properties are discussed below). Thus if we conclude that the population of all female heights is well described by a normal density (just as chi-square histograms are well described by a chi-square density), we can compute the probability of a height falling in an interval, such as $p(66'' \leq \text{height} \leq 69'')$, by simply finding the area under the curve from 66″ to 69″.

Table 8.5 **Batting Average Frequencies for 263 Baseball Players**

Batting average*	Experimental probability	Theoretical probability	Obtained player count	Expected player count
Under 0.195	0.000	0.010	0	2.543
0.195 to 0.205	0.008	0.013	2	3.480
0.205 to 0.215	0.027	0.026	7	6.842
0.215 to 0.225	0.061	0.046	16	11.982
0.225 to 0.235	0.080	0.071	21	18.691
0.235 to 0.245	0.084	0.099	22	25.971
0.245 to 0.255	0.137	0.122	36	32.146
0.255 to 0.265	0.137	0.135	36	35.441
0.265 to 0.275	0.160	0.132	42	34.807
0.275 to 0.285	0.103	0.116	27	30.450
0.285 to 0.295	0.061	0.090	16	23.729
0.295 to 0.305	0.046	0.063	12	16.471
0.305 to 0.315	0.046	0.039	12	10.184
0.315 to 0.325	0.023	0.021	6	5.609
0.325 to 0.335	0.019	0.010	5	2.752
0.335 to 0.345	0.004	0.005	1	1.203
0.345 to 0.355	0.004	0.002	1	0.468
0.355 to 0.365	0.004	0.001	1	0.162
Over 0.365	0.000	0.000	0	0.068

*Note: "0.195 to 0.205" means at least 0.195 and less than 0.205.

Source: Lorraine Denby, American Statistical Association Graphics Section Poster Session data set (1988).

Let's look a little more closely at the process. We will look at a baseball example with the goal of introducing continuous probability distributions: the 1985 batting averages of 263 major league baseball players, as displayed in Table 8.5. The batting average is the proportion of times at bat in which the player hits the ball well enough to get on base. An average major league player will have about four hundred at bats in a year and get one hundred or so hits. Typical batting averages are in the 0.240 to 0.300 range. The data we have go from 0.200 to 0.357. Figure 8.4 is a relative frequency histogram for the 263 players constructed from the raw data of the 263 batting averages. How would this distribution be modeled theoretically? It is not practical to write down the probability for each possible batting average: Even if we rounded to three digits, as is typical, we would have to specify hundreds of probabilities (one for a 0.100 batting average, one for a 0.101 batting average, and so on). In addition, the observed batting average of a particular player is surely different from his true underlying batting average, for the observed batting represents an element of luck (chance) in addition to the true ability. Because of these considerations, we can

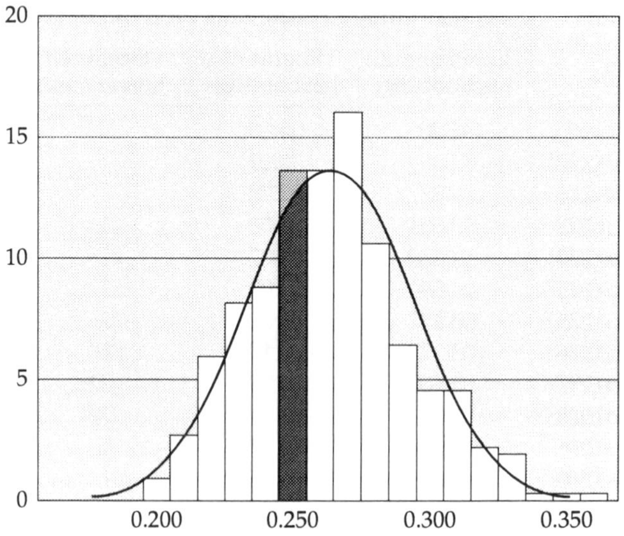

Figure 8.4 **Batting averages of 263 major league baseball players in 1985.**

imagine a smooth density with the general shape of the relative frequency histogram that represents the distribution of the "true" batting averages. Thus we seek a smooth curve approximating the relative frequency histogram. This idea is similar to that of the smooth curve we fit to the relative frequency polygons (which are essentially relative frequency histograms) of the chi-square data in Chapter 7.

Figure 8.4 has a so-called normal (bell-shaped) curve superimposed on the histogram. It does not conform exactly to the histogram, but it gives a good overall idea of the shape. This continuous distribution is in fact the most useful continuous distribution in statistics, partly because it describes so many data sets so well. Our goal is to use it to find probabilities, such as p(batting at least 0.300).

Theoretical probabilities can now be found by using areas under the curve. For example, the sixth block in the relative frequency histogram represents the players who hit between 0.245 and 0.255. There are 36 players with averages in that range, a proportion of $36/263 = 0.137$. The value 0.137 is also the area of that block in the histogram. The area between 0.245 and 0.255 under the curve of Figure 8.4 is the part that is shaded darkly. That area is the theoretical probability (according to the normal curve) of someone hitting between 0.245 and 0.255. That area turns out to be 0.122. (Later in this chapter you will see how to find this area.) Thus the theoretical probability is not exactly what we have observed from the players' data, but at least it is in the ballpark (no pun intended!). Indeed we expect some discrepancies between a player's observed batting average and that specified by the normal curve, which we view as representing his true ability. Column 3 of Table 8.5 gives the areas under the normal curve for all the blocks.

According to the normal curve, there is about a 0.010 probability (1%) that a player will hit less than 0.195. However, no player hit that low. It looks like the theoretical probability of hitting over 0.365 is 0, but actually it is 0.000259; the 0.000 appears in the table because we rounded off. The final two columns compare the actual counts (numbers of players) to the counts expected from the normal curve. Here we use our formula for expected count as (total number of players)(theoretical probability of the interval).

What is the chance that a player hits over 0.300? In the data, 33 players out of 263, or a proportion of 0.125, batted over 0.300 (28 people hit over 0.305, and, going back to the original data, one finds that 5 of the 12 who hit between 0.295 and 0.305 hit over 0.300). The theoretical probability is shaded in Figure 8.5. That area is 0.106, slightly less than the observed proportion.

The next section looks at perhaps the simplest theoretical continuous distribution, the uniform distribution, and then the subsequent section explains in more depth the normal curve that we introduced here.

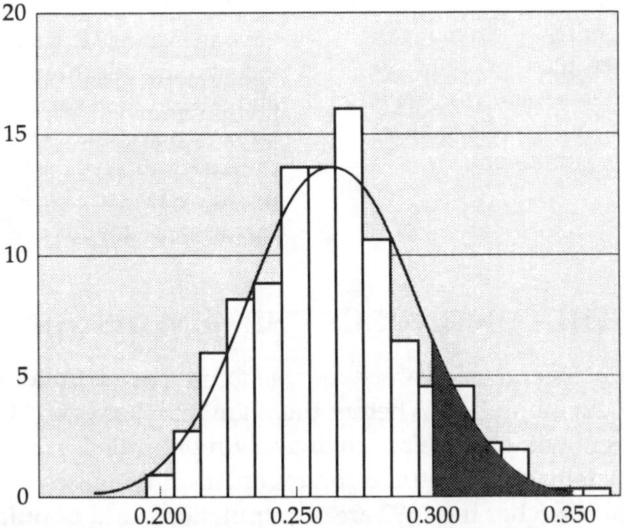

Figure 8.5 **p(batting over 0.300).**

SECTION 8.4 EXERCISES

1. What is the fundamental difference between a random variable that has a continuous distribution and a random variable that has a discrete distribution?

2. Each of the following situations describes a random variable. For each, tell if it is discrete or continuous.
   a. The length of the life of a battery
   b. The number of cars that pass by in an hour
   c. The number of games a team wins in a season
   d. The amount of rain that falls in a day
   e. The weight of a bag of apples

3. Refer to Table 8.5. Calculate each of the following theoretical probabilities:
   a. p(Someone is hitting between 0.215 and 0.225)
   b. p(Someone is hitting less than 0.195)
   c. p(Someone is hitting between 0.285 and 0.295)
   d. p(Someone is hitting between 0.245 and 0.265) (*Hint:* Add the corresponding probabilities.)

4. Another common continuous distribution is the exponential distribution. A random variable that has an exponential distribution can have any value greater than or equal to 0. The following table gives the probabilities for intervals from a standard (theoretical mean = 1) exponential distribution:

0 to 1	0.28
1 to 2	0.20
2 to 3	0.15
3 to 4	0.10
4 to 5	0.07
5 to 6	0.05
6 to 7	0.04
7 to 8	0.03
8 to 9	0.02
9 to 10	0.01
10 to 11	0.01
11 to 12	0.01
12 or greater	0.02

   a. Draw out a theoretical probability histogram using the above table. On that histogram draw a smooth curve that matches the shape of the histogram. (The curve you draw will be a rough approximation of the curve of the exponential distribution.)

   Calculate the following probabilities based on the table:
   b. p(exponential value between 1 and 2)
   c. p(exponential value between 4 and 6)
   d. p(exponential value of exactly 7)
   e. p(exponential value greater than 10)

5. Give an example of a random variable having a continuous distribution. Give a maximum and minimum value for the random variable. Draw a curve that in your opinion describes the distribution of that random variable (that is, draw the curve that mimics the relative frequency histogram).

8.5 THE CONTINUOUS UNIFORM DISTRIBUTION

Computers and calculators are able to use only a finite number of digits to represent a number. Whether there are 2 or 3 or even 100 digits, the actual numbers will only be approximately represented in general; that is, there will usually be round-off error, except in simple cases.

Table 8.6 has the 1992 areas, populations, and population densities (the average number of people per square mile) of the five U.S. cities with the highest populations. The fifth column rounds off the density to the nearest

Table 8.6 **Population Densities of the Five Most Populous Cities in the United States in 1992**

	Area (square miles)	Population	Density (people/square mile)	Density (rounded)	Round-off error
New York City	308.9	7,311,966	23,670.9809	23,671	−0.0191
Los Angeles	469.3	3,489,779	7436.1368	7436	0.1368
Chicago	227.2	2,768,483	12,185.2245	12,185	0.2245
Houston	539.9	1,690,180	3130.5427	3131	−0.4573
Philadelphia	135.1	1,552,572	11,492.0207	11,492	0.0207

integer. The round-off error is the difference between the actual value and the rounded value; for example, for New York, 23,670.9809 − 23,671 = −0.0191. (Of course, even these round-off errors have been rounded off to four digits, because the "actual number" itself had already been rounded off to four digits to the right of the decimal point.)

The round-off errors in the table are not very worrisome with regard to making practical use of the table. It would be hard to imagine that the missing 0.0191 person per square mile would be noticed when there are already over 23,000 per square mile. The problem in computer programs is that when making large numbers of complicated calculations, the round-off errors of the numbers involved often become compounded, so that many seemingly innocuous round-off errors can accumulate to produce a very significant error. Computer scientists often study the effects of round-off error by assuming that the round-off errors are distributed uniformly over the range from the largest possible negative to the largest possible positive value. In the case of Table 8.6, that would mean the round-off error is equally likely to be any number between −0.5 and +0.5.

Figure 8.6 is a relative frequency histogram of the observed round-off errors for population densities of 77 U.S. cities with populations over 200,000. The dark blocks represent the histogram for the observed round-off errors. The horizontal line at height 1 is the smooth curve (which does not curve, actually) that represents the (continuous) uniform distribution. The histogram fluctuates around the uniform curve.

The theoretical probabilities are found by looking at the areas under the curve in the different ranges. For example, the theoretical probability that a round-off error is between 0.3 and 0.4 is the area of the block in Figure 8.7. The height is 1, and the width is (0.4 − 0.3) = 0.1, so the area is 1 × (0.1) = 0.1. Because there are 77 cities, we expect 77 × (0.1) = 7.7 to have round-off errors between 0.3 and 0.4. We actually got only 4 such cities.

The last block in Figure 8.6, for which the round-off errors are between 0.4 and 0.5, seems to be unusually high for a uniform distribution. A chi-square test (viewed as a test of a 10-sided fair die that is tossed 77 times) can see whether that is true. See Table 8.7. The chi-square statistic turns

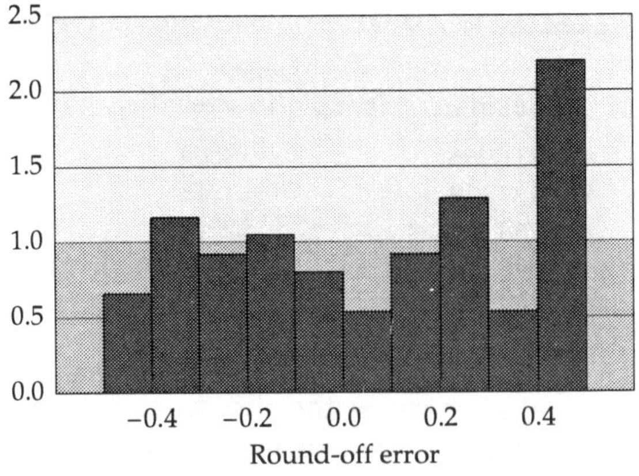

Figure 8.6 **Round-off error for population densities of 77 U.S. cities in 1992.**

out to be 17.16. There are nine degrees of freedom, because there are 10 categories and, as explained in Chapter 7, the number of degrees of freedom is thus 10 − 1. From the chi-square probability table (Table C) we obtain $p(\chi^2 \geq 16.92) = 0.05$. That is, 5% of the area is greater than or equal to 16.92 under the hypothesis of a fair 10-sided die. Thus $p(\chi^2 \geq 17.16)$ is less than 0.05, and we conclude that the uniform distribution does not quite fit the data. The culprit is clearly that last block. There are 17 cities there, when one would expect only 7.7. It would be interesting to try to find a reason why this last block is so large.

A statistic with a discrete uniform distribution can take on any integer value from the lowest to the highest possible value. A statistic with a **continuous uniform distribution** can take on any number at all from the

Table 8.7 **Chi-Square Table for Data of Figure 8.6**

Round-off error	Obtained (O)	Expected (E)	$(O - E)^2/E$
−0.5 to −0.4	5	7.7	0.95
−0.4 to −0.3	9	7.7	0.22
−0.3 to −0.2	7	7.7	0.06
−0.2 to −0.1	8	7.7	0.01
−0.1 to 0	6	7.7	0.38
0 to 0.1	4	7.7	1.78
0.1 to 0.2	7	7.7	0.06
0.2 to 0.3	10	7.7	0.69
0.3 to 0.4	4	7.7	1.78
0.4 to 0.5	17	7.7	11.23
Total	77	77	17.16

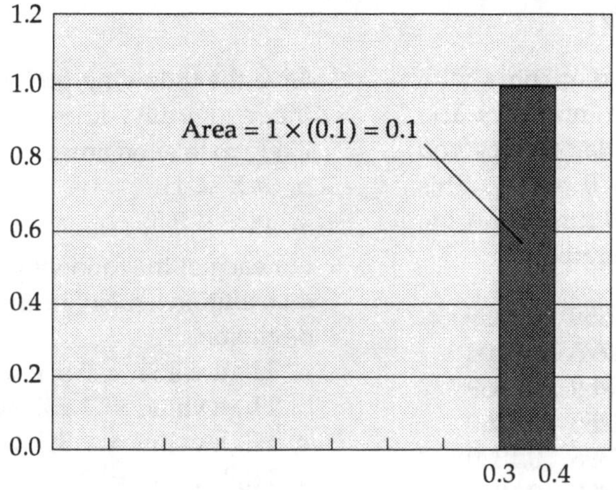

Figure 8.7 ***p*****(round-off error between 0.3 and 0.4).**

lowest to the highest possible value. The total area under the curve has to be 1, which means that the height has to be 1/(high value − low value). Figure 8.8 shows three continuous uniform distributions. The first two have widths of 1, so they have heights of 1. The third has width $6 - 3 = 3$, so its height has to be $\frac{1}{3}$.

The theoretical mean of a uniform distribution is (low value + high value)/2—right in the middle, as one would expect. The theoretical standard deviation is (high value − low value)/$\sqrt{12}$. For the round-off data, the actual average is 0.053 and the standard deviation is 0.310. Theoretically, the average should be about $(-0.5 + 0.5)/2 = 0$. The actual average is a little high, but not by much. The theoretical standard deviation is $(0.5 - (-0.5))/\sqrt{12} = 1/3.464 = 0.289$. So the actual standard deviation is also somewhat high, but not surprisingly so.

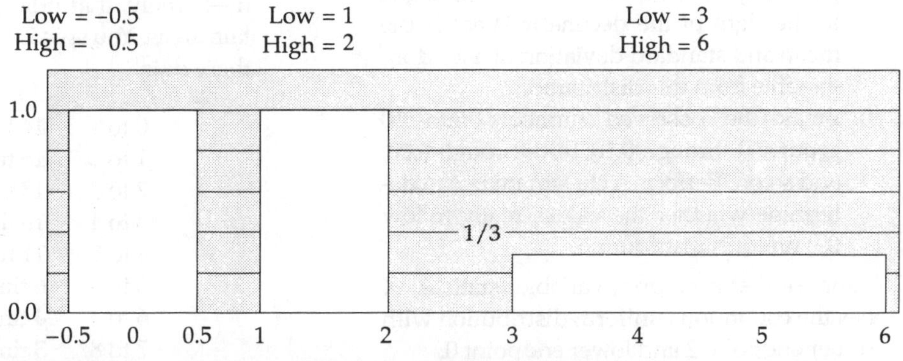

Figure 8.8 **Three continuous uniform distributions.**

SECTION 8.5 EXERCISES

1. Most scientific calculators today come with the ability to generate random numbers. You press a button, and the calculator gives you a random number between 0 and 1. This was done on a calculator 100 times, and the following numbers were generated:

0.2611	0.5247	0.5591	0.7859	0.4359
0.9090	0.4328	0.9389	0.4775	0.5404
0.7444	0.7275	0.1370	0.5759	0.5869
0.2847	0.0762	0.8667	0.3255	0.5060
0.6033	0.9665	0.6174	0.7396	0.1829
0.8156	0.0120	0.8709	0.6870	0.1810
0.1628	0.4085	0.3178	0.4318	0.7107
0.7162	0.9238	0.8096	0.4580	0.8141
0.0292	0.8796	0.9055	0.2570	0.1577
0.8230	0.9309	0.9878	0.1821	0.4076
0.6274	0.6446	0.6570	0.8063	0.0844
0.7330	0.4366	0.1622	0.9846	0.5486
0.9619	0.5447	0.7136	0.5774	0.6815
0.5262	0.9832	0.0542	0.3449	0.9968
0.7614	0.4203	0.8755	0.8861	0.9382
0.4191	0.6021	0.6149	0.7851	0.6420
0.5523	0.4077	0.6991	0.1097	0.4772
0.7795	0.6363	0.5749	0.5073	0.9759
0.8959	0.0876	0.3938	0.0035	0.7519
0.7539	0.4227	0.7432	0.4896	0.6945

   a. The calculator is supposed to generate a number between 0 and 1, with all numbers equally likely. What distribution is this (ignoring the rounding off to four digits to the right of the decimal)? What is the mean and standard deviation of a random variable from this distribution?

   b. Break the observed numbers into 10 groups: 0 through 0.10, 0.10 through 0.20, and so on. Perform a chi-square test to determine whether the values really follow the uniform distribution.

2. Suppose that a random variable (statistic) X has the continuous uniform distribution with upper endpoint 2 and lower endpoint 0.

   a. What is the theoretical mean and standard deviation of X?

   Give the following probabilities. Think about them logically, knowing that X is equally likely to take on any value between 0 and 2.

   b. $p(X < 1)$

   c. $p(X < 0.5)$

3. For each of the following continuous uniform distributions, calculate the mean and standard deviation.

   a. High value = 0.5, low value = 0

   b. High value = 1.25, low value = −1.5

   c. High value = 7.5, low value = 6.25

   d. High value = 20.3, low value = 13.6

4. A spinner that yields the values 0 to 10, equally spaced, is shown:

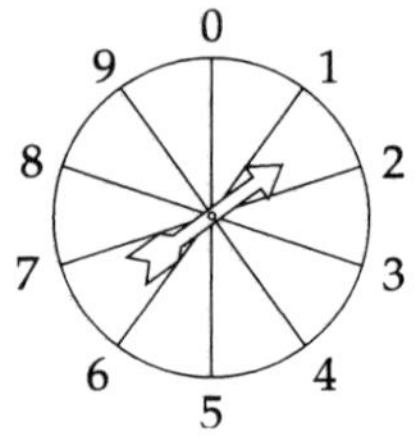

   a. Explain why when you spin the spinner, the exact value that comes up should have the continuous uniform distribution.

   b. What is the mean and standard deviation of this random variable?

   c. You try the spinner a few times, and you suspect that there is something wrong with it—namely, that it favors stopping in certain areas. You spin it 100 times, and collect these data:

0 to 1	14 times
1 to 2	15 times
2 to 3	13 times
3 to 4	16 times
4 to 5	11 times
5 to 6	6 times
6 to 7	4 times
7 to 8	3 times
8 to 9	8 times
9 to 10	10 times

Perform a chi-square test to test whether the spinner is really giving continuous uniform data.

5. Every time the phone rings, you check the exact position of the second hand of the analog clock on the wall. Explain how this is a continuous uniform variable. What are the upper and lower bounds?

8.6 THE NORMAL DISTRIBUTION

The most famous distribution is the **normal distribution.** It is also called the *Gaussian distribution,* after the famous mathematician Carl Friedrich Gauss. More colloquially, it is often referred to as the *bell-shaped curve.* As seen in Figure 8.4, it certainly is bell-shaped.

The normal curve fits an amazing number of situations. The reason for this lies in the central limit theorem, which is explained in Chapter 10. Figure 8.9 shows the relative frequency histogram of the heights of 102 women in a particular college statistics class. Figure 8.10 is the relative frequency histogram of the body temperatures of 130 people. The superimposed normal curves follow both histograms reasonably well.

The one in Figure 8.11 does not. This relative frequency histogram is one of the total number of dogs and cats that the 165 people in the

Statistics and Individual Differences

The French scientist A. Quetelet (1796–1874) first noted that if you measure the heights of a large group of people, the relative frequency polygon of these heights will resemble the bell-shaped curve of a normal distribution. He also studied the distributions of other characteristics, such as weight, chest girth, and arm length, and found that all of these follow nearly the same shape of distribution. This property of the distribution of physiological characteristics was found to occur so frequently that the British scientist Sir Francis Galton (1822–1911) coined the term *normal* to describe these distributions.

This property of most physiological characteristics has many practical applications. For example, the designer of an airplane cockpit must arrange it so that most pilots are comfortable and can reach all of the controls. Clearly, this requires knowledge of average heights, average arm lengths, and so on, as well as knowledge of the variability around these averages so that *most* pilots will be accommodated.

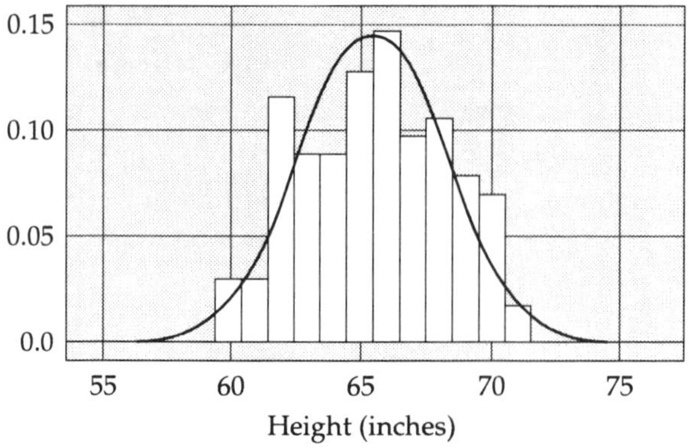

Figure 8.9 **Heights of 102 women.**

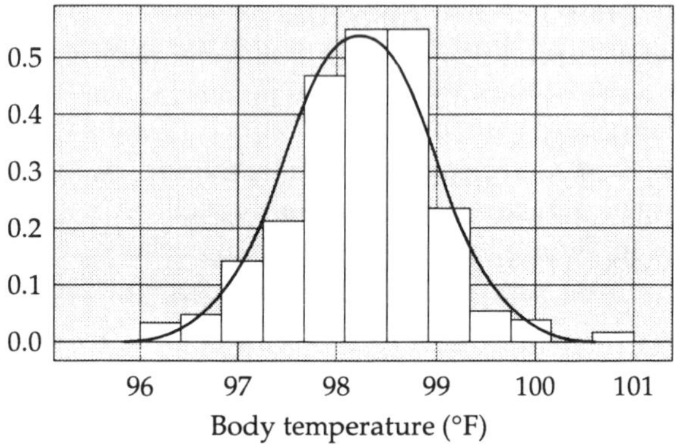

Figure 8.10 **Body temperatures of 130 people (data are from Allen L. Shoemaker, "What's Normal?—Temperature, Gender, and Heart Rate," *Journal of Statistics Education*, July 1996).**

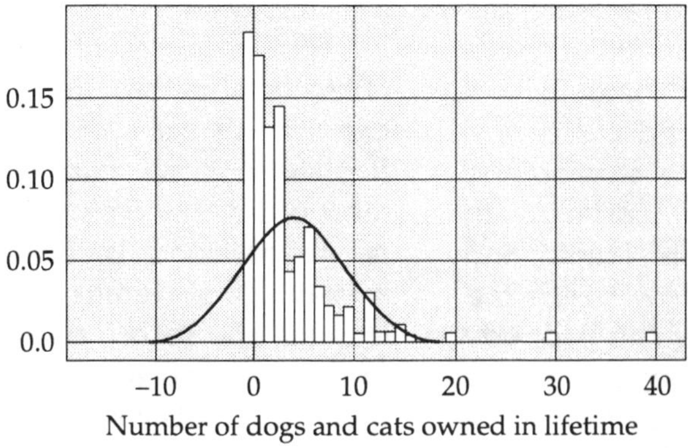

Figure 8.11 **Number of dogs and cats owned by 165 students.**

statistics class owned during their lives. The largest number was 40. Note that this is not a physiological measurement. We try to fit it with a normal curve. The normal curve tends to be centered around the bulk of the data, which in Figure 8.11 is in the range from 0 to 5. But in order to extend adequately out to the larger numbers, it also has to extend well below zero, because the normal curve is always symmetric in shape about its midpoint. The area to the left of 0 under the best-fitting normal curve in the graph is about 0.22, which means that 22% of the people should have fewer than 0 dogs and cats. One cannot have a negative number of pets, so the normal distribution is not appropriate here. The point here is not that there is a probability of fewer than 0 cats and dogs, for a probability of 0.01 of this, say, would be tolerable. Rather, the point is that there is the large probability of 0.22, indicating a serious error in fitting the relative frequency histogram of the data with the normal curve. We never seek a perfect fit of distribution to data, but we always seek a good fit! From our viewpoint that data represent a sample from a population, we expect the relative frequency histogram of the sample to fit the population distribution well.

If you look at the normal curves in the figures, you see that they all have the same shape but different centers (theoretical means) and spreads (theoretical standard deviations). For the batting averages (Figure 8.4), the best-fitting normal curve covers the data from about 0.200 to 0.360; for the heights (Figure 8.9), it covers the data from about 60 to 71 inches; for the body temperatures (Figure 8.10), it covers the data from about 96 degrees to 101 degrees. For any mean and standard deviation there is a normal curve, which is one reason the normal is so versatile. The three normal curves in Figure 8.12 have different means, -10, 0, and 10, but the same standard deviation of 5. The three normal curves in Figure 8.13, on the other hand,

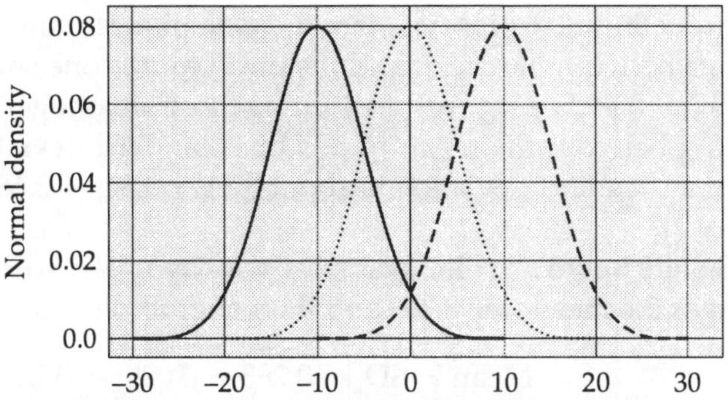

Figure 8.12 **Normal curves having a standard deviation of 5.**

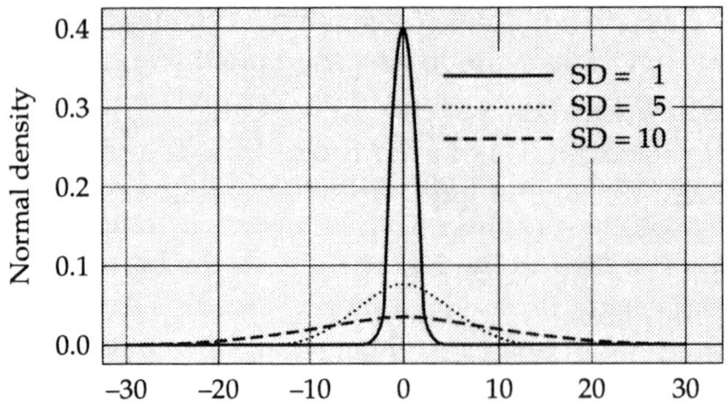

Figure 8.13 **Normal curves having a mean of 0.**

have the same mean of 0 but three different standard deviations: 2, 5, and 10. The one with standard deviation 2 is the tall narrow curve; the one with standard deviation 10 is the low flat curve.

These normal curves are symmetric around their mean; that is, the part of the curve to the right of the mean is a mirror image of the part of the curve to the left. The calculus-based formula for the theoretical mean shows that the mean of a distribution is the geometrical center of gravity, or balance point, of the density curve of the distribution (this is a property of all theoretical means of distributions). Thus in a symmetrical distribution the balance point, or theoretical mean, is located right at the theoretical median, namely, the point that has half the area on either side. Can you think up examples of curves in which the center of gravity (the mean) does not equal the theoretical median?

Because all normal curves have the same shape, the areas under normal curves will, if given in terms of distances from the mean in standard deviation units, be the same. For example, no matter what the mean and standard deviation are, the area, and hence the probability, between the mean and the mean plus one standard deviation is about 0.34. By symmetry, the area between the mean and the mean minus one standard deviation is also about 0.34. See Figure 8.14. Putting those areas together, the probability of being between the mean minus one standard deviation and the mean plus one standard deviation is about $0.34 + 0.34 = 0.68$—that is, approximately $\frac{2}{3}$.

Table 8.8 looks at the examples we have seen so far. For the batting averages the mean was 0.263 and the standard deviation was 0.029. So

$$\text{Mean} - \text{SD} = 0.263 - 0.029 = 0.234$$
$$\text{Mean} + \text{SD} = 0.263 + 0.029 = 0.292$$

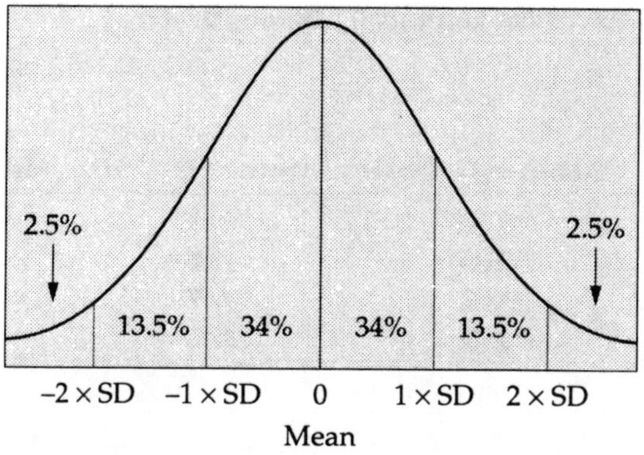

Figure 8.14 **Standard normal curve percentages.**

It turns out that 178 of the 263 players had batting averages between 0.234 and 0.292, which is a proportion of 0.68, exactly what the normal curve would indicate. In the case of the heights, the mean minus one standard deviation and the mean plus one standard deviation are 62.81 and 68.31, respectively. A proportion of 0.66 of the women had heights in that range—very close to the normal curve's value of 0.68. Similarly, 90 of 130, or 69% of the people measured, had temperatures between 97.52 and 98.98. Those examples have histograms that look reasonably normal. The dogs and cats histogram did not look normal. Going down one standard deviation from the mean gives −1.15, an impossible number of pets. And 89% of the people are in the range from the mean minus the standard deviation to the mean plus the standard deviation, which is a lot more than the normal distribution would predict. This is simply more evidence that the normal distribution does *not* fit the pet data well.

According to the normal curve in Figure 8.14, going from the mean minus two standard deviations to the mean plus two standard deviations

Table 8.8 **The (Mean ± SD): 67% Rule Empirically Demonstrated**

	Mean	SD	Mean − SD	Mean + SD	Actual proportion lying between (mean − SD) and (mean + SD)
Batting averages	0.263	0.029	0.234	0.292	178/263 = 0.68
Heights	65.56	2.75	62.81	68.31	67/102 = 0.66
Temperatures	98.25	0.73	97.52	98.98	90/130 = 0.69
Dogs and cats	4.03	5.18	−1.15	9.21	147/165 = 0.89

Table 8.9 **The (Mean ± 2SD): 95% Rule Empirically Demonstrated**

	Mean	SD	Mean − (2 × SD)	Mean + (2 × SD)	Actual proportion between (mean − 2 × SD) and (mean + 2 × SD)
Batting averages	0.263	0.029	0.205	0.321	251/263 = 0.95
Heights	65.56	2.75	60.06	61.06	99/102 = 0.97
Temperatures	98.25	0.73	96.79	99.71	123/130 = 0.95
Dogs and cats	4.03	5.18	−6.33	14.39	159/165 = 0.96

should yield 13.5% + 34% + 34% + 13.5% = 95% of the data. How does that work in the examples? For the batting averages,

$$\text{Mean} - 2 \times \text{SD} = 0.263 - 2 \times 0.029 = 0.205$$

$$\text{Mean} + 2 \times \text{SD} = 0.263 + 2 \times 0.029 = 0.321$$

Of the 263 players, 251 had batting averages between 0.205 and 0.321, which is 251/263 = 95%. Table 8.9 shows this and the other examples. This time, even in the case of the dogs and cats data, about 95% of the data are in the range. But this was merely a lucky chance occurrence, because the dogs and cats data do not follow a normal distribution and hence cannot be expected to follow this 95% rule.

The conclusion is that if the histogram of a set of data approximately follows a normal curve, then just by knowing the mean and the standard deviation of the data, one can estimate the proportion of the data in the range from the mean minus to the mean plus the standard deviation, or from the mean minus to the mean plus twice the standard deviation. In fact, we can do much more along these lines, as in the following.

The *z* Score One normal curve is the most used of them all. It is called the **standard normal curve,** and it is the one with mean 0 and standard deviation 1. Figure 8.15 shows the standard normal curve.

Any kind of normal data (that is, data whose relative frequency histogram is bell-shaped) can be turned into standard normal data (bell-shaped distribution having a mean of 0 and standard deviation of 1) by standardizing the observations, which means for each observation subtracting the mean of the observations and then dividing by the standard deviation of the observations. As an example, the data columns in Table 8.10 show the calories per hot dog in 20 brands of beef hot dogs. The hot dogs averaged 156.85 calories, and the standard deviation was 22.07 calories. The *z* columns give the corresponding *z* scores:

$$z \text{ score} = \frac{\text{data value} - \text{mean}}{\text{standard deviation}}$$

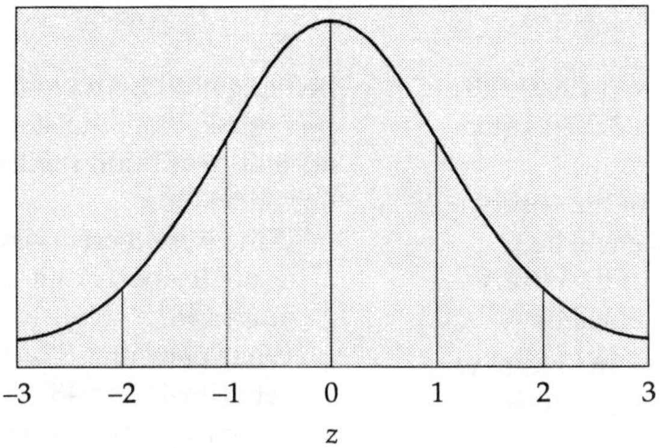

Figure 8.15 **Standard normal curve.**

The first brand of hot dog has 186 calories per hot dog. Its z score is $(186 - 156.85)/22.07 = 1.32$. What that means is that this brand has 1.32 standard deviations more than the average number (157) of calories per hot dog. There is one brand with 111 calories per hot dog. Its z score is $(111 - 156.85)/22.07 = -2.08$, which means it has 2.08 standard deviations fewer calories per hot dog than the average. Note that this brand of hot dog falls outside the 95% range of mean $\pm$ 2 standard deviations. In summary, the z score for an observation is the number of standard deviations above or below the average. Very often scientists will report data as z scores instead of reporting raw data. This is especially the case if the original scale is rather arbitrary or it is important to see at a glance where each individual observation ranks relative to the rest of the data.

The z scores now conform to the standard normal curve in that the average of the z scores is 0, the standard deviation of the z scores is 1, and the shape of the histogram is, provided the original histogram also was, bell-shaped.

One important use of z scores comes in the next section, where we use the normal curve to estimate proportions for any set of normal-looking data, no matter what the mean and standard deviation are, using just one table of probabilities, namely, the table for the standard normal (Table E).

Table 8.10 **Hot-Dog Calorie Data and *z* Scores**

Datum	z	Datum	z	Datum	z	Datum	z	Datum	z
186	1.32	184	1.23	175	0.82	141	−0.72	131	−1.17
181	1.09	190	1.50	148	−0.40	153	−0.17	149	−0.36
176	0.87	158	0.05	152	−0.22	190	1.50	135	−0.99
149	−0.36	139	−0.81	111	−2.08	157	0.01	132	−1.13

Sources: Davis S. Moore and George P. McCabe, *Introduction to the Practice of Statistics*, (New York: Freeman, 1989); and *Consumer Reports*, June 1986, pp. 366–367.

SECTION 8.6 EXERCISES

1. A random variable is assumed to have the normal distribution with mean 5.0 and standard deviation 1.5.
   a. State a range in which 68% of the observations of this random variable will lie.
   b. State a range in which 95% of the observations of this random variable will lie.
2. Another random variable is assumed to have the normal distribution, this time with mean −3.5 and standard deviation 0.5.
   a. State a range in which 68% of the observations of this random variable will lie.
   b. State a range in which 95% of the observations of this random variable will lie.
3. A random variable was observed 100 times. The observations were as follows:

43.50164	42.2318	39.81531	47.38623
42.17374	46.89344	41.78741	43.56211
47.51862	48.0677	42.98898	43.80182
44.4694	44.96988	50.89797	46.89443
49.31204	43.53012	43.29784	40.79547
44.50963	45.09659	43.77065	44.01106
41.58814	47.52947	45.71922	41.97859
44.30996	41.21647	41.54672	47.93914
53.50043	42.12499	44.59925	42.79054
41.1815	44.73116	44.26008	48.66678
43.72759	43.92343	45.75363	44.49416
48.56094	40.97019	44.42216	40.48248
40.81974	52.83752	44.54217	43.50889
46.58111	42.42954	40.9955	41.08607
46.51913	49.37738	39.9309	40.68009
41.1561	45.81967	48.91718	48.65868
45.62537	41.13073	40.75161	44.94595
44.88452	42.17687	47.00198	39.40947
41.70193	51.1321	45.50033	42.92722
48.42776	41.49603	49.58127	45.72352
45.43842	45.58613	50.40565	41.54957
41.4995	44.5927	47.77691	45.28159
48.26565	45.38359	43.03819	43.87346
47.51046	42.97241	47.79572	44.24591
47.30188	49.5567	42.30571	43.61332

   Before gathering the data, the researcher hypothesized that the data would follow the normal distribution with mean 45 and standard deviation 3.
   a. If the hypothesis is correct, what percentage of the observations should lie between 42 and 48?
   b. What percentage of the values really lie between 42 and 48?
   c. If the hypothesis is correct, what percentage of the observations should lie between 39 and 51?
   d. What percentage of the values really lie between 39 and 51?
   e. Does the researcher's hypothesis seem reasonable?
4. A random variable X has a normal distribution with mean 10 and standard deviation 2. Calculate the following probabilities:
   a. $p(X < 10)$
   b. $p(X > 10)$
   c. $p(8 < X < 10)$
   d. $p(6 < X < 12)$
5. The following 50 observations were made of the weights (kg) of 12-year-old boys:

40.96808	33.54756	37.64961	26.82471
42.99247	44.39386	32.04574	31.02483
38.53904	34.20286	31.54772	34.66573
44.8122	39.69208	40.73274	38.73469
37.06111	38.15901	34.85135	33.01202
44.57769	42.96117	49.89561	50.05673
42.11047	49.99663	54.18064	35.13623
41.06182	37.14643	39.58158	41.42845
48.64398	32.82423	41.62885	38.07567
44.61002	40.61515	38.93994	44.10412
36.13034	52.2878	29.71083	35.03201
50.87048	45.82631	34.87447	49.1575
31.00795	41.86156		

   The mean of this set of data is 40, and the standard deviation is 6.43.

a. What percentage of the observations fall within one standard deviation of the mean?
b. What percentage of the observations fall within two standard deviations of the mean?
c. Do these data seem to follow the normal distribution?

6. A random variable X has the standard normal distribution.
   a. What is the mean of X?
   b. What is the standard deviation of X?
   c. What is the probability of X being between -1 and 1?

7. A random variable has the normal distribution with mean 3 and standard deviation 2. Convert the following observations into standard units:
   a. 3.234
   b. 5.193
   c. 1.401
   d. -0.0184

8. Repeat Exercise 7, this time for a normal random variable having mean 6 and standard deviation 3.
   a. 5.290
   b. 2.816
   c. 8.791
   d. 10.271

9. Repeat Exercise 7, this time for a normal random variable having mean -5 and standard deviation 4.
   a. -4.823
   b. -5.972
   c. -11.732
   d. 1.672

8.7 USING THE NORMAL-CURVE TABLE

Since the normal curve is so useful, tables have been prepared that provide the area under the curve corresponding to given values of z. Recall that we used similar tables to find chi-square probabilities. Remember from the previous section how you can change any set of scores to z scores. If you can assume that the data you are working with can be smoothed to a normal curve—in other words, if their histogram is bell-shaped—then you can use a table of normal-curve areas for the standard normal to find probabilities. Let's see how to use a table of normal-curve areas.

A normal-curve (probability) table deals with z scores and areas (probabilities). It provides the area up to z. Parts of the normal-curve table are given in this chapter in Tables 8.11 and 8.12. A more complete table is given in Table E. The z scores are provided by the left column and the top row of the table (see Tables 8.11 and 8.12). The top row supplies the second digit to the right of the decimal point. The areas appear within the body of the table. Since the total area under the normal curve for z scores is 1.0, all areas given in the table are less than 1.0.

Before we use a normal-curve table, we have to make a decision: Are we going to the table with a z score or with a probability? If we go to the table with a z score, then we want a probability from the table. If we go to the table with a probability (area), then we want to read a z score from the table.

Table 8.11 **Sampling of Standard Normal Probabilities**

z	0	1	2	3	4	5	6	7	8	9
1.3	.9032	.9049	.9066	.9082	.9099	.9115	.9131	.9147	.9162	.9177
1.4	.9192	.9207	.9222	.9236	.9251	.9265	.9279	.9292	.9306	.9319
1.5	.9332	.9345	.9357	.9370	.9382	.9394	.9406	.9418	.9429	.9441
1.6	.9452	.9463	.9474	.9484	.9495	.9505	.9515	.9525	.9535	.9545
1.7	.9554	.9564	.9573	.9582	.9591	.9599	.9608	.9616	.9625	.9633
1.8	.9641	.9649	.9656	.9664	.9671	.9678	.9686	.9693	.9699	.9706
1.9	.9713	.9719	.9726	.9732	.9738	.9744	.9750	.9756	.9761	.9767
2.0	.9772	.9778	.9783	.9788	.9793	.9798	.9803	.9808	.9812	.9817
2.1	.9821	.9826	.9830	.9834	.9838	.9842	.9846	.9850	.9854	.9857
2.2	.9861	.9864	.9868	.9871	.9875	.9878	.9881	.9884	.9887	.9890

Table 8.12 **Sampling of Standard Normal Probabilities**

z	0	1	2	3	4	5	6	7	8	9
−1.0	.1587	.1562	.1539	.1515	.1492	.1469	.1446	.1423	.1401	.1379
− .9	.1841	.1814	.1788	.1762	.1736	.1711	.1685	.1660	.1635	.1611
− .8	.2119	.2090	.2061	.2033	.2005	.1977	.1949	.1922	.1894	.1867
− .7	.2420	.2389	.2358	.2327	.2296	.2266	.2236	.2206	.2177	.2148
− .6	.2743	.2709	.2676	.2643	.2611	.2578	.2546	.2514	.2483	.2451
− .5	.3085	.3050	.3015	.2981	.2946	.2912	.2877	.2843	.2810	.2776
− .4	.3446	.3409	.3372	.3336	.3300	.3264	.3228	.3192	.3516	.3121
− .3	.3821	.3783	.3745	.3707	.3669	.3632	.3594	.3557	.3520	.3483

Example 8.1

Using the table of normal-curve areas, find

$$p(z < 1.65)$$

Solution

We are asked to find the (theoretical) probability that a z score picked at random from a normally distributed population is less than 1.65.

Since z is a continuous variable (for example, it may assume all values in the interval 1.0–2.0), we know that $p(z < 1.65) = p(z \leq 1.65)$ because $p(z = 1.65) = 0$. Thus finding $p(z < 1.65)$ is the same problem as finding $p(z \leq 1.65)$. Thus for a continuous density like the normal, whether $<$ or $\leq$ appears is irrelevant.

Figure 8.16 is a picture of what is going on. A z score of 1.65 is shown on the graph. The shaded area to the left of $z = 1.65$ represents the probability that a random z is less than 1.65 and is the quantity we seek from the table. If we had the formula for the standard normal curve and were armed with integral calculus, we could in principle carry out the integration ourselves to evaluate these areas. In actuality, everybody uses normal tables for those purposes because such integration is not easy by standard calculus methods.

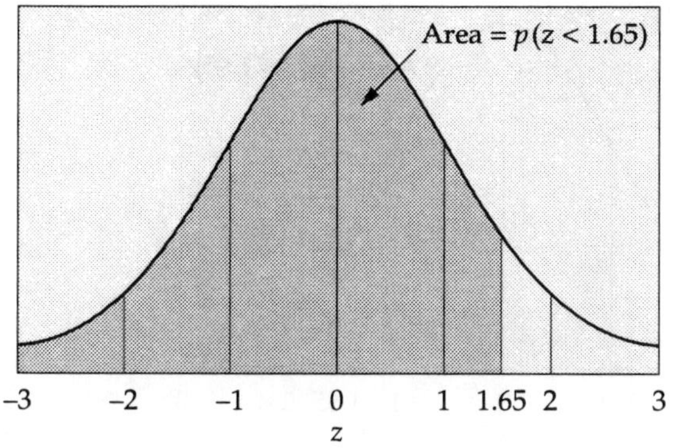

Figure 8.16 **$p(z < 1.65)$.**

In Table 8.11 we find the value of 1.6 in the left column. Then we look along the top row of the table for the column headed 5. The column headings are the hundredths digit for the z score. So for $z = 1.65$ we need column 5. Look in the body of the table for the number at which the row for $z = 1.6$ and column 5 intersect. We see the number is 0.9505. This is an area. The area of 0.9505 is for the shaded region to the left in Figure 8.16. So

$$p(z < 1.65) = 0.9505$$

Example 8.2

Using the normal-curve table, find

$$p(z > 1.65)$$

Solution

In this example we are to find the probability of getting a random z greater than a given value of 1.65. But our table gives areas (probabilities) for z less than (to the left of) given z values. Figure 8.17 shows a graph of what we are to find. For z values greater than a given value, we need the area to the right of the given z. How can we find this area? Conveniently, the total area under a probability distribution for z scores is 1.0. So we know that the area to the right of $z = 1.65$ is 1 minus the area under the curve to the left of $z = 1.65$. Recall from Section 5.6 that $p(\text{event}) = 1 - p(\text{complement of the event})$. Therefore

$$p(z > 1.65) = 1 - p(z < 1.65) = 1 - 0.9595 = 0.0495$$

or 0.05 when rounded to the nearest hundredth.

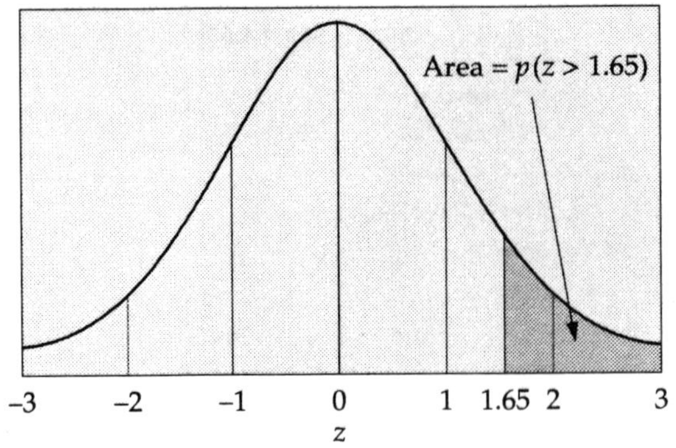

Figure 8.17 **p(z > 1.65).**

Example 8.3

Suppose we have drawn a z at random from a normal population. Suppose also that $p(z < \text{obtained value}) = 0.2$. What value of z was drawn?

We can state this problem as

$$p(z < ?) = 0.20$$

Solution

The problem is depicted in Figure 8.18. Here we are given an *area* (a probability) and need to find a z. So we will go to the table with the area of 0.20 and return with a corresponding z.

We look through the table (Table 8.12) and find 0.2005, which when rounded is closest to our given area of 0.20. We look along the row in which 0.2005 appears and find, in the far left column, a z of −0.8. Then we look at the top of the column in which 0.2005 appears and find the digit in the hundredths place for z. There we find a 4. So we read from the table a z of −0.84 and have, approximately,

$$p(z < -0.84) = 0.20$$

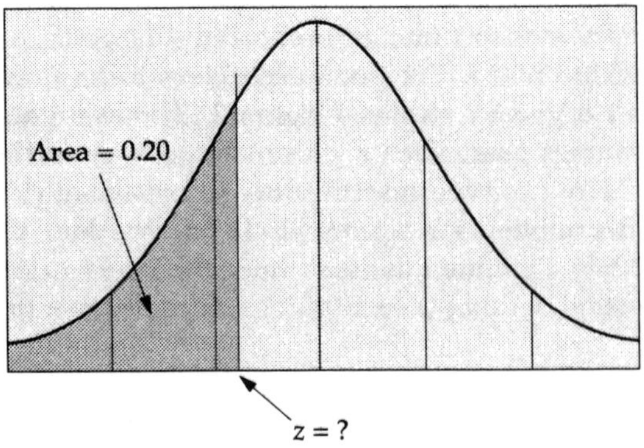

Figure 8.18 **p(z < ?) = 0.2.**

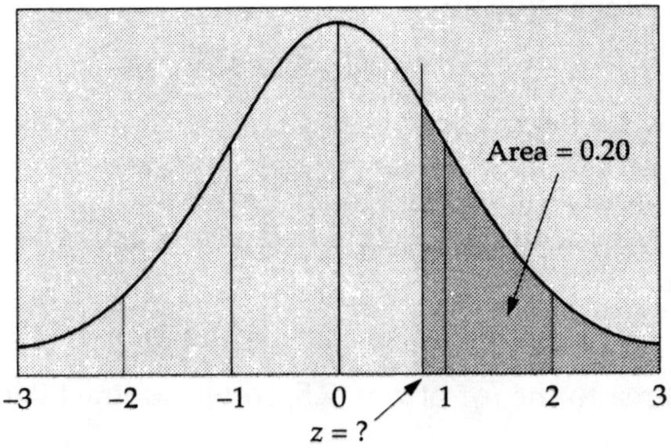

Figure 8.19 **$p(z > ?) = 0.2$.**

Example 8.4

Solve

$$p(z > ?) = 0.20$$

Solution

This is nearly identical to Example 8.3, except that we want to find a value of z such that a random z has probability 0.20 of being greater than that value. The graph is as shown in Figure 8.19. Notice that the difference between this graph and the one for Example 8.3 is that the given area of 0.20 is on the *right* since the random z is greater than the z we are to find.

We cannot go directly to the table with our given area, since Table E gives only areas to the left of a given value of z. So we must find from the given information what the area to the left of the required z is.

Since the total area under the standard normal curve is 1.0, the area we take to the table is

$$1 - 0.20 = 0.80$$

We find in the body of Table E that the value closest to 0.80, when rounded, is 0.7995. The z corresponding to an area of 0.7995 is 0.84—that is, approximately,

$$p(z > 0.84) = 0.20$$

Compare the value obtained in Example 8.4 with that obtained in Example 8.3. The fact that the answers differ only by a sign can be very useful in using normal-curve tables. The normal distribution is symmetrical (it has the same shape on both sides of the mean, or $z = 0$). So, for example, if we want to find

$$p(z > 2.5)$$

we can look at the table for

$$p(z < -2.5)$$

and find that it is 0.0062. So

$$p(z > 2.5) = 0.0062$$

Another way in which we could find $p(z > 2.5)$ is to find directly from Table E that

$$p(z < 2.5) = 0.9938$$

This is the area to the *left* of $z = 2.5$, so we subtract 0.9938 from 1.0000 to obtain

$$p(z > 2.5) = 1.000 - 0.9938 = 0.0062$$

SECTION 8.7 EXERCISES

Assume that z is produced by a normal probability model. In every case, first draw a simple sketch of the given information and the area or z value you are to find.

1. Use the table of normal-curve areas to find each value.
   a. $p(z < 1.96)$ b. $p(z < -1.96)$
   c. $p(z < 1.0)$ d. $p(z < -1.0)$
   e. $p(z < 0.5)$ f. $p(z < -0.5)$
   g. $p(z < 0)$
2. Use the table of normal-curve areas to find each value.
   a. $p(z > 1.96)$ b. $p(z > -1.96)$
   c. $p(z > 1.0)$ d. $p(z > -1.0)$
   e. $p(z > 0.5)$ f. $p(z > 0)$
3. Use the table of normal-curve areas to find each value.
   a. $p(z < 0.68)$ b. $p(z > 0.68)$
   c. $p(z < -0.68)$ d. $p(z > -0.68)$
4. Use the table of normal-curve areas to find the missing value.
   a. $p(z < ?) = 0.95$ b. $p(z < ?) = 0.90$
   c. $p(z < ?) = 0.98$ d. $p(z < ?) = 0.66$
   e. $p(z < ?) = 0.50$
5. Use the table of normal-curve areas to find the missing value.
   a. $p(z < ?) = 0.05$ b. $p(z < ?) = 0.25$
   c. $p(z < ?) = 0.01$ d. $p(z < ?) = 0.10$
   e. $p(z < ?) = 0.5$
6. Use the table of normal-curve areas to find the missing value.
   a. $p(z > ?) = 0.95$ b. $p(z > ?) = 0.90$
   c. $p(z > ?) = 0.99$ d. $p(z > ?) = 0.05$
   e. $p(z > ?) = 0.01$ f. $p(z > ?) = 0.10$
7. Find the missing value. But first decide whether the missing value is positive, negative, or zero.
   a. $p(z < ?) = 0.68$ b. $p(z < ?) = 0.16$
   c. $p(z < ?) = 0.80$ d. $p(z < ?) = 0.20$
   e. $p(z > ?) = 0.20$ f. $p(z > ?) = 0.15$
   g. $p(z > ?) = 0.68$ h. $p(z > ?) = 0.98$
   i. $p(z > ?) = 0.5$
8. Find the missing value. But first decide whether the missing value is positive or negative.
   a. $p(z > ?) = 0.40$ b. $p(z < ?) = 0.88$
   c. $p(z > ?) = 0.93$ d. $p(z < ?) = 0.25$
9. Find the desired value. First decide whether you are going to the table of normal-curve areas with a z value or a probability (that is, an area).
   a. $p(z < 2.38)$ b. $p(z > -1.65)$
   c. $p(z < ?) = 0.005$ d. $p(z > ?) = 0.975$

Graphing Calculator Exercises

In the following exercises use the program NORMAL. Do not just run the program to turn out an answer, but pay particular attention to the shape of the normal curve and the amount shaded by certain z scores. You should begin to develop an intuitive sense for about how much of the curve is shaded for a particular z-score range.

10. Find the following:
a. $p(z < 1.5)$ b. $p(z < 0)$ c. $p(z < -0.75)$
d. $p(z > 2.2)$ e. $p(z > -0.67)$ f. $p(z > 0.25)$

11. Find the following:
a. $p(-1 < z < 1)$ b. $p(-2 < z < 2)$ c. $p(-3 < z < 3)$
d. $p(-1.5 < z < 0.73)$ e. $p(0.45 < z < 1.75)$ f. $p(-1.1 < z < -0.25)$

8.8 APPLICATIONS OF THE NORMAL CURVE

If a set of real data, such as test scores or weights of persons or things, can be assumed to be generated by a normal probability model, the table of standard normal-curve areas can be applied to solve the problem even when the data are not standard normal.

Example 8.5

The lifetime of a certain type of battery has been found to be normally distributed with a mean of 200 hours and a standard deviation of 15 hours (this sort of information could be gathered, for example, by keeping records at the factory on the lifetimes of samples of batteries over a long period of time). What proportion of these batteries can be expected to last less than 220 hours?

Solution

Probabilities can be thought of as proportions, so this is a probability problem. The assumed distribution of battery lifetimes is shown in Figure 8.20. The shaded region shows the batteries that have a lifetime of less than 220 hours. We can calculate this area using the table of normal-curve areas if we know the z value corresponding to 220 hours.

$$z = \frac{220 - 200}{15} = \frac{20}{15} = 1.33$$

We now know the value of z that corresponds to 220 hours. Using this value, we can get the area of the shaded region in Figure 8.20 from the table of normal-curve areas (again, assuming that the battery lives are generated by a normal probability model). That is, we can find

$$p(z < 1.33)$$

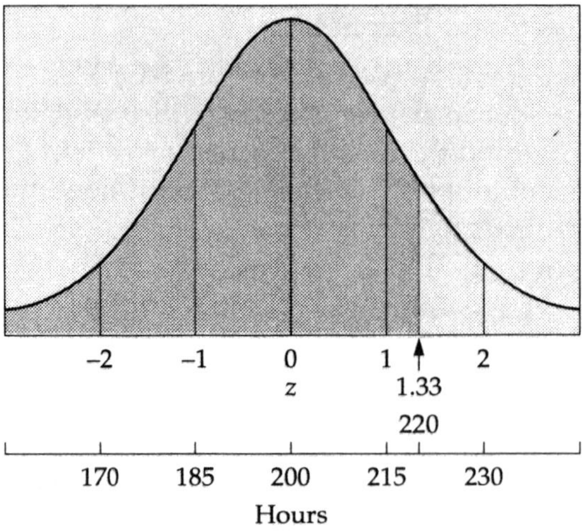

Figure 8.20 **p(battery life < 220 hours).**

From the table, we find the area to be 0.9082, so the proportion of batteries with lifetimes shorter than 220 hours (which corresponds to a z of 1.33) is expected to be 0.91 (rounded). That is, about 91% of the batteries can be expected to last less than 220 hours.

Example 8.6

What proportion of the batteries from Example 8.5 can be expected to last more than 220 hours?

Solution

Again, assuming the battery lifetimes to be distributed normally, we have the graph shown in Figure 8.21. The shaded region represents the proportion of batteries with lifetimes greater than 220 hours.

Since the z corresponding to 220 hours is 1.33, we want to find

$$p(z > 1.33)$$

We know from the table that

$$p(z < 1.33) = 0.91$$

Once again we use the fact that $p(\text{event}) = 1 - p(\text{complement of the event})$. So

$$p(z > 1.33) = 1.0 - 0.91 = 0.09$$

We expect about 0.09 (or 9%) of the batteries to last more than 220 hours.

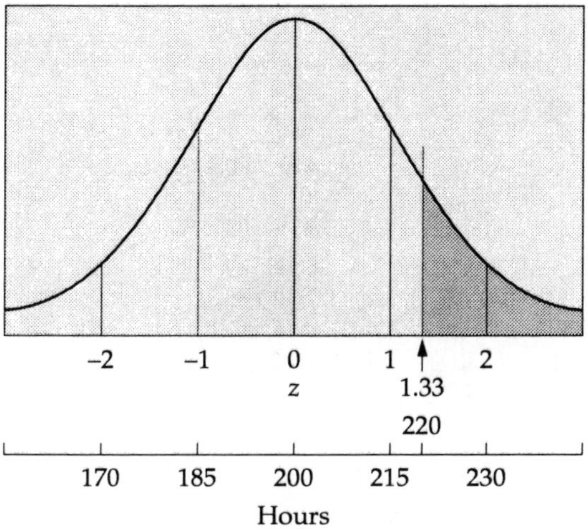

Figure 8.21 **p(battery life > 220 hours).**

Example 8.7

The SPWEHIQS Club (Society of People with Extremely High IQs) requires people to take an intelligence test as a condition for joining the club and restricts membership to the top 5% as measured by this test. Suppose the scores on the intelligence test have a mean of 100 and a standard deviation of 12 and are normally distributed for very large groups of people. What is the lowest score on this test that would be acceptable for admission to SPWEHIQS?

Solution

Figure 8.22 shows the distribution of the scores of the intelligence test. The figure indicates that 5%, or a proportion of 0.05, of the total area is in the shaded region under the curve. According to the normal-curve table (Table E), this area corresponds to a z score of 1.65.

A z score here is computed as

$$z = \frac{\text{IQ} - 100}{12}$$

Thus a z score can be changed to an intelligence test score by a scaling factor of 12 (the standard deviation of the test scores) and a centering factor of 100 (the mean of the test scores).

So the test score corresponding to a z score of 1.65 is

$$\text{IQ} = 12(1.65) + 100 = 19.8 + 100 = 119.8$$

Rounding this score to 120, we find that a score of 120 would be the lowest acceptable for admission to the club.

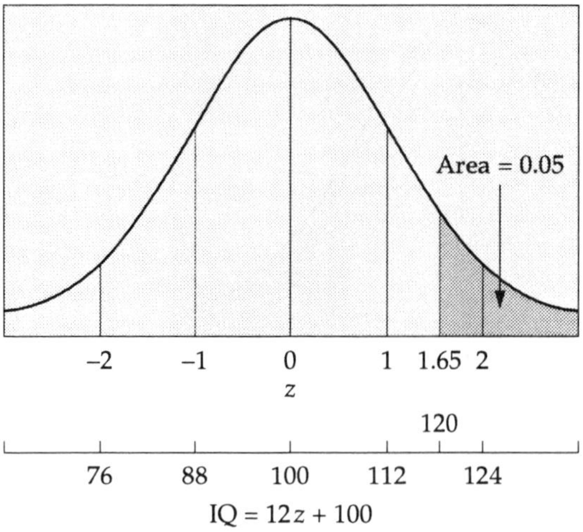

Figure 8.22 **$p(\text{IQ} > ?) = 0.05$.**

Example 8.8

A certain insect species has a mean length of 1.2 centimeters and a standard deviation of 0.12 centimeters. If there are estimated to be 1000 of these insects in a terrarium, how many would be expected to be less than 1 centimeter in length? Assume that the lengths are normally distributed.

Solution

Figure 8.23 is a sketch of the distribution of the insect lengths. The shaded part of the graph begins at the 1-centimeter mark and includes the region to the *left* of this point (since the problem specifies insects having lengths *less* than 1 centimeter).

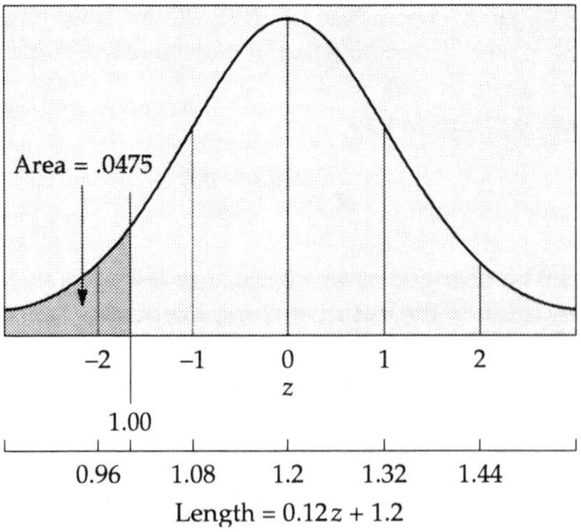

Figure 8.23 **$p(\text{insect length} < 1 \text{ cm})$.**

The z score corresponding to 1 centimeter is found by the transformation

$$z = \frac{1 - 1.2}{0.12} = -\frac{0.2}{0.12} = -1.67$$

According to Table E, the area to the left of $z = -1.67$ is 0.0475. So a proportion of 0.0475 (4.75%) of the 1000 insects are expected to be less than 1 centimeter long. That is, we expect

$$0.0475 \times 1000 \approx 48 \text{ insects}$$

to be less than 1 centimeter in length.

SECTION 8.8 EXERCISES

For each of the following problems, if the exact value is not available in the table, use the closest one.

1. The heights of a group of male students follow the normal distribution with mean 70 inches and standard deviation 3.1 inches.
   a. What percentage of the students would you expect to be shorter than 68 inches?
   b. What percentage of the students would you expect to be taller than 73.5 inches?
   c. What height are 31% of the students shorter than, and what height are 69% of them taller than?

2. To help ensure that boxes of its bananas weigh at least 40 pounds upon arrival at their destination, a packing plant might adopt this rule: Pack boxes to have a weight of 41.5 pounds of bananas with a maximum permissible range of 3 ounces above or below 41 pounds, 8 ounces (that is, pack boxes to have at least 41 pounds, 5 ounces, but no more than 41 pounds, 11 ounces, of bananas). With this rule and the shrinkage in travel, the distribution of box weights upon arrival may be assumed to be approximately normal with a mean of 41 pounds and a standard deviation of 4 ounces.

   Suppose 30 million boxes of bananas are packed a year. Given this packing plant rule and the assumption about the distribution of box weights upon arrival, how many boxes would be expected to weigh less than 40 pounds upon arrival? (*Note:* Tables tell us that for a standard normal z, $p(z < -4.00) = 0.000032$.)

3. A car manufacturer is producing a piston for its engines. The piston is supposed to have a diameter of 5.3 inches, but because of variability, the diameter of a piston actually follows a normal distribution with a mean of 5.3 and a standard deviation of 0.01. If a piston is more than 0.025 inch away from the needed size of 5.3 inches, the piston is rejected.
   a. What percentage of the pistons would you expect to be too large?
   b. What percentage of the pistons would you expect to be too small?

4. Recall the situation presented in Exercise 3. Suppose instead, because of a mechanical problem with the machine producing the pistons, the pistons have diameters that follow a normal distribution with a mean of 5.29 inches. The standard deviation is still 0.01. The specifications require the diameter of the piston to be 5.3 plus or minus 0.025.
   a. What percentage of the pistons would you expect to be too large in this situation?
   b. What percentage of the pistons would you expect to be too small?

5. A set of final exam scores has a mean of 52 and a standard deviation of 6. The scores are normally distributed. If a teacher wants to assign a grade of A to the top 15% of the scores, what score should be the lowest A? If the bottom 15% are to be Fs, what test score should be the highest F?

6. A pipette is a precise instrument used in science to dispense an exact amount of liquid. The pipette is carefully calibrated, but, as with any measurement, it will not dispense the same amount each time. The error in amount is well described by the normal distribution.

   Suppose a scientist is using a pipette known to dispense amounts with a standard deviation of 0.05 microliter (a microliter is a millionth of a liter). The scientist is able to set the pipette to dispense varying amounts of liquid.

   a. If the pipette is set to dispense 200 microliters, what percentage of the time will it dispense greater than 200.075 microliters?
   b. If the pipette is set to dispense 175 microliters, what percentage of the time will it dispense less than 174.90 microliters of liquid?
   c. The scientist wanted to set the pipette to dispense 200 microliters, but she accidentally set it to 199. Is there any chance that the pipette will dispense as much as 200 microliters?

Graphing Calculator Exercises

In the following exercises use the program NORMAL.

7. Suppose the height of men in the U.S. population is normally distributed with a mean of 69 inches (5 feet, 9 inches) and a standard deviation of 3 inches. In the past, metropolitan police forces would not accept candidates below a certain height, say 65 inches. What percentage of the male population would be rejected on the basis of this height requirement? (Compute the z score and try to guess an answer before you use your calculator.)

8. In certain manufacturing processes it is acceptable for a product to be a little larger, but not much smaller than a specified value. Suppose a machine makes washers with an average inner diameter of 1.2 centimeters and a standard deviation of 0.04 centimeter. The acceptable range for inner diameters is $1.12 < d < 1.4$. If we assume that the machine produces washers that are normally distributed, what percentage of washers are expected to be acceptable? What percentage are expected to be defective? In a package of 1000 washers, how many, on average, are expected to be defective?

9. In Exercise 7 we assumed that the height of adult males in the present U.S. population is normally distributed with a mean of 69 inches and a standard deviation of 3 inches. Suppose in the year 1800 their mean height was 66 inches (3 inches shorter) while the standard deviation remained the same at 3 inches. What percentage of the present male population is expected to be over 6 feet tall? What percent of the male population in 1800 would you expect to have been over 6 feet tall?

10. Suppose a standardized test produces results that are normally distributed in the population. If the average test score is 100 with a standard deviation of 12, what would you have to score to be in the upper 2% of those testing? From your experience in using the program NORMAL to do Exercises 10 and 11 in Section 8.7, guess at an appropriate z score. Check it with the program and see how close you came. Repeat the process a few times until you have a pretty good z score, and then use it to find the actual test score.

8.9 OTHER CONTINUOUS DISTRIBUTIONS (OPTIONAL)

Statisticians have invented a plethora of distributions. We have seen three of the most important continuous distributions so far—the chi-square, the normal, and the uniform—and we will see another very important continuous distribution, called *Student's t*, in Chapter 11. We consider two more important continuous distributions in this subsection: the exponential and the arcsin.

The Exponential Distribution

Recall the histogram of the dog and cat data in Figure 8.11. We tried to fit a normal curve to the relative frequency histogram of the numbers of dogs and cats people had, but the normal did not represent the data very well. The histogram starts fairly high at 0 pets, and then generally decreases as the number of pets increases. The **exponential distribution** often handles such situations quite well. It can also be useful for such diverse things as income data and lifetimes of lightbulbs. Indeed, if something is doomed to fail eventually (like a lightbulb) but does not wear out (its chance of failing in a given fixed length of time does not increase with age) then the exponential distribution can be shown to be the appropriate distribution.

The mean of the number of dogs and cats owned was 4.03. In Figure 8.24 the relative frequency histogram for the dogs and cats example is superimposed with an exponential curve whose mean is 4.03. This curve follows the histogram much better than the normal curve, although it is a little too high at 0 pets, and decreases to 0 a little too soon to capture the people with a large menagerie (30 or 40 pets).

We now consider an unusual distribution whose density is low in the middle and high on the ends.

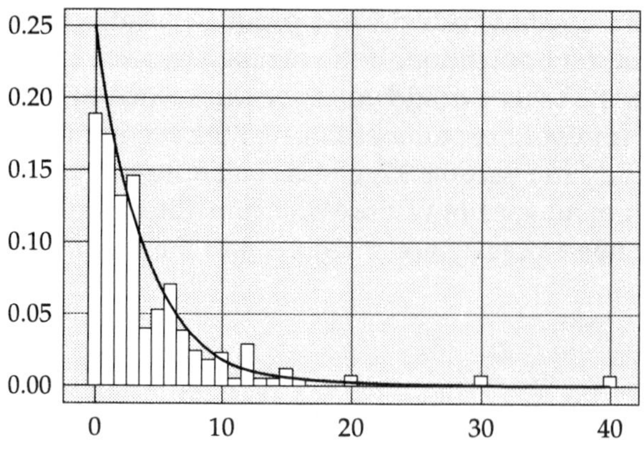

Figure 8.24 **Relative frequencies of dog and cat ownership and fitted exponential density.**

The Arcsin Distribution

During the major league baseball season, a team plays 162 games. Consider a mediocre team, that is, one that wins half its games. How much of the season will it spend with its winning proportion over 0.500 (which means having won more than it has lost so far), and how much of the season will it spend under 0.500?

Naively, we would expect it to spend about as much time above 0.500 as below, as well as some time having won exactly half. To explore the question theoretically, we modeled a season of 162 games by 162 flips of a fair coin and simulated the experiment on the computer. Note that we are back to carrying out the five-step method to explore a probability distribution we don't know. After each flip, we find the proportion of games won so far, and count 1 if it is larger than 0.500, and 0 if it is less than 0.500. If it is exactly 0.500, we count $\frac{1}{2}$. At the end of the season (162 flips), we average those counts, telling us what proportion of the time the team was over 0.500 (plus half the time they were exactly 0.500).

An example is in Table 8.13. Of the 162 flips, 82 were heads, a proportion of $82/162 = 0.506$. But we can see that because the team started out winning its first two games, it stayed above or at 0.500 for the first 14 games. As of game 15, the team had seven wins; this was the first time it was below 0.500. Overall, the team spent 103 days over 0.500, and 13 at exactly 0.500, so its average count is $103/162 = 0.6358$. Thus even though it just barely won more than half its games, 63.58% of the time the team was actually over 0.500.

Table 8.13 **Ahead of or behind 0.500: Game by Game**

Trial	Heads (1) or Tails (0)	Number of heads so far	Proportion of heads so far	Count
1	1	1	1.000	1.0
2	1	2	1.000	1.0
3	0	2	0.667	1.0
4	1	3	0.750	1.0
5	0	3	0.600	1.0
6	1	4	0.667	1.0
7	0	4	0.571	1.0
8	0	4	0.500	0.5
9	1	5	0.556	1.0
10	1	6	0.600	1.0
11	1	7	0.636	1.0
12	0	7	0.583	1.0
13	0	7	0.538	1.0
14	0	7	0.500	0.5
15	0	7	0.467	0.0
16	0	7	0.438	0.0
17	0	7	0.412	0.0
18	1	8	0.444	0.0
19	0	8	0.421	0.0
20	0	8	0.400	0.0
⋮	⋮	⋮	⋮	⋮
153	1	79	0.516	1.0
154	0	79	0.513	1.0
155	0	79	0.510	1.0
156	0	79	0.506	1.0
157	1	80	0.510	1.0
158	1	81	0.513	1.0
159	0	81	0.509	1.0
160	0	81	0.506	1.0
161	0	81	0.503	1.0
162	1	82	0.506	1.0

We then simulated 1000 such 162-game seasons. Table 8.14 gives the results. We see that for 152 of the seasons, the team spent at most 5% of the time above 0.500. Similarly, for 156 of the seasons this team spent at least 95% of the time over 0.500. Thus $(152 + 156)/1000$, or about $\frac{3}{10}$, of the teams spent 95% or more of their time above or spent 95% or more time below 0.500. Are you surprised? Hopefully, yes! In contrast, in relatively few seasons did teams spend very close to half the time over and half the time under 0.500.

Table 8.14 **Proportion of Time above 0.500 in Season**

Proportion of time above 0.500	Observed count	Experimental probability
0.00–0.05	152	0.152
0.05–0.10	45	0.045
0.10–0.15	43	0.043
0.15–0.20	54	0.054
0.20–0.25	42	0.042
0.25–0.30	31	0.031
0.30–0.35	28	0.028
0.35–0.40	27	0.027
0.40–0.45	25	0.025
0.45–0.50	39	0.039
0.50–0.55	31	0.031
0.55–0.60	31	0.031
0.60–0.65	35	0.035
0.65–0.70	34	0.034
0.70–0.75	30	0.030
0.75–0.80	47	0.047
0.80–0.85	35	0.035
0.85–0.90	47	0.047
0.90–0.95	68	0.068
0.95–1.00	156	0.156
	1000	

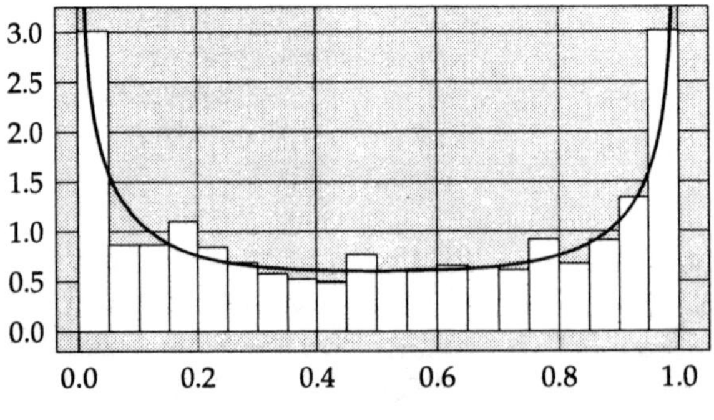

Figure 8.25 **Simulated proportion of time above 0.500 for a baseball team of 0.500 ability and fitted arcsin density.**

Figure 8.25 graphs the relative frequency histogram. It certainly does not look like a normal curve! It does follow the so-called **arcsin** law, which is the smooth curve superimposed on the histogram. In more advanced courses, one can derive this distribution using theoretical reasoning based on assumptions, as one does with the four binomial assumptions. The arcsin law is a theoretical description of the proportion of time one spends over 0.500. It describes this set of data very well. This type of distribution is called U-shaped, for obvious reasons. Also, the result is a shock to our intuition in that a surprisingly large proportion of these mediocre teams will be either ahead of 0.500 for *most* of the season or behind 0.500 for *most* of the season.

SECTION 8.9 EXERCISES

1. Suppose the number of calls that come into an office during a period of time follows the Poisson distribution. More specifically, the theoretical mean is 0.75 calls per minute. Suppose now, though, the secretary keeps track of the time between calls. These are the data collected for the times between the first 50 calls:

0.002144	0.370075	4.073172	1.109358
1.325636	5.990878	0.917768	0.60038
0.703133	0.571503	0.594772	5.103429
2.549596	0.267843	2.052013	0.833352
0.880532	0.144781	1.053058	1.558251
0.875297	0.365745	1.99231	0.112271
0.131846	0.432968	0.254621	0.255284
0.496691	2.866313	0.530326	0.888111
0.981161	1.125645	3.32624	0.194681
0.471359	0.116685	0.467262	1.883154
0.697345	2.050501	0.209754	1.095721
0.213761	4.492431	0.475652	2.259617
0.839236	6.701594		

   a. Group the data into groups of width 0.5. Draw a relative frequency histogram.
   b. What distribution does this seem to be?
   c. Find the average of the values. Compare the average you calculated to (1/theoretical mean) from the Poisson distribution. Are they close to each other?

2. Suppose you are playing a game in which you get one point if you win a round and you lose one point if you lose a round. You start with zero points (it is possible to go negative). You want to estimate the proportion of the time you are winning (have a positive number of points).
   a. Simulate this game by flipping a coin continually. After each flip, record the outcome. A head indicates a win, and a tail a loss. Also record whether you are ahead or behind at this point in the game. Play 50 rounds.
   b. What proportion of the time were you ahead?
   c. If you were to repeat this process of playing 50 rounds many times, what would you expect to be the distribution of the proportion of the time you were ahead?

CHAPTER 8 REVIEW EXERCISES

1. For each of the following distributions give the value of the theoretical mean and standard deviation:

a. Binomial distribution with $n = 15$ and $p = 0.35$
b. Poisson distribution with mean $= 4.3$
c. Discrete uniform distribution with upper limit 5 and lower limit 0

2. Repeat Exercise 1 for each of the following distributions:
   a. The standard normal distribution
   b. The continuous uniform distribution with upper limit 6 and lower limit 3
3. Convert the following values from the normal distribution with mean 5 and standard deviation 2 into standard units:
   a. 4.392
   b. 6.921
   c. 8.936
   d. 0.0638
4. Calculate the following probabilities, where z has the standard normal distribution:
   a. $p(z < -1.43)$
   b. $p(z > 0.97)$
   c. $p(z < -2.10)$
   d. $p(z > 1.75)$
5. Replace the question marks with the correct values from the standard normal distribution:
   a. $p(z > ?) = 0.1131$
   b. $p(z < ?) = 0.3557$
   c. $p(z > ?) = 0.0250$
   d. $p(z > ?) = 0.0104$
6. Consider the following situation: An engineer keeps track of how often an unacceptable part comes off the production line. He hypothesizes that there are 2.5 bad parts per hour.
   a. What distribution does the number of bad parts in an hour have? Explain.
   b. What is the theoretical mean and standard deviation, based on the engineer's hypothesis?
   c. Suppose, instead, we addressed the problem this way: The engineer knows that 1000 parts come through an hour, and he suspects that 0.0025 of them are defective. He is interested in finding out the probability of no defectives being produced in a given hour. What distribution would he use in this case? Explain.
7. When Jill leaves home in the morning, there is a 40% chance that the traffic light she comes to first will be green. Based on a five-day work week, calculate the following information:
   a. What is the mean number of times she will come to the light when it is green? What is the standard deviation?
   b. What is the probability that on all five days she will reach the light when it is green?
   c. What is the probability that she will reach the light when it is green on exactly four out of the five days?

d. What is the probability that she will reach the light when it is green on exactly three out of the five days?

8. A random variable is hypothesized to have the Poisson distribution with mean 1.3. The variable is observed 100 times, with the following distribution of results, along with the theoretical proportions:

Value	Observed proportion	Theoretical proportion
0	0.23	0.27
1	0.41	0.35
2	0.21	0.23
3	0.08	0.10
4	0.05	0.03
5	0.02	0.01

a. Draw a relative frequency histogram for both the observed and the theoretical proportions. Do they seem to be similar?
b. Calculate the value of the chi-square statistic based on these data.
c. Does the chi-square test give the same result as part (a)?

9. Describe the difference between the continuous and discrete uniform distributions. Give one example in which the discrete uniform applies, and another in which the continuous uniform applies.

10. For each of the following random variables, tell whether it is uniform or discrete:
a. The amount of time you have to wait in line
b. The number of people who are in front of you when you get into line
c. The total number of cars that pass by in an hour
d. The average speed of the cars that pass by in an hour

11. A random variable is suspected to follow the normal distribution with mean 6 and standard deviation 0.5. Once the random variable is observed many times, it is calculated that 53% of the observations lie between 5.5 and 6.5. Furthermore, 83% of the observations lie between 5 and 7. Do these observations of the random variable seem to give evidence for or against the case that the random variable has the normal distribution? Explain.

12. Convert the following variables from the normal distribution with mean -6.5 and standard deviation 2.4 into standard units:
a. -2.381
b. -12.905
c. -0.0964
d. Consider the value from part (b). What is the probability of seeing a random variable from this distribution smaller than that?

13. Calculate the following probabilities from the normal distribution specified:
a. $p(X < 4.0)$, where X has mean 5.0 and standard deviation 1.0
b. $p(X > 6.5)$, where X has mean 5.5 and standard deviation 0.75
c. $p(X > -5.44)$, where X has mean -7 and standard deviation 0.6

14. An engineer is designing a pedestrian walkway over an expressway. He knows that the heights of semis passing underneath follow the normal distribution with a mean of 14.5 feet and a standard deviation of 0.5 feet.
    a. If he builds the walkway with a clearance of 15.5 feet, what percentage of the trucks will not fit?
    b. How much clearance should the engineer build under the walkway if he wants to be sure that 99.9% of the trucks will fit?
15. A large city is conducting a campaign to increase the amount of carpooling in an effort to reduce pollution. The statistician involved with the project has figured that a good goal for the project is for the number of people in cars during rush hour to follow a discrete uniform distribution with limits 1 and 6.
    a. If the goal were met, what would be the mean and standard deviation of the number of people in a car?
    b. The statistician took a sample by sitting in a toll booth on an expressway during rush hour. This was the distribution she saw:

       42 cars with 1 person
       31 cars with 2 people
       12 cars with 3 people
       18 cars with 4 people
       12 cars with 5 people
       5 cars with 6 people

       Use the chi-square test to determine whether the goal has been met.
16. Recall the situation in Exercise 14. The engineer determined how high the walkway should be, but now he wants to figure out how much weight the walkway should be able to hold. The walkway must be large enough that 50 people can be on it at once. The engineer figures that the weight of 50 people follows the normal distribution with a mean of 7500 pounds and a standard deviation of 175 pounds.
    a. If the walkway is built to carry a weight of 7815 pounds, what percentage of the time will 50 people overload it?
    b. Suppose the engineer makes some mistakes in his calculations and has the bridge built to support only a weight of 7400 pounds. What percentage of the time will 50 people on the walkway overload it now?
17. A gambler is playing a game in which he has a 20% chance of winning. He decides to play the game seven times.
    a. What is the probability that he will lose all seven games?
    b. What is the probability that he will win exactly two games?
    c. What is the probability that he will win exactly one game?
    d. How many games will he win on average?
18. Calculate the following probabilities. The variable X has the specified normal distribution:
    a. $p(X > 9.5)$, where X has mean 8.4 and standard deviation 0.5
    b. $p(X < -6.7)$, where X has mean -6.0 and standard deviation 0.4
    c. $p(X > 0)$, where X has mean -0.4 and standard deviation 2

8

COMPUTER EXERCISES

These exercises use the program Five Step. Review the use of this program in the Computer Exercises of Chapter 4, if you wish.

1. TWO FLIPS OF A COIN

Recall the example in Section 8.1 that derived the binomial probability for the number of heads in two flips of a fair coin. Start the Five Step program, and create the box that mimics a coin flip—that is, has one 1 and one 0. Define a trial to be drawing $n = 2$ without replacement, and define the statistic to be "Sum." Run 500 trials. Notice the histogram after each set of 100 trials.

a. Does the histogram appears as it should—that is, with three bars, the middle being about twice as high as the others?
b. Choose "Frequency Table" from the Summary Statistics menu. What are the observed numbers of the possible values? (Note that "-0.5 to 0.5" means 0, "0.5 to 1.5" means 1, and "1.5 to 2.5" means 2.) What is the observed proportion of each number (out of 500)? Are the observed proportions close to the theoretical values of $\frac{1}{4}$, $\frac{1}{2}$, and $\frac{1}{4}$?

2. THREE LOTTERY TICKETS

Now consider the example in Section 8.1 in which you buy three lottery tickets and the chance of winning each day is 0.25. Change the box so that it represents one lottery ticket, and a 1 means win and a 0 means lose. (Thus there will be one 1 and three 0s.) Now we sample three with replacement, and again look at the sum. Run 500 trials.

Look at the frequency table, and find the proportions of trials for the possible values of the sums, 0, 1, 2, and 3. Are these sample proportions close to the theoretical ones of $\frac{27}{64}$, $\frac{27}{64}$, $\frac{9}{64}$, and $\frac{1}{64}$?

3. SEVEN DAYS OF LOTTERY TICKETS

Now move to the example in Section 8.1 in which you buy one lottery ticket each day for a week. The box represents one day, so again it has one 1 and three 0s. Now we sample seven with replacement, and again look at the sum. We also wish to find the chance that a week has seven winners, so set the "Statistic of interest" to be "Stat = a" with "a = 7."

Run 500 trials.

a. Write down the theoretical mean and standard deviation for the number of winning tickets in a week. How close are those to the sample average and sample standard error of the statistic, respectively?

b. Write down the sample proportions obtained for the possible values 0, 1, ..., 7. What was the maximum number of winners in any week? How do these proportions compare with the theoretical ones in Table 8.1?

c. How many of the sums had the value 7? If none, try 500 more trials. Are there any weeks with seven winners yet? If not, try more trials, until either you find a week with seven winners, or you give up. How many trials (weeks) did you perform? How many years does that translate to? How many times did you get seven winners?

4. ROLL OF A DIE

Set up the box to mimic a roll of a fair six-sided die. A trial will consist of one roll, so define the trial to be "One draw." That will automatically set $n = 1$ and the statistic to be the draw. Set the probability of interest to be blank ("blank" is a choice).

a. From Section 8.3, write down the average and standard deviation for the values on one roll. Are they the same as the values given at the bottom of the Five Step screen?

b. Run 500 trials. After each set of 100, notice the histogram. Is it getting closer to being uniform (all blocks being the same height)?

c. From the frequency table, find the proportion of times each value came up. Are these proportions close to $\frac{1}{6}$?

d. Compare the sample mean and sample standard deviation of the statistic to the theoretical mean and standard deviation. Are they close to what they should be?

e. Run 500 or more additional trials, and repeat questions (b), (c), and (d). Are the obtained values getting closer to the theoretical ones?

5. THE NORMAL DISTRIBUTION

We wish to set up a box to approximate the population of body temperatures as in Figure 8.10. We will use the normal distribution with an average of 98.25 and a standard deviation of 0.73, as in the sample of 130 people. To set this box up, press "Choose Box Model" (after pressing "Start Again," if necessary), and in the Type menu, choose "Named Distribution." A different screen will appear. Select the circle next to "Normal," and type in 98.25 for the average and .73 for the standard deviation. Press the Histogram button. The screen should look like this:

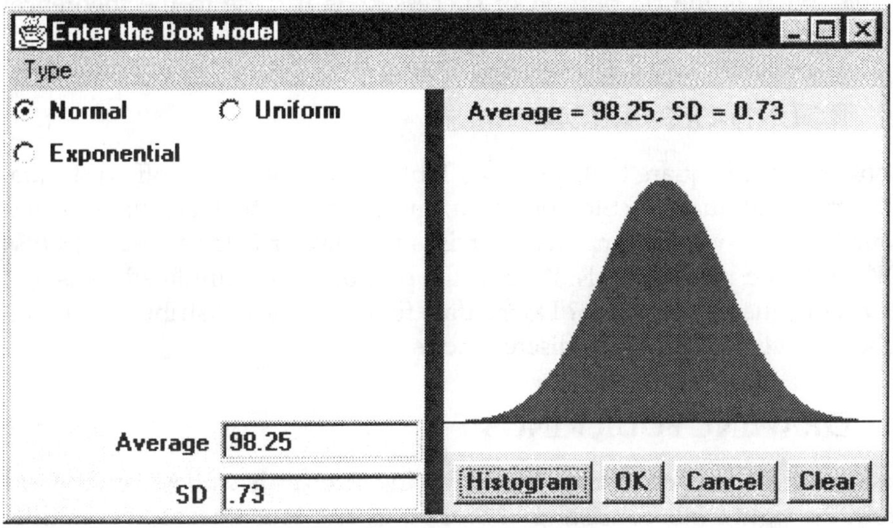

If it does, then press OK. Define the trial to be "Draw n with replacement," where $n = 1$. Set the statistic to be "Sum." (The sum of one draw is just the draw itself, of course.) We are going to draw a number of tickets from this box, and we wish to see how many are between average − SD and average + SD. So for the probability of interest, choose "a < Stat < b," where $a = 97.52$ and $b = 98.98$. (See Table 8.8.)

Run 500 trials.

a. Does the histogram look somewhat like a normal curve?

b. Compare the average and the standard error (which in this case is the standard deviation) of the statistic (which in this case is the draw) with the average and standard deviation of the box.

c. What proportion of draws were between average − SD and average + SD? How close is it to the theoretical value of 0.68?

d. Now press "Start Again," and change the statistic to be "Z." Where you see "mu," enter the mean, 98.25, and where you see "sigma," enter the standard deviation, 0.73. Now the statistic for each draw will be the standardized value: (value of draw − average)/(standard deviation). These statistics should have mean 0 and standard deviation 1. Set the probability of interest to be "a < Stat < b" with $a = -1$ and $b = 1$. Run 500 trials.

   i. Does the histogram look reasonably normal?

   ii. How close are the average and standard error to 0 and 1, respectively?

   iii. What is the proportion of successes? Is it close to the theoretical probability?

6. ROUND-OFF ERROR

Look at the chi-square table given as Table 8.7 for the round-off error data. Start the Chi-Square Six Step program, and fill in the obtained and expected counts. (Review the Computer Exercises in Chapter 7, if necessary.) Press OK, and then run 200 trials. What is the proportion of simulated trials that have chi-squares over 17.16? Do the data fit the uniform distribution? If not, where do you see the main discrepancies?

7. DRAWING FOUR KINGS

In Section 8.1 we noted that when drawing five cards *without replacement* from the eight cards consisting of four kings and four queens, the number of kings obtained is *not* distributed as a binomial variable. It is not binomial because the draws are not independent. If the draws are made *with* replacement, then the number of kings drawn *is* distributed as a binomial variable. How does the difference between drawing with and drawing without replacement manifest itself?

a. Set up the box with four 0s (the queens) and four 1s (the kings). Define a trial to be drawing five without replacement, and define the statistic to be the sum. Thus the statistic will be the number of kings in the five draws (the distribution of the number of kings is called the *hypergeometric* distribution and is discussed in Chapter 13). Run 500 trials. What are the values of the following?

   i. The average of the draws

   ii. The standard error of the statistic

   iii. The maximum number of kings drawn

iv. The minimum number of kings drawn

v. The proportion of each number of kings drawn (from the frequency table in the Summary Statistics menu)

b. Press "Start Again," and change the trial to be drawing five with replacement. Keep the statistic the sum. Now the distribution of the sum will be binomial. Run 500 trials of this experiment. Give the same quantities as above:

i. The average of the draws

ii. The standard error of the statistic

iii. The maximum number of kings drawn

iv. The minimum number of kings drawn

v. The proportion of each number of kings drawn

c. Are the averages in the two experiments about the same, or is one larger than the other?

d. Are the standard errors about the same, or is one larger?

e. How do the maxima and minima in the two experiments compare?

f. How does the proportion for each value compare between the two experiments?

g. In general, in which experiment is the distribution more spread out?

9

MEASUREMENT

Errors, like straws, upon the surface flow.
He who would search for pearls must dive below.

John Dryden

OBJECTIVES

After studying this chapter, you will understand the following:

- *How random error is inherent in the measurement process*
- *The existence of very large measurement errors called outliers*
- *How to improve measurement accuracy by using the average of a set of measurements*
- *The standard error as the typical error size associated with an average*
- *The distinction between precision and accuracy*

9.1 MEASUREMENTS VARY 448

9.2 MEASURING CAREFULLY 452

9.3 STANDARD ERROR 456

9.4 SYSTEMATIC ERROR (BIAS) 461

9.5 ACCURACY 463

9.6 AN APPLICATION OF MEASUREMENT: THE ACCELERATION DUE TO GRAVITY 467

CHAPTER REVIEW EXERCISES 470

COMPUTER EXERCISES 473

Measuring g

Objects fall to earth because of the pull of gravity. How to measure the acceleration due to gravity, g, has been studied for hundreds of years. Various experiments have been designed to measure g accurately, such as the timing of the fall of a steel bar using a photoelectric cell.

One well-known way to measure gravity is to use the relationship between the length of a pendulum and how long it takes to swing back and forth. You may have noticed that short pendulums swing back and forth in a short amount of time and long pendulums swing back and forth in a long amount of time. What you probably haven't realized, unless you learned it in a physics class, is that the stronger the pull of gravity g is, the less time the pendulum will take to swing back and forth. Put the same pendulum on Jupiter, which is much more massive than the Earth, and the pendulum will take much less time to complete one swing back and forth than on Earth.

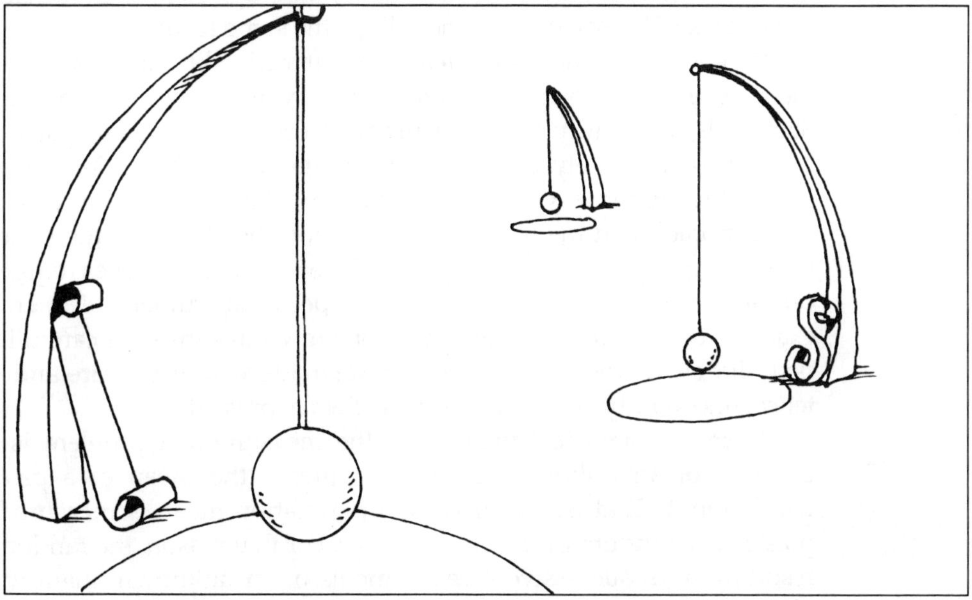

In this chapter we will see how statistical ideas can be used to help us accurately measure the acceleration due to gravity, g. Moreover, in general we will see the essential role statistics plays in helping people measure things *accurately.* We will discover that achieving measurement accuracy requires a blend of good science and good statistical practice. Such measurement accuracy is vital to the progress of science and technology, as well as in the affairs of ordinary life.

9.1 MEASUREMENTS VARY

A college statistics class of 30 students measured the height of a stack of textbooks as a group project whose goal was for the students to experience some of the practical and statistical aspects of real-world measurement problems. They divided into teams of two people, and each team made one measurement. Here are their results, in centimeters:

24.1	24.7	23.6	23.8	22.3
22.8	23.6	23.5	24.0	23.7
24.4	23.6	23.8	23.8	24.3

The results may surprise you, at first, for although each team measured the same stack of books, they did not all get the same results.

Whenever we measure, measurement variability and hence measurement error must creep in. Interestingly, even at the National Bureau of Standards, variability and measurement error are an inescapable reality! Measuring accurately involves care and good judgment. You have to be sure of the proper way to use the measuring device, such as a ruler, caliper, or electronic instrument. You have to know how to read the scale of the measuring device you are using. Good measuring techniques require much care and training and, in many cases, special instruments. Moreover, the exact set of rules for what and how one measures must be carefully specified. But even when measurements are made with great care and competence, uncertainty and variability are always present.

From the statistical viewpoint, the measurement problem is the estimation of an unknown population mean (the mean of a probability distribution). That is, the unknown population mean is measured in the presence of random measurement error. For that reason, the random errors resulting from successive measurements of an unknown quantity can be viewed as a random sample from the large (conceptually infinite) population of all possible measurement errors. This population of measurement errors has a population (theoretical) standard deviation that, if known, provides us with the size of a typical measurement error. For example, if the standard deviation of the measurement error population is 0.1, then

0.1 is the typical size of the error in measurement that is likely if we make one measurement. What we mean by "typical" can be quantified. According to the *normal law of error,* which is discussed below, most measurement data have the shape of the normal density. Therefore, as detailed in Section 8.6, a two-thirds rule and a 95% rule apply to most measurement data sets: if we had available many measurements of the unknown quantity, then about two-thirds of the time our measurement would fall within ± 1 SD (0.1 here) of the true, unknown population mean, and about 95% of the time our measurement would fall within ± 2 SD (0.2 here) of the true, unknown population mean.

The measurements of the height of the stack of books are presented in the stem-and-leaf plot in Table 9.1. Let's describe this data set using the summary statistics we have used before. The mean and median of these data are as follows:

$$\text{Mean} = 23.7 \text{ centimeters}$$
$$\text{Median} = 23.8 \text{ centimeters}$$

The range of the data is easily found from the table. The smallest measurement recorded is 22.3 centimeters, and the largest is 24.7 centimeters. So

$$\begin{aligned} \text{Range} &= \text{largest} - \text{smallest} \\ &= 24.7 - 22.3 = 2.4 \text{ centimeters} \end{aligned}$$

Using the 4th and 12th measurement magnitudes, the interquartile range is $24.1 - 23.6 = 0.5$ centimeter. The standard deviation of these 15 measurements is 0.59 centimeter.

We can describe the data in another way. Most of the data are concentrated near the middle of the set. That is, most of the measurements are close to a value of 23.7 (or 23.8) centimeters. The farther away from this value, the fewer measurements there are. And there are no measurements really far from this value. Values in a data set that are exceptionally far away from mean are called *outliers.* A widely used rule is that a data point is an outlier if its distance from the sample mean is more than 3 times the sample

Table 9.1

Stem	Leaf
21	
22	3, 8
23	5, 6, 6, 6, 7, 8, 8, 8
24	0, 1, 3, 4, 7
25	

Key: "22 3" stands for 22.3 centimeters.

THE
NORMAL
LAW OF ERROR
STANDS OUT IN THE
EXPERIENCE OF MANKIND
AS ONE OF THE BROADEST
GENERALIZATIONS OF NATURAL
PHILOSOPHY ◆ IT SERVES AS THE
GUIDING INSTRUMENT IN RESEARCHES
IN THE PHYSICAL AND SOCIAL SCIENCES AND
IN MEDICINE AGRICULTURE AND ENGINEERING ◆
IT IS AN INDISPENSABLE TOOL FOR THE ANALYSIS AND THE
INTERPRETATION OF THE BASIC DATA OBTAINED BY OBSERVATION AND EXPERIMENT

Figure 9.1 **The bell-shaped curve. (Source: W. J. Youden, *Experimentation and Measurement* [Washington, D.C.: National Science Teachers Association, 1962], p. 55.)**

standard deviation. Here that would be greater than $23.7 + 3(0.59) = 25.47$ or less than $23.7 - 3(0.59) = 21.93$. According to this often-used convention, none of these class measurements are outliers. It is common statistical practice in measurement to remove outliers because they can significantly distort the measurement process. Indeed, it is likely that a measurement outlier is the result of some sort of procedural error, such as misplacing a decimal point or grossly misusing a piece of equipment.

The shape of this data set is roughly similar to that of a theoretical distribution we studied in Chapter 8, namely, the normal distribution. In fact, the normal distribution is central in the statistical study of measurement. Its importance is due to the *normal law of error:* if a relative frequency polygon or relative frequency histogram is drawn of very many measurements, in most cases the graph will look like a normal curve, especially after outliers have been removed. Figure 9.1 provides an amusing and perhaps informative example of this bell-shaped curve, which we studied carefully in Chapter 8.

In measuring an object, we expect that exceptionally large measurement errors (either too large or too small) will occur very infrequently. As stated above, exceptionally large or small errors, called outliers (farther than three standard deviations away from the mean, according to our rule of thumb), do occasionally occur, even in well-designed, repeated measurement procedures, such as those the National Bureau of Standards would follow. But most measurements cluster around the middle of the set of measurements.

If we average measurements together, the positive and negative measurement errors will tend to cancel each other out, thus usually resulting in a value close to the correct measurement we are seeking, except when the measurements are biased, as when a badly calibrated grocer's scale adds one ounce to the weight of each meat purchase. It is standard practice to improve accuracy by averaging multiple measurements.

SECTION 9.1 EXERCISES

1. The diameter of a human hair (in micrometers) was measured in a laboratory as follows:

49	51	48	48	49	47	50	48	49
49	51	49	48	47	51	50	51	49
48	49	52	49	52	52	53	47	53
47	51	52						

Find the mean and standard deviation of the measurements. Draw a histogram of the data. Do the data appear to be normally distributed?

2. A state inspection office tested 50 gasoline pumps around the state for measurement errors. Five gallons of gasoline (according to the meter) were pumped, and the error was measured. The results, in cubic inches, are as follows (minus signs mean less than 5 gallons):

2	−4	−6	−3	0
−4	−1	0	0	−2
−1	−1	−4	2	−4
−1	−2	−4	−3	1
−2	−4	−2	−1	−4
−2	−6	−2	−1	−5
0	−3	−4	−3	−7
−5	−1	−4	−4	2
−2	−3	−3	1	1
−4	−3	1	−3	−3

Construct a histogram of the errors. Do the data appear to be normally distributed? What percentage of pumps had an error of more than 5 cubic inches?

3. The compressive strength of cement is the amount of squeezing pressure it can withstand before crumbling. The following data are values of the compressive strength (in pounds) of the same batch of cement as measured by 50 laboratories. What percentage of labs reported cement strengths 300 pounds or more greater than the average reported value?

4466	3914	4084	4135	4084
4154	4123	4030	4120	3626
3771	3889	4180	4141	4524
3810	3777	4072	3871	4181
3948	4081	4334	4046	4130
3676	3888	4126	4128	4058
3589	4135	4018	3675	4251
3842	4409	4134	4017	3888
3922	4300	4447	4228	3907
3825	4172	4005	4241	4339

Graphing Calculator Exercises

Exercises 4–6 require the program OUTLIER.

4. Here are the heights (in inches) of the boys on a soccer team:

65	68	72	64	67	69	67	82	70	71
70	71	68	69	73	64	70	66	68	69

a. Clear L1 in your calculator and enter these values.
b. Can you guess if there are any outliers in this data set?
c. Run the program OUTLIER and record the results.
d. Are there any outliers in this data set? What are they?
e. Trace the box box plot provided by the program.
f. The full name for a box plot is a "box-and-whisker" plot. What in the shape of the plot would lead you to suspect the presence of an outlier?

5. The table below presents the "winning percentages" (ratios of wins to games played, or winning proportions) of the major league baseball clubs as of April 26, 1997.

Team	Percentage	Team	Percentage	Team	Percentage
Cleveland	0.524	Seattle	0.636	Florida	0.550
Minnesota	0.500	Texas	0.579	Montreal	0.526
Kansas City	0.474	Oakland	0.500	New York	0.381
Milwaukee	0.444	Annaheim	0.500	Philadelphia	0.350
Chicago	0.333	Houston	0.667	Colorado	0.737
Baltimore	0.684	Pittsburgh	0.500	San Francisco	0.737
Boston	0.524	St. Louis	0.350	Los Angeles	0.579
Toronto	0.474	Cincinnati	0.286	San Diego	0.474
New York	0.455	Chicago	0.150		
Detroit	0.391	Atlanta	0.750		

a. Clear L1 in your calculator and enter the winning percentages.
b. Can you guess if there are any outliers in this data set?
c. Run the program OUTLIER and record the results.
d. What is the mean winning percentage? Would you always get this value? Why?
e. Are there any outliers in this data set?
f. Trace the box plot and look at the extreme values.

6. Find a data set or make up your own and test it using the OUTLIER program. Give two reasons why outliers may appear in a data set.

9.2 MEASURING CAREFULLY

Now let's look at another set of data, which was obtained as follows. The statistics class of Section 9.1 was given some scientific training on how to measure accurately. They practiced reading the scale on the ruler to the nearest millimeter. They placed the stack of books on a sturdy table and put a heavy weight on it to make sure that excess space between and within books was eliminated. They made sure that the stack of books was straight and that the ruler was held snugly against the books, and so on. Such development of techniques to ensure measurement accuracy is essential to good technology and science. In a serious technological setting the measurement protocols developed will be far more involved and sophisticated than in our simple statistics classroom example, but the basic approach will be the same—namely, a thorough study and standardization of all the steps going into the measurement process. This, it should be emphasized, is merely good science, and has nothing specifically to do with the field of statistics.

Table 9.2

Stem	Leaf
22	
23	4, 5, 6, 7, 7, 8, 9, 9, 9
24	0, 1, 1, 1, 2, 2
25	

Key: "23 4" stands for 23.4 centimeters.

To resume our example of the statistics class: after they were instructed on how to measure the height of the stack of books more accurately, the students measured the stack of books and obtained the following results (in centimeters):

23.6	23.7	24.2	23.9	23.9
24.1	24.2	23.9	24.1	23.4
23.7	24.0	23.8	23.5	24.1

The data are summarized in the stem-and-leaf plot of Table 9.2 (recall from Section 1.2 that we could split each of these stems into two stems to get a better idea of the shape of the data, if we wished). Notice how this set of data compares with the previous set:

	First set	Second set
Median	23.8	23.9
Mean	23.7	23.9
Range	2.4	0.8
Mean deviation	0.41	0.21
Interquartile range	0.5	0.4
Variance	0.35	0.06
Standard deviation	0.59	0.24

We see that the second set of data has considerably less variation in it than the first set. The standard deviation of the second set (0.24) is less than one-half of the standard deviation of the first set (0.59). This is quite an improvement! The extra scientific care taken in measuring is reflected in the smaller standard deviation for the second set of data. In general, careful measurements can be obtained by using high-precision instruments, such as calipers or micrometers, and by carefully training the technicians or scientists making the measurements.

In the next three sections we will carefully study what statistics can contribute to the goal of accurate measurement. No matter how good our scientific measurement techniques are, good statistical practice can improve upon whatever level of measurement accuracy is inherent in the physical measurement process. This potential for improvement becomes important when the measurement error standard deviation is large relative to the level of measurement accuracy we seek. For example, if measuring to the nearest inch suffices in the context of the particular measurement problem being considered and the standard deviation is 0.2 inch, then we can ignore statistical issues and simply make a single measurement and be confident of achieving the level of accuracy we require. However, if we wish to be accurate to the nearest 0.1 inch and the standard deviation is 0.4, then we are compelled to make multiple measurements and use statistics to turn these multiple measurements into an accurate estimate of the true length we wish to accurately measure. The next sections address this latter case, when statistics becomes important.

SECTION 9.2 EXERCISES

1. Recall the Key Problem (estimating the acceleration of gravity). Ten students measured the length of a pendulum and obtained the results in the table below.

Student	Measurement obtained (centimeters)
1	175.2
2	179.0
3	170.0
4	165.1
5	171.5
6	175.8
7	172.0
8	175.3
9	165.5
10	175.8

Find the mean, range, and mean deviation of the measurements. If a good estimate of the length of a pendulum were required, what value would you use?

2. In a statistics class, pairs of students measured the length (in inches) of a chalkboard. Their results are shown in the following table.

Team	Measurement
Steve–Scott	223
Heather–John	241
Mark–Art	239
Janine–Wendy	249
J. B.–Art	223
Karen–Kristin	224
Mike–Jim	238

What is the mean of the measurements? How much variation is in the measurements?

3. The table below shows two sets of 10 measurements made to find the diameter of a human hair (in micrometers) being examined in a crime laboratory. The first set was made by using a usual method, involving calipers. The second set (on the same strand of hair) was made with a new method involving laser beams. Find the mean and standard deviation of each set of measurements. Which set indicates less variability in measuring?

Method A					*Method B*				
85	79	68	62	70	70	77	75	74	76
76	74	70	75	74	76	72	76	75	73

4. A laboratory is trying to decide which of two scales to purchase. Both are claimed to be high-precision instruments. Ten people weigh a mineral sample with scale A, and then with scale B, using great care with each scale. The results are shown in the table below.

Scale A					*Scale B*				
74	77	79	77	75	80	77	75	78	68
75	74	77	77	79	77	74	69	74	78

Which scale appears to have less error? Explain.

5. The value of gravity has been calculated by a laboratory in Ottawa, Canada, using two methods. See Figure E1.

You read the graph by adding the mid-values of the intervals as third- and fourth-place decimals to 980.6100. For example, in the upper graph, one measurement of g was 980.6105 cm per second per second (see the left-most bar).

Which of the methods, A or B, do you think gives measurements with less variability? Why?

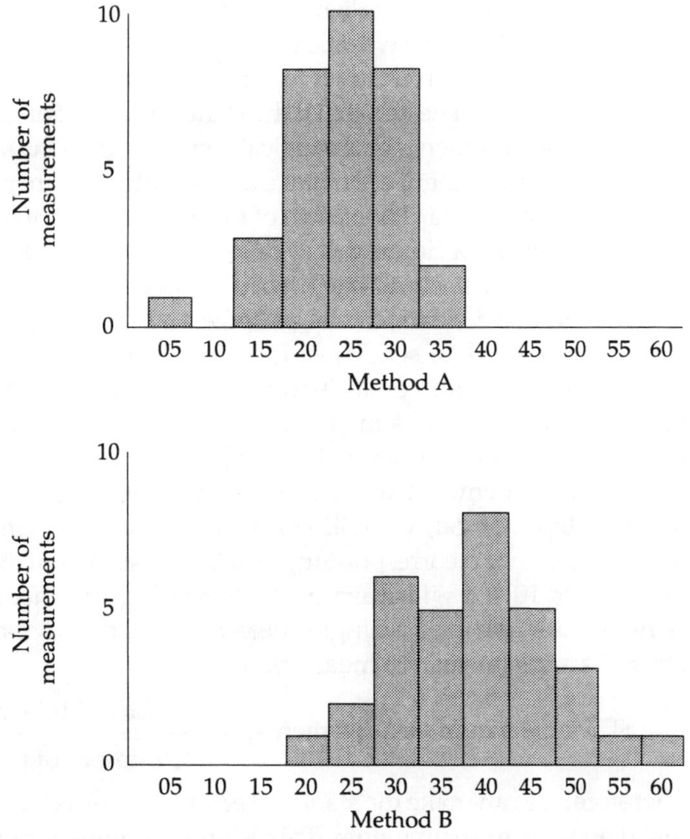

Figure E1 **Two histograms of measurements of gravity (mid-values of intervals are third- and fourth-place decimals to be added to 980.6100 cm/sec^2).**

9.3 STANDARD ERROR

Let's go back to our problem of measuring the stack of textbooks. We have talked about error of measurement. We have recognized that all measurements contain error.

Careful scientific training and accurate instrumentation can make a great difference, as discussed above. What can we do statistically to further reduce the error of measurement, after we have been as careful scientifically as we can be? There is a statistical way to improve accuracy. The idea we use is one we used in Chapter 4 on expected value, where we averaged observations via the five-step method to estimate an expected value. That is, rather than using only one measurement, we use the average (sample mean, sometimes referred to as just the *mean*) of a set of measurements. We intuitively expect that the sample mean of several measurements should with high probability be closer to the correct answer than what we would obtain if we relied on a single measurement.

In Chapter 10, when we study the central limit theorem and the problem of the estimation of an unknown population mean, we will learn why this intuition is correct. The key is that the standard deviation of the sample mean of a set of measurements is dramatically smaller than the standard deviation of a single measurement if we make a reasonable number of measurements. The theoretical standard deviation of the mean of a set of data is so important that it is given a special name. It is called the **standard error of the mean** or, more briefly, the **standard error** (often abbreviated SE). Remember that the standard deviation tells us the size of the typical variation of a population or a data set. Thus, individual measurements have a typical amount of variation as quantified by their standard deviation. By contrast, the typical variation of a mean of a set of measurements, as quantified by the standard error, can be much less. Hence we clearly favor using a sample mean of measurements rather than using a single measurement. Below, via the five-step method, we will study an example of standard error that is much less than the corresponding single-measurement standard deviation. In Chapter 10 we will learn a result relating the theoretical population standard deviation of a single measurement to the theoretical standard error of a sample mean of measurements:

$$\text{SE (of a mean of measurements)} = \frac{\text{SD (of a single measurement)}}{\sqrt{\text{number of measurements}}}$$

Therefore, we can make the standard error as small as we wish, provided we make enough measurements. This is an enormously powerful and useful statistical result.

To illustrate, we consider the Key Problem for this chapter, measuring the acceleration due to gravity, g. Finding a value for g depends on having values for L, the length of the pendulum, and T, the time the pendulum takes to make one complete swing back and forth. In order to obtain a more accurate measure for T, we make 50 measurements of the time it takes the pendulum to make a swing, using good scientific measurement techniques. These 50 measurements add to 132.2 seconds. Then the average, or mean, of these 50 observations is

$$T = \frac{132.2}{50} = 2.6 \text{ seconds}$$

(The value has been rounded to 2.6 seconds, since we have the original measurement of 132.2 only to the nearest 10th of a second.) As discussed above, this estimate of 2.6 seconds based on 50 measurements will have a theoretical standard error of

$$\text{SE (estimate of } T) = \frac{1}{\sqrt{50}}\text{SD (single measurement of } T) = 0.14 \times \text{SD}$$

which is a dramatic reduction in the variability of measuring T from the variability observed in using a single measurement. We return to this application in Section 9.6.

We now return to the class project of measuring the height of the stack of books. Because of what we have just learned about the advantages of replication of measurements to improve accuracy, the class measured the stack of books on 25 occasions and on each occasion found the mean of the set of measurements. They did so after learning the improved techniques of measuring discussed in Section 9.2. These *means* of the class's observations are recorded in Table 9.3, which shows 25 means of sets of 30 measurements each. These means also vary, since they, too, have error. But they vary much less than the original measurements given in Table 9.2. In particular, the range of $24.2 - 23.4 = 0.8$ in Table 9.2 has been reduced to $24.0 - 23.8 = 0.2$ in Table 9.3.

We have just learned that the standard error of the mean is an excellent indication of the accuracy we can achieve by using a mean of measurements instead of just one measurement. The standard error of the mean based on these 25 measurement means (of 15 measurements each) is about 0.06 (as

Table 9.3

23.9	24.0	23.9	23.9	23.9
24.0	23.9	24.0	23.8	23.9
23.9	23.9	23.9	24.0	23.8
24.0	23.8	23.9	23.9	23.9
23.9	23.8	23.9	23.9	24.0

computed from Table 9.3). That is, we have computed the standard deviation of the 25 means. It is interesting to compare this computed standard error of 0.06 with what is suggested by the above formula. If we substitute the estimated single-measurement standard deviation of 0.24 from Section 9.2, we can use the formula to get an estimated standard error:

$$\text{Estimated SE} = \frac{0.24}{\sqrt{25}} = 0.048$$

They are indeed relatively close. This topic will be taken up in Chapter 10.

The idea of the standard error of a statistic, such as that of the mean, is a very powerful one because it enables us to make some statement about how precise we believe our measuring statistic to be. For example, if we take an opinion poll, we will want to know the precision of our estimate of what proportion of people have a certain opinion. The standard error of estimate tells us this. It is often reported in the media. For example, it might be reported that 54% favor something with an error of ±3%. This reported 3% is the standard error!

In our examples in this section, instead of relying on the Chapter 10 formula for the standard error, we have computed the standard error of the mean by repeatedly observing a sample mean of measurements, as is consistent with the five-step method. In the case of an opinion poll we would have to repeat the poll many times and see how the results of the polls varied from one another in order to obtain the standard error. In the example of the height of the stack of books, we made 25 sets of 15 measurements each—375 measurements, which is a lot of measurements—and found the standard deviation of these 25 means, that is, the sample standard error. The purpose of repeating the basic sampling experiment 25 times would be for the class to discover for themselves empirically, by computing the resulting standard error, the great reduction in measurement error that occurs when we use a mean of measurements (in this case, a mean of 15 measurements). In an actual measurement problem we would take only *one* sample, such as a single set of measurements of the height of the stack of books or one opinion poll (for example, in actual practice the Gallup pollsters take one poll using anywhere from 500 to 3000 sampled people). In Chapter 10 we will learn that we can not only calculate the mean of a set of measurements but also, as was briefly introduced above, use our sample to help obtain a good estimate of the standard error of that mean. That is, besides giving us the estimate itself, information from the sample helps tell us how precise our estimate is. This is a major triumph of statistics.

For now, however, we will in the five-step tradition calculate standard errors using many samples, often simulated, so that we can become clear about the meaning and behavior of the very important standard error. Students in statistics courses that are not data-based often think that computing an average of a set of observations eliminates all variation. Our experience of collecting many samples and computing the standard error of the means of the samples certainly shows us the incorrectness of this impression, in that the variation of a mean, although often greatly reduced from that of a single observation, is *never* eliminated. The standard error is never zero, no matter how many measurements are averaged!

Statistics and Chemistry

About 100 years ago a supposed "error" of measurement led to the discovery of the rare gas argon. Chemists had long been able to remove the oxygen, carbon dioxide, and moisture from air. The remaining gas, it was believed, was solely nitrogen.

In 1890 the British scientist Lord Rayleigh did a study in which he compared nitrogen obtained from the atmosphere with nitrogen produced by heating the chemical ammonium nitrate. He filled a bulb of carefully determined capacity with each gas in turn under standard conditions (sea-level pressure and a temperature of 0° Celsius). The weight of the filled bulb minus its weight with the gas removed gave the weight of the gas itself. The study produced these results:

Weight of nitrogen from atmosphere	2.31001 grams
Weight of nitrogen from ammonium nitrate	2.29849 grams
Difference	0.01152 gram

Lord Rayleigh was faced with a problem: was the difference, which was very small, due to measurement error or to a real difference in the weights of the different kinds of nitrogen?

He applied the statistical technique of this chapter: he repeated the study several times and finally concluded that the difference was greater than could be explained away by measurement error and was therefore due to differences in the gases he was measuring. Further investigation led Lord Rayleigh to conclude that the nitrogen from the air contained other gases heavier than nitrogen. Under this assumption, he soon succeeded in isolating the rare gas argon. This gas proved to be one of a family of gases in the earth's atmosphere whose existence at the time was not even suspected. For his discovery Rayleigh was awarded a Nobel Prize in 1904. Statistical reasoning was a vital ingredient (in addition to some inspired scientific research, of course) in this crucial scientific triumph.

SECTION 9.3 EXERCISES

1. Here are the means of 10 sets of measurements of the weight of a newly minted penny (in decigrams). Each set contained 16 individual measurements using the same scale. Draw a stem-and-leaf plot of these data. What is the standard error of these data? What would you believe to be the true weight of the penny?

29.43	28.43	29.25	29.81	29.43
29.37	29.62	28.56	29.43	29.43

2. Exercise 1 was repeated with a different weighing scale. Find the standard error for this set of measurements.

29.43	29.68	28.37	29.50	30.18
30.62	29.12	29.06	30.12	29.62

Which scale do you believe to be more precise?

3. A laboratory is asked to find out how much radiation is produced by a sample of waste water from a nuclear power plant. A team of scientists takes 16 measurements and reports the mean of the measurements. This is repeated 25 times to give these 25 means:

50.93	49.50	49.25	48.43	51.68
48.18	47.25	48.87	50.37	49.93
50.00	50.81	48.06	50.25	48.62
49.81	49.68	51.18	51.81	49.93
50.06	49.00	48.62	49.25	47.87

Put these data into a stem-and-leaf plot and draw a graph of the results. What are the mean and estimated standard error of the data? What would you estimate the true amount of radiation to be?

4. The laboratory in Exercise 3 decided to take 64 measurements at a time and report the mean of those for the sample of waste water. Then this was repeated 25 times to give these results:

50.08	49.16	49.56	49.76	49.72
49.88	50.08	49.60	49.96	49.84
49.08	48.84	51.00	49.44	49.84
50.20	49.00	49.56	48.20	49.24
49.84	48.32	49.45	49.92	49.32

Put these data into a stem-and-leaf plot and draw a graph of the results. What are the mean and estimated standard error of the data? How do you interpret the differences between the sets of data in Exercises 3 and 4?

5. Ten students in the Netherlands were involved in an experiment to calculate the value of g, the acceleration of gravity, using a pendulum. One of the values needed to calculate g is T, the time for the pendulum to make a complete swing back and forth. In order to obtain precise measurements for T, the students measured the time required for the pendulum to swing back and forth 50 times. The table below lists the students' measurements of $50T$.

Student	$50T$ (seconds)
1	132.5
2	133.7
3	130.8
4	129.0
5	131.1
6	132.8
7	131.6
8	132.1
9	128.9
10	132.8

What is a precise estimate for T?

9.4 SYSTEMATIC ERROR (BIAS)

Sometimes when we are measuring, a systematic error will creep into our results. For example, suppose we are using a scale that overweighs. That is, the scale systematically shows a weight that is more (maybe several ounces more) than the actual weight of the object we are weighing. In that case our measurements may be very carefully obtained (and hence their standard deviation may be small), but they are **biased**—the true weight is consistently lower than what is given by the scale. Thus the average of many such measurements, in spite of having a very small standard error, will display (a possibly large) measurement error due to bias.

Consider the standard deviation of a set of numbers. Suppose a bias of 15 is added to each number to form a new set of numbers. Is there more spread or variation in this new set of data? Certainly not! Indeed, a little algebra can convince us that the new standard deviation of the biased measurements is the same as the old. First, note that

$$\text{average of } (x + 15)\text{'s} = (\text{average of } x\text{'s}) + 15$$

Then the key reason is seen to be that the deviation values used to compute the standard deviation are unchanged:

$$\begin{aligned} x - \text{average of } x\text{'s} &= (x + 15) - (\text{average of } x\text{'s}) + 15 \\ &= x + 15 - \text{average of } (x + 15)\text{'s} \end{aligned}$$

As we recall from Chapter 2, the standard deviation (the square root of the variance) is computed by adding up the squares of the deviation values, which, as the above equation shows, are unchanged by the addition of 15 to each original value. Thus the sum of the deviation values squared for the x's is the same as that for the the $(x + 15)$'s. It then follows that the variances and the standard deviations are also the same.

Suppose, for example, a newly minted penny is weighed on an unbiased scale (it does not measure consistently high or low) on 10 occasions, with these results (in decigrams):

Stem	Leaf
30	5, 6, 6
31	0, 0, 0, 2, 3, 4, 5

Key: "30 5" stands for 30.5 decigrams.

$$\text{Mean} = 31.0 \text{ decigrams}$$
$$\text{Standard deviation} = 0.4 \text{ decigram}$$

If we then use another scale with a constant error of +2 decigrams to weigh the penny, we would expect to get these results:

$$\text{Mean} = 33.0 \text{ decigrams}$$
$$\text{Standard deviation} = 0.4 \text{ decigram}$$

The mean is increased by the constant error of 2 decigrams, but, as expected, the standard deviation of measurements remains the same.

It thus becomes necessary to distinguish precision from accuracy, which is defined below in Section 9.5. The **precision** of a set of measurements is simply their standard deviation (either the theoretical standard deviation of the population of measurements or the estimated standard deviation of the set of measurements). If a sample mean of measurements is being used to estimate the true value of the quantity being measured, then the standard error of the mean becomes the precision of the mean. Note, however, that if a constant error, or bias, is present in a set of measurements, as with a defective scale, the precision of measurement will not be affected. The mean of the weights given by the scale will be increased by the constant error, but its standard error (that is, the precision of the mean) will remain the same.

How, then, can we check a measuring device for bias? We *cannot* do it statistically. The sample mean and the standard error of the mean are both totally silent on this vital issue. One way to assess bias is to compare the measurement it gives with that given by a device known to be free of bias. For example, the time tone given by the Canadian Broadcasting Corporation from the Dominion Observatory in Ottawa provides an accurate reference point for correct time for the entire country.

Another procedure for determining bias is to use more than one measuring device. A laboratory analyzing the amount of impurity in a drug might use two scales. If the two scales agree, that is a good indication (but not a proof) that both are free of bias. Both could, of course, have the same systematic error, but this is an unlikely situation and can usually be ignored.

SECTION 9.4 EXERCISES

1. A thermometer with a known constant error of +0.5° F (which measures too high on average) is used to give 10 measurements of the temperature of a greenhouse.

80.9	79.7	79	80.6	78.9
78.7	79.5	81.1	79.2	80.4

Find the mean and standard deviation of these temperatures. Then correct the mean for the constant error of the thermometer.

2. A second thermometer is used to measure the temperature of the greenhouse of Exercise 1 and gives these results on 10 occasions

(assume that the greenhouse temperature remains constant):

81.7	81.9	81	81.2	80.2
80.7	81.3	79.7	81.1	80.7

Which of the two thermometers appears to be more precise (have less error)? Does it appear that the second thermometer is biased? *Hint:* Recall that the thermometer in Exercise 1 is biased by 0.5° F.

3. A chemist is not sure whether a certain scale is biased, so she uses two scales when weighing the amount of precipitate found in a flask. Her measurements are in the following table. Which of the scales appears to have less error? What do you believe to be the true weight of the precipitate? Why?

Scale A				
49.2	48.0	51.5	52.5	51.9
50.6	46.5	52.0	49.0	49.5
50.4	50.5	49.5	55.4	48.4
Scale B				
54.1	50.5	49.6	48.0	50.2
52.3	48.6	51.2	49.8	48.5
49.6	52.0	51.1	51.2	48.0

4. Twenty measurements are made of the length of an insect (in millimeters):

41	38	37	38	40	39	39	40	42	39
38	37	42	40	38	42	35	40	39	38

What is the estimated true length of the insect? How could the precision of these measurements be improved?

9.5 ACCURACY

After all this discussion about measuring an object, it is reasonable to ask, What is the *real*, or *actual*, height of the stack of books? What is the actual weight of the penny? What is the true value of the acceleration due to gravity in Ottawa?

The actual length of an object can only be estimated. We can estimate more and more precisely by using a sample mean of measurements, but how do we know that bias has been acceptably reduced or eliminated?

Accuracy is defined to be the degree of closeness of a measurement to the true value. Accuracy is *not* the same as precision, which merely tells us that our measurements will all be quite close to each other (and that if we repeatedly constructed sample means of observation and used them as our measurement, they would be even closer to each other) but if bias exists the measurements may be unnacceptably far from the true value.

In understanding the various contributions to measurement error from a statistical viewpoint, the following basic formula is very useful:

$$\text{Measurement} = \text{true value} + \text{bias} + \text{random error (random noise)}$$

Here the "true value" and the "bias" are fixed numbers; the "random error" has a theoretical mean of 0, and its standard deviation is the standard deviation of the population of measurement errors, which, because the true value and the bias are not random, is the same as the standard deviation of the population of measurements.

Let's go back to the problem of measuring a stack of textbooks. When the class first measured the stack, there was a lot of variation in their measurements (see Section 9.1). The range of the lengths was 2.4 centimeters and the standard deviation was 0.59 centimeter. We had the feeling that the class was not very precise in measuring the height of the stack of books. So, as discussed in Section 9.2, they measured again, being much more careful. This time the range of measurements was only 0.8 centimeter and the standard deviation was reduced to 0.24 centimeter.

We decided that if we wanted to estimate the height of the stack of books even more accurately, we should use the mean, or average, of a set of measurements. We were more confident about using the mean as an estimate of the true height than we were in using any of the individual measurements because its standard error is smaller than the standard deviation of a single measurement.

When the class used more care in measuring, the standard deviation of the measurements became smaller. That is, the individual measurements had less variation and we concluded we had more confidence in their accuracy. But being careful and hence achieving high precision when measuring, on the one hand, and being accurate, on the other, are two somewhat different ideas because of possible bias, as we now see. We actually don't learn whether there is bias in the classroom measuring project. Consider the example of the lab technician who is carefully measuring the length of a pendulum. The required length should be in centimeters and could be obtained by means of a meter stick. But suppose the technician inadvertently (and carelessly) uses a yardstick (marked in inches) instead, believing it to be a meterstick (which is 39.37 inches long, whereas the yardstick is 36 inches long). The measurements may have a rather low standard deviation (indicating high precision), but they will indeed be biased, indicating very low accuracy! Moreover, using sample means will further improve precision but will not alter the bias that is preventing good accuracy.

So we see that there are two components of accurate measurements. First, they have a small standard deviation: they are precise. Second, they are relatively free from bias. Steps should be taken to see if there is bias in a measurement process and, if there is, to remove it. Note that low precision produces low accuracy because of lots of variability, but high precision guarantees high accuracy only if we can be confident of low bias. This is captured in the equation

$$\text{Measurement} = \text{true value} + \text{bias} + \text{random error}$$

And keep in mind that the "measurement" in the above equation may in fact be a sample mean of measurements and thus possess a standard error less than the standard deviation of the individual measurements.

SECTION 9.5 EXERCISES

1. A statistics class took 20 sets of 25 measurements of the length of a chalkboard to give a total of 20 means. The means of those sets (in inches) are as follows:

240.4	238.5	237.8	241.0	238.6
236.6	240.0	238.5	241.3	241.1
240.4	241.6	241.1	240.7	240.0
241.6	238.8	238.6	240.5	240.2

Construct a stem-and-leaf plot of these measurements. Estimate the standard error of the measurements by finding the standard deviation of the 20 mean measurements. Estimate the true length of the chalkboard from these data.

2. A safety inspector is examining the strength of sewing thread. She instructs her laboratory assistants to take 25 measurements of the strength of the thread and find the mean of those measurements. This is repeated 20 times, giving the following 20 means. What is the estimated true strength of the thread? What is the standard error of the measurements?

49.48	49.28	49.72	50.12	49.48
49.80	49.68	49.68	49.80	49.96
49.52	50.28	48.84	49.84	50.48
49.16	49.48	48.84	48.52	49.08

3. A cartographer (map maker) is measuring the area of a lake (in square miles), which he has drawn on a map. (This is done by using a device called a *planimeter*.) Since he is interested in the accuracy of his drawing, he is willing to take many measurements and to be very careful. So he takes 25 measurements and finds the mean. Then he repeats this for a total of 20 means as given below. What is the estimated true area using the planimeter? What is the standard error of his measurements?

39.08	38.52	39.64	39.80	39.00
39.28	38.40	39.40	39.36	39.76
39.16	39.00	39.68	39.00	38.60
40.04	38.60	38.60	38.92	39.44

4. Scientists estimating the acceleration of gravity on another planet take 36 measurements and average the results. Then they repeat this, taking additional groups of 36 measurements, until they have 16 means, as given below (in centimeters per second per second):

63.9444	64.4444	63.8333	63.7778
65.1667	64.7778	64.3889	65.1389
64.5833	63.2500	64.8056	64.3056
64.500	64.6111	64.9444	64.5278

What is the estimated value of gravity on that planet? What is the standard error?

5. Two classes measure the length of a field known to be exactly one kilometer long. They obtain these results:

Class I		*Class II*	
1.11	1.11	0.97	0.99
1.12	1.09	1.02	1.02
1.10	1.10	1.01	1.02
1.10	1.11	1.04	0.96

Which class has produced results with the smallest standard error? Which class seems to have more bias in its results? Assuming that bias is small, which set of data is more accurate?

Graphing Calculator Exercises

In the exercises below you will need the program MEASURE.

6. Suppose a scientist uses a certain measuring process that produces measurements that are normally distributed with a mean of 12 cm and a

standard deviation of 0.30 cm. If she were to take 30 such measurements, they would be expected to spread out three standard deviations above and below the mean (that is, between about 11.10 cm and 12.90 cm).

Individual measurements contain uncertainty (that is, error). But suppose she were to take, at random, five measurements and average them. How would the mean of these five measurements compare with the original measurements? For example, is this mean of five measurements expected to contain less error than any of the 30 original measurements? The program MEASURE can help us answer these questions.

a. Run MEASURE and enter 12 for the mean and 0.30 for the standard deviation. Record the sample mean and standard deviation and then press [ENTER].

(The program MEASURE now takes five random measurements from this specified distribution and finds their mean. This is repeated 25 times. The program now displays the mean and standard deviation of the 25 means.)

b. Record the mean of the means and the standard deviation of the means. How does the mean of the means compare with the mean of the original measurements? How does the standard deviation of the means compare with the standard deviation of the original measurements?
c. Press [ENTER] to look at the box plots provided. Compare the variation of the original measurements (plot 1–P1) with those produced by taking the average of five measurements (plot 2–P1).
d. How does taking the average of five measurements affect the precision of the measurement?

7. Suppose a properly calibrated measuring device produces measurements with a mean of 20 and a standard deviation of 0.15. But something has gone wrong, and you now have measurements with a mean of 21 and a standard deviation of 0.15. This would be like someone sneaking a weight onto a scale before you weigh your item. In this case, there is now a bias of +1. Use the program MEASURE to simulate this.

a. Run MEASURE and enter 21 for the mean and 0.15 for the standard deviation. Record the sample mean and standard deviation and then press [ENTER].
b. Record the mean of the means and the standard deviation of the means. How does the mean of the means compare with the mean of the original measurements? How does the standard deviation of the means compare with the standard deviation of the original measurements?
c. Press [ENTER] to look at the box plots provided. Compare the variation of the original measurements (plot 1–P1) with those produced by taking the average of five measurements (plot 2–P1).
d. Does taking five measurements and averaging them do anything to remove the bias?

e. Does taking five measurements and averaging them improve the precision of the measurements?
f. Does taking five measurements and averaging them improve the accuracy of the measurements?

9.6 AN APPLICATION OF MEASUREMENT: THE ACCELERATION DUE TO GRAVITY

We return to the pendulum method for estimating g, the acceleration due to gravity. As we learned, a pendulum can be used to measure the value of g. You may have noticed that short pendulums swing back and forth quickly and long pendulums swing back and forth slowly. The relationship between the length, L, of a pendulum and the time, T, it takes to go back and forth once (a cycle) is given in physics by the formula

$$T = 2\pi\sqrt{\frac{L}{g}}$$

Since the relationship between L and T involves g, we can solve this equation for g:

$$g = 4\pi^2\frac{L}{T^2}$$

We know the value of π, 3.14159 (approximately). So if we have accurate values for L and T, we can obtain a good estimate for g.

The data that follow were obtained by students at a university in the Netherlands. Ten students obtained measures of L and T for each of five pendulums (labeled A, B, C, D, E). The lengths are reported in Table 9.4.

Table 9.4 **Length *L* (in Centimeters)**

	Pendulum				
Student	**A**	**B**	**C**	**D**	**E**
1	175.2	151.5	126.4	101.7	77.0
2	179.0	150.0	125.0	100.0	75.0
3	170.0	149.3	124.8	100.4	76.4
4	165.1	149.8	125.0	100.0	75.0
5	171.5	150.0	124.9	100.0	75.0
6	175.8	—	125.0	99.7	75.0
7	172.0	150.0	125.0	100.0	75.0
8	175.3	149.9	125.0	100.0	75.0
9	165.5	150.0	125.0	100.0	75.0
10	175.8	150.7	125.8	100.8	74.6

Table 9.5 **Values of 50*T* (in Seconds)**

	Pendulum				
Student	**A**	**B**	**C**	**D**	**E**
1	132.5	123.4	112.8	101.2	88.2
2	133.7	122.3	111.3	99.8	85.8
3	130.8	133.5	113.1	100.5	87.6
4	129.0	133.8	112.2	100.0	86.8
5	131.1	122.6	111.9	100.1	86.8
6	132.8	—	112.0	100.1	86.8
7	131.6	122.8	112.0	100.2	86.8
8	132.1	122.2	111.8	100.0	86.5
9	128.9	122.7	113.0	100.3	86.9
10	132.8	123.0	112.3	100.7	86.7

Some remarks can be made about these measurements. For example, students 2 and 7 did not report any 10ths in their results. Are these accurate results, or did these students simply prefer to avoid decimal fractions? If the latter is true, these students should certainly have been trained to report their measurements to the nearest 10th. We will consider the students' length values in more detail below.

It is difficult to measure the time of just one swing of a pendulum, even if it moves slowly. So the students determined the time required for 50 complete swings back and forth, in seconds. Later, each value was divided by 50 to get T. For now, let's call the time required for 50 swings $50T$. The measured value of $50T$ for each student and for each pendulum is given in Table 9.5.

Now we can use our formula to calculate the value of g, based on each student's measurements for L and T. (Remember that Table 9.5 is for 50 times T!) For example, for pendulum A and student 1, we have

$$L = 175.2 \text{ centimeters}$$

$$T = \frac{132.5}{50} = 2.65 \text{ seconds}$$

$$\begin{aligned} g &= 4\pi^2 \frac{L}{T^2} \\ &= 4(3.14159)^2 \frac{(175.2)}{(2.65)^2} \\ &\approx (39.4783)\left(\frac{175.2}{7.0225}\right) \\ &\approx 984.9 \text{ centimeters per second per second} \end{aligned}$$

Table 9.6 Acceleration *g* Due to Gravity (in Centimeters per Second per Second)

	Pendulum				
Student	A	B	C	D	E
1	984.9	981.9	980.4	980.1	976.9
2	988.3	989.8	995.9	990.0	1005.5
3	980.7	981.9	980.2	981.1	982.6
4	979.2	980.4	980.0	987.0	982.5
5	984.8	984.9	984.5	985.0	982.5
6	983.8	—	983.5	982.0	982.5
7	980.2	981.7	983.5	983.0	982.5
8	991.5	990.7	987.0	987.0	989.3
9	983.1	983.3	983.5	983.0	980.2
10	983.8	983.1	984.5	981.1	979.5

Similarly, a value for g can be estimated for each of the other pendulums and for each of the other students—a total of 49 values of g, as given in Table 9.6. (There is one missing value—student 6 did not report results for pendulum B. Missing values are very common in science and a major problem for scientists and statisticians alike. What does a professional statistician do if a number needed from a table is missing? One approach is to provide a clever guess for the missing number based on a carefully reasoned extrapolation from other numbers in the table. For example, we could use the average of the other nine students' estimates of pendulum B to obtain a value for pendulum B for student 6.)

We now comment a little more about the values obtained by the students. To make the job easier, we will translate each of the values 980 points to the left (subtract 980 from each value in the table: a little like the standardization of data to produce a z score). The result is shown in Table 9.7. Consistent with the approach of this chapter, each student computed a mean of his or her five measurements, given in the right-hand column. The value of g is known by scientists in the Netherlands to be 981.3 centimeters per second per second. This compares with the combined student estimate of $980 + 4.163 \approx 984.2$ obtained by averaging the students' means to obtain their *grand mean*. This estimate is a tad high! But we notice that the values produced by students 2 and 8 are clearly outliers by any reasonable criterion (check whether 14.08 and 9.10 are outliers by computing $3 \times \text{SD}$, using the SD for the other eight student means). If we average the other eight means, we get $980 + 2.31 = 982.31$. This is much closer to the true g of 981.3. As this suggests, it is good practice to eliminate measurement outliers before using them to estimate the true value of a measured quantity.

Table 9.7

Student	Pendulum A	B	C	D	E	Average
1	4.9	1.9	0.4	0.1	−3.1	0.84
2	8.3	9.8	15.9	10.9	15.5	14.08
3	0.7	1.9	0.2	1.1	2.6	1.30
4	−0.8	0.4	0.0	7.0	2.5	1.82
5	4.8	4.9	4.5	5.0	2.5	4.34
6	3.8	—	3.5	2.0	2.5	2.95
7	0.2	1.7	3.5	3.0	2.5	2.18
8	11.5	10.7	7.0	7.0	9.3	9.10
9	3.1	3.3	3.5	3.0	0.2	2.62
10	3.8	3.1	4.5	1.1	−0.5	2.40
Mean	4.03	4.19	4.30	4.02	4.40	4.163*

*The value 4.163 is a mean of means, often called the *grand mean.*

SECTION 9.6 EXERCISES

1. Table 9.7 presents student estimates of the value of gravity, g. To obtain the actual estimate, add 980, as detailed in the discussion of Table 9.7. Redo the table by removing student 2 from the analysis. What effect will the elimination of student 2 have on the analysis? Is the accuracy of the student estimate improved?

2. Consider pendulum A in Table 9.7. Suppose we decide that any observation more than 2 SD away from 4.03 is to be considered an outlier. Are any of the 10 observations to be considered outliers?

CHAPTER REVIEW EXERCISES

1. A weight that is known to weigh exactly 10 pounds is weighed independently 50 times on the same scale. The measurements are listed below. Find the mean and standard deviation of these measurements. Draw a histogram of the data. Do the data appear to be normally distributed?

10.26	10.03	10.13	10.02	10.03	10.11	10.11	10.09	10.11	9.96
10.06	10.00	10.08	10.02	9.91	10.06	10.00	9.98	9.93	10.10
10.04	10.16	10.08	10.10	9.95	10.00	10.04	9.97	10.12	9.89
10.10	10.13	9.96	10.12	10.23	10.01	9.98	9.97	10.08	9.84
9.92	10.15	9.90	9.84	9.92	9.99	9.98	9.92	9.94	9.97

2. Suppose the scale in Exercise 1 is known to weigh items 0.1 pound heavy (a 10-pound item would read 10.1 pounds on the scale). Correct the estimated mean weight in Exercise 1 to reflect this information.

3. Twenty independent measurements, listed below, were taken to determine the length of a board (in inches). Find an estimate of the true value of the length.

12.04	11.92	12.15	11.89	11.95
11.94	11.97	12.03	11.86	11.76
12.02	11.95	11.91	12.03	11.85
12.00	12.10	12.16	11.82	11.97

4. Suppose the last three values in Exercise 3 were found to be measured very inaccurately. Find a new estimate of the true value of the length of the board by dropping those three values. Does your estimate change?

5. In the process of making computer chips, the chips must be heated in an oven at 1000° F. A factory has two ovens for this purpose. Twenty independent measurements of the temperature are taken within each oven. The measurements are listed below. Which oven has a more precise temperature? Which oven has a mean temperature closest to 1000°?

Oven 1

1006.46	1009.14	1020.33	994.83	1010.09
985.18	1017.71	1048.59	1040.32	990.01
965.25	1015.65	1000.23	1002.20	965.15
1004.90	1039.27	970.77	995.59	969.81

Oven 2

1041.91	1057.92	1037.62	1051.69	1060.53
1046.45	1045.51	1063.86	1043.75	1049.81
1044.74	1056.30	1058.64	1067.49	1050.35
1037.82	1059.99	1027.94	1048.98	1053.83

6. Twenty-five individual measurements were taken of the mass in grams of a chemical precipitate, and the mean measurement was recorded. The process was repeated 25 times, resulting in the following sample means:

18.05	17.94	18.00	17.99	18.03
18.07	17.96	18.07	17.94	17.96
18.01	17.96	18.01	17.97	17.96
18.01	18.12	17.95	18.00	18.06
17.94	17.98	17.97	18.04	18.00

What is a precise estimate of the weight of the chemical precipitate? What is the error of this estimate?

7. One measurement of the length of a bone, 34.36 cm, was made by a student. Unknown to her, the professor already knows the exact measurement of the bone, 35 cm. Is the student's measurement biased, or is it just randomly off? What other experiments could you ask the student to perform to better answer this question?

8. To determine the capacity of the fuel tank in a car, the fuel tank was filled to capacity 25 times and the mean amount of gallons of gasoline in the tank was recorded. The process was repeated 20 times, resulting in the following sample means:

23.3	23.4	23.7	23.6	23.5	23.2	23.1	23.7	23.6	23.4
23.6	23.5	23.6	23.8	23.7	23.2	23.7	23.0	23.3	23.5

What is a good estimate of the fuel capacity of the tank? What is the standard error of this estimate?

9. Suppose the manufacturer of the fuel tank in Exercise 8 listed the capacity of the tank as 23.5 gallons. Is your method of determining the fuel tank capacity biased? Are there any other explanations for the difference in the two estimates of fuel tank capacity?

9

COMPUTER EXERCISES

These two exercises are based on experiments for measuring the speed of light. Both data sets are found in "Do Robust Estimators Work with Real Data?" by Stephen M. Stigler (*Annals of Statistics,* vol. 5 [1997], pp. 1055–1078) and can be accessed through the Data Set and Story Library (http://lib.stat.cmu.edu/DASL/).

1. A. A. MICHELSON

The first data set contains 100 estimates of the speed of light made by A. A. Michelson in 1879. The actual observations, in kilometers per second, are 299,000 plus the values in the data set. The "true" speed of light in a vacuum is 299,792.5 km/sec, but Stigler explains that by using some corrections of Michelson's, the "true" value relative to the given data is 734.5. (These experiments and the ones in the next exercise were based on a method developed by the French physicist Foucault. They involve bouncing light off distant mirrors.)

Start the Data program (See the Computer Exercises of Chapter 1), and load the data set MICHELSN.DAT. Choose Histograms from the Data menu, and select the Statistics box. Create the histogram for velocity with 100 bins.

a. What is the average of the measurements? The median?

b. Does it appear that there is systematic bias? If so, in which direction?

c. Is there random error? If so, what is an estimate of the size of the random errors?

d. Are there any outliers? If so, how many? Are they generally too large or too small?

2. SIMON NEWCOMB

The second data set has 66 measurements of the speed of light made by Simon Newcomb in 1892. The estimates were based on measuring the time it took for light to travel from Fort Myer on the Potomac River to a mirror at the base of the Washington Monument and back again. Newcomb's actual measurements, in millionths of a second, are 24.8 plus 0.001 times the values in the data set. Stigler says that the true value, in the units used in the data set, is 33.02. The data set LIGHT.DAT contains the estimated speed of light for the 66 trials. Load this data set into the Data program, go to Histograms, check the Statistics box, set the number of bins to 100, and choose the variable "Speed_of_Light" (it is the only variable).

a. What is the average of the measurements? The median?

b. Does it appear there is systematic bias? If so, in which direction?

c. Is there random error? If so, what is an estimate of the size of the random errors?

d. There are two very low outliers. Remove them by following these steps:

   i. Close the Histogram screen, and select "Input/Modify Data" from the Data menu.

   ii. On the resulting screen, select the "Use current data set: Speed of Light" circle, and press OK.

   iii. In the first column of the Data area, find the observation number for one of the negative values.

   iv. Go to the Options menu, and choose "Delete Observation." Type in the observation number you found in part (iii). (Do not type in the value of the observation, but the number of it. For example, number 1 is 28, number 2 is 22, and so on.)

   v. Repeat parts (iii) and (iv), where in (iii) you find the other negative number.

   vi. Click on OK, and then select No when asked whether to save the data.

Now go back to the histogram screen, and answer these questions:

e. What is the average of the measurements? The median? How do these compare with the values in part (a)?

f. Does it appear there is systematic bias? If so, in which direction?

g. Is there random error? If so, what is an estimate of the size of the random error? Compare this to the value in part (c).

Part 3

ESTIMATION AND HYPOTHESIS TESTING

10 *E*STIMATION 476

11 *H*YPOTHESIS TESTING 522

12 *C*ORRELATION AND REGRESSION 566

13 *M*ORE PROBABILITY WITH APPLICATIONS 616

10

ESTIMATION

Everybody believes in the central limit theorem, the experimenters because they think it is a mathematical theorem and the mathematicians because they think it is an experimental fact.

G. Lippmann (1845–1921)

OBJECTIVES

After studying this chapter, you will understand the following:

- *The distinction between a population parameter and a statistic used to estimate it*
- *Two types of estimation: point estimation and confidence interval estimation*
- *The standard error as the typical estimation error size*
- *The central limit theorem for the probability law of a sample average*
- *How to obtain confidence intervals by means of the central limit theorem*
- *Point estimates for the population variance and the population median*

10.1 PARAMETERS AND STATISTICS 478

10.2 POINT ESTIMATES AND CONFIDENCE INTERVALS 481

10.3 STANDARD ERRORS OF ESTIMATES 488

10.4 CENTRAL LIMIT THEOREM 491

10.5 CONFIDENCE INTERVALS FOR THE POPULATION MEAN 496

10.6 CONFIDENCE INTERVALS FOR THE POPULATION PROPORTION 501

10.7 CONFIDENCE INTERVALS FOR THE DIFFERENCE BETWEEN TWO POPULATION MEANS 503

10.8 POINT ESTIMATES FOR THE POPULATION VARIANCE 505

10.9 POINT ESTIMATES FOR THE POPULATION MEDIAN 509

CHAPTER REVIEW EXERCISES 511

COMPUTER EXERCISES 513

Key Problem

Here is some information summarized from an October 12, 1986, *St. Louis Post Dispatch* article entitled "Majority Opposes Stadium, Poll Shows." It was found that 45% of St. Louis County residents opposed a new indoor football stadium (28% supported it, and 27% were undecided). The poll interviewed 301 registered voters in the county. The article states, "The county poll is accurate within plus or minus 5.7 percentage points [giving a (39.3%, 50.7%) interval] at a confidence level of 95%. That means if the survey were taken 100 times, the results for the [random] group of respondents would each vary no more than 5.7 percent in either direction [from the true population percentage opposing the stadium about] 95% of these times." (The material in the square brackets was added to help make the quote more understandable.)

The *St. Louis Post Dispatch* is to be commended for attempting to communicate an important statistical concept: confidence intervals for a population parameter. The Key Problem for this chapter is to understand how to construct confidence intervals such as the 95% interval (39.3%, 50.7%) for those members of the population of county registered voters who oppose the stadium construction and to understand clearly why we call such an interval a 95% confidence interval.

10.1 PARAMETERS AND STATISTICS

In Chapter 6 we discussed samples and populations. A *sample* is a collection of people or things that is chosen from a larger collection of people or things and is believed to represent that larger collection. The larger collection is called a *population*. In good statistical practice, the sample is chosen randomly from the population. Thus, every member of the population has an equal chance of appearing in the sample. Therefore, if the sample size is at all large, we can expect that the sample will be well representative of the population.

Example 10.1

Twelve hundred people selected from across the country are interviewed in order to determine the popularity of the U.S. president. The sample in this case is the 1200 people, and the population is all the people in the country.

Example 10.2

Ten 1-milliliter containers are filled from a swimming pool and tested for the density of *Escherichia coli* bacteria. The sample is the 10 containers of water taken from the swimming pool, and the population is all the water in the swimming pool.

In each of these examples, we are trying to determine an important characteristic of a population. This characteristic, which is useful in describing the population, is called a *parameter*. In Example 10.1 the parameter is the proportion of people in the country who approve of the president. Likewise, in Example 10.2 the parameter is the density of bacteria in the entire swimming pool.

As you might expect, we usually cannot measure a parameter directly. For example, we cannot actually count the number of people in the country who approve of the president. We would not be able to contact every person in the country, and even if we could, the process would be extremely expensive. Likewise, it is not practical or even possible to measure the density of bacteria in an entire swimming pool. Even when the population in question is not as large as the population of the United States or the amount of water in a swimming pool, the cost to survey an entire population is usually too expensive. Therefore it is not practical to survey the population to determine the value of a population parameter of interest.

If we cannot determine the parameter directly from the population, then, how can we get an accurate estimate of the parameter? We take a random sample of the population and calculate an estimate of the parameter from the sample instead. The estimate of the parameter that comes from a sample is called a *statistic*. Unlike the population parameter, a statistic can be calculated directly from the information in the sample. In Example 10.1 the statistic would be the proportion of people in the sample of 1200 who approve of the president. In Example 10.2 the statistic would be the average density of bacteria in the 10 small containers of water taken from the swimming pool. These statistics, or *estimates* as we will often call them, will estimate the values of the population parameters.

To reiterate: Statistics—that is, estimates—come from samples and can be calculated directly. Parameters come from populations and are usually unknown and hence are estimated by statistics. In order to distinguish between parameters and statistics, different symbols are used. We use our ordinary English alphabet when referring to statistics computed from sample values. For example,

Sample mean $= \overline{X}$

Sample variance $= S^2$

Sample standard deviation $= S$

Sample correlation coefficient $= r$

We use the Greek alphabet when referring to population values (parameters). For example,

Population mean $= \mu$, pronounced "mu"

Population variance $= \sigma^2$, pronounced "sigma squared"

Population standard deviation $= \sigma$, pronounced "sigma"

Population correlation coefficient $= \rho$, pronounced "rho"

Because of our experience with the five-step method, we understand that if the sample size (the number of trials in step 4) is sufficiently large, then the statistic of interest (our estimate) from step 2 will, with high probability, be close to the unknown population parameter of interest. Thus, we have already been estimating population parameters in Chapters 4 and 5! In particular, we have found that $\overline{X}$ is an estimate of μ, S^2 is an estimate of σ^2, S is an estimate of σ, and r is an estimate of ρ.

SECTION 10.1 EXERCISES

1. For each of the following, identify the sample, statistic population, and parameter involved.
   a. The high temperature at Los Angeles International Airport on August 27 was 73° F.
   b. In a poll of 64 freshmen at a large college, it was found that 65% were in favor of lowering the legal drinking age from 21 to 18.

c. Based on one-cubic-centimeter samples at 10 sites, the pollen count for Urbana, Illinois, was 228 grams per cubic meter of air on August 25.

d. A container full of corn taken from a truck unloading at a grain elevator was found to have 35% moisture.

2. A poll of 20 people at a shopping center was taken to rate the taste of a new cola drink. The results (1 = terrible to 5 = terrific) were as follows:

4	3	3	4	4	3	2	3	5	3
4	2	4	3	3	2	3	4	4	3

What is the mean preference score of the 20 people in the poll? What is the range of preference scores in the poll? What parameter could the mean preference score of the poll estimate?

3. A random sample of 25 college students is surveyed about their political views. Five of them have liberal views. What parameter could this statistic be used to estimate?

4. In the newspaper account below, what samples and population are involved? What factors could affect how well the samples represent the population?

> By measuring annual tree rings in groves scattered across the Western states, scientists here are seeking to define the likelihood of prolonged and extreme drought for each of nine river basins in that region.
>
> At least 10 trees are sampled at each site. A hollow drill extracts a pencil-sized core of wood without seriously harming the tree. The core is then studied under magnification to record climate-induced variations in tree-ring width back to the time when the tree began growing. Trees on well-drained slopes are preferred since they respond quickly to a drought.
>
> An objective of the project, funded in part by $286,000 from the National Science Foundation, is to determine whether there is evidence for what some hydrologists call a "Noah effect." This would be a weather extreme beyond known precedent, like the 40 days of rain that, according to Genesis, flooded the world in the days of Noah.
>
> A variant, known as the "Joseph effect," would be a condition—such as a drought—far more prolonged than any on record. The reference, again from Genesis, is to the seven years of famine predicted by Joseph.
>
> It is assumed that, if radical departures from normal behavior have taken place in the past, they may occur again and the frequency of occurrence may be estimated.
>
> It is known from other studies that major changes can occur. For example, during the "climatic optimum" 6,000 years ago, when regional temperatures rose 2 to 4 degrees Fahrenheit, the western grasslands moved east through Iowa and Illinois into Indiana and Ohio. (W. Sullivan, "Ancient Tree Rings Tell a Tale of Past and Future Droughts," *New York Times*, September 2, 1980)

5. Refer to Table E1. Identify the sample and population. Give a statistic from the report. Give the parameter it is estimating.

6. Give examples from newspaper or magazine article in which sample data (statistics) are used (or assumed to be used) to estimate population data (parameters).

Table E1 **Freshman Characteristics Responses of 188,000 Students Who Entered College in the Fall of 1982**

	Age		
16 or younger	0.1%	21	1.3%
17	2.5%	22	0.6%
18	74.2%	23–25	0.4%
19	18.9%	26–29	0.1%
20	1.8%	30 or older	0.0%
	Residence		
With parents or relatives	19.4%	Fraternity/sorority house	5.0%
Other private residence	25.8%	Other campus housing	3.8%
College dormitory	43.8%	Other	2.2%
	Reasons for choosing college		
Suggestions of relatives or friends	13.8%	Financial assistance	16.7%
Teacher's advice	4.0%	Special programs	25.5%
Advice of guidance counselor	7.7%	Low tuition	20.6%
Advice of former student	14.9%	Other	18.1%
Good academic reputation	53.5%		
	Religious preference		
Jewish	3.0%	Roman Catholic	38.9%
Protestant	33.7%	Other	17.2%

Political views	**Men**	**Women**
Liberal or left	20.6%	20.9%
Middle of the road	55.9%	63.7%
Conservative	22.0%	14.9%
Far right	1.5%	0.6%
Attitudes toward government		
Government isn't protecting consumer	64.4%	73.1%
Government isn't controlling pollution	74.9%	82.2%
Government should discourage the use of energy	74.5%	80.8%
Military spending should be increased	47.9%	29.9%

Source of Information: A. W. Astin, *The American Freshman: National Norms for Fall, 1982* (Los Angeles: Cooperative Institutional Research Program of the American Council of Education and the University of California at Los Angeles, 1982).

10.2 POINT ESTIMATES AND CONFIDENCE INTERVALS

Point Estimates

Suppose we want to estimate the proportion of Americans who approve of the president. In the previous section we took a random sample of size 1200 from the population and used the proportion of the people in the *sample* who approved of the president to estimate the proportion of the people in the *entire country* who approve of the president. The proportion of people in the sample who approve of the president is an example of what is called a **point estimate.** It is called a point estimate because it is a single number that estimates a population parameter, here the proportion of the people in the country who approve of the president.

Example 10.3 Suppose that out of 1200 people in the sample, 615 approve of the president. The point estimate of the proportion of the people in the country who approve of the president is $615/1200 = 0.5125$.

Example 10.4 Suppose that in the example of the 10 containers of sampled water, the sample average of the 10 observed *E. coli* bacteria densities is $1500/\text{cm}^2$. The point estimate of the density of *E. coli* bacteria in the swimming pool is $1500/\text{cm}^2$.

Example 10.5 It was desired to find the population mean and variance of the ages of all the students enrolled at a particular community college. It was impossible to survey all the students, so a random sample of 100 students was taken and the mean and the variance of the ages of the students in the sample were calculated. The sample mean was 20.25 and the sample variance was 5.05. The sample mean age, $\overline{X} = 20.25$, is a point estimate of the population mean age, μ. The sample variance of the ages of the students, $S^2 = 5.05$, is a point estimate of the population variance, σ^2.

In summary, a point estimate for a population parameter of interest is a statistic computed from the sample. It is believed to be an effective estimate of the unknown population parameter. Often, what formula to use for the statistic is obvious. For example, for the population parameter μ, which is the mean over a (presumably large) population, it seems that the sample mean will be a good estimate. However, in more complex situations the choice of an estimate of a population parameter is not always clear. Statisticians may find that either they have no idea how to use the sample data to estimate the population parameter of interest, or they may in fact have several equally plausible competing estimates to select from. A relatively simple example of the latter case is the estimation of σ^2. Let n denote the sample size. Many statisticians would use $S^2 = \text{sum}(X_i - \overline{X})^2/n$, while many others would use $\text{sum}(X_i - \overline{X})^2/(n-1)$. Here X_i denotes the *i*th observation of the random sample. A solid argument can be made for either estimate. In fact, a significant amount of statistical theory is devoted to finding the "best" estimate of a population parameter.

Let's consider an example in which the formula for a good estimate is not obvious.

Example 10.6

Suppose the waiting time for a train that goes to the parking area from a particular terminal at Chicago's O'Hare International Airport obeys a continuous uniform distribution on the time interval $[0, T]$, where T is an unknown population parameter. Thus a train always arrives within T minutes, but it is equally likely to arrive at any moment during this waiting time. Suppose we interview five randomly selected passengers and find that their waiting times were (in minutes) 1.4, 4.7, 4.2, 5.1, and 2.1. We want to estimate the parameter T. Clearly the sample mean is a poor choice for an estimate of T. What formula should we use? A widely used technique for producing an estimate when we have none in mind, called *maximum likelihood estimation*, leads us to an estimate that is equal to the maximum of the five observed times, which is 5.1 minutes. This choice of the maximum of the observed values as our point estimate is not obvious. Even though it might not be the "best" choice for an estimate of T, it seems clearly better than $\overline{X}$. We will not consider maximum likelihood estimation in this book.

A point estimate reports one single value that we estimate to be the true value of the population parameter. However, point estimates have the important limitation of not informing us how much the estimate is likely to be in error. Whenever we estimate a population parameter, we lack total accuracy. Thus our estimate will almost always be different from the actual population parameter. So the estimate will almost always have some amount of error associated with it. By simply reporting the point estimate of a parameter, we have essentially ignored the important issue of the likely error size associated with the estimate.

Example 10.7

Reconsider Example 10.5. A second random sample, this time of 200 students, was taken. Suppose, to keep our explanation simple, the point estimate of the population mean is the same in both cases, namely 20.25. The variance of the second sample mean would be much lower than the variance of the first mean, because the second sample is larger. So even though the two point estimates of the population mean are the same, the second one is surely more accurate, because it is based on more information—namely, twice as many observations.

Confidence Intervals

To improve on point estimates, statisticians usually report an interval of values that they believe the parameter is highly likely to lie in. Usually the point estimate is the middle point of the interval and the endpoints of the interval communicate the size of the error associated with the estimate (recall that point estimates ignore this error) and how "confident" we are that the population parameter is in the interval. The intervals are called **confidence intervals.** Typical confidence levels used in practice for confidence intervals are 90%, 95%, or 99%, with 95% occurring most frequently in applications. In Examples 10.1 and 10.3, if we are given a 95% confidence interval for the proportion of the population that approves of the president, which can be shown to be the interval (0.48, 0.54), we say we are 95% confident that the population proportion is contained in the confidence interval. In Examples 10.2 and 10.4, given a 90% confidence interval for the number of bacteria in the swimming pool, (1490, 1510) say, we say we are 90% confident that the density of bacteria in the swimming pool is contained in the confidence interval. (We will learn how to calculate confidence intervals for different population parameters later in the chapter.)

What does it really mean to state a 95% confidence interval for the unknown population proportion approving of the president? Although 95% sounds impressive, we cannot be satisfied unless we understand what it means. Theoretically, it means that the probability is 0.95 that such a confidence interval, which will be random because the sample it is formed from is random, will contain (surround) the unknown proportion in the population approving of the president.

Our experimental view of probability based on the five-step method will help us more clearly and deeply understand what the probability of 0.95 means practically. Just as we do simulations over and over in the five-step method, imagine that a statistician does the sampling experiment of Example 10.1 over and over—1000 times, say—and each time computes a 95% confidence interval from the 1200 sampled people. Now we can find the experimental probability given by the proportion of the 1000 confidence intervals that actually covers the true fraction of the population favoring the president. The 95% confidence means that this experimental probability of a confidence interval including the true value will be close to 0.95. (Below, by way of example, we will produce 100 such confidence intervals by use of the five-step method and calculate the experimental probability that a confidence interval covers the true population value.)

Of course, in a real application there will only be one random sample and hence only one such confidence interval, such as the (0.48, 0.54) interval of Examples 10.1 and 10.3. But the statistician obtaining this one sample knows, because of the experimental probability viewpoint, that this confidence interval is very likely to be correct in the sense that it contains the true value (since about 95% of such confidence intervals would cover the true population proportion). In the case of the (0.48, 0.54) interval of Examples 10.1 and 10.3, we know it is very likely that the true proportion of people favoring the president lies between 0.48 and 0.54.

Now in light of this insight into how to interpret the confidence interval percentage, let's return to the Key Problem. The *St. Louis Post Dispatch* explained the concept of its reported confidence interval this way: "[A 95% confidence interval] means if the survey were taken 100 times, the results for the [random] group of respondents would each vary no more than 5.7 percent in either direction from the true population percentage opposing the stadium [about] 95% of these times." The *Post Dispatch* quote is a bit roundabout and hence forces us to go through a slightly tricky piece of logic (draw yourself a picture if needed). If the interval, which extends 5.7% in either direction from its midpoint, indeed varies no more than 5.7% in either direction from the true population mean about 95% of the time, then about 95% of these intervals must contain the true population percentage as desired. The *Post Dispatch* could have more simply and more directly told its readers that such an interval can be expected to contain the true population parameter about 95% of the time.

In other words, just as explained above in the presidential popularity example, the *Post Dispatch* is pointing out that if you take 100 random samples from the same population and calculate the confidence interval for the population proportion for each sample, about 95% of the confidence intervals will include the true population proportion. That is, you will be correct in your claim that the unknown population proportion is in the interval computed using the sample for about 95 of the 100 samples.

Using our five-step method, we now simulate 100 confidence intervals and determine how many of them contain the true population parameter. Suppose that for the Key Problem the true proportion of people in the county who are opposed to the stadium is 50% (remember that the parameter value is never known to the statistician). In that case the probability that a person in the sample will be opposed to the stadium is the same as the probability of heads seen in flipping a fair coin: 0.5. Our goal is to obtain 100 samples of 301 people and calculate the confidence interval for each of these 100 samples (we will learn to compute such confidence intervals in Section 10.6 below). Each sampling is the same as flipping a coin 301 times, recording the number of heads, and calculating the confidence interval for the proportion of heads in the sample. We then repeat the process 100 times. The 100 confidence intervals obtained from this process are represented in Figure 10.1.

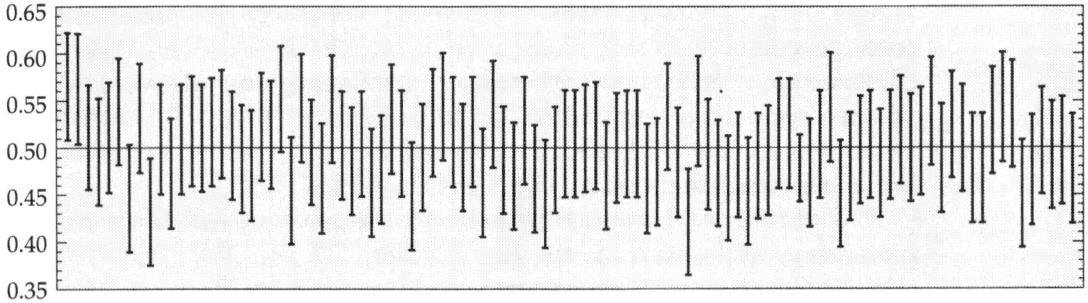

Figure 10.1 **One hundred simulated 95% confidence intervals for the Key Problem assuming a 50/50 population split.**

The line across the middle of the graph represents the true population proportion, 0.50. The 100 confidence intervals are the vertical lines on the graph. If the confidence interval covers the 0.50 line, then we say that the true population proportion is contained in the interval. Likewise, if the confidence interval does not cover the 0.50 line, then we say that the true population proportion is not contained in the interval. For the graph of Figure 10.1, we see that 96 out of the 100 confidence intervals cover the 0.50 line. So for the 100 confidence intervals, 96% of them (about 95%, as expected) contain the true population parameter, 0.50.

Thus, this example of the five-step method clearly illustrates how we are to correctly interpret a 95% confidence interval. As already discussed above, it is a sort of statisticians' success rate or batting average. If a statistician constructs 100 95% confidence intervals during a year's work, then, as our five-step simulation confirms, we can expect about 95% of them to be "hits": cases in which the population parameter is contained in the interval. Similarly we would expect about 5% of them to be "outs," or misses: cases in which the population parameter is not contained in the interval. Compared with baseball, in which a batting average of 0.300 is considered great, a 0.950 confidence interval coverage rate for statisticians is what is usually required. Batting 0.300 is never guaranteed in baseball, but in fact a statistician can guarantee a 95% confidence interval, as we shall see.

Confidence intervals have two basic characteristics that we need to understand. First, given the same set of data, a 95% confidence interval is wider than a 90% confidence interval, and a 99% confidence interval is wider than a 95% confidence interval. Thus, the higher the confidence level we require, the wider the interval we are forced to accept! Of course, a very wide interval is of little use to the scientist who has sought statistical advice. Thus there is no "free lunch" in specifying a 99% confidence instead of a

95% confidence, because the price paid is a wider interval. Here are the 90%, 95%, and 99% confidence intervals for the proportion of people in the country who approve of the president in Example 10.1:

90%: (0.49, 0.53)

95%: (0.48, 0.54)

99%: (0.47, 0.55)

As you can see from this example, the higher the level of confidence, the wider the confidence interval needs to be in order to contain the population proportion with the specified confidence.

The second characteristic of confidence intervals is that, given the same confidence level, a wider confidence interval is associated with data that have less information. Suppose in the situation of Example 10.1 a sample of 2400 people was taken from the population and the number of people in the sample that approved of the president was 1230. The point estimate of the population proportion would be the same: 1230/2400 = 0.5125. However, a 95% confidence interval for the population proportion based on this sample can be shown to be (0.492, 0.542). This is smaller than the 95% confidence interval for the original sample, which can be shown to be (0.484, 0.541). Clearly the original sample of 1200 is less informative. Indeed, the result is a wider interval (of length 0.057 compared with 0.04) for the same confidence level of 95%.

In summary, point estimates provide only a single number to estimate the value of a population parameter. By contrast, confidence intervals give a range of values that we reasonably expect will contain the population parameter. Again, a 95% confidence interval means that if we were to take a large number of samples (like 100 or 1000) of equal size from the same population and calculate a confidence interval for the population parameter, about 95% of the confidence intervals would contain the true value of the population parameter.

SECTION 10.2 EXERCISES

1. Suppose we want to estimate the proportion of a city's residents who drive to work. What is a good choice for the point estimate of this proportion?
2. Suppose, instead, we want to estimate the "average" number of miles people living in a city drive to work. What are two possible choices for the point estimate of this "average"?
3. Explain the meaning of a 99% confidence level.
4. Suppose you want a confidence interval for a population proportion. You want to be as accurate as possible, so you select a 100% confidence level. What would your confidence interval have to be?

5. Which confidence interval, when based on the same data, is wider: an 80% or an 85% confidence interval?
6. Suppose the confidence intervals in Exercise 5 are not calculated from the same data. Which is wider, the 80% confidence interval or the 85% confidence interval? Explain your answer.
7. Suppose two 95% confidence intervals are calculated using two different samples from the same population. Which sample data contains more error: the one associated with the wider or with the more narrow confidence interval?

10.3 STANDARD ERRORS OF ESTIMATES

Obviously it is very important to be able to calculate a quantity that measures the error associated with an estimate. One important way to measure this error is called the standard error (SE) of an estimate. We were introduced to this concept in Section 9.3, when we learned about the standard error of the sample mean. Recall that the size of the standard error is interpreted as the size of a typical error—that is, the typical distance between the sample estimate and the population parameter.

There are two distinct ways of finding the standard error of a particular sample estimate, such as the standard error of the sample proportion of Example 10.1 or the standard error of the sample mean of Example 10.2 or Example 10.5. The first method is to repeatedly simulate the drawing of a random sample from the population, following our five-step approach, and each time compute the estimate from the sample. Then you compute the standard deviation of this collection of estimates. This is an estimate of the needed standard error. The second method is to use a theoretical formula for the standard error of the estimate, as in fact we already did for the standard error of the sample mean of a set of measurements in Section 9.3. Both methods can be useful in practice. The first method can always be used, while the second, and more convenient, method can only be used when a theoretical formula for the standard error is available, as is often not the case.

We illustrate the *first method* by means of an actual statistical problem. As we saw in Chapter 9, we can find the standard error of a sample mean by computing the standard deviations of a large number of sample means that were sampled from the specified population by means of simulation using the five-step method. For example, if we need to determine the standard error of a sample mean, we can generate 100 random samples from the population and calculate the corresponding 100 sample means. The standard deviation of these 100 sample means is then an estimate of the standard error of the sample mean.

Let's consider the problem of estimating the mean of a population of heights of adult males. Because taking repeated random samples from a real population requires an enormous amount of work, we will, as we have often done before, *simulate* via the five-step method the taking of random samples from a population using a box model or a random number table. But this requires that we know the general shape of the population, especially that we know its spread. To illustrate, suppose we know the population has a normal distribution with a population standard deviation of 3 inches. Suppose our goal is to find the standard error of $\overline{X}$, the sample mean of a random sample of size 25. We simulate 10 such random samples of size 25 using a computer program to simulate normally distributed observations with a population standard deviation σ of 3 inches and an arbitrarily chosen population mean μ of 68 inches. (Here we are free to pick any μ we wish for this simulation because our goal is to find the standard error of $\overline{X}$, and this standard error does not depend on where the population is centered.) We chose $\mu = 68$ for convenience, even though this is surely not the true population mean, which is unknown and whose estimate is our ultimate goal. Then we obtain the following values, in inches, for the 10 sample means by sampling 10 sets of 25 random normal statistics, each having a mean $\mu = 68$ and a standard deviation $\sigma = 3$:

68.3 68.9 67.4 68.0 67.3 68.2 67.6 68.4 68.8 67.1

The standard deviation of these numbers is then our estimated standard error of the mean, which is 0.6.

It is perhaps interesting to note that Table B.5 in Appendix B presents simulated standard normal values—that is, normal variables with a mean of 0 and a standard deviation of 1. Let Z be such a value. Then we can generate a $\mu = 68$, $\sigma = 3$ normal value X from our population by the formula $X = \sigma Z + \mu$. That is, Table B.5 allows us to generate random normal values for any population μ and σ. Here we have used the fact from Section 8.6 that the equation $Z = (X - \mu)/\sigma$ converts a normal X into a standardized normal Z. Then we merely solve for X.

While this version of the first method works well as an instructional tool, there is usually a large problem with it in practice. It requires that the general shape of the probability law of the population be known, in particular that its spread, which is usually specified by the population standard deviation, be known. Usually this is not the case. One approach that gets around this problem is known as "bootstrapping." We will study bootstrapping in some detail in Chapter 13; here we only introduce it. When we do not know the general shape (probability law) of the population, the bootstrap method uses the shape of the sample to provide us with an estimate of the unknown shape of the population. In other words, the shape of the sample data supplies an estimated population distribution that, because the sample is a random sample, should be shaped approximately like the unknown population distribution. We then sample with replacement from this population created from the sample in order to repeatedly simulate obtaining the estimate whose standard error we seek.

Let's revisit the adult male height example and see how a bootstrap approach would work. We are given the 25 observations composing the random sample. The bootstrap approach simply declares these 25 observations to be our population to be repeatedly sampled from, 25 times *with replacement*. In this manner we can calculate as many $\overline{X}$'s as we need. Suppose we obtain 10 $\overline{X}$'s in this manner, and to simplify our explanation, suppose they are the same 10 $\overline{X}$'s as we obtained when sampling from a normal population. Then we compute the standard deviation of these 10 numbers and thus obtain the same value for our estimate for the standard error of the mean. The *key difference* is that here we have sampled not from a computer-generated normal population (such as that of Table B.5) but rather from the actual numbers forming the real sample. Note that we did not need to assume that we were sampling from a normal population, nor indeed did we assume *anything* about the shape or spread of the population distribution. Thus this method of bootstrapping to find the standard error is always an option in actual applied problems.

Can we succeed in estimating the standard error without doing any simulations? Yes: In many cases proven theories enable us to calculate standard errors of estimates. (The details of some of these theories will be discussed in Chapter 13.) For example, as we stated in Chapter 9, the standard error of a sample mean is $\sigma/\sqrt{n}$, where σ is the population standard deviation and n is the sample size. The standard error of a sample proportion is $\sqrt{p(1-p)/n}$, where p is the population proportion and n is the sample size. For example, in the Key Problem the standard error is estimated to be $\sqrt{(0.45)(0.55)/301} = 0.029$. Here we have substituted the sample proportion $\hat{p} = 0.45$ in the formula for the population proportion p ($\hat{p}$ is read "p hat"). We know that since $\hat{p}$ is a reasonable estimate of p, $\hat{p}$ will be close to p and therefore this estimate of the true standard error should be close to the true standard error.

This is an example of our *second method* of determining standard errors for estimates, and clearly it is the method we would like to use when it is applicable. It usually does require us to estimate parameters such as σ and p in the above standard error formulas, but reasonable estimates of these parameters exist and are easy to calculate. In summary, the second method is to use a theoretical formula for the standard error of our estimate, possibly requiring us to estimate a population parameter by using the random sample. Note that this method requires no repeated actual or simulated random samples from the population; the single obtained random sample suffices. However, we often need the standard error of an estimate for which there does not exist an easy method 2 formula like the two illustrated above. (The sample median, used to estimate the population median, is one such estimate.) In such situations we have to use the first method. Methods 1 and 2 are both important in statistical practice for determining the standard errors of estimates.

SECTION 10.3 EXERCISES

1. When is it necessary to use the bootstrap method to find the standard error of an estimate?
2. Complete the following sentence: The bootstrap method replaces having to know the probability law of the ________________ with the discrete uniform probability law of the ________________.
3. Explain how to find an estimate of the standard error of a sample mean using the bootstrap method.
4. Suppose the population standard deviation of a sample of 50 observations is 5. What is the standard error of the sample mean of the 50 observations from this population?

10.4 CENTRAL LIMIT THEOREM

A theorem is a statement of some mathematical property or principle that can be proven by logical arguments. The central limit theorem is a very important theorem in statistics. We will not prove it in this book, but we will look at examples that demonstrate what it tells us and its great usefulness.

Central Limit Theorem

Suppose we are taking random samples from a population with mean μ and standard deviation σ. Then as the sample size n increases, the shape of the distribution of the sample means $\overline{X}$—that is, the shape of the relative frequency histogram of these sample means—becomes more and more like the shape of a normal density whose mean is μ and whose standard deviation is $\sigma/\sqrt{n}$. Equivalently, $(\overline{X}-\mu)/(\sigma/\sqrt{n})$ becomes more and more like the standard normal density tabulated in Appendix E.

The theorem tells us that if the sample size n is large enough—the general rule of thumb being $n \geq 20$—then we can assume that the theoretical probability distribution of $\overline{X}$ is very close to the probability distribution of an appropriately centered and spread out normal distribution. Thus, referring to the quote at the beginning of the chapter, we are treating the central limit theorem as an "experimental fact." It is in reality both an experimental fact and a mathematical theorem. (See Figures 8.12 and 8.13 for some graphs of normal densities.) Thus we can use the normal distribution to calculate the probabilities involving $\overline{X}$, using the table in Appendix E to obtain numerical answers. Moreover, this normal distribution of $\overline{X}$ has mean μ and standard deviation $\sigma/\sqrt{n}$, where μ and σ are the population mean and standard deviation and n is the sample size used to obtain $\overline{X}$.

Example 10.8

Using a box model of size 200, we constructed a nonnormal population having a population mean of 63.5 and a population standard deviation of 12.0. We then randomly drew 100 samples of size 4, 16, and 36 *with replacement* from this population, and we calculated the sample mean for each of the 100 samples at each sample size. The resulting sample means are presented in the stem-and-leaf plots in Tables 10.1, 10.2, and 10.3. From these stem-and-leaf plots we constructed the relative frequency histograms of Figures 10.2, 10.3, and 10.4.

Let's make some observations about the results of our sampling as we look at these stem-and-leaf plots and the associated relative frequency histograms.

1. The mean of the 100 sample means is in all three cases close to the population mean of 63.5. For example, the mean of the 100 sample means of size 4 is 63.3, as can be computed from Table 10.1. The mean of the sample means is actually the mean of $4(100) = 400$ observations. As we know from the five-step method as it applies to estimating expected values (Chapter 4), a mean based on 400 observations should indeed be very close to the population mean of 63.5. Therefore even without knowing the central limit theorem, we would expect the mean of the sample means to be close to the population mean.
2. As the sample size increases, the distribution of the sample means looks more and more like a normal density. In other words, the shape of the curve suggested by the relative frequency polygon corresponding to the histogram looks more and more like the bell shape of a normal density as the sample size for each $\overline{X}$ increases (although the shape remains only roughly similar).
3. As the sample size increases, the variation among the sample means decreases. More specifically, as the sample size increases, the standard error (the standard deviation of the sample means) becomes smaller, as recorded in Tables 10.1, 10.2, and 10.3.
4. The calculated value of the standard error found by computing the standard deviation of the 100 sample means for each sample size is close to the theoretical value of $\sigma/\sqrt{n}$. The calculated standard error from Table 10.1 is 5.7, compared with $\sigma/\sqrt{n} = 6.0$, while the calculated standard error from Table 10.3 is 2.0, compared with $\sigma/\sqrt{n} = 2.0$—an exact (and lucky) match.

The central limit theorem is a very powerful tool in statistical analysis. In the present context we will use the central limit theorem to calculate confidence intervals for population means, for proportions, and for the difference between two means.

Table 10.1 **Sample Means of 100 Samples of Size 4 (Population Values: $\mu = 63.5$, $\sigma = 12.0$)**

Stem	Leaf	Frequency
49		0
50		0
51	8	1
52	7	1
53	1,5,6	3
54		0
55	2,3,3,5,6,7	6
56	4,6,6,6,9	5
57	1,1,1,3,8,9	6
58	1,3,5,8,8,9	6
59	1,1,3,4,5,5,6,8	8
60	3	1
61	3	1
62	2,2,5,5,8	5
63	2,2,4,5,5,5,7,9	8
64	0,0,0,2,4,7,9,9	8
65	1,1,3,3,4,7	6
66	1,3,5,5,6,6,9,9	8
67	0,2,7,9	4
68	0,2,3,5	4
69	3,5,5,6,6,8	6
70	8	1
71	1,2,4,5,6,6,8,9	8
72		0
73	2,9	2
74	1	1
75	3	1
76		0
77		0
	Total	100

Sample values:
$n = 4$
mean = 63.3
standard error = 5.7

Table 10.2 **Sample Means of 100 Samples of Size 16 (Population Values: μ = 63.5, σ = 12.0)**

Stem	Leaf	Frequency
54		0
55		0
56	9	1
57	7	1
58		0
59	0,2,2,5,6,6,7,9	8
60	0,2,5,7,7,7	6
61	1,2,5,8,9,9,9	7
62	0,1,3,3,4,5,5,6,7,8,8	11
63	2,2,2,2,3,3,3,3,3,4,4,4,4,4,5,6,6,7,7,8,8,9,9,9,9	25
64	2,2,3,4,4,5,6,6,7,8	10
65	0,0,0,3,3,3,5,7	8
66	0,1,2,4,4,5	6
67	0,4,4,5,6,7	6
68	2,3,3,4,5,6,6,9	8
69	5	1
70	2	1
71		0
72	1	1
73		0
74		0
	Total	100

Sample values:
n = 16
mean = 63.9
standard error = 2.9

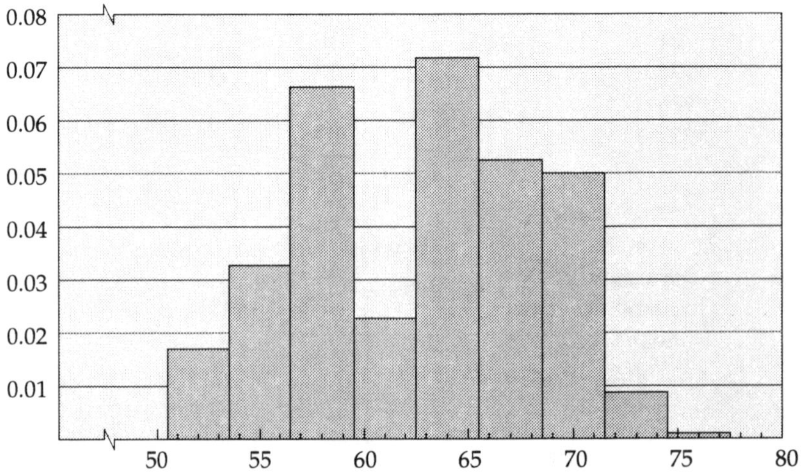

Figure 10.2 **Relative frequency histogram of sample means of 100 samples of size 4.**

Table 10.3 **Sample Means of 100 Samples of Size 36 (Population Values: μ = 63.5, σ = 12.0)**

Stem	Leaf	Frequency
56		0
57		0
58	1,6,9	3
59	7,8,9	3
60	0,0,3,6,7,7,7,8,8,9	10
61	0,1,2,4,5,8,9	7
62	0,0,2,2,2,2,4,5,6,6,7,7,7,8,8,8	16
63	1,1,1,1,2,3,3,4,4,5,5,7,8,9	14
64	0,0,1,1,2,2,2,3,3,4,4,4,4,4,4,5,5,6,6,6,7,8,8	23
65	0,0,0,2,3,3,4,4,4,4,5,5,8,9	14
66	0,3,4,5,5,8,8	7
67	1,2,4	3
68		0
69		0
	Total	100

Sample values:
n = 36
mean = 63.4
standard error = 2.0

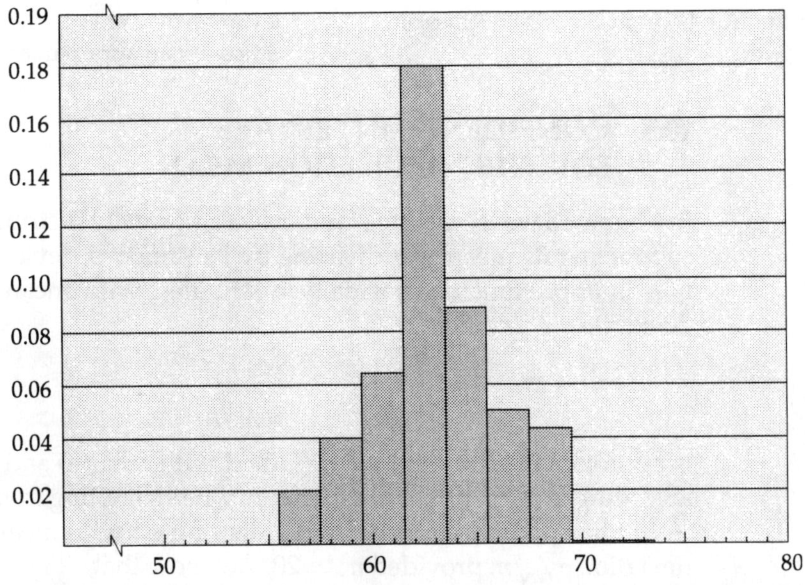

Figure 10.3 **Relative frequency histogram of sample means of 100 samples of size 16.**

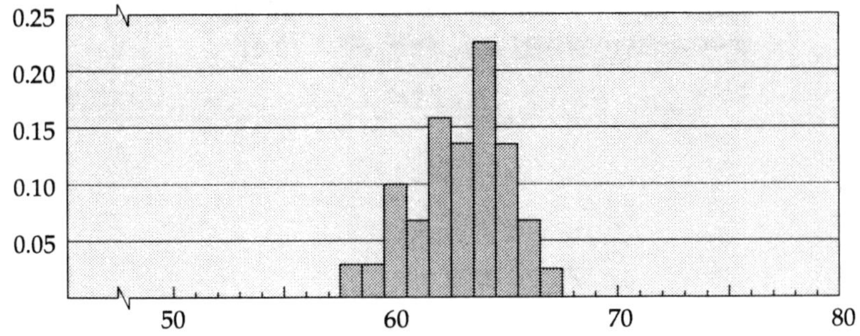

Figure 10.4 **Relative frequency histogram of sample means of 100 samples of size 36.**

SECTION 10.4 EXERCISES

1. Suppose the mean weight of all male Americans is 175 pounds with a standard deviation of 15 pounds. What are the estimated theoretical mean and standard deviation of a sample mean from this population based on 50 observations? How is the sample mean distributed?
2. Suppose the sample mean in Exercise 1 is based on 10 observations instead of 50. Can you still apply the central limit theorem to answer the questions in Exercise 1? Explain your answer.
3. Which relative frequency histogram will look more like a normal distribution—the histogram of sample means based on 20 observations per mean or the histogram based on 40 observations per mean? (Assume that the samples are taken from the same population).
4. What are the mean and standard deviation of a sample mean of 40 observations based on a population with mean 0 and standard deviation 1? What is the distribution of the sample mean approximately like?

10.5 CONFIDENCE INTERVALS FOR THE POPULATION MEAN

In Chapter 8 we learned to standardize a normally distributed statistic X by subtracting its mean and dividing by its standard deviation. In particular, if $\overline{X}$ is approximately normally distributed with mean μ and standard deviation $\sigma/\sqrt{n}$, then

$$\frac{\overline{X} - \mu}{\sigma/\sqrt{n}}$$

is approximately normally distributed with mean 0 and variance 1. Since according to the central limit theorem, when sampling from any population, $\overline{X}$ will be approximately normally distributed with mean μ and standard deviation $\sigma/\sqrt{n}$, provided $n \geq 20$, we know that

$$\frac{\overline{X} - \mu}{\sigma/\sqrt{n}}$$

will be approximately normally distributed with a mean of 0 and a variance of 1—that is, it will have a standard normal distribution.

Now we will use this knowledge to find a 95% confidence interval for a population mean μ. First we find two points that produce a standard normal distribution area of 0.95. We will choose the points so that the area is centered, leaving an area of 0.025 in each tail. See Figure 10.5 to visualize this area. The two points that produce this area are -1.96 and 1.96, as the table in Appendix E tells us. More precisely, we find from the table that $p(z < 1.96) = 0.975$ and $p(z < -1.96) = 0.025$.

Therefore, if $\overline{X}$ is approximately normal with mean μ and standard error $\sigma/\sqrt{n}$, as the central limit theorem tells us it must be when $n \geq 20$, then

$$p\left(-1.96 < \frac{\overline{X} - \mu}{\sigma/\sqrt{n}} < 1.96\right) \approx 0.95$$

Using algebra, we can solve for μ through the following steps. We start with the event that we know has an approximate probability of 0.95 when $n \geq 20$:

$$-1.96 < \frac{\overline{X} - \mu}{\sigma/\sqrt{n}} < 1.96$$

This is the same event (expressed differently) as

$$-1.96\frac{\sigma}{\sqrt{n}} < \overline{X} - \mu < 1.96\frac{\sigma}{\sqrt{n}}$$

This fact we obtained by multiplying both sides of the preceding left-hand inequality by the positive number $\sigma/\sqrt{n}$, and likewise for the right-hand

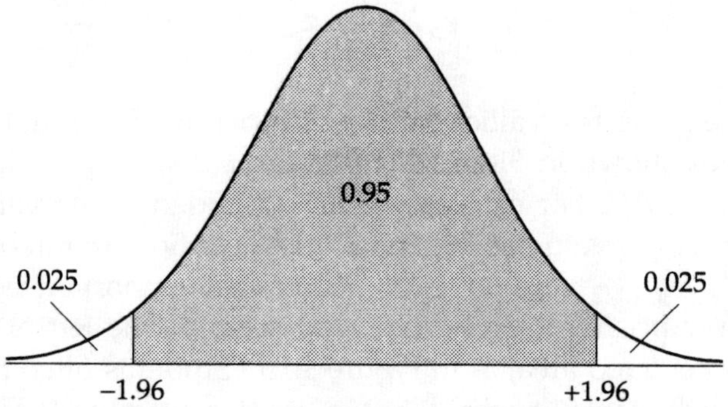

Figure 10.5 **An area of 0.95 under a standard normal curve.**

inequality. Next we subtract $\bar{X}$ from both sides of the two inequalities, yielding

$$-\bar{X} - 1.96\frac{\sigma}{\sqrt{n}} < -\mu < -\bar{X} + 1.96\frac{\sigma}{\sqrt{n}}$$

Now note that $-7 < -5$ while $7 > 5$. That is, if we multiply both sides of an inequality by a *negative* number (here -1), we change its direction. We therefore multiply our above inequalities by -1, yielding

$$\bar{X} + 1.96\frac{\sigma}{\sqrt{n}} > \mu > \bar{X} - 1.96\frac{\sigma}{\sqrt{n}}$$

Rearranging a bit, we obtain

$$\bar{X} - 1.96\frac{\sigma}{\sqrt{n}} < \mu < \bar{X} + 1.96\frac{\sigma}{\sqrt{n}}$$

Now this is the *same event* as we began with, so it must have the *same probability* of approximately 0.95. That is,

$$p\left(\bar{X} - 1.96\frac{\sigma}{\sqrt{n}} < \mu < \bar{X} + 1.96\frac{\sigma}{\sqrt{n}}\right) \approx 0.95$$

as desired! Thus we have convinced ourselves that we have a 95% confidence interval for the population mean μ.

Intervals for other levels of confidence can be similarly computed. In general terms, for whatever α the statistician wants to use (not just $\alpha = 0.05$), a $100(1 - \alpha)\%$ confidence interval is obtained by the following equation:

$$\bar{X} - z_{\alpha/2}\frac{\sigma}{\sqrt{n}} < \mu < \bar{X} + z_{\alpha/2}\frac{\sigma}{\sqrt{n}}$$

where $z_{\alpha/2}$ is the value from the Appendix E normal distribution table and is as shown in Figure 10.6; that is $p(z < z_{\alpha/2}) = 1 - \alpha/2$ and $p(z < -z_{\alpha/2}) = \alpha/2$. For example, a 95% confidence interval corresponds to a $z_{\alpha/2} = z_{0.05/2}$ value of 1.96, and a 90% confidence interval corresponds to a $z_{\alpha/2} = z_{0.10/2}$ value of 1.645. (Study this important point carefully. The table in Appendix E gives the *entire* area to the left of $z_{\alpha/2}$, whereas we want a centered area, as in Figure 10.5.) Students often find this confusing because the tabled probability we use (for example, 0.975) differs from the confidence probability of 0.95. The point to keep firmly in mind is that if there is to be 0.95 in the middle, then each tail area must be 0.025.

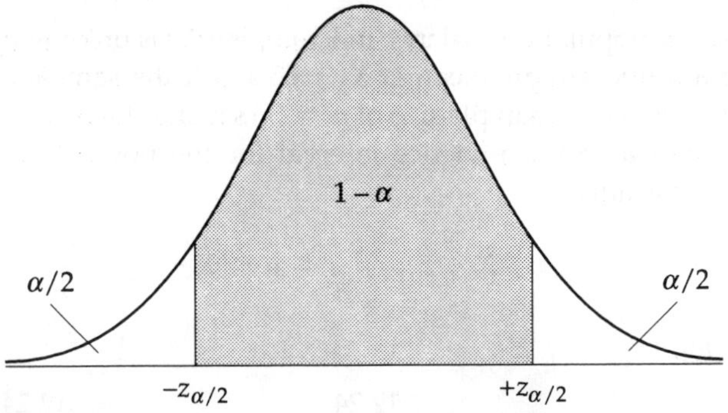

Figure 10.6 **An area of 1 − α under a standard normal curve.**

Example 10.9

Suppose for a sample of size 36 from a population with variance 144.0, the sample mean was 63.4 We can find a 90% confidence interval for the population mean, μ, using the following formula:

$$\overline{X} - z_{\alpha/2}\frac{\sigma}{\sqrt{n}} < \mu < \overline{X} + z_{\alpha/2}\frac{\sigma}{\sqrt{n}}$$

Thus

$$63.4 - 1.645\frac{12.0}{\sqrt{36}} < \mu < 63.4 + 1.645\frac{12.0}{\sqrt{36}}$$

yielding

$$60.11 < \mu < 66.69$$

Here we used the fact from Appendix E that p(standard normal variable $\leq$ 1.645) $=$ 0.95 to obtain the 1.645 value.

In the example above, the value of the population standard deviation was known to the statistician doing the problem. As we have seen, this will not generally be the case. However, we can still calculate the confidence interval for the population mean by substituting the sample standard deviation S into the formula in the place of the unknown population standard deviation, using the fact learned from the five-step method that the sample standard deviation should be close to the unknown population standard deviation σ. Because this estimate of the population standard deviation adds a further approximation to the normal approximation of the distribution of $\overline{X}$, the rule of thumb used by most statisticians requires $n \geq 30$ (rather than $n \geq 20$) in order to use the central limit theorem to obtain a confidence interval for μ when S is substituted for σ.

Example 10.10 Suppose the population variance in Example 10.9 is unknown. For a sample of size 36, suppose the sample mean is $\overline{X} = 63.4$ and the sample standard deviation is $S = 12.24$. Since the sample size of $n = 36$ satisfies the rule of thumb that $n \geq 30$, we can find a 95% confidence interval for the population mean, μ, using the following formula:

$$\overline{X} - z_{\alpha/2}\frac{S}{\sqrt{n}} < \mu < \overline{X} + z_{\alpha/2}\frac{S}{\sqrt{n}}$$

This yields

$$63.4 - 1.96\frac{12.24}{\sqrt{36}} < \mu < 63.4 + 1.96\frac{12.24}{\sqrt{36}}$$

and hence

$$59.40 < \mu < 67.40$$

Thus we can always find a confidence interval for a population mean, provided the sample size is large, even if σ is not known.

SECTION 10.5 EXERCISES

1. If possible, calculate a 95% confidence interval for the population mean based on a sample of size 40 with a sample mean of 4.5 and a sample standard deviation of 1.46. If it is not possible, explain your reasons.
2. If possible, calculate a 90% confidence interval for the population mean based on a sample of size 10 with a sample mean of 4.5 and a sample standard deviation of 1.46. If it is not possible, explain your reasons.
3. Suppose the population variance for the weight of all male Americans is 15 pounds. Find a 95% confidence interval for the population mean weight based on a sample size of 50 men with a sample mean of 178.9.
4. The popularity of a television show is found by taking a random sample of 100 households around the country. The mean rating (percentage of viewers watching) for a particular show was 31, and the standard deviation was 4.3. Find a 99% confidence interval for the population mean rating.
5. A set of 64 independent measurements of the length of a microorganism is made. The mean value of the measurements is 27.5 micrometers, with a standard deviation of 3.2 micrometers. Find a 90% and a 99% confidence interval for the true length of the microorganism.
6. A random sample of 40 students found that the mean study time was 19.5 hours per week with a standard deviation of 4.05 hours. Find a 90% confidence interval for the population mean study time. Explain the meaning of your confidence interval.
7. Thirty cars were tested to determine the fuel economy of a certain model. The sample fuel economy was 19.45 miles per gallon, with a standard deviation of 1.14 miles per gallon. Find a 95% confidence interval for the population mean fuel economy of this car model. Explain the meaning of your confidence interval.

10.6 CONFIDENCE INTERVALS FOR THE POPULATION PROPORTION

Recall the Key Problem for this chapter. We need to be able to construct a 95% confidence interval for the proportion p in the population of St. Louis County registered voters who oppose the construction of the football stadium. In this section we will learn how to solve this problem.

From the central limit theorem we know that the distribution of a sample mean will be approximately normal with mean μ and standard deviation $\sigma/\sqrt{n}$, when $n \geq 20$. But a sample proportion is just the sample mean of data whose only values are 0 and 1. This trick of recognizing that a sample proportion is really a sample mean is one you need to learn—it is used often in statistics. For example, in the Key Problem, out of the 301 registered voters sampled, 135 opposed the stadium and thus had a value of 1. On the other hand, 166 were not in this category and thus had a value of 0. So the sample proportion $\hat{p} = 0.45$ is just the mean of 135 1s and 166 0s. Since a sample proportion is a special kind of sample mean, we can apply the central limit theorem directly to a sample proportion as well. Therefore the distribution of the sample proportion, $\hat{p}$, is approximately normal with mean equal to the population proportion, p, and with standard deviation equal to $\sigma/\sqrt{n}$, where σ is the population standard deviation corresponding to randomly sampling one voter. So just as before, we can find a $100(1-\alpha)\%$ confidence interval for the population proportion by using the formula

$$\hat{p} - z_{\alpha/2}\frac{\sigma}{\sqrt{n}} < p < \hat{p} + z_{\alpha/2}\frac{\sigma}{\sqrt{n}}$$

where $z_{\alpha/2}$ is the value from the normal table, exactly as in Figure 10.6. The rule of thumb for when it is acceptable to use the normal approximation here is different from the $n \geq 20$ and $n \geq 30$ rules for the sample mean. The rule of thumb here is that both $np \geq 5$ and $n(1-p) \geq 5$ must hold. Therefore the closer p is to 0 or 1, the larger n must be to compensate. For example, if $p = 0.5$, we would feel comfortable using the normal approximation for the distribution of the sample proportion as long as $n \geq 10$. On the other hand, if $p = 0.10$, then we would need a sample size of $n \geq 50$ before we would feel comfortable using the normal approximation for the distribution of the sample proportion. We can apply the rule in practice by substituting $\hat{p}$ for p: $n\hat{p} \geq 5$ and $n(1-\hat{p}) \geq 5$ are then required.

The value of σ presents a problem we must solve. We cannot have unknown quantities in the formulas for the ends of a confidence interval because then we would not be able to obtain a numerical interval, as is required in practice. It can be shown that the formula for σ in a binomial problem like this is $\sqrt{p(1-p)}$, where p is the population proportion. But p is unknown. However, the sample proportion $\hat{p}$ is a good estimate for p, and therefore it seems reasonable to substitute $\hat{p}$ for p in the formula $\sqrt{p(1-p)}$. So a $100(1-\alpha)\%$ confidence interval for the population proportion is

$$\hat{p} - z_{\alpha/2}\frac{\sqrt{\hat{p}(1-\hat{p})}}{\sqrt{n}} < p < \hat{p} + z_{\alpha/2}\frac{\sqrt{\hat{p}(1-\hat{p})}}{\sqrt{n}}$$

Example 10.11

Pepsi and Coca-Cola have recently been in competition to have exclusive rights for their products on various college campuses and in companies across the nation. Suppose a college takes a random sample of 100 students on campus and finds that 60 students would prefer Pepsi to Coca-Cola. Find a 95% confidence interval for the population proportion of students who favor Pepsi.

Solution

We have just learned to use

$$\hat{p} - z_{\alpha/2}\frac{\sqrt{\hat{p}(1-\hat{p})}}{\sqrt{n}} < p < \hat{p} + z_{\alpha/2}\frac{\sqrt{\hat{p}(1-\hat{p})}}{\sqrt{n}}$$

This yields

$$0.6 - 1.96\sqrt{\frac{0.60(0.40)}{100}} < p < 0.6 + 1.96\sqrt{\frac{0.60(0.40)}{100}}$$

and thus

$$0.504 < p < 0.696$$

We are therefore 95% confident that a majority, indeed at least 50.4% of the student population, prefers Pepsi.

SECTION 10.6 EXERCISES

1. In a random sample of 64 people in a city, 37.5% were in favor of lowering the drunk-driving blood alcohol level from 0.10 to 0.08. Find a 90% confidence interval for the population proportion in favor of lowering the drunk-driving blood alcohol level from 0.10 to 0.08. Is it very likely that a majority of the population is in favor of the change?
2. In a random sample of 2000 U.S. citizens over the age of 18, 53.5% approve of the president. Find a 99% confidence interval for the population proportion who approve of the president. If the president were to run for reelection soon after this poll was taken, would the president be very likely to win reelection?
3. In a random sample of 2000 residents of Illinois, 75% believe in the fairness of the jury system. Find a 90% confidence interval for the proportion of people in the state of Illinois who believe the jury system is fair.

4. In a random sample of 2000 residents of Rhode Island, 75% believe in the fairness of the jury system. Find a 90% confidence interval for the proportion of people in the state of Rhode Island who believe the jury system is fair.
5. Your answers for Exercises 3 and 4 should be the same. The size of the populations are very different, but the size of the samples are the same. What can you conclude about the effect of the size of the population and the size of the sample on the width of the confidence interval?
6. Look at the formula for the standard error of the sample proportion, $p(1-p)/\sqrt{n}$. What value(s) of p will give the standard error its largest value? What value(s) of p will give the standard error its smallest value?
7. Suppose you believe the value of the population proportion is very close to 50%. How large should your random sample be to ensure that the width of a 95% confidence interval is no larger than 8%?
8. In a random sample of 20 college freshmen, 75% were in favor of lowering the drinking age from 21 to 18. If it is possible, find a 90% confidence interval for the proportion of college freshmen on campus who favor lowering the drinking age from 21 to 18. If it is not possible, explain your answer.

10.7 CONFIDENCE INTERVALS FOR THE DIFFERENCE BETWEEN TWO POPULATION MEANS

This type of problem occurs often in practice because we often wish to compare two things, such as two treatments or two teaching methods. Suppose we want to compare the mean population incomes for two cities. We can take a sample of size n from City A and a sample of size m from City B. We can then calculate the sample mean income of City A and the sample mean income of City B. From the central limit theorem we know that both the sample means, $\overline{X}$ and $\overline{Y}$, are approximately normally distributed with means μ_X and μ_Y and standard deviations $\sigma_X/\sqrt{n}$ and $\sigma_Y/\sqrt{m}$, respectively, provided that $n \geq 20$, and $m \geq 20$. Building on this fact, in Chapter 13 we will learn that the difference between two independent sample means, $\overline{X} - \overline{Y}$, is also approximately normally distributed, with mean $\mu_X - \mu_Y$ and standard deviation $\sigma_X/\sqrt{n} + \sigma_Y/\sqrt{m}$. So, reasoning exactly as in Section 10.5, from the central limit theorem we can construct a $100(1-\alpha)\%$ confidence interval for the difference between two population means, $\mu_X - \mu_Y$, through use of the formula

$$(\overline{X} - \overline{Y}) - z_{\alpha/2}\left(\frac{\sigma_X}{\sqrt{n}} + \frac{\sigma_Y}{\sqrt{m}}\right) < \mu_X - \mu_Y < (\overline{X} - \overline{Y}) + z_{\alpha/2}\left(\frac{\sigma_X}{\sqrt{n}} + \frac{\sigma_Y}{\sqrt{m}}\right)$$

As in the case of confidence intervals for one population mean, we are unlikely to know the values of the population standard deviations for City A and for City B. Just as in Section 10.5, we can substitute the sample standard deviations, denoted S_X and S_Y, for both cities A and B into the equation in place of the population standard deviations. The $100(1-\alpha)\%$ confidence interval for the difference of two population means, $\mu_X - \mu_Y$, then becomes

$$(\overline{X}-\overline{Y}) - z_{\alpha/2}\left(\frac{S_X}{\sqrt{n}} + \frac{S_Y}{\sqrt{m}}\right) < \mu_X - \mu_Y < (\overline{X}-\overline{Y}) + z_{\alpha/2}\left(\frac{S_X}{\sqrt{n}} + \frac{S_Y}{\sqrt{m}}\right)$$

This confidence interval is widely used when $n \geq 30$ and $m \geq 30$.

Example 10.12 Suppose in City A, out of a sample of 100 people, the mean income is $35,000 with a standard deviation of $2000. In City B, out of a sample of 400 people, suppose the mean income is $40,000 with a standard deviation of $1500. We find a 95% confidence interval for the difference between the two population means:

$$(\overline{X}-\overline{Y}) - z_{\alpha/2}\left(\frac{S_X}{\sqrt{n}} + \frac{S_Y}{\sqrt{m}}\right) < \mu_X - \mu_Y < (\overline{X}-\overline{Y}) + z_{\alpha/2}\left(\frac{S_X}{\sqrt{n}} + \frac{S_Y}{\sqrt{m}}\right)$$

This yields

$$(35{,}000 - 40{,}000) - 1.96\left(\frac{2000}{\sqrt{100}} + \frac{1500}{\sqrt{400}}\right) < \mu_X - \mu_Y < (35{,}000 - 40{,}000) + 1.96\left(\frac{2000}{\sqrt{100}} + \frac{1500}{\sqrt{400}}\right)$$

and hence

$$-\$5539 < \mu_X - \mu_Y < -\$4461$$

Thus we are confident that there is a real difference in mean income between the two populations of at least $4000.

SECTION 10.7 EXERCISES

1. Find the standard error of the difference of two sample means if the size of both samples is 50 and the population standard deviations are 10 and 4.
2. In a random sample of size 64, the sample mean was 50.3 with standard deviation 2.5. In another random sample of size 64, the sample mean was 49.25 with standard deviation 3.1. Find the standard error of the difference of the sample means.
3. For Exercise 2, find a 90% confidence interval for the difference of the population means.
4. Scientists want to study the effect of exercise on the amount of weight loss. One hundred people are divided into two equal groups. Both groups follow the same diet plan, but the first group also completes an exercise program. In the exercise group, the mean weight loss over three months was 25.2 pounds with a standard deviation of 10 pounds. In the nonexercise group the mean weight loss was 20.4 pounds with a standard deviation of 6.3 pounds. Find a 99% confidence interval for the difference of the population means.

5. On a college campus, a professor wanted to test the effect on learning of using a computer program to teach calculus. In his first class of 43 students, he taught using the computer as an instructional aid. In his second class of 35 students, he taught without using the computer. On the final exam the computer class scored a mean of 78.32 with a standard deviation of 8.07 while the noncomputer class scored a mean of 80.41 with a standard deviation of 8.53. Find a 99% confidence interval for the difference of the population means. Is it likely that the computer teaching made a difference in student learning?

6. Fifty independent measurements of the weight of a chemical compound were made on each of two scales. On the first scale, the mean of the 50 measurements was 19.45 grams with a standard deviation of 0.49 grams. On the second scale the mean of the 50 measurements was 18.42 grams with a standard deviation of 0.27 grams. Find a 95% confidence interval for the difference between the mean measurements of the two scales. Is it likely that the two scales weigh objects the same?

7. In a random sample of size 75, the mean SAT verbal score was 549.45 with a standard deviation of 21.12. In a random sample of size 75 the mean SAT math score was 539.25 with a standard deviation of 20.91. Find a 90% confidence interval for the difference between population mean verbal and math SAT scores.

10.8 POINT ESTIMATES FOR THE POPULATION VARIANCE

While population means and proportions are two of the most widely used parameters to report point estimates and confidence intervals for, other parameters are important as well. We have already looked at one such parameter, the population variance (and its square root, the population standard deviation). Information from a sample can be used to estimate the variance of the population from which the sample was taken. If a sample has a large variance, for example, we would expect the variance of the population to be large as well. Similar statements apply to the population and sample standard deviations.

To look at some of the properties of the sample, we have taken 100 samples each of size 4 from a normal population with mean 50 and variance 100, and we have calculated the variance for each sample. The results are listed in the stem-and-leaf plot in Table 10.4. The sample variances are given in the table to the nearest whole number. For example, the largest sample variance was 222, and the smallest variance was 1. The range of these variances is quite large: from 1 to 222. The mean of the 100 sample variances is 76.2—a slight surprise because $\sigma^2 = 100$.

We repeated the above process, but we increased the sample sizes to 16 in Table 10.5 and to 36 in Table 10.6. With a sample size of 16, the range is much smaller than before: from 42 to 148. The mean value of the 100 sample variances for a sample size of 16 is 93.9. (Recall that the population variance is 100 in all three cases.) With a sample size of 36, the range is even smaller—from 60 to 135—and the mean of the 100 sample variances is 97.0. The results of the three sets of samples are summarized in Table 10.7.

Table 10.4 **Sample Variances of 100 Samples of Size 4 (Population Values: μ = 50.0, σ^2 = 100.0)**

Stem	Leaf	Frequency
0	1,5,6	3
1	0,0,1,1,2,2,3,3,3,9,9	11
2	0,4,4,6,7,7,7,9	8
3	1,4,4,4,8,9	6
4	0,2,3,4,6,7,7,8,8	9
5	1,1,1,2,5,6	6
6	1,3,4,7,8,9	6
7	0,1,3,6,8,9	6
8	0,1,3,4,4,7,8	7
9	1,2,3,7,9	5
10	1,4,5,5,7,7,7,9	8
11	5,6,7	3
12	1,2,3	3
13	0,0,0,4,4,5,5,6,6	9
14	1,3	2
15	8	1
16	0,0,4	3
17	0,7,7	3
18		0
19		0
20		0
21		0
22	2	1
23		0
24		0
	Total	100

Sample values:
n = 4
Mean of sample variances = 76.2

We see from Table 10.7 that as the sample size increased from 4 to 36, there was less variation in the sample estimates of the population variances. In addition, the range of the estimates went down from 221 to 75. This indicates that the variance estimates became closer together as the sample size increased. Therefore, an increase in sample size has reduced the amount of variation in the estimates of the population variance.

We can also see from Table 10.7, particularly for the small sample sizes, that the mean value of the sample variances was considerably less than the population variance of 100. In fact, if we repeated this analysis several times, we would usually see the same result. So for small sample sizes the

Table 10.5 **Sample Variances of 100 Samples of Size 16 (Population Values: $\mu = 50.0, \sigma^2 = 100.0$)**

Stem	Leaf	Frequency
2		0
3		0
4	2	1
5	1,4,4,5,6,8,8,9	8
6	0,4,7,8,8,9,9,9	8
7	0,1,1,2,3,4,7,8,8,9,9	11
8	1,1,2,2,3,3,4,5,5,5,7,7,8,9	14
9	0,1,1,3,3,4,5,6,6,6,9,9,9	13
10	1,1,1,1,1,2,2,2,3,3,3,4,5,5,7,7,8,9,9	19
11	0,0,0,1,2,3,4,4,4,5,5,6,6,6,9	15
12	1,1,4,6	4
13	0,1,3	3
14	1,1,3,8	4
15		0
16		0
	Total	100

Sample values:
$n = 16$
Mean of sample variances = 93.9

Table 10.6 **Sample Variances of 100 Samples of Size 36 (Population Values: $\mu = 50.0, \sigma^2 = 100.0$)**

Stem	Leaf	Frequency
4		0
5		0
6	0,3,6,7,7,7	6
7	3,4,5,7,7,9,9,9	8
8	0,0,1,2,3,4,4,5,5,6,6,7,7,8,8,8,9,9,9	19
9	0,0,0,0,1,2,2,2,2,3,3,3,5,5,5,5,7,8,9,9	20
10	0,1,1,1,1,2,2,2,2,3,3,3,4,4,4,5,5,5,6,6,8,8,8,8,9	25
11	1,1,1,2,2,4,4,4,5,6,8,8,8,9	14
12	0,2,3,3,5,5,7	7
13	5	1
14		0
15		0
	Total	100

Sample values:
$n = 36$
Mean of sample variances = 97.0

Table 10.7

	Sample size		
	$n = 4$	$n = 16$	$n = 36$
Mean of sample variances	76.2	93.9	97.0
Range of sample variances	221.0	106.0	75.0

sample variance tends to underestimate, on average, the true population variance. That is, many statisticians using the sample variance would find if they pooled their individual sample variances by averaging that they on average would be quite low. Therefore we say that the sample variance is a *biased estimate* of the population variance.

It is possible to change the formula for the sample variance to make the statistic an unbiased estimator for the population variance. The formula we have been using for the sample variance is

$$S^2 = \frac{\text{sum of } (X - \overline{X})^2}{n}$$

Suppose we divide by $n - 1$ instead of n. Then new simulations (which we do not present) would show that the new value for the sample variance is no longer biased for the population variance. That is, even when $n = 4$, we would find the average of the 100 sample variances to be quite close to $\sigma^2 = 100$.

Generally, the sample variance we have been using in this text so far (that is, dividing by n) is quite sufficient for estimating the population variance, even though it is biased (especially for small sample sizes). Actually, the somewhat large bias of the biased estimate is compensated by a smaller standard error (a highly desirable property). Therefore, although the biased version tends to be too small on average, it also tends to be less variable than the unbiased estimate. Because of this lower variability, many statisticians use our biased sample variance. We will use this biased sample variance. If you are using a Texas Instruments graphing calculator, note that it divides by n.

SECTION 10.8 EXERCISES

1. Below is a random sample of 15 blood cholesterol levels taken from men at a local hospital. Use these data to find an estimate of the population variance.

164	271	262	287	175	215
247	291	326	202	189	335
305	242	215			

2. Below is a random sample of the compressive strength of a batch of concrete. Use these data to find an estimate of the population variance.

4466	3914	4084	4135	4084
4125	4123	4030	4120	3626
3771	3889	4180	4141	4524
3810	3777	4072	3871	4181

10.9 POINT ESTIMATES FOR THE POPULATION MEDIAN

We have seen that sometimes the population median is a better parameter for characterizing the center of a population because it is less affected by extreme population values. When this is indeed judged to be the case, clearly a reasonable estimate of the population median is the sample median. We do not have a simple formula for its standard error, which is absolutely central to the sample median's use as an estimate of the population median if we wish to assess the accuracy of estimation. However, as discussed in Section 10.3, we can use the bootstrap method to calculate an estimate of the standard error of the sample median. Indeed, this is a perfect application of the bootstrap method.

Example 10.13 A random sample of size 100 was drawn from a normal population. Its sample median was 9.1. We now replace the theoretical population (which the statistician does not know) by our observed random sample of size 100. That is, this is our (100-ticket) box model for step 1 of the five-step method. Next we draw 100 samples of size 21 with replacement from our new "population." Thus step 2 consists of a drawing of one sample of size 21. The statistic of interest of step 3 is the sample median of these 21 draws. In step 4 we repeat this process 100 times, each time obtaining a sample median. This is bootstrap sampling. Then in step 5 we compute the standard deviation of the statistic of interest, and this standard deviation will be our estimate of the standard error of the median. This is different from our usual step 5, which has been to compute the mean of the statistic of interest, rather than the standard deviation. The stem-and-leaf plot displays the 100 sample medians. We wish to find an estimate of the standard error of the sample median.

Stem	Leaf
6	8
7	1 1 4 4 4
7	6 6 6 6 6
8	0 0 0 1 1 1 1 1 1 1 1 1 1 1 1 1 1 2 2 2 3 3 3 3 3 3
8	6 6 6 6 6 6 6 6 6 8 8 8 8 8 8 8 8 8 8 8 8
9	0 0 0 0 3 3
9	6 6 6 6 6 6 6 6 6 8 8 8 8 9 9
10	0 0 0 0 0 0 0 0 0 1 1 1 1 3 3
10	5 5 7 7
11	1
11	
12	
12	5

Key: "12 5" stands for 12.5

An estimate of the standard error of the sample median is obtained by calculating the standard deviation of the 100 sample medians, which in this case is equal to 1.005. In particular, invoking the central limit theorem for the sample median, which is shown to be true in advanced statistics, we can get a 95% confidence interval for the sample median of

$$9.1 - 1.96(1.005) \leq \text{population median} \leq 9.1 + 1.96(1.005)$$

That is, $7.13 \leq$ population median ≤ 11.07.

SECTION 10.9 EXERCISES

1. Find the standard error of the following sample medians using the bootstrap method:

4124.0	4123.0	3871.0	4051.0
3901.5	3993.0	4122.5	4084.0
3999.0	3849.5	4123.0	4152.5
4084.0	4084.0	4121.5	3901.5
3903.5	4072.0	4072.0	3914.0
3849.5	4121.5	4057.0	4078.0
4084.0	3972.0	4078.0	4103.5
4124.0	4084.0	4130.5	3914.0
3972.0	3914.0	4295.5	4102.0
4084.0	3959.5	4078.0	3977.5
4078.0	4007.0	4102.0	4057.0
4017.0	4121.5	4127.5	4030.0
3993.0	3959.5	4084.0	3959.5
4123.0	4082.5	3889.0	3993.0
4122.5	4104.5	4135.0	4096.0
4157.5	3980.5	4138.0	4078.0
4057.0	4078.0	4084.0	4125.0
3972.0	4097.5	4084.0	4078.0
4121.5	4102.0	4102.0	4078.0
4084.0	4102.0	4057.0	3880.0
3999.0	4084.0	4120.0	4130.5
4121.5	4017.0	4120.0	4135.0
3972.0	4057.0	3977.5	4098.5
4098.5	4120.0	3999.0	4096.0
4084.0	4160.5	4180.5	4120.0

2. Find the standard error of the following sample medians using the bootstrap method:

228.5	254.5	267.0	279.0
238.5	164.0	279.0	254.5
242.0	267.0	279.0	266.5
252.0	215.0	252.0	242.0
274.5	215.0	244.5	242.0
244.5	232.0	231.0	262.0
228.5	252.0	254.5	254.5
266.5	228.5	254.5	247.0
202.0	228.5	279.0	228.5
222.0	283.5	289.0	274.5
262.0	259.0	164.0	238.5
247.0	252.0	244.5	243.0
252.0	228.5	195.5	231.0
228.5	279.0	228.5	215.0
228.5	281.0	242.0	218.0
247.0	254.5	242.0	242.0
267.0	228.5	231.0	251.0
262.0	298.0	287.0	271.0
215.0	283.5	208.5	271.0
215.0	267.0	266.5	228.5
244.5	215.0	208.5	228.5
256.5	238.5	222.0	215.0
228.5	252.0	244.5	266.5
244.5	244.5	215.0	279.0
254.5	215.0	256.5	230.0

CHAPTER REVIEW EXERCISES

1. Explain the meaning of a 90% confidence level.
2. Which relative frequency histogram will look more like a normal curve—one based on a sample mean of 20 observations or one based on a sample mean of 25 observations? (Assume that the samples are taken from the same population.)
3. What are the mean and standard deviation of a sample mean of size 100 taken from a population with mean value 65.25 and standard deviation 10.52? What shape does the relative frequency histogram of this sample mean have?
4. Explain how to estimate the standard error of a sample variance by using the bootstrap method.
5. In a random sample of 100 people, 70 are regular viewers of the television show *Seinfeld*. Find the standard error of this sample proportion.
6. In a random sample of 100 teenage boys, the mean amount of time spent watching television was 3.6 hours with a standard deviation of 0.9 hours. In a random sample of 100 teenage girls, the mean amount of time spent watching television was 4.2 hours with a standard deviation of 1.2 hours. Find the standard error of the difference between the two sample means.
7. See Exercise 5. Find a 95% confidence interval for the population proportion of people who watch *Seinfeld* on a regular basis.
8. See Exercise 6. Find a 99% confidence interval for the difference in the population mean television hours between boys and girls.
9. Fifty measurements of the length of a bone were made. The mean of the measurements was 34.4 centimeters with a standard deviation of 0.25 centimeters. Find a 90% confidence interval for the true bone length. Is it likely that the bone is 35 centimeters long?
10. In a random sample of 250 residents of a large city, 48% of the people in the sample would favor a raising the bar entrance age from 19 to 21. Find a 95% confidence interval for the population proportion of people who would favor raising the age. If a referendum were held, is it likely that more than one-half of the people would vote to raise the bar entrance age?
11. In a random sample of 50 women, the mean height was 64.45 inches with a standard deviation of 3.5 inches. Find a 90% confidence interval for the true mean height of women.
12. In a random sample of 100 men, the mean number of alcoholic drinks consumed per week was 5.4 with a standard deviation of 1.25. In a random sample of 100 women, the mean number of alcoholic drinks consumed per week was 3.2 with a standard deviation of 1.85. Find a 99% confidence interval for the difference between the two population means. Is it likely that in the population men and women consume the same amount of alcoholic drinks per week?
13. Complete the following table, which summarizes the information in this chapter on confidence intervals.

Parameter	Population variance	Standard error	Lower bound of confidence interval	Upper bound of confidence interval
Mean	Known			
Mean	Unknown			
Proportion	Unknown			
Difference between two means	Known			
Difference between two means	Unknown			

14. Find the standard error of the following sample medians by using the bootstrap method.

Stem	Leaf
232	0
232	
233	0,0,0,0
233	5,5,5,5,5
234	0,0,0,0,0,0,0,0,0,0,0,0,0,0,0,0
234	5,5,5,5,5,5,5,5,5
235	0,0,0,0,0,0,0,0,0,0,0,0,0,0,0,0,0
235	5,5,5,5,5,5,5,5,5,5,5,5,5,5,5,5,5,5,5,5,5
236	0,0,0,0,0,0,0,0,0,0,0,0,0,0,0,0,0,0
236	5,5,5,5,5,5
237	0,0

Key: "237 0" stands for 23.70.

10

COMPUTER EXERCISES

1. NORMAL DATA

The objective in this problem is to compare the sample average and sample median as estimators of the box (population) average and box (population) median. We will assume that the tickets in the box follow the normal distribution, so the average and median in the box will be the same. Start the Five Step program, and press the Choose Box Model button. When the Enter Box Model screen opens, select "Named Distribution" from the Type menu. (See Computer Exercise 2 of Chapter 8.) Select the Normal circle, type in 72 for the average and 3 for the standard deviation, and then press OK. This box represents a population of heights of men that average 72 inches and have a standard deviation of 3 inches.

	Average of statistics		*Standard error*		*Probability*	
Estimate	$n = 25$	$n = 100$	$n = 25$	$n = 100$	$n = 25$	$n = 100$
Average						
Median						

a. Set up the program to draw 25 with replacement. Choose the statistic as Average, and the probability of interest to be "a < Stat < b," with $a = 71.5$ and $b = 72.5$. This probability is the chance that the estimate of the average is within 0.5 of the box average, which we have set to be 72. Run 100 trials. Under "Average of statistics" in the above table, fill in the upper left cell (where the estimate is the average and $n = 25$) with the average of the statistics. Under "Standard error," fill in the corresponding cell with the standard error of the statistic. Under "Probability," fill in the experimental probability.

b. Press "Start Again." Repeat part (a), but with $n = 100$. Fill in the upper-right cells.

c. Press "Start Again." Repeat part (a), but select the median to be the statistic, and have $n = 25$. Fill in the lower-left cells.

d. Press "Start Again." Repeat part (c), with the median as statistic but with $n = 100$. Fill in the lower-right cells.

e. Compare the four averages of statistics to the box average, 72. Are they reasonably close? Compare each estimate's error with its corresponding standard error.

f. Compare the standard error of the average for $n = 25$ to that for $n = 100$. Which is smaller? Is the smaller more than half as large, about half as large, about one-fourth as large, or about one-tenth as large?

g. Compare the standard error of the median for $n = 25$ with that for $n = 100$. Which is smaller? Is the smaller more than half as large, about half as large, about one-fourth as large, or about one-tenth as large?

h. Find the theoretical standard error of the average for $n = 25$. How close is that to the standard error found from the program? Do the same for $n = 100$.

i. Compare the standard error of the average for $n = 25$ to the standard error of the median for $n = 25$. Which is smaller? Is the smaller more than half as large, about half as large, about one-fourth as large, or about one-tenth as large?

j. Compare the standard error of the average for $n = 100$ to the standard error of the median for $n = 100$. Which is smaller? Is the smaller more than half as large, about half as large, about one-fourth as large, or about one-tenth as large?

k. Look at the table of probabilities. Which estimator was more likely to be within 0.5 of 72:

   i. The average with $n = 25$ or the median with $n = 25$?

   ii. The average with $n = 100$ or the median with $n = 100$?

   iii. The average with $n = 25$ or the average with $n = 100$?

   iv. The median with $n = 25$ or the median with $n = 100$?

l. Overall, which is a better estimate of the average for this situation, the average or the median?

2. A BOX

Now create the box with one 1, one 11, two 12s, three 13s, two 14s, one 15, and one 25. Find the average and standard deviation of the box. What is the median in the box? Is it the same as the average?

Repeat Exercise 1, but with this box, and set the probability as the chance that the statistic is between 12.5 and 13.5. Be sure to compare the estimators to the box average for this box, not to 72.

Taking into account this exercise and the previous one, can you say whether the average or the median is better?

3. CONFIDENCE INTERVALS: THE KEY PROBLEM

Exercises 3 and 4 use the Box Models program. To start it, click on the Box Model icon:

This screen should appear:

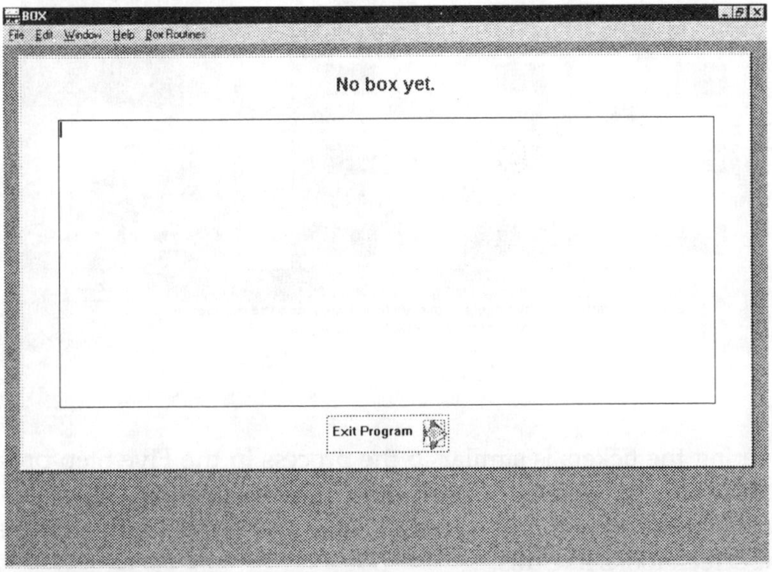

It says "No box yet" because you have not yet entered a box. To do so, choose the "Create/Modify a box" selection in the Box Routines menu:

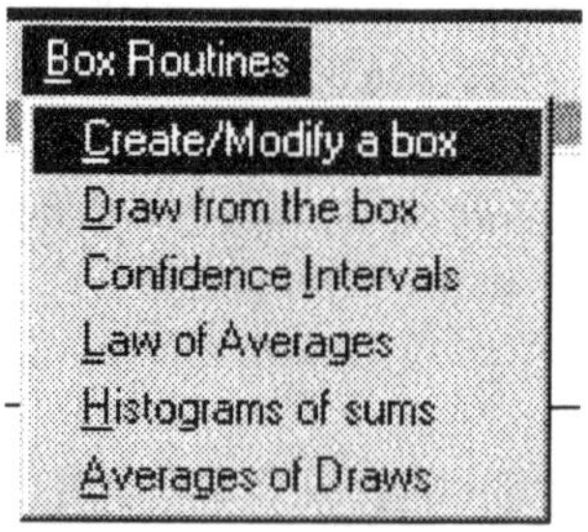

A screen allowing you to input the tickets should appear:

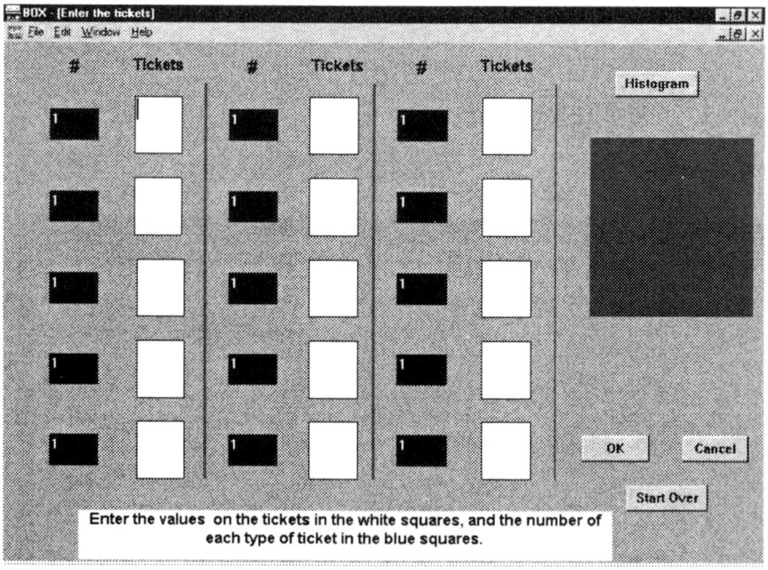

Entering the tickets is similar to the process in the Five Step program, except that now "#" means "Count" and "Tickets" means "Values." Therefore, to create a box with one 1, two 2s, and one 3, type in the numbers so that the screen looks like this:

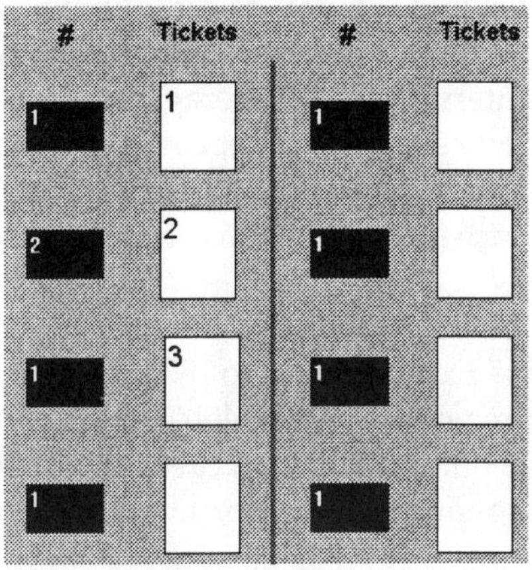

Once the tickets and counts have been entered, press OK. You will be returned to the opening screen.

In the Key Problem, a sample of 301 was taken to estimate the percentage of registered voters in St. Louis County who opposed a new stadium. In the text, to illustrate confidence intervals for proportions, it was assumed that the actual proportion in the entire county that opposed the stadium was $\frac{1}{2}$. Thus the sample is analogous to 301 flips of a fair coin.

a. Create the box to mimic a fair coin (one 0 and one 1) in the Box Model program, as described above. Then select "Confidence Intervals" from the Box Routines menu. You will be confronted with the following:

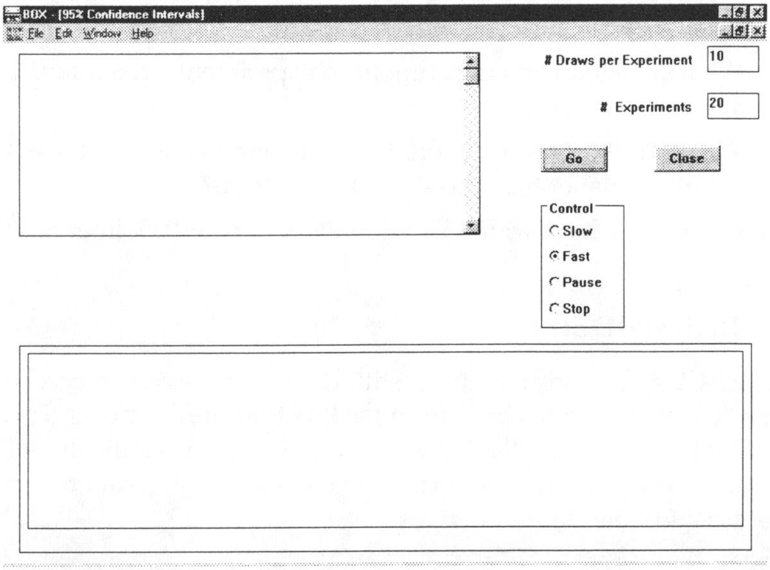

The "# Draws per Experiment" is the number of draws the program will take from the box. After making the draws, it will calculate a 95% confidence interval for the average in the box, which in this case is $\frac{1}{2}$. The whole process will be repeated, the number of times being indicated by "# Experiments." The way it is set up in the picture, there will be 20 confidence intervals, each one calculated from a sample of 10 draws.

Change "# Draws per Experiment" to 301, the actual number in the sample. Change "# Experiments" to 100. Press Go. Watch the confidence intervals. In the area on the upper left, the actual intervals are printed. In the graph at the bottom, the intervals are plotted. The blue line in the middle of the plot represents the box average of $\frac{1}{2}$.

i. What proportion of the intervals cover the blue line? Is the proportion close to 0.95?

ii. What changed from experiment to experiment—the interval or the average in the box?

iii. What are the widths of the last three intervals? (The width is the upper limit minus the lower limit.)

iv. Now set the number of experiments to 500, and press Go. What proportion of intervals cover $\frac{1}{2}$ this time?

b. Now suppose that instead of a sample size of 301, the pollsters took a sample of 1200 people. Set "# Draws per Experiment" to 1200, and set the number of experiments to 500. Press Go.

i. What proportion of the intervals cover the blue line? Is the proportion close to 0.95?

ii. What changed from experiment to experiment—the interval or the average in the box?

iii. What are the widths of the last three intervals? Are these widths larger or smaller than those in part (iv) of (a)?

iv. Is it better to have wide 95% intervals or narrow 95% intervals? Why?

4. THE SIX-SIDED DIE

Press the Close button if you are still at the confidence interval screen. Choose "Create/Modify a box" from the Box Routines menu, and press the Start Over button to clear the values. Create the box to mimic the rolling of a fair six-sided die, so that there is one each of 1, 2, 3, 4, 5, and 6. Press OK, and go back to Confidence Intervals.

a. For each number of draws per experiment in the table below, set the number of experiments to 500, press Go, and fill in the proportion of intervals that cover the box average (3.5) successfully.

Number of draws per experiment	Proportion of successes
2	
3	
5	
10	
15	
20	
25	
50	
100	

b. How many draws per experiment are needed for the proportion of successes to be near 0.95? If the number of draws is too small, is the chance that the interval covers 3.5 smaller or larger than 0.95?

5. CENTRAL LIMIT THEOREM: COIN TOSSES

The central limit theorem in Section 10.4 suggests that if a trial consists of drawing (with replacement) a reasonably large sample from a population, and the statistic is the sum of the draws, then the histogram of the sums from many trials should look somewhat like a normal curve. The number of draws per trial needed depends on the box itself. Sometimes only a few are needed, while other times a great many are necessary. Exercises 5–7 explore these numbers. The Five Step program will be used.

a. Start the Five Step program, and set up the box to have one 0 and one 1. Set the statistic to be the sum, and the trial to be drawing n with replacement. For each choice of the number of draws in the table below, run 500 trials and look at the resulting histogram of sums. Fill in the Normal? column with your assessment of how much the histogram looks like a normal curve. Choose from the following:

Not at all

Somewhat

Quite a bit

Exactly

Number of draws	Normal?
1	
2	
3	
5	
10	
25	
50	
100	

b. Repeat part (a), but with the box containing one 0 and three 1s.

Number of draws	Normal?
1	
2	
3	
5	
10	
25	
50	
100	

c. Repeat part (a), but with the box containing one 0 and 99 1s.

Number of draws	Normal?
1	
2	
3	
5	
10	
25	
50	
100	

d. For the three boxes above, which needed the fewest draws per trial for the histogram to look reasonably normal? Which needed the most?

6. NORMAL DATA

If the population one is sampling from is already normal, then the sum of the draws should be normal, too. Set up the box to be Normal with an average of 72 and a standard deviation of 3 as in Exercise 1. Fill in the table as for the previous problem, with each entry based on 500 trials:

Number of draws	Normal?
1	
5	
25	
100	

What do you notice?

7. OTHER BOXES

a. Now create the box with one 1, two 2s, three 3s, two 4s, and one 5. Repeat the previous problem, but with this box. Fill in the table:

Number of draws	Normal?
1	
5	
25	
100	

b. Change the box to have one 1, two 2s, three 3s, two 4s, one 5, three 10s, six 11s, and three 12s. Repeat the previous problem, but with this box. Fill in the table:

Number of draws	Normal?
1	
5	
25	
100	

c. Which of these two boxes needed fewer draws for the histogram to look normal? Why?

11

HYPOTHESIS TESTING

11.1 A BOOTSTRAP HYPOTHESIS TEST OF THE POPULATION MEAN 524

11.2 THE z TEST OF THE POPULATION MEAN 535

11.3 MAKING A WRONG DECISION 540

11.4 THE SIGN TEST 542

11.5 THE t TEST 547

11.6 COMPARING TWO POPULATION MEANS 550

11.7 THE z TEST FOR A HYPOTHESIS ABOUT A POPULATION PROPORTION 556

CHAPTER REVIEW EXERCISES 559

COMPUTER EXERCISES 561

The probability that this is a mere work of chance is, therefore, considerably less than $(1/2)^{60}$ .... Hence this coincidence must be produced by some cause, and a cause can be assigned which affords a perfect explanation of the observed facts.

G. Kirchoff (1824–1887)

OBJECTIVES

After studying this chapter, you will understand the following:

- *Using the bootstrap method to test a hypothesis about the population mean*
- *Using a normal* z *test of a population mean*
- *The two kinds of erroneous decisions possible when testing a hypothesis*
- *The sign test concerning a population median, using simulations and a* z *test*
- *The* **t** *test concerning a population mean when the sample size is small and the population is normal*
- *How to compare population means using the bootstrap method and a normal* z *test*
- *The normal* z *test of a hypothesized population proportion*

Key Problem

Is 98.6 Normal?

It is generally accepted that the normal body temperature of adult humans is 98.6 degrees Fahrenheit. The figure below is a histogram of 130 randomly sampled people's temperature readings from a large population of adults.* Notice that most of the histogram is to the left of 98.6, the supposed normal temperature. The average of the 130 readings is 98.25, which is less than 98.6. The question is whether the sample average of 98.25 is close enough to 98.6 to ascribe the difference to mere chance variation. That is, is such a difference likely to show up in another random sample of 130 people from a population with a mean temperature of 98.6? Then there would be no evidence that the population mean temperature is less than 98.6. Or is the difference of $98.25 - 98.6 = -0.55$ so large that it is not likely to show up in another random sample of 130 people from a population with a mean temperature of 98.6 degrees? Then the evidence would be strong that the mean population temperature is indeed less than 98.6.

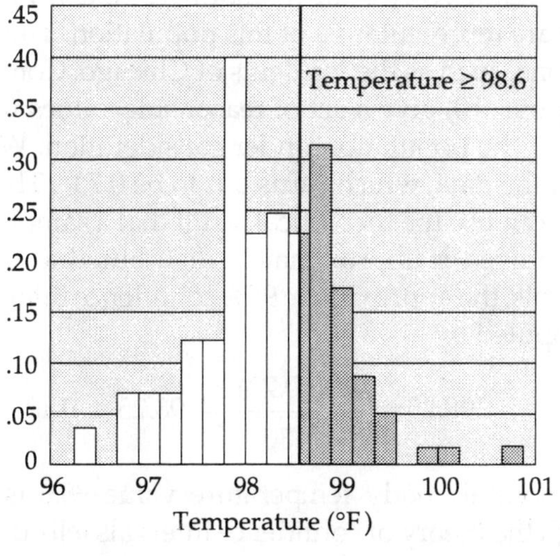

*Data are from Allen L. Shoemaker, "What's Normal?—Temperature, Gender, and Heart Rate," *Journal of Statistics Education,* July 1996.

11.1 A BOOTSTRAP HYPOTHESIS TEST OF THE POPULATION MEAN

In Chapter 6 we looked at sets of data and asked whether the data could have arisen from a binomial model with a certain population probability of success such as that given by a hypothesized drug cure rate. If the observed proportion of successes was too large, we rejected the hypothesized model. In Chapter 7 we looked at a set of data and asked whether it could have arisen from a particular hypothesized model such as that of a six-sided fair die or a many-sided loaded die. The chi-square statistic was calculated. If it was improbably large, we concluded that the data did not come from the hypothesized model. If the chi-square statistic was not too large, we concluded that the data may very well have come from that model. That is, we "rejected" or "accepted" the null hypothesis on the basis of the above considerations. The six-step decision-making process of Chapters 6 and 7 can be used for many other statistical purposes. Which of two new drugs, if either, is more effective? Are husbands more likely to be older than their wives? Is a particular hypothesized population blood pressure average correct for 60-year-olds? This chapter shows how to formally assess such hypotheses. The chapter is best viewed as a continuation of Chapter 6.

Let us consider the Key Problem. We will assume that the body temperature readings constitute a random sample from a population of adults and hence are representative of this population. The population is taken to be a large one, such as the residents of Chicago. Confidence intervals (discussed in Chapter 10) give a set of reasonable values for the unknown theoretical mean of the population under consideration. We need the standard deviation of the data, which turns out to be 0.73°. Thus, as explained in Chapter 10, we can use the theoretical result that a sample mean when appropriately standardized is approximately distributed as a standard normal variable. From this, the approximate 95% confidence interval for the population mean is computed by

$$98.25 \pm \frac{1.96(0.73)}{\sqrt{130}} = 98.25 \pm 0.13 = (98.12, 98.38)$$

The "normal" body temperature value 98.6 is not in that interval, even though the theory of confidence intervals tells us the interval covers the true population average temperature about 95% of the time. This suggests that the population mean is lower than 98.6—that it is not just chance causing the observed mean to be so low as 98.25.

Another approach to this question is to formally set out a null hypothesis:

H_0: The population mean is 98.6.

and then use the data to see if the hypothesis holds up or whether the population average temperature is in fact lower. The H_0 is the null hypothesis: H is for *hypothesis* and 0 is for *null*. This is the hypothesis-testing approach of this chapter. The idea is the same as in Chapter 6, where we asked whether the data conformed closely enough to a given hypothesized model. The null hypothesis represents the status quo: it is believed that the mean of the population is 98.6. It is usually the hypothesis of "business as usual" or "nothing of interest here." We then look at the data, which have a mean of 98.25. We have two choices:

1. It is plausible that the difference between the observed mean and 98.6 could be due to chance in a population with a mean of 98.6. We thus *accept* the null hypothesis, meaning that the evidence tells us either that the null hypothesis is true or that there is not enough evidence to say it is false. Thus "accept" does not *necessarily* mean we have strong evidence that the null hypothesis is true.
2. It is not plausible that the difference between the observed mean and 98.6 is due to chance in a population with a mean of 98.6. We thus *reject* the null hypothesis and conclude that the true population average temperature is less than 98.6.

How far does the sample average have to be from 98.6 before we are compelled to reject the null hypothesis? In particular, what is the chance that if the null hypothesis is true, we could observe a sample mean as low as or lower than 98.25? To decide, we take the six-step hypothesis-testing approach of Chapter 6.

A key step in our six-step simulation approach to statistical hypothesis testing is step 1: making a realistic choice of the model to be randomly sampled from. One cannot be effective as a statistician without understanding how to realistically specify the model generating the data to be analyzed. Because we are hypothesis testing, the specified model of step 1 must satisfy the null hypothesis as well as be a realistic model in terms of shape and spread for the problem at hand. In Chapters 6 and 7 supplying such a model was fairly straightforward. For example, in Chapter 7 we often used a fair many-sided die null hypothesis model, and in the Chapter 6 problem about community attitudes toward raising the driving age, a fair coin null hypothesis model sufficed. In this section something more subtle is often called for. When understood, this new approach to building the null hypothesis model of step 1 will seem most reasonable. The approach is called *bootstrapping* the observed data. We were introduced to bootstrapping in Section 8.3 as a method of approximately obtaining the standard error of an estimate. Bootstrapping's central idea is to use the shape of the data to supply a good estimate of the unknown model population that we wish to sample from. The name *bootstrapping* comes from the cliché of pulling one's self up by one's own bootstraps—that is, climbing upward with no assistance other than one's own body, clearly a feat not literally possible. In a statistical context it means making a statistical inference using only the data to produce the model: that is, we do not make the usual specification of a model (such as a binomial, a normal, or a uniform distribution), which is usually arrived at independently of the data. The statistical bootstrap has a valid justification and indeed often works well in actual applications. From the viewpoint of this book, it is a special version of our six-step method of hypothesis testing in which the model of step 1 is entirely determined by the observed data and the null hypothesis. Bootstrap methods are beginning to be heavily used in modern statistical practice; hence the method you will learn here is part of the modern statistical arsenal. Let us return to the Key Problem.

1. Choice of a Model (Definition of the Population): We must choose a realistic model—a population—that conforms with the null hypothesis. In particular, the population must represent the body temperatures of the large set of adults actually being sampled from, but with the null hypothesis being true and hence having a population average of 98.6. But simply saying the population mean is 98.6 does not define the population, for many populations have a theoretical mean of 98.6. What is the population standard deviation? What does the shape of the population histogram (that is, the theoretical probability distribution of the temperature of a randomly sampled person—see the introduction to Chapter 8) look like? For example, we could decide it is unwise to assume a normal shape for our population distribution, even though many would take this approach. Indeed, we will instead presume that the shape of the population histogram or theoretical distribution looks like that of the sample relative frequency histogram in Figure 11.1, except that it is shifted so that its theoretical mean is 98.6. (Note that the rectangles in Figure 11.1 are 0.2 in width, and keep in mind that the probability of an interval is given by the area, not the height, of the rectangle.) Since on average the data points are $98.6 - 98.25 = 0.35$ lower than the hypothesized value, we simply add 0.35 to all the data points.

Now we replicate each such data point many times to create a large population of adults satisfying the null hypothesis. (Example 11.1 below exhibits this procedure more explicitly.) This invented population that characterizes H_0 being true then has a theoretical mean of 98.6 but otherwise is shaped like the data. The advantage of this approach is that we have not needed to assume any particular theoretical distributional shape, but have let the data alone determine our estimate of the shape. An approach that makes no assumptions about the particular shape (for example, a bell shape) of the population distribution is called **nonparametric** because it is free of

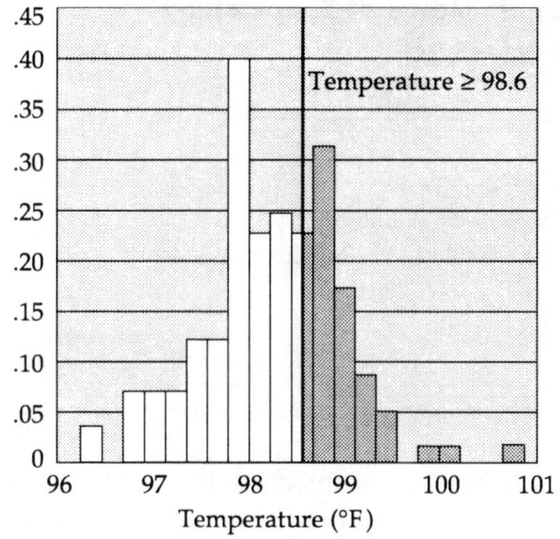

Figure 11.1 **Adult temperature data.**

restrictive assumptions that are usually given by parameters that specify the shape of the population distribution. The nonparametric approach is very powerful because the user takes no risk of being deceived by assuming an incorrect shape for the population histogram. The nonparametric, data-driven approach we are embarking on is one version of what statisticians call the *nonparametric bootstrap.*

2. Definition of a Trial (Sample): A trial consists of randomly choosing 130 readings from the population, sampling without replacement (noting that the actual data set was selected from the real population without replacement).

3. Definition of a Successful Trial: Because we are concerned with the average temperature of 130 people, the statistic is the average of the (new) 130 readings sampled from the population. The trial is a success if the average is less than or equal to 98.25.

4. Repetition of Trials: We perform the sampling 100 times, each time obtaining a new mean. The stem-and-leaf plot in Table 11.1 contains these means. The average of the means is 98.6048—very close to the null hypothesis value, as expected because the population mean is 98.6. The standard deviation of the means can be computed: it is 0.0606.

5. Estimation of the Probability of the Obtained Average or Less (Probability of a Successful Trial): We want to know the chance of obtaining a value as low as or lower than 98.25 from this null hypothesis population designed to have a population mean of 98.6. It turns out that all the simulated sample averages in step 5 are higher than 98.25: they range from 98.45 to 98.75. Thus we estimate the probability to be 0.

Table 11.1 **Sample Means of Temperatures from 100 Samples of Size 130**

Stem	Leaf
984	567
985	1222223334444444
985	55555667777778888889999
986	0000000000111112222222333344444
986	5555555566677778889
987	00014
987	55

Key: "984 567" stands for 98.45, 98.46, 98.47 degrees.

6. Decision: If the null hypothesis is true, the chance that the sample mean is as low as 98.25 is estimated to be 0, much less than the conventional value of 0.05 for an unlikely event. Thus there is strong evidence that the null hypothesis does not appear plausible—the observed value of 98.25 cannot be ascribed to chance under the null hypothesis. We reject the null hypothesis, believing the evidence to be very strong. That is, the evidence is strong that for the population from which the data were sampled, the average temperature is *not* the "normal" value of 98.6 degrees.

Since it is a totally new idea to form a nonparametric bootstrap null hypothesis population, we now examine the crucial step 1 in detail. To understand how to construct the null hypothesis population model from the sample data, let's simplify the body temperature problem by assuming that the sample is of 5 people rather than 130. Suppose their observed temperatures are 97.3, 97.5, 98.4, 98.6, and 99.2. The goal is to invent a large population (to be sampled from without replacement) shaped like these data but with a mean of 98.6, thereby satisfying the null hypothesis. The mean of the five points is 98.2. To start building the large population, we have to add something to each point so that their mean is 98.6. Because $98.6 - 98.2 = 0.4$, we have to add 0.4. We then have 97.7, 97.9, 98.8, 99.0, 99.6. (Check that these five have mean 98.6.) We now create the desired realistically large population by replicating each of these five points many times. Suppose the large population to be modeled is of size 5000. This choice of 5000 is arbitrary as long as it is large: for instance, 500 would be fine, too. Then we replicate each point 1000 times so that the total in our invented large population is 5000. The result is a box model of size 5000 that satisfies the null hypothesis and, we believe, is shaped approximately like the real population distribution of all people's temperatures. (See the following table.)

This invented population is our best guess of what the real population looks like in terms of centering, spread, and overall distributional shape if the null hypothesis is true. If the sample of size 5 is reasonably representative

97.7	97.7	97.7	97.7	97.7	97.7	97.7	97.7	97.7	97.7	...	97.7
97.9	97.9	97.9	97.9	97.9	97.9	97.9	97.9	97.9	97.9	...	97.9
98.8	98.8	98.8	98.8	98.8	98.8	98.8	98.8	98.8	98.8	...	98.8
99.0	99.0	99.0	99.0	99.0	99.0	99.0	99.0	99.0	99.0	...	99.0
99.6	99.6	99.6	99.6	99.6	99.6	99.6	99.6	99.6	99.6	...	99.6

of the unknown population (spread, shape, and so on), then our population created from the data should be (roughly) shaped like the unknown true population distribution we would use in step 1 *if* we knew it.

Now our plan is to repeatedly sample five numbers without replacement from this invented population. Because the population size (5000) is large relative to the planned sampling size (5), we know from Section 5.7 that the probability of a successful trial is almost unaffected if we instead sample with replacement, so we will sample with replacement. But if we sample with replacement, the probability law of the five-observation sample average needed for step 5 is the same if we use the shifted original (five-member) sample as our step 1 population instead of the population of 5000 formed from the 1000 replications. By switching to sampling with replacement to form our simulated samples, then, we can avoid all the effort of creating a large invented population. We can rather merely take the basic five shifted null hypothesis values, 97.7, 97.9, 98.8, 99.0, and 99.6 to be the entire null hypothesis population, and repeatedly randomly sample five observations from this five-member population *with replacement*. That is, we randomly choose one, record its value, and then put it back. We then randomly choose another, which could be the same as the first, record its value, and then put it back. We do this five times to obtain each step 2 sample. Such random sampling with replacement from the actually observed data set is *bootstrap sampling*.

In summary, we can bootstrap-sample repeatedly (that is, with replacement from the actual observed sample, but translated to make the null hypothesis true) as an acceptable substitute for repeatedly sampling without replacement from the large invented population we would have created to be shaped like the actual observed sample but translated to make the null hypothesis true.

Now imagine the above bootstrap sampling with replacement using as our box model the original 130 temperature readings shifted by $+0.35 = 98.6 - 98.25$. That is, we now return to the original data set of 130 body temperature measurements. We sample 130 observations with replacement 100 times. A statistician could use such a nonparametric bootstrap procedure in this situation in order to verify the choice of the distribution of the sample average under the null hypothesis for use in step 5. Indeed, in cases in which the population distributional shape is not known, the above bootstrap approach is very appealing and would be used by many professional statisticians when the sample size is too small for application of the central limit theorem—that is, well under 30 (and hence certainly in the case of 5). But when the sample size used to compute the sample mean is large, as 130 is in the example, most statisticians would appeal to the central limit theorem of Chapter 10 because it tells us that the distribution of these sample means will be well approximated by the normal distribution *regardless* of the shape of the distribution of temperatures in the population. This normal approach to carrying out step 5 is developed in Section 11.2.

In Example 11.1 below we illustrate the bootstrap approach as a special case of our six-step method of hypothesis testing.

Example 11.1 Are Husbands More Educated Than Their Wives?

Data on 177 Illinois husband-wife couples from the 1989 Current Population Survey* yielded the comparison of attained educational levels presented in the relative frequency histogram of Figure 11.2. The distributions look reasonably similar, although it appears that more of the husbands go through two years of college (14 years of education) and more wives have but one year of college (13 years). The average for the 177 husbands is 12.89, and that for the 177 wives is 12.65. The husbands average 0.24 year more of education. Could that be due to chance, or

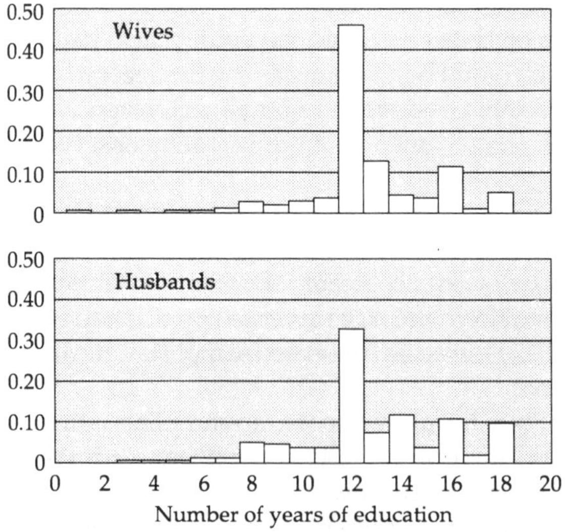

Figure 11.2 **Years of education data for married couples.**

*From data disk accompanying D. Freedman, R. Pisani, R. Purves, and A. Adhikari, *Statistics* (New York: Norton, 1991).

are husbands more educated than their wives on average in the population we are sampling from?

Because we are interested in differences, the variable we look at is

Difference = husband's education − wife's education

for each couple. Figure 11.3 shows the relative frequency histogram of the 177 differences. Over 30% of the couples have the same educational level. One wife (with 14 years) has 11 more years than her husband (with 3). Otherwise, the largest difference is seven years. The average of these differences is the 0.24 we saw above. Is it plausible that these data could be a sample from a population with a difference in number of years of education exactly 0? We wish to test the following null hypothesis:

H_0: The average difference in years of education for the population of Illinois husband-wife couples is 0.

Let us proceed to the six steps, using the nonparametric bootstrap method of this section.

1. Choice of a Model (Definition of the Population): We seek a null hypothesis population. It will be our best estimate of the Illinois husband-wife population satisfying the null hypothesis. Consistent with the bootstrap approach, we will use the sample to create this population. Because in the sample the observed difference is 0.24 year, we subtract 0.24 from each of the 177 differences in the data to produce a null hypothesis population of differences whose mean is exactly 0. That is, we shift all the differences by the same amount so that they can be viewed as a step 1 population for which the null hypothesis holds. Again we have the choice of creating a large realistic population by replicating each of the 177 members many times and then repeatedly sampling 177 differences without replacement. But once again we will more simply let the 177 translated differences be our entire null hypothesis population and produce our large number of size 177 sampled means by bootstrap-sampling with replacement between draws.

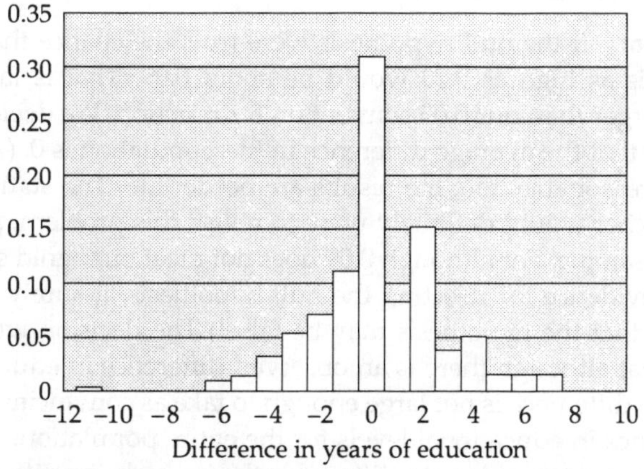

Figure 11.3 **Differences in years of education for married couples.**

***Table 11.2* Sample Means of Education Differences from 100 Samples of Size 177**

−6	1
−5	
−4	64
−3	84433331
−2	88877777554433
−1	988865554221110000
−0	877777653322221110
0	1113388889
1	01223345556667789
2	0024589
3	1334
4	0

Key: "−6 1" stands for −0.61.

2. Definition of a Trial (Sample): A trial consists of randomly choosing 177 differences (husband's years of education minus wife's years of education) from the population of 177 by sampling with replacement.

3. Definition of a Successful Trial: The statistic of interest is the average of the 177 differences sampled from the invented population. A trial is a success if the observed average of differences is as large as or larger than 0.24.

4. Repetition of Trials: We perform the sampling 100 times. The means of the 100 samples are shown in the stem-and-leaf plot in Table 11.2.

5. Estimation of the Probability of the Obtained Average or More (Probability of a Successful Trial): We want to know the chance of obtaining an average of differences as large as or larger than 0.24. From Table 11.2, we can count 9 that are 0.24 or above, so the probability is estimated to be 0.09.

6. Decision: If the null hypothesis were true, the chance that the sample mean difference is as high as 0.24 would be about 0.09. That is fairly small, but it is certainly larger than our 0.05 convention for rejection. We decide to *accept* the null hypothesis that the average difference in the population is 0. (A tentative rejection would be reasonable, too; the results are borderline. The statistically knowledgeable researcher would definitely want to revisit this problem, perhaps with a new and larger sample. For although 0.09 does not meet our "gold standard" for strong statistical evidence for rejecting the null hypothesis, it surely raises our scientific suspicions that the hypothesis may be false!) The decision we make here in step 6 means that although there is an observed difference in educational levels in the sample, the difference is not large enough to take as convincing evidence that there is a difference in educational levels for the entire population. That does not mean we are convinced there is not a difference; there may be no difference, or there could be a small one.

In Chapter 7, where we did chi-square testing, we learned that statisticians can, in fact, bypass the simulations of the six-step method and appeal to the method of chi-square density. In contrast, the bootstrap simulation method, which *does not* bypass the simulations of the six-step method, is often the professional statistician's method of choice when the sample size is small (say, under 30).

SECTION 11.1 EXERCISES

1. Suppose a sample of 100 heights of men has a mean of 70.53 inches and a standard deviation of 3.22 inches. Explain how to test the hypothesis that the mean height of men in the population is 69 inches.
2. Suppose for Exercise 1 we have replicated 100 sample means each from a sample size of 100 taken from the invented population with a mean height of 69 inches. They are recorded in the following stem-and-leaf plot. Is the mean height of the men in the population really 69 inches?

Stem	Leaf
681	2
682	2
683	08
684	2
685	01117
686	233459
687	01112334444559
688	003357778899
689	1122333445555678
690	001233459
691	000113333799
692	2223339
693	1234444478
694	1
695	
696	26
697	5

Key: "697 5" stands for 69.75 inches.

3. Suppose 100 small samples were taken from a certain batch of concrete. The mean compression strength of the samples was 4129.58 pounds with a standard deviation of 164.12 pounds. Explain how to test the hypothesis that the mean compression strength of the batch of concrete is 4200 pounds.
4. For Exercise 3, we replicated 100 sample means each from a sample of size 100 taken from the invented population with mean 4200 pounds. They are recorded in the following stem-and-leaf plot. Is the compression strength of the batch of concrete really 4200 pounds?

Stem	Leaf
416	778
417	2
417	557789
418	0022244
418	5567779
419	00001122334
419	55667777899999
420	11122333334444
420	5556777888999
421	011144
421	5666699
422	0012
422	5667
423	
423	57
424	
424	6

Key: "424 6" stands for 4246 pounds.

5. In a random sample of 100 husbands, the mean age was 46.62 years with a standard deviation of 4.40 years. The mean age of the 100 wives of the husbands was 41.88 years with standard deviation of 3.41 years. Explain how to test whether the difference between the ages of husbands and wives is 0.

6. Refer to Exercise 5. The following is a frequency table of 100 simulated differences taken from an invented population with a mean difference of 0. Test whether the difference between the ages of the husbands and wives is 0.

Difference (years)	Frequency
−1.4	2
−1.3	0
−1.2	1
−1.1	1
−1.0	0
−0.9	2
−0.8	1
−0.7	1
−0.6	3
−0.5	5
−0.4	7
−0.3	3
−0.2	10
−0.1	5
0.0	8
0.1	9
0.2	5
0.3	9
0.4	7
0.5	5
0.6	1
0.7	8
0.8	0
0.9	2
1.0	3
1.1	1
1.2	0
1.3	0
1.4	1

7. In a random sample of 200 high school seniors who took the SAT test, the mean SAT verbal score was 544.655 with a standard deviation of 36.36, and the mean SAT math score was 531.655 with a standard deviation of 47.10. Explain how to test whether the difference between SAT verbal and math scores is 0.

8. Refer to Exercise 7. The 100 simulated differences shown in the following stem-and-leaf plot were taken from the invented population with a mean difference of 0. Test whether the difference between SAT verbal and math scores is 0.

Stem	Leaf
−16	5
−15	
−14	
−13	0
−12	4
−11	
−10	
−9	2
−8	976
−7	9852
−6	98732
−5	9966621
−4	87653210
−3	77410
−2	97554
−1	55333321000
−0	665
0	011268
1	02358999
2	1578
3	3356888
4	1245
5	244469
6	0
7	37
8	24
9	0235
10	
11	2

Key: "11 2" stands for 1.12.

11.2 THE z TEST OF THE POPULATION MEAN

Introduction

In Chapter 10 we learned that according to the central limit theorem, sample means approximately follow a normal density if the number of observations n used to compute each such mean is at least 20. In the hypothesis-testing problem of Figure 11.1, we note that the relative frequency histogram of the sample data looks reasonably like a normal density. Although not discussed in Chapter 10, this resemblance implies that sample means will approximately follow a normal density even when the number of observations n used to compute each such mean is much less than 20. Indeed, the closer the shape of the original data is to normal (in turn suggesting an approximately normal population distribution), the smaller n can be with the means still approximately obeying a normal distribution.

Whenever we believe that the sample means approximately obey a normal distribution because of the central limit theorem, either because n is large or because we have evidence that the population distribution is itself approximately normal, then to test hypotheses about the population mean we can use the normal table (Appendix E) rather than do 100 (or more) bootstrap simulations of sample means.

Recall that we made an analogous decision to use the chi-square table to test hypotheses in Chapter 7 because the histograms of the simulated chi-square statistics appeared to be similar to the smooth chi-square density. That decision was also supported by a theory of the same kind as the central limit theorem, and hence using the chi-square density to compute probabilities was justified.

At step 5, then, instead of counting the proportion of the bootstrap-simulated means that are too big or too small, we will estimate this proportion using the normal density of Chapter 8. But in order to use the normal curve, we need a mean and a standard deviation for standardization purposes. Consider the six-step solution to the Key Problem presented in Section 11.1. In step 4 the 100 bootstrap-simulated means have a mean of 98.6048 and a standard deviation of 0.0606. Thus we do have the needed mean and standard deviation. Now we can substitute the following for step 5:

New Step 5. Estimation of the Probability of the Obtained Average or Less (the Probability of a Successful Trial): Recall from Sections 8.6 and 8.7 that in order to compute a probability for a statistic that obeys a normal distribution, we must first standardize the statistic by subtracting an appropriate mean and dividing by an appropriate standard deviation. The most straightforward way to do this is to simply subtract the sample average of the statistic and divide by the sample standard deviation, as was done in the hot-dog example of Table 8.10. Here the obtained mean is viewed under the null hypothesis to be the result of sampling from the null hypothesis population and obtaining 100 bootstrap-sampled means. Thus we can standardize the obtained mean of 98.25 by centering on (subtracting) the mean of the 100 bootstrapped means and normalizing (dividing) by the standard deviation of these 100 means. That is, we obtain the z statistic:

$$z = \frac{\text{obtained mean} - \text{mean of simulated means}}{\text{standard deviation of simulated means}}$$
$$= \frac{98.25 - 98.6048}{.0606} = -5.85$$

Now we use the normal table (Appendix E) to find the probability that a standard normal variable is less than -5.85. Appendix E shows that this probability is essentially 0—that the chance that a simulated mean is less than or equal to the obtained mean 98.25 is essentially 0. (Note that using the regular step 5 based on bootstrap simulations also gave the same answer of 0.)

Step 6. Decision: We make the same decision as in the old step 6 of Section 11.1 because it is very unlikely to obtain a z lower than -5.85 from the invented null hypothesis population. We have to reject the null hypothesis and declare that the population mean is not 98.6.

The new step 5, which took advantage of the fact that sample means follow a normal distribution, did not save much time, because in order to find the standardizing mean and standard deviation of the means we still had to simulate the 100 means from the invented null population, and doing so required $100 \times 130 = 13{,}000$ simulated observations. But the good news is that we can deduce what mean and standard deviation should be used to standardize 98.25, based only on the actual 130 obtained data points—that is, without doing the time-consuming 100 bootstrap simulations (needing 13,000 body temperatures).

In order to deduce what mean and standard deviation to use, imagine obtaining thousands of bootstrap means (instead of 100, as we did above), each consisting of 130 observations with replacement from the null hypothesis population. We know that the population mean of the null hypothesis population (the average of the 130 temperatures that constitute the null hypothesis population) is exactly 98.6. Now consider the mean of the thousands of imagined bootstrapped means. This is the "grand mean" of a huge number of temperatures sampled with replacement from the population of 130. In Chapter 4 we learned that such an observed mean based on a large number of observations will in fact be very close to the population mean (that is, the theoretical expected value of Chapter 4). Thus

Mean of thousands of simulated means

$$\approx \text{null hypothesis population mean} = 98.6$$

Thus, we should clearly center 98.25 at 98.6—that is, subtract 98.6 from 98.25.

Now recall that in Chapter 9 we learned that the standard deviation of a sample mean, called the *standard error of the mean*, is given by

$$\text{SD(mean)} \overset{\text{definition}}{=} \text{SE(mean)} = \frac{\text{population SD}}{\sqrt{\text{sample size}}}$$

But we can compute the population standard deviation exactly, because we know the 130 temperatures that make up the null hypothesis population. Now if we were to take the approach of Chapter 4 to estimate SE(mean), we would take the average of many standard deviations of means (100 of which yielded 0.0606 above). But if we instead used thousands of such simulated means, the standard deviation of all these means would have to be very close to the theoretical standard deviation of the mean, which as we saw above is given by the population standard deviation divided by $\sqrt{130}$.

We have already seen that the standard deviation of the data is 0.73, which is also the null hypothesis population standard deviation. Now the null hypothesis population was obtained from these data merely by shifting each number by -0.24. Shifting all the data points by the same quantity clearly leaves the spread of the data unchanged. In particular, the null hypothesis population will have the same standard deviation of 0.73. Thus

$$\text{SD of thousands of sampled means} \approx \frac{\text{sample SD}}{\sqrt{\text{sample size}}} = \frac{0.73}{\sqrt{130}} = 0.0642$$

Thus we have the two quantities needed to standardize 98.25. Note that they are quite close to the sampled values 98.6084 and 0.0606 that we used before.

These results show that in the equation for z in step 5 we can replace the 98.6048 by the population mean under H_0, which is 98.6, and we can replace the 0.0606 by the standard deviation of the obtained data divided by $\sqrt{130}$, which we have seen is 0.0642. The new standardized z is then

$$z = \frac{\text{obtained mean} - \text{mean under } H_0}{(\text{obtained SD})/\sqrt{n}}$$

$$= \frac{98.25 - 98.6}{0.0642} = -5.45$$

The probability of having a standard normal less than -5.45 is essentially 0, so we again conclude that the null hypothesis is not plausible, and we *have* saved a lot of time.

The z Test

We now have an alternative approach to testing the null hypotheses in Section 11.1, one that requires *no* bootstrap sampling. It works well when our interest is in making inferences about the mean of the population and when the sample size is reasonably large—say, $n \geq 30$—or when the original data set is approximately shaped like a normal density. The basic change is in steps 4 and 5. We will illustrate by repeating the example of the husbands' and wives' years of education using this normal-curve-based approach.

The null hypothesis is exactly the same:

H_0: The average difference in education for the population of Illinois husband-wife couples is 0.

The six steps are next.

1. Choice of a Model (Population): The population represents the number of years of education of husband-wife couples in Illinois, shifted so that the average difference in education in the population is 0—that is, so that H_0 holds. Since we will not be doing bootstrap sampling from a null hypothesis population, we do not need to explicitly build a null hypothesis box model.

2. Definition of a Trial (Sample): A trial would consist of randomly choosing 177 differences without replacement from the large created null hypothesis population. Again, we will not be doing simulation trials, so we can skip this step.

3. Definition of a Successful Trial The trial is a success if the average of the 177 differences is larger than 0.24. We will use this 0.24 in step 5, but we will not be doing the simulation trials of steps 2 through 4.

4. Repetition of Trials: Instead of actually bootstrap-sampling 100 times from the null hypothesis population and finding the means, we use our new z-test approach, which bypasses bootstrap sampling. In particular, we need to standardize the observed sample mean of 0.24. The centering is, as we learned, at the theoretical mean of the null hypothesis population, namely, 0. The standard deviation that we will divide by is given by

$$\frac{\text{SD of data}}{\sqrt{\text{sample size}}} = \frac{2.58}{\sqrt{177}} = \frac{2.58}{13.304} = 0.194$$

(We omit details showing that the data standard deviation is 2.58.) Thus we do not repeat the sampling of 177 students 100 times!

5. Estimation of the Probability of the Obtained Average or More (Probability of a Successful Trial): We want to know the chance of obtaining an average difference as large as or larger than 0.24. Because we are using the normal curve with mean 0 and standard deviation 0.194, we have to standardize:

$$z = \frac{0.24 - 0}{0.194} = 1.24$$

The area under the normal curve below 1.24 is 0.8925, so the area above 1.24 is 0.1025. By standardizing z and using a normal distribution table, then, we have avoided the process of simulating trials to obtain the experimental probability of success.

6. Decision: If the null hypothesis is true, the chance that the sample mean difference is as high as 0.24 is estimated to be about 0.1. (Compare that with the 0.09 we found in Section 11.1 using the bootstrap approach.) Again, we will decide to accept (barely) the null hypothesis that the average difference in the population is 0.

We now are in the same place with z testing as with doing chi-square testing using a chi-square table. Namely, *no* simulation is required in order to carry out the z test. This z test is heavily used in statistics. It applies whenever the null hypothesis concerns the population mean and whenever the sample size is fairly large ($n \geq 30$ is the convention usually followed by professional statisticians) or the population itself is known to be approximately normal in shape and hence n can be small.

You might ask whether the bootstrap method of Section 11.1 is ever needed. The answer is a definite yes. Whenever the shape of the population cannot be assumed to be normal and the size of the sample is well less than 30, the method of Section 11.1 is the one many statisticians would use to test a hypothesis about the population mean. This situation occurs often in statistical applications. Indeed, this bootstrap approach is becoming a keystone of modern statistics, because computer power is inexpensive and widely available, thus allowing fast and inexpensive simulations whenever needed.

Between the bootstrap method of Section 11.1 and the z test of Section 11.2 we have two methods from which we can always choose one for any test involving a hypothesis concerning a population mean. Thus you are empowered to test a hypothesis about a population mean in any setting. Sometimes both methods are appropriate, and one can use both and compare their answers.

SECTION 11.2 EXERCISES

1. The popularity of a television show is found by taking a random sample of 100 households around the country. The producers of a particular show believe the rating for their show is 33. For the sample of 100 households, the mean rating was 31 with a standard deviation of 4.3. Are the producers correct?
2. It is believed that the length of a particular microorganism is 25.5 micrometers. A set of

64 independent measurements of the length of the microorganism is made. The mean value of the measurements is 27.5 micrometers with a standard deviation of 3.2 micrometers. Is the length of the microorganism really 25.5 micrometers?

3. The chancellor of the University of Illinois at Urbana believes that undergraduates study, on average, 20 hours per week. A random sample of 40 students found the mean study time was 19.5 hours per week with a standard deviation of 4.05 hours. Is the chancellor correct?

4. For Exercise 1 in Section 11.1, use the z test to determine whether the true population mean is 69 inches. Is your answer the same as before?

5. For Exercise 3 in Section 11.1, use the z test to determine whether the true population mean is 4200 pounds. Is your answer the same as before?

6. A car manufacturer believes that the fuel economy of one of its models is 19.8 miles per gallon. In a random sample of 30 cars, the mean fuel economy was 19.45 miles per gallon with a standard deviation of 1.14 miles per gallon. Is the car manufacturer correct?

Graphing Calculator Exercises

In the following exercises use the z test and the normal curve program called NORMAL.

7. Victor C. Rental claims he can always tell when the economy is in a slump, because during times of economic depression, video rentals go up. From prior observation he knows that a family rents an average (mean) of 10 videos per month with a standard deviation of 5 videos. This month he randomly polled 25 families and found they rented a mean of 12 videos per month. Do you think that Victor has a valid claim for an economic downturn?

8. After extensive testing, the manufacturer of a new light-bulb claims that the bulb has a mean lifetime of 1500 hours with a standard deviation of 100 hours. Suppose you purchase 16 light bulbs and test them. You find the group of 16 to have an average life of 1475 hours. Can you disprove the manufacturer's claim? (Assume that the 16 light bulbs are chosen at random.)

11.3 MAKING A WRONG DECISION

In Section 11.1 we decided that the average body temperature in the population was less than 98.6 and that the difference in years of education between the husbands and wives could be 0. Were we correct? It is impossible to know without actually surveying the entire populations in question (usually a practical impossibility), but we certainly could have made a mistake. There are two types of errors one can make:

- **Type I error:** rejecting the null hypothesis when the null hypothesis is true

- **Type II error:** accepting the null hypothesis when the null hypothesis is not true

Unfortunately "type I error" and "type II error" are not well-chosen terms in the sense of being easy to memorize, but nonetheless they are what statisticians say!

As we have already stated, it is fairly standard practice in statistical work to want the probability of making a type I error to be smaller than some small probability, such as 0.10 or 0.05 or 0.01. Usually 0.05 is used. The way to guarantee that your chance of making a type I error is less than 0.05, say, is to reject the null hypothesis only when the probability is less than 0.05 that the null hypothesis model could yield a result as extreme as or more extreme than your data. In the body temperature example in Sections 11.1 and 11.2, we figured that the chance of obtaining an average temperature as low as or lower than the data's 98.25 was essentially 0. Because 0 is well below 0.05, it is safe to reject the null hypothesis that the average body temperature in the population is 98.6. By contrast, in the educational levels example of Section 11.2, the chance of obtaining an average difference between husband and wife as large as or larger than the data's 0.24 was 0.09 or 0.10, depending on which method one relies on. Because these values are greater than 0.05, we have to "accept" the null hypothesis that the husband-wife difference is 0, knowing we are not sure that the hypothesis is in fact true but lacking strong evidence that it is not.

If you are very conservative, you might decide that you want the chance of a type I error to be very small, such as 0.0001 (1/100th of a percent). Then you could reject a null hypothesis only if there is overwhelming evidence against it. It is interesting to note that such conservatism is used in our legal system. The null hypothesis is that the person on trial is innocent ("innocent until proven guilty"), and one rejects the null hypothesis—declares the person guilty—only if the evidence against innocence is "beyond a reasonable doubt."

The consequence of requiring such a small chance of type I error is that you will often incur a large chance of type II error. That is, even when the null hypothesis is *not* true, you will often still accept it. This is especially true when the sample size is small. (There may be many jury trials in which people who are probably guilty are not declared guilty.) In this sense it is important to understand that to the statistician, "accepting" the null hypothesis does not mean that the null hypothesis is believed to be true (although it may be), but rather that sufficiently strong evidence for rejecting the null hypothesis is *lacking*.

There is always this trade-off: trying to decrease the chance of one type of error increases the chance of the other type. The only way to decrease both types is to collect more data and hence have more information. How to balance the chances of the two types of errors depends on the particular situation. In the legal example, the idea is that it is much worse to convict an innocent person (type I error) than release a guilty one (type II error). When researchers are developing new drugs to treat diseases, the null hypothesis is that the drug is no better than conventional treatment. In such cases, it is bad to declare a new drug useless when it can be of help to people (type I error), but it is also bad to declare a new drug useful when it is not (type II error), because of possible side effects, expense, and replacement of effective treatments. Sometimes statisticians will choose a type I error value different from 0.01, 0.05, or 0.1 because of information they may have about the relative damage resulting from a type I error as compared with a type II error.

SECTION 11.3 EXERCISES

1. What are some common values for the probability of a type I error?
2. Suppose you lower the probability of a type I error from 0.05 to 0.01. What happens to the probability of a type II error?
3. What must you do to decrease the probability of both type I and type II errors?
4. In a court of law in the United States, a conviction for a crime requires "guilt beyond a reasonable doubt." What level of certainty should be required for reasonable doubt? Should you be 99% certain of a person's guilt to convict? Explain your answer.

11.4 THE SIGN TEST

The test statistics so far in this chapter have been based on the mean and have dealt with hypotheses concerning the population mean of measured data, as in the example concerning the number of years of education. But when one says that 98.6 degrees is "normal," does one imply that it is the population average? The population median? Something else? Statisticians often form hypotheses about the median when they wish to study the location of the center of a population. Indeed, one could argue that the population median is really the better indicator of the center of the population, as discussed in Chapter 2.

There is a simple way to test whether the population median in the example of the body temperatures is 98.6. Start by defining the null hypothesis:

$$H_0\text{: The population median is 98.6.}$$

Table 11.3 is the stem-and-leaf plot of the 130 observations of human body temperature. If the null hypothesis is true, we would expect about half of the data to be less than 98.6, and about half to be greater than 98.6. We actually obtain 81 values less than 98.6, 39 values greater than 98.6, and 10

Table 11.3 **Body Temperatures of 130 Randomly Sampled People**

96	34
96	7789
97	0111222344444
97	556666777888888899999
98	00000000000111222222222233333444444444
98	555666666666677777777888888888899
99	000001112223344
99	59
100	0
100	8

Key: "96 34" stands for 96.3, 96.4 degrees.

exactly equal to 98.6. The distribution looks a little lopsided—that is, there are too many data below 98.6. To assess whether 81 below and 39 above are too many to ascribe to chance, we have to proceed through the six steps. The procedure we are going to use only looks at the values above or below 98.6, not at the 10 that are exactly 98.6. The data will therefore be the 120 observations that are not exactly 98.6.

1. Choice of a Model (Population): We build our null hypothesis model in a manner similar to the way we did it in Section 11.1. Again, we apply the useful trick of sampling with replacement from a small population constructed directly from the sample. But instead of setting up this null hypothesis population model by specifying the population mean, we specify the population median instead. That is, we assume a population of 120 observations having a median of 98.6. More precisely, we require 60 observations to be less than 98.6 and 60 observations to be greater than 98.6, thus making 98.6 the median.

Below we will define a successful trial on the basis of the number of the 120 sampled observations from the null hypothesis population that are less than 98.6.

The probabilistic behavior (distribution) of the statistic of interest is in fact the same for every null hypothesis population having median 98.6. Hence we do not need to specify the shape of the null hypothesis population as long as we obtain a median of 98.6, which implies that

$$p(\text{sampled observation} < 98.6) = \frac{1}{2}$$

which is the only fact that matters. Because of this we can simulate the decision-making statistic of interest used in step 5 in a particularly simple manner similar to that of the simulation trials of Chapters 6 and 7.

2. Definition of a Trial (Sample): A trial consists of randomly choosing, with replacement, 120 readings from the null hypothesis population of 120 that we would construct. But as remarked above, the null hypothesis implies that the probability of a value being less than 98.6 is $\frac{1}{2}$, so our sampling procedure is the same as flipping a fair coin 120 times. That is, instead of sampling from the 120 null hypothesis population observations, we simulate 120 flips of a fair coin and count the number of heads, a head signifying a temperature less than 98.6. (Recall from Chapter 8 that the distribution of this statistic is binomial with 120 trials and with a probability of a success on any one trial being $\frac{1}{2}$.) Note also that the independence of different tosses of a coin implies we are sampling with replacement, because sampling with replacement means that the sampled values of different draws are independent.

3. Definition of a Successful Trial: Because we are concerned with whether the median is 98.6, the statistic is the number of readings less than 98.6. We really do not care about the values of the readings, only whether they are less than or greater than 98.6. Therefore a trial is a success if 81 tosses are heads, corresponding to 81 values below 98.6.

4. Repetition of Trials: We perform the sampling 100 times, each time obtaining a new count for the number (out of 120) less than 98.6 (that is, each time counting the number of heads in the 120 tosses). Since we need $100 \times 120 = 12{,}000$ coin flips, we simulate them on a computer. Here are the numbers of heads obtained in the 100 trials:

3	99
4	012233334444555555666666666666777777777788888889999999999999
5	0000000000000111112222233333333444455566666677899999
6	01235

Key: "3 9" stands for 39 tosses.

5. Estimation of the Probability of the Obtained 98.6 or Less (Probability of a Successful Trial): We want to know the chance of obtaining 81 or more values less than 98.6, that is, the chance of obtaining 81 or more heads in 120 flips of a fair coin. All the trials in step 4 came up with fewer than 81 heads. The most was 65. Thus we estimate the chance that one obtains 81 or more readings below 98.6 if the median is 98.6 to be 0.

6. Decision: The chance that if the null hypothesis is true, one obtains 81 or more readings less than 98.6 has been estimated to be 0. Thus the null hypothesis does not appear at all plausible: the observed value cannot be ascribed to chance. Thus, as in Section 11.1, we reject the null hypothesis that a temperature of 98.6 is "average." More precisely, we conclude that the median temperature of the population is less than 98.6.

The test we just performed is called the **sign test** because we count the number of observations less than 98.6—that is, the number of observations that yield a negative sign after subtraction of the hypothesized median of 98.6. It has the advantage over the test based on the sample average in that we do not have to specify the null hypothesis population in any detail: we only required that it have a median of 98.6. For the test in Section 11.1, we had to specify not only the average of the population but also the shape of the histogram of the population (which we obtained from the data via bootstrapping). For the sign test, all we need is the hypothesized median, because all we consider is whether the observations are below or above the median. Statisticians often use the median test, although ignoring the size of the observations, as the sign test does, often seems to throw away valuable information.

We can also modify step 5, as we did in Section 11.2 and in the case of chi-square testing, in order to avoid doing simulations. When the null hypothesis is true, the statistic we generate is binomial with 120 trials and with the chance of a success on any one trial being $\frac{1}{2}$, as mentioned in step 3. We then know from Chapter 8 that the theoretical mean of the statistic is $120(1/2) = 60$ and the theoretical standard deviation is $\sqrt{120(1/2)(1/2)} = 5.48$. Thus we can standardize the observed data to obtain another kind of z statistic:

$$z = \frac{\text{number observed} - \text{number expected under } H_0}{\text{standard deviation under } H_0}$$

$$= \frac{81 - 60}{5.48} = 3.82$$

We can look that up in the normal table because as we saw in Chapter 10, when the number of trials is large, the standardized binomial statistic is approximately standard normal. The probability of getting 81 or more observations below 98.6 is therefore (approximately) the chance that a standard normal statistic is above 3.82, which is not even in the table! That is, the chance is essentially 0, the same chance estimated using the simulations. Professional statisticians always bypass simulations when they do sign tests and use the z-statistic approach given above.

SECTION 11.4 EXERCISES

1. A feed company claims that cattle switched to its feed should gain a median of 100 pounds. To test that claim, a farmer gave the feed to 17 cattle. The amount of weight gained is given below:

62	83	90	101	104	106
109	109	109	127	143	187
204	205	209	266	277	

Use these data and the following stem-and-leaf plot to test the feed company's claim.

Number of Heads in 17 Flips of a Coin

Stem	Leaf
0	44455555555
0	66666666777777777777777
0	888888888888889999999999
1	00000000000000000111111111111111
1	222222233
1	5

Key: "1 5" stands for 15.

2. Some researchers believe that the median number of mistakes rats should make in completing a certain maze is 8. The numbers of mistakes made by a random sample of 14 rats in completing the maze are given below:

7	8	8	8	8	9	11
12	12	14	14	14	14	17

Use these data and the following frequency table to test the researchers' hypothesis.

Number of Heads in 10 Flips of a Coin

Number of Heads	Frequency
0	0
1	1
2	4
3	7
4	15
5	33
6	24
7	9
8	7
9	0
10	0

Key: "0 8" stands for 8.

3. The owners of an exercise club believe the median value of the amount of weight lost in a program during six months should be 29 pounds. The numbers of pounds lost by a sample of 40 people over six months are given below:

3	4	5	5	5	6	7	7	7	10
10	10	10	10	11	12	15	15	18	18
19	20	20	23	24	24	25	28	29	30
33	38	38	40	42	42	44	49	50	55

Use these data and the following stem-and-leaf plot to test the owners' hypothesis.

Number of Heads in 39 Flips of a Coin

Stem	Leaf
1	2
1	4455555
1	66666666666777777777
1	888888888899999999999999
2	00000000000011111111111
2	222222222222333333
2	444455
2	6

Key: "2 6" stands for 26.

4. For Exercise 1, calculate the probability using the binomial distribution instead of the stem-and-leaf plot.
5. For Exercise 2, calculate the probability using the binomial distribution instead of the stem-and-leaf plot.
6. For Exercise 3, calculate the probability using the binomial distribution instead of the stem-and-leaf plot.

11.5 THE *t* TEST

The *z* test, which is based on the average, requires that the stem-and-leaf plot of the simulated sample means follows the normal curve reasonably well. Often that is the case, but if the sample size is too small (5, for example), then using the normal tables often gives inaccurate answers. This is true even if the population has a normal distribution, in which case the sample means can be shown to exactly follow the normal curve, even if the sample size is quite small. There is an alternative test, called the *t* test, that is better than the *z* test if both the following conditions hold:

1. The distribution of the values in the population is close to the normal curve.
2. The sample size is small.

The first condition is important. If the original data do not have a normal shape—at least approximately—the *t* test will not necessarily be better than the *z* test and should *not* be used. (They may both be bad. It is better then to either use the sign test for the median or perform the actual bootstrap simulation approach of Section 11.1 to assess the population mean.) If the sample size is small—say, less than 30—the *t* test will be better than the *z* test (if condition 1 holds). If the sample size is large, the *t* test and the *z* test will give practically the same answer as to whether to accept or reject the null hypothesis, so either is fine to use. Hence most statisticians would use the *z* test because it is so easy to use.

The *t* statistic is almost exactly the same as the *z* statistic. There is only a slight adjustment to be made. Instead of using

$$\frac{\text{Standard deviation of obtained sample}}{\sqrt{n}}$$

in the denominator, we use

$$\frac{\text{Standard deviation of obtained sample}}{\sqrt{n-1}}$$

so that

$$t = \frac{\text{obtained mean} - \text{hypothesized mean}}{\text{Standard deviation of obtained sample}/\sqrt{n-1}}$$

Student's Example

In 1908, W. S. Gossett proposed the *t* test to adjust for the fact that the standard deviation of the obtained sample is a somewhat inaccurate estimate of the standard deviation of the population, at least for small sample sizes.* The z test requires that the sample standard deviation be a quite accurate estimate of the population standard deviation. (Gossett published under the pseudonym "Student," because the brewery he worked for, Guinness, did not want its research to be published. It was perhaps concerned that competitors would become aware of Gossett's results and its implications for their quality control work!) Gossett illustrated the t test on two treatments designed to increase sleep time. The treatments were dextro hyoscyamine hydrobromide and laevo hyoscyamine hydrobromide. There were 10 patients, and for each the number of additional hours of sleep gained after taking the drug was measured. The results are presented in Table 11.4.

The 0.7 under *dextro* for patient A is the number of hours of sleep after taking dextro hyoscyamine hydrobromide minus the number of hours of sleep after taking no treatment. For this patient the drug helped a little. The 1.9 under *laevo* is the number of hours of sleep after taking laevo hyoscyamine hydrobromide minus the number of hours of sleep after taking no treatment; the laevo sleep was almost two hours more than the sleep with no treatment. The difference column gives the additional hours using the laevo treatment minus the additional hours using the dextro treatment. Thus for patient A, the laevo treatment added 1.2 more hours of sleep than the dextro.

There are three questions:

Does the dextro treatment help increase sleep?
Does the laevo treatment help increase sleep?
Does the laevo treatment increase sleep more than the dextro?

Table 11.4 **Sleep Gains for Two Drugs**

Patient	Dextro	Laevo	Difference (laevo − dextro)
A	0.7	1.9	1.2
B	−1.6	0.8	2.4
C	−0.2	1.1	1.3
D	−1.2	0.1	1.3
E	−0.1	−0.1	0.0
F	3.4	4.4	1.0
G	3.7	5.5	1.8
H	0.8	1.6	0.8
I	0.0	4.6	4.6
J	2.0	3.4	1.4

*Student, "The Probable Error of a Mean," *Biometrika*, vol. 6 (1908), pp. 1–25.

We will only address the first question. The null hypothesis is that the dextro treatment does not have any effect—that is, that the average increase in hours of sleep for the population from which the 10 patients were sampled is 0:

H_0: The population mean increase from dextro in hours is 0.

The obtained average for the dextro data column is 0.75, and the sample standard deviation SD of the number of hours of additional sleep using dextro is 1.70. We go through the six steps, much as in Section 11.2, but at steps 4, 5, and 6 we use the *t* test and the *t* table (Appendix F) instead of the *z* test and the normal distribution table (Appendix E).

1. Choice of a Model (Population): The population is the (very large) set of all people who could have been chosen for the experiment. It is presumed that for this null hypothesis population the average increase in hours of sleep when using the dextro treatment is 0.

2. Definition of a Trial (Sample): A trial consists of randomly choosing 10 people from the population without replacement. Note that the results would be almost the same if we sampled with replacement, because the population is large compared with the sample size of 10.

3. Definition of a Successful Trial: One statistic of interest is the average of the 10 observations made in step 2. We will calculate the *t* statistic from this first statistic. We need $SD/\sqrt{n-1} = 1.70/\sqrt{9} = 0.57$. For the sample,

$$t = \frac{\text{obtained mean} - \text{mean under } H_0}{SD/\sqrt{n-1}}$$

$$= \frac{0.75 - 0}{0.57} = 1.32$$

A trial is a success if the simulated t is greater than 1.32.

4. Repetition of Trials: Instead of actually sampling 100 times from the population and finding the means, we imagine what the results of such sampling would be. We can skip to step 5 and use a *t* table instead of doing simulations to evaluate $p(t > 1.32)$.

5. Estimation of the Probability of the Obtained Average or More (Probability of a Successful Trial): We want to know the chance of obtaining an average increase in number of hours of sleep as large as or larger than 0.75. This will equal the chance that when we sample from a population with a mean of 0, the *t* statistic is as large as or larger than 1.32. We have to look to the *t* table given by Appendix F. As with the chi-square tables, we need to know the number of degrees of freedom (df). Recall that

n denotes our sample size. For this t test the number of degrees of freedom is $n - 1$, which in this case is 9. Looking in the t table, we see that for df $= 9$, the number 1.38 is in the column under 0.10. What this means is that the chance that the t statistic is above 1.38 is 0.1. We want the chance that the t statistic is above 1.32, which must be slightly more than 0.1.

6. Decision: We found in step 5 that the probability of obtaining from the null-hypothesis population a t statistic of 1.38 or greater is 0.1. Thus we decide to accept the null hypothesis: we are not convinced that the dextro treatment does any good.

SECTION 11.5 EXERCISES

1. For the data in Table 11.4, test whether the mean population increase in hours due to the laevo drug is 0. The mean of the 10 observations is 2.33, and the standard deviation is 1.90.
2. For the data in Table 11.4, test whether the mean population difference in hours of sleep between laevo and dextro is 0. The mean of the 10 differences is 1.58, and the standard deviation is 1.17.
3. A farmer believes the corn yield from his land will have a mean of 50 bushels per acre. In a random sample of eight plots, the farmer finds a mean yield of 49.25 bushels per acre with standard deviation of 3.25 bushels per acre. Test the farmer's claim.
4. Suppose the producer of cattle feed in Exercise 1 of Section 11.4 claims that the median weight gain will be 160 pounds, not 100 pounds. Use the data from the problem to test the company's claim.

11.6 COMPARING TWO POPULATION MEANS

The Bootstrap Approach

Until now we have tested hypotheses for one population. But the most common statistical problem is to compare two population means. That is the topic of this section.

A small study was conducted several years ago to determine whether taking the drug LSD has deleterious effects on one's chromosomes. Four users of LSD and four controls (nonusers) were studied. A sample of cells was taken from each person, and the percentage of cells in which there was chromosomal breakage was recorded. Here are the data:

Controls:	3.3	4.8	6.4	7.1	Mean = 5.40	SD = 1.47
Users:	0.9	2.6	3.4	11.5	Mean = 4.60	SD = 4.08

That is, the first control had breakage in 3.3% of the studied cells, the second control had 4.8% breakage and so on. Looking at the mean breakages, we see that actually the users had a slightly smaller amount of breakage. Is that difference due to chance? Because there is such a small sample, one would be inclined to say yes. But how can one formally test the hypothesis?

What, precisely, is the null hypothesis? It is that the mean breakages in the two *populations* are equal, the populations being the population of users and the population of controls:

$$H_0\text{: The mean breakage in the control population} \\ = \text{the mean breakage in the user population}$$

We will focus on the difference between sample means of the controls and the users in the sample: $5.40 - 4.60 = 0.80$. Is this difference small enough to be compatible with the null hypothesis? What is the chance of getting a difference that large if the two population means are the same? Let's turn to the six steps. There are some modifications because we have two populations to worry about instead of just one.

1. Choice of a Model (Two Populations): We have two populations now, the controls and the users. The null hypothesis insists only that the two populations have the same mean; beyond that, we merely expect the populations to have the same distributional shapes as the samples. We have no reason to assume that the populations are normal in shape, so we will not try a t-test approach. Instead, we will use a bootstrap approach, as introduced in Section 11.1.

We invent two populations. The control population will be the invented population obtained by replicating the control data:

Invented control population				
3.3	3.3	3.3	···	3.3
4.8	4.8	4.8	···	4.8
6.4	6.4	6.4	···	6.4
7.1	7.1	7.1	···	7.1

The mean of that invented control population is still 5.40. In order for our invented user population to have that same mean, we have to add 0.80 to each value in the sample of users before replicating, so that the new values are $0.9 + 0.8 = 1.7, 2.6 + 0.8 = 3.4, 3.4 + 0.8 = 4.2$, and $11.5 + 0.8 = 12.3$. These new numbers have a mean of 5.40, as desired. Thus we now have two populations with the same population mean of 5.40. Moreover, we have created two populations whose shapes closely approximate the unknown shapes of the true populations. This is true because the two sample histograms that form our created populations are in fact good estimates of the unknown population histograms.

Invented user population				
1.7	1.7	1.7	···	1.7
3.4	3.4	3.4	···	3.4
4.2	4.2	4.2	···	4.2
12.3	12.3	12.3	···	12.3

2. Definition of a Trial (Sample): A trial consists of randomly choosing without replacement four persons from the control population and four from the user population. (We could have avoided the creation of the large null hypothesis populations and instead sampled with replacement from small populations—review Section 11.1 if needed to understand this assertion.)

3. Definition of a Successful Trial: The statistic of interest is the difference between the means of the four controls and four users sampled in step 2. A trial is a success if the difference is 0.8 or more.

4. Repetition of Trials: We repeat steps 2 and 3 100 times. Thus we obtain 100 simulated differences in means. Table 11.5 gives the stem-and-leaf plot of the 100 simulated differences.

5. Estimation of the Probability of the Obtained Difference in Means or More (Probability of a Successful Trial): The difference in means obtained from the original data was 0.8. If we look at the stem-and-leaf plot of step 4, we see that 36 of the simulated differences were 0.8 or more.

Table 11.5 **One Hundred Simulated Mean Differences for Sleep Example**

−6	4
−5	
−4	310
−3	8887652210
−2	9754222
−1	65433220
−0	976666444333222222221100
0	122223344456689
1	022333378888
2	000122456788899
3	002448
4	
5	0

Key: "−6 4" stands for −6.4.

6. Decision: From step 5 we estimate a 0.36 chance that under the null hypothesis, the difference in the means will be as large as or larger than what we obtained. That is, it is not at all unusual to see a difference of 0.8 when the populations have the same mean. Thus we accept the null hypothesis.

Now is a good time to reiterate that accepting the null hypothesis is not the same as believing that the null hypothesis is actually true. All we have done is shown that the null hypothesis and the data are compatible. It could very well be that the reason we do not see much difference in the means is that we have too small a sample. This analysis does not prove that LSD has no effects on chromosomal breakage. Nor does it prove that LSD does effect chromosomal breakage. We cannot conclude much of anything! In fact, the actual means suggest that LSD could actually retard chromosomal damage—something that is very unlikely to be true. Perhaps the real problem is the *very* small sample size of four from each population used!

Larger Sample Sizes: Another *z* Test

If the sample sizes are fairly large, we can find a *z* statistic to test a hypothesis involving the equality of two population means. Let's consider an example. A beginning statistics class (STAT 100) at the University of Illinois at Urbana-Champaign had 104 students, 64 women and 40 men. The average percentage on homework assignments among the women was 78.56, and that among the men was 75.04. Thus the women did on average 3.52 percentage points better than the men. Was this due to chance, or do women in general perform better on the homework? The standard deviations were 25.08 for the men and 19.54 for the women.

We first have to decide what the populations we are trying to compare are. The students in the class are not a random sample from the entire population of the United States, nor even of the university. It does seem reasonable to think of these people as a random sample of all the people who take STAT 100 now or will do so in the near future. We will go on that supposition. The null hypothesis is that the average percentage scores on homework are the same for the populations of women and of men:

H_0: The mean homework score for the population of women
= the mean homework score for the population of men

We will follow the six steps, except that we will introduce a new *z* statistic to replace the simulations. That is, we will develop a *z* test for comparing the means.

1. Choice of a Model (Populations): The two populations are the women who take STAT 100 and the men who take STAT 100. The null hypothesis is that the two populations have the same mean. We invent two populations so that they have a common mean. The population of men is a large replication of the sample of 40 men. Their mean is 75.04, so to create a population of women with the same mean, we have to subtract 3.52 from each woman's score before replication, and then replicate the new data as often as we did for the male population (maybe 500 or 1000 times for each original observation).

2. Definition of a Trial (Sample): A trial consists of randomly choosing 64 students without replacement from the women and 40 without replacement from the men. (Again, we could have replaced replication and sampling without replacement with no replication and sampling with replacement.)

3. Definition of a Successful Trial: The statistic is the difference between the mean of the 40 men and that of the 64 women sampled in step 2. A trial is a success if the observed difference exceeds 0.76.

4. Repetition of Trials: Instead of actually sampling, we appeal to theory, which tells us what the average and standard deviation of such sampled differences would be if we sampled very many times. The average of very many differences would be approximately 0 (the difference implied by the null hypothesis), and the sample standard deviation of very many differences would be

SD of very many differences in means

$$\approx \sqrt{\text{theoretical variance of difference in means}}$$

$$= \sqrt{\frac{(\text{theoretical SD of men})^2}{\text{number of men}} + \frac{(\text{theoretical SD of women})^2}{\text{number of women}}}$$

This formula is based on a result that is fully developed in Chapter 13. Here we give a brief partial justification. First note that by *theoretical SD of men* we mean the variance of the male population created in step 1, and similarly for the theoretical standard deviation of the women. Chapter 13 will show that the theoretical variance of the sum or difference of two independent statistics is the sum of the individual variances. Thus

$$\begin{aligned}
&\text{Theoretical variance of (average of men} - \text{average of women)}\\
&\quad = \text{theoretical variance of average of men} + \text{theoretical variance of average of women}\\
&\quad = \frac{\text{theoretical variance of male population}}{\text{number of men in sample}} + \frac{\text{theoretical variance of female population}}{\text{number of women in sample}}\\
&\quad = \frac{(\text{theoretical SD of men})^2}{\text{number of men in sample}} + \frac{(\text{theoretical SD of women})^2}{\text{number of women in sample}}
\end{aligned}$$

The women's and men's theoretical standard deviations in the formula are those obtained from the women's and men's data, because the two populations are built from these data and hence the population standard deviations are the same as the standard deviations of the data. Thus for our data

$$\text{SD of the differences in means} = \sqrt{\frac{(25.08)^2}{40} + \frac{(19.54)^2}{64}}$$
$$= \sqrt{21.69} = 4.66.$$

Now we can define the z statistic, using the fact that the subtracted 0 is the expected difference under the null hypothesis:

$$z = \frac{\text{Difference in means} - 0}{\text{SD of difference in means}}$$
$$= \frac{3.52}{4.66} = 0.76$$

5. Estimation of the Probability of the Obtained Difference in Means or More (Probability of Success): We wish to calculate the chance that the difference in means is equal to or exceeds 3.52, which is approximately the chance that a standard normal exceeds $z = 0.76$, so all we need to do is look up 0.76 in the normal table (Appendix E). The area to the right of 0.76 is 0.7764, so the area to the left is 0.2236. That is, there is approximately a 0.22 chance of seeing a difference of 3.52 just by chance.

6. Decision: The chance calculated in step 5, 0.22, shows that it is not unusual to see such a difference when the null hypothesis that the populations have the same means is true. Thus we accept the null hypothesis.

SECTION 11.6 EXERCISES

1. To test the effectiveness of the new weight-loss drug Redux, 40 women were split into two groups: group A, the control group, who took a placebo drug, and group B, the experimental group, who took the real drug, Redux. The amount of weight lost by each member of the groups over a six-month period is given below. Explain how you would test the claim that the population mean weight loss of the two groups is the same. (*Hint:* You cannot use the z test because the sample sizes are not large enough.)

Group A:	3	4	5	6	7
	10	11	12	15	18
	19	20	23	24	25
	30	33	38	40	42
Group B:	5	5	7	7	10
	10	10	10	15	18
	20	24	28	29	38
	42	44	49	50	55

2. The following is a stem-and-leaf plot of 100 simulated differences between the population means of the amount of weight lost by the two groups in Exercise 1 (group A − group B). (Group A was used as the control group in the simulation.) Using the stem-and-leaf plot, test the claim that the difference between the population means is 0.

Stem	Leaf
−11	32
−10	54
−9	0
−8	20
−7	74322
−6	840
−5	955431
−4	987443
−3	87630
−2	88775441
−1	885410
−0	993
0	00126789
1	001356679
2	02233589
3	00335
4	02244478
5	45679
6	5799
7	234
8	0

Key: "8 0" stands for 8.0.

3. Scientists want to study the effect of exercise on the amount of weight loss. One hundred people are divided into two equal groups. Both groups follow the same diet plan, but the first group also follows an exercise program. In the exercise group the mean weight lost over three months was 25.2 pounds with a standard deviation of 10 pounds. In the nonexercise group the mean weight lost was 20.4 pounds with a standard deviation of 6.3 pounds. Test the claim that the difference between the population means is 0.

4. On a college campus, a professor wanted to test the effect on learning of using a computer program to teach calculus. In his first class of 43 students, he taught using the computer as an instructional aid. In his second class of 35 students, he taught without using the computer. On the final exam the computer class scored a mean of 78.32 with a standard deviation of 8.07 while the noncomputer class scored a mean of 80.41 with a standard deviation of 8.53. Test the claim that the difference of the population means is 0.

5. Fifty independent measurements of the weight of a chemical compound were made on each of two scales. On the first scale the mean of the 50 measurements was 19.45 grams with a standard deviation of 0.49 gram. On the second scale, the mean of the 50 measurements was 18.42 grams with a standard deviation of 0.27 gram. Test the claim that the difference between the population means is 0.

6. In a random sample of size 75, the mean SAT verbal score was 549.45 with a standard deviation of 21.12. In a random sample of size 75 the mean SAT math score was 539.25 with a standard deviation of 20.91. Test the claim that the difference between the population means is 0.

11.7 THE z TEST FOR A HYPOTHESIS ABOUT A POPULATION PROPORTION

In Chapter 6 we tested hypotheses concerning population proportions using our six-step decision method. Such problems are important because just as we may want to test a hypothesis about a population mean or about the difference between two population means, we may want to test a hypothesis about a proportion of a binomial population (in which case the number of successes follows a binomial distribution—see Chapter 8) or the difference in the proportions of two binomial populations. Here we consider the case in which the number of observations from the population is at least 20 and it is therefore accurate to approximate a standardized sample proportion by a standard normal distribution via the central limit theorem, which was thoroughly treated in Chapter 10. This enables us to bypass doing simulations in the six-step method, just as we did both in chi-square testing and in testing the population mean in Sections 11.2, 11.5, and 11.6 above. We will consider an example. We will not put it in the six-step framework, as we did above, although we could do so.

Example 11.2

The standard drug is known to produce a cure 60% of the time. A new drug is produced and tested on 36 people believed to be a random sample from those afflicted with the disease. Of these, 30 are cured. Is there evidence that we can reject the null hypothesis

$$H_0\text{: The new drug is no better.}$$

This null hypothesis is

$$H_0\text{: Success probability} p = 0.6 \text{ for the new drug}$$

Although this is just like some of the examples of Chapter 6, the point here is that it can be done without doing simulations using the central limit theorem approach of Section 11.2. Recall from Section 10.6 that the observed proportion $\hat{p}$ of successes in a binomial experiment, when standardized, follows the central limit theorem. That is,

$$z = \frac{\hat{p} - p}{\sqrt{p(1 - p)/n}} \approx \frac{\hat{p} - p}{\sqrt{\hat{p}(1 - \hat{p})/n}}$$

is approximately standard normal.

We are using the fact that the theoretical mean of $\hat{p}$, a binomial random variable divided by n, is $(np)/n = p$, and $\text{Var}(\hat{p}) = \text{Var}(\text{binomial random variable}/n) = (1/n^2)np(1 - p) = p(1 - p)/n \approx \hat{p}(1 - \hat{p})/n$. This is discussed in Chapters 8 and 13.

Under H_0, $p = 0.6$. Therefore

$$z = \frac{30/36 - (0.6)}{\sqrt{(0.6)(0.4)/36}} = 2.84$$

Thus we seek

$$p(z \geq 2.84) = 1 - p(z < 2.84)$$
$$= 1 - 0.9977 = 0.0023$$

This is less than 0.05, and hence we reject the null hypothesis and conclude that the new drug is better. Of course, this hypothesis-testing problem could have been solved by simulations, as in Chapter 6.

In Chapter 13 we will consider the important two-population hypothesis

H_0: Population 1 success probability = population 2 success probability

For example, we could ask which of two manufacturing methods produces the smallest proportion of defectives or whether, as would be stated in the null hypothesis, they are the same in this regard.

SECTION 11.7 EXERCISES

1. In a random sample of 64 people in a city, 37.5% were in favor of lowering the drunk-driving blood alcohol level from 0.10 to 0.08. Test whether the population proportion is equal to 50%.
2. In a random sample of 2000 U.S. citizens over the age of 18, 53.5% approve of the president. Test whether the population proportion is equal to 50%.
3. In a random sample of 2000 residents of Illinois, 75% believe in the fairness of the jury system. Test whether the population proportion is equal to 80%.
4. In a random sample of 60 college freshmen, 75% are in favor of lowering the drinking age from 21 to 18. Test whether the population proportion is equal to 70%.
5. In a random sample of 100 people, 70 are regular viewers of the television show *Seinfeld*. Test whether the population proportion is equal to 75%.

Graphing Calculator Exercises

6. A roulette wheel has 38 slots—18 black, 18 red, and 2 green. You suspect that the wheel is improperly balanced and not returning truly random results. In 152 turns of the wheel, red comes up 80 times. Can you decide whether the wheel is biased?

7. You have a new commemorative coin made to be tossed at the beginning of the Super Bowl. You want to test it to see whether it is a fair coin. You have your referees toss the coin 500 times, and they report that it comes up heads 278 times. Is it likely that the coin is fair ?

CHAPTER REVIEW EXERCISES

1. Explain the six steps of simulating a hypothesis test for a population mean.
2. A weight that is known to weigh exactly 10 pounds is weighed independently 50 times on the same scale. The values obtained are listed below. Test whether the scale is unbiased. (Test whether the population mean of the weight readings is 10.)

10.26	10.03	10.13	10.02	10.03	10.11	10.11	10.09	10.11	9.96
10.06	10.00	10.08	10.02	9.91	10.06	10.00	9.98	9.93	10.10
10.04	10.16	10.08	10.10	9.95	10.00	10.04	9.97	10.12	9.89
10.10	10.13	9.96	10.12	10.23	10.01	9.98	9.97	10.08	9.84
9.92	10.15	9.90	9.84	9.92	9.99	9.98	9.92	9.94	9.97

3. A state inspection office kept records of the error in 50 gasoline pumps tested around the state. Five gallons of gasoline (according to the meter) were pumped, and the error was measured by an inspection. The results, in cubic inches, are as follows:

2	−4	−6	−3	0
−4	−1	0	0	−2
−1	−1	−4	2	−4
−1	−2	−4	−3	1
−2	−4	−2	−1	−4
−2	−6	−2	−1	−5
0	−3	−4	−3	−7
−5	−1	−4	−4	2
−2	−3	−3	1	1
−4	−3	1	−3	−3

Use the binomial distribution to test whether the population median error is 0.

4. In a random sample of 100 teenage boys, the mean amount of time spent watching television was 3.6 hours with a standard deviation of 0.9 hour. In a random sample of 100 teenage girls, the mean amount of time spent watching television was 4.2 hours with a standard deviation of 1.2 hours. Test whether the difference between the population means is 0.
5. In a random sample of eight people, the mean amount of sleep per night is 7.45 hours with a standard deviation of 1.15 hours. Test whether the population mean amount of sleep per night is 8 hours.
6. In a random sample of four people, the mean body temperature was 98.25 with a standard deviation of 0.73 degrees. Test whether the population mean temperature is 98.6 degrees.
7. Is your answer to Exercise 6 different from the answer for the Key Problem in Section 11.2? If your answer is different, explain why.

8. Fifty measurements of the length of a bone were made. The mean of the measurements was 34.4 centimeters with a standard deviation of 0.25 centimeter. Test whether the true length of the bone is 35 centimeters.
9. In a random sample of 250 residents of a large city, 48% of the people in the sample would favor raising the bar entrance age from 19 to 21. Test whether the true population proportion is equal to 50%.
10. It is generally believed that the mean height of U.S. women is 65 inches. In a random sample of 50 women, the mean height was 64.45 inches with a standard deviation of 3.5 inches. Test whether the population mean height of U.S. women is 65 inches.
11. In a random sample of 100 men, the mean number of alcoholic drinks consumed per week was 5.4 with a standard deviation of 1.25. In a random sample of 100 women, the mean number of alcoholic drinks consumed per week was 3.2 with a standard deviation of 1.85. Test whether the difference between the population means is 0.
12. In a random sample of 300 people, 58% reported owning at least one cat. Test whether the true population proportion of people who own cats is 60%.

11

COMPUTER EXERCISES

1. ESP

Eighty-two people in a statistics class tried to guess the suits of eight cards randomly chosen (without replacement) from a regular deck of cards. If a person is guessing randomly, he or she should guess about two of the eight correct. (Why?) The table below shows how many people guessed 0, 1, and so on, correctly:

Number correct	Number of people
0	8
1	29
2	33
3	7
4	4
5	0
6	1

No one got more than six correct. The average number correct is 1.68293. That number is less than one would expect. Is it enough less to suspect that the class actually has negative ESP? We will test the null hypothesis that everyone is randomly guessing.

Start the Five Step program. Set up the box so that it looks like the data, but with an average of 2. Because the average in the data is 1.68, it is 0.31707 below the expected value under the null hypothesis. Therefore 0.31707 must be added to each value so that there are 8 tickets with the value 0.31707, 29 with the value 1.31707, and so on. The "Enter the Box Model" screen should look like the following:

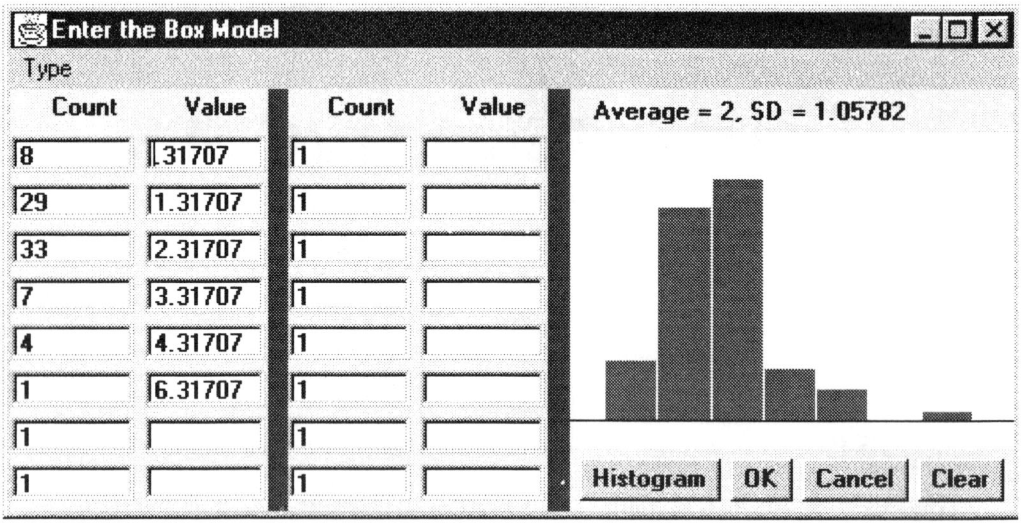

(The screen was widened somewhat so that the entire values could be read.) Press OK.

A trial is to draw 82 with replacement. The statistic is the average. We wish to find the chance that one obtains an average as small as or smaller than the value 1.61707, obtained when the box actually has an average of 2. Thus the probability of interest is "Stat <= a," where $a = 1.61707$.

a. Run 500 trials. What is the average of the statistics? The standard error? What is the standard deviation of the box?

b. What proportion of the statistics were less than or equal to 1.61707?

c. Do you accept or reject the null hypothesis? What do you conclude?

d. How close is the average in part (a) to 2?

e. The standard error of the statistic for many many simulations should be around SD/$\sqrt{82}$, where SD is the standard deviation of the box. What is the value in this case? How close is it to the standard error in part (a)?

f. What is the value of the z statistic, using the standard error from part (e)? What is the area under the normal curve to the right of that number? Do you accept or reject the null hypothesis?

2. TEMPERATURES: THE SIGN TEST

Recall Section 11.4, in which we tested to see whether the population median for the body temperature example could be 98.6. This problem is to perform the median test using the Five Step program. Set up the box model to have one 0 and one 1. We draw 120 with replacement, and the statistic is the sum. The probability of interest is that of obtaining 81 or more 1s.

a. Run 100 trials. How many were successes?
b. Find the maximum number of 1s in any trial. (Look in the Statistics part of the Summary Statistics menu.)
c. Do you accept or reject the null hypothesis?

3. STUDENT'S EXAMPLE

Recall Student's example in Section 11.5, and the data in Table 11.4. This problem is to test the three hypotheses mentioned in the example:

a. Does the dextro treatment help increase sleep?
b. Does the laevo treatment help increase sleep?
c. Does the laevo treatment help increase sleep more than the dextro? (Is the difference positive?)

For each hypothesis, perform the following steps:

i. Find the average of the 10 observations for the relevant variable.
ii. Subtract that average from each observation, so that the average of the new observations is 0.
iii. In the Five Step program, create the box model with these new observations (one of each of these values). Press the Histogram button in the "Enter the Box Model" screen to make sure the average is 0. Note down the standard deviation of the box.
iv. Set up the program so that the trial is to sample 10 with replacement, the statistic is the average, and the probability of interest is "Stat >= a," where a is the average of the data obtained in part (a).
v. Perform 500 trials. Find the proportion of trials that were successes. Do you accept or reject the null hypothesis? What do you conclude?
vi. Write down the standard error of the statistic. How close is it to (SD in box)/$\sqrt{10}$? Perform the t test. Does your conclusion differ from that in part (v)?

4. MAKING ERRORS

Recall Section 11.3. Consider the following situation. A lottery company has a machine that randomly chooses 1 of 10 table-tennis balls marked 0, 1, 2, ..., 9. Every day the company tests whether the machine is working properly by having the machine choose a ball 100 times. If the results do not look random enough, the company declares the machine "out of control."

One test the company does is to count the number of 0s that are chosen and declare the machine out of control if there are 16 or more 0s.

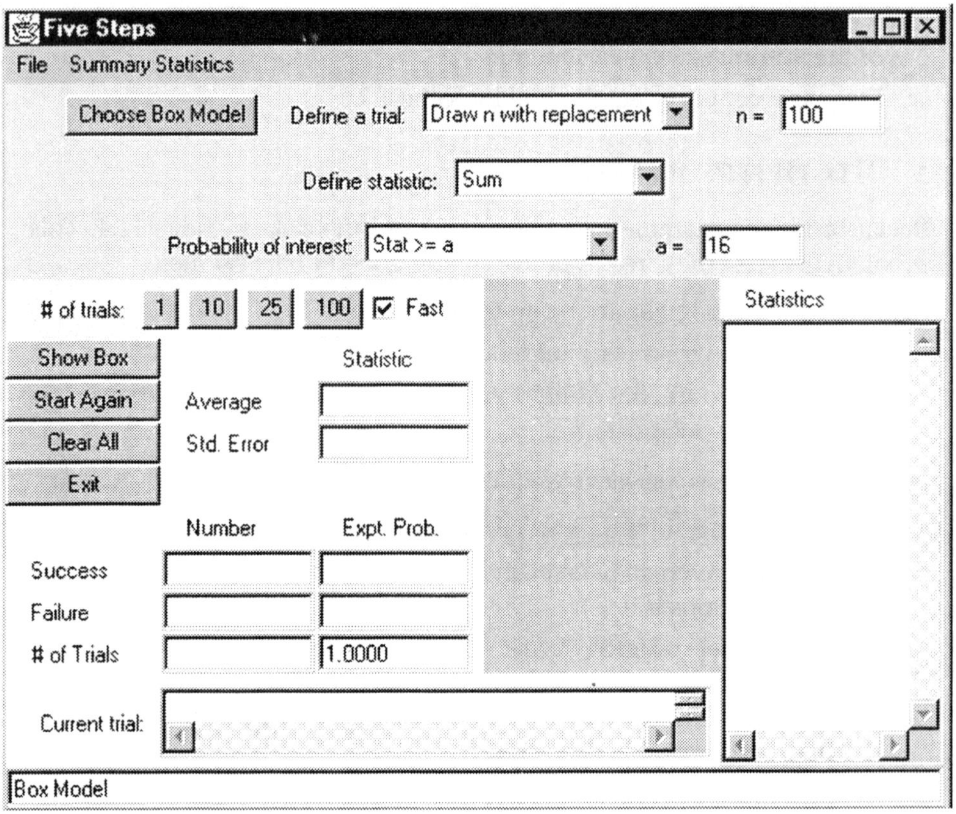

a. What is the chance this test declares the machine out of control even though it is not? (This error is a type I error, or a false positive.) To answer the question, set up the box model with one 1 and nine 0s, a trial being to draw 100 with replacement, the statistic being the sum, and the probability of interest being the probability that the statistic is greater than or equal to 16. Run 500 trials. What is your answer?

b. Suppose the machine is out of control. What is the chance that the company successfully declares it to be out of control? What is the chance it does not (that is, makes a type II error)?

   To answer these questions, run 100 trials, but change the box to reflect the machine not working correctly. Do this experiment for each of the following boxes:

   i. The chance of a 0 is actually 0.11, so the box has 11 1s and 89 0s.

   ii. The chance of a 0 is actually 0.13, so the box has 13 1s and 87 0s.

iii. The chance of a 0 is actually 0.15, so the box has 15 1s and 85 0s.

iv. The chance of a 0 is actually 0.20, so the box has 2 1s and 8 0s.

v. The chance of a 0 is actually 0.07, so the box has 7 1s and 93 0s.

Fill in the estimated chances of type II errors:

Chance of a 0	Experimental probability of a type II error
0.11	
0.13	
0.15	
0.20	
0.07	

c. How effective is this test in discovering whether the machine is out of control?

12

CORRELATION AND REGRESSION

Like mother, like daughter.

OBJECTIVES

After studying this chapter, you will understand the following:

- *How to test whether a straight line is needed to fit a set of data*
- *Reasons why a straight line can fail to fit a set of data*
- *Examples of curved regressions*
- *How to test for a linear relationship using the sample correlation*
- *How to use two variables to explain the variation in a third variable*
- *The multiple correlation coefficient as a measure of how well a curved line fits the data*
- *How to assess a trend when the relationship is not necessarily linear*

12.1 IS THE REGRESSION LINE WORTHWHILE? 568

12.2 DOES A STRAIGHT LINE WORK? 574

12.3 NONLINEAR RELATIONSHIPS 581

12.4 USING THE CORRELATION COEFFICIENT AS A TEST STATISTIC 588

12.5 TWO EXPLANATORY VARIABLES 591

12.6 THE MULTIPLE CORRELATION COEFFICIENT 595

12.7 ANOTHER TEST FOR TREND: KENDALL'S TAU 597

CHAPTER REVIEW EXERCISES 605

COMPUTER EXERCISES 610

Key Problem

Crickets, Anyone? (Or, "Did you hear what the temperature is today?")

Crickets make their chirping sounds by rapidly sliding one wing over the other. They do this very quickly about four times to produce a single chirp. Scientists have shown that crickets move their wings faster in warm temperatures than in cold temperatures. Data from one experiment designed to show the relation between wing speed and ambient temperature are shown in Table 12.1. The pulse rate provided in the table is a measure (obtained electronically) of the cricket wing speed. The table gives the pulse rate and the temperature recorded on 15 different days. A scatter plot of the data is given in Figure 12.1.

In this chapter we will learn ways of finding the relationship between two measures. For a certain cricket pulse rate, for example, we will be able to calculate the corresponding temperature to be expected. Hence, instead of a thermometer, crickets, anyone?

Table 12.1 **The Cricket Data**

Pulses per second	14	20	16	20	18	17	16	15	17	15	16	15	17	16	17
Temperature (°F)	76	89	72	93	84	81	75	70	82	69	83	80	83	81	84

Source: George Pierce, *The Songs of Insects* (Cambridge, Mass.: Harvard University Press, 1949), pp. 12–21.

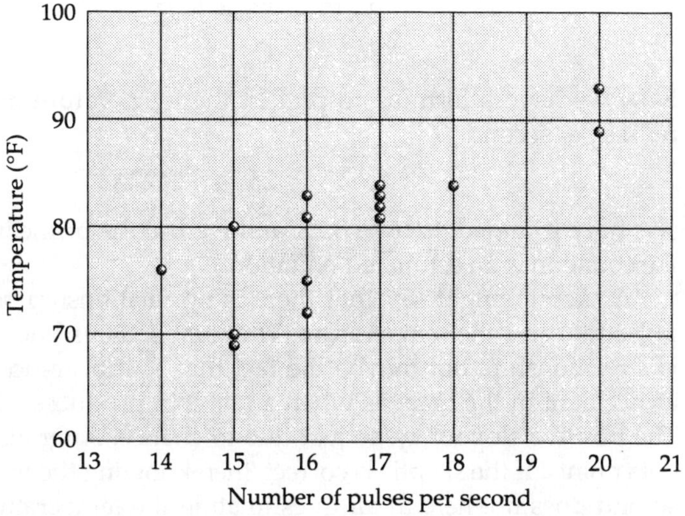

Figure 12.1 **Cricket data.**

One focus of this chapter is to begin to learn how statisticians decide whether and how strongly two variables, like height and weight, are related. A second, related focus is to begin to learn methods of estimating the relationship, as in the equation $Y = 2.1X + 3 +$ (random noise), which expresses an underlying straight-line relationship.

Chapter 3 introduced scatter plots for visualizing the relationship between two variables and discussed a number of ways to fit straight lines to scatter plots. It also introduced the Pearson correlation coefficient as a measure of how well the regression line fits the data or, alternatively, how strong the linear relationship between the two variables is. In this chapter, for a given scatter plot we focus on the least squares line, that is, the line that has the least mean square error (a concept developed in Chapter 3). One can fit such a "best" line to any scatter plot. Of course, even the best line may not make much sense, may be relevant only for certain values of the variables, or may be useless for prediction. We take up these concerns in this chapter. In particular, we consider cases in which other regression approaches than those presuming straight lines are appropriate.

12.1 IS THE REGRESSION LINE WORTHWHILE?

Consider the cricket chirp data from the Key Problem. We consider the least squares line. It turns out that its slope is 3.216. Recall that the least squares line goes through the point of averages, the average X, number of pulses per second, being 16.60, and the average Y, temperature, being 80.13. So

$$\begin{aligned} 80.13 &= 3.216(16.60) + C \\ 80.13 &= 53.39 + C \\ 26.74 &= C \end{aligned}$$

Now we have a formula to predict the temperature from the number of pulses per second:

$$Y' = 26.74 + 3.22X$$

(We have rounded 3.216 to 3.22, and we use the prime in Y' to denote that the equation is an estimated equation.)

A skeptic might say that there is no relationship between the cricket pulse rate and the temperature. The least squares line suggests that there is a relationship, but maybe the fact that Y appears to increase with X is an accident in the data, as when a fair coin produces 8 heads in 10 tosses. The positive slope may be just due to chance. After all, there are only 15 data points! If the skeptic is correct, then knowing the number of pulses per second does not help at all in estimating the temperature. In terms of the line, that means that if we were able to take a very large number of (X, Y) data points, we would discover that the true slope is 0. We now have to think in terms of a population that we are sampling from, as we did in the previous chapter, but a population whose members have values for two variables, X and Y.

The population is conceptual and consists of all possible (cricket, temperature) pairs. For each pair i, there is an X_i (pulses per second) and a Y_i (temperature). We assume that the 15 observed pairs are a random sample from all possible pairs and hence should be typical of the population. The skeptic is saying the following about this population: There is no relationship between the X's and the Y's; that is, the number of pulses per second has nothing to do with the temperature. We make this skeptic's claim the null hypothesis:

$$H_0\text{: The } Y_i\text{'s are independent of the } X_i\text{'s.}$$

Now we wish to see if this null hypothesis holds up under an analysis of the data. From the scatter plot in Figure 12.1, it certainly looks as if the null hypothesis is false. We have to create a population that satisfies the null hypothesis, and by repeatedly sampling 15-point data sets from it, determine what the chance is that even though the population slope is zero (because of no relationship) we see a relationship as strong as the one given by the least squares slope of the data. Note that the careful specification of a null hypothesis population was step 1 of our six-step approach in Chapters 6, 7, and 11. We will use the slope as the statistic of interest to test our null hypothesis. If the null hypothesis is true, the slope is 0. We proceed to the usual six-step method for hypothesis testing.

1. Choice of a Model (Definition of the Population): Just as in Chapter 11, we want the data to guide us in setting up the null hypothesis population. That is, we want the null hypothesis population to look as much like the data as possible but with no relationship between the X's and Y's. One way to achieve this is to construct the population to include, as equally likely, all possible pairings of the observed X's with the observed Y's. For example, look at the first observation in Table 12.1: $X = 14$ and $Y = 76$. To start constructing the invented population, we use that observation and we add every other Y to that X, to obtain these 15 observations:

$$(14,76),\quad (14,89),\quad (14,72),\quad (14,93),\quad \ldots,\quad (14,84);$$

We do the same with the other values of X:

$$\begin{array}{cccccc} (20,76), & (20,89), & (20,72), & (20,93), & \ldots, & (20,84); \\ (16,76), & (16,89), & (16,72), & (16,93), & \ldots, & (16,84); \\ & & & \vdots & & \\ (17,76), & (17,89), & (17,72), & (17,93), & \ldots, & (17,84). \end{array}$$

Now we have $15 \times 15 = 225$ observation pairs. Note that any X is just as likely to occur with any Y in this set of 225 pairs. Under the null hypothesis model, any X is equally likely to occur (that is, occurs randomly) with any Y. The creation of the basic 225 pairs to build our null hypothesis population is what makes the test a randomization hypothesis test. Thus X is truly independent of Y. We replicate the 225 pairs as many times as we wish in order to create a large population satisfying H_0 to be sampled from without replacement. (Or, as you might recall from Chapter 11, we could simply sample with replacement from the original 225 pairs.) Note that this large population has independent X's and Y's, as required by the null hypothesis, and yet has the distribution of X's and the distribution of Y's of the original sample. We thus have a null hypothesis population with X unrelated to Y and a distribution of X that estimates the population X distribution, and likewise of Y. Therefore our null hypothesis population attempts to capture the shape of the X distribution and that of the Y distribution in the actual population. If we fit a least squares line to all the points in this large invented population, we would find that the slope is 0. In fact, the least squares line would be

$$Y' = 80.13,$$

where the 80.13 is the average of the Y's. That is, the line is flat.

2. Definition of a Trial (Sample): A trial consists of randomly choosing 15 observations, each of which is an (X, Y) pair, from this large invented population satisfying H_0. The sampling is without replacement.

3. Definition of the Test Statistic and Definition of a Successful Trial: Once the 15 points are obtained, we find the least squares line and note down the slope. The first sampling we find is the following:

X:	16	16	16	17	14	17	17	15	17	16	15	16	16	15	17
Y:	75	69	81	81	76	81	83	75	84	81	80	69	75	81	75

The scatter plot of these points is in Figure 12.2. Note that some points have occurred more than once; this is not surprising, because both X and Y have repeated values in the original data of Table 12.1. Its least squares line can be calculated to be $Y' = 1.33X + 56.4$. Thus the first slope we record is 1.33. The trial is a success if its slope is greater than or equal to 3.22, the slope of the obtained data.

4. Repetition of Trials: We do step 3 100 times. It is definitely helpful to have a computer do the calculations for you. Table 12.2 contains the 100 slopes in a stem-and-leaf plot. The mean of these slopes is 0.177, and the standard deviation is 1.164.

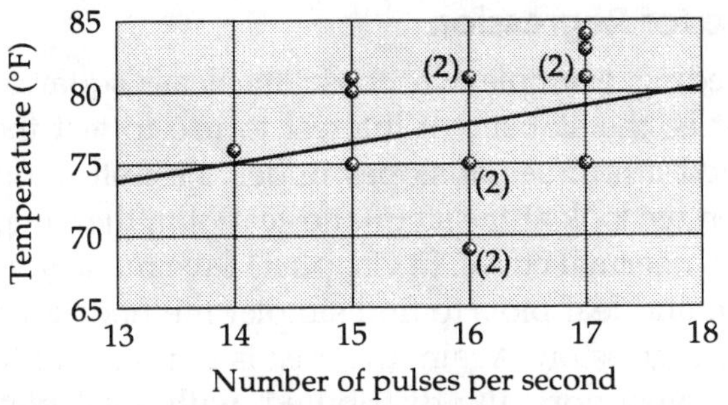

Figure 12.2 **One randomization of the cricket data.**

5. Estimation of the Probability of Obtaining a Slope As Large As or Larger Than the Observed Slope (Probability of a Successful Trial): In the original data, the slope was calculated to be 3.22. What is the probability of getting a slope that high or higher by chance if the *X*'s and *Y*'s are independent? Looking at the stem-and-leaf plot of the slopes in Table 12.2, we see that all are below 3.22. The largest one is 2.9. So we estimate the chance at 0%.

6. Decision: Step 5 shows that it is very unlikely to get a slope as high as 3.22 by chance under the null hypothesis. Therefore, we have to reject the null hypothesis—that is, we must conclude that *X* and *Y* are dependent, and that the number of pulses per second and the temperature are indeed related. Moreover, their relationship can be described by a straight line.

It is standard terminology to say that *X* and *Y* are *correlated.*

Table 12.2 **One Hundred Simulated Slopes from Null Hypothesis Population**

−2	7
−2	0
−1	97655
−1	443322211110
−0	9999888777665
−0	443222221100
0	011222333334444
0	5555666677789
1	0000111112334
1	555677
2	0011344
2	69

Key: "−2 7" stands for −2.7.

A z Statistic for Regression

In the preceding example we chose the least squares slope as the *test statistic* (that is, the statistic of interest to use to test the null hypothesis) and simulated a number of slopes under the null hypothesis population model. When we look at the stem-and-leaf plot, the sampled slopes appear to be close to a normal curve. In Chapter 11 we managed to exploit normal-looking stem-and-leaf plots to find simpler methods for testing. We can do the same for regression. Again, the idea is to note that the sampled slopes are approximately normally distributed, with a calculated mean of 0.177 and a standard deviation of 1.164. The obtained slope was 3.22, so the z statistic would be $(3.22 - 0.177)/1.164 = 2.61$. Again, as in Chapter 11, we can figure out what the mean of the slopes would be if we were to sample over and over from the null hypothesis population, and (approximately) the standard deviation of the slopes if we were to sample over and over. Then we do not need to sample slopes via simulation trials and can—a big advantage, as we know—simply base the inference on the observed data without any simulation. Clearly, the mean of the slopes is 0, the value in the null hypothesis. Let n denote the sample size ($n = 15$ in our sample). The standard deviation of the estimated slope can be shown by advanced statistical reasoning to be approximately

$$\frac{\text{SD of } Y}{\text{SD of } X}\frac{1}{\sqrt{n}}$$

when in fact the population slope is 0, as it is when our null hypothesis, H_0, is true.

For the cricket data, the standard deviation of X is 1.67, the standard deviation of Y is 6.49, and hence the standard deviation of the slope is $(6.49/1.67)(1/\sqrt{15}) = 1.003$. The final z statistic is then

$$z = \frac{\text{obtained slope} - 0}{\text{SD of slopes}} = \frac{3.22 - 0}{1.003} = 3.21$$

Now we have a new step 5.

New Step 5. Estimation of the Probability of Obtaining a Slope As Large As or Larger Than the Observed (Probability of a Successful Trial). We seek the area under the normal curve to the right of the z value 3.21. (The fact that the computed z is close to the slope of 3.22 in this case is just coincidental.) The value is not in Appendix E; the table only yields $p(z \leq 3.0) = 0.999$ and hence $p(z > 3.0) = 0.001$. This means the probability is less than 0.001 and hence much less than 0.05 and even 0.01.

Step 5 again shows that we are very unlikely to get a slope as high as 3.22 by chance under the null hypothesis. Hence in step 6 we strongly reject the null hypothesis that temperature and number of pulses per second are independent.

SECTION 12.1 EXERCISES

1. Using Table 12.2, determine what would be the conclusion of the hypothesis test of

   H_0: The X's are independent of the Y's.

   in the example of the cricket data if the slope of the least squares regression line were as follows:
   a. 1.1
   b. −2.0
   c. 2.1

2. Suppose a set of data is analyzed with the following characteristics:

   Standard deviation of X: 3.20

   Standard deviation of Y: 3.43

   n: 10

   Obtained slope: −0.09

   Do a hypothesis test of whether the X's are independent of the Y's.

3. In Chapter 3 we performed regression analysis on the following set of data regarding diamond sizes and prices:

Size (X)	Price (Y)
0.17	$353
0.16	328
0.17	350
0.18	325
0.25	642
0.16	342
0.15	322
0.19	485
0.21	483

   a. Calculate the standard deviation of X and the standard deviation of Y.
   b. The slope of the least squares regression line in this case is 3368. Perform a hypothesis test to determine whether size and price are independent.

4. Fill in the blanks in this statement: If X and Y are ____________________, then we assume that the slope of the least squares regression line would be ____________________ if we used the entire population to calculate this slope.

5. The following table shows the 1977 prices of fishing and hunting licenses for residents of 10 states (compare with today's prices!).

State	Fishing license (X)	Hunting license (Y)
Alabama	$9.50	$16.00
Delaware	8.50	12.50
Georgia	9.00	10.00
Kansas	15.50	15.50
Maryland	10.00	15.50
Michigan	10.35	13.35
New Jersey	16.50	22.00
Pennsylvania	17.00	12.75
South Carolina	10.00	12.00
Texas	13.00	19.00

   The slope of the least squares regression line relating X to Y is 0.568. The standard deviation of X is 3.27, and the standard deviation of Y is 3.56. Perform a hypothesis test of whether the price of a fishing license is independent of the price of a hunting license.

6. What characteristic of the stem-and-leaf plot of the slopes (such as Table 12.2) enables us to use the normal curve to test whether the slope is significantly different from 0?

7. Is there a relationship between the sign of the correlation coefficient and the sign of the slope of the regression line for every set of data in two variables?

Graphing Calculator Exercises

8. a. Consider the following scatter plots. See if you can guess their correlation. Then use a ruler to visually estimate the least squares line of best fit.

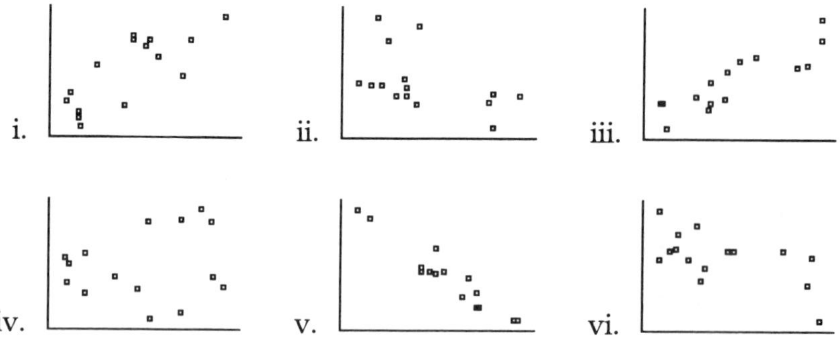

b. Use the program CORRSAMP to generate your own scatter plots with the following correlations. When using the program, record the regression lines given.
   i. $r = 0.95$
   ii. $r = 0.30$
   iii. $r = -0.65$
   iv. $r = 0$
   v. $r = -0.77$

9. Refer to the cricket data from Section 12.1:

Pulses per second:	14	20	26	20	18	17	16	15	17	15	16	15	17	16	17
Temperature (°F):	76	89	72	93	84	81	75	70	82	69	83	80	83	81	84

Do you think the pulse rates are random, or is there a relation between the pulse rate and the temperature? To answer this question, use the program BOTSTRAP. (See Section 1.19 in the TI Graphing Calculator Supplement for hints on how to use this program.)

12.2 DOES A STRAIGHT LINE WORK?

Suppose we have a scatter plot in which we know that X and Y are related. One can find the best line to fit the scatter plot, but that does not necessarily mean that the line fits the data at all well. Even if a line fits the data well, one has to be careful about the range of the observations for which it is relevant. We will look at some cautionary examples in this section that deal with these and related concerns.

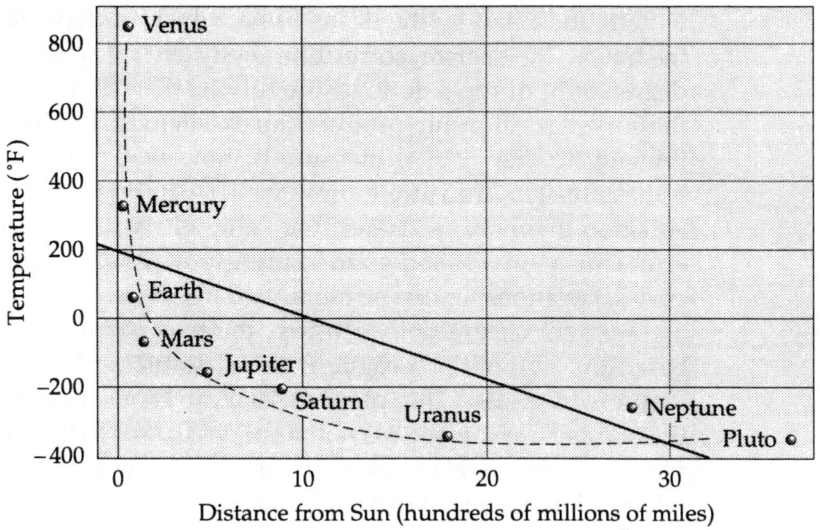

Figure 12.3 **Temperature as a function of distance from Sun.**

Curvature

Two variables can be highly related, but not in a straight line. Figure 12.3 plots Y, the average temperature in degrees Fahrenheit, against X, the average distance from the Sun in hundreds of millions of miles, for the nine planets. Earth is about 92,900,000 miles from the Sun, so its X is 0.929. (See Table 12.3 for the actual data.)

Clearly, the farther a planet is from the Sun, the colder it is. The line shown is the least squares line. It loosely tracks the points, but one can see that the points are not close to the line; rather, they clearly have a curved

Table 12.3 **Temperature as a Function of Distance from Sun**

	X = Distance (hundreds of millions of miles)	$Z = 1/X$ = 1/(distance)	Y = Temperature (°F)
Mercury	0.3596	2.781	332
Venus	0.6720	1.488	854
Earth	0.9290	1.076	59
Mars	1.4150	0.707	−67
Jupiter	4.8330	0.207	−162
Saturn	8.8670	0.113	−208
Uranus	17.8200	0.056	−344
Neptune	27.9300	0.036	−261
Pluto	36.6400	0.027	−355

pattern, indicated by the dashed line, which we have roughly sketched in the figure. The Pearson correlation coefficient $r = 0.62$, which measures the degree of fit of the data to a straight line, is well less than 1, even though the curved relationship produces an excellent fit. It is ingrained in statistical thinking to view r as a measure of how well Y can be predicted by X. But r is really a measure of how well Y can be predicted by a *straight-line* equation involving X. Hence the value of r will be very deceiving if it is erroneously interpreted as measuring how good an underlying (possibly curved) relationship can be found to fit the data.

For another example, consider the two plots in Figure 12.4. They are based on 1986 Major League Baseball statistics. For each of 263 players, Figure 12.4*a* plots Y, the player's salary in thousands of dollars, against X, the number of years the player has played in the Major League. Figure 12.4*b* plots the average salary for each value of X to make it easier to see what the relationship between years and salaries really is.

The least squares lines in both graphs have positive slope, suggesting that overall, the longer the players have been playing, the higher their salaries. However, even though there is a lot of variability in the data, the straight line seems to be a misleading indication of what happens. Rather, as the second graph shows, salaries rise very quickly over the first 5 to 10 years but then seem to level off and eventually decline. They do not proceed upward in a steady line. The dashed line, which was informally arrived at, suggests the curved relationship that appears to hold.

Section 12.3 gives some ideas of what one can do when there is curvature in the relationship between X and Y.

Outliers

Imagine a scatter plot in which X is body weight (in kilograms) and Y is brain weight (in grams) for a number of animal species. You would probably expect there to be positive correlation: larger animals would tend to have larger brains. Figure 12.5 shows such a scatter plot for a particular set of 27 animal species. Most of them (21) are near the $(0, 0)$ point, even though the graph does not clearly show this fact.

Notice that the least squares regression line (the line shown in the figure) is practically flat, with a slight negative slope. The correlation coefficient for these points is $r = -0.01$, suggesting a very weak relationship between X and Y. In fact, body weight and brain weight are slightly negatively correlated, which seems bizarre. But along the left side of the plot, the points go up very quickly. The problem is that the large number of points near $(0, 0)$ and the three points farthest to the right combine to produce a flat regression line. The three points with the biggest X—the highest body weight—are dinosaurs. They were huge, and they had relatively tiny brains. Their brains

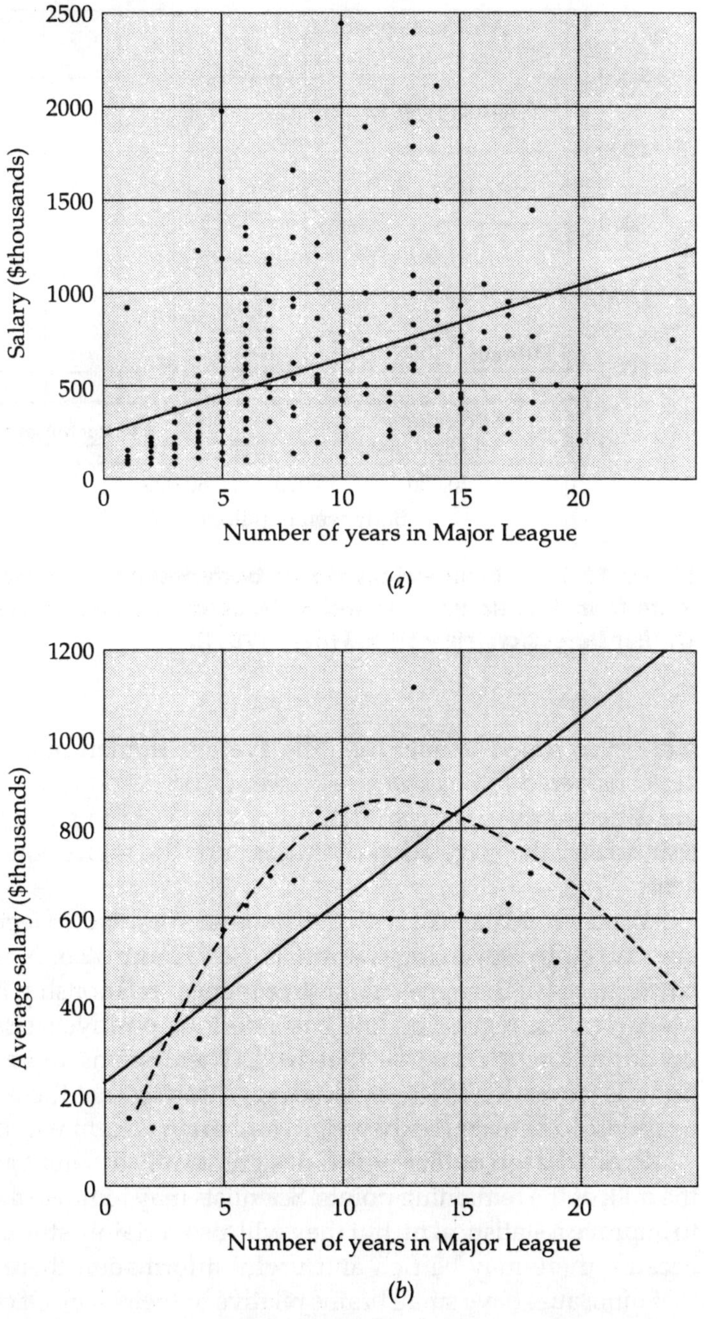

Figure 12.4 **Player's salary versus number of years in Major League Baseball. (data from Lorraine Denby, American Statistical Association Data Expo data set, http://lib.stat.cmu.edu/data=expo/1988.html).**

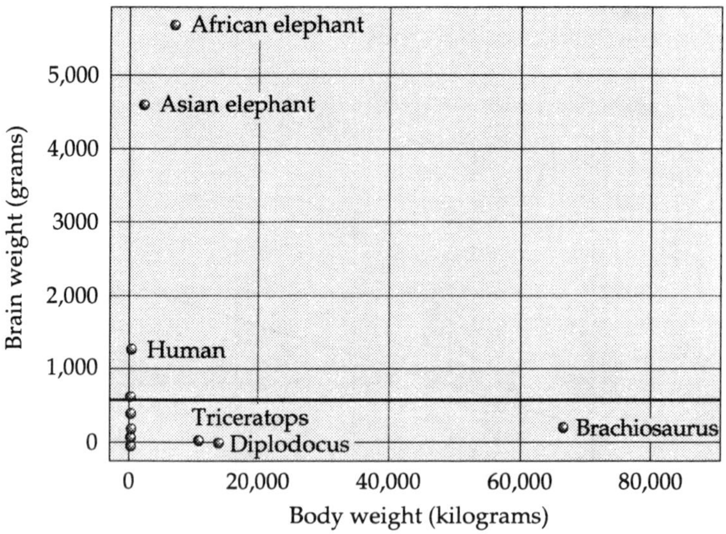

Figure 12.5 **Brain weight versus body weight of animal species. (data from P. J. Rousseeuw and A. M. Leroy, *Robust Regression and Outlier Detection* [New York, Wiley, 1987]).**

were even much smaller than the average human brain. These dinosaurs are considered to be *outliers,* a concept first introduced in Chapter 9. They are different enough from the bulk of the data to skew the regression, so that neither the correlation coefficient nor the regression line makes much sense.

We can redo the plot with the dinosaurs omitted, in fact then considering only currently surviving mammals. See Figure 12.6. Now the correlation coefficient is 0.93, suggesting a strong linear relationship between X and Y. Moreover, the regression line has a definite positive slope. Still, the points do not follow the line all that well. There seems to be some curvature. But this regression is less misleading, showing that there is a very positive relationship between body weight and brain weight without the dinosaurs.

Recall that an outlier is a point significantly different in its location from the bulk of the remaining points. Scientists may remove one or more outliers to improve a statistical fit, but they will also certainly study outliers carefully because there may be rich and useful information there (such as the fact that dinosaurs have small brains relative to their size). Of course, sometimes an outlier is simply the result of an experimental or measurement error, as we saw in Chapter 9, on measurement. But an outlier can be an accurately measured point that simply has a different scientific cause than the majority of the data (such as dinosaurs versus mammals).

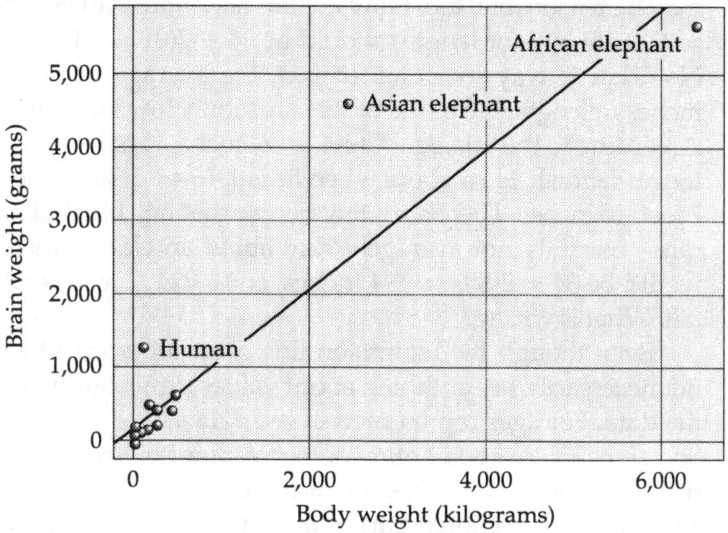

Figure 12.6 **Brain weight versus body weight of mammals.**

Interpolation and Extrapolation

Table 12.4 gives the average height in inches of boys at ages 8 through 14. If we plot the points, we see they are exactly on a straight line, so the correlation coefficient is 1. The least squares regression line is

$$\text{Average height} = 34 + 2(\text{age})$$

What would you guess is the average height of boys who are 10.5 years old? According to the regression line,

$$\text{Average height} = 34 + 2(10.5) = 55 \text{ inches}$$

That seems reasonable. How about the average height of boys who are 7 years old? We can plug in the value, and get $34 + 2(7) = 48$ inches. That

Table 12.4 **Average Height versus Age of Boys**

Age	Average height (inches)
8.0	50
9.0	52
10.0	54
11.0	56
12.0	58
13.0	60
14.0	62

may be reasonable, too. How tall are new-born babies? Their age is $X = 0$, so their height (or length) should be $34 + 2(0) = 34$ inches. Is that correct? No. That is way too long. In fact, the average newborn boy is only 20 inches in length. Go in the other direction. How tall will a boy be at age 20? According to the line, $34 + 2(20) = 74$ inches. That seems a little tall, but not too outlandish. How about when he is 30? $34 + 2(30) = 94$ inches tall. That is 7 feet, 10 inches. There are a few people that tall, but that height is extremely rare—certainly not average! What about an elderly man? An 80-year-old would be $34 + 2(80) = 194$ inches, or 16 feet, 2 inches tall. Nobody is that tall. What is wrong?

Even though the regression line gives a perfect fit to the data, it does not necessarily say anything about values significantly outside the range of the data. For ages represented in the data set—8, 9, ..., 14—the regression line gives a very good notion of the average height (in fact, contrary to what usually happens in regression, a perfect notion). For other ages between 8 and 14 (with fractional parts), the regression line is surely just as useful. Using the line for values *between* values actually represented in the data is called *interpolation*. Interpolation is usually reasonable, although if there are large gaps in the data, it may not work very well within those gaps.

Using the regression line for values *outside* the range of the data—called *extrapolation*—is much more risky. Using it a little outside the data is usually acceptable. For example, age 7 is only one year from age 8, and the regression line looks as if it gives a reasonable value. But the regression line fails completely at age 0, or 30, and especially at age 80. This is because people's growth rates are not constant. If people continued to grow at the rate they grew as adolescents, people would be very tall indeed. For general regression problems, one must be cautious when extrapolating. Do not do it unless you have reason to believe the trend outside your range of data is the same as the trend for your data. Supposedly Mark Twain used regression with extrapolation to predict that at some time in the twenty-first century the length of the Mississippi River will become negative. He was merely poking fun at the dangerous practice of undue extrapolation.

Summary

The last three subsections pointed out some pitfalls in regression analysis. The most effective way to avoid many of the pitfalls is to plot the data and look at the plot carefully. Does a systematic relationship that is not a straight line underlie the random jiggle of the data? Are there outliers—points not belonging with the majority of the points? The numerical calculations that produce the correlation coefficient and regression line may seem to be working, but unless one actually looks at the data, one could be misled. The same holds with interpolation and extrapolation. Even if the regression looks fine for the data you have, you have to be careful when using it to make predictions concerning values of X that you do not have. Generally, interpolation is safe and extrapolation is not, but even these broad statements have to be taken with a grain of salt. Even interpolation is dangerous when there is a large gap in the X values used to find the regression curve. Be sure to use your common sense. And graph the data!

SECTION 12.2 EXERCISES

1. State a definition of an outlier. How can an outlier affect a regression line?
2. What is the difference between extrapolation and interpolation? With which of these two procedures should one be more careful?
3. List potential concerns one should consider when inspecting a scatter plot of two variables.
4. The following table lists the 1997 population (in millions of people) and the size (in thousands of square feet) of nine states:

State	Population (X)	Land area (Y)
Alaska	0.55	570
Florida	13.0	54
Georgia	6.5	58
Illinois	11.5	56
Kentucky	3.7	40
Kansas	2.5	82
Michigan	9.3	57
Mississippi	2.6	47
New Hampshire	1.1	9

   a. Make the scatter plot for all nine states. What problem do you see with one of the points?
   b. Remove the point that was causing the problem and make another scatter plot. Does there appear to be a linear relationship now?
5. A least squares regression line is fit to a set of data whose X values range from 5 to 10. The scatter plot of the data shows a linear relationship, and there are no outliers. The equation for the least squares regression line is $Y' = 4.2X + 8.4$. Predict the value of Y for each of the following X's:
   a. $X = 4.3$
   b. $X = 1.2$
   c. $X = 16.5$
   d. Which of the three predictions you made in parts (a) through (c) do you feel will be the least reliable? Why?

12.3 NONLINEAR RELATIONSHIPS

Reciprocal Relationships

As we saw in Section 12.2, points can exhibit a distinct relationship pattern that is not a straight line. There are times when a judicious rescaling of either X or Y can improve the regression substantially by turning a nonlinear (curved) relationship into a linear (straight line) one. Consider the data in Figure 12.3. The points seem to lie near a curve, but what curve? We will try taking the reciprocal of the distance X, that is, $Z = 1/X$. The reciprocals are given in Table 12.3. Figure 12.7 plots Y versus Z, and the least squares

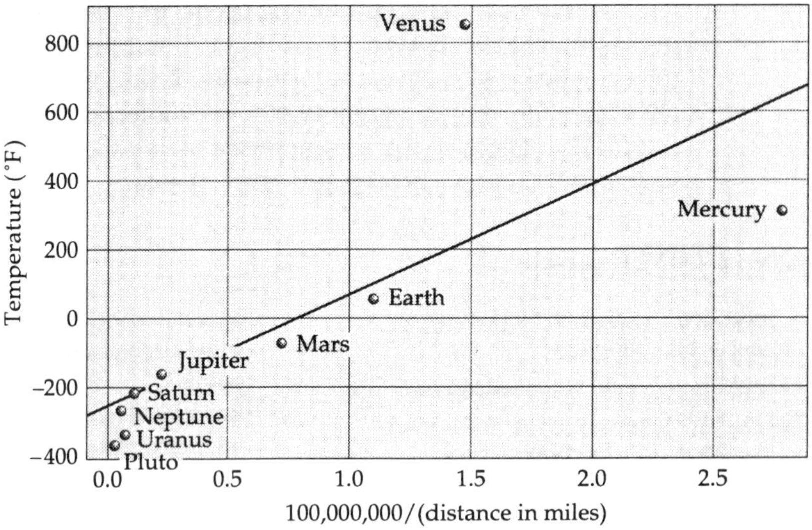

Figure 12.7 **Temperature versus reciprocal of distance from Sun.**

regression line, which turns out to be

$$Y = -246.58 + 318.47\,Z$$

The correlation between Z and Y is $r = 0.76$, which represents an increase from the X and Y correlation of 0.62. Venus is a possible outlier. Notice how the planets far away from the Sun now have very small values of Z, and vice versa. To go back to putting Y in terms of X, we now substitute for Z. Because $Z = 1/X$, we have $X = 1/Z$. The new *nonlinear* equation relating X and Y is thus

$$Y = -246.58 + 318.47/X$$

The equation is called nonlinear because it does not describe a straight line. Figure 12.8 plots the original X versus Y and draws the above regression equation with $X = 1/Z$. Notice how much better it follows the points than the regression line in Figure 12.3. Unlike the informally guessed relationship given by the dashed line in Figure 12.3, the curve of Figure 12.8 is the result of a formal least squares line (Figure 12.7). The technique of changing the scale of X or Y or both to reduce the problem to a straight-line regression problem for the new variables is a commonly used statistical practice.

Quadratic Regression

The regression model using Z instead of X is conceptually no different from the other regression models we have considered in this chapter. That is,

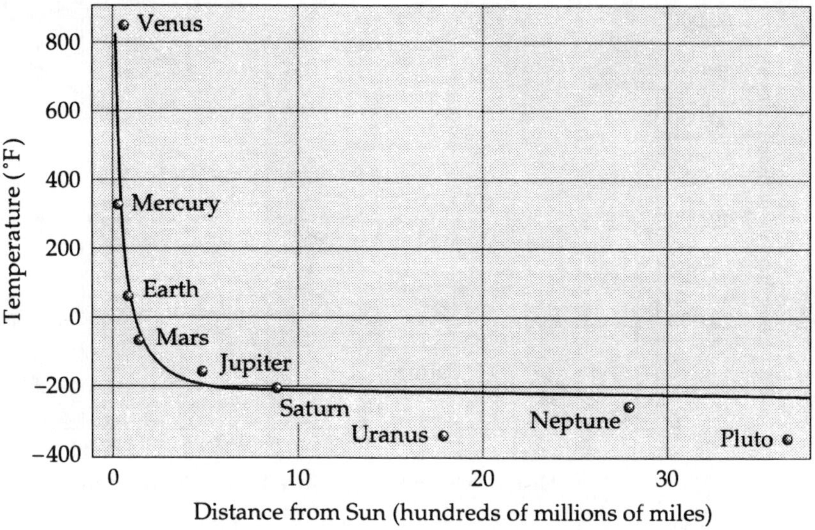

Figure 12.8 **The fitting equation of Figure 12.7 for temperature versus distance.**

we have a linear relationship between Y and Z, just as we did between Y and X earlier. Thus the trick of rescaling was used to allow us to continue to depend on least squares straight-line regression. In this subsection we will introduce a somewhat more complicated model, one that forces us to completely abandon straight-line regression. We will again look at the planets, but now the Y will be the length of the planet's year, that is, how much time in Earth days the planet takes to orbit the Sun. The plot and the regression line are in Figure 12.9.

The correlation coefficient is 0.99, so the regression line seems to be explaining the pattern well. However, a close look at the graph (remember our advice to graph the data) suggests there is definite systematic curvature in the data. Recall that we addressed the curvature in Figure 12.3 by using the reciprocal of the distance, obtaining the fit in Figure 12.8. For the plot of years, we will instead try a *quadratic* function, that is, try to fit

$$\text{Length of year} = a + b \cdot (\text{distance}) + c \cdot (\text{distance})^2$$

Here a, b, and c are unknown parameters of this quadratic equation model. The intercept of the curve with the vertical is a, and b and c control the shape of the curve. Such a function gives a parabola when plotted. (For example, in Figure 12.10 the simple parabola $Y = 1 + 2X + X^2$ is graphed.) Notice now that there are two slopes: one for X = distance, and one for $X^2 = (\text{distance})^2$. The data are given in the first three columns of Table 12.5.

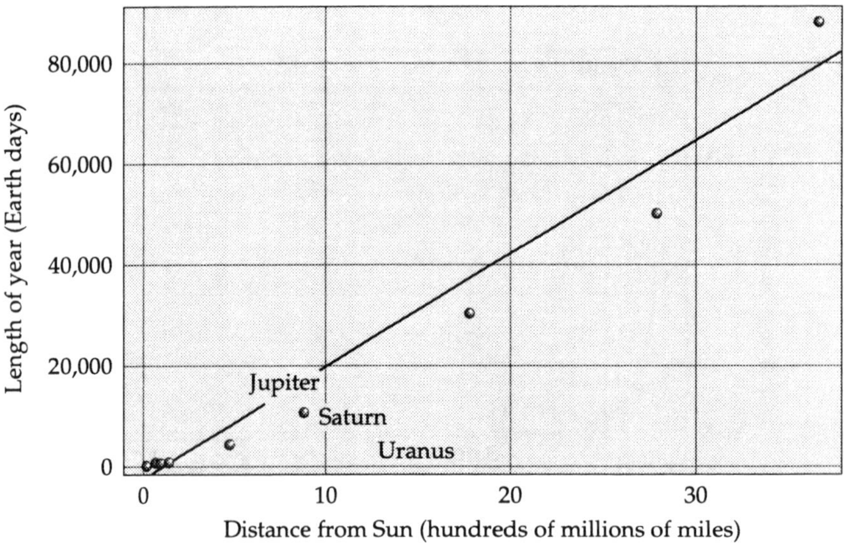

Figure 12.9 **Planetary year as a function of distance from Sun.**

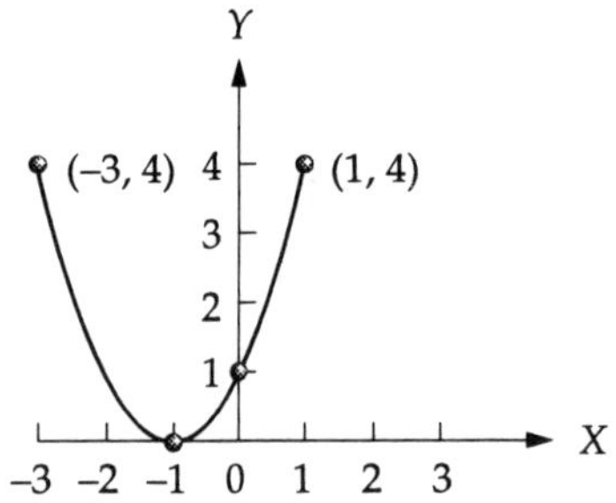

Figure 12.10 **The parabola $Y = 1 + 2X + X^2$.**

Table 12.5 **Tabulation of Planetary Year (Y) versus Distance (X) from Sun**

	Y	X	X^2	Y'	$Y - Y'$	$(Y - Y')^2$
Mercury	88.00	0.360	0.129	−428.00	516.00	266,264.6
Venus	224.70	0.672	0.452	−91.36	316.06	99,897.5
Earth	365.26	0.929	0.863	191.44	173.82	30,213.3
Mars	687.00	1.415	2.002	740.70	−53.70	2884.3
Jupiter	4332.60	4.833	23.358	5138.20	−805.60	648,997.0
Saturn	10,759.20	8.867	78.624	11,532.48	−773.28	597,972.1
Uranus	30,685.40	17.820	317.552	30,381.74	303.65	92,207.5
Neptune	60,189.00	27.930	780.085	59,387.53	801.46	642,346.1
Pluto	90,465.00	36.640	1342.490	90,943.42	−478.42	228,885.6

One can find the least squares estimates of the b and c slopes using a procedure similar to that of Chapter 3. We will not explore the details of how to do this. Computers are quite adept at finding such best slopes for the best-fitting least squares parabola, so we will let the computer do it. We obtain the following equation:

$$Y' = -805.8 + 1036X + 40.06X^2$$

Table 12.5 contains Y', $Y - Y'$, and $(Y - Y')^2$ for this equation; Y is the observed value and Y' is that predicted by the equation. The mean of the squared errors (average of the $(Y - Y')^2$ values) is 289,962. Compare this to the mean of the squared errors for the straight line in Figure 12.9, which is the much larger 21,110,278. The average squared error is a good way to quantify how closely the curve fits the data. The plot of the above estimated parabola for Y' is shown in Figure 12.11.

The new parabolic curve follows the points extremely well, but there are still some problems. Look at the predicted values of the length of the year for Mercury and Venus. They are negative! Also, Earth is predicted to have a year of length 191.44 days. The first two are impossible, and the third is from our perspective as "Earthlings" a poor prediction. The predictions for the planets farther out are much better, at least relative to their years' lengths, and this is what makes the fit of the curve to the data look so good visually.

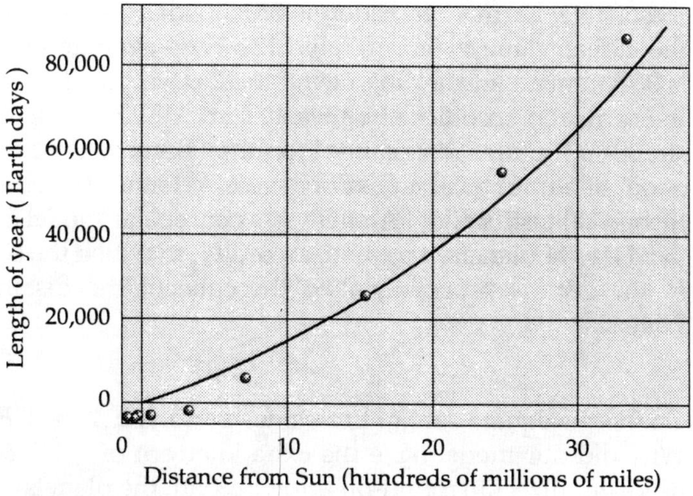

Figure 12.11 **Planetary year as a quadratic function of distance from Sun.**

Table 12.6 **Tabulation of Planetary Year (Y) versus $X\sqrt{X}$, Where X Denotes Distance**

	Y	X	$X\sqrt{X}$	Y'	$Y - Y'$	$(Y - Y')^2$
Mercury	88.00	0.360	0.216	86.24	1.755	3.081
Venus	224.70	0.672	0.551	222.97	1.724	2.972
Earth	365.26	0.929	0.895	363.50	1.759	3.093
Mars	687.00	1.415	1.683	684.81	2.189	4.790
Jupiter	4332.60	4.833	10.625	4331.83	0.764	0.584
Saturn	10759.20	8.867	26.404	10767.48	−8.281	68.582
Uranus	30685.40	17.820	75.225	30679.99	5.403	29.187
Neptune	60189.00	27.930	147.607	60202.15	−13.154	173.017
Pluto	90465.00	36.640	221.786	90457.15	7.841	61.487

A Better Fit

It turns out that in the above example an even better model can be had by using $X^{1.5}$, which is the same as $X\sqrt{X}$. This model, involving $Z = X\sqrt{X}$, is again like the one using $Z = 1/X$ in that there is only one slope to worry about. The least squares line is

$$Y' = -1.71 + 407.87X\sqrt{X}$$

Table 12.6 gives the statistics for this regression. Look at the errors, $Y - Y'$. They are all quite small, usually around 1 or 2, with the largest being −13.154, for Neptune. Even the values for the planets close to the Sun are reasonable; Earth's prediction is 363.501 days, which is much better than the earlier value given in Table 12.5. The average of the squared errors is 38.53, much smaller than even the 289,962 for the quadratic fit. In fact, the correlation coefficient between Y and $X\sqrt{X}$ is calculated as 1 to many significant figures. You cannot top that. Because the fit is so spectacularly good, one might ask an astrophysicist whether the above equation has a theoretical justification! Actually, this particular model is basically Kepler's third law of planetary motion, according to which the square of the length of the year is proportional to the cube of the distance from the sun. That is,

$$Y^2 = BX^3$$

Taking the square root of both sides gives you $Y = \sqrt{B}X\sqrt{X}$, or $\sqrt{B}X^{1.5}$. Why did our model have the extra intercept of −1.71? First, the data we used are subject to round-off error. Second, the planets orbit in ellipses, not exact circles, so the distance from the Sun, X, is not a constant. Kepler's law takes account of the ellipticity in the formula, and hence that X varies throughout the year, a fact that we have ignored.

SECTION 12.3 EXERCISES

1. Look at the scatter plot given in Figure 12.7.
   a. What problem does there seem to be with the data point associated with Venus?
   b. Remove the point corresponding to Venus from the data set. Calculate the variance of Z and the covariance between Z and Y.
   c. Using the quantities you found in part (b), determine the slope of the least squares regression line (recall the formula from Chapter 3 relating the slope to the covariance and the variance of X).
   d. Find the equation for the regression line, recalling that the line must pass through the point (X, Y).
   e. Do you think that the correlation coefficient for this line will be higher than that for the data set including Venus?
   f. Calculate the variance of Y. Now recall the formula for the correlation coefficient given in Chapter 3. You should be able to calculate the correlation coefficient with values you have calculated so far in this exercise. Is the correlation coefficient higher than it was when Venus was included?
2. Suppose you found other planets and wanted to estimate the length of their years. Give your best estimate for the following distances from the sun:
   a. 0.743
   b. 2.54
   c. 10.23
   d. How close do you expect these estimates to be to the true values?
3. Consider the following set of data:

X	Y	X	Y
0.83	6.12	0.88	6.14
0.70	7.98	1.05	5.40
0.77	6.09	0.48	9.60
0.52	8.83	0.72	7.66
0.56	7.82	0.60	7.20

   a. Construct a scatter plot of the data. Does there seem to be a linear relationship?
   b. Try transforming the X variable by letting $Z = 1/X$. Now construct a new scatter plot after replacing X with Z.
   c. Does there seem to be a linear relationship in this case?
4. Consider the following set of data:

X	Y	X	Y
0.58	−0.04	3.07	14.25
1.79	5.55	3.44	16.37
2.07	5.94	3.91	22.33
2.53	10.07	4.45	27.52
2.80	11.81	4.99	36.24

   a. Construct a scatter plot of the data. Notice that it appears to be a parabola.
   b. Create a new variable Z that equals X^2. Now construct a new scatter plot by replacing X with Z. Does there seem to be an improvement in the shape of the plot?
5. Suppose we performed a transformation on a set of data where we let $Z = 1/X$. The least squares regression equation turned out to be $Y' = 5.6Z + 7.8$.
   a. What would be the equation relating Y' to X?
   b. Would you refer to this as a linear or a nonlinear equation? Explain.

12.4 USING THE CORRELATION COEFFICIENT AS A TEST STATISTIC

In Section 12.1 we showed how to test whether the straight-line regression was worthwhile, that is, whether the slope is significantly different from 0. If the X and Y are independent, then the slope is 0. Similarly, if X and Y are independent, the correlation coefficient *for the population* of all possible (X, Y)'s is 0. This section shows how to use the correlation coefficient as a test statistic—that is, to use it as a statistic to decide whether to accept or reject a null hypothesis.

Using 165 students in an introductory statistics course at the University of Illinois at Urbana-Champaign for the data, for each student we have X equal to the number of brothers the student has and Y equal to the number of sisters the student has. The correlation coefficient is 0.0487, which is very small. From this, can we conclude that the correlation coefficient in the population is 0—that is, that the number of brothers is independent of the number of sisters?

As in Section 12.1, we start with the null hypothesis that X and Y are independent:

$$H_0\text{: The } X_i \text{ are independent of the } Y_i \text{ in the population.}$$

What is the real population actually being sampled from? We will take it to be the entire undergraduate student body at Illinois. Can the statistics students be assumed to be a random sample from that population? Not really, but probably with respect to the issue of brothers and sisters, it is not a bad assumption. Now we are ready for the six steps.

1. Choice of a Model (Definition of the Population): We put together an invented population exactly as in step 1 of Section 12.1, one that satisfies the null hypothesis. Thus, for each X we create 165 pairs, matching that X with each possible Y. We have 165×165 data points for which the number of brothers is obviously uncorrelated with the number of sisters. Once again this is a randomization hypothesis test.

2. Definition of a Trial (Sample): A trial consists of randomly choosing 165 observations without replacement, each one of which is an (X, Y) pair from the invented population of step 1.

3. Definition of the Test Statistic and Definition of a Successful Trial: In Section 12.1 this step had us calculate the slope from the sample we found in step 2. Here we calculate the correlation coefficient. A successful trial occurs if the correlation coefficient is greater than or equal to 0.0487.

4. Repetition of Trials: We perform steps 2 and 3 100 times, yielding 100 sampled correlation coefficients. The stem-and-leaf plot appears in Table 12.7.

Table 12.7 Simulated Correlation Coefficients for Null Hypothesis Population of Independent X's and Y's

−1	433221100
−0	999888776655
−0	44443333332221111100
0	000011111122222222223333344444
0	55567788888999
1	00333334
1	5677

Key: "−1 4" stands for −0.14.

5. Estimation of the Probability of Obtaining a Correlation Coefficient As Large As or Larger Than the Observed (Probability of a Successful Trial): In the original data the correlation coefficient was 0.0487. What is the chance that, when the null hypothesis that X and Y are independent is true, the correlation coefficient will be that high or higher? From step 4 we find 23 of the samples had a correlation coefficient over 0.0487. (The three values that are 0.05 in the stem-and-leaf plot are actually 0.0454, 0.0478, and 0.0481, which are slightly less than 0.0487.) Thus we estimate the chance to be 0.23.

6. Decision: Step 5 shows that it is quite plausible to have a correlation coefficient as high as 0.0487 when X and Y are independent. Therefore we have to accept the null hypothesis—that is, conclude that there is not enough evidence to convince us that X and Y are not independent.

A *t* Test for the Correlation Coefficient

Once again we can avoid the need for simulation by using a well-known theoretical statistical result. Suppose we know in our null hypothesis model of step 1 that the populations of X's and Y's are each normally distributed and the X's are independent of the Y's. It is shown in advanced textbooks that if r is the sample correlation coefficient based on a sample of (x, y) pairs of size n from the population, then

$$T = \frac{r\sqrt{n-2}}{\sqrt{1-r^2}}$$

is distributed exactly as a t distribution with $n - 2$ degrees of freedom. Thus by calculating T, one can use the t table (Table F) for step 5 instead of simulating. In our example, $n = 165$, so we actually use the normal table because, as you might recall, the normal table and the t table are indistinguishable when the number of degrees of freedom in the t table exceeds 30. Therefore

$$T = (0.0487)\frac{\sqrt{165-2}}{\sqrt{1-0.0487^2}} = 0.62$$

Now we have a new step 5:

New Step 5. Estimation of the Probability of Obtaining a *T* Statistic As Large As or Larger Than the Observed (Probability of a Successful Trial): Using T, we find that the area under the normal curve to the right of the value 0.62 is 0.27, or 27%.

Step 5 again shows it is not at all unusual to get a correlation coefficient as large as 0.487 when X and Y are independent, so in step 6 we accept the null hypothesis.

We should note that the assumption about the population made in step 1 fails, because for the real population the numbers of brothers and sisters are not normally distributed. However, with the large sample size of 165, the approximation can be proven reasonable via a type of central limit theorem for correlation coefficients. Note that the z test we have used for T here produces a result very similar to that of the six-step simulation approach.

SECTION 12.4 EXERCISES

1. Using Table 12.6, estimate the probability of seeing a correlation coefficient as large as or larger than the following values:
   a. 0.17
   b. 0.07
   c. 0.13

2. Under what conditions does the test statistic

$$T = \frac{r\sqrt{n-2}}{\sqrt{1-r^2}}$$

   have the t distribution? How many degrees of freedom will it have?

3. For each of the following sets of conditions, test whether the X's are independent of the Y's in the population.
   a. There are 10 data points, and $r = 0.2$.
   b. There are 8 data points, and $r = 0.65$.
   c. There are 6 data points, and $r = -0.43$.

4. Repeat Exercise 3 for the following sets of conditions:
   a. There are 12 data points, and $r = -0.1$.
   b. There are 13 data points, and $r = 0.4$.
   c. There are 7 data points, and $r = -0.7$.

5. Suppose we conducted two experiments. In one we had 7 data points, and in the other we had 14 data points. In both cases the sample correlation coefficient was 0.50.
   a. For each of the two cases described above, would we conclude that the X's and Y's are independent or not independent?
   b. What does your result from part (a) tell you about what happens as you increase the number of data points you are working with? Does this seem reasonable?

12.5 TWO EXPLANATORY VARIABLES

We will look at three variables regarding the 50 states in 1977.* The variable Y is the average life expectancy of the people in each state. The two variables we will use to try to "explain" Y are X, the murder rate, and Z, the high school graduation rate. Clearly, neither variable can be expected to do a good job, but both may be expected to be of some predictive power. In Figure 12.12 are plots of Y versus X and Y versus Z, and their individual regression lines.

The least squares lines and the corresponding mean square errors for the two regressions are

$$Y' = 72.97 - 0.2839X \qquad \text{mean square error} = 0.6892$$

$$Y' = 65.74 + 0.0968Z \qquad \text{mean square error} = 1.167$$

respectively. Thus we see that the murder rate (X) explains the life expectancy (Y) a little better than does the high school graduation rate (Z), judging from the mean square errors. Rather than looking at X and Z separately, however, one might be able to explain more about the variation in Y by looking at them simultaneously. Can we do this? How do we plot all three variables?

Imagine a level desk with the X and Z axes painted on the desk top and the Y axis pointing straight up. Then an (X, Z) point is found using the axes on the desk in the usual way, and the point is plotted floating in the air above the (X, Z) point, how far up being determined by the value of Y. For example, for Alabama, $X = 15.1$, $Z = 41.3$, and $Y = 69.05$. On the desk, find the $(15.1, 41.3)$ point, and then place the point floating 69.05 above the desk. Without something to keep the points in the air, this procedure is not really practical. If one wanted a physical representation, one could use little helium balloons, tied down at the points on the desk with the lengths of the strings equal to the Y value, or one could build small flagpoles, stuck in the desk at the (X, Z) points, and being as high as the Y values. Figure 12.13 gives a computer plot of what the 50 points would look like. The point for Alabama is labeled on the plot.

To see how X and Z explain Y, instead of a line we will fit a *plane* to the points. Think of taking a large piece of cardboard and inserting it among the floating points in such a way that the points are reasonably close to the cardboard. The cardboard is a plane. The equation of the plane gives how high the plane is above each (X, Z) point on the desk. The equation has the form

$$Y' = \text{intercept} + (\text{slope } 1)X + (\text{slope } 2)Z$$

*Data are in S-PLUS statistical software, MathSoft, Inc.

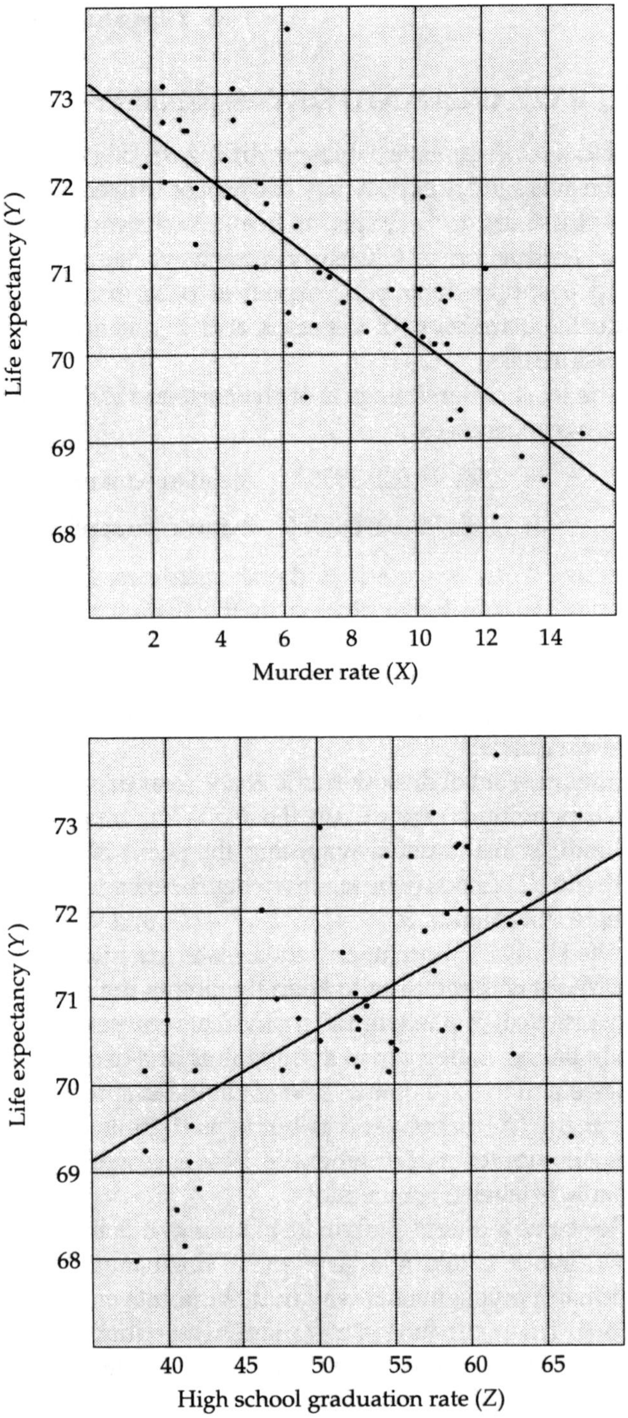

Figure 12.12 **Life expectancy as function of murder rate and high school graduation rate.**

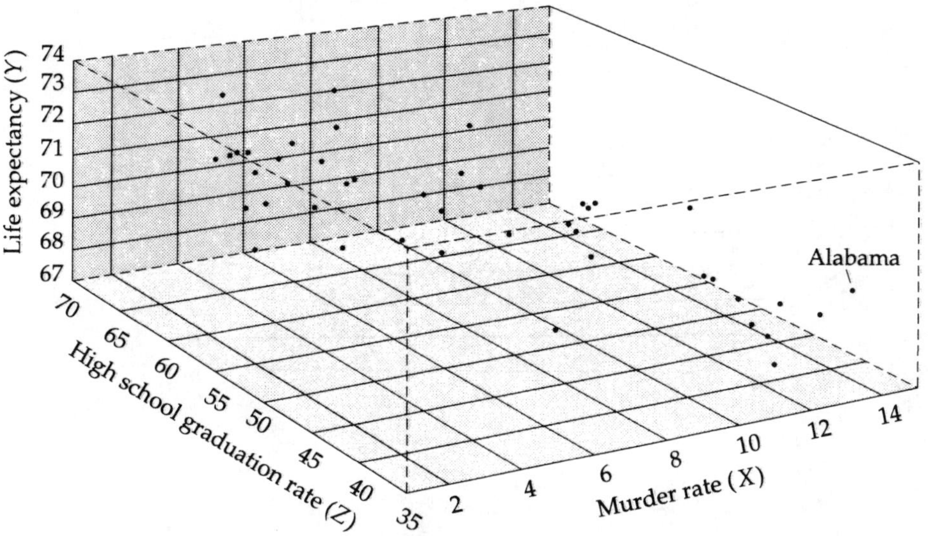

Figure 12.13 **Life expectancy as joint function of graduation rate and murder rate.**

Thus there are two slopes, one for each variable. (We also had two slopes in the quadratic regression problem in Section 12.3.) To find the best values for the slopes, we use least squares—that is, try different values for the intercept and two slopes and pick the values that minimize the sum of squares of the $(Y - Y')$'s. Geometrically, we are finding the plane such that the sum of the squared vertical distances between it and all the points is a minimum. Again, we let the computer perform this task. The resulting equation is

$$Y' = 70.30 - 0.2371X + 0.0439Z$$

Notice that the slopes in this equation are not the same as the slopes for the individual equations, but they are somewhat similar. In Figure 12.14 the plane is drawn among the points.

One can see how the plane tilts. The higher the murder rate (Z), the lower the life expectancy (Y), but the higher the high school graduation rate (X), the higher the life expectancy.

How well does the plane fit? The mean square error turns out to be 0.5954, which is smaller than either of the variables produced on its own, although not greatly lower than the one for the murder rate. It does seem we are doing a little better predicting using both variables than either one alone.

Caution: Just as we discussed in Section 12.2, one has to worry about curvature, outliers, and extrapolation and interpolation when doing regression with two variables. In fact, it can be even more difficult to detect such problems because the graphs are harder to read.

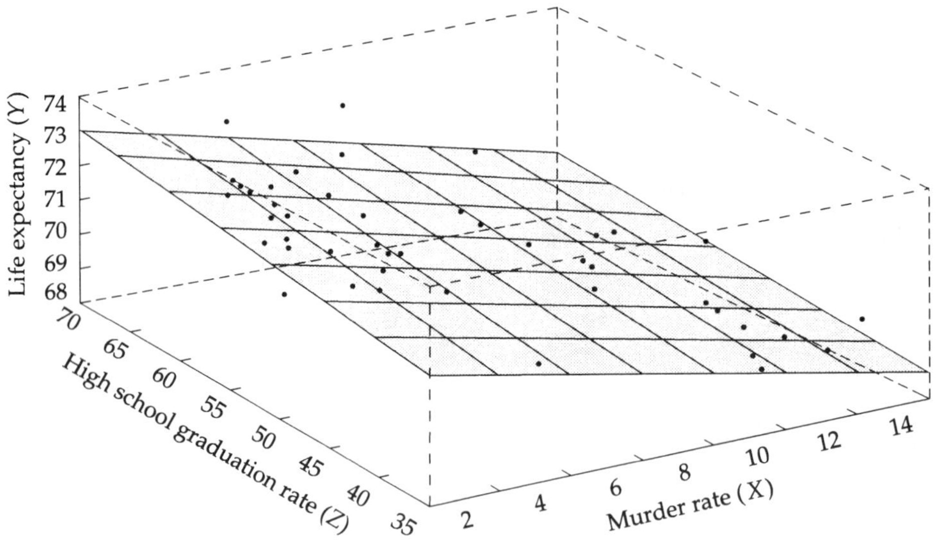

Figure 12.14 **Least-squares best-fitting plane for data of Figure 12.13.**

The technique presented above is called **multiple linear regression.** In many applications Y is predicted linearly from many more variables than the two of the example. This technique is of enormous importance in the social sciences, where human behavior of some sort (quantified by Y) is imperfectly predicted by several measured variables. For example, in personality theory, one approach uses 16 measured personality variables (extrovertedness, friendliness, and so on) to predict human behavior, such as overt aggressive behavior in various social settings.

SECTION 12.5 EXERCISES

1. Suppose a least squares regression plane is fit to a set of data consisting of the variables X, Y, and Z. The equation is found to be $Y' = 12.1 - 4.2X + 6.9Z$. For each of the following, respond with one of the following answers: "Y' increases," "Y' decreases," or "It is impossible to say whether Y' increases or decreases."
   a. You increase X, but hold Z constant.
   b. You increase Z, but hold X constant.
   c. You increase both X and Z.
2. Suppose we have the same equation for a plane as in Exercise 1: $Y' = 12.1 - 4.2X + 6.9Z$. Assume that this plane is based on values of X ranging from 0 to 10 and values of Z ranging from 100 to 200. Make predictions of the values of Y at the following points:
   a. $X = 5.6, Z = 123$
   b. $X = 2.3, Z = 270$
   c. $X = 12.4, Z = 90$
   d. Which of these three estimates will most likely be the best? Which of the three will most likely be the worst? Justify your answers.

3. Refer to the example pertaining to the 50 states and their life expectancies discussed at the beginning of Section 12.5. Determine the prediction of life expectancy (Y'), based on three different equations:
   a. The equation relating murder rate to life expectancy
   b. The equation relating high school graduation rate to life expectancy
   c. The equation relating both murder rate and high school graduation rate to life expectancy
   d. Which of these three equations does the best job of predicting the life expectancy?
   e. What does this teach regarding predictions on individual locations and using more variables?

12.6 THE MULTIPLE CORRELATION COEFFICIENT

The correlation coefficient is a useful measure of the linear relationship between two variables. In a regression context, another measure is the **multiple correlation coefficient,** often denoted R^2. It is useful because it can be used to measure the linear relationship between Y and X, or between Y and a pair of variables (X, Z), or even between Y and any number of explanatory variables. It can also be used in nonlinear regressions such as the quadratic regression of Section 12.3. It is based on comparing the mean square error of the estimated regression equation for Y', which we have been using repeatedly, to the usual variance of the Y variable, which can be thought of as the mean square error obtained by predicting Y only by the average of the Y's and thus ignoring all use of explanatory variables. If the regression equation (possibly nonlinear or involving multiple explanatory variables $X_1, X_2, \ldots$) for Y' explains Y very well, then the mean square error should be small relative to the variance. If the regression does not really help explain Y, its mean square error will be close to the variance of Y.

The multiple correlation coefficient R^2 is defined in such a way that it stays between 0 and 1. A value of 1 means the regression explains the Y perfectly, and a value of 0 means the regression does no good at all. Here is the formula:

$$R^2 = 1 - \frac{\text{mean square error}}{\text{variance of } Y}$$

You see that if the regression is perfect, the mean square error will be 0, and thus R^2 will be $1 - 0 = 1$. If the regression is no good, the mean square error will be close to the variance, so the ratio (mean square error/variance) will be about 1, and R^2 will be about $1 - 1$, or 0.

Let us look at the life expectancy example from the previous section. The standard deviation of Y is 1.329, so the variance of Y is $(1.329)^2 = 1.766$. The mean square error for the regression with just X (murder rate) was 0.6892.

Thus

Murder rate: $$R^2 = 1 - \frac{0.6892}{1.766} = 0.6097$$

When we used just the high school graduation rate, the mean square error was 1.167, so

HS graduation rate: $$R^2 = 1 - \frac{1.167}{1.766} = 0.3392$$

We would say that murder rate alone "explains" about 61% of the variation in life expectancy, while the high school graduation rate alone only explains about 34% of the variation. What about using both at the same time? The mean square error using both was 9.5954, so

Murder rate and HS graduation rate: $$R^2 = 1 - \frac{0.5954}{1.766} = 0.6629$$

We see that now about 66% of the variation can be explained. That is about 5% more than by using the murder rate alone. It looks as if two variables are somewhat better than one.

It turns out that when just one variable is being used (that is, in the case of straight-line regression), R^2 is the same as the square of the ordinary correlation coefficient between Y and the explanatory variable. In the above example, the correlation coefficient between X and Y is -0.7808, which when squared gives 0.6097, the value of R^2. Similarly, the correlation between Y and Z is 0.5822, which when squared is 0.3390. This should be equal to the R^2, but there has been some round-off error. The advantage of R^2 over the set of individual correlation coefficients is most evident when there are several explanatory variables, because then R^2 actually gives information not found in the individual correlation coefficients. Indeed one can find examples in which R^2 is close to 1 (and hence approximately 100% of the variation is explained) and yet all the correlation coefficients are far less than 1. Whenever a multiple regression is carried out, it is standard statistical practice to report R^2.

SECTION 12.6 EXERCISES

1. What do values of R^2 close to 1 indicate? What about values close to 0?
2. For each of the following sets of conditions, state what the value of R^2 would be:
   a. Mean square error = 3.41, variance of $Y = 6.78$
   b. Mean square error = 0.941, variance of $Y = 2.31$
   c. Mean square error = 9.83, variance of $Y = 10.49$
   d. Mean square error = 4.71, variance of $Y = 15.02$
3. What is the relationship between the correlation coefficient and R^2 when we have only one explanatory variable (the simple regression case)?

12.7 ANOTHER TEST FOR TREND: KENDALL'S TAU

In 1969 the U.S. government instituted a draft lottery for choosing young men to be drafted into the military. Numbers from 1 to 366 were randomly assigned to the days of the year, including February 29. Young men born in 1950 or earlier received the number corresponding to their birthday. They were then to be drafted into the military in the order given by these numbers. After the numbers were assigned, some people thought they detected a trend: namely, that people born later in the year tended to have lower lottery numbers. Figure 12.15 is a plot of the lottery data, X being the day of the year (1 = January 1, 2 = January 2, ..., 366 = December 31) and Y being the lottery number assigned to that day. On the face of it, there does not appear to be much trend in this graph.

However, instead of looking at individual points, one could look at the average lottery number for each month. That is what Table 12.8 provides. The plot of the monthly averages is in Figure 12.16. Now there is a distinct trend down. Is it significant, or could it just be due to chance? With our knowledge of statistics, we know this is the key question!

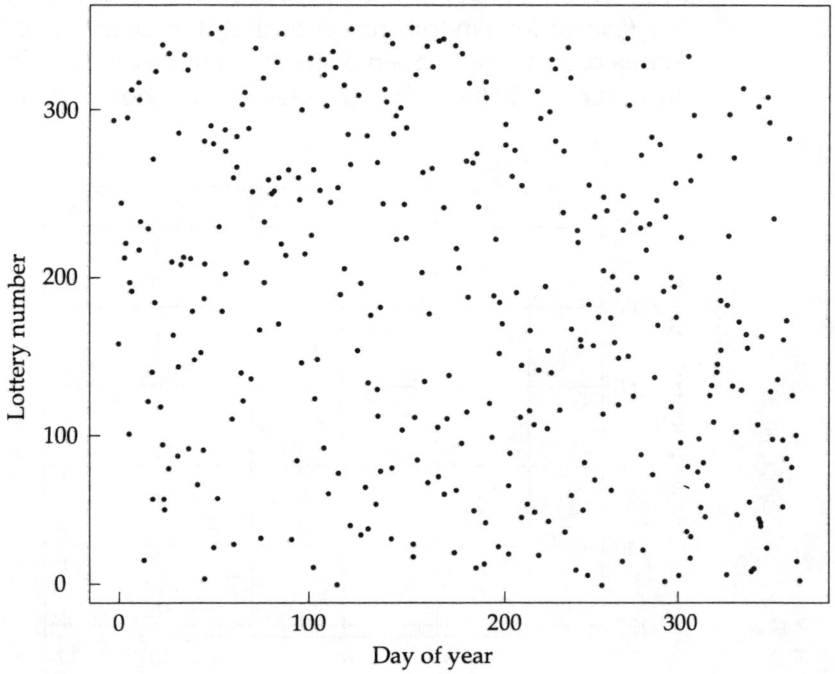

Figure 12.15 **Draft lottery number versus birthday (data are from Stephen E. Fienberg, "Randomization and Social Affairs: The 1970 Draft Lottery," *Science*, Jan. 22, 1971, pp. 155–261).**

Table 12.8 **Average Assigned Lottery Number for Month *X***

Month	Month's number, X	Average lottery number, Y
January	1	201.2
February	2	203.0
March	3	225.8
April	4	203.7
May	5	208.0
June	6	195.7
July	7	181.5
August	8	173.5
September	9	157.3
October	10	182.5
November	11	148.7
December	12	121.5

We could use the regression procedures already introduced, but instead we will use a procedure called **Kendall's tau.** *Tau* refers to the Greek letter τ, which we will use below. This is a nonparametric procedure because it does not make strong assumptions, such as that the data lie in a straight line (except for random noise) or that the observations have been drawn from a particularly shaped distribution or box model. Since we really have no reason to believe that the trend here should be a straight line, this

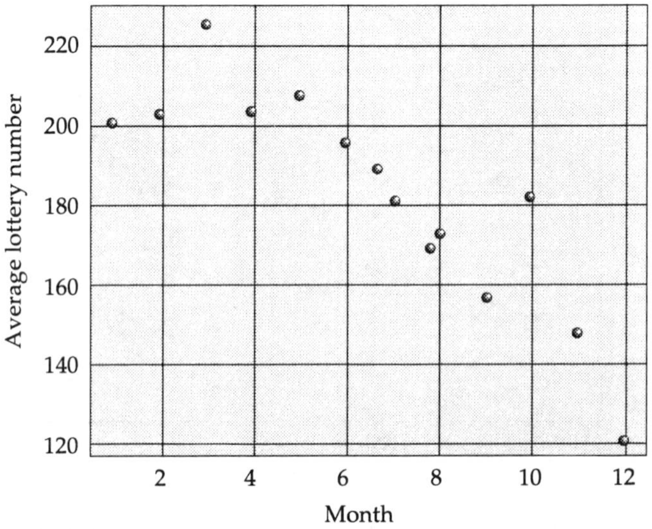

Figure 12.16 **Average lottery number versus birth month.**

nonparametric approach seems a wise choice. Kendall's tau is simpler in concept than the linear regression procedures above, and as just stated it can be used to test nonlinear trends as well as linear ones. The basic idea is to connect all pairs of points and then count how many of these line segments have negative slope and how many have positive slope. Clearly, if a large proportion have negative slopes, this suggests that the null hypothesis of no downward trend is likely not defensible. Figure 12.17 is the same plot as Figure 12.16, except that now there are lines connecting all pairs of points.

There are a lot of line segments: a total of 66. Most of them slope downward, but a number slope upward (such as two of the segments in the upper left). Counting carefully reveals that there are 55 segments with negative slopes and 11 segments with positive slopes. Kendall's tau is defined to be the difference between the number of positives and the number of negatives, divided by the total number of segments:

$$\tau = \frac{\text{number of positive slopes} - \text{number of negative slopes}}{\text{number of segments}}$$

For the draft lottery data,

$$\tau = \frac{11 - 55}{66} = -0.67$$

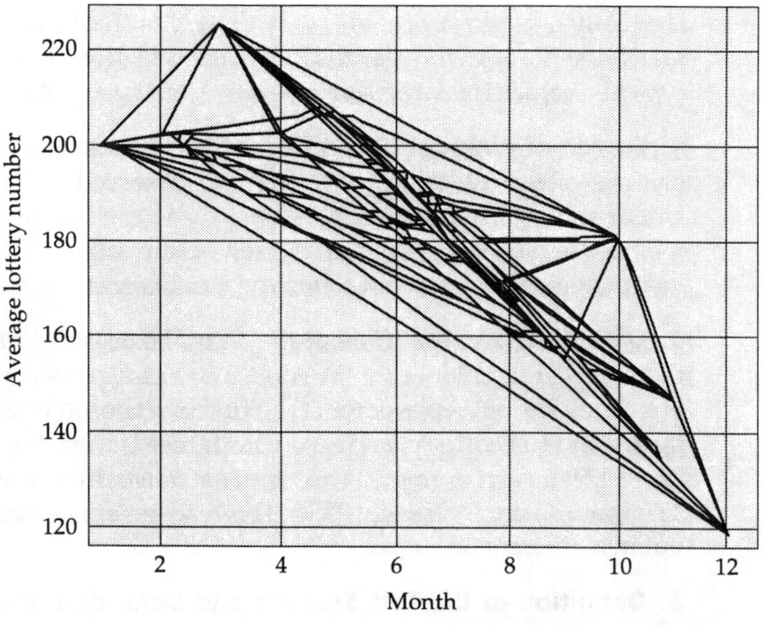

Figure 12.17 **Points of Figure 12.16 joined by line segments.**

The variable τ is read somewhat like a correlation coefficient. If $\tau = 1$, then the points trend exactly up (although it may not be in a straight line); if $\tau = -1$, then the points trend exactly down; and if $\tau = 0$, there is no particular trend up or down. The -0.67 for the draft lottery data shows a fairly strong trend down.

Note: If a line segment has a slope of 0, it is counted as neither positive nor negative. If two points share the same value of x, then there is no line segment between the points, so it is not counted as a positive or a negative and is not counted as a segment in the denominator of τ.

Using τ as a Test Statistic

The number $\tau = -0.67$ appears to provide impressive evidence of a trend, but could it be just because of chance that it is that far below 0? The null hypothesis we wish to test is that the average lottery numbers Y are independent of the month number X:

$$H_0: Y \text{ is independent of } X.$$

If H_0 is true, there should be no trend either up or down: τ should be around 0. The way we build a step 1 population from the data to formally test this hypothesis is to note that if the hypothesis is true, then any rearrangement of the given set of Y's in Table 12.8 is as plausible as any other. That is, for the 12 given X's and the 12 given Y's, if there is no trend then a particular Y is equally likely to appear with any of the X's. That is, we should be able to match the Y's with the months (X's) in any randomly chosen order to get a typical τ value. This procedure we now detail using the six-step method.

1. Choice of a Model (Definition of the Population): The population consists of the 12 values of Y that we observed: 201.2, 203.0, ..., 121.5. Under the null hypothesis, these are to be randomly assigned to the 12 months, because any order is as likely as any other if X and Y are indeed independent. Thus again we employ a randomization approach.

2. Definition of a Trial (Sample): A trial consists of randomly assigning the 12 Y's to the 12 months. We do this by randomly choosing the Y's, one at a time, without replacement. The first one chosen is assigned to January, the second to February, and so on, until there is just one left for December. Table 12.9 has an example. Thus the first Y we drew was 208.0, then came 121.5, and so on, and finally 225.8. The Y values are the same as in Table 12.8, but in a different order.

3. Definition of the Test Statistic and Definition of a Successful Trial: Now that we have the new arrangement, we proceed to calculate Kendall's tau for it. Figure 12.18 shows the points of the new arrangement and their

Table 12.9 **Randomly Assigned Lottery Number for Month *X* Using Data of Table 12.8**

Month	Month's number, *X*	Average lottery number, *Y*
January	1	208.0
February	2	121.5
March	3	181.5
April	4	157.3
May	5	195.7
June	6	148.7
July	7	201.2
August	8	182.5
September	9	203.7
October	10	173.5
November	11	203.0
December	12	225.8

connecting segments. This time the trend is not so obvious, and there are many segments sloping up and many sloping down. Counting shows that 44 of the slopes are positive and 22 are negative. Thus

$$\tau = \frac{44 - 22}{66} = 0.33$$

This τ is nearer 0 than the -0.67 of the obtained data. Such a simulated τ test statistic produces a successful trial if $\tau \leq -0.67$.

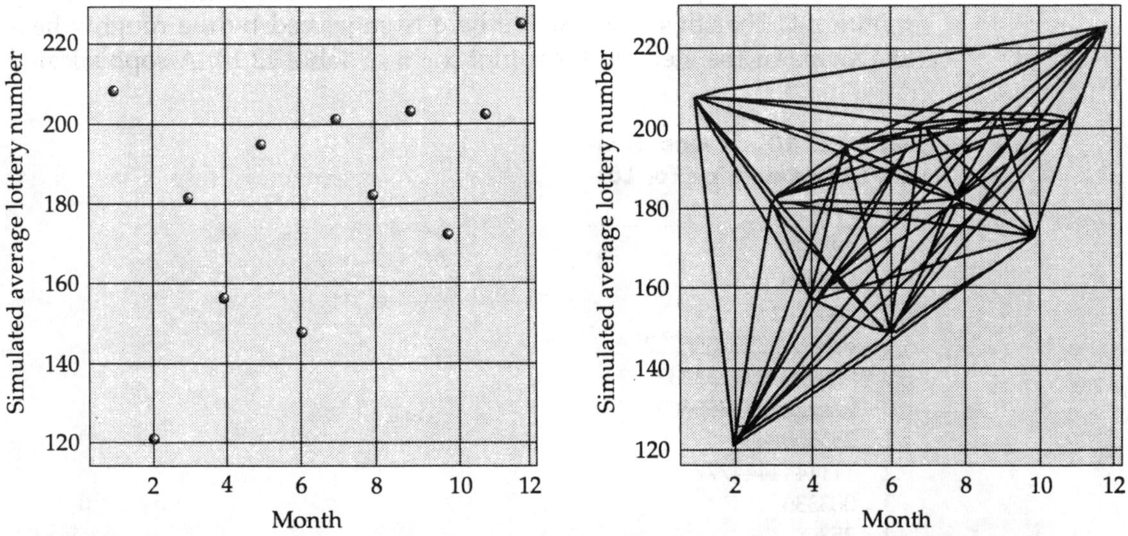

Figure 12.18 **Data of Table 12.9 randomly assigned and joined by line segments.**

4. Repetition of Trials: We do steps 2 and 3 100 times. Each time, we end up with a new τ based on a new random assignment of months to the given Y's. Table 12.10 shows the stem-and-leaf plot for the simulated τ's. The mean of the τ's is -0.012, and the standard deviation is 0.2218.

5. Estimation of the Probability of Obtaining a τ As Small As or Smaller Than the Observed (Probability of a Successful Trial): From the data, the τ was -0.67. Looking at all the simulated τ's, we see that the lowest is -0.45. Thus we estimate the chance of getting a τ as small as or smaller than -0.67 to be about 0% if the X and Y are independent.

6. Decision: Step 5 shows that it is very unlikely to get a τ as low as -0.67 by chance under the null hypothesis. Therefore we have to reject the null hypothesis—that is, conclude that X and Y are dependent and that there is a trend to the average lottery numbers among the months.

From the above analysis, we have to conclude that the 1969 draft lottery was *not* completely random, and that indeed, on average, people born later in the year did have lower lottery numbers to an extent not explained by chance. The personnel in the U.S. government also decided the lottery was not completely random. The following year a better randomization process was used, and there did not appear to be any trend.

A z Test for Kendall's Tau

As for many of the other hypothesis tests we have carried out by simulations based on bootstrapping or randomization, one can also use a test based on the normal approximation to see whether the observed τ is statistically significant. That this is legitimate here is suggested by the roughly bell-like shape of the stem-and-leaf plot for τ of Table 12.10. A sophisticated

Table 12.10 **One Hundred Simulated τ's for Lottery Data**

−4	52222
−3	993300
−2	74444411
−1	88888888555555222
−0	9999996633333333
0	zzzzzzzz33366669
1	2555588888
2	1111444444777
3	003336
4	258

Key: "−4 5" stands for −0.45.

probability argument can be given to justify the assumption that these sampled τ's are approximately normally distributed.

The key to the test is to figure out what the mean and the standard deviation of τ are when the null hypothesis is true. The formulas are given below. They are based on the assumption that the X's are independent of the Y's. It is clear that under the null hypothesis,

$$\text{Population mean of } \tau = 0$$

However, deriving the standard deviation of τ is beyond the scope of this book. The value is

$$\text{SD of } \tau = \sqrt{\frac{2(2n+5)}{9 \cdot n(n-1)}}$$

where n is the number of pairs, 12 in our case. Then the z statistic is

$$z = \frac{\tau - \text{mean of tau}}{\text{SD of tau}}$$

For our example, $n = 12$; hence

$$\text{SD of } \tau = \sqrt{\frac{2 \cdot (2 \cdot 12 + 5)}{9 \cdot 12 \cdot 11}} = \sqrt{0.04882} = 0.2210$$

Note that the standard deviation of the simulated τ's was 0.2218, very close to the theoretical value. Now the z statistic:

$$z = \frac{-0.67 - 0}{0.2210} = -3.03$$

Now we can finish up with the new step 5.

New Step 5. Estimation of the Probability of Obtaining a Slope As Large As or Larger Than the Observed (Probability of a Successful Trial): Looking in the normal table (Appendix E), we see that the chance of z being less than -3.03 is 0.0012, or about 1/10 of a percent. Thus in step 6 we conclude that the chance of getting a τ below -0.67 is so small that we have to rule out chance. Again, we strongly reject the null hypothesis that X and Y are independent.

We have made considerable progress in this chapter on one of the major problems in statistics, namely, discovering relationships between variables in the presence of random noise. We can use linear regression if it is appropriate to assume a straight-line relationship. Moreover, we can test for the presence or absence of such a relationship using the correlation coefficient. We can solve some nonlinear relationship problems by rescaling to turn them into ordinary linear regression problems. We have learned about curvilinear and multiple regression and the use of the multiple correlation coefficient R^2 to measure how good the fit is in these cases. Finally, we have learned to test for the presence of either an upward or a downward trend via Kendall's tau, even if the trend is curved rather than linear. Sometimes we used our often-used six-step simulation approach, and sometimes we used the often-used normal approximation approach.

SECTION 12.7 EXERCISES

1. Consider the five points in the following scatter plot:

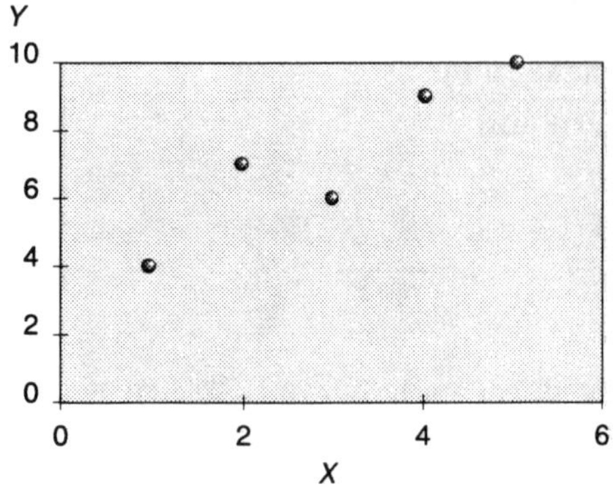

   a. Calculate τ for this set of data.
   b. Use the z test for Kendall's tau to test the hypothesis that the X's are independent of the Y's.

2. For each of the following sets of conditions, calculate Kendall's tau:
   a. Seven segments have positive slopes, 14 segments have negative slopes, and a total of seven points appear on the scatter plot.
   b. Twenty-one segments have positive slopes, 15 segments have negative slopes, and a total of nine points appear on the scatter plot.
   c. Thirteen segments have positive slopes, 32 segments have negative slopes, and a total of 10 points appear on the scatter plot.

3. For each of the sets of conditions in Exercise 2, test the hypothesis that X and Y are independent.

4. Consider the six points in the following scatter plot:

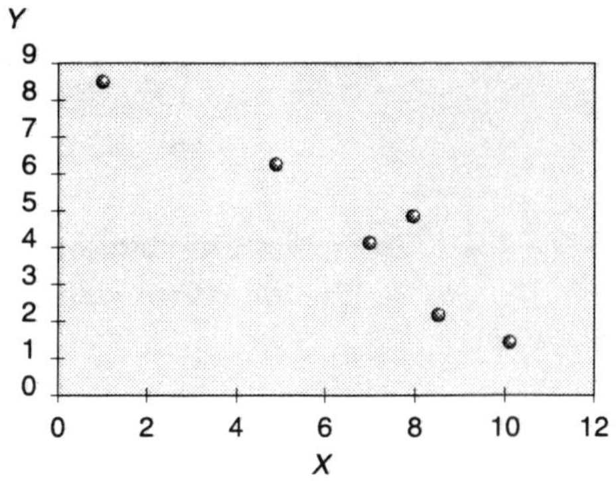

   a. What is the value of Kendall's tau for this set of data?
   b. Perform the hypothesis test to determine whether X is independent of Y.

5. Describe the similarity in the interpretation of the Kendall's tau statistic and the Pearson correlation coefficient.

6. State two advantages of the use of Kendall's tau to test for trend over the use of the standard method with the least squares regression line.

CHAPTER REVIEW EXERCISES

1. Recall the following set of data regarding the return and risk for different types of investments, first presented in Chapter 3.

	Return (X)	Risk (Y)
Common stocks	9.8	15.8
Long-term corporate bonds	6.9	11.1
Long-term treasury bonds	6.4	10.9
Short-term treasury bonds	6.4	2.9
Residential real estate	10.6	10.7
Farm real estate	9.9	8.5
Business real estate	8.7	4.9

   a. Make a scatter plot of the data. Does a straight-line fit seem reasonable in this case?
   b. The actual equation for the least squares line in this case is $Y' = 2.10+0.854X$. The standard deviation of X is 1.80, and the standard deviation of Y is 4.30. Perform a test to determine whether X is independent of Y.

2. Using the data and least squares equation given in Exercise 1, predict the risk for the following values of the return:
   a. Return = 7.5
   b. Return = 5.2
   c. Return = 15.4
   d. Which of these three estimates is the most reliable? Why?

3. Look back at Exercise 1 once again. The covariance between X and Y is 2.37.
   a. What is the value of r, the Pearson correlation coefficient? Recall the formula from Chapter 3.
   b. Using this value of r, perform another test as to whether X is independent of Y.

4. Consider the following data concerning the population (in millions) of the state of Michigan over the years:

Year (X)	Population (Y)	Year (X)	Population (Y)
1840	0.212	1920	3.668
1860	0.749	1940	5.256
1880	1.636	1960	7.823
1900	2.420	1980	9.262

a. Make a scatter plot of the data. Does a straight line seem to be the best fit?
b. Try transforming Y by letting Z equal the square root of Y. Plot Z versus X. Does there seem to be more of a straight line now?

5. The following table gives the number of calories, saturated fat (in grams), and cholesterol (in milligrams) in three-ounce portions of different types of red meat:

Meat	Saturated fat (X)	Cholesterol (Z)	Calories (Y)
Veal top round	1.0	88	127
Pork tenderloin	1.4	67	133
Beef top round	1.4	71	153
Pork sirloin chop	1.5	78	156
Pork loin roast	2.4	66	160
Lamb leg	2.3	78	162
Pork loin chop	2.5	70	165
Beef tenderloin	3.2	71	272

When fitting the least squares regression line to these data, you come up with $Y' = -12.62 + 54.38X + 0.98Z$.

a. As you increase the amount of saturated fat and cholesterol, what will happen to the number of calories? Explain why this is logical.
b. Predict the number of calories in a serving of red meat having 2.1 grams of fat and 72 milligrams of cholesterol. Was this estimation interpolation or extrapolation?
c. The mean square error for this regression equation is 595.1, and the variance of Y is 2024. What is the value of R^2, the multiple correlation coefficient?

6. Return to the population data given in Exercise 4.
   a. What is the value of Kendall's tau for the original, untransformed data?
   b. Now calculate Kendall's tau using the transformed data (using Z instead of Y).
   c. What important property of using Kendall's tau as a test for trend do these results show?

7. The table on the following page gives three variables for each baseball team in the American League Western Division for the 1987 season: the number of runs scored, the number of runs allowed, and the number of wins.
   a. Make a scatter plot of Y versus X and a scatter plot of Y versus Z.
   b. Predict what the signs of the coefficients for X and Z would be in the multiple regression model used to fit these data. Explain how you made your predictions.

Team	Number of runs scored (X)	Number of runs allowed (Z)	Number of wins (Y)
Athletics	800	620	104
Twins	759	672	91
Royals	794	648	84
Angels	714	771	75
White Sox	631	757	71
Rangers	637	735	70
Mariners	664	744	68

8. Assume that you have 15 data points ($n = 15$). For each of the following values of r, the Pearson correlation coefficient, give the value of the T statistic and then determine whether we would accept or reject the hypothesis that X is independent of Y.
   a. $r = 0.5$
   b. $r = 0.4$
   c. $r = 0.3$
   d. $r = 0.2$
   e. Judging from the results of parts (a) through (d), if $n = 15$, between what two values is the cutoff point for r where we would decide to reject the hypothesis that X is independent of Y?

9. Explain the difference between interpolation and extrapolation.

10. For each of the following sets of conditions, calculate the value of the z statistic for testing the hypothesis that X is independent of Y.
    a. Obtained slope is 4.62, standard deviation of X is 0.29, standard deviation of Y is 1.57, $n = 9$.
    b. Obtained slope is 1.42, standard deviation of X is 4.36, standard deviation of Y is 44.60, $n = 14$.
    c. Obtained slope is 12.36, standard deviation of X is 4.71, standard deviation of Y is 62.45, $n = 10$.
    d. In which of the above three cases would you conclude that X is independent of Y?

11. In a group of seven boys, two skills are measured: the amount the boy is able to bench-press (in pounds), and the number of push-ups the boy is able to complete consecutively. The results are in a table on the following page.
    a. Make a scatter plot of the relationship between X and Y.
    b. What do you notice in your scatter plot?
    c. True or false: Outliers should always be discarded from the data set and ignored as freak occurrences.

Boy	Number of push-ups (X)	Weight (in pounds) bench pressed (Y)
1	12	150
2	15	175
3	9	140
4	13	155
5	12	250
6	6	135
7	13	160

12. You have a set of data with an X and a Y variable. After calculating the Pearson correlation coefficient, you perform a significance test and find that the correlation is not significantly different from zero. True or false: You should conclude that there is no relationship between X and Y. Explain.

13. The following table gives two variables for each of 10 countries: first, thousands of residents per doctor, and second, the life expectancy for a resident of that country.

Country	Thousands of residents per doctor	Life expectancy (years)
Argentina	0.37	70.5
Brazil	0.684	65
Canada	0.449	76.5
France	0.403	78
Pakistan	2.364	56.5
Peru	1.016	64.5
Philippines	1.062	64.5
Poland	0.48	73
United States	0.404	75.5
Vietnam	3.096	65

a. Make a scatter plot of the data. Describe the pattern in the data.
b. Suggest a way to transform the X variable to get this into a straight line. Try your suggestion by making a new scatter plot. (*Hint:* Look back in the chapter to a data set that had a similar original pattern.)

14. Refer to the original (untransformed) data given in Exercise 13.
a. Calculate Kendall's tau for this set of data.

b. Is this value significant? (That is, is there a trend in the data?)
c. Should you be concerned about the fact that the data do not follow a straight line? Explain.

15. Jan operates a hot-dog stand in a park. She suspects that there is a relationship between the temperature on a given day and the number of hot dogs she sells that day. She begins to keep track of the data, and she sees the following results:

Day	Temperature (X)	Number of hot dogs sold (Y)
1	52	67
2	48	61
3	43	49
4	44	54
5	56	65
6	61	75
7	63	72
8	62	77
9	55	64
10	51	60

a. The variance of X is 51.83, and the covariance between X and Y is 54.6. What is the slope of the least squares regression line?
b. Test whether X is independent of Y. The standard deviation of X is 7.20, and the standard deviation of Y is 8.90.
c. Do these data support Jan's idea that on hotter days she sells more hot dogs?

12

Computer Exercises

The following problems will provide some practice in determining whether a straight line is reasonable for certain scatter plots.

1. Airport Data

The data set AIRPORT.DAT contains data on 135 large and medium-sized U.S. airports.* The variables are

Number of scheduled departures (Scheduled Depart)
Number of performed departures (Depart)
Number of enplaned passengers (Passengers)
Tons of freight (Freight)
Tons of mail (Mail)

Load the data set in the Data program, and select the Regression option from the Data menu.

a. Do the regression with X = Depart and Y = Passengers. Do you see any outliers? Is there any curvature?

b. Often one can see outliers and curvature better by looking at the residuals. Let X be the original X variable, but let Y be the variable of errors, $Y - Y'$. Press the Save Residuals button. A new set of variables, $R(3.2)$, will appear in the variable lists. These are the values of $Y - Y'$, where the 3 means Y is the third variable (Passengers) and the 2 means X is the second variable (Depart). Now keep X = Depart, but choose $Y = R(3.2)$, and press Update. On the vertical axis this scatter plot shows how far the points are from the original regression line.

---

*The data were collected by Larry Winner, Department of Statistics, University of Florida, from U.S. Federal Aviation Administration and Research and Special Programs Administration, "Airport Activity Statistics" (1990). In *Journal of Statistics Education* Data Archives http://www2.ncsu.edu/ncsu/pams/stat/info/jse/datasets.index.html.

i. Which point has the largest positive residual? Click the mouse on it. What is the name of the airport? How would you characterize this airport?

ii. Which point has the largest negative residual? Click the mouse on it. What is the name of the airport? How would you characterize this airport?

c. Sometimes functions of the data produce a better fit. This part will look at the square roots of Departures and Passengers. To find these, close the regression screen, and select "Create New Variables" from the Data menu. You should see the following screen:

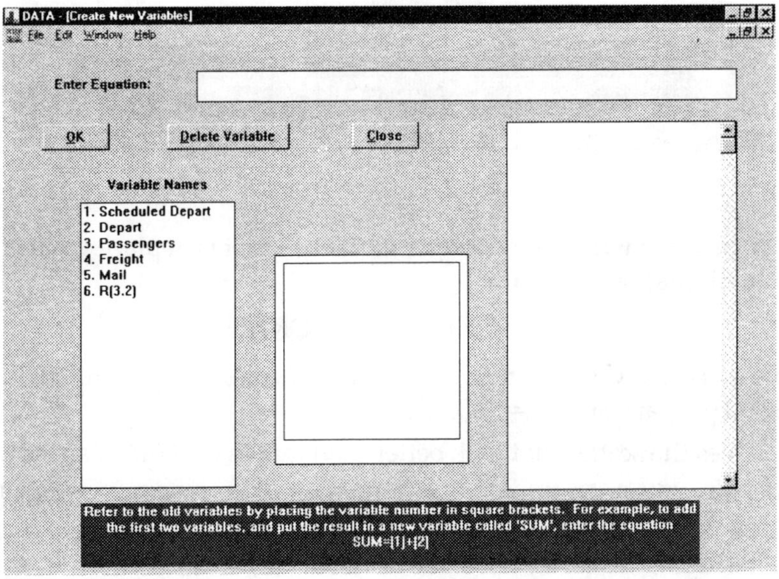

You can create new variables from the existing ones by entering an equation in the Enter Equation area. The program is fairly picky about how you enter the equation. To find the square root of Depart, which is variable 2, type

$$\text{SQRT(Depart)} = \text{SQRT([2])}$$

in that area. It is important that the 2 be in square brackets. Then press OK. The program will create a new variable called *SQRT(Depart)* by taking the square roots of all the values in Depart. It lists the new data at the right, plots a histogram of the data, and adds the new variable to the variable list. The screen should look like this:

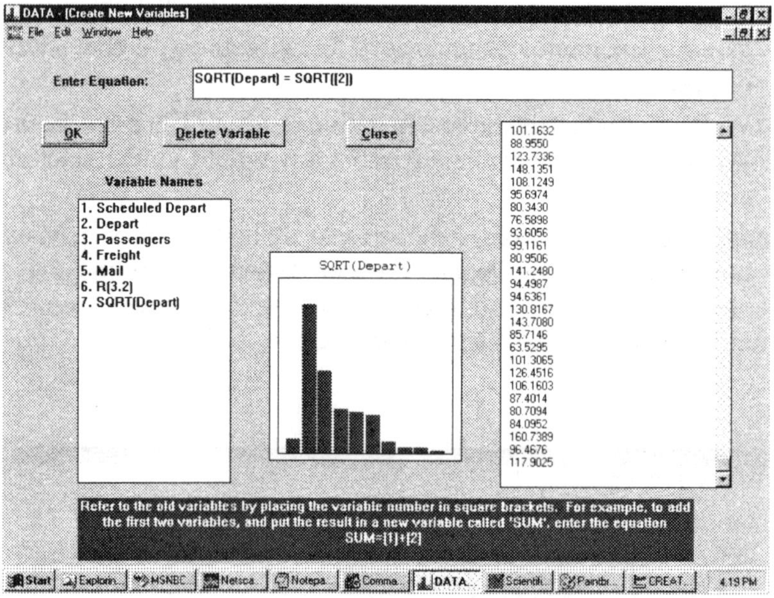

Do the same with the Passenger variable—that is, replace what is in the Enter Equation area with

$$\text{SQRT(Pass)} = \text{SQRT([3])}$$

Now press Close, and go to the regression screen. Plot X = SQRT(Depart) and Y = SQRT(Pass).

i. Does the scatter plot look better than that in part (a)? If so, how?

ii. Press "Save Residuals," and then plot the residuals (which will be $R(8.7)$) versus X = Depart. What are the highest and lowest residuals? Are they the same as in part (b)?

d. Now create the scatter plot with X = Depart and Y = Freight.

i. Are there any outliers? Is there curvature?

ii. Press "Save Residuals," and plot these new residuals versus X. Does there seem to be curvature? Which airport has the largest positive residual? Is it one of the airports in part (b)? How would you characterize this airport?

e. Now close the regression screen, go to the Create New Variables screen, and create the variable

$$\text{SQRT(Freight)} = \text{SQRT([4])}$$

(because Freight is the fourth variable). Close this screen, and go back to the regression screen.

i. Create the plot with X = SQRT(Depart) and Y = SQRT(Freight).

ii. Is there still some curvature, or does the scatter plot appear to fall along a straight line?

iii. Press "Save Residuals," and plot Y = the new residuals versus X = SQRT(Depart). Is there curvature?

2. BIRTH RATES

The data set BIRTHRT.DAT contains the number of births per 10,000 23-year-old women in the United States for the years 1917 to 1975.* Load the data set into the Data program, and go to the regression screen. Create the plot with X = Year and Y = Birthrate.

1. Is there curvature?
2. Select the regression line box and press Update. Does the regression line give a good idea of the pattern in the data?
3. What phenomena explain the dip and peak in the birthrates?

3. PLANETS

Load the PLANETS.DAT data set into the Data program. These are the data of Table 12.5.

a. Go to the regression screen, and plot X = Distance versus Y = Year. (See Figure 12.9.) There is some curvature. Press "Save Residuals," and plot Y = the residuals (R(3.1) in this case) versus X = Distance. Is the curvature clearer?

†b. Press Close, and go to "Create New Variables" in the Data menu. Create the logarithm in base 10 ($\log_{10}$) of the variables Distance and Year. That is, enter the formula

log 10(Distance) = log 10([1])

press OK, enter the formula

log 10(Year) = log 10([3])

press OK, and then press CLOSE.

Go to the regression screen, select the Regression Line square, and plot X = log 10(Distance) and Y = log 10(Year).

---

*P. F. Velleman and D. C. Hoaglin, *Applications, Basics, and Computing of Exploratory Data Analysis* (Belmont, Calif.: Wadsworth, 1981).

† This question uses the logarithm in base 10. It can be skipped if the student is not familiar with logarithms.

i. How close are the points to a straight line?

ii. What is the correlation coefficient?

iii. What is the regression equation?

According to the formula in Section 12.3, the relation between Distance and Year should be

$$\text{Distance}^2 = B \times \text{Year}^3$$

Take $\log_{10}$ of both sides in that equation, and rewrite it to obtain

$$\log_{10}(\text{Distance}) = m + b \times \log_{10}(\text{Year})$$

What is the value of b in that equation? How does it compare with the slope in the regression line equation of part (ii)?

c. Now plot X = Diameter and Y = Day to see whether larger planets take a longer time to rotate.

i. Is the correlation positive or negative?

ii. Do the data follow a straight line, or is there curvature?

iii. Now close the regression screen, and go to "Create New Variables." Create the variable

$$1/\text{Diameter} = 1/[4]$$

Close the screen, and go back to the regression screen. Now plot X = 1/Diameter and Y = Day. How would you describe this plot? What is the correlation? What are the two points above the line? What do they have in common?

4. NONRESIDENT TUITION IN THE BIG TEN

The data set NONRES.DAT compares the Big Ten schools on the percentage of out-of-state students they have and the tuition these students pay. The total undergraduate enrollment is also included. The data are for the 1994–95 academic year. The Big Ten has eleven schools(?!). All are public except Northwestern.

Load the data set into the Data program, and go to the regression screen. Check the Regression Line box.

a. Create the regression with X = NR Tuition (the tuition for nonresident, or out-of-state, students) and Y = % Nonresident (the percentage of undergraduates who are nonresident).

i. What is the correlation? Is that surprising, or would you expect that the higher the tuition for nonresidents, the smaller the percentage of nonresidents?

ii. Which school has the highest nonresident tuition? Which has the highest percentage of nonresident students? Why do you think this is?

iii. What is the regression equation? According to this equation, what would be the percentage of nonresident students if tuition were free—that is, if $X = 0$? Is this a reasonable answer? Why or why not?

b. Now create the regression with X = Undergrads (the total number of undergraduates) and Y = NR Tuition.

i. What is the correlation? Which school has the fewest undergraduates?

ii. Close the regression screen, go to "Input/Modify Data," select "Use current data set: Out of Staters," and press OK. Delete Michigan and Northwestern. (See Computer Exercise 2 of Chapter 9.) Press OK, press NO when the program asks whether to save data to a file, and go back to the regression screen. Redo the regression with X = Undergrads and Y = NR Tuition. What is the correlation now? Is it the same as in part (i)? How do you explain the difference?

13

MORE PROBABILITY WITH APPLICATIONS

We are so poorly equipped . . . to grasp the working of random processes and patterns in nature.

S. J. Gould

OBJECTIVES

After studying this chapter, you will understand the following:

- *The basic laws of theoretical probability computations*
- *Formulas for the binomial, Poisson, and hypergeometric probability distributions*
- *Elemental conditional probability*
- *Expected value as computed theoretically*
- *A test of the equality of two population proportions*
- *A chi square test of the independence of two categories*

13.1 SOME BASIC RULES OF PROBABILITY 618

13.2 THREE IMPORTANT DISCRETE PROBABILITY MODELS 624

13.3 CONDITIONAL PROBABILITIES 632

13.4 EXPECTED VALUE AS A THEORETICAL CONSTRUCT 635

13.5 HYPOTHESIS TESTING OF THE EQUALITY OF TWO POPULATION PROPORTIONS: AN APPLICATION OF THE BINOMIAL DISTRIBUTION 643

13.6 CHI-SQUARE TEST FOR PROBABILISTIC INDEPENDENCE 646

CHAPTER REVIEW EXERCISES 652

COMPUTER EXERCISES 657

Key Problem

In a certain area, if any day in July is chosen at random, the probability of rain occurring on that day is 0.10. The probability of rain on any two consecutive days is known to be 0.05. If it is raining today, what is the theoretical probability that it will rain tomorrow?

To answer this problem, we need to develop some more theory concerning theoretical probability, in particular theory regarding outcomes, events, independence, and conditional probability. These topics, along with others, will be discussed in this chapter.

13.1 SOME BASIC RULES OF PROBABILITY

In Chapter 8 we discussed how probability distributions are used to assign a theoretical probability to each possible outcome of an experiment. In this section we will discuss this topic further.

The term **sample space** refers to the set of all possible outcomes of an experiment. Remember that an *experiment* is to be widely interpreted as any random, or chance, data-generation situation. It is not an experiment in the sense of the physical, biological, and social sciences. We will let S denote the sample space. For example, if we flip a coin, then $S = \{\text{heads, tails}\}$. If we toss a die, $S = \{1, 2, 3, 4, 5, 6\}$.

There are two kinds of sample spaces: *discrete* and *continuous*. The feature that distinguishes a discrete sample space from a continuous sample space is that in the discrete case the sample space S can be written as a sequence. The sequence could be very long, even infinite (as for the Poisson distribution, where $S = \{1, 2, 3, \ldots\}$), but all of its outcomes can be listed one after another. Conversely, consider writing out all the possible values in the case of the normal distribution (which, as you may recall from Chapter 8, is a continuous distribution). Even if we consider just writing the values a normally distributed variable could take between 0 and 1, it is somewhat intuitive that a sequential listing of this sort, which would have to include every number between 0 and 1, is not possible. A formal mathematical proof can verify this intuition.

In a **discrete probability model** a probability is assigned to each of the outcomes in the sample space S. For example, we know that for the flip of a coin, $S = \{\text{heads, tails}\}$. If the coin is fair, we assign a probability to the two outcomes in the sample space as follows:

$$p(\text{heads}) = 0.5$$
$$p(\text{tails}) = 0.5$$

In the example of the toss of a die, $S = \{1, 2, 3, 4, 5, 6\}$, and we make the assignments

$$p(1) = \frac{1}{6} \qquad p(2) = \frac{1}{6} \qquad \ldots$$

These are both examples in which each outcome is equally likely, as discussed in Section 5.5.

Note that in both these examples, if one trial is performed, at least one of the outcomes must occur. For example, if a coin is tossed, either heads or tails must come up. When a set of outcomes has this property, it is said to be *exhaustive*. Since the set S is defined so as to contain every possible outcome, it is exhaustive. Therefore when you build the sample space, you want to be sure to list all possible outcomes that have some chance of occurring.

Also notice that in both examples, only one of the outcomes can occur in one trial. For example, you cannot obtain a head and a tail on the same flip of a coin. When outcomes have this property, they are said to be *mutually exclusive.* In a set of mutually exclusive outcomes, only one of them can occur on a single trial. Besides being exhaustive, the outcomes of a sample space (in either the discrete or the continuous case) must be mutually exclusive. In order to construct a sample space, you must ensure that its outcomes are both exhaustive and mutually exclusive. Thus when you build the sample space (in either the discrete or the continuous case), you must be careful to specify outcomes that cannot simultaneously occur. In what follows we will assume that every set of outcomes composing a sample space is exhaustive and mutually exclusive.

In the discrete case, whenever you have a list of outcomes that are both exhaustive and mutually exclusive, the sum of the probabilities of those outcomes must be 1. This is the case because at least one of the outcomes has to occur (since they are exhaustive), but more than one cannot occur (since they are mutually exclusive).

Generally, any set of outcomes is called an **event.** Once a trial is carried out and an outcome results, one can then say whether a specified event has occurred. For example, let's toss a die: then $S = \{1, 2, 3, 4, 5, 6\}$. Let A be the event that the die comes up an even number, or $A = \{2, 4, 6\}$. If a trial results in outcome $\{4\}$, we then say that A has occurred. Let $B = \{1, 5, 6\}$. If a trial results in outcome $\{3\}$, then B has not occurred. Clearly, an event can be described in words, as in the case of A above—"The die is even"—or it can be described by specifying its outcomes.

The definition of mutual exclusivity certainly applies to events just as it does to outcomes. Also note that an outcome is an event, too. For example, $A = \{1\}$ in the die example is clearly an event. S is also an event, although an uninteresting one in the sense that $p(S) = 1$ and hence S is sure to occur regardless of which outcome the trial produces.

Example 13.1

Consider tossing a die. Let

$$A = \text{event that it comes up a 1}$$
$$B = \text{event that it comes up a 2}$$

The events A and B are mutually exclusive since only one, not both, can occur in one trial.

Example 13.2

Again, consider a die. Let

$$A = \text{event that it comes up a number greater than 3}$$
$$B = \text{event that it comes up a number less than 5}$$

In this case the events are not mutually exclusive, because if you roll a 4, both of the events occur. Event A occurs since 4 is greater than 3, but event B also occurs since 4 is less than 5.

One important goal in any probability problem is to be able to compute $p(A)$ for any choice of event A, given the probabilities of all the possible outcomes. Here we give a general rule for discrete sample spaces that holds even if the outcomes are not equally likely. The fundamental rule is a simple one. Let a denote an outcome in A. Then

$$p(A) = \text{sum of } p(a)\text{'s}$$

where the sum is over every outcome a in A. You will use this formula quite often. A special case yields

$$1 = p(S) = \text{sum of } p(a)\text{'s}$$

where the sum is over every a in S.

Example 13.3

Let $S = \{1, 2, 3, 4\}$ and let the probability distribution be as given here:

a:	1	2	3	4
$p(a)$:	1/12	2/12	3/12	6/12

(This could model a loaded four-sided die, for example.) This table tells us the probabilities of the four possible outcomes. Let A be the event that the die is even. Then

$$p(A) = p(2) + p(4) = \frac{2}{12} + \frac{6}{12} = \frac{8}{12} = \frac{2}{3}$$

Combining Events to Form New Events

Consider events A and B. The event that either event A *or* event B occurs is denoted $A \cup B$, and the event that both event A *and* event B occur is denoted $A \cap B$. We use the set theory symbols for union ($\cup$) and intersection ($\cap$) because they represent these same relationships when we express these events as sets. For example, the outcomes corresponding to A or B (or both) occurring are precisely the outcomes in the set $A \cup B$.

Example 13.4

Consider the tossing of a die. Let A be the event that the die comes up a 1, and let B be the event that the die comes up a 2. Then

$A \cup B$ is the event that the die comes up a 1 or a 2.

$A \cap B$ is the event that the die comes up both a 1 and a 2.

Of course, this second event is impossible and hence $p(A \cap B) = 0$, since the events A and B are mutually exclusive. We are applying the obvious rule that $p(\text{impossible event}) = 0$. By an impossible event we simply mean an event, such as $A \cap B$ here, that contains no outcomes and hence can never occur.

Let C be the event that the die comes up greater than 3, and let D be the event that the die comes up smaller than 5. Then

$C \cup D$ is the event that the die comes up greater than 3 or it comes up smaller than 5.

$C \cap D$ is the event that the die comes up both greater than 3 and smaller than 5.

Note that the event $C \cup D$ occurs every time: every roll is either greater than 3 or less than 5. Thus $C \cup D = S$, and hence $p(C \cup D) = 1$. The event $C \cap D$ occurs only when the die comes up a 4, since 4 is the only number greater than 3 and smaller than 5.

Basic Rules of Probability for Events

Using the concepts just developed, we can introduce a few rules of probability that apply in every theoretical probability problem. Thus you are always free to use one of these rules, if it helps you solve a probability problem.

1. $$0 \le p(A) \le 1 \quad \text{for any event } A$$
2. $$p(S) = 1$$
3. $$p(A \cup B) = p(A) + p(B) - p(A \cap B)$$
4. $$p(A \cap B) = 0 \quad \text{if the events } A \text{ and } B \text{ are mutually exclusive}$$
5. $$p(A) = 1 - p(\text{not } A)$$

Rule 1 simply states that the probability of any event is between 0 and 1. We know this must be true since we interpret the probability as the proportion of the time an event occurs in a very large number of trials. Rule 2 is true because one outcome in the sample space must occur on every trial, since the sample space is defined as the set of all possible outcomes, and hence the proportion of the time S occurs is always 1 regardless of how many trials we carry out. Rule 3 is often useful, especially when it is known that A and B are independent and therefore $p(A \cap B) = p(A)p(B)$, as discussed in Section 5.5. Rule 4 should be obvious since if two events are mutually exclusive, it is impossible for both to occur on the same trial. Therefore $P(A \cap B)$ must equal 0. Rule 5 was first introduced in Section 5.5 of Chapter 5, on probability.

Example 13.5

Let $p(A) = \frac{1}{2}$ and $p(B) = \frac{1}{2}$. Suppose A and B are independent. Find $p(A \cup B)$.

Solution

By rule 3,

$$\begin{aligned} p(A \cup B) &= p(A) + p(B) - p(A \cap B) \\ &= p(A) + p(B) - p(A)p(B) \\ &= \frac{1}{2} + \frac{1}{3} - \frac{1}{6} \\ &= \frac{4}{6} = \frac{2}{3} \end{aligned}$$

Thus rule 4 can be used whenever we know $p(A)$ and $p(B)$ and know that independence holds.

Rules 3 and 4 can be combined to create one of the most important results of this section. You will note that if the events A and B are mutually exclusive, then

$$p(A \cup B) = p(A) + p(B)$$

since rule 4 tells us that $p(A \cap B)$ necessarily equals 0. What the above statement says is that if two events are mutually exclusive, the probability that either of them happens on a trial is the sum of the probabilities of their happening individually. This rule will turn out to be very useful in cases when we know the events to be mutually exclusive. When people say that probabilities are "additive," it is precisely this rule they have in mind. Keep in mind that it *only* holds provided A and B are exclusive; otherwise, rule 3 is needed.

So far, in all but one of the examples we have presented, the outcomes of the experiment had the same probabilities. In the example of the coin, it either came up heads or tails, each with probability $\frac{1}{2}$. It is important to note that the basic outcomes need not always have the same probabilities, as long as the total probability of all the basic outcomes sums to 1, because of the exclusivity and exhaustiveness of the outcomes of a sample space. Consider the following example of unequal probability outcomes.

Example 13.6 A die is loaded in such a way that the theoretical probability of obtaining each of the sides is as follows:

Side	Probability
1	0.10
2	0.20
3	0.15
4	0.25
5	0.10
6	0.20
Total	1.00

Note that the sample space S is still the same as in the example of a fair die ($S = \{1, 2, 3, 4, 5, 6\}$), and the sum of the probabilities of all the sides is 1, as required for the sample space in a discrete probability model. All of the rules presented above can be used.

Let A be the event that the die comes up a 1, and let B be the event that the die comes up a 2. Then

$$p(\text{die comes up either a 1 or a 2}) = p(A \cup B) = p(A) + p(B)$$

since A and B are mutually exclusive. Hence

$$p(\text{die comes up either a 1 or a 2}) = 0.10 + 0.20 = 0.30$$

Consider other examples illustrating the use of the rules:

$$p(A \cap B) = 0$$

$$p(\text{number showing is less than or equal to 5}) = 1 - p(\text{number showing equals 6})$$
$$= 1 - 0.2 = 0.8 \qquad \text{(rule 5)}$$

$$p(\text{number showing is even}) = p(2) + p(4) + p(6) = 0.20 + 0.25 + 0.20 = 0.65$$

SECTION 13.1 EXERCISES

1. A four-sided die is to be rolled. What are all the possible events based on the four outcomes of this experiment?
2. In a table of random digits, you look at single digits one at a time. Consider each digit you look at as one trial of a random experiment.
   a. What is the sample space S of the possible outcomes of this experiment?
   b. Assign a probability to each of the outcomes you listed as part of S.
3. Return to the situation introduced in Exercise 2, in which you are picking random digits from a random number table. Let A be the event that you pick a digit larger than 4. Let B be the event that you pick the digit 7.

a. Are the events A and B mutually exclusive? Explain.
b. What are $p(A)$ and $p(B)$?
c. What is $p(A \cap B)$?
d. What is $p(A \cup B)$?

4. Let A be the event that it rains today. Let B be the event that it rains tomorrow. Write out in words what the following events are:
a. $A \cap B$
b. $A \cup B$

5. After shuffling a standard deck of cards, you draw the top card. Let A be the event that the card is a heart, let B be the event that the card is a diamond, and let C be the event that the card is a spade.
a. What must event D be for the entire set of events A, B, C, and D to be mutually exclusive and exhaustive? (Exhaustiveness simply means $A \cup B \cup C \cup D = S$.)
b. What is $p(A \cup B)$?
c. What is $p(A \cap B)$?
d. Explain why the model here is a discrete probability model. *Hint:* Can you list all the outcomes that make up the sample space?

6. In your class there are 100 students, 45 of whom are women. In addition, you know that 75 people in the class own their own automobiles. Finally, you know that 30 of the women in the class own automobiles. You pick a student at random from the class. Let A be the event that the student is a woman, and let B be the event that the student owns a car.
a. What are $p(A)$ and $p(B)$?
b. Write out in words the event $A \cap B$.
c. What is $p(A \cap B)$?
d. What is $p(A \cup B)$? Look at rule 3.
e. Look at your answer to part (b). Write out the complement of the event $A \cap B$?
f. What is the probability of the complement of $A \cap B$? Use rule 5.

7. In a box there are 10 red marbles, 5 blue marbles, 3 green marbles, and 2 white marbles. You draw one marble from the box.
a. What are the four possible outcomes that make up S, the sample space, in this case?
b. What is the probability of each of the outcomes listed in part (a)?
c. What is p(draw a red or a blue marble)?
d. What is p(do not draw a red or blue marble)?

8. Consider the following loaded die, with the probability of rolling each side:

a:	1	2	3	4	5
$p(a)$:	0.10	0.20	0.40	0.15	0.15

What are the following probabilities?
a. p(rolling a 1, a 2, or a 3)
b. p(rolling a 3, a 4, or a 5)
c. p(rolling a 2 or a 5)

9. Consider two events A and B that you know are independent.
a. If $p(A) = 0.5$ and $p(B) = 0.2$, what is $p(A \cup B)$?
b. If $p(A) = 0.1$ and $p(B) = 0.4$, what is $p(A \cup B)$?

13.2 THREE IMPORTANT DISCRETE PROBABILITY MODELS

In Section 13.1 the idea of a discrete probability model was introduced. In the examples of that section the probability of each basic outcome of the experiment was explicitly stated. In the example of the fair die, each side was given a probability of $\frac{1}{6}$. Discrete probability models (that is, "discrete distributions") used commonly by statisticians, such as the Poisson, binomial, and hypergeometric distributions, all use a formula to assign a probability to each basic outcome. In Chapter 8, examples concerning theoretical probabilities for the binomial and Poisson distributions were introduced, but the formulas were not given. Here we will learn the formulas.

Poisson Distribution

From Chapter 8 recall the data concerning the Prussian army and the number of soldiers killed by horses. The data gave the number of units (of a total of 280) in which a certain number of people had died in a particular year:

Number of soldiers kicked to death	Number of units	Proportion
0	144	0.514
1	91	0.325
2	32	0.114
3	11	0.039
4	2	0.005
5 or more	0	0

The theoretical probabilities can be computed using the following formula, the probability law for the Poisson distribution:

$$p(x) = \frac{e^{-\lambda}\lambda^x}{x!} = \frac{\lambda^x}{e^{\lambda}x!}$$

Here e, sometimes called *Euler's number*, is a famous number in mathematics used in problems involving continuously compounded interest, the natural logarithm, and so on. For practical purposes, $e = 2.718$ suffices in most problems.

There are two other parts to this formula that need to be explained. First, for a positive integer x, the notation $x!$ is read as "x factorial" and is given by

$$x! = x(x-1)(x-2)(x-3)\cdots(2)(1)$$

For example, $3! = 3 \times 2 \times 1 = 6$, and $4! = 4 \times 3 \times 2 \times 1 = 24$. We define $0! = 1$, which is a property sometimes needed.

Second, the value λ in the formula stands for the theoretical mean, or in the current example, the theoretical mean number of soldiers kicked to death per unit. This is what is referred to as a *parameter* for the distribution.

The Poisson distribution, like most other distributions used by statisticians, is actually a family of distributions, and its parameter (for some distributions, parameters) specifies exactly which distribution we are working with. In each application where the Poisson model fits the data well, a different choice of λ will be appropriate.

In the case of the Prussian soldier data, $\lambda = 0.714$ provides a probability model that fits the data best. That is, we assume that the theoretical average number of soldiers kicked to death per unit is 0.714. Plugging this value into the Poisson formula, we come up with the following theoretical probabilities for the different values:

$$\begin{aligned} p(0 \text{ deaths in a unit}) &= 0.490 \\ p(1 \text{ death in a unit}) &= 0.350 \\ p(2 \text{ deaths in a unit}) &= 0.124 \\ p(3 \text{ deaths in a unit}) &= 0.030 \\ p(4 \text{ deaths in a unit}) &= 0.005 \\ p(5 \text{ deaths in a unit}) &= 0.001 \\ &\ldots \end{aligned}$$

We could continue calculating these probabilities forever, for 6 deaths, 7 deaths, and so on. For example, the theoretical probability that 10 deaths occur in a unit is a small number (0.00000000465), very close to 0. But still it is greater than 0, so that event is possible. The sample space in the case of a Poisson-distributed random variable is any number $0, 1, 2, 3, 4, \ldots$. Note that the theoretical and obtained proportions are very close. Note also that for practical purposes we can ignore the probabilities for large integers because they are all approximately 0.

In general, recall from Chapter 8 that the Poisson distribution can be used in any situation in which we are given a rate of occurrence—that is, when we know that a certain event occurs, on average, a given number of times in a given time period, provided the three conditions of Section 8.2 hold. Crudely, these conditions are that the event occurrence rate is stationary over time, that events occur independently of each other, and that events occur one at a time. Any situation in which one of these three conditions is seriously violated should not be considered to be well described by a Poisson distribution.

In the case of the Prussian army data, we used the Poisson distribution with a theoretical mean of 0.714 deaths due to horses *per* unit. Other examples are the number of phone calls arriving per hour and the number of customers entering a store per minute. In ecology one often studies the population of a certain organism per square mile, or per square meter, as when a biologist counts crab holes on an ocean beach.

Once we can arrive at a numerical value for the theoretical mean λ (which is often estimated from the data), the Poisson distribution formula can be used to give the theoretical probability that the event occurs any given number of times in the time interval. For example, if we concluded that the theoretical mean number of phone calls arriving in an hour is 5.5, we could calculate the probability that we will receive exactly 3 (or any other integer) phone calls in an hour.

The Binomial Distribution

Another distribution first introduced in Chapter 8 is the binomial distribution. The formula for the theoretical probabilities was not given for this distribution either. However, we did learn how to compute the theoretical binomial probabilities in some simple situations. Recall that the binomial distribution deals with situations that have two outcomes per trial, referred to as *success* and *failure* (these terms are used merely for convenience; they do not necessarily imply success or failure in any real sense). Recall the four conditions of Section 8.1 that must hold if we are to use the binomial distribution to model a situation: there must be two possible outcomes per trial, a fixed number of trials (not a random number of trials), the same probability of success in every trial, and independence of outcomes for different trials. If these conditions hold, then the probability of x successes in n trials is as follows:

$$p(x) = \binom{n}{x} p^x (1 - p)^{n-x} \quad \text{where } x = 0, 1, \ldots, n$$

Here p, the parameter of the binomial distribution, is the probability of success on one trial. The sample space for a binomial random variable is the set of every integer between 0 and n, inclusive.

There is yet another new notation in this formula that needs to be explained. The notation

$$\binom{n}{x}$$

is read as "n choose x." We sometimes also call this the number of combinations of size x selected from n items. This notation is related to the factorial, just introduced, through the following definition:

$$\binom{n}{x} = \frac{n!}{x!(n-x)!}$$

Here $0! = 1$ is needed when $x = 0$ or $x = n$.

Example 13.7

On any given day, the probability that Jill will make it to class on time is 75%. Assume that her attendance behaviors on different days are independent. What is the probability that she will make it on time on exactly four days in a five-day school week?

Solution

Because the four assumptions necessary for using the binomial distribution hold in this case, we will use the binomial distribution to solve this problem. In this case $n = 5$, since there are five "trials." The probability of success, or of Jill's making it to class on time, is 0.75, so $p = 0.75$.

$$p(4 \text{ successes in 5 trials}) = \binom{5}{4}(0.75)^4(0.25)^1 = 0.3955$$

What is the probability that she will make it to class on time all five days?

$$p(5 \text{ successes in 5 trials}) = \binom{5}{5}(0.75)^5(0.25)^0 = 0.2373$$

Thus Jill will come late to class at least one day a week in about 74% of the weeks in the term. This is because

$$\begin{aligned} p(\text{fewer than 5 successes}) &= 1 - p(5 \text{ successes}) \\ &= 1 - 0.2373 \\ &\approx 0.74 \end{aligned}$$

The Hypergeometric Distribution

Now we will introduce a new discrete distribution, the **hypergeometric distribution.** We will introduce it through an example from quality control, which is perhaps its main area of application.

Suppose a manufacturer ships 50 units of its product to a store, and 10 of the units are defective. If an individual comes along and buys seven of these units (choosing them at random), what is the probability that exactly two of them will be defective?

This is the type of situation that can be modeled using the hypergeometric distribution. In general, the hypergeometric distribution is used when you have a total of n elements and this total of n is divided into two distinct groups, which we will label, for convenience, "defectives" and "nondefectives," even if the context of the problem is not one of defective versus nondefective objects.

The number of defectives is denoted a, and the number of nondefectives is denoted b. Note that $a + b$ must equal n. If r draws are made without replacement, as was the case when the seven units were purchased in the example, the probability of getting exactly x elements from the group of defectives is as follows:

$$p(x \text{ defectives out of the } n \text{ draws}) = \frac{\binom{a}{x}\binom{b}{r-x}}{\binom{n}{r}}$$

This is the function for assigning probabilities given by the hypergeometric distribution. It can be used for any value of x such that $0 \leq x \leq n$, though of course $x \leq a$ is a practical constraint, too: there cannot occur more sampled defectives than exist in the population.

Continuing the example of the defective units:

$$p(\text{exactly 2 defectives}) = \frac{\binom{10}{2}\binom{40}{5}}{\binom{50}{7}} = 0.297$$

Likewise,

$$p(\text{exactly 1 defective}) = \frac{\binom{10}{1}\binom{40}{6}}{\binom{50}{7}} = 0.384$$

$$p(\text{none defective}) = \frac{\binom{10}{0}\binom{40}{7}}{\binom{50}{7}} = 0.187$$

Finally, we can conclude, based on the additivity of probabilities of exclusive events, discussed in the previous section, that

$$\begin{aligned} p(1 \text{ defective or less}) &= p(0 \text{ defectives}) + p(1 \text{ defective}) \\ &= 0.187 + 0.384 = 0.571 \end{aligned}$$

The hypergeometric distribution has important applications in the field of quality control. As seen in the above example, the distribution can be used to model situations in which we are interested in finding out how many defectives one can expect when drawing units from a larger group of a known defective rate. In statistics one often infers from the number of defectives observed in a random sample from the population something about the proportion of defectives in the population. In this manner one can assess whether a manufacturing process is in control in the sense of producing, on average, an acceptably small percentage of defectives. The hypergeometric distribution is the probability backbone of this theory.

More generally, the hypergeometric distribution can be used whenever we are looking at one large group that can be divided into two subgroups on the basis of a certain characteristic and we sample randomly without replacement. Using the hypergeometric distribution, we can determine the probability of pulling out a certain number from the first group when we take a sample of a specific size and are sampling without replacement.

The use of formulas such as these we have introduced for the Poisson, binomial, and hypergeometric distributions greatly simplifies the task of coming up with probabilities for certain outcomes. The five-step method was our original method for the calculation of probabilities of this type, but of course in that case we were only obtaining estimates of the probabilities—accurate estimates if we had a large enough number of trials, of course. Now we have formulas that give us the exact theoretical probabilities and are also much simpler to use and less time consuming.

The disadvantage, though, is that in order to compute probabilities involving the statistic of interest, we must logically deduce, from the original model given in the first two steps of the five-step method, which theoretical probability distribution describes the statistic of interest. In the case of the Poisson distribution, there are three conditions that must be met, and in the case of the binomial there are four. The five-step method is much more general, allowing one to estimate probabilities for almost any situation in which we are able to set up a model and decide what constitutes a trial and what the statistic of interest is. We can therefore use the five-step method when the statistic of interest is neither binomial, Poisson, or hypergeometric!

SECTION 13.2 EXERCISES

1. What are the following expressions equal to?
   a. 4!
   b. 2!
   c. 5!
   d. 0!
2. What are the following expressions equal to? *Hint:* "5 choose 2" is the same as $\binom{5}{2}$.
   a. (5 choose 2)
   b. (4 choose 1)
   c. (2 choose 2)
   d. (4 choose 2)
3. A secretary is working in an office where telephone calls arrive at a rate of 2.3 per minute throughout the day.
   a. Explain why the number of telephone calls arriving per minute follows a Poisson distribution.
   b. What is the value of the parameter λ, the theoretical mean?
   c. What is the probability that the secretary does not receive any calls in a minute?
   d. What is the probability that the secretary receives one call in a minute?

e. What is the probability that the secretary receives three or more calls in a minute? (*Hint:* Recall that $p(A) = 1 - p(\text{not } A)$.)

4. In a box there are five doughnuts, and Alice takes two of them at random. Out of the five doughnuts, three are cream-filled, and two are not.
   a. Explain why the number of cream-filled doughnuts Alice obtains follows a hypergeometric distribution.
   b. What are the values of n, a, b, and r in the hypergeometric formula in this case?
   c. What is the probability that Alice obtains exactly one cream-filled doughnut in the two she takes out of the box?
   d. What is the probability that both of the doughnuts Alice pulls out are cream-filled?

5. During rush hour, a certain intersection sees an average of 5.1 cars every minute. Use the Poisson distribution to model the number of cars arriving at the intersection per minute.
   a. What is the value of λ, the theoretical mean?
   b. What is the probability that exactly two cars arrive during a given minute?
   c. What is the probability that exactly four cars arrive during a given minute?

6. Refer to the situation given in Exercise 5.
   a. How many cars would you expect every two minutes on average?
   b. Using your result from part (a), what is the probability that there will be exactly 7 cars in a two-minute period?
   c. What is the probability that there will be exactly 12 cars in a three-minute period?

7. A certain stock has a 60% chance of going up and a 40% chance of going down on any given day. Assume that each day is independent of any other (so whatever happened yesterday does not effect what happens today).
   a. Explain why the number of lines the stock goes up in five days follows a binomial distribution. What is a "success" in this case?
   b. What is the value of p, the probability of success on one trial?
   c. What is the probability that in five days, the stock goes up on every day?
   d. What is the probability that in five days, the stock goes up on exactly four days?

8. In a group of four used cars at an auction, two have previously been involved in serious accidents. A car dealer will buy two of the cars chosen at random. Use the hypergeometric distribution to model the number of cars purchased that have been in accidents.
   a. What are the values of n, a, b, and r in this case?
   b. What is the probability that the dealer picks both of the cars that have been in accidents?
   c. What is the probability that the dealer picks exactly one of the cars that have been in accidents?

9. There are six multiple-choice questions on a test, and each one has five choices. Paul has no idea as to what the correct answers are, so he guesses on each one. Use the binomial distribution to model the number of correct answers.
   a. What is the value of n, the number of trials, and the value of p, the probability of success?
   b. What is the probability that Paul guesses none of the six questions correctly?
   c. What is the probability that Paul guesses exactly two of the question correctly?
   d. What is the probability that Paul gets very lucky and guesses all six questions correctly?

Graphing Calculator Exercises

10. On a 10-question true-false test, you flip a coin to decide your answer. What is the probability that you would get 7 out of 10 questions correct? What is the probability that you would get 7 out of 10 or better correct?

11. In a multiple-choice test with four options (a, b, c, and d), if you were to just guess randomly, what is the probability that you get 7 out of 10 correct? What is the probability that you get 7 out of 10 or better correct?

12. What do the results of Exercises 10 and 11 tell you about why test-makers often prefer multiple-choice questions over true-false questions?

13. Michael Jordan is a career 88% free-throw shooter. What is the probability that he will make at least 7 out of 10 free throws in a game? What is the probability that he will make 4 or fewer out of 10?

14. John Doe is a career 60% free-throw shooter. What is the probability that he will make at least 7 out of 10 free throws in a game? What is the probability that he will make 4 out of 10 or fewer?

15. Suppose that in a shipment of tulip bulbs, it is expected that out of every 100 bulbs, 8 will not germinate. If you order one dozen tulip bulbs, what is the probability that two or fewer will not germinate? What is the probability that all will germinate?

16. A geode is a stone that when split open will sometimes exhibit a crystal. Suppose a rock shop tells you that out of every 20 geodes, 5 do not exhibit crystals. If you purchase eight geodes, what is the probability that one is "defective"?

17. Fiber optic cable contains an average of one flaw per 4000 feet. If one assumes a Poisson distribution, what is the distribution of X, the number of flaws in a 16,000-foot reel, and what is the probability of four or fewer flaws?

13.3 CONDITIONAL PROBABILITIES

In Section 5.6 the idea of a conditional probability was informally introduced in the framework of estimating probabilities by the five-step method. Now we want to focus on how theoretical conditional probabilities can be calculated. Recall that a conditional probability is simply the probability of an event's occurrence given that another event is known to have occurred. This is written $p(A \mid B)$, read as "the probability of event A occurring given event B occurs." The key formula we will use in the calculation of theoretical conditional probabilities is the following definition of the conditional probability:

$$p(A \mid B) = \frac{p(A \cap B)}{p(B)}$$

This expression relates the conditional probability $p(A \mid B)$ to the probability that both A and B occur, $p(A \cap B)$, and the probability that B occurs, $p(B)$.

Example 13.8

Suppose we are tossing a fair six-sided die. Let A be the event that the die comes up a 2, and let B be the event that the die comes up an even number. In this case, clearly $p(A) = \frac{1}{6}$ and $p(B) = \frac{1}{2}$. Now we want to calculate $P(A \mid B)$, or, in words, What is the probability of getting a 2, if you know that you got an even number?

We will proceed theoretically, using the above conditional probability formula. According to the formula, $p(A \cap B)$ equals the probability that the die comes up a 2 and comes up an even number. Note, however, that the event $A \cap B$ is the same as the event A, since if event A occurs, event B must occur. So $p(A \cap B) = p(A) = \frac{1}{6}$. As stated above, it is clear that $p(B) = \frac{1}{2}$, since there are three even numbers on a six-sided die. So, using the formula for conditional probability,

$$p(A \mid B) = \frac{p(A \cap B)}{p(B)} = \frac{p(A)}{p(B)}$$

$$= \frac{1/6}{1/2} = \frac{1}{3}$$

This answer seems intuitively correct, of course, but the point is that we have derived it theoretically! Thus knowing B has occurred increases the chance of A (from the value $p(A) = \frac{1}{6}$ to the value $p(A \mid B) = \frac{1}{3}$). In real life, our assessment of the likelihood of an event is often altered by the knowledge that another event has occurred.

Example 13.9

Recall the Key Problem introduced at the beginning of this chapter. In a certain area, if any day in July is chosen at random, the probability of rain on that day is 0.10. The probability of rain on any two consecutive days is known to be 0.05. If it is raining today, what is the probability that it will rain tomorrow?

Solution

Let A be the event that it rains today, and let B be the event that it rained yesterday. We are interested in calculating $p(\text{rains today} \mid \text{rained yesterday})$, or $p(A \mid B)$.

We know that $p(B) = 0.10$. We also know that $p(A \cap B) = p(\text{rains both today and yesterday}) = 0.05$, since this is the probability of rain on two consecutive days. Therefore

$$p(A \mid B) = \frac{p(A \cap B)}{p(B)} = \frac{0.05}{0.10} = 0.5$$

So the probability is 0.5 that it will rain today, given that it rained yesterday. If it rains on one day it is much more likely (0.5 compared with 0.1) to rain on the next. Thus rain on successive days is not a case of independence (note that this is merely common sense, because weather systems tend to persist).

Independence

Another important concept discussed in Chapter 5 is that of independence. The intuitive idea is that two events are independent if the occurrence of one does not affect the probability of the other. This fits in very well with the concept of conditional probability, since we are investigating the probability of one event given that another has happened. If the information that the event B has occurred does not affect the probability of the event A's occurrence, we would write this as $p(A \mid B) = p(A)$. This turns out to be a good definition for independence:

The events A and B are independent if $p(A \mid B) = p(A)$.

Look at this in terms of the formula for conditional probability that we introduced above. There we stated that $p(A \mid B) = p(A \cap B)/p(B)$. If the events A and B are independent, this formula becomes $p(A) = p(A \cap B)/p(B)$. Multiplying both sides by $p(B)$, we come up with $p(A)p(B) = p(A \cap B)$. This was our definition of the independence of A and B in Chapter 5. Thus we now see that the Chapter 5 definition comes from $p(A \mid B) = p(A)$, which is clearly what we really mean by *independence.*

Example 13.10

Consider a standard deck of 52 playing cards. Let A be the event that the card you pull out is a spade, and let B be the event that the card you pull out is a king. It seems that events A and B should be independent of each other since whether you obtain a king should have no effect on whether you obtain a spade. We can check whether independence holds by using our definition above.

Clearly, $p(A \mid B) = 1/4$, since one of the four kings is a spade. We also know that $p(A) = 13/52 = 1/4$. Therefore $p(A \mid B) = p(A)$, and independence indeed holds.

Note also that $p(A \cap B) = 1/52$ and that $p(A) \times p(B) = (1/4)(1/13) = 1/52$. Once again, this verifies that independence holds, according to the equivalent definition used in Chapter 5.

If the events A and B are mutually exclusive, then $p(A \mid B) = 0$. This is true since, as we noted before, $p(A \cap B) = 0$ when A and B are mutually exclusive. This fact should come as no surprise, though. If it is impossible for both events A and B to occur on the same trial (as in the definition of mutual exclusivity), then the probability of A occurring given that B occurs is logically 0. Note that events that are mutually exclusive cannot be independent because $p(A) > 0$ and yet $p(A \mid B) = 0$. Since exclusivity suggests unrelatedness, which in ordinary English suggests independence, students sometimes find this result a little disconcerting. But it certainly makes sense. Indeed, $p(A \mid B) = 0$ means that there is much dependence between A and B!

SECTION 13.3 EXERCISES

1. Suppose you read single digits from a random number table that has digits from 0 to 9. Let A be the event that a number is a less than 5. Let B be the event that a number is even.
   a. What are $p(A)$ and $p(B)$?
   b. Write out the event $A \cap B$ in words.
   c. What is $p(A \cap B)$?
   d. Write out the event $A \mid B$ in words.
   e. What is $p(A \mid B)$?
2. The events A and B are independent, with $p(A) = 0.5$, and $p(B) = 0.9$.
   a. What is $p(A \mid B)$?
   b. What is $p(A \cap B)$?
   c. What is $p(B \mid A)$?
3. The events A and B are mutually exclusive, with $p(A) = 0.3$, and $p(B) = 0.6$.
   a. What is $p(A \cap B)$?
   b. What is $p(A \mid B)$?
   c. What is $p(A \cup B)$?
   d. Are A and B independent? Argue according to the rule that $p(A \mid B) = p(A)$ if A and B are independent.
4. One hundred customers shopped during one hour at a certain grocery store. The numbers of customers who purchased milk and eggs were as follows:

Bought eggs	23
Bought milk	42
Bought both eggs and milk	18

   Suppose you choose a customer at random from those 100 who shopped during the hour. Let A be the event that that customer bought eggs, and let B be the event that the customer bought milk.
   a. What are $p(A)$, $p(B)$, and $p(A \cap B)$?
   b. Write out the event $A \mid B$.
   c. What is $p(A \mid B)$?
   d. What is $p(B \mid A)$?
   e. Why can you say that during this hour the purchasing of milk and the purchasing of eggs were not independent of each other?
5. Assume that birthdays are randomly scattered throughout the year (that is, people are equally likely to be born on any day of the year). The following list shows the number of days in each month, ignoring leap year:

   January, March, May, July, August, October, December: 31 days each

   April, June, September, November: 30 days each

   February: 28 days

   a. What is p(born in January)?
   b. What is p(born in August | born in a month with 31 days)?
   c. What is p(born in a month with 31 days | born in March)?

13.4 EXPECTED VALUE AS A THEORETICAL CONSTRUCT

In Chapter 4 the five-step method was used to estimate the expected value of a statistic, as in the following example: Suppose a company sends a shipment of 20 units to a store, and five of the units are defective in some way. A customer randomly buys three of the units. How many defectives can he expect to get in his group of three?

With the five-step method, one approach to this problem is to use a box model. We use 20 tickets to represent the 20 units, and we assign five of the units to be defective (labeled 0 to 4, say). Then we draw three tickets without replacement from the box to simulate the purchase of three units. The statistic of interest is the number of tickets corresponding to defective units in the set of three. We then repeat this procedure many times, and the average of all the statistics of interest provides the estimate of the expected value (step 5).

We can use a random number table to simulate draws from the box model, but we cannot allow a number to appear more than once in the three draws, because we are drawing without replacement. Let the numbers 00 through 19 represent the 20 units. Since we are drawing without replacement, we must ignore any number that has already appeared, thereby obtaining three distinct numbers. Let the numbers 00 through 04 indicate defective units and 05 through 19 indicate nondefective units. Thus we would ignore the numbers 20 through 99. Then the series 04 76 07 04 93 18 becomes D I N I I N, with D meaning defective, N nondefective, and I ignore. The third I resulted because the number 04 had already appeared once.

The five-step method, of course, gives us only an estimate of the expected value. As you may suspect, there is a way to obtain the theoretical expected value, using the theoretical probability ideas we have developed. Continuing with our example, the theoretical probability that the customer purchases a certain number of defectives in his set of three is as follows:

Number of defectives	Theoretical probability
0	0.40
1	0.46
2	0.13
3	0.01

These probabilities were obtained from the hypergeometric distribution. This can be seen by referring back to the formula given in Section 13.2 for the hypergeometric distribution and letting $n = 20, a = 5, b = 15$, and $r = 3$. We want to find the probability that the customer draws out 0, 1, 2, and 3 defectives out of the stock of units.

$$p(0 \text{ defective in } 3) = \frac{\binom{5}{0}\binom{15}{3}}{\binom{20}{3}} = 0.40$$

$$p(\text{1 defective in 3}) = \frac{\binom{5}{1}\binom{15}{2}}{\binom{20}{3}} = 0.46$$

$$p(\text{2 defectives in 3}) = \frac{\binom{5}{2}\binom{15}{1}}{\binom{20}{3}} = 0.13$$

$$p(\text{3 defectives in 3}) = \frac{\binom{5}{3}\binom{15}{0}}{\binom{20}{3}} = 0.01$$

Our goal is to find the theoretical expected value of defectives in the set of three that the customer purchases. One way to think about the above theoretical probabilities is to realize that if we were to perform the five-step method for this problem, we would, in the long run, obtain 0 defective 40% of the time, 1 defective 46% of the time, 2 defectives 13% of the time, and 3 defectives 1% of the time. Note that these four outcomes make up the entire sample space S for this problem, and that these four probabilities add up to 1. The final step in the five-step method would be to average all of these outcomes from the repeated trials. When we compute this average, though, we would be averaging (approximately) a list of numbers that was 40% 0's, 46% 1's, and so on. So, if we had performed a total of N trials, then $(0.40)(N)$ of them would be 0 in the long run, and likewise $(0.46)(N)$ would be 1, and so on. Our average for a very large number N of trials would therefore be calculated as follows:

Average from five-step method

$$= \frac{(0)(0.40)(N) + (1)(0.46)(N) + (2)(0.13)(N) + (3)(0.01)(N)}{N}$$

Note that in this formula we could factor the N out of the numerator and it would cancel the N in the denominator. This leads to the definition of the theoretical expected value as a weighted average of the four outcomes. In other words, our expected number of defectives is as follows:

$$\begin{aligned} E(\text{number of defectives}) &= (0)(0.40) + (1)(0.46) + (2)(0.13) + (3)(0.01) \\ &= 0.75 \end{aligned}$$

Thus the theoretical expected number of defectives is 0.75.

In general, in the discrete case the formula for the expected value of a statistic X, or as it is called in probability theory, a random variable, is as follows:

$$E(X) = \text{sum of } xp(x)$$

where the summation is over every value of x in the sample space of the experiment. As in the above example, this is just the average of all the possible values of X weighted by how likely they are. This is a fundamental formula in probability and statistics. There is a similar formula for the continuous case, but it requires the use of calculus. Note that now we have the theoretical analogue to the experimental expected value of Chapter 4. Whenever the $p(x)$'s are known, we can just use this formula to find an expected value, and we will not need the five-step method of Chapter 4. Of course, it may be mathematically difficult to evaluate the sum in the equation for $E(X)$.

Thus we have a theoretical substitute for the five-step method of Chapter 4—namely, use of the formula for $E(X)$ given above.

Properties of Expected Values

Next we will list some properties of theoretical expected values. Consider a random statistic X. Let $\text{Var}(X)$ denote the theoretical variance of the random variable X. The first property we will discuss is the following:

$$\text{Var}(X) = E[(X - \mu)^2]$$

That is,

$$\text{Var}(X) = \text{sum of } (x - \mu)^2 p(x)$$

over all values of x. Here μ denotes the theoretical mean of the variable X. This formula relates the variance of X to an expression involving the theoretical mean. Recall from Chapter 2 that the variance is a measure of the spread in the data. The more the spread, the farther away on average you would expect a single observation to be from the theoretical mean. This idea is what is being expressed by $E[(X - \mu)^2]$. To become clear on exactly what the above formula tells us to compute, we consider an example.

Example 13.11 Assume that X takes on four possible values, 1, 2, 3, and 4, each with probability $\frac{1}{4}$. The theoretical mean in this case is $\mu = 2.5$.

$$E[(X - 2.5)^2] = \left[(4 - 2.5)^2 \frac{1}{4}\right] + \left[(3 - 2.5)^2 \frac{1}{4}\right] + \left[(2 - 2.5)^2 \frac{1}{4}\right] + \left[(1 - 2.5)^2 \frac{1}{4}\right]$$

$$= 1.25$$

Thus the theoretical variance of X is 1.25. We could find it approximately by the five-step method, letting $(X - \mu)^2$ be the statistic of interest.

If you were to multiply a random variable by a constant, the expected value of the resultant random variable is likewise multiplied by the same constant. More generally, we have our second property of expected values:

$$E(a + bX) = a + bE(X)$$

where a and b are any constants we choose.

Recall that in the formula for the Poisson distribution discussed in Section 13.2 there is the parameter λ, the theoretical mean rate of occurrence per unit time interval. It turns out that the theoretical expected value of a random variable that has a Poisson distribution with mean rate of occurrence λ is also λ:

$$E(\text{Poisson random variable}) = \lambda$$

Example 13.12 Prussian Soldier Data

Let's return once again to the Prussian army data concerning the number of men killed by horses. We determined in Section 13.2 that the theoretical probabilities according to the Poisson distribution having $\lambda = 0.714$ are as follows:

Number of deaths in a unit	Theoretical probability
0	0.490
1	0.350
2	0.124
3	0.030
4	0.005
5	0.001
6	0.000
⋮	⋮

Using this table, we determine the theoretical expected value of the number of deaths per unit as

$$\begin{aligned} E(\text{number of deaths per unit}) &= (0)(0.490) + (1)(0.350) + (2)(0.124) \\ &\quad + (3)(0.030) + (4)(0.005) + (5)(0.001) + \cdots \\ &= 0.713 \end{aligned}$$

Note that this agrees with the value that we assigned to the parameter λ, the theoretical mean. Thus the equation $E(\text{Poisson random variable}) = \lambda$ has been demonstrated in this case.

Recall the formula given for the binomial distribution in Section 13.2. Let n be the number of trials, and let p be the probability of success. That the expected value of a random variable with a binomial distribution is np seems plausible. If one performs n trials, and on each trial the probability of success is p, then one would expect that there would be np successes in those trials. For example, one expects $20(\frac{1}{2}) = 10$ heads in 20 tosses of a fair coin. That is our fourth rule:

$$E(\text{Binomial random variable}) = np$$

We also have

$$\text{Var(binomial random variable)} = np(1 - p)$$

This formula for the variance is not intuitive, but can be derived mathematically.

Properties of the Variance

Above we noted the defining relationship between the theoretical expected value and the theoretical variance of a random variable. Here are a few important properties concerning the theoretical variance:

1. $\text{Var}(aX + b) = a^2\text{Var}(X)$, where a and b are constants we choose
2. $\text{Var}(X + Y) = \text{Var}(X) + \text{Var}(Y)$, when X and Y are independent
3. $\text{Var}(X - Y) = \text{Var}(X) + \text{Var}(Y)$, when X and Y are independent
4. $$\text{Var(sample average)} = \frac{\text{population variance}}{n}$$

   and hence

4′. $$\text{SD(sample average)} = \sqrt{\frac{\text{population variance}}{n}}$$

This last property has an important relationship to the idea of estimation accuracy as introduced in Chapter 9 and further discussed in Chapter 10. In both chapters the relationship between the estimation accuracy of a statistic and its variance was established. In particular, property 4′ was stated. In this case, we see that as we increase the sample size n, the variance of the sample average estimate of the population mean μ will decrease correspondingly. Of course, practically, property 4 (and 4′) is useless unless we know or can accurately estimate the population variance (or its square root, the standard deviation).

Example 13.13 Two separate investigators are interested in finding out the average amount of time that college students spend viewing television. The following table compares the two investigators' results:

	Number of students asked	Mean number of hours per week	Sample variance
Investigator 1	100	10.1	2.3
Investigator 2	1000	10.5	2.5

When these researchers present their respective results, which has a more accurate measure of the average number of hours college students spend watching television during the week?

Solution

By property 4, the estimated variance of Investigator 1's estimate of the average is

$$\text{Var(mean number of hours viewing TV)} = \frac{2.3}{100} = 0.023$$

or SE $= 0.15$. For Investigator 2 the variance estimate is

$$\text{Var(mean number of hours viewing TV)} = \frac{2.5}{1000} = 0.0025$$

or SE $= 0.05$. Note that we are merely using the variance of the sample mean number of hours of television viewing per week because we are using the sample variances as stand-ins for the unknown population variances. Clearly, by surveying more students, the second researcher has obtained a more accurate estimate of the population mean number of hours the students spend watching television. Indeed, the estimated standard error is reduced by a factor of 3.

SECTION 13.4 EXERCISES

1. A random variable has five possible values. These values and their corresponding theoretical probabilities are in the following table. What is the theoretical expected value (that is, the mean μ) of the random variable?

Value	Probability (p)
0	0.10
1	0.20
2	0.35
3	0.20
4	0.15

2. A gambler has a 20% chance of winning \$5, a 30% chance of breaking even, and a 50% chance of losing \$5 in a certain game.
   a. Write out the three possible outcomes of the game, and their corresponding probabilities.
   b. How much can the gambler expect to win in one play of the game?
3. You are conducting an experiment to determine the average number of children there are per household in your neighborhood. Explain using statistical reasoning why asking more people will increase the accuracy of your estimate of the average.
4. Use the properties of expected values and of the variance to evaluate the following expressions:
   a. $E(4X)$, where $E(X) = 2$
   b. $\text{Var}(S - T)$, where S and T are independent and $\text{Var}(S) = 9$ and $\text{Var}(T) = 3$
   c. $\text{Var}(5Y)$, where $\text{Var}(Y) = 4$
5. Before taking a course, a student feels her chances of obtaining certain grades are as follows:

Grade	Probability (p)
A (4.00)	0.10
A− (3.67)	0.15
B+ (3.33)	0.35
B (3.00)	0.25
B− (2.67)	0.10
C+ (2.33)	0.05

   What is the student's theoretical expected grade point on the exam?
6. A certain type of flower bulb is sold with the guarantee that 80% of the bulbs will eventually flower. A customer purchases 15 of these such bulbs.
   a. Explain why the binomial distribution fits this situation. What is the value of n, the total number of trials, and p, the probability of a success on one trial?
   b. What is the theoretical expected value of the number of bulbs that will eventually bloom? This should be a simple calculation.
7. **Expected value of a hypergeometric random variable.** In this exercise you will determine the expected value of a random variable that has the hypergeometric distribution.

   In a store bakery there are six loaves of bread. A customer comes to buy bread, not knowing that three of the loaves are older than the other three, and not quite as fresh. The customer purchases three loaves without looking at the dates on the labels.
   a. What are the values of n, a, b, and r in the hypergeometric probability function in this case?
   b. What are the possible values of the number of old loaves the customer buys?
   c. Using the probability function, calculate the probability of each of the possible values you found in part (b).
   d. Finally, find the theoretical expected value of the number of old loaves the customer buys.
8. Find the variance of the random variable X, assuming that X takes on the values 1, 2, 3, 4, and 5 with equal probabilities. (*Hint:* Refer to Example 13.11. Find the value of μ first.)

13.5 HYPOTHESIS TESTING OF THE EQUALITY OF TWO POPULATION PROPORTIONS: AN APPLICATION OF THE BINOMIAL DISTRIBUTION

Recall that we tested a hypothesis concerning a single population proportion in Chapter 11—namely, a hypothesis about a drug's cure proportion in the population. An even more widely occurring statistical problem is that of testing the equality of two population proportions.

The Drugtech Pharmaceutical Company is researching the effectiveness of a new drug designed to prevent an individual from coming down with the common cold. In a preliminary study, 100 people are chosen. Half are given the drug, and the other half a placebo (a pill with no active ingredients). It can be argued in each case that the number of "successes" (subjects coming down with colds) is a binomial random variable. The drug and the placebo possibly have different rates of infection (p) in the population, because the placebo should yield the "natural" rate of cold contraction, while if the drug is at all effective, its population infection rate should be less.

After a period of six months, the researchers interview the individuals to find out if they have come down with a cold while they were taking the medication (either the drug or the placebo). The results are as follows:

Treatment	Number given treatment	Number who came down with cold
Drug	50	15
Placebo	50	25

We are interested in determining whether the drug had a positive effect in preventing colds. This is an example of hypothesis testing, as explained in Chapter 11. In this case the null hypothesis would be "the drug had no effect in preventing colds," as contrasted with "the drug had a positive effect in preventing colds."

To address this problem in a statistical way, we begin by letting p_x be the probability that a randomly sampled person on the drug develops a cold, and likewise we let p_y be the probability that a randomly sampled person on the placebo develops a cold. These are binomial probabilities of success that apply to the placebo-taking population and the drug-taking population. Restating our null hypothesis H_0 in terms of these variables,

$$H_0 : p_x = p_y$$

How are we going to be able to test whether there is a significant difference between p_x and p_y, or whether the difference is just due to chance? Our goal is to answer this question and thus determine whether we should accept or reject the null hypothesis.

The key to solving this problem is to recall the central limit theorem, which was introduced in Chapter 10 and was used in Section 11.7 to solve hypothesis-testing problems involving one proportion when there is a large number of trials (such as the 50 in each group here). We will use the central limit theorem to bypass the simulations of the six-step hypothesis testing method of Chapter 11. Without the central limit theorem, we have no information about what sort of distribution the difference between the proportion $\hat{p}_x$ developing a cold in the treatment sample and the proportion $\hat{p}_y$ developing a cold in the placebo (control) sample should have. (Recall that the hat ^ over the population parameter means "estimate of.") That is, we do not know the probability distribution of $\hat{p}_x - \hat{p}_y$ when the null hypothesis model is true. Thus we have no idea how negative $\hat{p}_x - \hat{p}_y$ should be before we believe we have strong evidence that the drug is effective. That is, without the null hypothesis distribution of $\hat{p}_x - \hat{p}_y$, we have no way of determining whether a value is significantly large or could easily have occurred by chance under the null hypothesis.

Using the central limit theorem, we will test the hypothesis below, which you will note is the same as the hypothesis first introduced above:

$$H_0 : p_x - p_y = 0$$

First, what are our estimates for the values of p_x and p_y? All we have to use is the data that the company gathered, and it seems logical that p_x would be estimated as $\hat{p}_x = 15/50 = 0.3$ and p_y would be estimated as $\hat{p}_y = 25/50 = 0.5$. So using these values, we can estimate $\hat{p}_x - \hat{p}_y = -0.2$. This value is certainly not 0, but is its difference from 0 significant enough to be unlikely under the null hypothesis and so enable us to conclude that the drug is effective? The difference could be just due to chance when in fact H_0 is true—that is, when the drug is no more effective than the placebo.

Just as $\hat{p}_x$ obeys the central limit theorem, it can be shown that $\hat{p}_x - \hat{p}_y$ obeys the central limit theorem in the sense that the distribution of $\hat{p}_x - \hat{p}_y$ is bell-shaped. But to use this application of the central limit theorem, we have to determine the value of $\text{Var}(\hat{p}_x - \hat{p}_y)$. From the previous section, we know that $\text{Var}(\hat{p}_x - \hat{p}_y) = \text{Var}(\hat{p}_x) + \text{Var}(\hat{p}_y)$, since $\hat{p}_x$ and $\hat{p}_y$ are independent. Hence we need $\text{Var}(\hat{p}_x)$ and $\text{Var}(\hat{p}_y)$. Since the number of cures is a binomial random variable, we can compute $\text{Var}(\hat{p}_x)$ as follows:

$$\begin{aligned}\text{Var}(\hat{p}_x) &= \text{Var}\left(\frac{\text{Number of successes}}{50}\right)\\ &= \frac{1}{(50)^2}50p_x(1-p_x) \approx \frac{1}{50}\hat{p}_x(1-\hat{p}_x)\end{aligned}$$

Here we have used property 1 of the variance with $a = 1/50$ and $b = 0$. We have also used the rule that Var(binomial random variable) $= np(1-p)$, which was stated just after the fourth property of expected values discussed in Section 13.4. We have also used the fact that $\hat{p}_x \approx p_x$, which is true because the sample mean is close to its theoretical mean.

By the same reasoning we have

$$\text{Var}(\hat{p}_y) \approx \frac{1}{50}\hat{p}_y(1-\hat{p}_y)$$

We therefore estimate $\text{Var}(\hat{p}_x) = (0.3)(0.7)/50 = 0.0042$ and $\text{Var}(\hat{p}_y) = (0.5)(0.5)/50 = 0.005$.

Finally, we are able to state, with the use of the central limit theorem, that if the null hypothesis is true, then

$$\frac{\hat{p}_x - \hat{p}_y - 0}{\sqrt{\text{Var}(\hat{p}_x - \hat{p}_y)}}$$

has the standard normal distribution. We have subtracted $p_x - p_y = E(\hat{p}_x - \hat{p}_y)$ in the numerator, which is 0 when the null hypothesis is true. Again, we will use property 3 of the variance, namely, that $\text{Var}(\hat{p}_x - \hat{p}_y) = \text{Var}(\hat{p}_x) + \text{Var}(\hat{p}_y)$ because $\hat{p}_x$ and $\hat{p}_y$ are clearly independent. Plugging in the estimates we have obtained, the value for this statistic is

$$\frac{\hat{p}_x - \hat{p}_y}{\sqrt{\text{Var}(\hat{p}_x - \hat{p}_y)}} = \frac{-0.2}{\sqrt{0.0042 + 0.005}} = -2.09$$

We go to our table and note that the probability of seeing a value of -2.09 or smaller on the standard normal curve is less than 0.02. Thus, we reject the null hypothesis that $p_x - p_y = 0$. This experiment supports the drug company's claim that the drug is helping prevent the individuals who take it from getting the cold.

This problem has shown how a hypothesis test can be performed on data that are supposed to follow the binomial distribution. In our case, the number of people who came down with the cold in each of the two groups (those who took the drug and those who took the placebo) has a binomial distribution. Note that a "success" in this case occurs when someone comes down with the cold. We were testing whether the two samples were coming from the same binomial distribution, or in other words, whether the value of p was the same in each case.

SECTION 13.5 EXERCISES

1. Refer to the drug test discussed in Section 13.5. Suppose 20 of the individuals (instead of 15) who had been taking the drug came down with a cold during the period of the study.

What would have been the conclusion of the statistical analysis in that case?

2. A study is performed to compare the voting behaviors of women and men during an election. An exit poll was carried out, and the results were as summarized in the following table:

	Total number polled	Number who voted for candidate A
Men	45	21
Women	51	36

Let p_x denote the proportion of men who voted for candidate A, and let p_y denote the proportion of women who voted for candidate A.

a. What are your estimates of p_x and p_y?

b. Using the formula given in the section, determine an estimate of the value of $\mathrm{Var}(\hat{p}_x - \hat{p}_y)$.

c. Using the values you found in parts (a) and (b), find the value of the statistic used to test whether there is a significant difference.

d. Is there a significant difference in how the women and men voted in this race?

3. In both 1978 and 1990, a survey was taken of college freshmen asking whether they opposed the death penalty. The results were as follows:

Year	Number opposing death penalty
1978	165
1990	105

Suppose in both cases the survey polled 500 college freshmen. Use the method described in this chapter to determine whether there is a significant difference between the percentages opposing the death penalty.

4. In another survey similar to that in Exercise 3, college freshmen were asked in both 1970 and 1990 what their political orientation was. The following table shows how the percentage of students who replied "liberal/far left" changed:

Year	Number saying they were "liberal/far left"
1970	190
1990	125

Once again, assume that 500 students were surveyed in each case. Is there a significant difference between the percentages?

5. Conduct your own poll of students at your school as to whether they agree with the death penalty. Using the method described in this section, determine whether there is a significant difference between the attitudes of the students on your campus and the students asked in the 1990 survey (where 21% said that they opposed the death penalty).

13.6 CHI-SQUARE TEST FOR PROBABILISTIC INDEPENDENCE

In this section we will introduce a new hypothesis test that utilizes the chi-square test first considered in Chapter 7. In that chapter the tests investigated were tests whether a set of data fit a certain discrete model (as given by a box model or loaded many-sided die). For example, we tested whether a die was "fair." We performed those tests by comparing the number of data in each given category with the number expected under the model. Earlier in Chapter 13 (in Section 13.3) as well as in previous chapters, we carefully studied the concept of theoretical independence. Here we will see how the chi-square statistic can be used to test for independence in actual data. If you have a good imagination and are interested in sociological issues, you can see there are an enormous number of interesting hypotheses regarding independence of traits in human populations. For example, are presidential election preferences independent of gender?

Example 13.14 At a swimming club in Charlottesville, Virginia, a study was conducted into whether people who swim frequently experience more dental enamel erosion than those who do not. The (real) data from the study are summarized in the following table. The scientific (and health) issue is whether frequent exposure to chemically treated water (such as chlorinated water) causes erosion of tooth enamel.

	Evidence of enamel erosion	
	No	**Yes**
Frequent swimmer	383	69
Infrequent swimmer	286	9

Source: "Erosion and Dental Enamel among Competitive Swimmers—Virginia," *Morbidity and Mortality Weekly Report,* vol. 32, no. 20 (July 22, 1983).

The null hypothesis in this case is as follows:

H_0 : Whether a person has dental erosion is independent of whether he or she is a frequent swimmer.

We build a second table, this one with column and row totals, along with a grand total (747) for the table:

	Evidence of enamel erosion		
	No	**Yes**	**Total**
Frequent swimmer	383	69	*452*
Infrequent swimmer	286	9	*295*
Total	*669*	*78*	*747*

Remember from Chapter 7 that we needed expected frequencies to solve chi-square problems. So to begin, we ask the question, If the null hypothesis is indeed true, what would our expected frequencies in the table be? To answer this question, we first require that in a table of *expected* frequencies for our two-way table the column and row totals (such as the total number having enamel erosion in the combined category of frequent and infrequent swimmers) must be the same as in the above table of *observed* frequencies. It is not the numbers that fall into any of the four marginal quantities (such as 78, the column total of all those experiencing enamel erosion, and 669, the column total of all those not experiencing enamel erosion) that is in question. The issue is whether or not swimming frequency and enamel erosion (as given by the four interior numbers, 383, 69, 286, and 9) are independent of each other, and hence we focus on the four interior numbers, not on the marginal totals (452, 295, 669, 78).

What does the theoretical law of independence tell us in this case? It states that if frequency of swimming and enamel erosion are independent of each other, then

$$p(\text{Person is a frequent swimmer and has no enamel erosion}) \\ = p(\text{Person is a frequent swimmer}) \times p(\text{Person has no enamel erosion})$$

Using the fixed column totals, we obtain p(Person is a frequent swimmer) as $452/747 = 0.6050870$ and p(Person has no enamel erosion) $= 669/747 = 0.8955823$. Thus, p(Person is a frequent swimmer and has no enamel erosion) $= 0.605 \times 0.896 = 0.5419052$, assuming that swimming frequency and enamel erosion are independent (that is, under H_0).

Likewise, we can calculate these probabilities using independence:

$$p(\text{Person is a frequent swimmer and has enamel erosion}) = 0.0631818$$
$$p(\text{Person is an infrequent swimmer and has no enamel erosion}) = 0.3536771$$
$$p(\text{Person is an infrequent swimmer and has enamel erosion}) = 0.0412359$$

Knowing that there are a total 747 of people, if our assumption of independence is true, then the number of people who are both a frequent swimmer and have no enamel erosion is expected to be $747 \times 0.5419052 = 404.80$. We now have a third version of the table, this time with the needed expected values under the null hypothesis of independence in parentheses alongside the actual data:

	Evidence of enamel erosion		
	No	**Yes**	**Total**
Frequent swimmer	383 (404.80)	69 (47.20)	*452*
Infrequent swimmer	286 (264.20)	9 (30.80)	*295*
Total	*669*	*78*	*747*

That there are differences between the expected and actual frequencies is obvious, but as always when testing hypotheses, the question is, Is the difference significant, or easily explainable by chance error? Note how similar this situation is to the original chi-square testing situation. Recall that

$$\chi^2 = \text{sum of } \frac{(O - E)^2}{E}$$

where O denotes observed frequencies (such as the value 383 in our example) and E denotes expected frequencies (such as 404.80). Once again, we are testing how well our observed data fit our model, so we calculate the chi-square statistic as follows:

$$\chi^2 = \frac{(383 - 404.80)^2}{404.80} + \frac{(69 - 47.20)^2}{47.20} + \frac{(286 - 264.20)^2}{264.20} + \frac{(9 - 30.80)^2}{30.80}$$
$$= 28.47$$

Before we can look up this value in a chi-square table, though, we need to determine the number of degrees of freedom for the chi-square statistic. When we just had rows in computing a chi-square, as we did in Chapter 7, the number of degrees of freedom was the number of rows minus 1. Here we have two dimensions in the table: rows and columns. Recall that the row and column totals are fixed at their current amounts (both the expected and the actual number of frequent swimmers sum to 452, ignoring slight rounding error). If one of the four items in the table were specified, then the other three could be determined because these row and column totals are fixed. For example, if we knew that the number of frequent swimmers with no enamel erosion was 400, we would necessarily know that the number of frequent swimmers with enamel erosion was $452 - 400 = 52$. We would also know that the number of infrequent swimmers without enamel erosion was $669 - 400 = 239$. Finally, we could also say that the number of infrequent swimmers with enamel erosion was $78 - 52 = 26$. Thus, we can see that there is only one degree of freedom in this chi-square statistic. One of the four values completely determines the table. This provides some intuitive evidence for the correct rule, namely that the number of degrees of freedom in a table such as this is given in the following formula:

Number of degrees of freedom

$$= (\text{number of rows} - 1) \times (\text{number of columns} - 1)$$

Looking in Table C in the row for one degree of freedom, we see that the probability of obtaining a chi-square as large as 28.47 or larger is essentially 0. We conclude that the model of the null hypothesis does not fit the data. Thus, we reject the idea that enamel erosion and frequency of swimming are independent. Judging from these data, there does seem to be a dependence between the two variables: frequent swimming is likely positively associated with dental erosion.

This test for independence is a useful tool, and it can also be used in cases in which there are more than two rows and two columns in the table. Of course, in those cases the number of degrees of freedom will be greater than 1, but the procedure of the test will remain the same as what we just carried out. The test for independence can be used to assess whether there are relationships between many apparently related categorical variables, such as religious preference and political party preference, to mention just one interesting example. Try to imagine interesting hypotheses about independence you could test using chi-square testing.

SECTION 13.6 EXERCISES

1. Tell how many degrees of freedom there would be in a chi-square statistic based on a table having
   a. Four rows and three columns
   b. Three rows and three columns
   c. Five rows and two columns
2. A survey asked 320 high school boys whether they were right- or left-handed and whether they participated in sports. The results were as follows:

	Participate in sports?		
	Yes	**No**	**Total**
Right-handed	86	194	280
Left-handed	22	18	40
Total	108	212	320

   a. Using the row and column totals given in the table, what are the values for
      i. p(Student is left-handed)
      ii. p(Student is right-handed)
      iii. p(Student participates in sports)
      iv. p(Student does not participate in sports)
   b. To determine the expected frequencies in each of the four positions in the table, begin by calculating the following probabilities, assuming independence between the two factors (use the law of theoretical independence):
      i. p(Student is right-handed and participates in sports)
      ii. p(Student is left-handed and participates in sports)
      iii. p(Student is right-handed and does not participate in sports)
      iv. p(Student is left-handed and does not participate in sports)
   c. Calculate the expected frequencies for the four cells in the table, using the probabilities you found in part (b).
   d. Calculate the chi-square statistic and test whether the claim of independence is legitimate.
3. On March 25, 1997, the weather forecast for 98 North American cities was compiled and two factors were looked at: the prediction for high temperature and the prediction for the sky conditions (cloudy, clear, or precipitation). The results are summarized in Table E1.
   a. Write out the table of the expected values, assuming that the forecast for temperature is independent of the forecast for sky conditions.
   b. Perform the chi-square test of independence. How many degrees of freedom are there in this case? What is your conclusion?

Table E1

	Clear	Cloudy	Precipitation	Total
Over 60 degrees	24	28	11	63
50–59 degrees	4	7	7	18
Less than 50 degrees	2	12	3	17
Total	30	47	21	98

Table E2

	Passed course	Failed Course	Total
Liked professor	38	3	41
Disliked professor	7	9	16
No opinion	15	4	19
Total	60	16	76

4. In a class with 76 students, a semester-end survey was conducted to determine the students' feelings about the professor. The results of that survey and the numbers of students passing or failing the course are summarized in Table E2.
   a. Write out the table of expected values, assuming that whether a student likes the professor is independent of whether the student passes.
   b. Perform the chi-square test for independence. What is your conclusion?
5. Suppose a survey regarding the major fields of study of 500 college students and their political party preferences had the results found in Table E3.
   a. Write out the table of expected values, assuming that the two factors are independent.
   b. What is the value of the chi-square statistic? What is your conclusion?
6. Suppose a car dealership interested in whether men or women prefer different colors obtained the following data from 400 single men and women who purchased a car:

	Men	Women	Total
Black	56	29	85
White	29	27	56
Red	72	32	104
Blue	52	40	92
All others	44	19	63
Total	251	149	400

   a. Write out the table of expected values, assuming the two factors are independent.
   b. What is the value of the chi-square statistic? What is your conclusion?

Table E3

	Democratic	Republican	Other	Total
Engineering	72	31	5	108
Science	90	22	2	114
Business	62	61	5	128
Fine Arts	30	3	1	34
Liberal Arts	64	48	4	116
Total	318	165	17	500

CHAPTER REVIEW EXERCISES

1. What are the definitions of the terms *exhaustive* and *mutually exclusive* as applied to probability?
2. Assume that X is a Poisson random variable with theoretical mean λ. Find the following theoretical probabilities:
   a. $p(X = 0)$ when $\lambda = 2.1$
   b. $p(X = 2)$ when $\lambda = 1.3$
   c. $p(X = 6)$ when $\lambda = 3.7$
   d. $p(X = 4)$ when $\lambda = 1.9$
3. Assume that X is a binomial random variable collected on n trials with probability of success p. Find the following theoretical probabilities:
   a. $p(X = 1)$ when $n = 4$ and $p = 0.45$
   b. $p(X = 2)$ when $n = 3$ and $p = 0.95$
   c. $p(X = 4)$ when $n = 6$ and $p = 0.35$
   d. $p(X = 0)$ when $n = 5$ and $p = 0.25$
4. Evaluate each of the following expressions, using properties or rules given in this chapter.
   a. 4!
   b. (5 choose 3)
   c. $E(-4X)$, where $E(X) = 9$
   d. $\text{Var}(-2Y)$, where $\text{Var}(Y) = 5$
   e. $p(A \cap B)$, where the events A and B are mutually exclusive
5. In a survey of members of the Biopharmaceutical Section of the American Statistical Association, individuals were asked their type of employer and what their current salary was. Out of 188 respondents who said that they were working in academia, 41 said that they made more than \$91 thousand a year. Of the 529 people who worked for pharmaceutical companies, 180 said that they made more than \$91 thousand a year.

   Using the method described in Section 13.5, test the hypothesis that the proportion of people making over \$91 thousand is the same for both those working in academia and those working for pharmaceutical companies.
6. Return to the salary survey described in Exercise 5. This time consider the following summary of data gathered:

	Less than \$60,000	Between \$60,000 and \$105,000	More than \$105,000	Total
Academia	95	66	27	188
Government	11	20	6	37
Pharmaceutical co.	106	296	127	529
Self-employed	11	21	16	48
Total	223	403	176	802

a. If we assume that type of employer and salary are independent of each other, write out the table with the expected values for the 12 categories.
b. How many degrees of freedom are there in this case?
c. What is the value of the chi-square statistic? What is your conclusion?

7. Explain why there is one degree of freedom in a table with two columns and two rows when performing the chi-square test for independence. Do not just quote the formula.

For Exercises 8 through 11, refer to the following situation: There are 10 poker chips in a bag. The following table describes the 10 chips:

Chip	Color	Letter on chip
1	Red	A
2	Red	B
3	Red	C
4	White	A
5	White	B
6	White	D
7	Yellow	A
8	Yellow	C
9	Yellow	D
10	Violet	C

Define the following events:

Event	Description
A	Draw a chip with the letter *A* on it.
B	Draw a chip with the letter *B* on it.
C	Draw a chip with the letter *C* on it.
D	Draw a chip with the letter *D* on it.
R	Draw a red chip.
Y	Draw a yellow chip.
W	Draw a white chip.
V	Draw the violet chip.

8. Write out in words what the following events are:
   a. $A \cup B$
   b. $R \cup D$
   c. $Y \cap B$
   d. $W \cap B$

9. What are the following probabilities?
   a. $p(A)$
   b. $p(A \cap R)$
   c. $p(A \cup R)$
   d. $p(C \cap D)$
   e. $p(B \cup W)$
10. What are the following conditional probabilities?
    a. $p(A \mid W)$
    b. $p(C \mid V)$
    c. $p(D \mid R)$
    d. $p(R \mid A)$
11. Once again, recall the data from the Prussian army regarding soldiers killed by being kicked by a horse. The following table provides the actual data along with the expected amounts under the assumption that the data follow the Poisson distribution:

Value	Actual number of units	Expected number of units
0	144	137.2
1	91	98.0
2	32	35.0
3	11	8.4
4	2	1.4
5 or more	0	0.3

Do the data really follow the Poisson distribution? Recall Chapter 7. There we performed tests of this exact type, comparing actual distributions to hypothesized distributions.

Perform the chi-square test for this set of data. Do you conclude that the data follow the Poisson distribution?

12. The table below shows the possible values for a random variable, and the corresponding theoretical probability for each value:

Value	Probability
0.0	0.03
0.5	0.08
1.0	0.14
1.5	0.18
2.0	0.27
2.5	0.30

a. What is the sample space S in this case?
b. What properties does this distribution have that make it a discrete probability model?
c. What is the theoretical expected value for this random variable?

13. In an actual advertisement, a drug company reported the results of studies as to the side effects of its new drug. One of the side effects investigated was "tingling." Some patients were given a placebo, and another group was given the actual drug. The table below shows the results:

Treatment	Number in group	Number who reported "tingling"
Placebo	112	4
Drug	114	9

Use the test described in Section 13.5 to determine whether there is a significant difference between these proportions.

14. Respond *true* or *false* to each of the following statements, and if a statement is false, explain why:
a. If X and Y are independent, then $\text{Var}(X - Y) = \text{Var}(X) - \text{Var}(Y)$.
b. If the events A and B are mutually exclusive, then $p(A \cup B) = p(A) + p(B)$.
c. If the events A and B are independent, then $p(A \mid B) = p(B)$.
d. The theoretical expected value of a variable having a binomial distribution is np, where n is the number of trials and p is the probability of a success on any one trial.

15. You go into a video store to rent three movies. You have already narrowed your search down to six choices. You are not aware of it, but four of the six videos you are considering have been reviewed favorably, and the other two have not. Suppose you choose three videos at random from the six. Note that this is an example of the hypergeometric distribution.
a. What are the values of n, a, b, and r for the hypergeometric formula in this case?
b. What is the probability that you will choose three movies that have been reviewed favorably?
c. What is the probability that you will choose exactly two movies that have been reviewed favorably?

16. A survey performed in 1995 revealed that 20% of adolescents aged 12 to 17 had smoked a cigarette within the past month. Suppose you choose seven children from this age group at random. This is an instance in which the binomial distribution can be used.
a. What is the theoretical expected number of children in this group of seven who would have smoked in the past month? Recall that there is a shortcut formula for the binomial case.

b. What is the probability that three of the students will have smoked in the past month?
c. What is the probability that exactly six of the students will have smoked in the past month?

17. In a group of 10 fuses, three are defective. You choose two fuses from the group.
a. What is the probability that you choose no defective fuses?
b. What is the probability that you choose exactly one defective fuse?
c. What is the probability that you choose exactly two defective fuses?
d. Note that in parts (a–c), you found the probabilities of all the possible outcomes of the experiment. Use those results to find the theoretical expected value of the random variable.

18. A certain island in the Caribbean sees an average of 1.1 hurricanes per year. Assume that the number of hurricanes seen per year follows the Poisson distribution.
a. What is the theoretical mean in this case?
b. What is the probability that there will be no hurricanes during a year?
c. What is the probability that there will be exactly two hurricanes during a year?

19. The 50 states were classified as to the number of national parks they had and their population. The classifications were made as follows:

Many parks: four or more national parks

Few parks: three or fewer national parks

Large population: more than 3 million residents

Small population: fewer than 3 million residents

The following table summarizes the data:

	Few parks	Many parks	Total
Small population	17	5	22
Large population	15	13	28
Total	32	18	50

Perform a chi-square test of independence to test whether the fact that a state has many parks or few parks is independent of whether the state has a large population. What is your conclusion?

13

COMPUTER EXERCISES

1. DEFECTIVE UNITS

Consider the example of the defective products discussed in connection with the hypergeometric distribution in Section 13.2. The Five Step program can be used to model this hypergeometric situation. Start the Five Step program. (See the computer exercises in Chapter 4 for an explanation of this program.) Create the box that mimics the 50 units shipped; that is, the box has 10 ones, representing the defective units, and 40 zeros, representing the nondefectives (thus $a = 10$ and $b = 40$). We wish to draw $n = 7$ without replacement. The chance that there are exactly two defectives (two 1s) in this sample is given in the text to be 0.297. Set the statistic to be Sum, and the probability of interest to be Stat $= a$ with $a = 2$. Run 500 trials.

a. What is the experimental probability of getting exactly two defectives? Is it close to 0.297?

b. The expected number of defectives in the sample is $7 \cdot (10/50) = 1.4$. What was the average of the 500 simulations? Was it close to 1.4?

2. DRAFT LOTTERY

In part (b) of Computer Exercise 4 of Chapter 6, about the draft lottery, we counted how many of the first six months had an average lottery number above the median lottery number of all 12 months. Under the null hypothesis that the drawings of lottery numbers are totally random, this number has a hypergeometric distribution, described in Section 13.2. The value of $a + b$ is the total number of months; the "defectives" are those months with average lottery number greater than the median, and hence $a = 6$; and the "nondefectives" have an average less than the median, so $b = 6$. The sample consists of the first six months, so $n = 6$. The number x is the number of months among the first six with average above the median.

a. What is the value of x? (See part (b) of Computer Exercise 4 of Chapter 6.)

b. Use the formula on page 629 to calculate the chance of obtaining $X \geq x$. How does that compare with the experimental proportion found in part (c) of Computer Exercise 4 of Chapter 6?

3. DRAWING FOUR KINGS

Look again at Computer Exercise 7 of Chapter 8. There are eight cards—four kings and four queens—and you draw five (a five-card hand) without replacement. The statistic X is the number of kings drawn. This X is hypergeometric with $a = b = 4$ and $n = 5$. (Why?)

a. Start the Five Step program, and run 500 trials of drawing five cards without replacement from the box with four 0s and four 1s. Look at the frequency table from the Summary Statistics menu. Fill in the obtained counts for each possible number of kings drawn in the obtained count column in the table below. (You can use your results from Computer Exercise 7 of Chapter 8, part (a), if you have them.)

Number of kings drawn	Theoretical probability	Expected count	Obtained count
0			
1			
2			
3			
4			

b. Now use the hypergeometric formula to find the theoretical probabilities of $x = 0, 1, 2, 3,$ and 4, and write them in the theoretical probability column.

c. Multiply each entry in the theoretical probability column by 500, and put the results in the expected count column. These are the numbers of hands in which you would expect to have the corresponding numbers of kings. Are the expected and obtained counts fairly close?

d. Now start the chi-square program. (See the computer exercises for Chapter 7 to review the use of this program.) Type in the expected and obtained counts from the above table. Press OK. What is the obtained chi-square?

e. Run 200 trials in the program. What proportion of the simulated chi-squares exceeded the obtained one? Do you think the hypergeometric distribution models the obtained counts adequately? Why or why not?

4. AVERAGE OF A NORMALLY DISTRIBUTED VARIABLE

Is the sample average a good estimate of the theoretical average? Start the Five Step program, and choose a normal distribution with an average of 0 and a standard deviation of 1. Set the sampling to draw three without replacement, and set the statistic to be "Average." Run 500 trials.

a. What is the average of the sample averages? How close is that to the theoretical average of 0?

b. Repeat the above, but set the number of draws to be 20. What is the average of the sample averages? How close is that to the theoretical average of 0? Is it closer than the estimate in part (a)?

c. Repeat the above, but set the number of draws to be 100. What is the average of the sample averages? How close is that to the theoretical average of 0? Is it closer than the estimate in parts (a) and (b)?

5. STANDARD DEVIATION OF A NORMALLY DISTRIBUTED VARIABLE

This problem repeats the previous one, but looks at the standard deviation instead of the average. Is the sample standard deviation a good estimate of the theoretical standard deviation? Start the Five Step program, and choose a normal distribution with an average of 0 and a standard deviation of 1. Set the sampling to draw three without replacement, and set the statistic to be the standard deviation, "SD." Run 500 trials.

a. What is the average of the sample standard deviations? How close is that to the theoretical standard deviation of 1?

b. Repeat the above, but set the number of draws to be 20. What is the average of the sample standard deviations? How close is that to the theoretical standard deviation of 1? Is it closer than the estimate in part (a)?

c. Repeat the above, but set the number of draws to be 100. What is the average of the sample standard deviations? How close is that to the theoretical standard deviation of 1? Is it closer than the estimate in parts (a) and (b)?

A

COMPUTATIONALLY GENERATED RANDOM DIGITS

Computers (and some calculators) can produce random digits rapidly and in very large quantities. Often it is not practical to require such devices to store long lists of random data. Therefore it is often better to generate such random digits only as they are needed.

Formulas have been invented to compute random digits. At first, you might think this is impossible, because if you *compute* a number by a formula then you will *know* what the next digit will be, and thus such digits cannot be random. This is true, but although these integers are generated deterministically, they nonetheless appear to be true random digits for all practical purposes. Digits produced by such formulas are therefore sometimes called *pseudo-random*. They are *pseudo*-random because they are not random. For most purposes, such pseudo-random digits usually work extremely well in that they appear in every observable way to be random and are very convenient to use.

Some such formulas are very complicated; others are surprisingly simple. You can usually find out from a user manual or a consultant the formula that a particular computer program uses for generating its random numbers.

There is one particularly easy method for producing pseudo-random numbers that you can use on a calculator or computer with little or no programming. It has some flaws and hence is not often used, but it shows us how a deterministic approach can produce digits that appear to be random. It is called the *mid-square method*, and it was suggested by the mathematician John von Neumann.

Take an arbitrary number of five (or more) digits (you could use the last five digits of your telephone number or social security number, for example). Suppose we start with 63537. This is not a part of your list of random digits—it is only the starting value for the procedure. We square this number:

$$(63537)^2 = 403\mathbf{69503}69$$

We now take the middle five digits, 69053, shown in boldface in the above equation. These are our first five "random" digits. Next we square these:

$$(69503)^2 = 48\mathbf{30667}009$$

Again, we take the middle five digits and square:

$$(30667)^2 = 940464889$$

We repeat this process as long as desired. We then write down the string of random digits. For our example, we get

69503 30667 04648

This string of digits may be regarded as a set of random digits. The advantage is that is has been mechanically produced. In this case, we used an ordinary calculator to do the job.

A caution is necessary about this method: after a while it can sometimes start generating all 0s.

Nobody uses this method for actual production of random number tables. We have presented it because it illustrates the idea that a simple computation done repeatedly can produce digits that appear to be random. In addition, you can certainly use it if you wish.

B

RANDOM NUMBER TABLES

Table B.1 **Random Number Table for Tossing a Fair Coin: {0, 1}**

01110	10000	00010	10111	00010	11001	10011	10001	10110
11101	00111	10111	10011	11001	11011	00000	00000	01111
11111	11000	01101	01101	00000	00101	01001	11100	11001
10001	00101	00001	01100	11101	10001	11101	10011	11110
00101	11011	10100	00110	11110	00110	10111	10011	10101
00111	10000	11111	00010	10000	10011	10111	01100	11001
01110	00100	00001	11100	11010	01011	01100	01100	11010
10011	11100	00111	10010	01011	11010	10011	10111	11010
10100	10001	01101	01100	11100	10011	00001	11110	00011
00111	00100	00110	00101	01011	01110	10110	11101	01111
01001	10001	00101	00000	10111	00001	10000	00000	00110
11011	01010	11100	01100	01011	11111	10000	00110	10100
11010	10101	00001	10011	01110	11101	01110	01111	11011
01010	00001	11110	11101	00111	01100	01001	01000	10110
01111	11111	10000	10000	01000	00101	00100	00110	00110
11010	11000	00111	01010	01111	11101	01001	10100	01001
01100	11011	00111	01011	10010	10000	00110	01011	01010
00111	01011	10011	10100	11111	11111	01011	10011	11000
01001	11011	10011	00100	10110	01010	00110	10111	11000
10010	10111	10001	10100	00010	01000	10011	01110	11000
01001	01001	00100	00100	10110	10000	11011	00001	00101
10000	00011	00111	00010	10111	01111	00100	00000	01111
00011	11110	01101	10100	00010	10010	00001	01100	11100
10001	10000	01011	00010	01010	10000	11000	11100	01010
00111	00111	00000	11100	00110	11001	10101	01001	00010
01010	01010	10011	00000	11001	10101	10111	10011	00010
10100	11111	10001	10011	01010	11011	00101	11010	10011
01000	01100	01011	10110	00101	10100	11000	10000	01000
11010	11001	00110	00100	11101	01000	11011	10011	11110
10101	00010	01001	00111	10111	10100	00110	01010	11111
10000	10110	10110	00110	10001	01110	01010	00110	11001
01011	01001	10111	01011	10000	11111	00000	10000	00001
00000	00000	01111	00101	11111	01100	01111	10011	10110
10110	01111	01111	01000	01010	01000	00110	01001	11000
11100	11100	10010	00101	00111	01111	01010	11000	00010
11100	10001	10001	01010	00110	10111	11001	11001	11111
00110	00100	00001	11100	10001	10101	01110	01010	11100
01001	10011	10011	10010	01011	00111	11110	01010	00110
10001	01100	10001	00011	00111	00001	11110	10111	10010
00010	10100	01101	00110	00111	00000	01010	01000	10101

Table B.2 Random Number Table for Rolling a Fair Six-Sided Die: {1,2,3,4,5,6}

66533	45332	24614	22231	26431	35541	12165	62116	16111
61261	22613	26252	14622	32262	33244	34614	13316	41136
61144	46631	56646	24544	36461	14612	21234	23335	16212
21341	66222	53246	24444	13311	44244	41643	54163	21243
15365	46135	23345	53331	46112	54655	65626	24216	11144
51612	21315	16156	56511	31516	26121	23151	66611	64242
31244	26623	33555	53333	24465	14566	54345	25532	43426
41452	65222	25316	44431	33141	64245	62514	45553	24234
25515	41332	25311	65143	61134	65443	45366	61566	63112
35226	33554	55613	54321	21434	22314	11416	64624	26565
41554	23265	43245	46443	26431	23326	51232	45243	54452
24356	54551	52551	64423	13134	25513	43312	43234	12543
56341	24435	44336	62665	26123	25311	55156	46545	45462
26452	16243	13225	66222	31311	41621	51634	63265	22422
41265	26326	23356	62315	32132	55212	26512	56646	42132
13355	53313	42141	56223	61515	22554	44363	42162	52564
22434	61332	13551	43314	45151	43646	15241	55526	61141
62616	22135	55522	62532	43455	14443	61544	55551	34455
53565	42466	22156	21551	13322	54653	62516	66132	36151
55634	65353	51535	66264	61332	11323	15613	44653	36232
64424	24211	63132	25346	41315	66415	33444	16552	56425
34324	62343	35623	22645	41232	35323	25646	52446	24233
44252	16335	35563	33652	12265	25133	56646	14452	45423
55613	35253	13233	26555	56412	23255	61144	36162	51462
34446	33554	16144	16156	23641	24226	52346	26545	55324
16263	13232	52551	32542	41514	52151	43454	32133	34634
34544	15413	14125	62245	13145	53112	52116	63164	42245
23235	36666	61125	44255	62636	53632	56521	26641	56613
32161	66226	66631	33144	12225	42633	61631	14334	45532
13423	54133	12612	15644	41562	36533	25533	22642	34621
66565	24634	23326	42543	55652	25144	12623	46551	42423
52544	45645	23641	64431	41455	21554	23525	26554	66453
21311	46163	14532	63152	12231	34425	41356	64213	62225
14163	41364	32442	55443	54236	45413	43531	12112	45356
63431	26466	35321	56544	25625	66235	56345	15116	65553
66142	14561	25443	53424	45123	32155	41112	45254	54661
56651	14313	51634	65631	45624	34654	32533	63314	11516
61641	44461	14462	35555	26443	61415	62343	22661	31154
25144	62266	63315	51561	26416	26626	46624	55541	64515
26423	66151	45645	44611	16624	56615	55414	36255	25265

Table B.3 Random Number Table for Rolling a Fair 10-Sided Die: {1, 2, 3, 4, 5, 6, 7, 8, 9, 0}*

32236	12683	41949	91807	57883	65394	35595	39198	75268
40336	50658	32089	78007	58644	73823	62854	31151	64726
88795	93736	22189	47004	48304	77410	78871	98387	44647
12807	65194	58586	78232	57097	01430	00304	32036	23671
65929	96713	94452	56211	85446	13656	32155	84455	38125
50339	82178	19650	41283	03944	13736	02627	41929	60613
73840	53838	90804	94332	63639	73187	87067	37557	29635
87062	66298	10731	40629	64955	08081	31443	72112	58006
48038	94580	55223	97799	10105	27952	62493	42176	69615
89830	54426	02692	21233	39553	33483	03141	90919	99219
72234	40065	24052	95658	98335	21125	45364	67989	32451
02833	78254	05112	95160	62546	85982	85567	27427	21436
20565	20846	63664	72162	75338	04022	77166	83339	99021
18090	91089	09799	75883	36480	37067	40933	65634	79883
11519	97203	70899	00697	84864	24470	07933	48202	15392
12732	61573	22852	32281	84871	13331	08947	09023	38248
26823	44530	84507	10396	12240	62603	13396	69378	37173
74622	88768	90819	54769	95306	07685	50369	13763	02205
58535	99062	55182	89858	67701	94838	37317	10432	75653
78551	56329	09024	81507	90137	19241	55198	74006	52851
41477	58940	04016	38081	45519	27559	92403	30967	86797
17004	22782	09508	37331	94994	67305	34040	91360	83009
36925	31844	12940	51503	24822	53594	72930	23342	88646
97569	75612	07237	92264	77989	09054	03863	83891	09041
35122	31549	19982	66024	68615	15959	40347	50052	35312
34358	17573	32838	68335	93497	17412	19850	98965	27357
09285	03384	61410	01932	26797	92577	42580	23354	38677
98363	76867	91821	26538	47181	50938	95676	45306	96725
91769	65764	52386	18551	22196	50282	19985	90730	95175
09393	34982	05654	51208	37731	33916	49063	76700	63094
33202	19891	95374	23650	64877	70661	53330	07223	03469
34111	18376	08231	95163	62837	56995	03022	93618	67560
04410	39026	99536	85083	34607	38979	47259	30921	90092
88209	36990	83284	50578	83549	19006	29501	04565	48865
89257	25109	26253	57523	99297	29901	29472	52817	66611
25581	36013	52215	47684	55094	93140	32969	05603	66922
31742	82956	36361	52786	79761	49819	41375	67628	81707
46012	07959	79667	95325	49142	99596	25691	85964	53568
00174	92988	49499	74089	99209	33816	51757	25031	35862
04861	63886	09763	03265	12748	77513	91010	15062	20270

*This 10-digit case is what is usually referred to when one refers to a *random number table*.

Table B.4 **Random Number Table for Rolling a Fair 12-Sided Die: {1, 2, 3, 4, 5, 6, 7, 8, 9, 0, *E*, *T*}**

7840T	9647E	42368	8309T	85516	24907	42T0E	T5120	31E34
E6072	29031	3111E	3TT34	94619	E53E6	E5125	ET57E	28T5T
161T7	381T7	700T0	97592	377E6	41843	19T19	8501T	58981
2T9E3	T6947	9T20E	9908E	57680	92044	92636	8E772	6566T
76E3T	E7914	T5562	77322	E6458	14867	42267	95466	4E236
6034T	E328E	1T860	522E4	03T45	13947	03829	51T2T	70723
94E40	6493E	T0905	T5432	8474T	831E9	99078	48668	3T746
E9173	773T9	10842	4073T	85T66	0E092	31554	82122	6E008
5E04E	E5690	76223	T98TT	20106	38T62	735T4	52288	7T726
9TE40	65538	037E3	31244	4T664	445E4	04251	T1T1T	TT21T
80805	7E200	42316	6T291	282E2	6TTT7	TT09T	39T8E	6T79T
45555	00461	5T561	1TT83	96145	22037	5T24E	53983	36055
69772	09193	1528T	3950E	T276T	068T5	7394E	7E006	56675
44647	21073	41228	58T2E	3T915	40E9E	E6E23	79389	E8714
T05E0	63503	4E014	02325	840T1	9T7T3	10014	1833T	20T0E
13943	82793	22E63	422E1	95E91	14342	0ET58	0T024	3E349
37E24	55640	E5608	103T7	23341	83804	143T7	7T399	49294
95723	EE97E	T092E	7588T	T6407	09896	6148T	13973	02316
6E647	ET083	662E3	ET96E	89802	T593E	38418	10432	96764
9E761	6843T	0E035	92608	T0148	2E352	762TE	95007	63E71
52598	7ET50	05018	3E091	5662T	2866T	E3504	40T79	E78T9
29015	328E3	0T609	49341	T5TT5	79306	45040	T1470	E300T
47T26	41E55	13T50	61604	35933	646T5	82E31	34453	EE757
T968T	86712	08338	T3275	99TET	0T064	04E73	E4ET2	0T041
46222	4265E	1TEE3	87035	7E826	26T6T	50458	61603	46413
3546E	18783	32E49	8E346	T45T9	19512	1TE60	T9T87	29468
0T2E6	044E4	72550	02T33	289T8	T3688	536E0	23365	4E788
TE484	97E78	T1921	27739	58192	60T39	E6797	56407	T7836
E187E	76985	633E7	1E662	221E8	60393	2TTE6	T0940	T6196
0T3T4	45TE3	06765	71319	39841	44T17	5T074	87801	830T5
44203	1E9E2	T6485	34870	85E99	90871	74440	09334	1357T
45211	2E397	0E231	T7183	73E48	67TT6	04032	E371E	79670
05520	4T027	TT638	E60E4	45809	39T8T	5837T	40T21	T00T2
E920E	48EE0	943E5	6069E	9475T	2T018	2T711	05686	59986
9T268	3710E	37263	68624	TE2T9	2TT01	2T593	73929	78721
366E1	47023	63316	59895	660T5	T4150	42T8E	06704	77E22
41853	E3T68	38382	639E7	8T972	5T91T	41497	7873E	92909
57022	08T7T	8T779	T7326	5T253	TT6T8	367T2	E6T85	6468E
00284	T2TE9	4T5TT	850ET	TT21T	33917	61869	26042	46973
05E71	74EE7	0T984	04276	13959	98614	T1010	16083	20290

Table B.5 **Random Number Table for a Standard Normal Variable**

−0.3713	−0.2478	−0.6191	0.0207	−0.3093	−0.6102	−1.5955	−0.4378	0.6066
2.1352	−0.8940	−0.0062	0.6737	−0.6091	−0.3083	−0.0850	1.7532	−1.3563
0.9692	−0.6652	−1.9856	−0.6354	0.5742	−0.7707	0.9028	−0.4957	2.0585
−0.1241	−0.1783	−1.0843	−0.7939	−0.1232	−1.4296	2.7796	0.4968	0.1914
0.0297	0.3037	−1.3847	0.0486	0.2966	0.5970	0.2056	0.1413	−0.1678
−0.3438	1.2274	−1.1718	0.8335	0.7018	0.0750	0.3924	−0.6794	−0.0852
1.1178	1.8089	1.3865	−0.4378	0.6506	1.2432	0.1845	−0.6945	−0.2513
0.7023	−0.1139	0.3233	1.6031	−0.5919	−1.2176	0.7271	−0.0229	−0.1501
0.4092	0.3912	−1.4844	−0.1497	−1.3398	−0.3652	−1.7329	−1.0306	0.9890
0.2007	−2.3895	−0.5677	−1.3207	−1.3434	0.4915	1.5763	0.6146	1.3502
1.0997	−1.4403	−0.6747	−0.3954	−1.5214	1.9471	−0.4529	0.1710	0.1409
−0.8716	−0.9046	0.1550	2.4269	−2.0841	1.2182	−1.5685	−0.0411	1.9245
0.0647	0.3757	0.0207	0.8082	1.3206	0.4904	−0.8063	−1.4134	1.7442
−0.5876	1.0762	0.1382	1.1575	−0.1087	−1.3951	0.0209	−0.3103	−0.1289
0.9306	0.5654	−0.7672	1.4689	0.2111	0.6171	−2.2243	−0.7565	0.8861
0.2412	−0.8951	−0.7605	0.1465	−0.0505	1.0706	−0.0549	−0.3443	0.4225
2.4332	−0.2278	−1.2614	−1.4321	0.1858	−0.2124	0.0483	1.0501	0.3017
−1.3169	−0.9054	−1.7658	0.5712	−0.5247	−0.1075	−1.0862	0.4623	0.0951
1.0249	−1.4323	−0.5404	−1.0173	−0.4266	−1.0423	−0.9953	1.1717	−0.7675
−0.9134	0.3134	−0.9818	0.6456	−1.5494	−1.5576	−1.9423	−1.0402	0.9997
2.3278	0.5612	−1.4350	0.4842	0.2576	0.3703	0.4033	0.1137	0.7600
1.0979	1.4661	1.0422	1.6997	0.3377	1.3285	−0.8753	1.2995	0.6081
−0.0024	−0.1728	0.4132	−0.0027	0.2905	−2.3368	−0.5928	−0.7066	−0.3341
−1.5959	−0.5691	0.6664	1.1476	0.0528	−0.2079	0.4760	−0.3523	−0.6874
−1.2178	0.0393	0.9418	0.9675	−0.0114	−0.3220	−0.9746	−0.0663	0.1718
0.5636	0.8843	1.2293	0.8789	−0.4601	−0.9147	0.6869	−0.7901	−0.9107
−1.8032	1.1091	0.3521	−0.0354	−1.3890	−0.0706	−1.7638	−0.1438	−0.3393
−0.9270	0.2753	−0.3444	1.2687	−0.0046	−1.3528	1.3998	−0.4565	−0.4570
0.8453	0.6641	−0.0013	0.0300	0.0875	0.1401	0.3621	0.9879	−0.5743
−0.1557	−0.1358	−0.1314	−0.6539	−0.9242	0.2399	0.0259	0.0477	−1.0546
−0.4832	−0.2027	1.2441	1.1753	−0.9627	0.6382	1.2822	−0.5381	1.3506
−1.0901	0.2983	−1.3574	−0.7218	−0.2277	0.0415	−0.4500	0.7018	0.4680
0.4859	2.1673	1.3597	−0.3802	0.5927	−2.0448	1.2126	0.9862	−0.0708
0.0313	−0.1177	−0.3527	−0.5585	0.5493	−0.7759	−1.0542	2.7646	1.4251
0.0114	−0.8899	1.1575	1.0503	0.5112	0.9398	0.1870	−1.0656	0.0948
−0.4968	−1.1686	−0.1628	−0.6773	0.8950	1.4502	0.3037	−1.9043	0.8748
2.1473	−2.6982	−0.3968	−0.6915	−1.3053	0.1680	0.9301	−0.1786	−0.2899
−1.6251	1.6827	0.5806	−1.5906	−0.9129	−1.1662	−0.9490	−1.3759	−1.6889
1.1831	−1.1058	−0.7076	0.7401	1.2219	1.1260	−0.6956	−0.2945	−0.6131
1.2256	−0.2552	1.4166	−3.4767	0.3159	−0.1370	0.0577	−1.1569	−0.2177

C

CHI-SQUARE PROBABILITIES

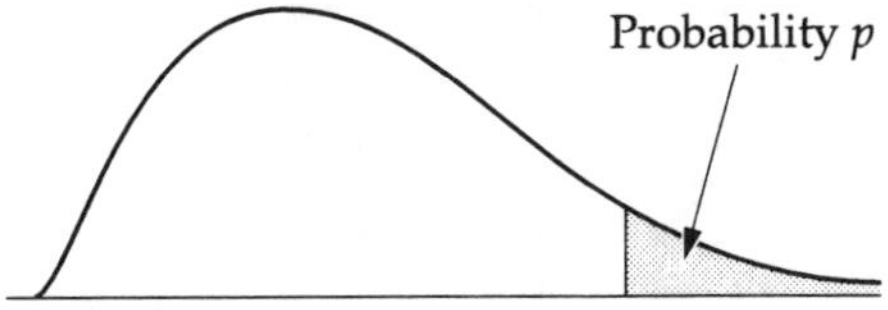

Table entry is the value of x such that $p(\chi^2 \geq x) = p$ for given p (column) and number of degrees of freedom df (row).

Table C **Right-Hand-Tail Chi-Square Density Areas**

	p								
df	**0.25**	**0.20**	**0.10**	**0.05**	**0.025**	**0.01**	**0.005**	**0.001**	**0.0005**
1	1.32	1.64	2.71	3.84	5.02	6.63	7.88	10.83	12.12
2	2.77	3.22	4.61	5.99	7.38	9.21	10.60	13.82	15.20
3	4.11	4.64	6.25	7.81	9.35	11.34	12.84	16.27	17.73
4	5.39	5.99	7.78	9.49	11.14	13.28	14.86	18.47	20.00
5	6.63	7.29	9.24	11.07	12.83	15.09	16.75	20.51	22.11
6	7.84	8.56	10.64	12.59	14.45	16.81	18.55	22.46	24.10
7	9.04	9.80	12.02	14.07	16.01	18.48	20.28	24.32	26.02
8	10.22	11.03	13.36	15.51	17.53	20.09	21.95	26.12	27.87
9	11.39	12.24	14.68	16.92	19.02	21.67	23.59	27.88	29.67
10	12.55	13.44	15.99	18.31	20.48	23.21	25.19	29.59	31.42
11	13.70	14.63	17.28	19.68	21.92	24.72	26.76	31.26	33.14
12	14.85	15.81	18.55	21.03	23.34	26.22	28.30	32.91	34.82
13	15.98	16.98	19.81	22.36	24.74	27.69	29.82	34.53	36.48
14	17.12	18.15	21.06	23.68	26.12	29.14	31.32	36.12	38.11
15	18.25	19.31	22.31	25.00	27.49	30.58	32.80	37.70	39.72
16	19.37	20.47	23.54	26.30	28.85	32.00	34.27	39.25	41.31
17	20.49	21.61	24.77	27.59	30.19	33.41	35.72	40.79	42.88

(continued)

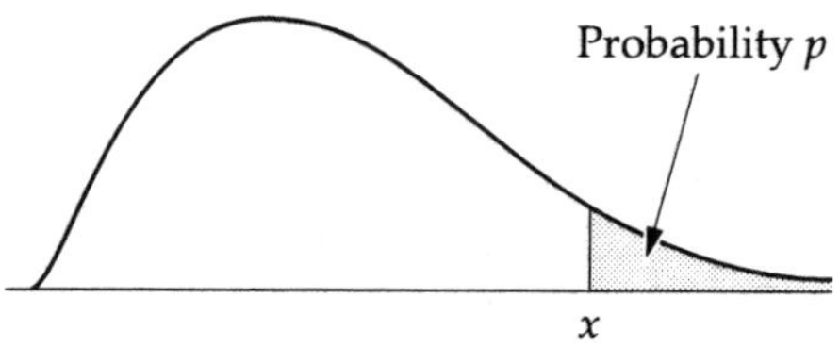

Table entry is the value of x such that $p(\chi^2 \geq x) = p$ for given p (column) and number of degrees of freedom df (row).

***Table C** (Continued)* **Right-Hand-Tail Chi-Square Density Areas**

	p								
df	**0.25**	**0.20**	**0.10**	**0.05**	**0.025**	**0.01**	**0.005**	**0.001**	**0.0005**
18	21.60	22.76	25.99	28.87	31.53	34.81	37.16	42.31	44.43
19	22.72	23.90	27.20	30.14	32.85	36.19	38.58	43.82	45.97
20	23.83	25.04	28.41	31.41	34.17	37.57	40.00	45.31	47.50
21	24.93	26.17	29.62	32.67	35.48	38.93	41.40	46.80	49.01
22	26.04	27.30	30.81	33.92	36.78	40.29	42.80	48.27	50.51
23	27.14	28.43	32.01	35.17	38.08	41.64	44.18	49.73	52.00
24	28.24	29.55	33.20	36.42	39.36	42.98	45.56	51.18	53.48
25	29.34	30.68	34.38	37.65	40.65	44.31	46.93	52.62	54.95
26	30.43	31.79	35.56	38.89	41.92	45.64	48.29	54.05	56.41
27	31.53	32.91	36.74	40.11	43.19	46.96	49.64	55.48	57.86
28	32.62	34.03	37.92	41.34	44.46	48.28	50.99	56.89	59.30
29	33.71	35.14	39.09	42.56	45.72	49.59	52.34	58.30	60.73
30	34.80	36.25	40.26	43.77	46.98	50.89	53.67	59.70	62.16
40	45.62	47.27	51.81	55.76	59.34	63.69	66.77	73.40	76.09
50	56.33	58.16	63.17	67.50	71.42	76.15	79.49	86.66	89.56
60	66.98	68.97	74.40	79.08	83.30	88.38	91.95	99.61	102.7
80	88.13	90.41	96.58	101.9	106.6	112.3	116.3	124.8	128.3

D

LINEAR INTERPOLATION

Interpolation can be used to help find a value that is not provided in a table. The procedure is called *linear* interpolation, since it assumes that the relations between the values involved are (or are nearly) linear—that is, they are straight-line relationships. This assumption gives very good estimates of desired values for the exercises in this book.

Example 1: Find $p(\chi^2_{10} \geq 20.0)$.

The chi-square table in Appendix C does not have an entry of 20.0 for a chi-square with 10 degrees of freedom. So interpolation must be used.

The chi-square values of 18.31 and 20.48 are entries in the table, with corresponding probabilities (areas) of 0.05 and 0.025, respectively (that is, $p(\chi^2_{10} \geq 18.31) = 0.05$ and $p(\chi^2_{10} \geq 20.48) = 0.025$). So $p(\chi^2_{10} \geq 20.0)$ is somewhere between 0.05 and 0.025. Note that $20.48 - 18.31 = 2.17$ and $20.0 - 18.31 = 1.69$. Now consider the following proportionality argument. The following table shows the proportions we are dealing with:

2.17 [1.69 [...]] d] 0.025

Chi-square	Probability (area)
18.31	0.05
20.0	$p(\chi^2_{10} \geq 20.0)$
20.48	0.025

The unknown probability $p(\chi^2_{10} \geq 20.0)$ is proportionally decreased from 0.05 by an amount d according to the proportionality equation

$$\frac{d}{0.025} = \frac{1.69}{2.17}$$

Thus

$$d = \frac{1.69}{2.17}(0.025) \approx 0.019$$

Thus
$$p(\chi^2_{10} \geq 20.0) \approx 0.05 - 0.019 \approx 0.031$$

Example 2: Find the value of a in the following:
$$p(\chi^2_{12} \geq a) = 0.03$$

The chi-square table in Appendix C does not have a column for a probability of 0.03. However, there are columns for 0.05 and 0.025, so we can use interpolation to find the required chi-square value a with 12 degrees of freedom.

$$\begin{array}{cc} \text{Chi-square} & \text{Probability (area)} \end{array}$$

$$2.31 \left[d \left[\begin{array}{cc} 21.03 & 0.05 \\ a & 0.03 \\ 23.34 & 0.025 \end{array} \right] 0.02 \right] 0.025$$

Thus the unknown a is proportionally increased from 21.03 by an amount d according to the proportionality equation
$$\frac{d}{2.31} = \frac{0.02}{0.025} = 0.8$$

So we have
$$d = 0.8(2.31) \approx 1.85$$

The unknown value a is thus
$$a \approx 21.03 + d \approx 21.03 + 1.85 = 22.88$$

Therefore
$$p(\chi^2_{12} \geq 22.88) \approx 0.03$$

E

NORMAL PROBABILITIES

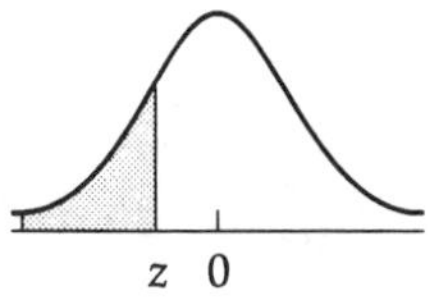

Table E **Standard Normal Probabilities to Left of Given *z***

t	0	1	2	3	4	5	6	7	8	9
−3.0*	.0013	.0013	.0013	.0012	.0012	.0011	.0011	.0011	.0010	.0010
−2.9	.0019	.0018	.0017	.0017	.0016	.0016	.0015	.0015	.0014	.0014
−2.8	.0026	.0025	.0024	.0023	.0023	.0022	.0021	.0021	.0020	.0019
−2.7	.0035	.0034	.0033	.0032	.0031	.0030	.0029	.0028	.0027	.0026
−2.6	.0047	.0045	.0044	.0043	.0041	.0040	.0039	.0038	.0037	.0036
−2.5	.0062	.0060	.0059	.0057	.0055	.0054	.0052	.0051	.0049	.0048
−2.4	.0082	.0080	.0078	.0075	.0073	.0071	.0069	.0068	.0066	.0064
−2.3	.0107	.0104	.0102	.0099	.0096	.0094	.0091	.0089	.0087	.0084
−2.2	.0139	.0136	.0132	.0129	.0125	.0122	.0119	.0116	.0113	.0110
−2.1	.0179	.0174	.0170	.0166	.0162	.0158	.0154	.0150	.0146	.0143
−2.0	.0228	.0222	.0217	.0212	.0207	.0202	.0197	.0192	.0188	.0183
−1.9	.0287	.0281	.0274	.0268	.0262	.0256	.0250	.0244	.0239	.0233
−1.8	.0359	.0351	.0344	.0336	.0329	.0322	.0314	.0307	.0301	.0294
−1.7	.0046	.0436	.0427	.0418	.0409	.0401	.0392	.0384	.0375	.0367
−1.6	.0548	.0537	.0526	.0516	.0505	.0495	.0485	.0475	.0465	.0455
−1.5	.0668	.0655	.0643	.0630	.0618	.0606	.0594	.0582	.0571	.0559
−1.4	.0808	.0793	.0778	.0764	.0749	.0735	.0721	.0708	.0694	.0681
−1.3	.0968	.0951	.0934	.0918	.0901	.0885	.0869	.0853	.0838	.0823
−1.2	.1151	.1131	.1112	.1093	.1075	.1056	.1038	.1020	.1003	.0985
−1.1	.1357	.1335	.1314	.1292	.1271	.1251	.1230	.1210	.1190	.1170
−1.0	.1587	.1562	.1539	.1515	.1492	.1469	.1446	.1423	.1401	.1379
−0.9	.1841	.1814	.1788	.1762	.1736	.1711	.1685	.1660	.1635	.1611
−0.8	.2119	.2090	.2061	.2033	.2005	.1977	.1949	.1922	.1894	.1867
−0.7	.2420	.2389	.2358	.2327	.2296	.2266	.2236	.2206	.2177	.2148
−0.6	.2743	.2709	.2676	.2643	.2611	.2578	.2546	.2514	.2483	.2451
−0.5	.3085	.3050	.3015	.2981	.2946	.2912	.2877	.2843	.2810	.2776
−0.4	.3446	.3409	.3372	.3336	.3300	.3264	.3228	.3192	.3156	.3121
−0.3	.3821	.3783	.3745	.3707	.3669	.3632	.3594	.3557	.3520	.3483
−0.2	.4207	.4168	.4129	.4090	.4052	.4013	.3974	.3936	.3897	.3859
−0.1	.4602	.4562	.4522	.4483	.4443	.4404	.4364	.4325	.4286	.4247
−0.0	.5000	.4960	.4920	.4880	.4840	.4801	.4761	.4721	.4681	.4641

*For any $z \leq -3$, the area is approximately 0; indeed, for any $z \leq -4$, the area is 0 to four decimal places.

(*continued*)

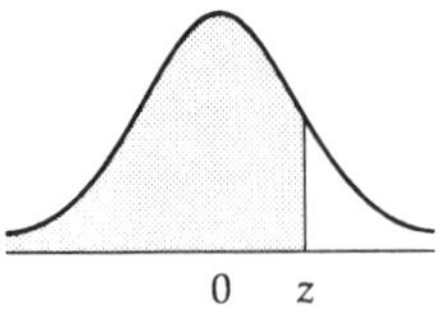

Table E (Continued) **Standard Normal Probabilities to Left of Given z**

z	0	1	2	3	4	5	6	7	8	9
0.0	.5000	.5040	.5080	.5120	.5160	.5199	.5239	.5279	.5319	.5359
0.1	.5398	.5438	.5478	.5517	.5557	.5596	.5636	.5675	.5714	.5753
0.2	.5793	.5832	.5871	.5910	.5948	.5987	.6026	.6064	.6103	.6141
0.3	.6179	.6217	.6255	.6293	.6331	.6368	.6406	.6443	.6480	.6517
0.4	.6554	.6591	.6628	.6664	.6700	.6736	.6772	.6808	.6844	.6879
0.5	.6915	.6950	.6985	.7019	.7054	.7088	.7123	.7157	.7190	.7224
0.6	.7257	.7291	.7324	.7357	.7389	.7422	.7454	.7486	.7517	.7549
0.7	.7580	.7611	.7642	.7673	.7704	.7734	.7764	.7794	.7823	.7852
0.8	.7881	.7910	.7939	.7967	.7995	.8023	.8051	.8078	.8106	.8133
0.9	.8159	.8186	.8212	.8238	.8264	.8289	.8315	.8340	.8365	.8389
1.0	.8413	.8438	.8461	.8485	.8508	.8531	.8554	.8577	.8599	.8621
1.1	.8643	.8665	.8686	.8708	.8729	.8749	.8770	.8790	.8810	.8831
1.2	.8849	.8869	.8888	.8907	.8925	.8944	.8962	.8980	.8997	.9015
1.3	.9032	.9066	.9049	.9082	.9099	.9115	.9131	.9147	.9162	.9177
1.4	.9192	.9207	.9222	.9236	.9251	.9265	.9279	.9292	.9306	.9319
1.5	.9332	.9345	.9357	.9370	.9382	.9394	.9406	.9418	.9429	.9441
1.6	.9452	.9463	.9474	.9484	.9495	.9505	.9515	.9525	.9535	.9545
1.7	.9554	.9564	.9573	.9582	.9591	.9599	.9608	.9616	.9625	.9633
1.8	.9641	.9649	.9656	.9664	.0671	.9678	.9686	.9693	.9699	.9706
1.9	.9713	.9717	.9726	.9732	.9738	.9744	.9750	.9756	.9761	.9767
2.0	.9772	.9778	.9783	.9788	.9793	.9798	.9803	.0808	.9812	.9817
2.1	.9821	.9826	.9830	.9834	.9838	.9842	.9846	.9850	.9854	.9857
2.2	.9861	.9864	.9868	.9871	.9875	.9878	.9881	.9884	.9887	.9890
2.3	.9893	.9896	.9898	.9901	.9904	.9906	.9909	.9911	.9913	.9916
2.4	.9918	.9920	.9922	.9925	.9927	.9929	.9931	.9932	.9934	.9936
2.5	.9938	.9940	.9941	.9943	.9945	.9946	.9948	.9949	.9951	.9952
2.6	.9953	.9955	.9956	.9957	.9959	.9960	.9961	.9962	.9963	.9964
2.7	.9965	.9966	.9967	.9968	.9969	.9970	.9971	.9972	.9973	.9974
2.8	.9974	.9975	.9976	.9977	.9977	.9978	.9979	.9979	.9980	.9981
2.9	.9981	.9982	.9982	.9983	.9984	.9984	.9985	.9985	.9986	.9986
3.0*	.9987	.9987	.9987	.9988	.9988	.9989	.9989	.9989	.9990	.9990

*For any $z \geq 3$, the area is approximately 1; indeed, for any $z \geq 4$, the area is 1 to four decimal places.

F

STUDENT'S *t* PROBABILITIES

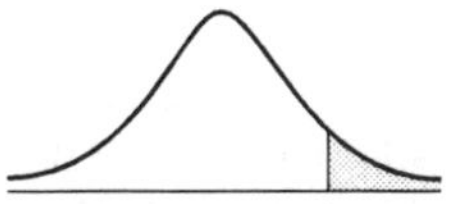

Table F **Right-Hand-Tail *t* Distribution Probabilities**

df	0.10	0.05	0.025	0.01	0.005
1	3.08	6.31	12.71	31.82	63.66
2	1.89	2.92	4.30	6.96	9.92
3	1.64	2.35	3.18	4.54	5.84
4	1.53	2.13	2.78	3.75	4.60
5	1.48	2.02	2.57	3.36	4.03
6	1.44	1.94	2.45	3.14	3.71
7	1.42	1.89	2.36	3.00	3.50
8	1.40	1.86	2.31	2.90	3.36
9	1.38	1.83	2.26	2.82	3.25
10	1.37	1.81	2.23	2.76	3.17
11	1.36	1.80	2.20	2.72	3.11
12	1.36	1.78	2.18	2.68	3.05
13	1.35	1.77	2.16	2.65	3.01
14	1.35	1.76	2.14	2.62	2.98
15	1.34	1.75	2.13	2.60	2.95
16	1.34	1.75	2.12	2.58	2.92
17	1.33	1.74	2.11	2.57	2.90
18	1.33	1.73	2.10	2.55	2.88
19	1.33	1.73	2.09	2.54	2.86
20	1.33	1.72	2.09	2.53	2.85
21	1.32	1.72	2.08	2.52	2.83
22	1.32	1.72	2.07	2.51	2.82
23	1.32	1.71	2.07	2.50	2.81
24	1.32	1.71	2.06	2.49	2.80
25	1.32	1.71	2.06	2.49	2.79
26	1.32	1.71	2.06	2.48	2.78
27	1.31	1.70	2.05	2.47	2.77
28	1.31	1.70	2.05	2.47	2.76
29	1.31	1.70	2.05	2.46	2.76
∞	1.28	1.64	1.96	2.33	2.58

Note: The last row of the table (df = ∞) gives values for the area to the right of t using the standard normal density of Table E. For example, the table shows that $p(z > 1.28) = 0.10$ and $p(z > 1.64) = 0.05$, as can also be seen in Table E.

G

BINOMIAL PROBABILITIES

Table G p(*x* successes in *n* attempts, with success probability *p*)

$n = 1$

	x	
p	0	1
0.10	0.90	0.10
0.25	0.75	0.25
0.50	0.50	0.50
0.75	0.25	0.75
0.90	0.10	0.90

$n = 2$

	x		
p	0	1	2
0.10	0.810	0.180	0.010
0.25	0.562	0.375	0.062
0.50	0.250	0.500	0.250
0.75	0.062	0.375	0.562
0.90	0.010	0.180	0.810

$n = 3$

	x			
p	0	1	2	3
0.10	0.729	0.243	0.027	0.001
0.25	0.422	0.422	0.141	0.016
0.50	0.125	0.375	0.375	0.125
0.75	0.016	0.141	0.422	0.422
0.90	0.001	0.027	0.243	0.729

$n = 5$

	x					
p	0	1	2	3	4	5
0.10	0.590	0.328	0.073	0.008	0.000	0.000
0.25	0.237	0.396	0.264	0.088	0.015	0.001
0.50	0.031	0.156	0.312	0.312	0.156	0.031
0.75	0.001	0.015	0.088	0.264	0.396	0.237
0.90	0.000	0.000	0.008	0.073	0.328	0.590

$n = 10$

	x										
p	0	1	2	3	4	5	6	7	8	9	10
0.10	0.349	0.387	0.194	0.057	0.011	0.001	0.000	0.000	0.000	0.000	0.000
0.25	0.056	0.188	0.282	0.250	0.146	0.058	0.016	0.003	0.000	0.000	0.000
0.50	0.001	0.010	0.044	0.117	0.205	0.246	0.205	0.117	0.044	0.010	0.001
0.75	0.000	0.000	0.000	0.003	0.016	0.058	0.146	0.250	0.282	0.188	0.056
0.90	0.000	0.000	0.000	0.000	0.000	0.001	0.011	0.057	0.194	0.387	0.349

Glossary

accuracy The extent to which a measurement is close to its true value.

average The sum of a set of numbers divided by the number of numbers in the set; same as the **mean** of the set of numbers.

bias Systematic error of measurement in repeated measurements, as distinct from random errors, which vary unsystematically between measurements.

bimodal data set A data set with two locations of high concentration.

binomial distribution The probability model for the number of successes x in a fixed number n of two-outcome ("success," "failure") trials that are independent and are of equal probability p; the distribution is computed as

$$p(x) = \binom{n}{x} p^x (1 - p)^{n-x} \qquad x = 0, 1, \ldots, n$$

bivariate data Data involving two variables, such as height and weight, or amount of smoking and a measure of health; often graphed in a scatter plot.

bootstrap Any inference method that approximates the unknown population (or model) distribution by the discrete uniform distribution on the sample data points and then repeatedly samples with replacement from the sample data to discover empirically the behavior of the statistic being used to make the inference; sometimes called the *nonparametric bootstrap*.

box-and-whisker plot A graphical way of displaying data that shows the median as center and the quartiles and extreme values as spread; also called a *box plot*.

box model A method of generating simulated data by randomly drawing a number from a box repeatedly either with or without replacement between draws, depending on the application. The box may contain any number of real numbers, some appearing more than once; the box containing 1, 2, 3, 4, 5, 6 simulates the throwing of a fair die.

categorical data Data that fall into a finite number of categories; the emphasis is on the frequency in each category (an example is the number of men and women in a college class).

causation A relationship between bivariate variables X and Y that holds when it is known that varying X causes a change in the value of Y; a correlation, even when large, between X and Y does *not* imply causation.

center of data A value that indicates the middle, or center, of a set of data. Also called *measure of central tendency*. Important measures of central tendency include the mean and the median.

central limit theorem The empirical and theoretical result that $\overline{X}$ has a distribution more and more like a normal distribution with mean μ and standard deviation $\sigma/\sqrt{\mu}$ as the sample size n increases, where μ and σ are the population parameters of the random sample that $\overline{X}$ is computed from.

chi-square density The theoretical curve used to calculate chi-square probabilities for chi-square hypothesis testing; also called the *smooth chi-square curve.*

chi-square statistic χ^2 = sum of $(O - E)^2/E$, where the sum is over all categories, O is the observed frequency, and E is the expected frequency.

chi-square table A table that provides numerical values for probabilities involving the chi-square density and is needed for chi-square testing; provided in Appendix C.

chi-square test A statistical hypothesis test for situations in which an observed frequency and a theoretical frequency expected under a null hypothesis are available for the possible categories and the data are from a large number of trials.

circle graph A graph for categorical data. The proportion of elements belonging to each category is proportionally represented as a pie-shaped sector of a circle. Sometimes called a **pie chart.**

complementary events Given an event A, the complement of A consists of all the outcomes not in A. For example, if A is the event that two of three children are boys, then not A is the event that there are either zero, one, or three boys. The equations $P(\text{not } E) = 1 - P(E)$ and $p(\text{not } E) = 1 - p(E)$ are important laws regarding probabilities of complementary events.

conditional probability The probability that one event will occur given that another has occurred. The conditional probability that event A will occur given that event B has occurred is given by $p(A \mid B) = p(A \cap B)/p(B)$.

confidence interval An interval computed from a random sample that is used to estimate some population parameter; the "confidence" probability that the confidence interval contains the parameter must be stated with the confidence interval.

confidence level The probability (typically 0.9, 0.95, 0.99) that the statistician's confidence interval contains the true, unknown population parameter.

continuous distribution A probability law for a continuous variable; the probability that the variable falls within any given interval is given by the area of the interval under a specified curve called a *density*. Examples include the normal and chi-square distributions.

correlation coefficient, *r* A measure of how close two variables of a scatter plot are to being perfectly linearly related; computed by dividing the covariance by the product of the two variables' standard deviations. Also called the *Pearson correlation coefficient.*

covariance A measure of how closely two variables of a scatter plot are linearly related. A closely related concept is **correlation.**

data Numerical information, usually about the real world.

descriptive statistics Techniques of describing or summarizing data that capture the essence of the data.

deviation The signed distance of a data point from the mean of the data.

distribution The probability law for a statistic of interest; for example, the height of a person chosen at random may follow the normal distribution.

equally likely outcomes A theoretical probability model, such as that for a fair die, in which every outcome has the same probability; also called *discrete uniform distribution* (see **uniform distribution [discrete]**).

error variance The mean squared error for the best-fitting least squares regression line; also called the **residual variance.**

estimation Using data to make an educated guess about the magnitude of a population or probability model parameter.

event A set of outcomes of interest in a probability model; for example, the event E that the number showing on a fair die is even is given as $E = \{2, 4, 6\}$. Usually the goal is to compute or estimate $p(E)$.

exclusive events Events with no outcomes in common.

expected value (experimental) The average value of a random quantity that has been repeatedly observed in replications of an experiment; possibly obtained through the five-step method.

experimental probability The estimated probability of an event; obtained by dividing the number of successful trials by the total number of trials. It is often obtained by applying the five-step method.

five-step method Basic simulation method of this book; the five steps are the choice of a model, the definition of a trial, the definition of the statistic of interest associated with the trial (often the number of successful trials), the repetition of trials, and the calculation of the average of the statistic of interest (often an experimental probability or probability estimate).

frequency interpretation of probability The interpretation of a theoretical probability as approximately predicting the proportion of occurrences of the event of interest (such as the proportion of heads being approximately $\frac{1}{2}$) occurring in a large number of trials of a random experiment (such as repeated tossing a coin).

frequency table A table giving the number of data points in a data set falling in each of a set of given intervals.

histogram A bar graph presenting the frequencies of occurrence of data points. Sometimes called a *frequency histogram.*

hypergeometric distribution A distribution used to assess the probability of a certain number of "defectives" in a random sample; given by

$$p(x) = \frac{\binom{a}{x}\binom{b}{r-x}}{\binom{n}{r}}$$

where a is the number of defectives in the population, b is the number of nondefectives in the population, r is the sample size, and $n = a + b$.

hypothesis testing Deciding which of two realities about a population, such as whether a coin is fair or biased, is true on the basis of a random sample from the population.

independence Property of two events that holds if the occurrence of one does not affect the probability of the occurrence of the other; see also **law of experimental independence; law of theoretical independence.**

inferential statistics Techniques used to draw conclusions from data.

interquartile range The difference between the third quartile and the first quartile of a set of data.

Kendall's tau test A particular hypothesis test used to detect a (possibly nonlinear) trend in bivariate (X, Y) data.

law of experimental independence The fact that if the events A and B are independent, then $P(A \text{ and } B) \approx P(A)P(B)$, where P denotes experimental probability.

law of theoretical independence The fact that if the events A and B are independent, then $p(A \text{ and } B) = p(A)p(B)$, where p denotes theoretical probability.

least mean error regression The method of locating the best-fitting regression line by minimizing the mean absolute error; a robust method that is preferred if outliers are suspected.

least squares regression The method of locating the best-fitting line to a scatter plot by minimizing the sum of the squared vertical distances between the points of the scatter plot; amounts to choosing the best slope among all lines passing through $(\overline{X}, \overline{Y})$.

linear relationship A relationship between two variables whose scatter plot is well fit by a straight line; the equation $Y = mX + C$ is often used, where m is the slope and C is the vertical axis intercept value.

line plot A line graph that orders the data along a real number line.

mean (sample) The sum of a set of numbers divided by the number of numbers in the set; same as the **average,** and often defined by $\overline{X}$.

mean (theoretical) The population or probability distribution mean. In the case of a discrete distribution given by $p(x)$, it equals the sum of $xp(x)$.

mean absolute error Also **mean error:** the average of the vertical distances between the points of a scatter plot and the fitted regression line.

mean deviation The measure of spread (variation) that is the average of the deviations of the data points from the sample mean; also called the *mean absolute deviation.*

mean square error The average of the squared vertical distances between the points of a scatter plot and a proposed fitted regression line; equals S_e^2 for the best-fitting least squares line.

median The number that appears in the middle of a set of numbers when they are arranged in order. By convention, the median is the average of the two middle numbers if the number of numbers is even.

median test A hypothesis test used to assess the null hypothesis that two populations have the same centers as measured by the population medians.

mode The number in a set of data that occurs the most frequently; a seldom used measure of the center of a data set.

model Also **probability model:** a mathematical set of probability rules, a random physical mechanism, a box model, or a random-number-based simulation for producing data that are as similar as possible to actual real-world data.

Monte Carlo method A simulation method for solving probability problems by repeatedly doing an experiment, such as tossing a coin repeatedly, rolling a die repeatedly, repeatedly drawing from a box model, or repeatedly choosing random digits.

multiple correlation coefficient, R^2 A measure of the degree of fit of the best-fitting curve to bivariate data when the regression is nonlinear or multiple linear; reduces to the ordinary squared correlation r^2 when the regression line is $Y = A + BX$.

negative relationship A relationship between two variables in which one decreases as the other increases; in the special case of a straight line, a negative relationship means that the slope of the regression line is negative.

negative slope The slope of a line on which the Y value decreases as the X value increases.

nonlinear regression Regression that seeks the relationship between bivariate (X, Y) pairs when the relationship is not linear, as in the case of $Y = A + BX + CX^2 +$ random error.

normal distribution Also called the *bell-shaped curve* or the *Gaussian curve:* the most widely used continuous distribution; often used to model biological measurements and errors of measurement. Probabilities are computed using the table of Appendix E.

null hypothesis The presumed model (such as that of a fair coin) in hypothesis testing. The data provide a measure of how weak or strong the evidence for or against this null hypothesis is; it is the model of step 1 if the six-step method is being used.

outlier A data value that is extreme relative to the rest of the data; farther than ± 3 standard deviations from the sample mean is the usual criterion.

parameter A number describing a characteristic of a population or a probability model, such as the theoretical mean, μ.

pie chart See **circle graph.**

point estimate An estimate of a population parameter that is one specific number (as opposed, say, to an interval).

Poisson distribution The probability law of the number of occurrences of a randomly occurring event in a fixed time interval when the rate of occurrence is fixed across the interval, separate occurrences are independent, and simultaneous occurrences are precluded; an example is the number of phone calls arriving in a given period of time. The distribution is computed as

$$p(x) = \frac{\lambda^x}{x! e^\lambda} \qquad x = 0, 1, \ldots$$

population The entire collection of objects or people under consideration for statistical study. Often the statistical goal is to use a random sample to make inferences about the population, such as about its center or spread; the population is modeled by a probability distribution.

positive relationship A relationship between two variables in which one increases as the other increases; in the special case of a straight line, a positive relationship means that the slope of the regression line is positive.

positive slope The slope of a line on which the Y value increases as the X value increases.

precision The degree of smallness of the variation of a set of measurements; high-precision measurements display little variation. Precision ignores the vital issue of measurement bias.

probability See **experimental probability; frequency interpretation of probability; theoretical probability.**

quartile The third quartile is the value above which $\frac{1}{4}$ of the data lie; the first quartile is the value above which $\frac{3}{4}$ of the data lie.

r Symbol for **correlation coefficient.**

random Not predictable, occurring by chance. For example, the outcome of tossing a fair coin is random.

random digits Also **random numbers:** digits (most commonly 0, 1, . . . , 9) that occur in equally likely and random fashion, as when produced by using a spinner having 10 equal sectors; often produced by a computer program.

random sample A set of data chosen from a population in such a way that each member of the population has an equal probability of being selected.

random walk The path taken along a line or in a two-dimensional rectangular grid by an object moving at random.

range The measure of spread (variation) that is the difference between the largest number and the smallest number in a set of data; not a robust measure of spread.

regression equation The equation of the regression line.

regression line A straight line used to estimate the relationship between two variables, based on the points of a scatter plot; often determined by a least squares analysis.

relative frequency (of an event) Same as the **experimental probability.**

relative frequency histogram A frequency histogram vertically scaled so that the sum of the areas of its rectangles is 1.

residual The distance, or error, between an actual data point and that predicted by the statistically inferred model; in the case of regression, the distance between the actual Y value and the regression line Y' value.

residual variance See **error variance.**

robust A statistic computed from a set of data is robust if it is not overly influenced by the location of a single number in the data set—a desirable property!

S_e^2 The average of the squared errors for the best-fitting least squares regression line, given by the average of the squared vertical distances between the scatter plot points and the best-fitting line. The subscript e denotes *error.*

S_Y^2 The sample variance of a set of data whose values are denoted by Y.

sample Same as a **random sample** unless it is known that the sample was not randomly selected. Nonrandom samples cannot be trusted!

sample variance See **variance (sample).**

scatter plot A graph of two-variable (bivariate) data in which each point is located by its coordinates (X, Y).

sign test A hypothesis test to assess a hypothesized value for a population median.

simulation Any method for generating data from a given probability model; methods used include box models, physical models such as coins or dice, and simulation based on random number generation.

six-step method A modification of the five-step method to do hypothesis testing in which a sixth decision-making step about the truth or falsity of the step 1 null hypothesis model is included.

slope The rate of change of the Y values with respect to the rate of change of the X values.

spread of data The degree to which data are spread out around their center. Measures of spread include the mean deviation, variance, standard deviation, and interquartile range. Also called **variation.**

standard deviation (sample) The square root of the sample variance. It is the most widely used measure of the amount of spread (variation) in a set of data.

standard deviation (theoretical) The standard deviation of the population or distribution, given by the square root of the theoretical variance.

standard error of a sample mean A measure of the typical variation of the sample mean from the population mean; given by $\sigma/\sqrt{n}$, where n is the number of observations used to compute the sample mean and σ is the population standard deviation.

standard normal distribution A normal distribution with a mean of 0 and a variance of 1.

statistic A piece of numerical information computed from a sample, such as $\overline{X}$.

statistical regularity The empirical (real-world) fact that the experimental probability becomes closer and closer to a number, called the **theoretical probability,** as the number of trials becomes large; this law is the foundation of statistical reasoning.

statistics (1) The science of gathering, describing, and drawing conclusions from data; (2) reported numerical information, such as in a newspaper.

stem-and-leaf plot A graphical display of data using certain digits (such as those in the 10s place) as *stems* and the remaining digits (such as those in the 1s place) as *leaves*. It is a special kind of histogram.

theoretical probability The true probability of an event; what the experimental probability approaches in a very large number of trials. In the special case of an equally likely outcomes probability model, the theoretical probability is obtained by dividing the number of outcomes in the event of interest by the total number of possible outcomes.

***t* test** A special hypothesis test about the population mean used when the population is known to be normally distributed, the sample size is small, and the population standard deviation is unknown.

two-sample problem Any inference problem about two populations in which the data consist of a random sample from each population.

type I (hypothesis test) error The error of incorrectly rejecting a null hypothesis when it is true.

type II (hypothesis test) error The error of incorrectly accepting a null hypothesis when it is false.

uniform distribution A probability model in which each number has the same chance of occurring; can be discrete or continuous.

uniform distribution (continuous) The continuous probability law in which

$$p(\text{variable in interval}) = \frac{\text{length of interval}}{\text{range of the variable}}$$

uniform distribution (discrete) A distribution in which each outcome has the same probability of occurrence; an example is the distribution computed by

$$p(x) = \frac{1}{n} \qquad x = 1, 2, \ldots, n$$

variable A quantity that varies, often randomly. For example, the weight of a randomly chosen member of a football team is such a variable. Variables are usually represented by letters.

variance (sample) A measure of the amount of spread (variation) in a set of data. It is the average of the squared distances of all the data values from their mean.

variance (theoretical) The variance of the population or distribution. In the case of a discrete distribution given by $p(x)$, it equals the sum of $(x - \mu)^2 p(x)$, where μ is the population mean.

variation See **spread.**

visual method of linear regression Visually choosing the line through $(\overline{X}, \overline{Y})$ that appears to best fit a scatter plot; useful for instructional purposes, but not usually used in practice.

without replacement Describes random drawings from a set of objects, or from a box model, in which each drawn object is not replaced before the next random drawing.

with replacement Describes random drawings from a set of objects, or from a box model, in which each drawn object is replaced before the next random drawing.

z score Also called a *standardized score:*

$$z \text{ score} = \frac{\text{data value} - \text{sample mean}}{\text{sample standard deviation}}$$

z test A hypothesis test about the population mean μ (or about two population means) using the central limit theorem result that $(\overline{X} - \mu)/(\sigma/\sqrt{n})$, where σ is the population standard deviation and n is the sample size, has an approximately standard normal distribution as long as n is large.

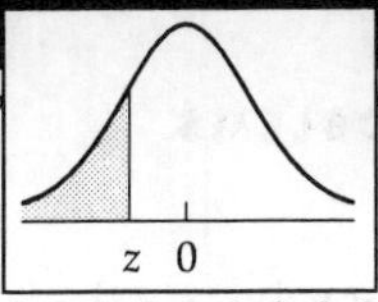

ANSWERS TO SELECTED PROBLEMS

CHAPTER 1

Section 1.1

1\. a.

1.	United States	44
2.	Russia	26
3.	Germany	20
4.	China	16
5.	France	15
6.	Italy	13
7.	Australia	9
7.	Cuba	9
7.	Ukraine	9
10.	South Korea	7
10.	Hungary	7
10.	Poland	7
13.	Spain	5
14.	Romania	4
14.	Netherlands	4
14.	Czech Republic	4
17.	Canada	3
17.	Bulgaria	3
17.	Brazil	3
17.	Japan	3
21.	Britain	1
21.	Belarus	1

b. Countries whose ranking went up:

Russia
France
Italy
Cuba
Ukraine
Hungary
Poland
Spain
Czech Republic
Japan

c. Countries whose ranking went down:

Germany
Australia
South Korea
Romania
Canada
Britain
Belarus

3\. New York City is only a very small part of the entire state of New York. The density of the state spreads the population of New York

City over the entire state. Hence, the density for the state is much lower than for the city.

5. a. The percentage responding true is a total of 58%. So yes, generally parents approve of their teenagers working in a fast-food restaurant.
   b. Answers will vary.
7. a. In a population of 1 million people, there are 1000 groups of 1000 people each. This gives $27 \times 1000 = 27{,}000$ millionaires.
   b. No, you would need the number of millionaires per 1000 people and the population of each state.

Section 1.2

1.

Stem	Leaf
2	7
3	1,1,3,3,5,5,6,7,7,7,8
4	2,2,3,6,6,7,8,9
5	0,1,1,1,2,5,6,7,7
6	5

3. 60, 61, 64, 64, 68, 71, 73, 75, 80, 82
5. a. 72
   b. 96
   c. 24

7.

Stem	Leaf
1	09,52,82
2	40,67
3	87,95
4	82
5	23
6	10,27

Section 1.3

1.

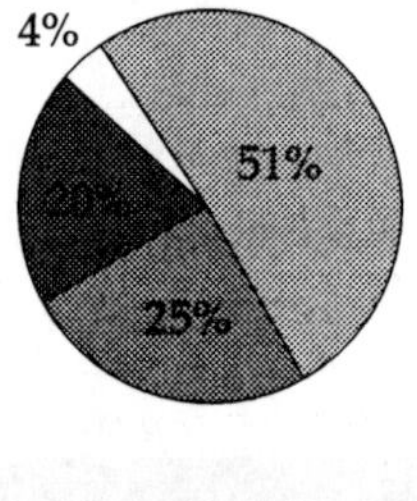

3.

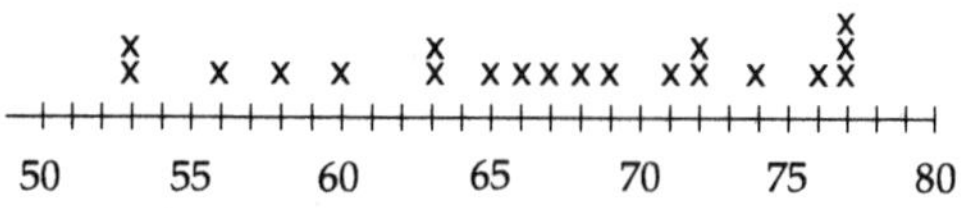

   a. Highest = 77, lowest = 53
   b. Range = 24
   c. The weather around the country was fairly mild on that day, with not a lot of variation.
5. Chicago and New York tie for buildings with largest number of stories (110). Austin has the smallest number (32).

Section 1.4

1. The main disadvantage is that there is a loss of the original data values. Two advantages are that histograms provide a clearer picture of the data and indicate the center, spread, and shape of the data.

3.

Stem	Leaf
7	2,3,4,5
7	9
8	0,2,3,4,4,5
8	7,8
9	0,1,2,3,5
9	6

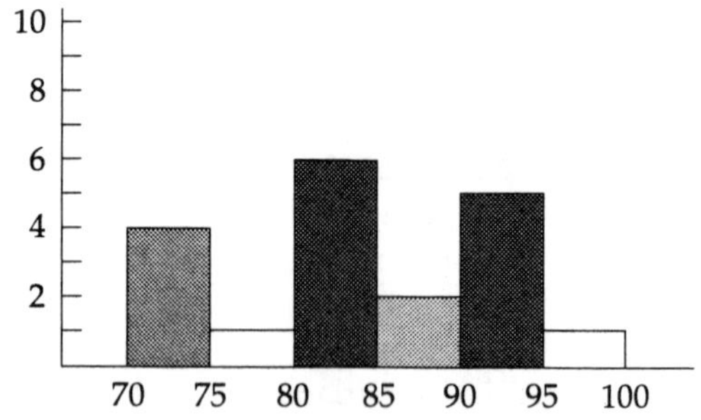

The histogram with more bars does not really give you a better picture of the data.

5. Answers will vary. Here is an example:

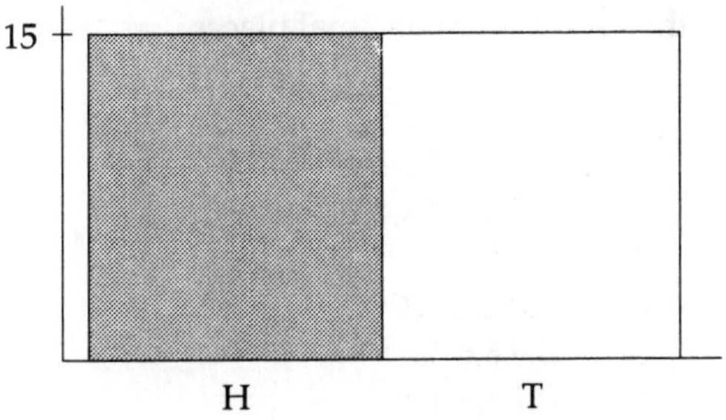

If you performed the experiment again, you would not expect to get the same results, because flipping a coin is a random event—that is, an event that will not happen the same every time.

7. a.

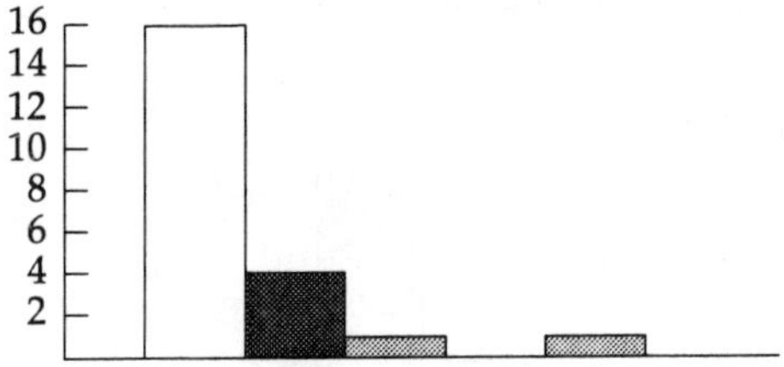

b.

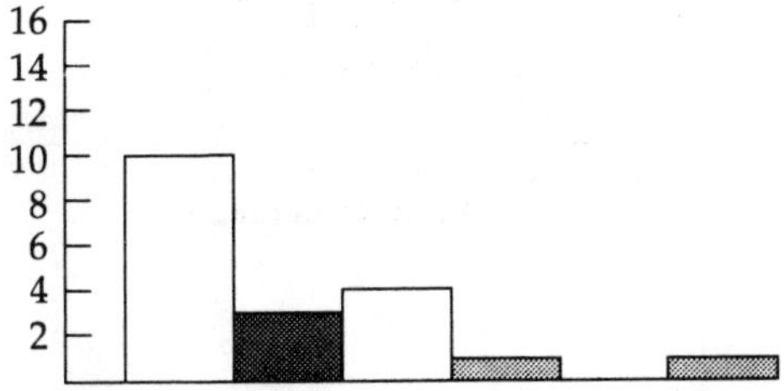

c.

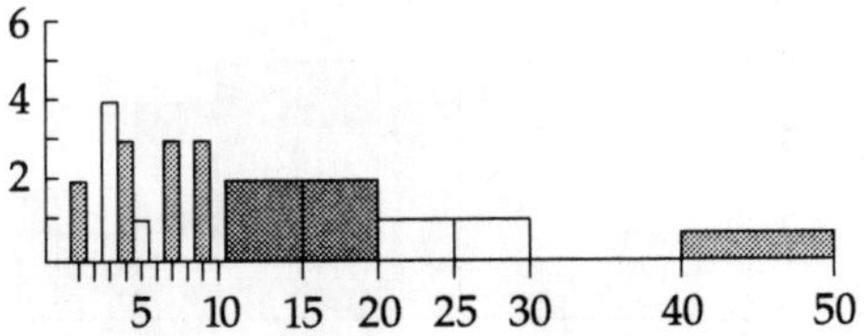

d. The histogram in part (b) is the best way to represent the data.

11. Check your answers with Table 1.10.

Section 1.5

1. a.

Number of children	f	Proportion
0	3	0.06
1	8	0.16
2	6	0.52
3	10	0.20
4	2	0.04
5	0	0.0
6	1	0.02

b.

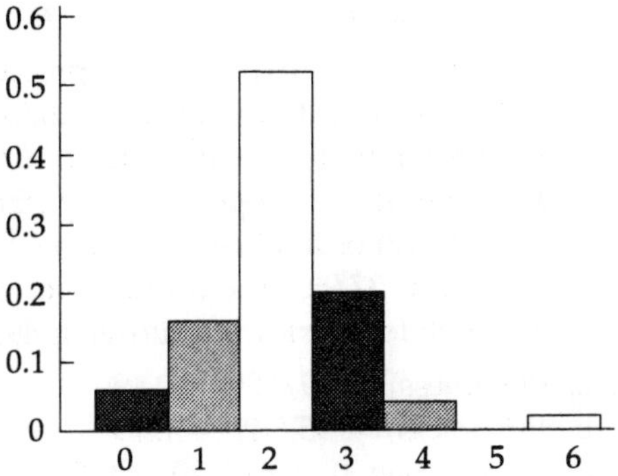

c. 0.74

3. a.

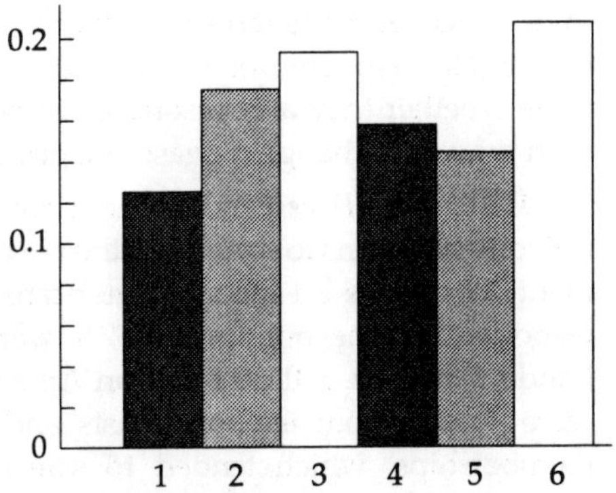

b. i. 0.492

ii. 0.35

iii. 0.541

5.

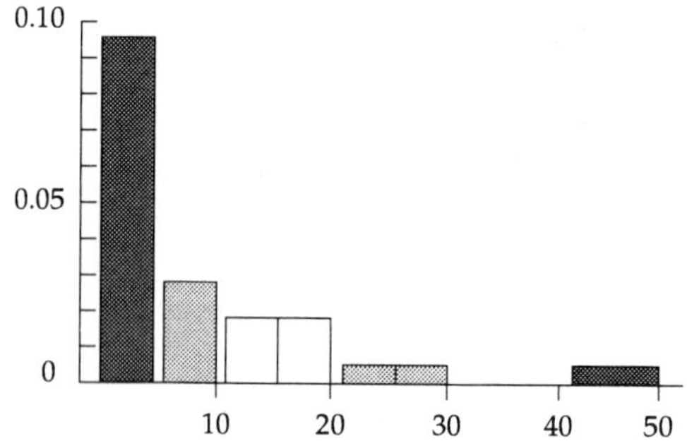

7. a. The number of households with two children or fewer is 37. The number with three or fewer is 47.
   b. The proportion (experimental probability) of families with two children is 26/50. The proportion with three children is 10/50.
   c. The cumulative proportion (experimental probability) of families with two or fewer children is 37/50. The cumulative proportion with four or fewer children is 49/50.

9. a. $P(\text{3 or less}) = 59/120 = 0.49$
   b. $P(\text{5 or less}) = 95/120 = 0.79$
   c. $P(\text{more than 3}) = 61/120 = 0.51$

Section 1.6

1. Answers will vary. You should check for things such as the heights and widths of the bars and the diameters of the circles and determine whether they are reasonable compared with the actual changes in gasoline prices.

3. In 1932 *Literary Digest* conducted a poll of 10 million Americans to determine the next president. Out of the 2.4 million Americans who responded to the poll, about 57% were for Landon. However, the 10 million Americans were chosen from telephone lists and club memberships, which tended to screen out poor voters. Those poor voters were much more likely to vote Democratic.

Chapter Review

1. Answers will vary. Generally, you should be thinking about using the "middle" number. The more recent results were obtained using more precise measurement instruments and thus have more decimal places.

3.

Stem	Leaf
0	0,0,1,2,3,4,4,4
0	7,9,9
1	0,0,1,2,2,2
1	5,6,6,6,6,7,8
2	0,1,1,3,4,4
2	5,6,8,8,9
3	
3	6

Most frequent: 16 years; lowest: 0 years (both Harrison and Garfield died in office); highest: 36 years (Herbert Hoover)

5.

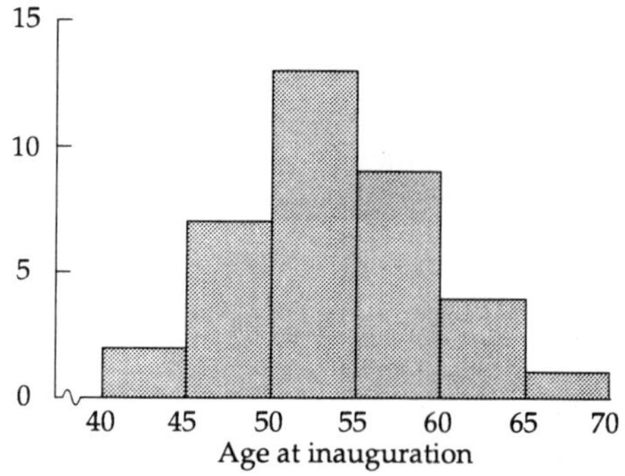

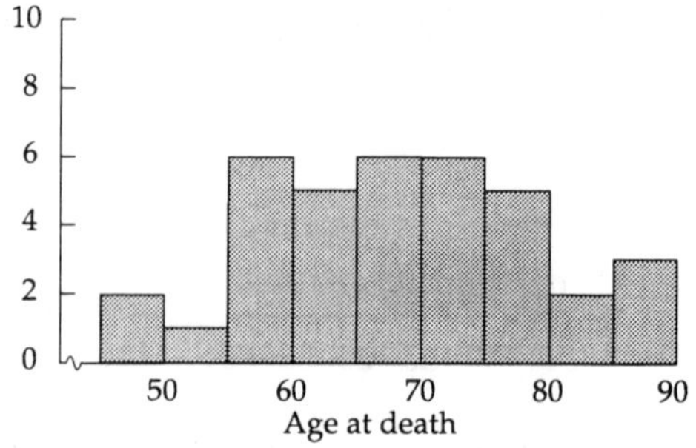

7. 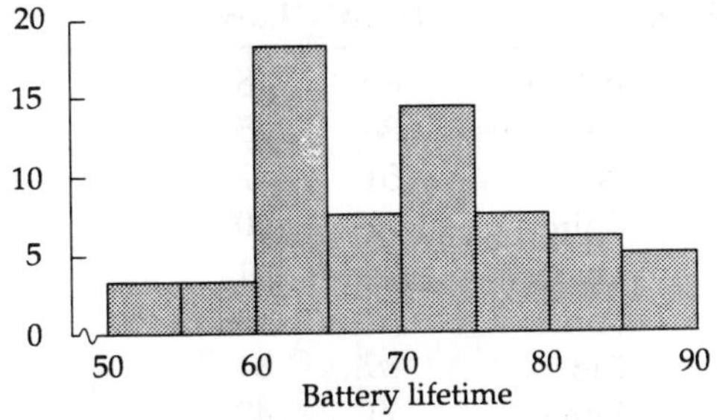

9. Answers will vary.
11. Answers will vary. Here is one example:

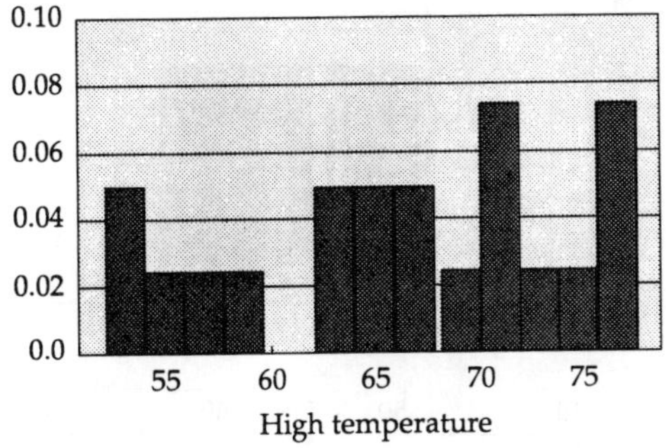

13. a. 0
    b. 21/36
    c. 10/36

CHAPTER 2

Section 2.1

1. a. Mean = 45.1; median = 46; the values are fairly equal.
   b. Mode = 27
   c. The mode is much lower than both the mean and the median.
3. a. Mean = 1.365077 million; median = 0.325 million
   b. The mean is highly influenced by the high salaries of Mark Grace and Ryne Sandberg.
   c. The median is a better measure. Only 9 players out of the 27 actually make more than the mean salary.
5. Mean = 118.1667
7. a. Mode = 3 (Sweden, Kazakhstan, China, Slovenia)
   b. Mean = 2.77, median = 1, mode = 0. The median is much better than the mean or the mode. Only 8 out of the 22 countries are above the mean, and 14 out of the 22 countries are above the mode.
9. The mode of a set of numbers can be thought of as fashionable because by the definition of mode, it is the most common number in the data set.
11. Answers may vary. We obtained these results for 10 tosses of three coins: 2, 2, 0, 1, 1, 1, 3, 2, 1, 2. Mean = 1.5 heads; median = 1.5 heads.

Section 2.2

1. a. Mean = 17.5; median = 17.5
   b. Mean = 636.25; median = 17.5

   The mean greatly increases when the 25 is changed to 2500, but the median stays the same.
3. a. Mean = \$23.64; median = \$28.00
   b. The mean will increase, but the median will stay the same.
5. The owner could resolve the dispute by including her salary in the calculations but, instead of computing the mean salary, computing the median or mode salary, which would both be \$17,000.
7. Mary's score on the third test must be greater than 85 for the three tests to have an average of 80.
9. a. Mean = 71.36; median = 71; mode = 70 or 72
   b. Any of them would be sufficient.

Section 2.3

1. a. Mean: 24.25; median: 23
   b. First quartile: 18.5; third quartile: 29
   c. Interquartile range: 29 − 18.5 = 10.5

d.

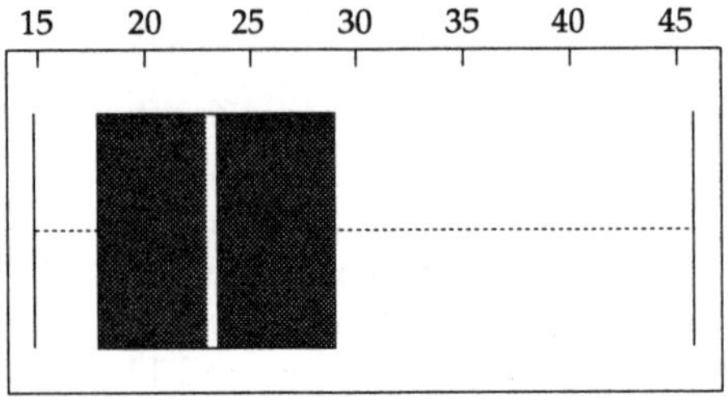

3. a. Chicago: mean = 23.821; median = 24
   Denver: mean = 20.607; median = 17
   b. Chicago: first quartile = 18, third quartile = 27.75, IQR = 9.75
   Denver: first quartile = 14, third quartile = 26, IQR = 12
   c.

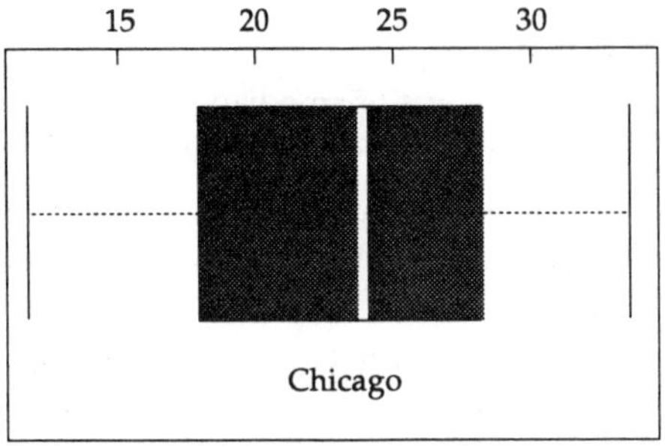

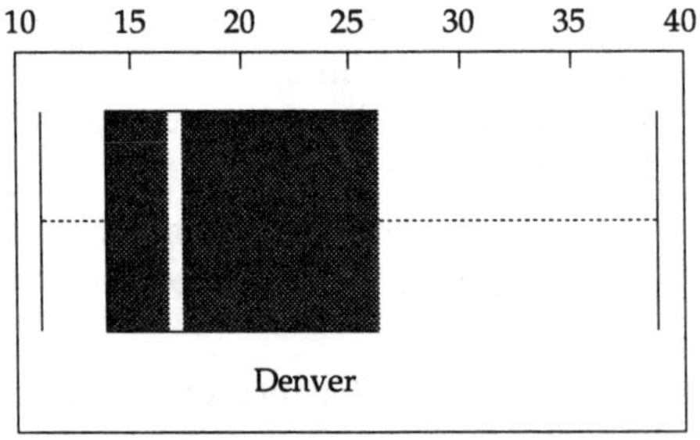

   d. On average, Chicago is a windier city than Denver. But Denver had days with higher wind gusts than Chicago.

5. a.

	High	Low
Athens	60	48
Barbados	86	75
Berlin	51	42
Cairo	77	60
Dublin	55	41
Paris	62	48
Rio	89	64
Rome	71	42
Seoul	62	50
Tokyo	77	62

   b.

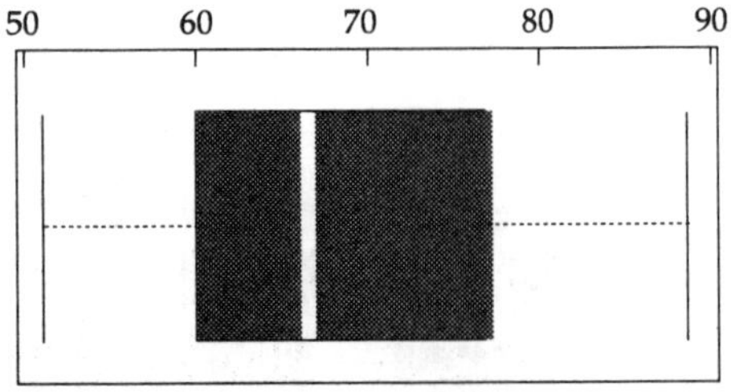

   c.

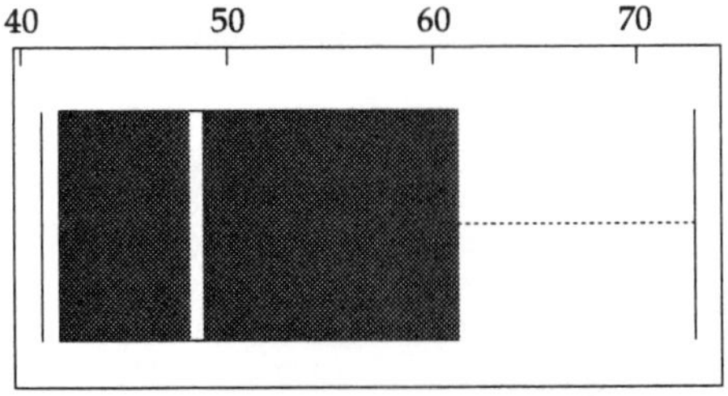

7. a.

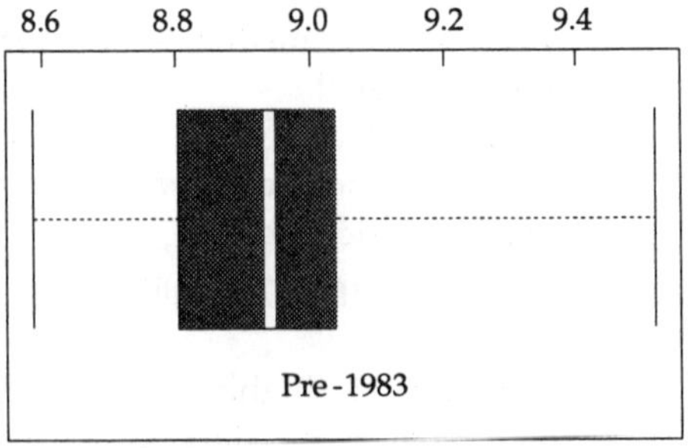

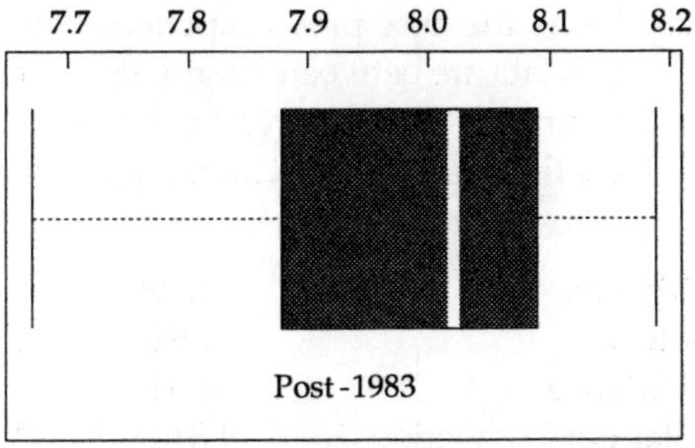

b. The post-1983 pennies are less dense than the pre-1983 pennies. The Treasury Department likely changed the amounts of metals going into pennies in 1983. (Answers will vary.)

9. In 1997: mean = \$322, median = \$287, range = \$216

Section 2.4

1. a. Mean = 77.6
   b. In order: 3.4, −17.6, −11.6, 7.4, 18.4
   c. The city closest to the mean temperature is Atlanta.

3. a.

4.61	3.03	2.22	2.13	2.01
1.59	1.18	1.01	0.33	−0.37

   b. Ryne Sandberg has the largest deviation score, 4.61.

5. Smallest deviation: Acura Integra, 0.75 miles per gallon. The city MPG for the Acura Integra is close to the average for these cars. Largest deviation: Geo metro, 21.75 miles per gallon. The Geo Metro has much better than average city MPG.

Section 2.5

1. Deviations −3, 1, 0, 1, 3; mean absolute deviation: 1.6.

3. Deviations:

21.75	8.75	6.75	4.75
4.75	1.75	1.75	0.75
−3.25	−3.25	−5.25	−5.25
−7.25	−8.25	−9.25	−9.25

   Mean absolute deviation: 6.375.

5. a. Use the program to store the deviations in L2:

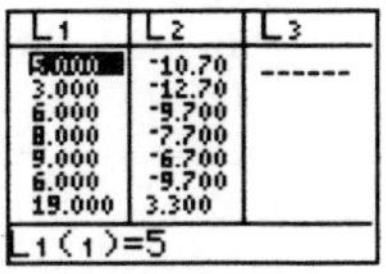

   b. Mean absolute deviation: 11.44 vibrations

7. a. California: 5.02° F
   b. Colorado: 7.59° F
   c. California. The smaller mean absolute deviation indicates that the locations have temperatures closer to their mean temperature (fewer extreme temperatures).

Section 2.6

1. 4

3. The variance of this set of numbers is 0. All the observations are the same, so they all equal the mean. Hence all the deviation scores are zero.

5. The variance is 9.5. The mean number of errors has decreased from 7 to 3, but the spread in the data is still the same.

7. The recommendation should be for Electronics Superstore. The mean for its parts is very close to 100 (100.1), but more importantly, the variance is very small (only 4.29). While Circuits R Us has a mean close to 100 (96.9), the variance of its parts is huge (74.89).

9. 63.44

Section 2.7

1. Variance = 10.4; standard deviation = 3.22

3. 2

5. The mean absolute deviation of 10, 15, 20, and 25 is 5; the mean absolute deviation of 10, 15, 20, and 2500 is 931.875. Changing 25 to 2500 makes less of a difference in the mean absolute deviation than it does in the variance or standard deviation. However, there is still a large increase in the mean absolute deviation.

7. Standard deviation $= \sqrt{61.22} = 7.82$ medals
9. a. 7.69° F
   b. 10.72° F
   c. California

Chapter Review

1. Mean = 77.8; median = 78
3. Low: mode = 0; high: mean = 2.2
5. a.

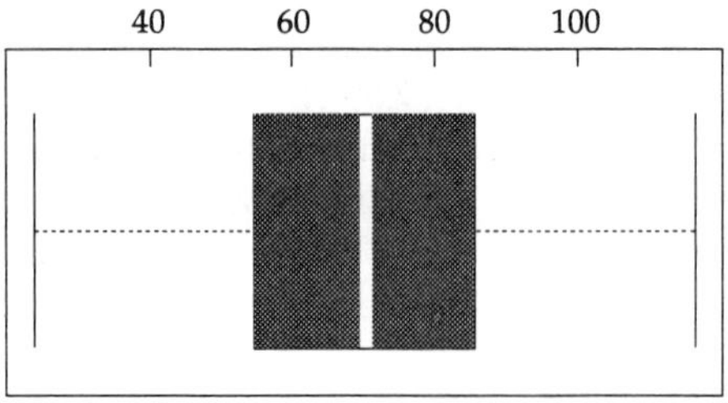

   b.

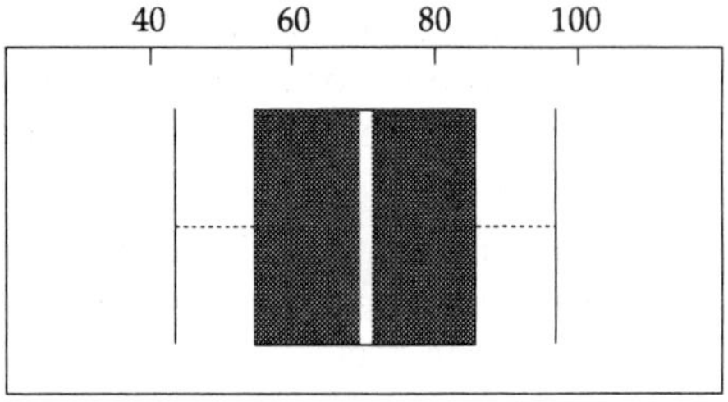

7. Median = 67.5; range = 77 − 53 = 24; interquartile range = 72.5 − 62.25 = 10.25.

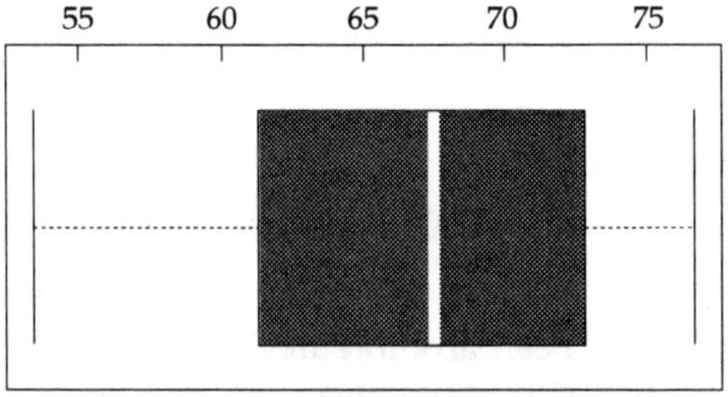

9. The mean is 58; the deviations are −8, −9, −4, 2, 21, 7, −13, and 4; the mean absolute deviation is 8.5.
11. Mean = 43.33; variance = 62.62; standard deviation = 7.91342
13. From the box plot it appears that 50% of the data are between 30 and 36, but the minimum of 20 could lower the mean value. Here are the summary statistics for this data set:

Median	32.15
Mean	31.92
Variance	18.54
Standard deviation	4.31
Range	38.3 − 19.4 = 18.9
Interquartile range	35.2 − 29.95 = 5.25
Mean absolute deviation	3.36

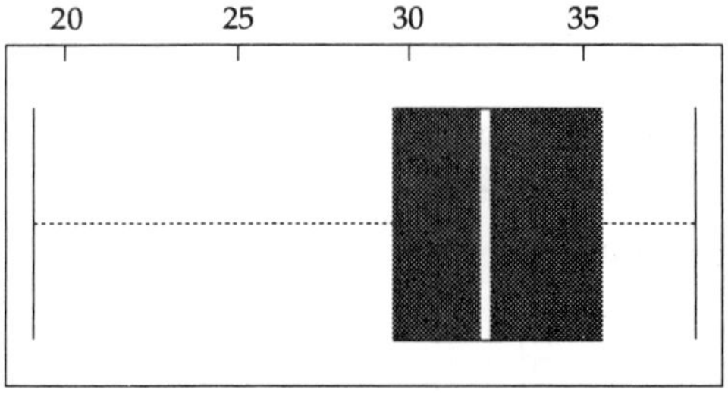

CHAPTER 3

Section 3.1

1. a. Amount of television viewing and grades in school
   b. Temperature and size of the copper bar
   c. Number of vaccinations and disease rate
   d. Inflation of tires and car mileage
3. One possible answer is age and degree of participation in active sports. In this case, the degree of participation goes down as age goes up—a negative relationship. Another possible answer is age and the amount of time spent watching sports on TV. Here, there is a positive relationship between the two variables.
5. a. About 41 miles per gallon
   b. About 18 miles per gallon
   c. Average is about 30 miles per gallon.
   d. As the weight increases, the mileage decreases.

9. a. As city mileage increases, so does highway mileage.
   b. The range for city mileage (47 − 15 = 32) is just slightly greater than the range for highway mileage (50 − 20 = 30).

11. a. Lowest life expectancy: Sudan (53 years)
    b. Largest number of persons per TV: India (44 persons)
    c. There appears to be a general tendency for countries with lower life expectancy to be those with more persons per TV. Generally, poorer countries are those with shorter life expectancy and less access to TV. But certainly less access to TV does not cause people to die sooner. Clearly poverty influences *both* access to TV and life expectancy!

Section 3.2

1. a.

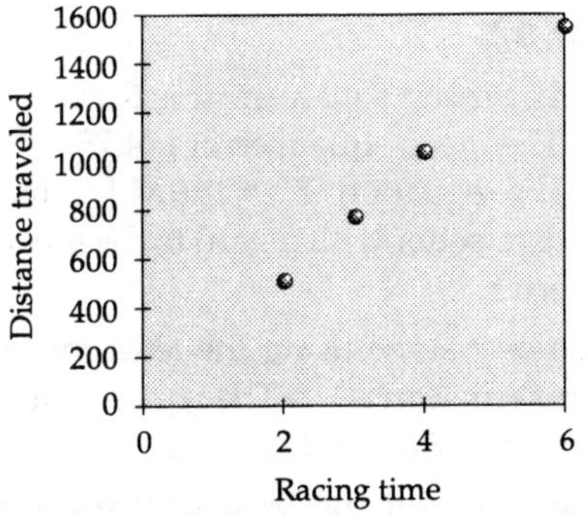

   b. The rule should approximately be distance traveled = 260 (racing time).
   c. The errors (residuals) are as follows: For racing time of 2, residual = 0. For racing time of 3, residual = −10. For racing time of 4, residual = 10. For racing time of 6, residual = 0.

3. The rule should approximately be cooking time = 1.8 + 0.25(weight).

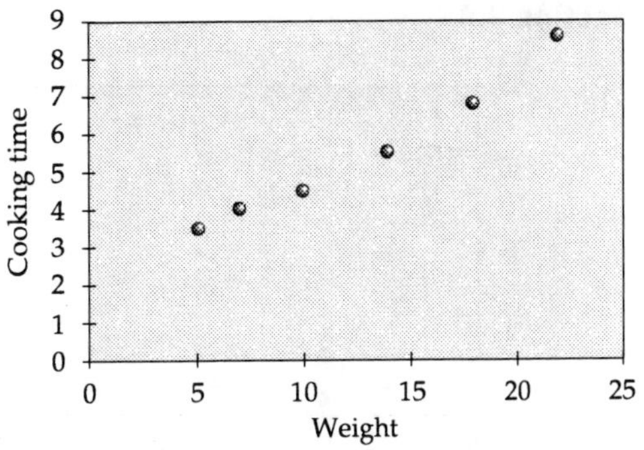

5. The rule should approximately be weight = −2 + 2(length).

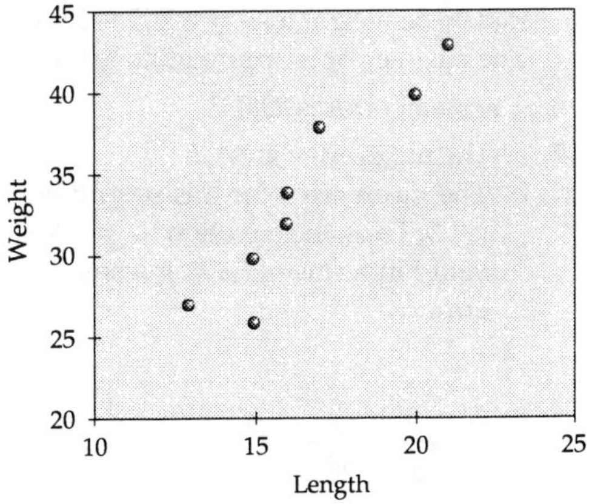

Section 3.3

1. a. $Y = 1.5X$
   b. $Y = 1.67 + 1.33X$
   c. $Y = 1.75 + 0.25X$
   d. $Y = 5 + 0.5X$

3. a. Circumference = 2 + 2.5(diameter), or a similar line
   b. The mean error should be around 2.

5. a. College score = 5.5 + 0.9(high school score), or a similar line
   b. The mean error should be around 6.

7. Answers will vary. A possible equation of best fit is $Y' = -1.5X + 67.7$.

Section 3.4

1. 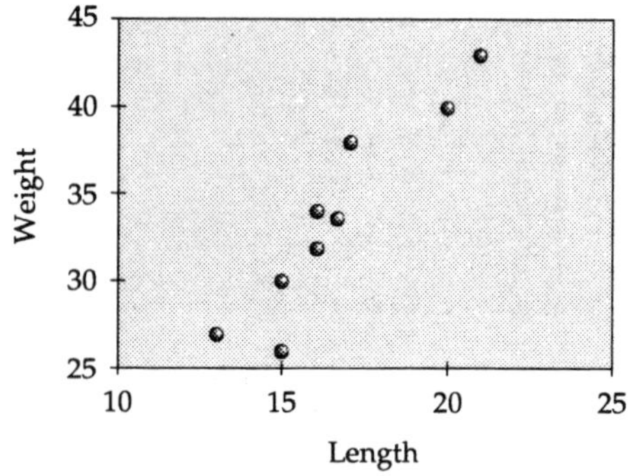

3. The equation for the line should approximately be weight loss = 6.505 + 1.5(weeks). The mean error is approximately 5.11.
5. The mean error is 2.8833.
7. a. The mean error is 3.125.
   b. The mean error for this second equation is 1.5. Thus, the equation $Y' = 2X + 0.5$ is better in terms of having a smaller mean error.

9. a.

X	Y	Y'	$\|Y - Y'\|$
12	20	20.2	0.2
10	12	16.2	4.2
10	18	16.2	1.8
8	10	12.2	2.2
7	12	10.2	1.8
6	14	8.2	5.8
6	6	8.2	2.2
5	7	6.2	0.8
4	3	4.2	1.2
2	1	0.2	0.8

   b. The mean error for this equation is 2.1.
11. The slope of the best-fitting regression line is 2.0. The equation for the line is $Y' = 0.5 + 2X$.
13. The mean error is 1.5.
15. Scatter plot with possible equation of line $Y' = -1.5X + 67.7$:

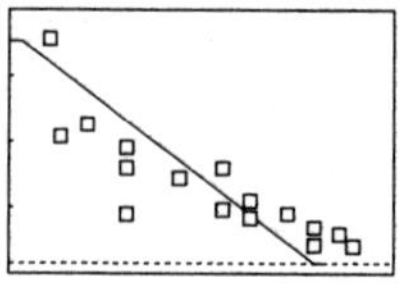

   The mean error using the above equation is 4.91. Here, the slope is negative.
17. The following is a plot of mean absolute error used to find the slope that gives the least mean error. The slope is 2.8 and the least mean absolute error is 3.09.

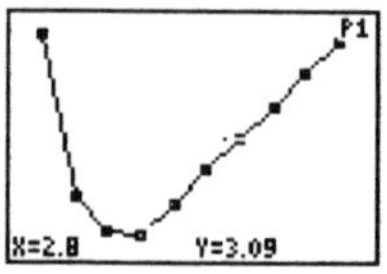

Section 3.5

1. a. The mean square error is 6.99.
   b. The mean square error is 4.75.
   c. The equation $Y' = 1.9X + 2.16$ fits the data better with regard to the mean square error.
3. The best slope using the least mean square error criterion is 1.8. The equation for the line is $Y' = -2.3 + 1.8X$.
5. The slope for the least square error line is 2.1. The equation for the line is $Y' = 2.1X - 1.1625$.
7. The mean square error for Y'_1 is 7.0; the mean square for Y'_2 is 4.75. Y'_2 has a smaller square error and is therefore a better fit.
9. The slope that gives the least square error is 1.4.

Section 3.6

1. a. The covariance is 13.41.
   b. Based on the table, the best slope is 2.1. Using the formula, the best slope is 2.15.

3. a. The covariance is −14.56. The variance of X is 101.33.
   b. The slope of the least squares best-fitting line is −0.16.
5. a. Strong negative
   b. Weak positive
   c. Strong positive
   d. Weak negative
7. −0.92
9. 0.93
11. The correlation is −0.47. You know the correlation is negative since the slope of the best-fitting least squares regression line is negative.
13. a. Scatter plot:

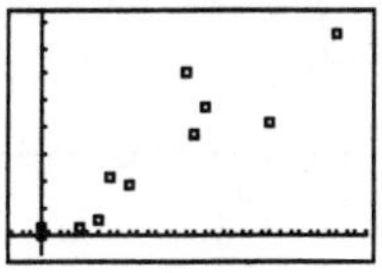

   b. A possible equation is $Y' = 2.32X + 1.12$.

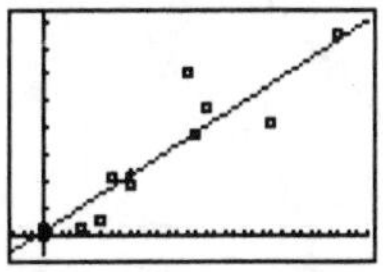

   c. The covariance between X and Y is 220.8. The standard deviation of X is 9.6. The standard deviation of Y is 24.5. The correlation is 0.936. The more you foul, the more likely you are to score points.
15. The covariance is −96.4. The variance of X is 182.8. The slope is −0.53.

Section 3.7

1. a.

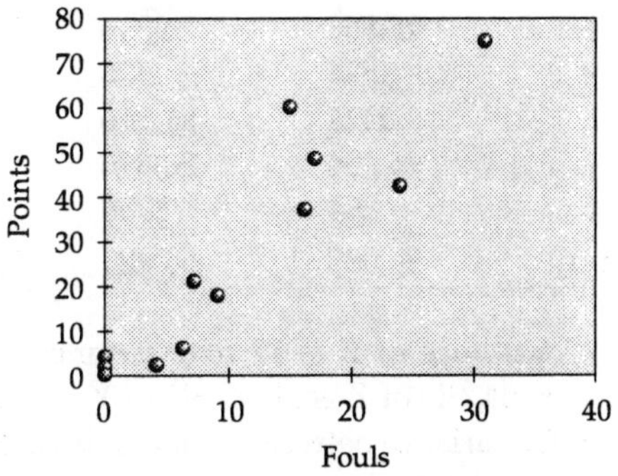

   b. The equation for the line is $Y' = 2.5X - 0.525$.
   c. The covariance is 220.82. The standard deviation of X is 10.00. The standard deviation of Y is 25.41. The correlation between X and Y is 0.94.
   d. If these data were interpreted causally, one would infer that the more a person fouls, the more points he or she will score.

Chapter Review

1. a. Height and number of blocked shots; close to 1
   b. Number of skips and score on the final exam; close to −1
   c. Family income and score on IQ test; close to 0
   d. Years of education and income; close to 1
   e. Number of people and number of farms; close to −1
3. a.

Student	X	Y	$X - \overline{X}$	$Y - \overline{Y}$
1	670	710	96	134
2	550	500	−24	−76
3	720	620	146	44
4	410	490	−164	−86
5	520	560	−54	−16
Sum	2870	2880	0	0

Student	$(X-\overline{X})(Y-\overline{Y})$	$(X-\overline{X})^2$	$(Y-\overline{Y})^2$
1	12,864	9216	17,956
2	1824	576	5776
3	6424	21,316	1936
4	14,104	26,896	7396
5	864	2916	256
Sum	36,080	60,920	33,320

b. Variance of X is 12,184, variance of Y is 6664, SD of X is 110.4, SD of Y is 81.63.
c. Covariance between X and Y is 7216.
d. $r = 0.80$
e. There is a strong positive relationship between verbal and mathematical scores on the SAT.

5. a. The largest squared error is for the Florida Marlins, 4489. The smallest is for the New York Mets, 9.
   b. False; the method of least mean error finds the line that minimizes the mean error, and that is not necessarily the same as the line that minimizes the mean square error.
   c. 849 runs

7. 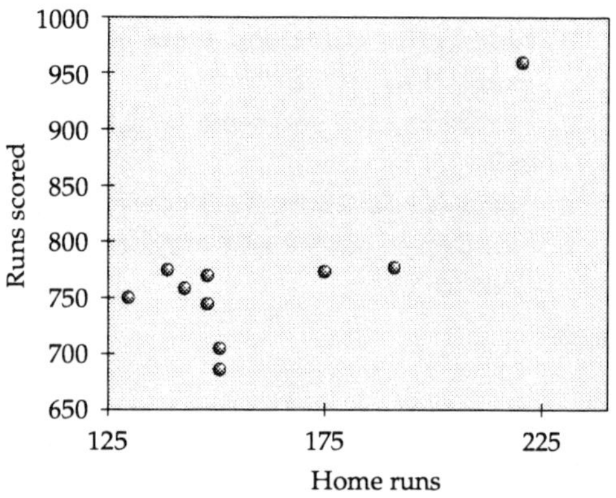

   a. The line must pass through (138, 759) and (191, 778). The equation for the line is thus $Y' = 0.36X + 709.32$. The lines are not very similar. This is because one observation (Colorado Rockies) is really throwing off the least squares best-fitting line.
   b. The method of median fit must give a larger mean squared error, because the regression line found with the method of least squares has the smallest mean squared error possible.

9. b. Recall that statistical correlation does not imply causation.

11. a. Method of least squares
    b. Informal visual method
    c. Method of least squares
    d. Method of least mean error

13. 2.90

15. a. 362
    b. 0.68
    c. $Y' = 0.68X + 58.88$
    d. 231.12 pounds.

17. The equation with the lowest mean square error is $Y' = 5.8X - 6.5$.

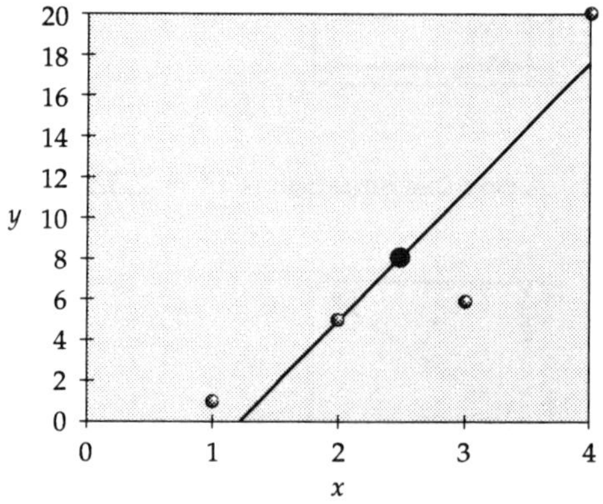

CHAPTER 4

Section 4.1

1. a. 18.4 boxes.
   c. Answer should be around 15.
3. Answer should be around eight boxes.
5. a. Answer should be around 10/3.
   b. A die would work well. Assign two of the sides to guessing correctly, and the other four sides to guessing incorrectly.
7. Answer should be around four days.
9. Answer should be around 23 boxes.

11. Answers will vary. The theoretical expected number of packs needed before a complete set of eight cards is 21.7.

Section 4.2

1. a. Answer should be around 5.
   b. Answer should be around 10.
3. a. Answers should be around 0.58 empty seat and 1.08 persons not seated.
   b. Answer should be around 0.25 empty seat and 1.10 persons not seated.
5. Answer should be around eight keys.
7. Answers will vary. We obtained these:
   a. 4.0 bags
   b. 1.7 bags
   c. 11.3 bags

Section 4.3

1. In this case, if it rains in one city, you know it rains in the other, so you can model this with only one coin: heads = rains in both, tails = does not rain in both.
3. Your friend's argument does not make sense because each flip of the coin is independent of the others. Thus, the probability of heads on the next flip does not depend on what has happened before and is still 50% for every flip.

Section 4.4

1. Average stopping point should be around zero, right where the man started.
3. The person should take an average of around 18 moves to reach one of the walls.
5. a. Should be around 19 degrees.
   b. Should be around 19 degrees.
   c. Should be around 24 degrees.
   d. Should be around 31 degrees.
7. Answers will vary. We obtained these:
   a. Average distance from zero is 2.27.
   b. Average distance from zero is 3.2.

Chapter Review

1. a. Flip the coin twice. There is a 25% chance it will come up heads both times.
   b. Pull a card at random from the deck. A club will turn up 25% of the time.
   c. Use only sides one through four of the die. Thus, there is a 25% chance of seeing a one. Roll again if it comes up a five or a six.
   d. Look only at digits 1 through 4. A 1 will appear 25% of the time.
3. Answer should be around −5.
5. Answer should be around 1 time.
7. a. Answer should be around 4.
   b. There should be either one family or none having seven girls.
9. Answer should be around 2.5 five-card hands.
11. Answer should be around 10 days.
13. a. Answer should be around 3.75 points.
    b. Answer should be around 5 points.
    c. He should concentrate on two-point shots.
15. Answer should be around nine packs.
17. a. Answer should be around 2.5 points.
    b. John should guess on all the problems, since if he does not, his score will be zero.
19. Answer should be around 3.77.

CHAPTER 5

Section 5.1

1. a. 0.48
   b. 0.47
   c. 0.486
3. Answer should be around 4/52 (0.077). This is expected since four of the 52 cards are aces.
5. 0.07
7. a. Answer should be around 1/36 (0.027).
   b. Answer should be around 1/6 (0.167).
   c. Answer should be around 1/12 (0.083).
   d. Answer should be around 5/36 (0.139).
9. With 180 trials, the estimated probabilities should each cluster more closely around 1/6.
11. As is shown later in this chapter (Section 5.5), the theoretical probabilities are as follows:
    a. $p(\text{2 heads}) = 0.25$
    b. $p(\text{exactly 1 head}) = 0.50$

c. $p(\text{1 or more heads}) = 0.75$
d. $p(\text{0 heads}) = 0.25$

Do your estimates based on 100 trials cluster more closely around these theoretical values than your answers to Exercise 10 did?

13. Answers will vary. Here are estimates based on 50 trials:
a. $P(\text{Sum equals 2}) = 1/50 = 0.02$
b. $P(\text{Sum equals 7}) = 8/50 = 0.16$
c. $P(\text{Sum equals 10}) = 9/50 = 0.18$
d. $P(\text{Sum is greater than 6}) = 35/50 = 0.7$

Section 5.2

1. Answer should be around 0.125.
3. Answer should be around 0.333.
5. Answer should be around 0.341.
7. Answers will vary. Here is a sample for 50 trials:

$$P(\text{3 boys}) = 5/50 = 0.10$$

The theoretical value is 0.125.

9. The following experimental probabilities are based on 50 trials. The theoretical values are in parentheses.
a. $P(\text{at least 2 girls}) = 35/50 = 0.70\ (0.688)$
b. $P(\text{exactly 3 girls}) = 8/50 = 0.16\ (0.250)$

Section 5.3

1. Answer should be around 0.41.
3. Answer should be around 0.12. You do not really have a reason to complain. There was a 12% chance that you would get two or more bad prints.
5. Answer should be around 0.671.
7. Answers will vary.
9. Here are sample outcomes of program LOAD for $p(\text{heads}) = 0.10$, 7 tosses, and 50 trials. Answers may vary.

$P(\text{0 heads}) = 17/50$

$P(\text{1 head}) = 20/50$

$P(\text{2 heads}) = 7/50$

$P(\text{3 heads}) = 3/50$

$P(\text{4 heads}) = 2/50$

$P(\text{5 heads}) = 1/50$

$P(\text{6 heads}) = 0/50$

$P(\text{7 heads}) = 0/50$

So, $P(\text{one or more bursts}) = 0.33$.

11. Here are sample outcomes of program LOAD for $p(\text{heads}) = 0.1$, 5 tosses, and 100 trials. Answers may vary.

$P(\text{0 heads}) = 61/100$

$P(\text{1 head}) = 27/100$

$P(\text{2 heads}) = 10/100$

$P(\text{3 heads}) = 1/100$

$P(\text{4 heads}) = 0/100$

$P(\text{5 heads}) = 1/100$

So $P(\text{two or more defectives}) = 0.12$

Section 5.4

3. a. The probability of getting a green on both lights should be around 0.111.
5. a. $P(\text{Both engines fail})$ approximately equals $P(\text{Engine 1 fails}) \times P(\text{Engine 2 fails})$.
b. $P(\text{Both engines fail})$ should be around 0.01.
7. In 50 trials using the program RANDLIST, six successes were obtained. So

$$P(\text{rain in San Juan and Chicago}) = 6/50 = 0.12$$

9. Select two lists. For best results, try first choosing three theoretical probabilities all at least 3/4.

Section 5.5

3. Each digit has a probability of 0.1. The probability of getting an odd digit is 0.5.
5. a. 1/2
b. 1/4
c. 1/13
d. 1/52
7. Results should be close to those of Exercise 6, but not exactly.

9. a.

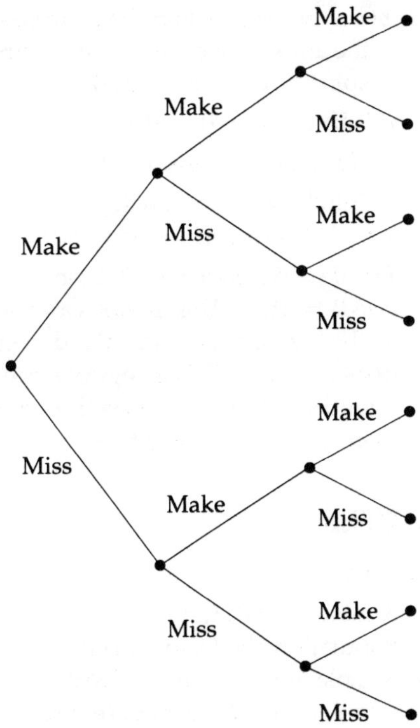

b. 1/8

c. p(makes all three) = p(makes first) × p(makes second) × p(makes third) = 1/2 × 1/2 × 1/2 = 1/8

d. 7/8

Section 5.6

1. The event not E is getting at least one head in the two flips. $P(E)$ is 0.20, and $P(\text{not } E)$ is 0.80.

3. a. Not A is the event where you do not get an ace.
   b. One possibility: Repeatedly turn over the first card in a deck of cards and keep track of how often it comes up an ace.

5. a. Not G is the event of there being no girls in a family of four children.
   b. Answer should be around 0.0625.
   c. $p(\text{not } G) = 0.0625$.

7. a. Answer should be around 0.52.
   b. 0.52
   c. 0.48

9. a. p(It does not rain any day) = 0.133
   b. The complement is, "It rains at least one day during the week."
   c. p(It rains at least one day) = 1 − 0.133 = 0.866

Section 5.7

1. a. 0.578
   b. 0.382
   c. 0.0675

3. a. It is sampling without replacement, since the man does not replace the wrong keys into his pocket after he tries them.
   b. 0.3

5. Answer should be around 0.444.

7. a. 1/24
   b. 1/4
   c. 1/4
   d. 18/24

Section 5.8

1. Answer should be around 0.427.

3. You need 13 people to be certain. This because if you have 12 or fewer, it is still possible that they were all born in different months.

5. a. Answer should be around 0.03.
   b. Answer should be around 0.12.
   c. Answer should be around 0.25.
   d. Answer should be around 0.41.

7. Answer should be around 0.28.

Section 5.9

1. Answer should be around 2/3 = 0.667.

3. Answer should be around 0.475.

5. a. Answer should be around 0.833.
   b. Answer should be around 0.667.
   c. Answer should be exactly 1.
   d. Answer should be around 0.5.
   e. Answer should be around 0.167.

Chapter Review

1. a. 0.12
   b. 0.88
   c. Answer should be around 0.27.

d. The probability in part (b) is the probability of the event that at least one of the students is left-handed, and the probability in part (c) is the probability of the event that exactly one of the students is left-handed.

3. a. People who marry each other will tend to have the same characteristics, beliefs, and so on.
   d. In about 38% of the couples, both husband and wife should be Democrats, and in about 38% of the couples, both husband and wife will be Republicans.

5. If you sample without replacement, once you sample an item, you cannot sample it again (that is, you do not replace it). When sampling with replacement, it is possible to sample an item more than once.

7. a. "None of the five retirees has volunteered during the past 12 months."
   b. p(None of the five retirees volunteered) = p(retiree 1 did not volunteer) $\times$ p(retiree 2 did not volunteer) $\times \cdots \times$ p(retiree 5 did not volunteer)
   c. 0.08
   d. 0.92

9. Answer should be around 0.30.

11. a. Answer should be around 0.67.
    b. Answer should be around 0.14.

13. a. 0.60
    b. Answer should be around 0.44.

15. b. 21/27
    c. 6/27

17. a. Each spot has a 1/10 chance of coming up.
    b. Each combination has a 1/100 chance.
    c. They are the same since Jan increases her amount if she gets a five or higher, given that she got a four on her first spin. This probability is 0.50.
    d. p(On the second spin she increases the amount she wins | On the first spin she won \$5) = 0.40

    p(On the second spin she increases the amount she wins | On the first spin she won \$6) = 0.30

19. a. P(Wife exercised) = 0.56

    P(Husband exercised) = 0.49

    P(Both husband and wife in the same couple exercised) = 0.40
    b. P(Wife exercised) $\times$ P(Husband exercised) = 0.27. This is not close to 0.40, so the law of experimental independence does not hold. This is logical since if one member of the couple exercises, the other member is more likely to exercise also.

CHAPTER 6

Section 6.1

1. a. Sample: the 40 students
   Population: the freshman class
   b. Sample: the few pieces of rock
   Population: rock from the entire river valley region
   c. Sample: the spoonful of soup
   Population: the pot of soup
   d. Sample: the water where Elaine put her toe
   Population: the water in which they will swim
   e. Sample: the sample from each truckload
   Population: each truckload of concrete
   f. Sample: the blood used for the blood test
   Population: the blood in Lisa's body

3. Answers will vary.

5. The sample is the group of people surveyed in the *Weekend Magazine* poll. The population is all Canadians.

7. The sample is the 130,000 households responding to the survey. The population is all households in the United States.

9. Answers will vary.

Section 6.2

1. a. 0.15
   b. 0.03

c. 0
d. 0.51
e. 0.10
f. 0

3. P(9 or more cures) = 0.04. Yes, one can conclude that the new drug is more effective than the old drug.
5. P(5 or fewer) = 0. Yes, one can conclude that the company has a gender bias at the executive level.
7. P(4 or fewer) = 0.02. Yes, one can conclude there is a problem with racial bias in that jury.
9. P(7 or fewer) = 0.13. No, one cannot conclude that the community is opposed to the amendment.
11. Answers will vary. We obtained these:
    a. P(3 or more heads) = 0.80
    b. P(4 or fewer heads) = 0.36
    c. P(6 or more heads) = 0.36
    d. P(8 or fewer heads) = 1.00
    e. P(9 or more heads) = 0
13. Toss 12 coins 50 times. Let heads represent a male juror and tails a female juror. Let H be the number of heads. Answers will vary. We obtained:

$$P(H \geq 9) = \frac{2}{50} = 0.04$$

Section 6.3

1. The median of the combined data is 109. Group I has seven observations above the median. P(7 or more above median) = 0.01, so we conclude that the new feed produces higher weight gain than the old feed.
3. The median of the combined data is 10. The farmers have one observation above the median. P(1 or less) = 0, so we conclude that the two groups have different attitudes toward food stamps.
5. Answers will vary. The probability of there being 10 cures out of 12 trials that we obtained was 0.1. This is somewhat unusual. There is some evidence that the new drug is more effective than the old one.
7. We obtained P(19 out of 20) = 0.02.

Chapter Review

Note: Answers obtained from the five-step method may vary.

1. The population is citizens of the United States over the age of 18 and registered to vote in the election.
3. Assume that the city is split 50/50. Flip a coin 25 times and record the number of heads. Repeat the process 100 times. From Table 6.3, the number of times there were 18 or more heads out of 25 flips of a coin was 3 out of 100. Since $0.03 < 0.05$, we conclude that the candidate can feel confident about receiving more than 50% of the vote.
5. Combine the first 20 random numbers, the second 20 random numbers, and so on. Use 1 as a success and 2, 3, 4, 5, 6, 7, 8, 9, and 0 as failures. The number of trials with six or more successes is $0/25 = 0$. Since $0 < 0.05$, we conclude that Josh's shooting has improved. Yes, the final answer is the same in both problems, but the conclusion is much stronger in Exercise 5.
7. The combined median of the 16 cars is 19.85. Out of the eight cars with the new motor, five had mileage greater than the median value. P(5 or more) = 32/100 = 0.32. No, we cannot conclude that the cars with the new engine have better fuel economy.

CHAPTER 7

Section 7.2

1. $D = 38$. Based on Table 7.4, the die is not fair.
3. $D = 24$. The phone book is not a very good place to find random digits. This value of D is rather high compared with the other values in the table.

5. Answers will vary. We obtained the following:

a.

Side	Observed	Expected
1	33	25
2	18	25
3	22	25
4	30	25
5	25	25
6	22	25

These results yield $D = 26$.

b.

Side	Observed	Expected
1	34	50
2	65	50
3	39	50
4	48	50
5	65	50
6	49	50

These results yield $D = 60$.

c.

Side	Observed	Expected
1	91	100
2	109	100
3	101	100
4	112	100
5	95	100
6	92	100

These results yield $D = 44$.

7. The value of D calculated from the phone book data is 24. Here are 30 D values obtained from rolling a fair 10-sided die 50 times (your frequencies may vary):

D	Frequency	D	Frequency
10	2	18	1
12	5	19	2
13	2	20	2
14	3	22	1
15	1	23	2
16	4	24	1
17	4		

Notice that a D of 24 occurred only 1 time in 30. This value of D is rather large. We are doubtful that the phone book data are random.

Section 7.3

1. a. 0.833
   b. 0.867
   c. 0.567
   d. 0.233
   e. 0.067

3. The chi-square statistic is 25.733. Using the stem-and-leaf plot for χ^2 in Exercise 3, we determine that people did prefer one color over another.

5. a. The chi-square statistic is 3.8.
   b. According to the stem-and-leaf plot for χ^2 in Exercise 5, the probability of seeing a value greater or equal to that is 0.28. Conclude that the die is fair.

7. a. 6.24
   b. 16.64
   c. 3.96

9. The calculated chi-square for the phone-book data from Exercise 7 in Section 7.2 is 16.4. The following figure is a histogram of 30 chi-square values obtained from 30 trials of rolling a 10-sided die 50 times. (Use the program RANDOM to produce each set of outcomes; your results may vary from these.)

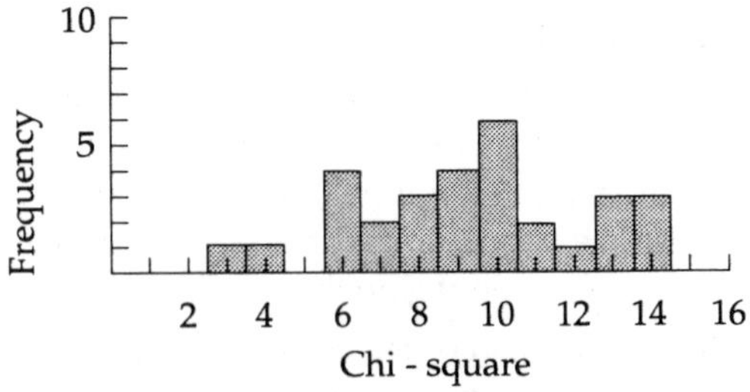

A chi-square value of 16.4 is large. It was not obtained in these 30 values. So we are doubtful that the data are random.

Section 7.4

1. a. 0.76
   b. 0.86

3. The chi-square statistic is 22.8. Using the table in Exercise 1, we find that the probability of seeing a value equal to or greater than 22.8 is 0. Conclude that some numbers are more preferred than others.
5. The chi-square statistic is 13.4. Using the table in Exercise 1, we find that the probability of seeing a value equal to or greater than 13.4 is 0.12. Conclude that the phone book is a satisfactory source of random digits.

Section 7.5

1. a. 0.13
   b. 0.02
3. a. 0.05
   b. 0.08
5. a. The chi-square statistic is 2.8.
   b. There are five degrees of freedom, since there are six categories.
   c. 0.73
   d. Conclude that the die is fair since the probability in part (c) is high.
7. There are nine degrees of freedom, since there are 10 possible categories.

Section 7.6

1. Estimated: 0.42; theoretical: 0.49
3. Estimated: 0.05; theoretical: 0.10
5. Estimated: 0.10; theoretical: 0.20
7. a. 0.05
   b. 0.01
9. a. 0.29
   b. 0.03
11. The chi-square statistic is 12.75. The probability of getting a chi-square this large or larger with five degrees of freedom is 0.03. Conclude that the die is loaded since this probability is less than 0.05.
13. The chi-square statistic is 23.19. The probability of getting a chi-square this large or larger with seven degrees of freedom is 0. Conclude that some fractions are more common than others. This is most likely due to the human tendency to round the fractions off.

Section 7.7

1. a. 5
   b. The chi-square statistic is 11.6. The probability of seeing a chi-square this large or larger with five degrees of freedom is 0.04. Conclude that people prefer certain types of crust over others, since this probability is less than 0.05.
   c.

Type	Expected number of persons
LMH	6.9
LHM	5.4
MLH	5.4
MHL	4.5
HLM	4.5
HML	3.3

   d. The chi-square statistic is 9.85. The probability of seeing this chi-square statistic or one larger with five degrees of freedom is 0.08. Conclude that this is a good model, and that people do prefer light crust.
3. The chi-square statistic is 9.69. The probability of seeing a chi-square this large or larger with three degrees of freedom is 0.02. Conclude that the results do not support the theory.
5. The chi-square statistic is 4.55. The probability of seeing a chi-square this large or larger with two degrees of freedom is 0.10. Conclude that the city council is representative of the actual population.
7. The chi-square statistic is 0.4. The probability of seeing a chi-square this large or larger with four degrees of freedom is 0.98. Conclude that the model predicts the thunderstorms well.

Chapter Review

1. The chi-square statistic is 6.6. With five degrees of freedom, the probability of seeing a chi-square this large or larger is 0.25. Conclude that the items are preferred equally.

3. a.

Type	Expected amount
Newspapers	10.1 tons
Books/magazines	6.7 tons
Office paper	4.8 tons
Corrugated cardboard	13.5 tons
Mixed paper	13.0 tons

b. The chi-square statistic is 0.727. The probability of seeing a chi-square this large or larger with four degrees of freedom is 0.95. Conclude that the distribution of wastepaper does follow the national average.

5. a. 2
b. 0.533
c. The probability of seeing a chi-square this large or larger with two degrees of freedom is 0.77. Conclude that the model did do a good job of predicting.

9. A larger chi-square means that the observed values are farther away from the expected values, suggesting that the die is loaded.

11. a. 10
b. The chi-square statistic is 1.4. The probability of seeing a chi-square this large or larger with three degrees of freedom is 0.71. Conclude that the cards were being drawn randomly from a fair deck.

13. a. You should have expected 27 of each size to be sold.
b. The chi-square statistic is 13.85. The probability of seeing a chi-square that large or larger with three degrees freedom is 0. Conclude that certain sizes are more likely to be bought than others.

15. a. For two degrees of freedom, 0.01; for four degrees of freedom, 0.04; for six degrees of freedom, 0.12
b. As the number of degrees of freedom increases, the probability of seeing 10.2 or larger increases.
c. The probability of seeing a chi-square greater than or equal to 10.2 with 100 degrees of freedom must be larger than 0.12, since the probabilities increase as the number of degrees of freedom increases. Therefore conclude that the expected and observed values have the same distribution.

17. The chi-square statistic is 1434.8. Clearly, we should conclude that this is not a good source of random numbers.

19. a. Nixon, 50.5%; Humphrey, 33.8%; Wallace, 15.7%
b. Nixon, 271.5; Humphrey, 181.8; Wallace, 84.7
c. The chi-square statistic is 21.32. The probability of seeing a chi-square this large or larger with two degrees of freedom is 0. Conclude that the popular vote and the electoral college vote did not follow the same distribution.

CHAPTER 8

Section 8.1

1. The number of games won does not have the binomial distribution since it is not reasonable to assume that the games are independent of each other. The team could be experiencing a streak of good playing, or bad playing. Thus, condition 4 is violated.

3. a, b.

Question number					
1	2	3	4	5	Probability
R	R	W	W	W	0.026
R	W	R	W	W	0.026
R	W	W	R	W	0.026
R	W	W	W	R	0.026
W	R	R	W	W	0.026
W	R	W	R	W	0.026
W	R	W	W	R	0.026
W	W	R	R	W	0.026
W	W	R	W	R	0.026
W	W	W	R	R	0.026

c. The probability that Joan guesses exactly two of the questions correctly is 0.26.

5. a. 0.80
   b. 0.33
   c. 0.4096
   d. Theoretical mean = 4; theoretical SD = 0.89

7. a. 32.5
   b. 3.37
   c. Drawing 18 or fewer chips is not very likely since 18 is over four standard deviations away from the mean. Drawing 29 or fewer chips is possible but not very likely since 29 is one or more standard deviations away.

9. 0.60 + 0.32 = 0.92

Section 8.2

1. The event we are concerned with is the scoring of runs. First, during certain times or situations during the game, the team is more likely to score than others. Second, nonoverlapping intervals are not necessarily independent of each other.

3. a. 0.4966
   b. 0.1404
   c. 0.0082
   d. 0.0905

5. The theoretical standard deviation is the square root of the theoretical mean. For the Prussian army data, the theoretical standard deviation is 0.834.

7. a, b.

Outcome	Obtained	Expected
0	53	33
1	153	113
2	138	193
3	103	219
4	225	186
5	214	126
6	68	72
7	47	35

   c. The chi-square statistic is 177.37. The number of degrees of freedom is 8 − 1 = 7.
   d. Conclude that the data do not follow the Poisson distribution.

Section 8.3

1. a. mean = 5.5, SD = 2.87
   b. mean = 6, SD = 1.41
   c. mean = 0, SD = 3.16
   d. mean = 10.5, SD = 2.87

3. a. The possible values are 0, 1, 2, and 3.
   b. The theoretical mean is 1.5 and the standard deviation is 1.12.
   c. The value of the chi-square statistic is 1.04. The number of degrees of freedom is 4 − 1 = 3. Conclude that the data likely follow the uniform distribution.

Section 8.4

1. A random variable that has a discrete distribution has as its possible values a set of numbers that could be listed. The possible values of a continuous distribution cannot be listed.

3. a. 0.046
   b. 0.010
   c. 0.090
   d. 0.257

Section 8.5

1. a. This is the continuous uniform distribution. The mean is 0.5; the standard deviation is 0.28.
   b. The chi-square test statistic is 15.8. The number of degrees of freedom is 10 − 1 = 9. The probability of seeing a chi-square this large or larger is 0.07. Conclude that the data do follow the uniform distribution, but just barely.

3. a. mean = 0.25, SD = 0.144
   b. mean = −0.125, SD = 0.794
   c. mean = 6.875, SD = 0.361
   d. mean = 16.95, SD = 1.93

5. You can assume that when a call arrives, the second hand is equally likely to be at any position on the clock. This is continuous, though, since the hand swings around the clock, not just counting off seconds as on a digital clock. The upper bound is 60, and the lower bound is 0.

Section 8.6

1. a. Between 3.5 and 6.5
   b. Between 2 and 8
3. a. 68%
   b. 61%
   c. 95%
   d. 97%
   e. The data seem to follow the normal distribution well.
5. a. 66% are within one standard deviation.
   b. 96% are within two standard deviations.
   c. The data seem to follow the normal distribution very well.
7. a. 0.117
   b. 1.10
   c. −0.800
   d. −1.51
9. a. 0.0443
   b. −0.243
   c. −1.68
   d. 1.67

Section 8.7

1. a. $p(z < 1.96) = 0.9750$

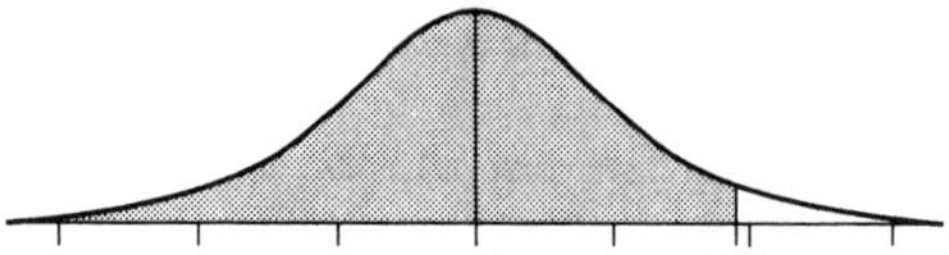

1.96

   b. $p(z < -1.96) = 0.0250$

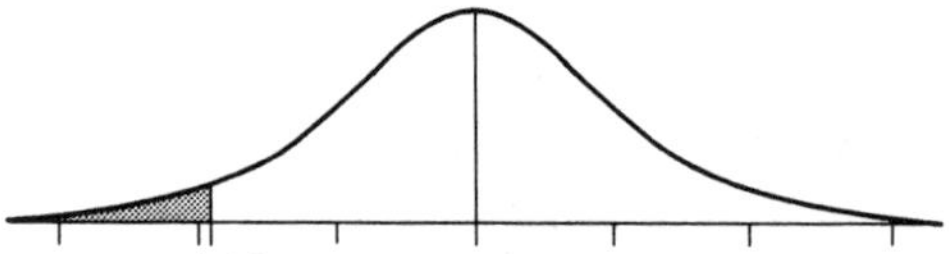

−1.96

   c. $p(z < 1.0) = 0.8413$
   d. $p(z < -1.0) = 0.1587$
   e. $p(z < 0.5) = 0.6915$
   f. $p(z < -0.5) = 0.3085$
   g. $p(z < 0) = 0.5000$

3. a. $p(z < 0.68) = 0.7517$
   b. $p(z > 0.68) = 1 - 0.7517 = 0.2483$
   c. $p(z < -0.68) = 0.2483$
   d. $p(z > -0.68) = 1 - 0.2483 = 0.7517$
5. a. $p(z < -1.64) = 0.05$
   b. $p(z < -0.67) = 0.25$
   c. $p(z < -2.32) = 0.01$
   d. $p(z < -1.2) = 0.10$
   e. $p(z < 0) = 0.5$

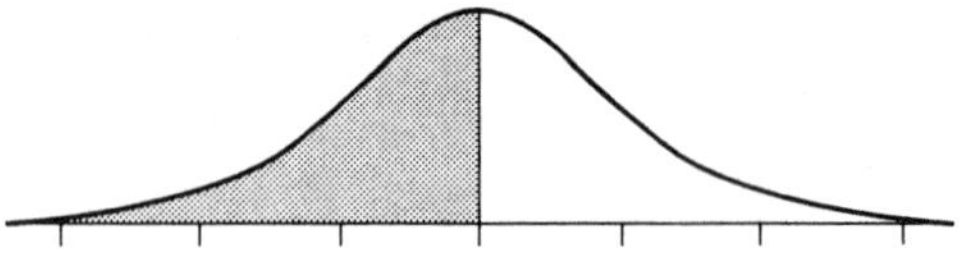

0

7. a. $p(z < 0.47) = 0.68$ (positive)

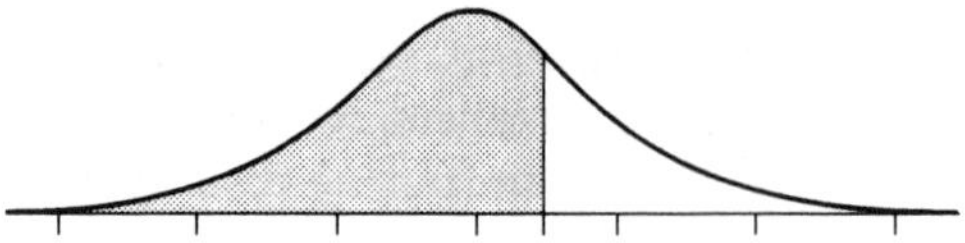

0.47

   b. $p(z < -0.99) = 0.16$
   c. $p(z < 0.84) = 0.80$
   d. $p(z < -0.84) = 0.20$
   e. $p(z > 0.84) = 0.20$ (use $1 - 0.20 = 0.80$)
   f. $p(z > 1.04) = 0.15$ (use $1 - 0.15 = 0.85$)
   g. $p(z > -0.47) = 0.68$ (use $1 - 0.68 = 0.32$)
   h. $p(z > -2.05) = 0.98$ (use $1 - 0.98 = 0.02$)
   i. $p(z > 0) = 0.5$
9. a. $p(z < 2.38) = 0.9913$

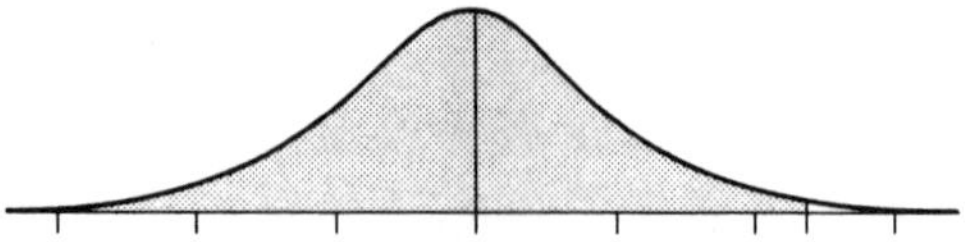

2.38

b. $p(z > -1.65) = 1 - 0.0495 = 0.9505$

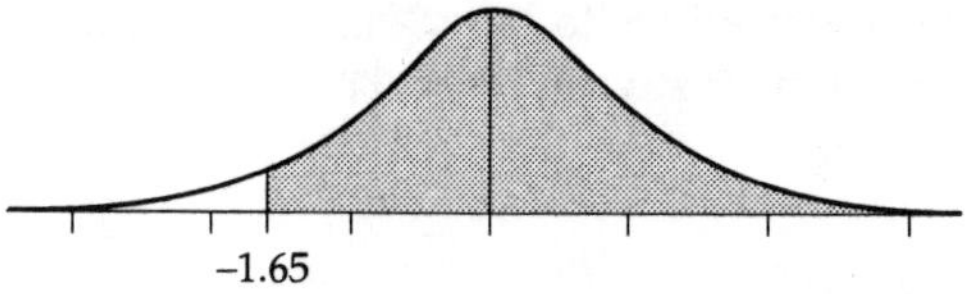

c. $p(z < -2.57) = 0.005$

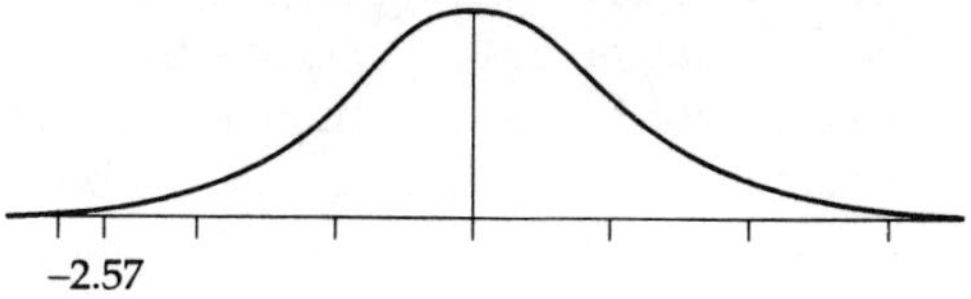

d. $p(z > -1.96) = 0.975$ (use $1 - 0.975 = 0.025$)

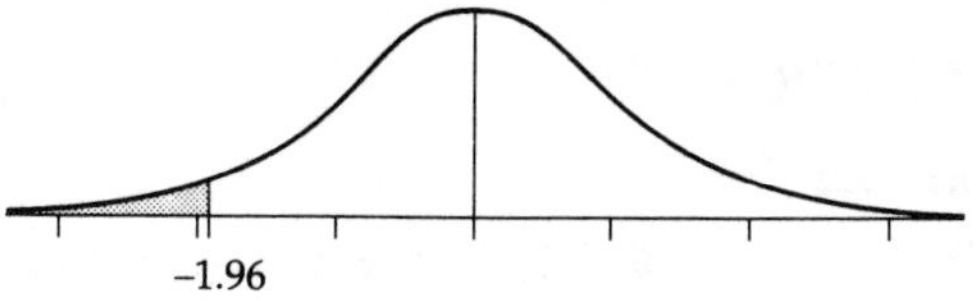

11. a. $p(-1 < z < 1) = 0.6827$

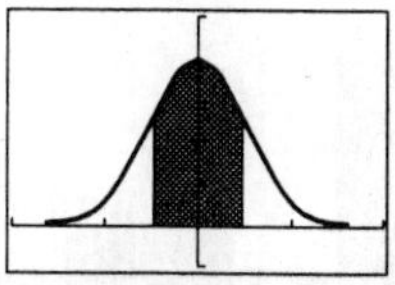

b. $p(-2 < z < 2) = 0.9545$

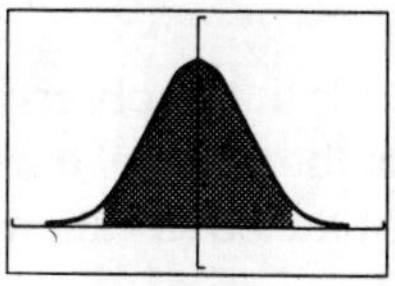

c. $p(-3 < z < 3) = 0.9973$

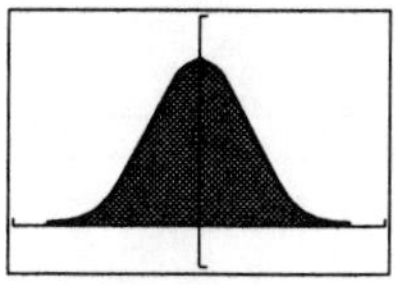

d. $p(-1.5 < z < 0.73) = 0.7005$

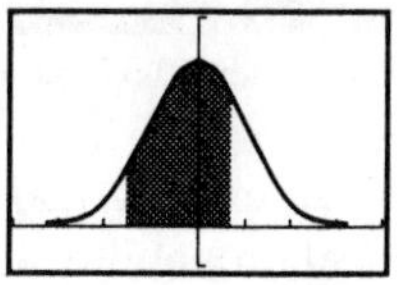

e. $p(0.45 < z < 1.75) = 0.2863$

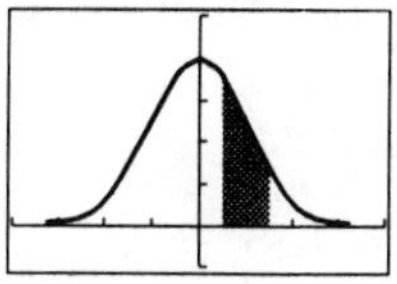

f. $p(-1.1 < z < -0.25) = 0.2656$

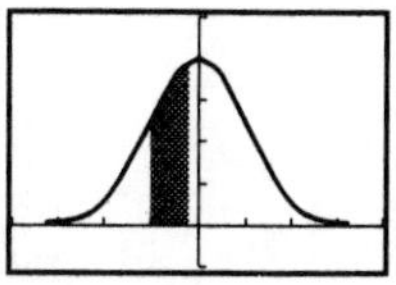

Section 8.8

1. a. 26%
   b. 13%
   c. 68.45 inches
3. a. 0.621%
   b. 0.621%
5. Lowest A is 58. Highest F is 46.
7. 0.0912, or about 9%
9. The present: 16%; the year 1800: 2%

Section 8.9

1. a.

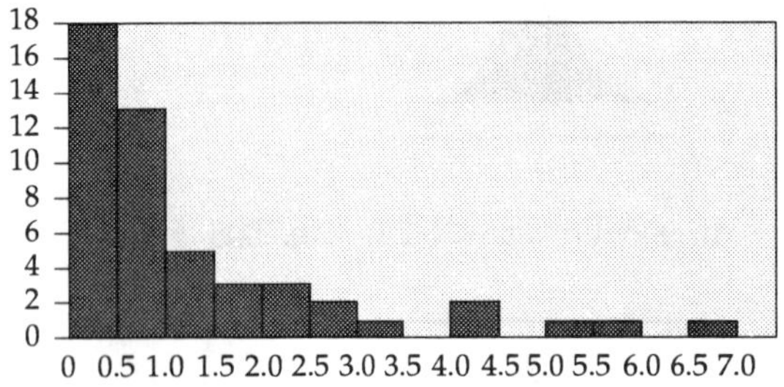

   b. This appears to be an exponential distribution. (A theoretical probability analysis will confirm this impression.)
   c. The average of the waiting times is 1.35. This is very close to the value of 1/(theoretical mean) = 1.33 from the Poisson distribution.

Chapter Review

1. a. mean = 5.25, SD = 1.85
   b. mean = 4.3, SD = 2.07
   c. mean = 2.5, SD = 1.71
3. a. −0.304
   b. 0.9605
   c. 1.968
   d. −2.47
5. a. 1.21
   b. −0.43
   c. 1.96
   d. 2.31
7. a. mean = 2, SD = 1.10
   b. 0.010
   c. 0.0768
   d. 0.2304
9. A discrete uniform variable can only take on integer values between its lower and upper limits. A continuous uniform variable can take on any value between its lower and upper limits.
11. The observations suggest the variable is not normally distributed, since in the case of the normal distribution you would expect 68% to be within one standard deviation (between 5.5 and 6.5). Also, you would expect 95% of the observations to be within two standard deviations (between 5 and 7).
13. a. 0.159
    b. 0.0913
    c. 0.0047
15. a. mean = 3.5, SD = 1.71
    b. The value of the chi-square statistic is 48.1. The probability of seeing a chi-square this large or larger is 0. Conclude that the data do not follow the discrete uniform distribution. The goal has not yet been met.
17. a. 0.21
    b. 0.275
    c. 0.367
    d. He will win 1.4 games on average.

CHAPTER 9

Section 9.1

1. The mean is 49.63333; the standard deviation is 1.816284. From the histogram, it appears that the data are fairly normally distributed.

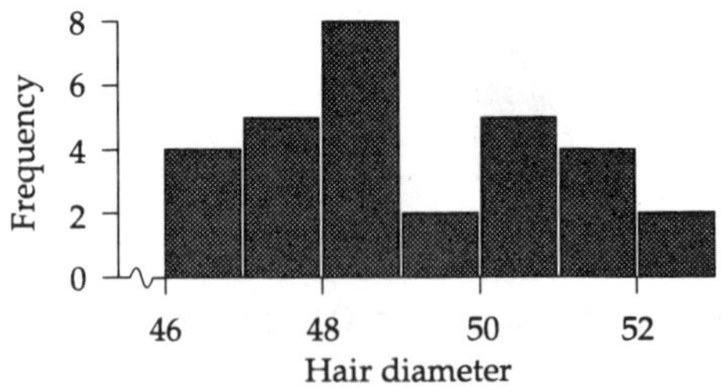

3. The mean is 4054.24. Of the 50 observations, 4 had deviations from the mean of more than 300 pounds, for 8% of the observations.
5. c. mean = 0.502
   SD = 0.142
   mean − 3SD = 0.075
   mean + 3SD = 0.929
   d. The mean winning percentage is 0.500. The 0.502 value recorded is due to the

rounding employed in reporting the winning percentages. Since there are no ties in baseball, there are the same cumulative numbers of wins and losses. For every team that wins, another loses, so we would expect the average winning percentage to always be 0.500.

e. No

f. 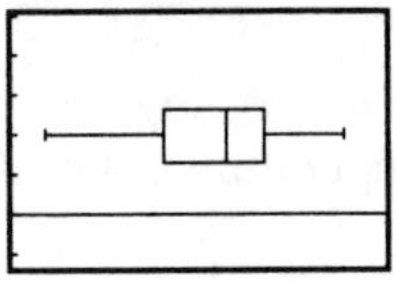

Section 9.2

1. The mean is 172.52; the range is 179.0 − 165.1 = 13.9. The mean absolute deviation is 3.7. A good estimate of the length of the pendulum would be the mean of the measurements, 172.52.
3. Method A: the mean is 73.3; the standard deviation is 5.984146. Method B: the mean is 74.4; the standard deviation is 2.059126. Method B has much less error associated with it than method A.
5. Method A appears to give measurements with less error. The histogram for method A covers values from 05 to 35, while that for method B covers values from 20 to 60. In addition, in method A there is only one observation at 5, the rest of the observations being contained within a 15-unit space, from 15 to 30.

Section 9.3

1.

Stem	Leaf
28	43
29	56,25,37,43,43,43,43
29*	62,81

The standard error is 0.4166821. The true weight is the mean of the data, 29.276.

3.

Stem	Leaf
47	9
48	1,2,2,4
48*	6,6,9
49	0,2,2,5
49*	7,8,9,9
50	0,1,2,4
50*	8,9
51	2
51*	7,8

The mean is 49.6136; the standard error is 1.099665. The estimate of the true amount of radiation present in the water is 49.6136.

5. A precise estimate of T is the mean of the 10 observations, 131.53.

Section 9.4

1. The mean is 79.8; the standard deviation is 0.8378544. Correct the mean for the +0.5° bias by subtracting 0.5; the corrected mean is 79.3.
3. Scale A: the mean is 50.33; the standard error is 2.087891. Scale B: the mean is 50.31; the standard error is 1.664879. So it appears that scale B has less error than scale A. Since the means of the two scales are almost equal, the true weight of the precipitate is likely around 50.32.

Section 9.5

1.

Stem	Leaf
236	6
237	8
238	5,5,6,6,8
239	
240	0,0,2,4,4,5,7
241	0,1,1,3,6,6

The grand mean is 239.865; the standard error is 1.364286. The estimate of the true length of the chalkboard is the mean, 239.865.

3. The mean is 39.164; the standard error is 0.464. The estimate of the true area of the drawing is the mean, 39.164.

5. Class I: the mean is 1.105; the standard error is 0.0087. Class II: the mean is 1.00375; the standard error is 0.0260. Class I has the smaller standard error, but class II has observations that are biased, probably by +0.1 meter. So class I has the more accurate results.
7. b. The means are fairly close in value, as the box plots suggest.
   c. Those produced by taking the average (mean) of five measurements have far less variability than the original measurements.
   d. No, it does nothing to remove the bias.
   e. Yes, precision has been improved.
   f. No, bias remains.

Section 9.6

1. Pendulum A mean: 3.56
   Pendulum B mean: 3.4875
   Pendulum C mean: 3.01
   Pendulum D mean: 3.26
   Pendulum E mean: 2.06

   Overall mean average is 3.06, for an estimate of the acceleration due to gravity in the Netherlands of 983.06 centimeters per second per second. This is a better estimate of the true value of 981.3 cm/sec/sec than was obtained when student 2 was included in the computations.

Chapter Review

1. The mean is 10.0258; the variance is 0.0085; the standard deviation is 0.092. Yes, the data appear to be normally distributed.
3. The estimate is the mean, 11.966 inches.
5. Oven 1: the mean is 1002.574; the standard error is 23.64074. Oven 2: the mean is 1050.256; the standard error is 9.6342. Oven 1 is closer to 1000 degrees, but oven 2 has a much smaller standard deviation.
7. One cannot be certain whether the student's findings are biased, or whether the student's finding is just the result of random error. To answer the question more precisely, the student would need to repeat the measurement several times.
9. No, your estimate is not likely to be biased. You are only off by 0.03 gallons. The error of your estimate is very likely just due to chance.

CHAPTER 10

Section 10.1

1. a. Sample: the air around the thermometer
      Statistic: 73° F
      Population: the air in Los Angeles
      Parameter: the temperature in Los Angeles
   b. Sample: the 64 college freshmen
      Statistic: 65%
      Population: the freshmen at the college
      Parameter: the percentage in favor of lowering the legal drinking age
   c. Sample: the air used to make the pollen count
      Statistic: 228 grams per cubic yard
      Population: the air in Urbana, Illinois
      Parameter: the pollen count
   d. Sample: the container of corn
      Statistic: 35% moisture
      Population: the truckload of corn
      Parameter: the moisture of the corn in the truck
3. The statistic is 5/25, or 20%, liberal. The parameter is the percentage of liberal students on campus.
5. The sample is the incoming college freshmen who responded to the survey. The population is all incoming college freshmen in the United States. One example of a statistic is the value 33.7% Protestant; the parameter it estimates is the percentage of incoming freshmen in the United States who are Protestant.

Section 10.2

1. The sample proportion
3. A 99% confidence level means that if we calculate 100 such confidence intervals, about 99 of them will contain the true population parameter of interest.

5. When based on the same data, an 85% confidence interval is wider.
7. When the confidence level of two intervals is the same, the data with the higher error are associated with the wider confidence interval.

Section 10.3

1. It is necessary to use the bootstrap method when the probability law of the population is unknown or when there does not exist an easy formula for the standard error of an estimate.
3. One possible way to proceed: Take a sample of size 100 from a population. Use the sample as the probability law of the population. Now take a number of smaller samples, such as 25, using this probability law. Repeat the process 100 times and record the value of the sample mean for each of the 100 samples. Calculate the standard deviation of the 100 sample means. This is an estimate of the standard error of the sample mean.

Section 10.4

1. The mean is 175 pounds; the standard deviation is $15/\sqrt{50} = 2.12$. The sample mean is approximately normally distributed.
3. When based on samples taken from the same population, the relative frequency histogram for the sample mean will look more like a normal distribution when the number of observations is higher. The answer is therefore 40.

Section 10.5

1. (4.04755, 4.95245)
3. (174.7423, 183.0577)
5. The 90% confidence interval is (26.84206, 28.15794). The 99% confidence interval is (26.46967, 28.53033).
7. The 95% confidence interval is (19.04206, 19.85794). We are 95% confident that the true population mean gas mileage is between 19.04206 and 19.85794 miles per gallon.

Section 10.6

1. The 90% confidence interval is (32.68%, 42.32%). Because the interval does not come close to including 50%, it is unlikely that a majority of the population is in favor of the change.
3. (74.31%, 75.69%)
5. The size of the population has no effect on the width of the confidence interval. Only the size of the sample affects the width of the confidence interval.
7. Width of CI is 2 $\mathrm{SE}(p)z_{\alpha/2} = 2p(1-p)/\sqrt{n}(1.96) = 2(0.25)/\sqrt{n}(1.96)$. We want width of CI to be smaller than 0.08: width $= 0.98/\sqrt{n} < 0.08$, so $\sqrt{n} > 0.98/0.08 = 12.25$ and hence $n > 150.06 > 5$. The sample size must be greater than 150.

Section 10.7

1. SE = 1.52
3. (0.231, 1.869)
5. The 99% confidence interval is (−6.97, 2.79). Since the confidence interval includes 0, it is not likely that the computer made much of a difference in student learning.
7. (4.555, 15.845)

Section 10.8

1. $S^2 = 2746.107$; $S = 52.40$

Section 10.9

1. The sample variance is 6907.968; the standard error is 83.11419.

Chapter Review

1. For a 90% confidence level, if 100 confidence intervals were calculated, about 90 of them would contain the true population parameter.
3. The mean is 65.25; the standard deviation is $10.52/\sqrt{100} = 1.052$.
5. $p = 70\%$; $\mathrm{SE} = (0.70)(0.30)/\sqrt{100} = 0.021$
7. (0.66, 0.74)

Parameter	Population variance	Standard error (true or estimated)	Lower bound of confidence interval	Upper bound of confidence interval
Mean	Known	$\frac{\sigma}{\sqrt{n}}$	$\bar{X} - z_{\alpha/2}\frac{\sigma}{\sqrt{n}}$	$\bar{X} + z_{\alpha/2}\frac{\sigma}{\sqrt{n}}$
Mean	Unknown	$\frac{S}{\sqrt{n}}$	$\bar{X} - z_{\alpha/2}\frac{S}{\sqrt{n}}$	$\bar{X} + z_{\alpha/2}\frac{S}{\sqrt{n}}$
Proportion	Unknown	$\sqrt{\frac{\hat{p}(1-\hat{p})}{n}}$	$\hat{p} - z_{\alpha/2}\sqrt{\frac{\hat{p}(1-\hat{p})}{n}}$	$\hat{p} + z_{\alpha/2}\sqrt{\frac{\hat{p}(1-\hat{p})}{n}}$
Difference between two means	Known	$\sqrt{\frac{\sigma_x^2}{n} + \frac{\sigma_y^2}{m}}$	$(\bar{X} - \bar{Y}) - z_{\alpha/2}\sqrt{\frac{\sigma_x^2}{n} + \frac{\sigma_y^2}{m}}$	$(\bar{X} - \bar{Y}) + z_{\alpha/2}\sqrt{\frac{\sigma_x^2}{n} + \frac{\sigma_y^2}{m}}$
Difference between two means	Unknown	$\sqrt{\frac{S_x^2}{n} + \frac{S_y^2}{m}}$	$(\bar{X} - \bar{Y}) - z_{\alpha/2}\sqrt{\frac{S_x^2}{n} + \frac{S_y^2}{m}}$	$(\bar{X} - \bar{Y}) + z_{\alpha/2}\sqrt{\frac{S_x^2}{n} + \frac{S_y^2}{m}}$

9. The 90% confidence interval is (34.34, 34.46). Since the confidence interval does not contain the value 35, it is not likely that the true length of the bone is 35 cm.
11. (63.64, 65.26)
13. See table at the top of the next page.

CHAPTER 11

Section 11.1

1. Subtract 70.53 − 69 = 1.53 from each observation in the sample. Now the data look like the sample data but have a mean of 69 inches. Sample from this invented population with replacement 100 times and record the sample mean. Repeat the process 100 times. Base your decision on the observed distribution of these sample means.
3. Subtract 4200 − 4129.58 = 70.42 from each observation in the sample. Now the data look like the sample data but have a mean of 4200 pounds. Sample from this invented population with replacement 100 times and record the sample mean. Repeat the process 100 times. Base your decision on the observed distribution of these sample means.
5. Find the differences between the 100 variables. Add or subtract a number so that the differences have a mean of 0. Sample with replacement from the differences 100 times and find the mean difference. Repeat the process 100 times. Base your decision on the observed distribution of these sample means.
7. Find the differences between the 200 variables. Add or subtract so that the differences have a mean of 0. Sample with replacement from the differences 100 times and find the mean difference. Repeat the process 100 times. Base your decision on the observed distribution of these sample means.

Section 11.2

1. z value = −4.65; p value = 0; no, the producers are incorrect.
3. z value = −0.781; p value = 0.22; yes, the chancellor is probably correct.
5. z value = −4.29; p value = 0; no, the mean is not 4200 pounds. The answer is the same.
7. Run NORMAL, and choose "ABOVE":

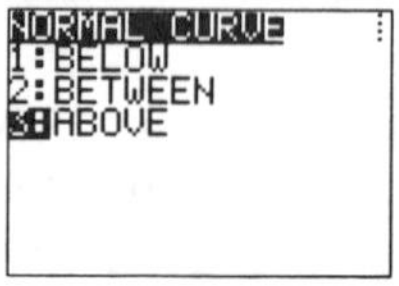

Compute

$$z = \frac{12 - 10}{5/\sqrt{25}} = 2$$

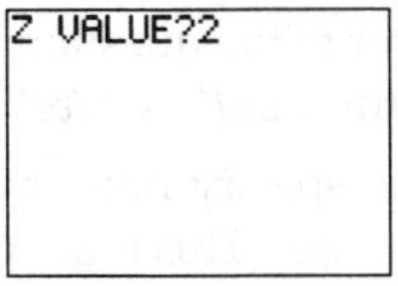

Press ENTER:

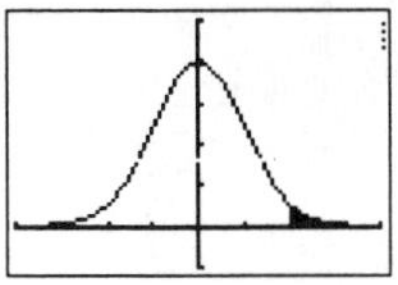

Press ENTER:

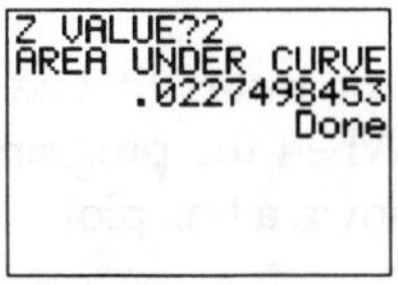

Since $P(z > 0) = 0.0227$, it is unlikely that this occurred by chance, and therefore we may consider Victor's claim plausible.

Section 11.3

1. 0.10, 0.05, 0.01, 0.005
3. Increase the sample size.

Section 11.4

1. There are 14 observations out of 17 greater than the supposed median of 100. In the stem-and-leaf plot there is one observation of 14 or more. The probability of getting 14 or more out of 17 observations above the median is therefore $1/100 = 0.01$. The feed company's claim is not correct. The median amount it reports is too low.
3. There are 11 observations out of 39 above the supposed median of 29. In the stem-and-leaf plot there are no observations of 11 or fewer. The probability of getting 11 or fewer out of 39 observations above the median is therefore $0/100 = 0$. The owners' claim is not correct. The median amount they report is too high.
5. Number above = 9, expected number = 5, z value = 2.72, p value = 0.003.

Section 11.5

1. t value = 3.68; p value = 0.0025; the laevo drug had an effect on sleep time.
3. t value = −0.611; p value = 0.280; the farmer's claim is correct.

Section 11.6

1. Find the mean of the two groups. Use group A as the control group. Since the difference between the two means is $23.8 - 19.25 = 4.55$, subtract 4.55 from each observation in group B so that the mean of group B is now the same as the mean of group A. Sample from both groups A and B 20 times with replacement and calculate the difference between the sample means.
3. z value = 2.87; p value = 0.002; population means are different.
5. z value = 13.02; p value = 0; population means are different.

Section 11.7

1. z value = −2; p value = 0.023; no, the population proportion is not 50%.
3. z value = −5.59; p value = 0; no, the population proportion is not 80%.
5. z value = −1.15; p value = 0.124; yes, the population proportion is 75%.
7. H_0: The coin is fair. This means we expect a mean of 250 with a standard deviation of $\sqrt{500(0.5)(0.5)} \approx 11.18$, and therefore $z = (278 - 250)/11.18 = 2.5$.

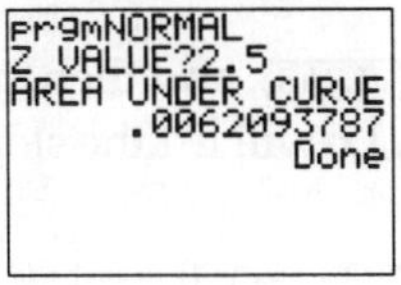

The probability that $z > 2.5$ is 0.006, or six in a thousand. We reject the null hypothesis and conclude that either the coin is weighted or there is something wrong in the tossing.

Chapter Review

1. Step 1: Choose a model.
   Step 2: Define a sample.
   Step 3: Define a successful trial.
   Step 4: Repeat the trials.
   Step 5: Estimate the probability of observing the obtained average or more (less).
   Step 6: Make a decision.
3. z value $= 4.72$; p value $= 0$
5. t value $= -1.265359$; p value $= 0.1231163$
7. In Section 11.1, the number of observations was 130, but in this problem, the number of observations is significantly lower.
9. z value $= -0.6324555$; p value $= 0.2635446$
11. z value $= 1.741035$; p value $= 0.04083868$

CHAPTER 12

Section 12.1

1. a. Conclude that X and Y are independent of each other.
   b. Conclude that X and Y are not independent of each other.
   c. Conclude that X and Y are not independent of each other.
3. a. The standard deviation of X is 0.031; the standard deviation of Y is 110.42.
   b. The z statistic is 2.85. Conclude that X and Y are not independent of each other.
5. The z statistic is 1.65. Conclude that X and Y are independent of each other on the basis of this evidence.
7. The sign of the correlation coefficient and that of the slope of the least squares best-fitting line will be the same.

   Now press [ENTER] [CLEAR], and then [STAT] [ENTER] to view the lists. L6 will list the slopes in order:

−2.3	−1.6	−1.3	−1.1	−1
−1	−.8	−.4	−.2	.1
.1	.2	.2	.5	.5
.7	.8	1.1	1.2	1.3
1.4	1.5	1.6	1.8	1.9

   Now consider the slope of the regression equation for the cricket problem ($m = 3.22$). We see that the probability of getting a slope of 3.22 or larger, by chance, is estimated to be zero. Thus, the answer to the question, "Is the regression line worthwhile?" is "Yes!"

9. BOTSTRAP asks you to specify how many slopes you want to compute. (This is also the number of random samples from the data you are going to draw.) Suppose we request 25:

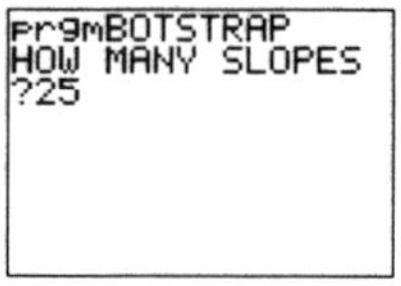

   The calculator now computes (it takes a while, so be patient). When the program is finished, you will be shown a box plot:

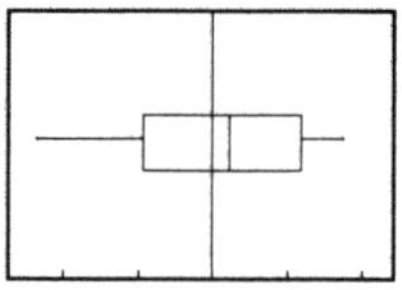

   Press [TRACE]:

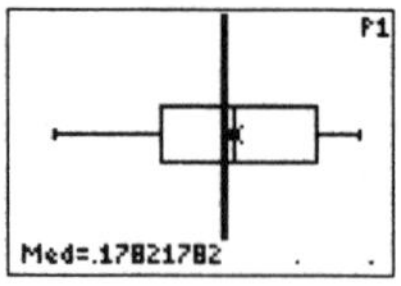

   Now press → to get the third quartile. Press → again for the maximum value:

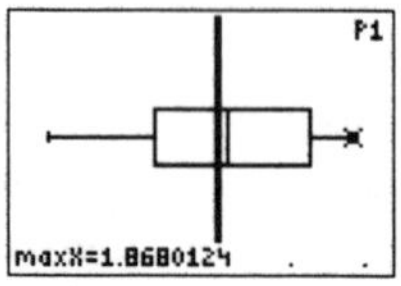

Press ← for the first quartile:

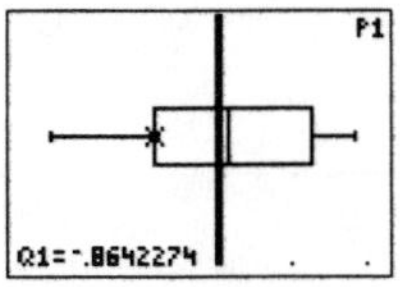

Finally, press ← again for the minimum value:

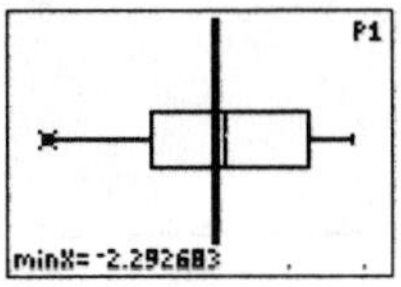

Section 12.2

1. An outlier is a point significantly different in location from the rest of the points. An outlier can be very influential in pulling the regression line in a certain direction.
3. When you look at a scatter plot, you should check whether there are any outliers, and also check whether the data really do have a linear relationship.
5. a. 26.46
   b. 13.44
   c. 77.7
   d. The prediction in part (c) is the least reliable, since it was extrapolated. The other two were interpolated.

Section 12.3

1. a. Venus appears to be an outlier.
   b. The covariance between Z and Y is 184.15. The variance of Z is 0.903.
   c. The slope of the least squares line is 232.976.
   d. The equation for the line is $Y' = 232.976\ Z - 271.445$.
   f. The new correlation is 0.96, much higher than that obtained when Venus was included, 0.76.
3. a. There does not seem to be a linear relationship between X and Y:

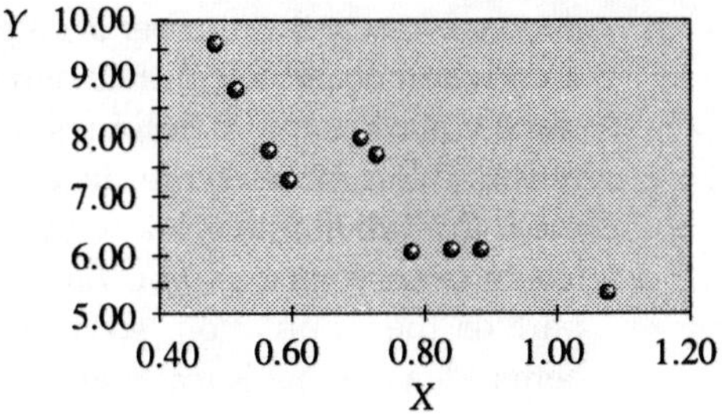

   b.

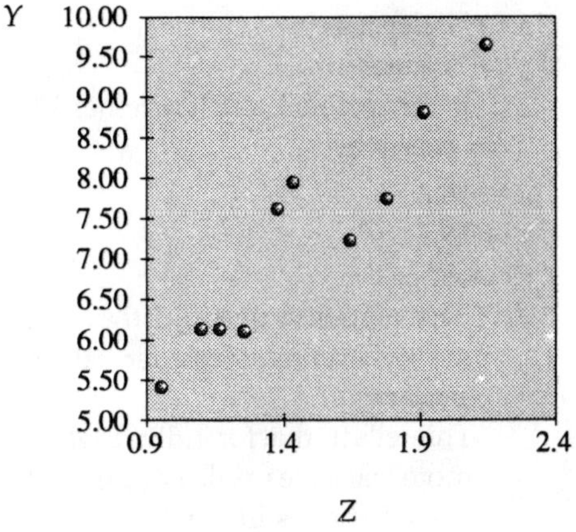

   c. There does now seem to be a linear relationship between Z and Y.
5. a. $Y' = 5.6/X + 7.8$
   b. This is a nonlinear equation since it does not describe a line.

Section 12.4

1. a. 0.02
   b. 0.22
   c. 0.10
3. a. $T = 0.577$; probability $= 0.29$; conclude that the X's and Y's are independent.
   b. $T = 2.095$; probability $= 0.04$; conclude that they are not independent.
   c. $T = -0.953$; probability $= 0.20$; conclude that they are independent.
5. a. In the case in which there are 7 data points, we would conclude that X and Y are independent; in the case in which there are 14 data points, we would conclude that they are not independent.

b. As you increase the number of data points, a correlation coefficient that remains the same distance from 0 becomes more significant. This makes sense since you have more information to work with, so less is left to chance (that is, you have a larger part of the population, so you have a better idea of what is really going on).

Section 12.5

1. a. Y' decreases.
   b. Y' increases.
   c. It is impossible to tell whether Y' increases or decreases.
3. a. 68.68
   b. 69.74
   c. 68.53
   d. The equation that did the best job was the one relating murder rate to life expectancy.
   e. This tells us that for individual cases, using more variables will not necessarily give you a better estimate.

Section 12.6

1. Values of R^2 close to 1 indicate that the regression is nearly perfect—that is, that the regression fits the data very well. Values of R^2 close to 0 indicate that the regression does not fit the data very well—that is, that it does not do any good.
3. The correlation coefficient squared equals R^2.

Section 12.7

1. a. The value of Kendall's tau is 0.8.
   b. The value of the z statistic is 2.25. The probability of seeing a standard normal random variable 2.25 or greater is 0.03. Conclude that there is a trend to the data.
3. a. z statistic is -1.04.
   b. z statistic is 0.638.
   c. z statistic is -1.69.

   In all three cases, conclude that X and Y are independent of each other.
5. In both cases, a value of 1 indicates a positive relationship between the variables, 0 indicates no relationship, and -1 indicates a negative relationship.

Chapter Review

1. a.

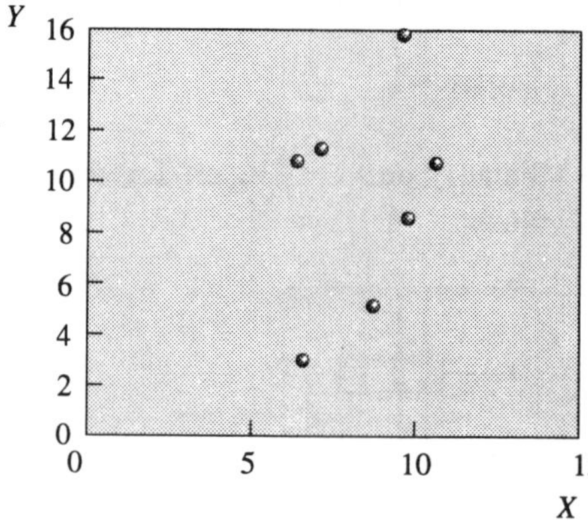

   There is definitely not a perfect linear relationship, but it does not seem too bad.
   b. The value of the z statistic is 0.945. The probability of seeing a z this large or larger is 0.25. Conclude that X is independent of Y.
3. a. 0.357
   b. The value of the T statistic is 0.855. The probability of seeing a t-distributed statistic with five degrees of freedom this large or larger is 0.22. Conclude once again that X is independent of Y.
5. a. The number of calories will increase. This is logical since meat with more fat and cholesterol also has more calories.
   b. 172.14 calories; interpolation
   c. 0.706
7. a.

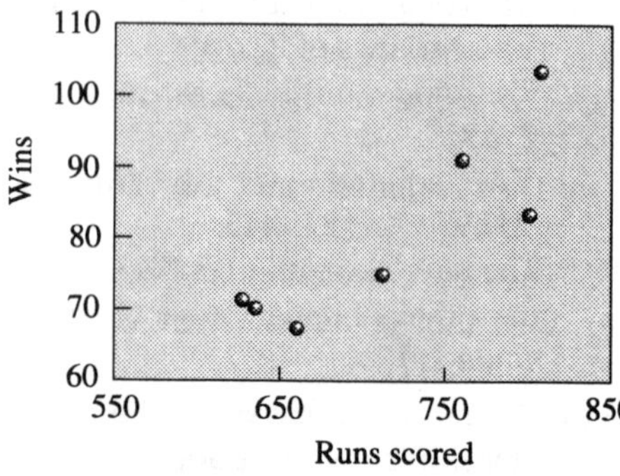

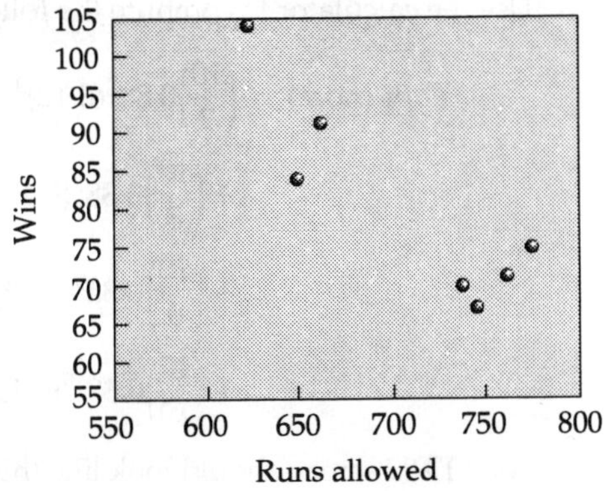

b. The sign of the coefficient of X should be positive, and the sign of the coefficient of Z should be negative. These predictions are based on the slopes of the lines in the scatter plots in part (a).

9. Interpolation is the estimation of a Y value using the regression equation when the X value you are using is between values of X that were part of the data when the regression line was created. Extrapolation is the estimation of a Y value when the X you are using is either larger than or smaller than all of the values of X you used when creating the regression equation.

11. a.

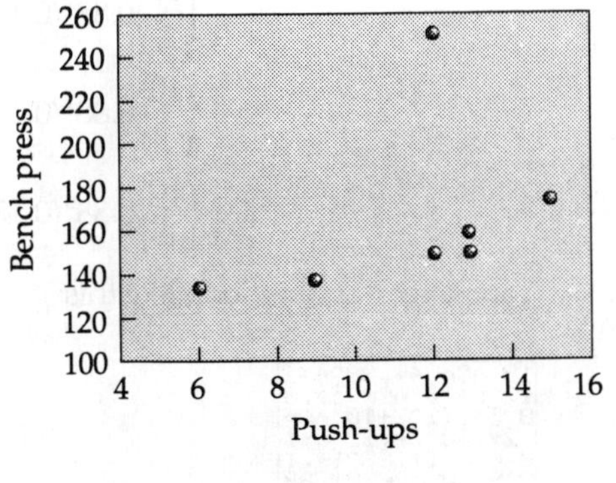

b. The one point appears to be an outlier.
c. False. A researcher should be interested in the cause of the outlier, and only after careful consideration remove the point from the data.

13. a.

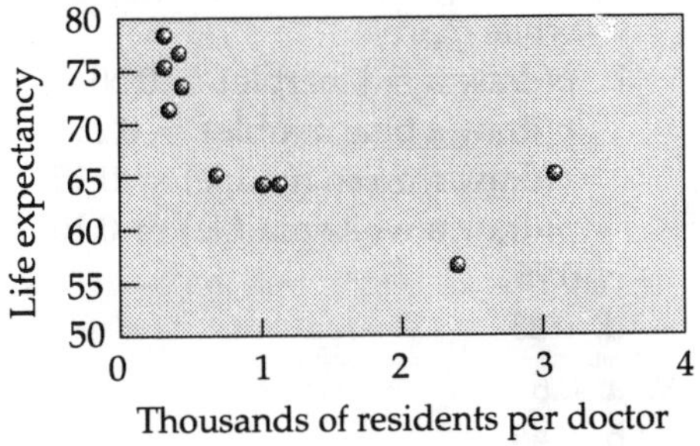

b. The reciprocal transformation (letting $Z = 1/X$) creates a more linear pattern.

15. a. 1.17
b. The z statistic is 3.0. Conclude that X is not independent of Y.
c. The data support Jan's idea.

CHAPTER 13

Section 13.1

1. {1}, {2}, {3}, {4}, {1, 2}, {1, 3}, {1, 4}, {2, 3}, {2, 4}, {3, 4}, {1, 2, 3}, {1, 3, 4}, {1, 2, 4}, {2, 3, 4}, {1, 2, 3, 4}

3. a. They are not mutually exclusive since they could both occur on the same trial. If event B occurs, then event A has necessarily occurred also, since 7 is larger than 4.
b. $p(A) = 5/10 = 0.5$, $p(B) = 1/10 = 0.10$
c. $1/10 = 0.10$
d. 0.50

5. a. Event D is the event that the card is a club.
b. 0.5
c. 0
d. The model is a discrete probability model since the total number of outcomes is finite.

7. a. The four possible outcomes in S are drawing a red marble, drawing a blue marble, drawing a green marble, and drawing a white marble.
   b. $p(\text{draw a red marble}) = 0.50$
      $p(\text{draw a blue marble}) = 0.25$
      $p(\text{draw a green marble}) = 0.15$
      $p(\text{draw a white marble}) = 0.10$
   c. 0.75
   d. 0.25
9. a. 0.6
   b. 0.46

Section 13.2

1. a. 24
   b. 2
   c. 120
   d. 1
3. a. The Poisson fits well because we are looking at a situation where an event is occurring at a given rate. In this case 2.3 calls arrive *per* minute.
   b. $\lambda = 2.3$
   c. 0.10
   d. 0.23
5. a. $\lambda = 5.1$
   b. 0.079
   c. 0.172
7. a. The binomial distribution fits since each day is a separate trial with two possible outcomes. A "success" in this case is the event that the stock goes up.
   b. 0.60
   c. 0.078
   d. 0.259
9. a. $n = 6, p = 0.20$
   b. 0.262
   c. 0.246
   d. 0.000064
11. $p(\text{7 out of 10 correct}) = 0.0031$; $p(\text{7 out of 10 or better correct}) = 0.0035$
13. Use the calculator to compute the following:

$$p(\text{7 or more}) = \binom{10}{7}0.88^7(0.12)^3 + \binom{10}{8}0.88^8(0.12)^2 + \binom{10}{9}0.88^9(0.12)^1 + \binom{10}{10}0.88^{10}(0.12)^0$$

Your TI-82 screen should look like this:

```
10 nCr 7*.88^7*.
12^3+10 nCr 8*.8
8^8*.12^2+10 nCr
 9*.88^9*.12^1+1
0 nCr 10*.88^10*
.12^0
      .9760611779
█
```

Therefore $p(\text{7 out of 10 or more}) = 0.9760$, or Jordan will be making 7 or more free-throws out of 10 about 98% of the time.

Similarly,

$$p(\text{4 or fewer}) = \binom{10}{4}0.88^4(0.12)^6 + \binom{10}{3}0.88^3(0.12)^7 + \binom{10}{2}0.88^2(0.12)^8 + \binom{10}{1}0.88^1(0.12)^9 + \binom{10}{0}0.88^0(0.12)^{10}$$

Your screen should look like this:

```
10 nCr 4*.88^4*.
12^6+10 nCr 3*.8
8^3*.12^7+10 nCr
 2*.88^2*.12^8+1
0 nCr 1*.88^1*.1
2^9+10 nCr 0*.88
^0*.12^10█
```

The result is 0.000406889, or about 0.04%.

15. Probability of two nongerminating bulbs is 0.1922. Probability of all germinating (0 defective bulbs) is 0.3454.

17. $\lambda = 16{,}000/4000 = 4$; $p(x) = 4^x e^{-4}/x!$. Use the method in Section 23 of Appendix H to produce the following:

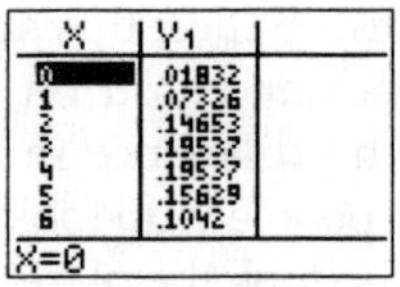

$$p(x \leq 4) = 0.01832 + 0.07326 + 0.14653 + 0.19537 + 0.19537 = 0.6288$$

Section 13.3

1. a. $p(A) = 0.5$, $p(B) = 0.5$
   b. The event $A \cap B$ is the event that the number is both even and less than 5.
   c. 0.3
   d. The event $A \mid B$ is the event that the number is less than 5, given that it is even.
   e. 0.6

3. a. 0
   b. 0
   c. 0.9
   d. No, they are not independent. Since A and B are mutually exclusive, $p(A \mid B) = 0$, but $p(A) > 0$.

5. a. 0.0849
   b. 0.143
   c. 1

Section 13.4

1. The expected value is 2.1.

3. You know that the variance of the estimate of the mean is σ^2/n. So as you increase n, the variance decreases. The variance can be looked at as a measure of the accuracy of the estimate.

5. Her expected grade point is 3.2495.

7. a. $n = 6, a = 3, b = 3, r = 3$
   b. The possible values are 0, 1, 2, and 3.
   c. p(0 bad loaves) $= 0.05$
   p(1 bad loaf) $= 0.45$
   p(2 bad loaves) $= 0.45$
   p(3 bad loaves) $= 0.05$
   d. The theoretical expected number of bad loaves is 1.5.

Section 13.5

1. In this case, the value of the test statistic would be -1.01. The probability of a random variable from the standard normal distribution being this small or smaller is 0.16. Conclude based on this evidence that there is not enough reason to reject the null hypothesis that the drug is not causing any improvement.

3. The value of the test statistic in this case is 4.31. The probability of seeing a standard normal random variable this large or larger is 0. Conclude that there is a significant difference between the groups.

Section 13.6

1. a. Six degrees of freedom
   b. Four degrees of freedom
   c. Four degrees of freedom

3. a. Table of expected values:

	Clear	Cloudy	Precip.	Total
Over 60°	19.3	30.2	13.5	63
50–59°	5.5	8.6	3.9	18
Less than 50°	5.2	8.2	3.6	17
Total	30	47	21	98

   b. The value of the chi-square statistic is 8.77. The probability of seeing a chi-square this large or larger with four degrees of freedom is 0.067. There is not enough evidence to reject the hypothesis that sky condition and temperature are independent.

5. a. Table of expected values:

	Dem.	Rep.	Other	Total
Engineering	68.69	35.64	3.67	108
Science	72.50	37.62	3.88	114
Business	81.41	42.24	4.35	128
Fine Arts	21.62	11.22	1.16	34
Liberal Arts	73.78	38.28	3.94	116
Total	318	165	17	500

   b. The value of the chi-square statistic is 38.98. The chance of obtaining a chi-square this large or larger with eight degrees of freedom is 0. Conclude that the two factors are not independent of each other.

Chapter Review

1. A set of events is exhaustive if one of them must occur on a single trial. Events are mutually exclusive if only one of them can occur on a single trial.

3. a. 0.299
   b. 0.135
   c. 0.095
   d. 0.237

5. The value of the test statistic is -3.35. The probability of seeing a random variable from the standard normal distribution this small or smaller is less than 0.001. Conclude that the proportions are not the same.

7. If you know all the data except for the value in a given column and a given row, you can determine that column and row by using the interdependencies in the data. That is why the formula for the number of degrees of freedom is (rows − 1)(columns − 1).

9. a. 0.3
   b. 0.1
   c. 0.5
   d. 0
   e. 0.4

11. The value of the chi-square statistic in this case is 2.456. The probability of seeing a chi-square this large or larger is 0.78. Conclude that the Poisson distribution does fit the data well.

13. The value of the test statistic is -1.16. The probability of seeing a random variable from the standard normal distribution this small or smaller is 0.12. Conclude based on this evidence that there is no difference in this side effect between the people who took the drug and the people who took the placebo.

15. a. $n = 6, a = 4, b = 2, r = 3$
    b. 0.2
    c. 0.6

17. a. 0.467
    b. 0.467
    c. 0.0667
    d. 0.60

19. Table of expected values:

	Few parks	Many parks	Total
Small population	14.1	7.9	22
Large population	17.9	10.1	28
Total	32	18	50

The chi-square statistic is 2.96. The probability of seeing a chi-square this large or larger with one degree of freedom is 0.085. Conclude that there is not enough evidence to reject the hypothesis that the two factors are independent of each other.

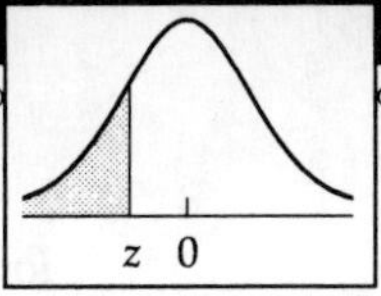

*I*NDEX

Absolute value, 84
Acceleration due to gravity, *g*, 447, 456–457, 467–469
Accuracy, 463–464
Arcsin distribution, 434–437
Average. *See* Mean

Bacon, Sir Francis, 3
Bell-shaped curve. *See* Normal distribution
Bias. *See* Measurement bias
Bimodal data, 65
Binomial distribution, 379–392, 627–628, 643–645, 674
  mean of, 388–402
  standard deviation of, 388–402
Binomial experiment, 381
Binomial probability, 234, 383–388
Birthday problem, 264–267
Bootstrap sampling. *See* Bootstrapping
Bootstrapping, 489–490, 509, 525–526, 529
  nonparametric, 527
Box model, 180, 235–239, 378
Box plot, 57, 74
Box-and-whisker plot. *See* Box plot

Center of a data set, 58–71
  mean, 58–59
  median, 22, 59–63
  mode, 63–65
  properties of, 69–70
Central limit theorem, 491–495, 644–645
Chi-square, 318–354
  density, 337, 379
  distribution, 335–344, 379
  frequency table, 325
  smooth curves, 335–344
  statistic, 325–327
  tables, 345–349, 667–668
  test for independence, 646–650
Circle graph, 20
Combining events to form new events, 620–621
Comparing population means, 550–555
Complementary events, 254–257
Conditional probability, 268–270, 632–634
Confidence interval, 484–487
  for the difference between two population means, 503–504
  for population mean, 496–500
  for population proportion, 501–502
Continuous distribution, 379, 403–411
Continuous uniform distribution, 408–412
  mean of, 411
  standard deviation of, 411
Correlation, 141–149, 568–604
  coefficient of, 142–148
  finding from the slope, 147
Covariance, 138–140

de Fermat, Pierre, 225
de Méré, Chevalier, 224
Decision making, 286–306
Degrees of freedom, 341–344, 549, 649, 655
Descriptive statistics, 11
Deviation values, 79–82, 114
  mean of, 83–85
Discrete distribution. *See* Discrete probability model
Discrete probability model, 618, 624–630
Discrete uniform distribution, 401–402
  mean of, 402
  standard deviation of, 402
Distribution. *See* Probability distribution

Equally likely outcomes, 246–251
Error of estimation, 114
Error variance, 134, 148
Estimation, 478–510
Euler's number, *e*, 625
Event, 619–622
Exclusive events, 621–622
Exhaustive set of outcomes, 618
Expected frequencies, 320
  unequal, 351–354
Expected value, 176–206
  properties of, 638–640
  as a theoretical construct, 635–641
Experiment, 618
Experimental probability, 31, 224–227
  property of complementary events for, 256
Explanatory variables, 591–594
Exponential distribution, 433
Extrapolation, 579–580

Federalist Papers, 3–4
Five-step method, 182
Frequency, 24, 189, 201, 318
  expected, 320, 647–649
  obtained, 319, 648–649
Frequency histogram. *See* Histogram
Frequency interpretation of probability, 184

Galton, Sir Francis, 413
Gaussian distribution. *See* Normal distribution
Goodness of fit, 141–142, 148–149
Gossett, W. S., 547–548
Grand mean, 469, 536

Hamilton, Alexander, 3–4
Herschel, William, 295
Histogram, 14, 24–30
  bimodal, 101
  relative frequency, 31-35
  split, 51–53
  unimodal, 101
How to Lie with Statistics, 37
Hughes, Howard, 3
Hypergeometric distribution, 261
Hypothesis testing, 289, 524–558
  for bivariate trend, 597–604
  of the equality of two medians, 304–306
  of the equality of two population proportions, 643–645
  for independence, 646–650
  of the median, 541–544
  of the population mean, 524–533, 535–539
  of a population proportion, 556–558
  randomization, 570

Independence, 180, 199, 241–244, 634
  chi-square test for, 646–650
  theoretical, 251–253
Inferential statistics, 11
Interpolation, 579–580
  linear, 348, 669–679
Interquartile range, 74, 80
Interval, 25

Jay, John, 3–4

Kendall's tau, 597–604
  use as a test statistic, 600–602
  z test for, 602–604
Kerrich, J. E., 226

Law of averages, 189, 278–280
Law of experimental independence, 243
Law of theoretical independence, 251
Law of statistical regularity, 227
Least mean error linear regression, 122–126
Least squares best-fitting line, 132, 133, 140–141
Least squares linear regression, 105
Line plot, 21–23
Linear interpolation, 348, 669–670
Linear regression, 116–135
  least mean error, 122–126
  least squares, 130–135
  multiple, 591–594
  visual method, 116–120

Madison, James, 3–4
Mean, 58–59
  of continous uniform distribution, 411
  of discrete uniform distribution, 402
  population, 389
Mean absolute deviation, 83–85
Mean absolute error, 119, 122–126
Mean error. *See* Mean absolute error
Mean square error, 132
Measurement, 448–470
Measurement bias, 352, 461–462

Median, 22, 59–63
Median test, 302–306
Mendel, Gregor Johann, 355, 369–374
Michel, John, 295
Michelson, A. A., 473
Mid-square method of random number generation, 660
Modal value. *See* Mode
Mode, 63–65
Modeling, 176–206
Monte Carlo method, 184
Multinomial probabilities, 318
Multiple correlation coefficient, 595–596
Multiple linear regression, 591–594
Multiplication rule for independence, 250
Mutually exclusive outcomes, 619

Newcomb, Simon, 474
Nonlinear regression, 581–586
Nonparametric approach, 526–527
Normal distribution, 413–431
  standard, 418–431, 671–672
Normal law of error, 449, 450
Null hypothesis, 292, 525

Obtained frequency, 319
Outliers, 16, 449–450, 576–578

Parameter, 396, 478–479
  population, 287
Pascal, Blaise, 224–225
Pearson correlation coefficient, 142–148
  calculating, 145–147
  t test for, 589–590
  use as a test statistic, 588–590
Point estimate, 481–483
  for the population mean, 482
  for the population median, 509–510
  for the population variance, 505–508
Poisson distribution, 356, 393–399, 625–627
  mean of, 396
  standard deviation of, 397
Population, 286, 478
Population means, comparing, 550–555
Population parameters, 287
Population proportion, 490, 501, 643–655
Precision, 462
Probability, 184, 224–270
  with applications to hypothesis testing, 616–650
  basic rules of, 618–623
  binomial, 234, 383–388
  conditional, 268–270, 632–634
  discrete model, 618, 624–630
  estimating, 229–234
  experimental, 31, 224–227
  frequency interpretation of, 184
  large, 301
  Poisson, 396
  multinomial, 318
  small, 301
  theoretical, 246–251
Probability distribution, 378–437
  arcsin, 434–437
  binomial, 379–392, 627–628, 643–645, 674
  chi-square, 335–344, 379
  continuous, 379, 403–411
  continuous uniform, 408–412
  discrete, 379
  discrete uniform, 401–402
  exponential, 433
  hypergeometric, 261, 628–630
  normal, 413–419, 671–672
  Poisson, 356, 393–399, 625–627
  Student's *t*, 673
Pseudo-random digits, 660–661

Quadratic regression, 582–586
Quartile, 74
Quetelet, A., 413

Raleigh, Sir Walter, 3
Random number table, 179–182, 662–666
Random sample. *See* Sample, random
Random walk, 202–206
Range, 18, 22, 73
Rayleigh, Lord, 459
Regression, 116–135, 568–604
  quadratic, 582–586
  z test for, 572
Regression equation, 116
Relationships, 105–154
  and causation, 153–154
  curvature in, 576–577
  linear, 111–114
  nonlinear, 581–586
  reciprocal, 581–582
Relative frequency histogram, 31-35
Relative frequency polygon, 32–35
Replacement, sampling with and without, 180, 258–262

Residual. *See* Deviation values
Robustness, 60, 69, 91

Sample, 286, 478
  random, 289
  using to decide on hypotheses about populations, 291–302
Sample proportion, 490, 501
Sample space, 618
Sampling with and without replacement, 180, 258–266
Scatter plot, 105–106
Shakespeare, William, 3
Sign test, 542–545
Six-step method, 300–301, 322–323, 329–334
Spread of a data set, 71–91
  interquartile range, 74, 80
  mean absolute deviation, 83–85
  range, 18, 22, 73
  standard deviation, 88–91
  variance, 86–87
Standard deviation, 88–91
  of binomial distribution, 640–641
  of continous uniform distribution, 411
  of discrete uniform distribution, 402
  of Poisson distribution, 397
  population, 389
Standard error, 456–459
  of an estimate, 488–491
  of the mean, 537
Standard normal curve, 418–431
  applications, 427–431
  table, 421–426
Statistic, 5, 478–479
  of interest, 183
Statistical inference, 5–6
Statistics
  defined, 5, 7
  descriptive, 11
  inferential, 11
  misusing, 37–41
Stem-and-leaf plot, 13-19
Student's *t*. *See t* test.
Successful outcome, 247, 627
Systematic error. *See* Measurement bias

t test, 547–550
  probability distribution, 673
Theoretical independence, 251–253
Theoretical probability, 246–254
  property of complementary events for, 257
Tree diagram, 248
Tukey, John, 14
Twain, Mark, 580
Two-sample problem, 306
Type I and type II errors, 540–542

Uniform distribution. *See* Continuous uniform distribution, Discrete uniform distribution

Variance, 86–87
  properties of, 640–641
Variation. *See* Spread of a data set
von Neumann, John, 660

Whisker, 57
z score, 318–319
z test
  for comparing two population means, 553–554
  for a hypothesis about a population proportion, 556–558
  for Kendall's tau, 602–604
  of the population mean, 535–540
  for regression, 572